Theilheimer's

Synthetic Methods

of Organic Chemistry

Vol. 72

Vol. 72 2008

Theilheimer's Synthetic Methods of Organic Chemistry

Editor Gillian Tozer-Hotchkiss, Wirral, UK

Assistant Editors Alan F. Finch, Cambridge, UK
Jill Entwistle, Berkhamsted, UK
Chris Hardy, Leeds, UK
Julian Hayward, Leeds, UK

Basel • Freiburg
Paris • London
New York • Bangalore
Bangkok • Shanghai
Singapore • Tokyo
Sydney

KARGER

digital chemistry

Deutsche Ausgaben

Vol. 1	1946	1. Auflage
	1948	2., unveränderte Auflage
	1950	3., unveränderte Auflage
Vol. 2	1948	
Vol. 3	1949	with English Index key
	1953	2., unveränderte Auflage
	1966	3., unveränderte Auflage
	1975	4., unveränderte Auflage
Vol. 4	1950	with English Index key
	1966	2., unveränderte Auflage

English Editions

Vol. 1	1948	Interscience Publishers
	1975	(Karger) Second Edition
Vol. 2	1949	Interscience Publishers
	1975	(Karger) Second Edition
Vol. 5	1951	with Reaction Titles Vol. 1-5 and Cumulative Index
	1966	Second Edition
Vol. 6	1952	1975 Second Edition
Vol. 7	1953	1975 Second Edition
Vol. 8	1954	1975 Second Edition
Vol. 9	1955	
Vol. 10	1956	with Reaction Titles Vol. 6-10 and Cumulative Index
	1975	Second Edition
Vol. 11	1957	1975 Second Edition
Vol. 12	1958	1975 Second Edition
Vol. 13	1959	1975 Second Edition
Vol. 14	1960	1975 Second Edition
Vol. 15	1961	with Reaction Titles Vol. 11-15 and Cumulative Index
Vol. 16	1962	
Vol. 17	1963	
Vol. 18	1964	
Vol. 19	1965	
Vol. 20	1966	with Reaction Titles Vol. 16-20 and Cumulative Index
Vol. 21	1967	
Vol. 22	1968	
Vol. 23	1969	
Vol. 24	1970	
Vol. 25	1971	with Reaction Titles Vol. 21-25 and Cumulative Index
Vol. 26	1972	
Vol. 27	1973	
Vol. 28	1974	
Vol. 29	1975	
Vol. 30	1976	with Reaction Titles Vol. 26-30 and Cumulative Index
Vol. 31	1977	
Vol. 32	1978	
Vol. 33	1979	
Vol. 34	1980	
Vol. 35	1981	with Reaction Titles Vol. 31-35 and Cumulative Index
Vol. 36	1982	
Vol. 37	1983	
Vol. 38	1984	
Vol. 39	1985	
Vol. 40	1986	with Reaction Titles Vol. 36-40 and Cumulative Index
Vol. 41	1987	
Vol. 42	1988	
Vol. 43	1989	
Vol. 44	1990	
Vol. 45	1991	with Reaction Titles Vol. 41-45 and Cumulative Index
Vol. 46	1992	
Vol. 47	1993	
Vol. 48	1994	
Vol. 49	1995	
Vol. 50	1996	with Reaction Titles Vol. 46-50
Vol. 51	1997	
Vol. 52	1997	
Vol. 53	1998	
Vol. 54	1998	
Vol. 55	1999	
Vol. 56	1999	
Vol. 57	2000	
Vol. 58	2000	
Vol. 59	2001	
Vol. 60	2001	
Vol. 61	2002	
Vol. 62	2002	
Vol. 63	2003	
Vol. 64	2003	
Vol. 65	2004	
Vol. 66	2004	
Vol. 67	2005	
Vol. 68	2005	
Vol. 69	2006	
Vol. 70	2006	
Vol. 71	2007	

Library of Congress, Cataloging-in-Publication Data

Theilheimer's synthetic methods of organic chemistry = Synthetische Methoden der organischen Chemie. –
Vol. 72 (2008) - Basel; New York: Karger, © 1982 -
v.
Continues: Synthetic methods of organic chemistry.
Editor: Gillian Tozer-Hotchkiss
1. Chemistry, Organic – yearbooks I. Tozer-Hotchkiss, Gillian II. Finch, Alan F. III. Theilheimer, William, 1914–2005
ISBN: 978-3-8055-8639-9
e-ISBN: 97-3-8055-8928-4

Distributed by S. Karger AG, Allschwilerstrasse 10, P.O. Box, CH-4009 Basel (Switzerland)
ISBN: 978-3-8055-8639-9
e-ISBN: 97-3-8055-8928-4

Contents

Preface

This is the first of the biannual volumes of *Theilheimer* for 2008 containing abstracts of new synthetic methods and supplementary data mainly from papers published in the literature up to December 2007.

For browsing purposes, abstracts are displayed according to the Systematic Classification (symbol notation: summary p. VIII) so that reactions of the same type and associated data appear together. For example, all deprotections appear in the early symbols (under HO↓↑, HN↓↑, HS↓↑); reduction of oxo compds., imines and carbon-carbon multiple bonds under the HC⇓ sections; C-defunctionalization under the HC sections; oxy-functionalization under the OC sections; aminations, nitrations, peptide coupling etc. under the NC sections; halogenation under the HalC sections; sulfurations under the SC sections; selenation, stannylation, phosphorylation, etc. under the RemC sections; syntheses involving C-C bond formation in the latter half of the book under the CC sections; and data on resolutions (Res) at the end. A list of reaction symbols and page references thereto is given in the Systematic Survey (p. XXI).

The displayed data are supported by the customary in-depth Subject Index (p. 399) and access to supplementary data can be made in the usual manner via the Supplementary References section, e.g. the reader interested in updates on the Biginelli synthesis (Synth. Meth. *55*, 337) will note from p. 453 that additional references can be found on p. 307 of this volume.

As usual, the volume contains a 'Reviews' section (p. 378), covering reviews published up to and including April 2008, and a 'Trends' section (p. XI) incorporating key developments in synthetic chemistry up to and including May 2008.

This volume marks the start of a new collaboration between Karger and Digital Chemistry (in place of Thomson Scientific). I would like to express my gratitude to Alan Finch for his considerable help in the selection and preparation of abstracts as well as his encouragement. We welcome Chris Hardy and Julian Hayward to the Theilheimer team and thank them and Jill Entwistle for their contributions. Thanks also go to Eliot Cartwright-Finch, Chloë Cyrus-Kent, Lucy Ranner and Andy Hotchkiss for their assistance and support.

June 2008 *G. Tozer-Hotchkiss*, Editor

Advice to the User

General Remarks

New methods for the synthesis of organic compounds and improvements of known methods are being recorded continuously in this series.

Reactions are classified on a simple though purely formal basis by symbols, which can be arranged systematically. Thus searches can be performed without knowledge of the current trivial or author names (e.g. 'Oxidation' and 'Friedel-Crafts reaction'). Once a reaction of interest has been located, further examples may be found by browsing under the same symbol.

Users accustomed to the common notations will find these in the subject index (see page 399). By consulting this index, use of the classification system may be avoided. It is thought that the volumes should be kept close at hand. The books should provide a quick survey, and obviate the immediate need for an elaborate library search. Syntheses are therefore recorded in the index by starting materials and end products, along with the systematic arrangement for the methods. This makes possible a sub-classification within the reaction symbols by reagents, a further methodical criterion. Complex compounds are indexed with cross reference under the related simpler compounds. General terms, such as synthesis, replacement, heterocyclics, may also be brought to the attention of the reader.

A brief review, *Trends and Developments in Synthetic Organic Chemistry* (see page XI), stresses highlights of general interest and calls attention to key methods too recent to be included in the body of the text.

The abstracts are limited to the information needed for an appraisal of the applicability of a desired synthesis. In order to carry out a particular synthesis it is therefore advisable to have recourse to the original papers or, at least, to an abstract journal. In order to avoid repetition, selections are made on the basis of most detailed description and best yields whenever the same method is used in similar cases. Continuations of papers already included will not be abstracted, unless they contain essentially new information. They may, however, be quoted at the place corresponding to the abstracted papers. These supplementary references (see page 449) make it possible to keep abstracts of previous volumes up-to-date.

Syntheses that are divided into their various steps and recorded in different places can be followed with the help of the notations such as *startg. m. f.* (starting material for the preparation of ...).

Method of Classification

Reaction Symbols. As summarized in the Systematic Survey (page XXI), reactions are classified firstly according to the bond formed in the synthesis, secondly according to the reaction type, and thirdly according to the bond broken or the element eliminated. This classification is summarized in the reaction symbol, e.g.

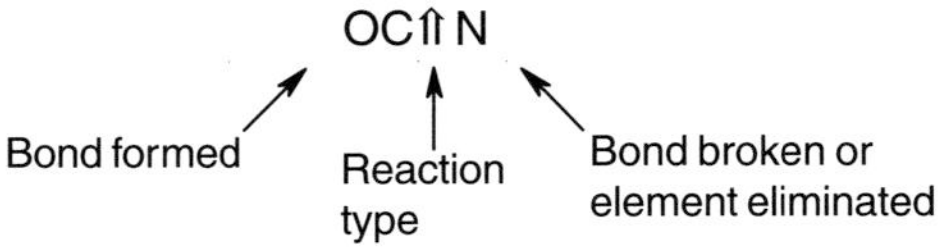

The first part of the symbol refers to the chemical bond formed during the reaction, expressed as a combination of the symbols for the two elements bonded together, e.g. HN, NC, CC. The order of the elements is as follows:

H, O, N, Hal (Halogen), S, Rem (Remaining elements), and C.

Thus, for the formation of a hydrogen-nitrogen bond, the notation is HN, not NH.

If two or more bonds are formed in a reaction, the 'principle of the latest position' applies. Thus, for the reduction

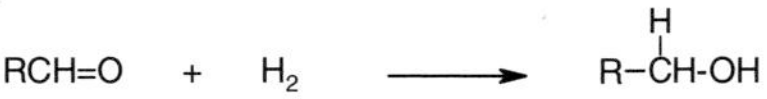

in which both hydrogen-oxygen and hydrogen-carbon bonds are formed, the symbol is HC⇓OC and not HO⇓OC.

The second part of the symbol refers to the reaction type. Four types are distinguished: addition (⇓), rearrangement (∩), exchange (⇅), and elimination (⇑), e.g.

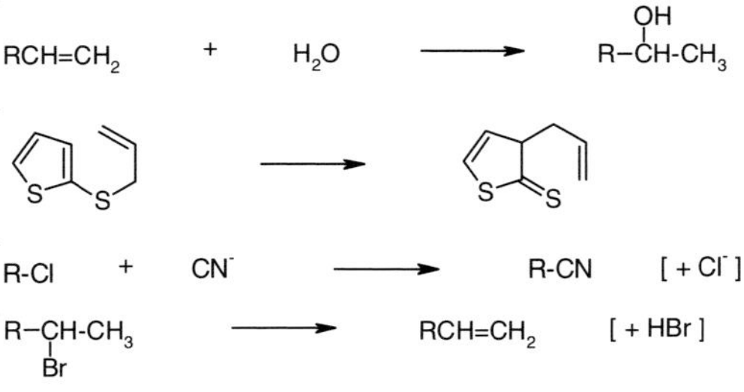

OC⇓CC

CC∩SC

CC⇅Hal

CC⇑Hal

Monomolecular reactions are either rearrangements (∩), where the molecular weight of the starting material and product are the same, or eliminations (⇑), where an organic or inorganic fragment is lost; bimolecular and multicomponent reactions are either additions (⇓), such as intermolecular Diels-Alder reactions, Michael addition and 1,4-addition of organometallics, or exchanges (↓↑), such as substitutions and condensations, where an organic or inorganic fragment is lost.

The last part of the symbol refers to the essential bond broken or, in the case of exchange reactions and eliminations, to a characteristic fragment which is lost. While the addition symbol is normally followed by the two elements denoting the bond broken, in the case of valency expansion, where no bonds are broken, the last part of the symbol indicates the atom at which the addition occurs, e.g.

$R_2S \longrightarrow R_2SO$ OS⇓S

$RONO \longrightarrow RONO_2$ ON⇓N

For addition, exchanges, and eliminations, the 'principle of the latest position' again applies if more than one bond is broken. However, for rearrangements, the most descriptive bond-breakage is used instead. Thus, for the thio-Claisen rearrangement depicted above, the symbol is CC∩SC, and not CC∩CC.

Deoxygenations, quaternizations, stable radical formations, and certain rare reaction types are included as the last few methods in the yearbook. The reaction symbols for these incorporate the special symbols El (electron pair), Het (heteropolar bond), Rad (radical), Res (resolutions), and Oth (other reaction types), e.g.

$R_2S{=}O \longrightarrow R_2S$ ElS⇑O

$R_3N + R'Cl \longrightarrow R_3N^+R'\ Cl^-$ Het⇓N

The following rules simplify the use of the reaction symbols:

1. The chemical bond is rigidly classified according to the structural formula without taking the reaction mechanism into consideration.

2. Double or triple bonds are treated as being equivalent to two or three single bonds, respectively.

3. Only stable organic compounds are usually considered: intermediates such as Grignard compounds and sodiomalonic esters, and inorganic reactants, such as nitric acid, are therefore not expressed in the reaction symbols.

Reagents. A further subdivision, not included in the reaction symbols, is based on the reagents used. The sequence of the reagents usually follows that of the periodic system. Reagents made up of several components are arranged according to the element significant for the reaction (e.g. $KMnO_4$

under Mn, NaClO under Cl). When a constituent of the reagent forms part of the product, the remainder of the reagent, which acts as a 'carrier' of this constituent, is the criterion for the classification; for example, phosphorus is the carrier in a chlorination with PCl_5 and sodium in a nitrosation with $NaNO_2$.

Trends and Developments in Synthetic Organic Chemistry 2008

The arrangement of this survey of the recent literature reflects the layout of the reviews section (page 378).

Firstly, for reviews on asymmetric catalysis see page 379. Recently it has been shown that mass spectrometric monitoring allows the rapid screening of chiral catalysts[1]; both copper catalysts and metal-free organocatalysts were tested for asymmetric Diels-Alder reactions, and the simultaneous screening of catalyst mixtures was examined, which has potential for high-throughput catalyst development. Catalytic asymmetric hydrogenation of 2,3,5-trisubstituted N-Boc-pyrroles proceeds with high enantioselectivity by using a ruthenium catalyst modified with a *trans*-chelating chiral ferrocenyldi(phosphine), PhTRAP[2]. Copper-catalyzed asymmetric N-substitution of acoxy-2-acetylenes may be achieved using chiral 2,6-bis(oxazolinyl)pyridine[3] or di(phosphine)[4] ligands. Iron-catalyzed asymmetric hydrosilylation of ketones proceed smoothly in the presence of inexpensive and environmentally benign $Fe(OAc)_2$ and (S,S)-Me-duphos[5]. Formal [4+3]-cycloadducts may be obtained from benzofuran-3-yl(diazo)acetates and 1,3-dienes with high diastereo- and enantio-selectivity under catalysis by the dirhodium tetracarboxylate, $Rh_2((R)\text{-}DOSP)_4$, reaction involving a tandem cyclopropanation-Cope rearrangement followed by stereoselective tautomerization[6]. Palladium-catalyzed asymmetric 4-vinyl-2-imidazolidone formation from 1,3-dienes (cf. *72*, 179) may now be performed with a chiral N-heterocyclic carbene as ligand[7].

Continuing with asymmetric *organo*catalysis, reviews of its progress from infancy to adolescence[8] and of hydrogen-bond-mediated asymmetric catalysis [by chiral (thio)ureas, diols, and phosphoric acids][9] have now been published. A chiral bicyclic guanidine catalyzes the phospha-Michael reactions of secondary phosphine oxides to nitroalkenes with high enantioselectivity[10]. A novel application of acetaldehyde as two-carbon nucleophile in proline-catalyzed asymmetric Mannich reaction with N-Boc-imines has been reported[11]. A diarylprolinol has been shown to be an outstanding catalyst in an asymmetric aldol reaction with formaldehyde[12]. Ring closure to decalin derivatives have achieved by oxidative dearomatization of phenols bearing a remote formyl group [with $PhI(OAc)_2$] followed by chiral 2-α-siloxy-pyrrolidine-catalyzed intramolecular Michael addition with desymmetrization[13]. Chiral N-carbo-*tert*-butoxy-2-hydroxy-1,2,3,4-tetrahydropyridines may be obtained by formal aza-[3+3]-cycloaddition of enals and N-Boc-enamines with a diphenylprolinol silyl ether as organocatalyst[14]. An asymmetric 3-component synthesis of 4-aryl-1,4-dihydropyridine-3-carbonyl compounds from cinnamaldehydes, primary amines and β-keto esters or β-diketones using a chiral 3,3'-diaryl 1,1'-binaphthyl-2,2'-diyl phosphoric acid has emerged[15]. Asymmetric organocatalysis using SOMO activation (effectively an umpolung of the enamine/iminium ion) has now been demonstrated in α-vinylation of aldehydes with potassium trifluoro(vinyl)-borates[16]. The importance of the N-aryl substituent for a family of chiral bicyclic triazolium salt catalysts has recently been discussed[17]. New double axially-chiral 4,4'-diaryl-2,2'-binaphthyl-3,3'-diyl hydrogen phosphates have been prepared and applied in asymmetric catalytic transfer-hydrogenation of 2- and 2,3-di-substituted quinolines[18].

For reviews on transition metal catalysis see page 380. Density functional theory calculations and catalyst-screening protocols have been used to find *non-precious*-metal alloy catalysts[19]; the process has been demonstrated for the selective hydrogenation of acetylene (with NiZn and $NiZn_3$ having catalyst selectivity and reactivity profiles similar to or better than PdAg catalysts) but potentially has wider applicability. Ketones may be obtained from thiolic acid esters and boronic acids under copper(I) catalysis (in the absence of palladium), C-C bond formation occuring via templating of the thiolate and the boronic acid at copper[20]. N,N'-Ditosyl-1,2,3,4-tetrahydropyrazines may be prepared by copper(I)-catalyzed reaction of an N,N'-ditosyl-1,2-diamine with α,β-acetylenebromides or dibromomethylene compounds; 7-membered or oxa-analogs may also be produced[21]. Gold-catalyzed alkylation with *o*-alkynylbenzoates (*72*, 126) has been extended to a glycosylation method[22] which is orthogonal to known procedures. α-Acoxy-α'-silylketones may be obtained from 1-acoxy-2-acetylenesilanes with $(Ph_3P)AuOTf$ in dioxane containing 1 equivalent of water via [3.3]-sigmatropic rearrangement[23]. Oxidative aromatic homo-coupling using a little $HAuCl_4$ and $PhI(OAc)_2$ as oxidant is reported to be mild and general for non-activated arenes[24].

An iron-based small molecule catalyst is highly active for oxidation of a variety of unactivated sp^3 C–H bond to alcohols with predictable selectivity using hydrogen peroxide[25], while γ-lactones are formed from acids. Iron(III) amine-bis(phenolate) complexes have been described as catalysts for the Kumada coupling of primary and secondary alkyl halides bearing β-hydrogens with aryl Grignard reagents[26]. Iron-catalyzed Sonogashira reactions using $FeCl_3$/dmeda have been reported, including an extension to benzofuran synthesis with *o*-iodophenol[27].

A ruthenium catalyst based on a dearomatized 2-(di-*tert*-butylphosphino-methyl)-6-(diethylaminomethyl)pyridine-type ligand facilitates efficient amide synthesis from alcohols and amines without coupling reagents and with formation of hydrogen as by-product[28]. N-Acyl-*p*-toluenesulfonamides may be obtained efficiently from aldehydes and phenyl(tosylimino)iodinane under catalysis by ruthenium(II) porphyrin complexes[29]. Amides may be obtained by hydration of a broad variety of nitriles in high yield and chemoselectivity under nearly neutral, ambient conditions using a Rh^{I}(OMe) species, the method complementing enzymatic ones[30]. Aldehydes may be obtained from α,β-ethylenecarboxylic acids via decarboxylative hydroformylation using a supramolecular catalyst [31].

Palladium nanoparticles have been supported on ionic liquid-functionalized, multi-walled carbon nanotubes to create catalysts that may be recycled up to 50 times[32]. $Pd(OAc)_2$-catalyzed aromatic methylation by dicumyl peroxide (as both methylating agent and hydrogen acceptor) has been shown to be applicable to various acetanilides and 2-phenylpyridine[33]. A microwave-assisted palladium/copper co-mediated direct 2-arylation of oxazoles with aryl bromides under ligandless conditions has been developed with good functional group tolerance, and applied to a two-step synthesis of 2,5-diaryloxazole alkaloids (including balsoxin and texaline) from commercially available starting materials[34]. A Pd(II)-catalyzed direct α-arylation of cyclic enaminones with potassium aryltrifluoroborates has been reported, using a catalytic amount of $Pd(OAc)_2$ and an excess of $Cu(OAc)_2$ in unpurified solvents[35].

Group 9 metal complex-catalyzed syntheses involving hydrogen transfer have now been reviewed[36]. Following on from the reaction with allenes (*72*, 462), iridium-catalyzed addition of 1,3-cyclohexadiene to benzyl alcohols under hydrogen autotransfer conditions or to aldehydes under transfer-hydrogenation conditions has been reported[37], as have analogous ruthenium-catalyzed reactions of acyclic 1,3-dienes[38]. Carboxylic acids have been shown to react with 1,1-dimethylallene to afford 1,1-dimethylallyl (reverse prenyl) esters in the first branch-regioselective allene hydrocarboxylation[39].

For reviews of metathesis see page 382, while an overview of ruthenium-indenylidene complexes in metathesis[40] and a description of the quest for the ideal olefin metathesis catalyst[41] are now available. A comparison of solvents for Grubbs II-catalyzed ring-closing metathesis indicates that acetic acid may be a useful solvent[42]. Phosphine ligands bearing an imidazolium fragment have been used to prepare ionophilic second generation Grubbs-type catalysts which show good catalytic activity in ring closing metathesis and may be stable up to eight cycles[43].

For reviews on Lewis and Brønsted acid/base and solid acid catalysis see page 383, while acid catalysis where neutralization is 'built in'[44], liquid-phase heterogeneous catalysis with phosphates[45] and Nafion-silica composite catalysts[46] have recently been discussed.

For reviews of N-heterocyclic carbenes as organocatalysts see page 383. N-Heterocyclic carbenes derived from N-mesityl-N-methyltriazolium salts are effective catalysts for the preparation of N-acyl-3-pyrazolidones from enals and N-acylazo compounds; reductive N-N bond cleavage may then afford β-amino acid derivatives[47].

Reviews on biocatalysis may be found on page 384. Chiral N-acylated benzylamines may be obtained in high yield and enantioselectivity from aromatic ketoximes under enzyme-metal cocatalysis with a Pd nanocatalyst and lipase[48]. A rapid and inexpensive assay (using an alcohol oxidase such as horseradish peroxidase as reporter enzyme) has been developed to determine the activity and selectivity of ketoreductases[49]. Enzymatic reactions may now be performed in water-in-ionic liquid microemulsions, wherein the catalytic activity of lipase PS is higher than in isooctane-based microemulsions[50].

For reviews on heterocyclic chemistry see page 384; one on the preparation, reactivity and importance of 4-thiazolidones has recently emerged[51].

Reviews of natural product synthesis and bioactive molecules may be found on page 386. A relatively short and efficient synthesis of (-)-oseltamivir has been developed, involving a novel palladium-catalyzed asymmetric allylation and chemo-, regio-, and stereo-selective aziridination as key transformations[52]. The first total synthesis of marine pyrrole-imidazole alkaloids of the axinellamine/palau'amine/massadine class is reported, featuring a new method for chemoselective oxidation without using protecting groups[53].

For reviews on carbocyclic chemistry see page 387. A method for construction of 7,5-fused ring systems has been reported through ring expansion via dibromocyclopropanation of bicyclic enoxysilanes, and applied to the synthesis of an intermediate of tricholomalide A[54].

For reviews on carbohydrate chemistry see page 388. An ionic liquid-supported glycosylation strategy is shown to produce linear oligomannan thioglycosides rapidly, avoiding the need for column purification after each glycosylation step[55].

For reviews on amino acids and peptides see page 389; an assessment of Bruce Merrifield and solid-phase peptide synthesis[56] has also come to light recently. A new, O-carbamyl-type, tyrosine side chain-protective group for Fmoc-based solid-phase peptide synthesis reportedly increases the solubility of synthetic peptides during purification, after which it may be cleaved via intramolecular cyclization under slightly basic conditions[57]. A series of 5-aryl-

3,4-(ethylenedioxy)thien-2-ylmethanols has been evaluated as carboxylic acid protecting groups for peptide synthesis, with lability to very low concentrations of trifluoroacetic acid (0.01-0.5%)[58].

For reviews on steroid chemistry see page 390, as well as a recent one on azasteroids[59].

For reviews on phosphorus chemistry see page 390, while ones on stabilized P-ylids[60] and 4-membered P-heterocycles[61] are now to hand.

For reviews on fluorine chemistry see page 390, plus a recent one on biomedicinal applications[62]. In the formation of fluorides from mesylates with CsF, the addition of imidazolium ionic liquids bearing an N-hydroxyalkyl group has a synergistic effect, increasing the nucleophilic reactivity of the fluoride anion but reducing the amount of olefin byproduct[63].The first highly enantioselective fluorination of *tert*-butyl methyl malonates is now available using N-fluorobenzenesulfonimide and catalytic amounts of $Zn(OAc)_2$ and a chiral bis(oxazolinyl)dibenzofuran[64]; one of the ester groups may subsequently undergo selective reduction.

For reviews on radical chemistry/cations see page 391, while recent advances in free radical C-C bond formation in aqueous media have also been discussed[65]. N-(Carbo-*tert*-butoxyamino) Hantzsch dihydropyridine has been shown to be an efficient precursor of aminyl radicals and may be used for thiol-catalyzed radical-transfer hydroamination of a variety of olefins including chiral enecarbamates[66]. Pre-formed transition metal-imidazolin-2-ylidene complexes are effective catalysts for atom transfer radical cyclization [ATRC] reactions[67].

'Click' chemistry is still to the fore, with several reviews in the cycloaddition section on page 393, while discussions of its role in medicinal chemistry are now available[68]. The 'Click' reaction has been used for the rapid and selective combinatorial synthesis of a family of pincer-type ligands (applicable in the Heck reaction with high catalytic efficiency)[69]. Its catalysis by a highly reactive and recyclable [tris(2-dioctadecylaminoethyl)amine)]-copper(I) bromide complex, stable under aerobic conditions, was reported recently[70]. A tandem Ferrier and 'Click' reaction with O^3-acylglycals, 1-alkynes and trimethylsilyl azide is reported to provide triazolyl-2,3-dideoxypyranosides readily[71]. The copper-free 'Click' reaction with strained alkynes has been applied in the non-invasive imaging of membrane-associated glycans in live developing zebrafish[72] (with work underway on the creation of tumor-imaging agents). Benzotriazoles may be prepared by cycloaddition with benzyne (generated with CsF)[73].

For reviews of rearrangement and isomerization reactions see page 394, while applications of O→N-acyl migration in medicinal chemistry[74] and

1,2-alkyl migration in pi-acid-catalyzed cascade reactions[75] have also been described. *trans*-γ,δ-Ethylenechlorides may be obtained from N-allyl-N-carbo-*tert*-butoxyhydrazones with $CuCl_2$ via a concerted [3.3]-sigmatropic rearrangement[76].

For reviews of reagents and auxiliaries see page 395; one on applications of metal hydrogen sulfates has recently been uncovered[77]. Arylacetic acids may be converted to aromatic nitriles or arylacetonitriles using NaN_3 and bis(2-methoxyethyl)aminosulfur trifluoride in the presence of ethyldiisopropylamine or triethylphosphine/triethylamine, respectively[78].

For reviews on polymer-based synthesis see page 396. Diversity-oriented synthesis using a solid-supported phosphonate has allowed the synthesis of a family of drug-like compounds, notably including gemmacin, which is active towards methicillin-resistant *Staphylococcus aureus* (MRSA)[79].

For reviews of multicomponent syntheses see page 396. A *catalytic* 3-component Ugi reaction using phenylphosphinic acid as the catalyst is now available[80].

For reviews of tandem, sequenced, domino and cascade reactions see page 396. Polydimethylsiloxane and polycyclooctene/polydicyclopentadiene thimbles may be used to site-isolate incompatible reagents (such as water and $LiAlH_4$, Grignard or cuprate reagents) for cascade reactions, the hydrophobic nature of the thimble preventing polar molecules such as water from diffusing across the membrane[81].

For reviews on reaction media and 'green chemistry' see page 397. The role of Giacomo Ciamician in the formation of early concepts of green chemistry (100 years ago) has been highlighted[82]. Concerning reactions 'on water', addition of a hydrophobic isonitrile to an aqueous suspension of a hydrophobic aldehyde affords a Passerini adduct, in which both the carbonyl and ester functions are derived from the aldehyde[83]. It has been shown that stable carbocations generated from alcohols and characterized by $E<-2.5$ (where E is Mayr's parameter) react readily with nucleophiles in pure water at 80°, without added Brønsted or Lewis acid[84].

Propylene carbonate may be used advantageously as solvent in the Rh-catalyzed asymmetric hydrogenation of functionalized olefins (such as acylaminoenoates) with self-assembling catalysts[85]; similarly, its use in Pd-catalyzed asymmetric C- and N-allylation with acoxy-2-ethylenes may result in improved enantioselectivity[86].

Ionic liquids as media for palladium-catalyzed reactions have now been reviewed[87], while recyclable trihexyl(tetradecyl)phosphonium hexafluorophosphate is recommended for preparation of aromatic thiolic acid esters from iodoarenes and mercaptans by Pd-catalyzed carbonylation[q27c1].

Reviews on photochemistry may be found on page 397, while an article covering photoaddition, photooxygenation, photoelectrocyclization and photoelectron transfer reactions has also come to light[89].

For reviews on methodology see page 398. Applications of microwave irradiation in combinatorial and diversity-orientated chemistry have been reviewed[90].

Trimethylaluminum-mediated amide bond formation has been developed as a safe, functional group-tolerant, high-throughput variant in a continuous flow microreactor, and applied to the synthesis of rimonabant and efaproxiral[91].

Finally, a survey of synthetic organic chemistry 2007 is also now available[92].

1 A. Teichert, A. Pfaltz, Angew. Chem. Int. Ed. Engl. *2008*, *47* (18), 3360-2; screening of palladium catalysts for asymmetric allylation, C.A. Müller, A. Pfaltz, ibid. 3363-6.

2 R. Kuwano, M. Kashiwabara, M. Ohsumi, H. Kusano, J. Am. Chem. Soc. *2008*, *130* (3), 808-9.

3 R.J. Detz, M.M.E. Delville, H. Hiemstra, J.H. van Maarseveen, Angew. Chem. Int. Ed. Engl. *2008*, *47* (20), 3777-80.

4 G. Hattori, H. Matsuzawa, Y. Miyake, Y. Nishibayashi, Angew. Chem. Int. Ed. Engl. *2008*, *47* (20), 3781-3.

5 N.S. Shaikh, S. Enthaler, K. Junge, M. Beller, Angew. Chem. Int. Ed. Engl. *2008*, *47* (13), 2497-501.

6 J.P. Olson, H.M.L. Davies, Org. Lett. *2008*, *10* (4), 573-6.

7 L. Xu, Y. Shi, J. Org. Chem. *2008*, *73* (2), 749-51.

8 A. Dondoni, A. Massi, Angew. Chem. Int. Ed. Engl. *2008*, *47* (25), 4638-60.

9 X. Yu, W. Wang, Chem. Asian J. *2008*, *3* (3), 516-32; review of chiral (thio)ureas in asymmetric organocatalysis s. Q.-H. Wu, Y.-J. Gao, Z. Li, J.-M. Wang, C. Wang, Chin. J. Org. Chem. *2007*, *27* (12), 1491-501.

10 X. Fu, Z. Jiang, C.-H. Tan, Chem. Commun. *2007* (23), 5058-60.

11 J.W. Yang, C. Chandler, M. Stadler, D. Kampen, B. List, Nature *2008*, *452* (7186), 453-5.

12 Y. Hayashi, T. Itoh, S. Aratake, H. Ishikawa, Angew. Chem. Int. Ed. Engl. *2008*, *47* (11), 2082-4.

13 N.T. Vo, R.D. Pace, F. O'Hara, M.J. Gaunt, J. Am. Chem. Soc. *2008*, *130* (2), 404-5.

14 Y. Hayashi, H. Gotoh, R. Masui, H. Ishikawa, Angew. Chem. Int. Ed. Engl. *2008*, *47* (21), 4012-5.

15 J. Jiang, J. Yu, X.-X. Sun, Q.-Q. Rao, L.-Z. Gong, Angew. Chem. Int. Ed. Engl. *2008*, *47* (13), 2458-62.

15 H. Kim, D.W.C. MacMillan, J. Am. Chem. Soc. *2008*, *130* (2), 398-9; for application to 1,4-addition to enals s. *72*, 596 and for a review in Science s. page 379; geospecific Wittig synthesis of unsaturated organotrifluoroborates in water s. G.A. Molander, R.A. Oliveira, Tetrahedron Lett. *2008*, *49* (7), 1266-8.

17 T. Rovis, Chem. Lett. *2008* (1), 2-7.

18 Q.-S. Guo, D.-M. Du, J. Xu, Angew. Chem. Int. Ed. Engl. *2008*, *47* (4), 759-62.

[19] F. Studt, F. Abild-Pedersen, T. Bligaard, R.Z. Sørensen, C.H. Christensen, J.K. Nørskov, Science *2007*, *320* (5881), 1320-2.

[20] J.M. Villalobos, J. Srogl, L.S. Liebeskind, J. Am. Chem. Soc. *2007*, *129* (51), 15734-5.

[21] Y. Fukudome, H. Naito, T. Hata, H. Urabe, J. Am. Chem. Soc. *2008*, *130* (6), 1820-21.

[22] Y. Li, Y. Yang, B. Yu, Tetrahedron Lett. *2008*, *49* (22), 3604-8.

[23] K. Sakaguchi, T. Okada, T. Shinada, Y. Ohfune, Tetrahedron Lett. *2008*, *49* (1), 25-8.

[24] A. Kar, N. Mangu, H.M. Kaiser, M. Beller, M.K. Tse, Chem. Commun. *2008* (3), 386-8.

[25] M.S. Chen, M.C. White, Science *2007*, *318* (5851), 783-7.

[26] R.R. Chowdhury, A.K. Crane, C. Fowler, P. Kwong, C.M. Kozak, Chem. Commun. *2008* (1), 94-6.

[27] M. Carril, A. Correa, C. Bolm, Angew. Chem. Int. Ed. Engl. *2008*, *47* (26), 4862-5.

[28] C. Gunanathan, Y. Ben-David, D. Milstein, Science *2007*, *317* (5839), 790-2.

[29] J.W.W. Chang, P.W.H. Chan, Angew. Chem. Int. Ed. Engl. *2008*, *47* (6), 1138-40.

[30] A. Goto, K. Endo, S. Saito, Angew. Chem. Int. Ed. Engl. *2008*, *47* (19), 3607-9.

[31] T. Smejkal, B. Breit, Angew. Chem. Int. Ed. Engl. *2008*, *47* (21), 3946-9.

[32] Y.S. Chun, J.Y. Shin, C.E. Song, S. Lee, Chem. Commun. *2008* (8), 942-4; cf. M.J. Park, S. Lee, Bull. Korean Chem. Soc. *2007*, *28* (11), 1925-6.

[33] Y. Zhang, J. Feng, C.-J. Li, J. Am. Chem. Soc. *2008*, *130* (10), 2900-1.

[34] F. Besselièvre, F. Mahuteau-Betzer, D.S. Grierson, S. Piguel, J. Org. Chem. *2008*, *73* (8), 3278-80; by direct arylation on water cf. S.A. Ohnmacht, P. Mamone, A.J. Culshaw, M.F. Greaney, Chem. Commun. *2008* (10), 1241-3.

[35] H. Ge, M.J. Niphakis, G.I. Georg, J. Am. Chem. Soc. *2008*, *130* (12), 3708-9.

[36] K. Fujita, J. Synth. Org. Chem. Jpn. *2008*, *66* (4), 322-31.

[37] J.F. Bower, R.L. Patman, M.J. Krische, Org. Lett. *2008*, *10* (5), 1033-5.

[38] F. Shibahara, J.F. Bower, M.J. Krische, J. Am. Chem. Soc. *2008*, *130* (20), 6338-9.

[39] I.S. Kim, M.J. Krische, Org. Lett. *2008*, *10* (3), 513-5.

[40] F. Boeda, H. Clavier, S.P. Nolan, Chem. Commun. *2008* (24), 2726-40.

[41] A. Michrowska, K. Grela, Pure Appl. Chem. *2008*, *80* (1), 31-43.

[42] C.S. Adjiman, A.J. Clarke, G. Cooper, P.C. Taylor, Chem. Commun. *2008* (24), 2806-8.

[43] C.S. Consorti, G.L.P. Aydos, G. Ebeling, J. Dupont, Org. Lett. *2008*, *10* (2), 237-40.

[44] J.P. Hallett, P. Pollet, C.L. Liotta, C.A. Eckert, Acc. Chem. Res. *2008*, *41* (3), 458-67.

[45] S. Sebti, M. Zahouily, H.B. Lazrek, J.A. Mayoral, D.J. Macquarrie, Curr. Org. Chem. *2008*, *12* (3), 203-232.

[46] Á. Molnár, Curr. Org. Chem. *2008*, *12* (2), 159-81.

[47] A. Chan, K.A. Scheidt, J. Am. Chem. Soc. *2008*, *130* (9), 2740-1.

[48] K. Han, J. Park, M.-J. Kim, J. Org. Chem. *2008*, *73* (11), 4302-4.

[49] M.D. Truppo, F. Escalettes, N.J. Turner, Angew. Chem. Int. Ed. Engl. *2008*, *47* (14), 2639-41.

[50] M. Moniruzzaman, N. Kamiya, K. Nakashima, M. Goto, Green Chem. *2008*, *10* (5), 497-500; for the preparation of chiral cyanohydrin acetates from aldehydes via lipase-catalyzed kinetic resolution in ionic liquids s. Z.-L. Shen, W.-J. Zhou, Y.-T. Liu, S.-J. Ji, T.-P. Loh, ibid. *10* (3), 283-6; for ionic liquid effects on the activity of monooxygenase P450BM-3 s. K.L. Tee, D. Roccatano, S. Stolte, J. Arning, B. Jastorff, U. Schwaneberg, ibid. *10* (1), 117-23.

[51] A.P. Liesen et al., Quim. Nova *2008*, *31* (2), 369-76.

[52] B.M. Trost, T. Zhang, Angew. Chem. Int. Ed. Engl. *2008*, *47* (20), 3759-61; review of oseltamivir phosphate synthesis s. M. Shibasaki, M. Kanai, Eur. J. Org. Chem. *2008* (11), 1839-50.

[53] D.P. O'Malley, J. Yamaguchi, I.S. Young, I.B. Seiple, P.S. Baran, Angew. Chem. Int. Ed. Engl. *2008*, *47* (19), 3581-3.

[54] S.-J. Min, S.J. Danishefsky, Tetrahedron Lett. *2008*, *49* (21), 3496-9.

[55] A.K. Pathak, C.K. Yerneni, Z. Young, V. Pathak, Org. Lett. *2008*, *10* (1), 145-8.

[56] A.R. Mitchell, Biopolymers *2008*, *90* (3), 175-84.

[57] K. Wahlström, O. Planstedt, A. Undén, Tetrahedron Lett. *2008*, *49* (23), 3779-81.

[58] A. Isidro-Llobet, M. Álvarez, F. Albericio, Tetrahedron Lett. *2008*, *49* (20), 3288-90.

[59] M. Ibrahim-Ouali, L. Rocheblave, Steroids *2008*, *73* (4), 375-407.

[60] A. Ramazani et al., Curr. Org. Chem. *2008*, *12* (1), 59-82.

[61] Z. Mucsi, I.G. Csizmadia, Curr. Org. Chem. *2008*, *12* (1), 83-96.

[62] K.L. Kirk, Org. Proc. Res. Dev. *2008*, *12* (2), 305-21.

[63] S.S. Shinde, B.S. Lee, D.Y. Chi, Org. Lett. *2008*, *10* (5), 733-5.

[64] D.S. Reddy, N. Shibata, J. Nagai, S. Nakamura, T. Toru, S. Kanemasa, Angew. Chem. Int. Ed. Engl. *2008*, *47* (1), 164-8.

[65] V.T. Perchyonok, I.N. Lykakis, Mini-Rev. Org. Chem. *2008*, *5* (1), 19-32.

[66] J. Guin, R. Fröhlich, A. Studer, Angew. Chem. Int. Ed. Engl. *2008*, *47* (4), 779-82.

[67] J.A. Bull, M.G. Hutchings, C. Luján, P. Quayle, Tetrahedron Lett. *2008*, *49* (8), 1352-6.

[68] A.D. Moorhouse, J.E. Moses, Chem. Med. Chem. *2008*, *3* (5), 715-23; G.C. Tron, T. Pirali, R.A. Billington, P.L. Canonico, G. Sorba, A.A. Genazzani, Med. Res. Rev. *2008*, *28* (2), 278-308.

[69] E.M. Schuster, M. Botoshansky, M. Gandelman, Angew. Chem. Int. Ed. Engl. *2008*, *47* (24), 4555-8.

[70] N. Candelon, D. Lastecoueres, A.K. Diallo, J.R. Aranzaes, D. Astruc, J.-M. Vincent, Chem. Commun. *2008* (6), 741-3; cf. use of a dinuclear copper(I) complex of tris(N-benzyltriazol-4-ylmethyl)amine, P.S. Donnelly, S.D. Zanatta, S.C. Zammit, J.M. White, S.J. Williams, ibid. (21), 2459-61.

[71] J.S. Yadav, B.V.S. Reddy, D.N. Charya, C.S. Reddy, Tetrahedron Lett. *2008*, *49* (16), 2649-52.

[72] S.T. Laughlin, J.M. Baskin, S.L. Amacher, C.R. Bertozzi, Science *2008*, *320* (5876), 664-7; for new insights and perspectives on the copper-free reaction s. J.-F. Lutz, Angew. Chem. Int. Ed. Engl. *2008*, *47* (12), 2182-4.

[73] F. Shi, J.P. Waldo, Y. Chen, R.C. Larock, Org. Lett. *2008*, *10* (12), 2409-12.

[74] M. Skwarczynski, Y. Kiso, Curr. Med. Chem. *2007*, *14* (26), 2813-23.

[75] B. Crone, S.F. Kirsch, Chem. Eur. J. *2008*, *14* (12), 3514-22.

[76] D.A. Mundal, J.J. Lee, R.J. Thomson, J. Am. Chem. Soc. *2008*, *130* (4), 1148-9.

[77] F. Shirini, M.A. Zolfigol, P. Salehi, M. Abedini, Curr. Org. Chem. *2008*, *12* (3), 183-202.

[78] C.O. Kangani, B.W. Day, D.E. Kelley, Tetrahedron Lett. *2008*, *49* (5), 914-8.

[79] G.L. Thomas, R.J. Spandl, F.G. Glansdorp, M. Welch, A. Bender, J. Cockfield, J.A. Lindsay, C. Bryant, D.F.J. Brown, O. Loiseleur, H. Rudyk, M. Ladlow, D.R. Spring, Angew. Chem. Int. Ed. Engl. *2008*, *47* (15), 2808-12.

[80] S.C. Pan, B. List, Angew. Chem. Int. Ed. Engl. *2008*, *47* (19), 3622-5; for reviews on amino acid chemistry see page 389.

[81] M.B. Runge, M.T. Mwangi, A.L. Miller II, M. Perring, N.B. Bowden, Angew. Chem. Int. Ed. Engl. *2008*, *47* (5), 935-9.

[82] A. Albini, M. Fagnoni, ChemSusChem *2008*, *1* (1-2), 63-6.

[83] N. Shapiro, A. Vigalok, Angew. Chem. Int. Ed. Engl. *2008*, *47* (15), 2849-52; for a review of

the Passerini reaction see Reviews section (page 395).

84 P.G. Cozzi, L. Zoli, Angew. Chem. Int. Ed. Engl. *2008*, *47* (22), 4162-6.

85 B. Schäffner, J. Holz, S.P. Verevkin, A. Börner, Tetrahedron Lett. *2008*, *49* (5), 768-771.

86 B. Schäffner, J. Holz, S.P. Verevkin, A. Börner, ChemSusChem *2008*, *1* (3), 249-53.

87 R. Singh, M. Sharma, R. Mamgain, D.S. Rawat, J. Braz. Chem. Soc. *2008*, *19* (3), 357-79.

88 H. Cao, L. McNamee, H. Alper, J. Org. Chem. *2008*, *73* (9), 3530-4.

89 N. Hoffmann, J.-C. Gramain, H. Bouas-Laurent, L'Actualité Chimique *2008*, *317*, 6-13.

90 Microwave reactions under continuous flow conditions s. I.R. Baxendale, J.J. Hayward, S.V. Ley, Comb. Chem. High Throughput Screening *2007*, *10* (10), 802-36; diversity-oriented synthesis and solid-phase organic synthesis under controlled microwave heating s. W.-M. Dai, J. Shi, ibid. 837-56; microwave-assisted reactions in heterocyclic compounds s. A. de la Hoz et al., ibid. 877-902; microwave irradiation as a tool for functionalization of fullerenes and carbon nanotubes s. F. Langa, P. de la Cruz, ibid. *10* (9), 766-82.

91 T. Gustafsson, F. Pontén, P.H. Seeberger, Chem. Commun. *2008* (9), 1100-2.

92 C. Tschierske, Nachrichten aus der Chemie *2008*, *56* (3), 269-94.

Systematic Survey

Abbreviations and Symbols

abs.	absolute
alc.	alcoholic
aq.	aqueous
ar.	aromatic
asym.	asymmetric
atm.	atmosphere(s)
compd(s).	compound(s)
deriv(s).	derivative(s)
e.e.	enantiomeric excess
eq(s).	equivalent(s)
E.	Example
F.e.s.	Further example(s) see
M	molar
prepn.	preparation
prim.	primary
s*72*	supplementary reference in Volume *72*
sec.	secondary
startg. m.f.	starting material for (the preparation of ...)
subst.	substituted
sym.	symmetrical
tert.	tertiary
v.i.	via intermediates
w.a.r.	without additional reagents
Y *	Yield
ϟ	Electrolysis
///	Irradiation
[\\\\]	Microwave irradiation
○	Ring closure
◌	Ring contraction
◌	Ring expansion
C	Ring opening
Ⓗ	Ring hydrogenation
←	'see title or reagent on the left half of the page'

* Yields in parentheses refer to the immediately preceding step of a multi-step reaction

Formation of H-O Bond

Uptake ⇓

Addition to Oxygen and Carbon HO ⇓ OC

Sodium azide *NaN_3*
p-Quinols from p-quinones ←
cf. *5*, 3; simple, mild, efficient method under neutral conditions using NaN_3 in the presence of water s. F. Algi, M. Balci, Synth. Commun. *2006*, *36* (16), 2293-7.

Exchange ↓↑

Sulfur ↑ HO ↓↑ S

Aluminum chloride *$AlCl_3$*
O-Desulfonation of aryl sulfonates $OSO_2R \rightarrow OH$
s. *21*, 9s*72*

Remaining Elements ↑ HO ↓↑ Rem

Copper(II) bromide *$CuBr_2$*
Selective O-de-*tert*-butyldimethylsilylation $OSiMe_2Bu\text{-}t \rightarrow OH$
cf. *23*, 6s*66*; *56*, 4s*71*; of alcohol/phenol derivs. with $CuBr_2$ in acetonitrile at room temp., leaving functional groups such as ketals, alkenes, ketones, *tert*-butyldiphenylsilyl, tetrahydropyran-2-yl, allyl and benzyl ethers unaffected, s. S. Bhatt, S.K. Nayak, Tetrahedron Lett. *2006*, *47* (47), 8395-9; alcohols with silica-supported sodium hydrogen sulfate as heterogeneous catalyst in methylene chloride s. B. Das, K.R. Reddy, P. Thirupath, ibid. 5855-7.

Lithium tetrahydridoaluminate *$LiAlH_4$*
Protection of alcohols OSi⋲ → OH
as chiral (1-methoxy-2,2,2-triphenylethyl)dimethyl silyl ethers
and asym. induction with the same

1. 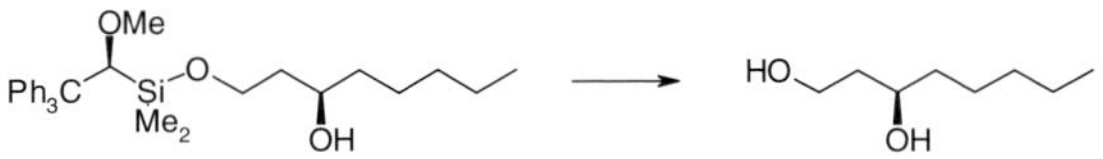

Cleavage of the protective group. 1.5 Eqs. 1 *M* $LiAlH_4$ in ether added to a soln. of the startg. protected alcohol in THF, and quenched with satd. aq. NH_4Cl after 40 min → (R)-octane-1,3-diol. Y 95% (e.e. >98%). α- and β-Hydroxyoxo compds. protected with such chiral silyl groups underwent **Grignard addition with asym. induction** with added $MgBr_2$, while the corresponding chiral α,β-ethylenecarboxylic acid silyl esters underwent **Diels-Alder reaction with asym. induction** in the presence of the same catalyst. Furthemore, the **absolute configuration of sec. alcohols and prim. amines** could also be determined via ^1H NMR analysis of such chiral silyl derivs. F.e.s. M. Campagna, M. Trzoss, S. Bienz, Org. Lett. *2007*, *9* (19), 3793-6.

N-Bromosuccinimide/methanol *NBS/MeOH*
O-Detrimethylsilylation s. *29*, 415s*72* $OSiMe_3 \rightarrow OH$

Vanadyl triflate *$VO(OTf)_2$*
Lewis acid-catalyzed O-de-*tert*-butyldimethylsilylation $OSiMe_2Bu\text{-}t \rightarrow OH$
s. *72*, 9

Silica-supported sodium hydrogen sulfate *$NaHSO_4$-SiO_2*
Selective O-de-*tert*-butyldimethylsilylation $OSiMe_2Bu$-*t* → OH
s. *23*, 6s*72*; *56*, 4s*72*

Chlorotris(triphenylphosphine)rhodium(I)/catecholborane ←
Preferential reductive cleavage of trialkylsilyl groups under mild conditions OSi≤ → OH

2. 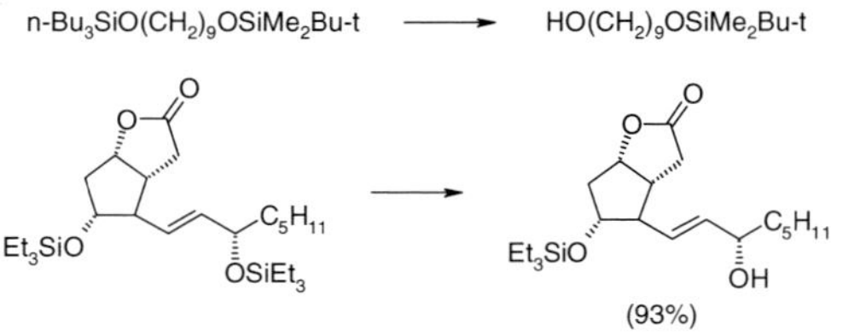

Preferential O-detributylsilylation. 3 Mol% $RhCl(PPh_3)_3$ added to a soln. of the startg. protected diol in dry THF, cooled to 0°, 3 eqs. catecholborane added slowly under N_2, the ice-bath removed after 10 min, and stirring continued for 18 h → product. Y 95%. The procedure is convenient for the preferential cleavage of triethylsilyl, triisopropylsilyl and tributylsilyl ethers, leaving *tert*-butyldimethylsilyl and *tert*-butyldiphenylsilyl ethers unaffected. Triethylsilyl ethers are also removed more quickly than triisopropylsilyl and tributylsilyl ethers, and triethylsilyl ethers at acyclic secondary sites are removed in preference to those at cyclic secondary sites (e.g. in prostaglandins). Significantly, halides and esters are left intact, as also are both isolated and conjugated double bonds (without associated hydroboration!), suggesting that catecholborane is 'tied up' in the catalytic cycle rather than in the form of an active hydridorhodium-borane hydroborating agent. F.e.s. P. Patel, C.-T. Chang, N. Kang, G.-J. Lee, W.S. Powell, J. Rokach, Tetrahedron Lett. *2007, 48* (30), 5289-92.

Carbon ↑ HO ↓↑ C

Irradiation ///
Protection of carbonyl groups as 6-methoxy-4,4-diphenyl-1,3-benzodioxanes C
Photochemical removal of the protective group s. *30*, 5s*72*

Microwaves (s.a. under $In(OTf)_3$, Me_3SiBr and TsOH) [\\\\]
Uncatalyzed cleavage of acetals in pure water under microwave irradiation $C(OR)_2$ → CO

3. 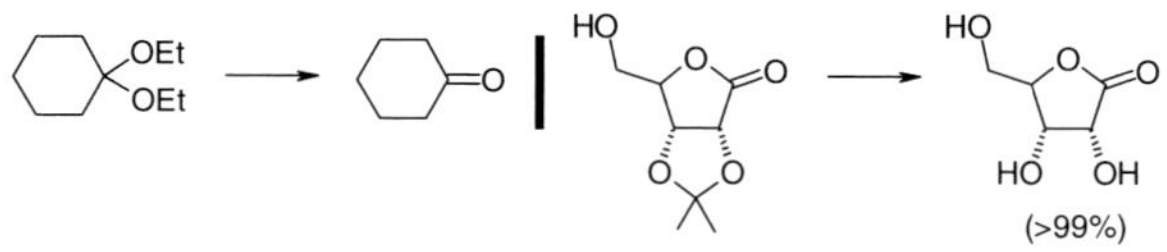

Cyclohexanone diethyl acetal (2 mmol scale) suspended in deionized water (pH 6-6.7) contained in a Teflon reaction vessel, the Teflon tube stopped, and the mixture stirred in a Synthos 3000 microwave synthesizer at 80° for 15 min under irradiation at a power setting of 1000 W → cyclohexanone. Y >99%. The procedure is *entirely 'green'*, and generally applicable to both aliphatic and aromatic acetals and ketals, notably including carbohydrate O,O-isopropylidene and O,O-benzylidene derivs., leaving a variety of functions (carbonyl, triple bonds, hydroxyl groups) and other O-protective groups (acetyl, *tert*-butyldimethylsilyl, benzyl, *p*-methoxybenzyl) unaffected. F.e.s. A. Procopio, M. Gaspari, M. Nardi, M. Oliverio, A. Tagarelli, G. Sindona, Tetrahedron Lett. *2007, 48* (49), 8623-7.

Sodium hydroxide *NaOH*
Carboxylic acids from their esters – Partial hydrolysis COOR → COOH
cf. *44*, 8; of a cyclic unsatd. diester with aq. NaOH and a little co-solvent (especially THF or acetonitrile) at 0° s. S. Niwayama, H. Wang, Y. Hiraga, J.C. Clayton, Tetrahedron Lett. *2007*, *48* (48), 8508-10.

Carboxylic acids and alcohols from carboxylic acid esters COOR → COOH + HOR
Saponification in non-aqueous media at low hydroxide ion concentration

4. $BuCH{=}CHCH_2CO_2Et \longrightarrow BuCH{=}CHCH_2CO_2H + HOEt$

A large excess of hydroxide ion and forcing conditions can be avoided by conducting saponification of esters in non-aqueous media. **E:** *2 eqs.* methanolic NaOH [affording a final concentration of 0.2 *N*] added to a soln. of ethyl oct-3-enoate *in ca. 9:1 methylene chloride/ methanol,* stirred for 2-3 min *at room temp.* when the sodium carboxylate started to precipitate out, stirring continued *for 30 min*, the solvents removed *in vacuo*, the residue taken up in water, the aq. soln. extracted with ether, the aq. layer cooled, acidified with dil. HCl (or $NaHSO_4$) to pH 2-3, and the free acid extracted with ethyl acetate or ether → oct-3-enoic acid (Y ca. 80-90%) and ethanol (Y ca. 80-90%). Reaction is generally applicable to both alkyl and aryl esters, incl. orthogonally protected amino acids (without racemization), triglycerides and terpenoids. Interestingly, there was no isomerization of the double bond on saponification of β,γ-ethylenecarboxylic acid esters, but hindered esters required harsher conditions and a longer reaction time. If an aqueous work-up is problematic, the acid can be isolated as the precipitated sodium salt. The same mechanistic features are presumed as those pertaining to aqueous saponification. F.e.s. V. Theodorou, K. Skobridis, A.G. Tzakos, V. Ragoussis, Tetrahedron Lett. *2007, 48* (46), 8230-3.

Potassium hydroxide s. under $(EtO)_2P(O)N_3$ *KOH*
Sodium iodide s. under Dimethyl sulfoxide *NaI*
Triethylamine s. under $(EtO)_2P(O)N_3$ *Et_3N*
Magnesium s. under Ti(OPr-i)$_4$ *Mg*

Magnesium bromide (s.a. under Ti(OPr-i)$_4$) *$MgBr_2$*
Selective Lewis acid-catalyzed O-detritylation of O-tritylhydroxylamine derivs. N-OTr → N-OH

5. 

Fmoc = carbo-9-fluorenylmethoxy

Cleavage of hydroxamic acid trityl esters. 5 Eqs $MgBr_2$ and the O-tritylhydroxylamine deriv. stirred in methylene chloride at room temp. for 40 min, the soln. poured into 1:1 ethyl acetate/ water, the organic layer washed with brine, dried, and evaporated → product. Y 86%. The method is successful using either $MgBr_2$ or BF_3-OEt_2 in methylene chloride or methanol (or mixtures thereof), providing mild, efficient and selective deprotection of O-tritylhydroxylamines in the presence of Fmoc-, Boc-, Cbz- and tosyl-protected amines in 74-93% yields. A single example of $ZnBr_2$-catalyzed cleavage of an acid-labile O-trityloxime is also given in 82% yield. F.e. incl. cleavage of O-trityl-N-hydroxyureas, -N-hydroxyisoureas and -N-hydroxysulfonamides, also preparation of substrates, and use of the methodology in an efficient synthesis of (-)-compactin T, s. S.-M. Yang, B. Lagu, L.J. Wilson, J. Org. Chem. *2007*, *72* (21), 8123-6.

Indium/ammonium chloride *In/NH_4Cl*
Reductive cleavage of 2,2,2-trichloroethyl esters $COOCH_2CCl_3$ → COOH
cf. *24*, 9s*68*; of aliphatic or aromatic esters with In/NH_4Cl in aq. THF, also monodechlorination of α- or β-aryl-derivs. to 2,2-dichloroethyl esters (cf. *19*, 93), s. T. Mineno, H. Kansui, T. Kunieda, Tetrahedron Lett. *2007*, *48* (29), 5027-30.

Dodecapotassium hexakis[N,N′-bis(2,3-dihydroxybenzoyl)-1,5-diaminonaphthalene]tetragallate/potassium carbonate ←

Cleavage of acetals in basic aqueous media under supramolecular ['nanozymatic'] catalysis $C(OR)_2 \rightarrow CO$

6. 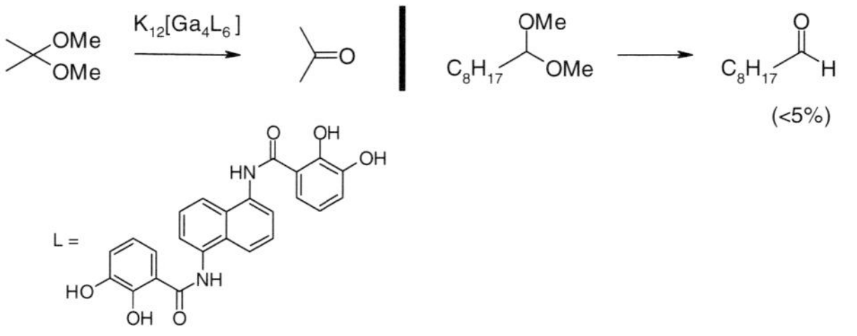

Well-defined supramolecular assemblies with the stoichiometry M_4L_6 (where M is central metal, and L = N,N′-bis(2,3-dihydroxybenzoyl)-1,5-diaminonaphthalene) serve as hosts to acetal molecules *of limited size*, facilitating their hydrolysis in a *basic* aqueous medium. **E:** Dodecapotassium hexakis[N,N′-bis(2,3-dihydroxybenzoyl)-1,5-diaminonaphthalene]tetragallate (20 mg), water (1 ml), buffered to pH 10 with K_2CO_3, and 20 eqs. of acetone dimethyl ketal introduced into a vial (in a glove box), the vial removed from the glove box, heated at 50° for 6 h in an oil bath, cooled to room temp., saturated with NaCl, and worked up → acetone. Y >95%. The naphthalene walls of the assembly provide hydrophobic cavities, permitting stabilization of cationic intermediates along the hydrolytic pathway. A large number of acetals and ketals were cleaved in this way, incl. 2-adamantanone dimethyl ketal, but acetals having a long side chain, e.g. nonanal dimethyl acetal, remained unchanged, as they cannot be accommodated within the host assembly. F.e.s. M.D. Pluth, R.G. Bergman, K.N. Raymond, Angew. Chem. Int. Ed. Engl. *2007, 46* (45), 8587-9.

Vinylboronic anhydride-pyridine complex s. under Pd(OAc)$_2$ ←

Aluminum silicate ←
Mesoporous aluminosilicates ←
Cleavage of acetals
cf. *29*, 9; *48*, 7s*54*; with mesoporous aluminosilicates, cleavage in acetone at 55°, also formation in (m)ethanol at room temp. (cf. *22*, 191), s. M.W.C. Robinson, A.E. Graham, Tetrahedron Lett. *2007*, *48* (27), 4727-31; with Al-silicate s. *28*, 141s*72*.

Montmorillonite *clay*
Cleavage of O,O-alkylidene derivs. C
cf. *5*, 12s*71*; of 5′-*O*-acetyl-2′,3′-isopropylidene/cyclohexylidene nucleosides with montmorillonite K10 in refluxing methanol s. M.K. Gurjar, D. Mondal, S.V. Ravindranadh, M.S. Chorghade, Synth. Commun. *2006*, *36* (16), 2321-7.

Indium(III) triflate/acetone *In(OTf)$_3$/Me$_2$CO*
Indium(III) triflate/acetone/microwaves *In(OTf)$_3$/Me$_2$CO/[\\\\]*
Cleavage of acetals under mild, neutral conditions $C(OR)_2 \rightarrow CO$

7. 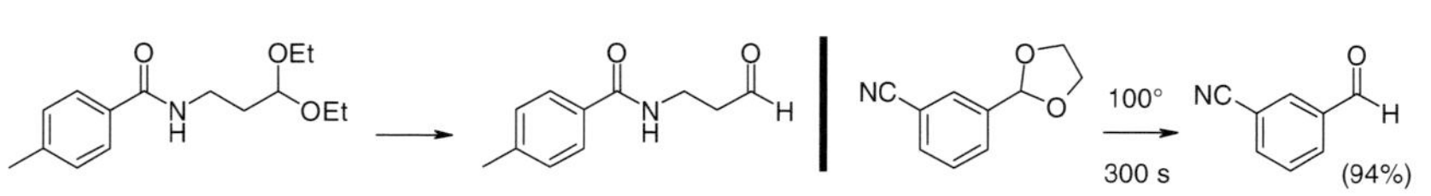

by transacetalation at low catalyst loading. *0.1 Mol%* In(OTf)$_3$ added to a soln. of N-(3,3-diethoxypropyl)-4-methylbenzamide in acetone, and stirred *at room temp.* for 2 h → 4-methyl-N-(3-oxopropyl)benzamide. Y 91% (after column chromatography). This inexpensive, simple and rapid conversion is clean and generally applicable to acetals and ketals up to the multi-gram scale (ca. 20 g) and in parallel synthesis, notably leaving N-Boc groups as well as TBS and THP ethers unaffected. Microwave enhancement (generally for 300-400 s at 100°) was applied for more stable substrates, such as 1,3-dioxolane derivs. No particular precautions to eliminate moisture and air are required. F.e. incl. cleavage of electron-rich, electron-poor and hindered substrates s. B.T. Gregg, K.C. Golden, J.F. Quinn, J. Org. Chem. *2007, 72* (15), 5890-3.

Boron fluoride *BF_3*
Selective Lewis acid-catalyzed O-detritylation N-OTr → N-OH
of O-tritylhydroxylamine derivs. s. *72*, 5

Aluminum chloride *$AlCl_3$*
O-Deacylation of phenolesters OAc → OH
cf. *61*, 7; nitro-directed O-deacylation with $AlCl_3$, also O-desulfonylation of phenol sulfonates (cf. *21*, 9s*54*), s. X. Ji, C. Li, Synthesis *2006* (15), 2478-82.

Ammonium cerium(IV) nitrate *CAN*
Cleavage of cyclic acetals C
cf. *53*, 3s*57*; chemoselective deprotection of diacetone sugars with 20 mol% CAN in moist acetonitrile at room temp. s. S. Jana, S.C. Roy, Indian J. Chem. *2007*, *46B* (4), 707-9.

Glycerol s. under $(EtO)_2P(O)N_3$ *$(HOCH_2)_2CH(OH)$*
Acetone s. under $In(OTf)_3$ *Me_2CO*

Enzymes ←
O-Deacylation OAc → OH
cf. *26*, 6; *42*, 6s*67*; regioselective hydrolysis of peracylated β-monosaccharides with immobilized lipases from *Thermomyces lanuginose, Aspergillus niger* and *Candida antarctica* B s. J.M. Palomo, M. Filice, R. Fernandez-Lafuente, M. Terreni, J.M. Guisan, Adv. Synth. Catal. *2007*, *349* (11-12), 1969-76; of ar. diolesters in *sec.* alcohols s. T. Miyazawa, M. Hamada, R. Morimoto, T. Murashima, T. Yamada, Tetrahedron Lett. *2007*, *48* (47), 8334-7.

Asym. enzymatic hydrolysis of carboxylic acid esters COOR or OAc → COOH or OH
s. *28*, 13s*70*; bis(2-hydroxyethyl)amine oxides with desymmetrization s. I. Suzuki, R. Kerakawauchi, T. Hirokawa, K. Takeda, Tetrahedron Lett. *2007*, *48* (39), 6873-6; 2-phthalimidoalcohols from α-bromoketones via reduction and enzymatic resolution of their acetates s. P. Gupta, B.A. Shah, R. Parshad, G.N. Qazi. S.C. Taneja, Green Chem. *2007*, *9* (10), 1120-5.

Acetic acid s. under HBr *AcOH*

2-(Diethylamino)ethanethiol *$Et_2NCH_2CH_2SH$*
O-Demethylation OMe → OH
cf. *23*, 13; odorless O-demethylation of phenolethers with 2-(diethylamino)ethanethiol·HCl/KOBu-*t* s. J. Magano, M.H. Chen, J.D. Clark, T. Nussbaumer, J. Org. Chem. *2006*, *71* (18), 7103-5.

Dimethyl sulfoxide-sodium iodide *DMSO-NaI*
O-Deallylation OC-C=C → OH
of ethers cf. *58*, 7; with a little DMSO-NaI (cf. *49*, 15s*70*) s. M. Nagaraju, A. Krishnaiah, H.B. Mereyala, Synth. Commun. *2007*, *37* (15), 2467-72; with low-valent titanium [$Ti(OPr\text{-}i)_4$/Me_3SiCl/Mg or $Ti(OPr\text{-}i)_4$/$MgBr_2$/Mg,], also O-depropargylation (cf. *56*, 21), incl. preferential cleavage in the presence of 2 eqs. ethyl acetate, s. M. Ohkubo, S. Mochizuki, T. Sano, Y. Kawaguchi, S. Okamoto, Org. Lett. *2007*, *9* (5), 773-6.

4,4′-Bis(dichloroiodo)biphenyl/tetraethylammonium bromide *$Cl_2IAr\text{-}ArICl_2/Et_4NBr$*
Selective O-debenzylation OBn → OH
cf. *49*, 8; of sugars with 4,4′-bis(dichloroiodo)biphenyl/Et_4NBr at room temp., leaving acetates, benzoates, and glycosidic linkages unaffected s. V.N. Telvekar, Synth. Commun. *2007*, *37* (16), 2647-50.

N,N′-Dibromo-N,N′-1,2-ethanediylbis(benzenesulfonamide) ←
O-Detetrahydropyran-2-ylation OThp → OH
s. *48*, 120s*72*

Triethylsilane s. under Pd-C *Et_3SiH*

Silica s. under Silica-sulfuric acid SiO_2

Titanium tetraisopropoxide/trimethylsilyl chloride (or magnesium bromide)/magnesium *Ti(OPr-*i*)$_4$/Me$_3$SiCl (or MgBr$_2$)/Mg*

O-De-allylation or -propargylation OC-C=C or OC-C≡C → OH

s. *58*, 7s*72*, *56*, 21s*72*

Trimethylsilyl bromide/microwaves *Me$_3$SiBr/*[\\\\]

O-Dealkylation of phosphorus(V) esters P-OR → P-OH

cf. *33*, 21s*38*; of phosphoric, phosphonic or phosphoromonoamidic acid esters with Me_3SiBr under microwave irradiation, scope and limitations, s. G.D.K. Kumar, D. Saenz, G.L. Lokesh, A. Natarajan, Tetrahedron Lett. *2006*, *47* (35), 6281-4.

Diethyl phosphorazidate/triethylamine/potassium hydroxide/glycerol ←

Cleavage of α-(arylhetero)carboxylic acids $ArXC\text{-}CO_2H \rightarrow ArXH$

8. 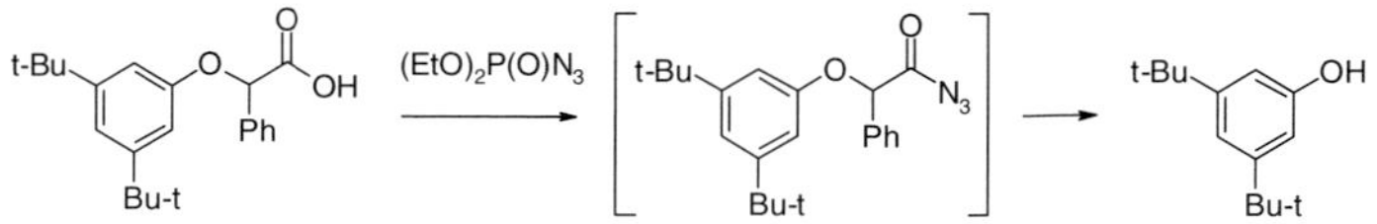

Phenols. 1.1 Eqs. diethyl phosphorazidate and 1.01 eqs. triethylamine added to a stirred soln. of the startg. carboxylic acid in dry toluene/DMF (10:1), the mixture stirred for 30 min, refluxed for 3 h, the intermediate acyl azide treated with 50% aq. KOH, ethanol and glycerol, the mixture refluxed for 2 h, cooled, acidified with citric acid (pH 5), extracted, and evaporated → product. Y 99%. The method is also successful for the preparation of **arylmercaptans** (isolated as the disulfides) **and arylselenols** from the corresponding α-functionalized carboxylic acids in yields of 65-93%. Nitrogen analogs, however, gave low yields in the case of aniline derivs. but moderate yields with more lipophilic substrates, such as *N*-carboxymethylcarbazole (66%) and an *N*-carboxymethylbenzamide (46%). Reaction proceeds **via Curtius rearrangement** of an intermediate acid azide, diethyl phosphorazidate being the most effective azide-transfer agent. The addition of glycerol during the hydrolytic work-up also improves yields. F.e.s. A. Spurg, S.R. Waldvogel, Synlett *2007* (19), 3042-4.

4-(Diphenylphosphino)benzoic acid s. under Pd(OAc)$_2$ ←

Vanadyl triflate *VO(OTf)$_2$*

Lewis acid-catalyzed cleavage of acid-labile O-protective groups ←

9. 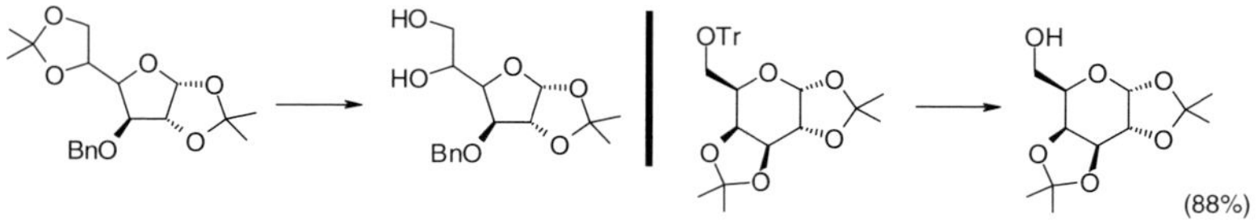

Regiospecific cleavage of carbohydrate O,O-isopropylidene groups. 20 Mol% $VO(OTf)_2$ added at room temp. to a soln. of the startg. protected carbohydrate in 3:5 methanol/methylene chloride, stirred at 55° for 17 h, cooled to room temp., ice-cold water and methylene chloride added, and the organic phase worked up with column chromatographic clean-up → product. Y 78%. The environmentally friendly catalyst is readily recoverable from the aqueous phase and retained its activity for subsequent cycles. The procedure is also suitable for the **cleavage of O,O-benzylidene groups, trityl ethers, and *tert*-butyldimethylsilyl ethers**, leaving a number of functions, O-, S- and N-glycoside linkages, and O-protective groups (notably BnO, BzO, LevO, OAc) unaffected. **Selective O-detritylation** is also possible with retention of O,O-isopropylidene groups, and O,O-benzylidene derivs. possessing electron-donating groups at the *para*-position can be cleaved under the same conditions. F.e.s. M.-C. Yan, Y.-N. Chen, H.-T. Wu, C.-C. Lin, C.-T. Chen, C.-C. Lin, J. Org. Chem. *2007*, *72* (1), 299-302.

Nafion-H ←
Silica-sulfuric acid *SiO_2-HSO_4*
O-Detritylation OTr → OH
cf. *23*, 15; rapid, mild 5′-O-de-tritylation, -monomethoxytritylation and -dimethoxytritylation of nucleosides using catalytic amounts of silica-sulfuric acid in acetonitrile at room temp. with retention of acid-sensitive hydroxyl protective groups such as *p*-methoxybenzyl, isopropylidene, cyclohexylidene, di(*p*-anisyl)methylidene, triisopropylsilyl and *tert*-butyldimethylsilyl s. A. Khalafi-Nezhad, A. Parhami, M.N.S. Rad, M.A. Zolfigol, A. Zare, Tetrahedron Lett. *2007*, *48* (30), 5219-22; selective cleavage of trityl ethers and terminal O,O-isopropylidene derivs. with Nafion-H in methanol at room temp. s. G.K. Rawal, S. Rani, A. Kumar, Y.D. Vankar, Tetrahedron Lett. *2006*, *47* (51), 9117-20.

p-*Toluenesulfonic acid/microwaves* *TsOH/[\\\\]*
Dodecatungstophosphoric acid *$H_3PW_{12}O_{40}$*
Aldehydes from acylals s. *51*, 55s*72*; *56*, 63s*72* $CH(OAc)_2$ → CHO

Heteropolyacid salts *$H_{14}[NaP_5W_{30}O_{110}]$*
Cleavage of tetrahydropyran-2-yl ethers s. *48*, 120s*72* OThp → OH

Hydrogen bromide/acetic acid *HBr/AcOH*
Cleavage of ethers OR → OH
with HHal/phenol cf. *28*, 15; cyclopentyl as protective group for phenols, O-decyclopentylation with 48% HBr in AcOH, s. J.M. Gajera, L.A. Gharat, A.V. Farande, Synth. Commun. *2007*, *37* (17), 2877-80.

Ammonium chloride s. under In *NH_4Cl*
Tetraethylammonium bromide s. under 4,4′-Bis(dichloroiodo)biphenyl *Et_4NBr*

Palladium-carbon/triethylsilane *Pd-C/Et_3SiH*
O-De-allylation or -benzylation ←
under mild, neutral conditions s. *28*, 45s*72*

Palladium(II) acetate/4-(diphenylphosphino)benzoic acid/vinylboronic anhydride-pyridine complex ←
Palladium-catalyzed O-deallylation with an improved work-up OC-C=C → OH

10.

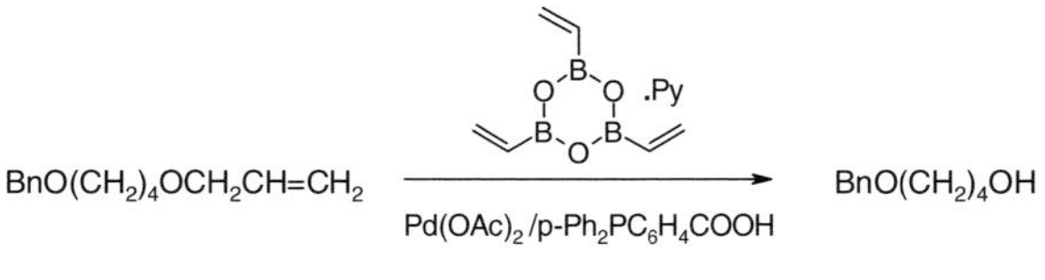

A chromatography-free, non-aqueous work-up has been devised for Pd-catalyzed O-deallylation, wherein both catalyst and non-volatile by-products can be easily removed with a polymer-supported diethanolamine. **E:** 2.3 µmol $Pd(OAc)_2$ in dry THF added under argon to a test-tube containing 0.226 mmol of the startg. allyl ether, 0.151 mmol vinylboronic anhydride-pyridine complex (as allyl scavenger) and 6.9 µmol 4-(diphenylphosphino)benzoic acid, the tube sealed with a screw cap, agitated at 80° for 12 h, cooled to room temp., 0.45 mmol polymer-supported diethanolamine and THF added, agitated again at room temp. for 12 h, filtered, the filtrate concentrated *in vacuo*, and the residue subjected again to the same purification → product. Y 96% (95.4% reduction in palladium). The procedure is extremely clean: volatile by-products (1,4-pentadiene, 1,4-butadiene and pyridine) are simply driven off, and the palladium catalyst, the *acidic* ligand (or its P-oxide) and the allylated boronic by-product are captured by the polymer-based aminodiol. F.e. and procedure s. H. Tsukamoto, T. Suzuki, M. Sato, Y. Kondo, Tetrahedron Lett. *2007, 48* (48), 8438-41.

Palladium(II) chloride *$PdCl_2$*
Cleavage of diphenylmethyl ethers s. *72*, 108 $OCHPh_2$ → OH

Formation of H-N Bond

Uptake ⇓

Addition to Nitrogen-Nitrogen Bonds HN ⇓ NN

Sodium dithionite $Na_2S_2O_4$
Reduction of nitrogen-nitrogen double bonds N=N → NHNH
with hydrazine cf. *9*, 41s*67*; with tin/hydrazine hydrate, ar. hydrazines, s. B. Dinesh, K. Abiraj, G.R. Srinivasa, D.C. Gowda, J. Chem. Res. *2006* (7), 422-3; with Na-dithionite, N,N′-diarylhydrazines, s. L.K. Sydnes, S. Elmi, P. Heggen, B. Holmelid, D. Malthe-Sørensen, Synlett *2007* (11), 1695-8.

Exchange ↓↑

Oxygen ↑ HN ↓↑ O

Microwave irradiation s. under Mo(CO)$_6$ [\\\\\]
Gold nanoparticles/zirconium dioxide Au/ZrO_2
Ar. amines from nitro compds. with retention of ar. chlorine $NO_2 \rightarrow NH_2$

11. 

p-Chloronitrobenzene (2 mmol) and 25 mg Au-nanoparticles-on-zirconia (average particle size ca. 4 nm) in ethanol (4 ml) heated at 150° under 1 MPa H_2 for 5 h → *p*-chloroaniline. Conversion 100%; selectivity 99.4%. The procedure is exceptionally clean, and the pre-reduced catalyst may be recycled four times with the same selectivity. Solvent, temperature and pressure effects were studied. F.e. incl. reduction of *o*-chloronitrobenzene and 2,5-dichloronitrobenzene s. D. He, H. Shi, Y. Wu, B.-Q. Xu, Green Chem. *2007*, *9* (8), 849-51.

Aluminum amalgam/lithium tetrahydridoaluminate or palladium-carbon ←
Sodium tetrahydridoborate/charcoal $NaBH_4/C$
Sodium tetrahydridoborate/antimony trifluoride $NaBH_4/SbF_3$
Samarium diiodide/water/tert. amines $SmI_2/H_2O/R_3N$
Amines from nitro compds.
cf. *11*, 67s*67*; rapid reduction of nitroalkanes with SmI_2/water/tert. amine, also of nitroalkenes, s. T. Ankner, G. Hilmersson, Tetrahedron Lett. *2007*, *48* (32), 5707-10; ar. amines with $NaBH_4$/charcoal in aq. THF at 50-60° s. B. Zeynizadeh, D. Setamdideh, Synth. Commun. *2006*, *36* (18), 2699-704; with $NaBH_4/SbF_3$ in wet acetonitrile s. B. Zeynizadeh, H. Ghasemi, J. Chem. Res. *2006* (8), 542-4; 2-step conversion of 1,2-nitramines to 1,2-diamines via reduction to the corresponding hydroxylamines with Al-amalgam then N-dehydroxylation with Pd-C/H_2 or $LiAlH_4$ (cf. *12*, 55) without loss of diastereoselectivity s. J.C. Anderson, H.A. Chapman, Synthesis *2006* (19), 3309-15; ar. amines with Fe-powder in AcOH/ethanol under ultrasonication, improved selectivity over use of $SnCl_2$, s. A.B. Gamble, J. Garner, C.P. Gordon, S.M.J. O′Conner, P.A. Keller, Synth. Commun. *2007*, *37* (16), 2777-86; with 3 eqs. each sulfur and $NaHCO_3$ in DMF at 130° s. M.A. McLaughlin, D.M. Barnes, Tetrahedron Lett. *2006*, *47* (51), 9095-7; with $Mo(CO)_6$/DBU (3 eqs.)/EtOH under microwave irradiation s. J. Spencer, N. Anjum, H. Patel, R.P. Rathnam, J. Verma, Synlett *2007* (16), 2557-9; with a nickel-iron mixed oxide prepared from a nickel-iron hydrotalcite precursor, chemoselectivity, s. Q. Shi, R. Lu, L. Lu, X. Fu, D. Zhao, Adv. Synth. Catal. *2007*, *349* (11-12), 1877-81; with a nanostructured, water-soluble, poly(diallyldimethylammonium chloride)-supported, platinum carbonyl cluster $\{[Pt_{30}(CO)_{60}]^{2-}\}$ derived catalyst, also hydrogenation of aldehydes and alkenes, s. P. Maity, S. Basu, S. Bhaduri, G.K.

Lahiri, ibid. 1955-62; with (acetoacetonato)bis(2-carboxyphenyl)diphenylphosphine)palladium(II) fluoroborate, also hydrogenation of aldehydes (cf. *23*, 51), alkenes and alkynes, s. P.U. Rani, P.M. Reddy, M. Sarangapani, V. Ravinder, J. Indian Chem. Soc. *2007*, *84* (2), 122-9.

Borane-tetrahydrofuran/chiral spiroborates ←
Prim. amines from alkoximes – Asym. reduction C=NOR → $CHNH_2$
cf. *25*, 15s*41-3*; of O-benzyloxime ethers with BH_3-THF and 10 mol% chiral spiroborates derived from 2-aminoalcohols (especially diphenylvalinol) and ethylene glycol s. X. Huang, M. Ortiz-Marciales, K. Huang, V. Stepanenko, F.G. Merced, A.M. Ayala, W. Correa, M. De Jesús, Org. Lett. *2007*, *9* (9), 1793-5.

Triethylsilane s. under Pd-C *Et_3SiH*
Zirconium dioxide s. under Au *ZrO_2*
Antimony trifluoride s. under $NaBH_4$ *SbF_3*

Sulfur/sodium hydrogen carbonate *$S/NaHCO_3$*
Molybdenum hexacarbonyl/1,8-diazabicyclo[5.4.0]undec-7-ene/microwaves ←
Iron/acetic acid *Fe/AcOH*
Nickel-iron oxide ←
Amines from nitro compds. s. *11*, 67s*72* NO_2 → NH_2

Palladium-carbon/triethylsilane *$Pd\text{-}C/Et_3SiH$*
Amines from nitro compds. under mild, neutral conditions s. *28*, 45s*72*

(Acetoacetonato)bis[(2-carboxyphenyl)diphenylphosphine]palladium(II) fluoroborate ←
Nanostructured, water-soluble, poly(diallyldimethylammonium chloride)-supported platinum carbonyl cluster complex ←
Via intermediates *v.i.*
Amines from nitro compds. s. *11*, 67s*72*

Nitrogen ↑ HN ↓↑ N

Triphenylphosphine *Ph_3P*
Amines from azides N_3 → NH_2
cf. *12*, 42s*68*; preferential Staudinger reduction of equatorial over axial azides in glycopyranoside and mannitol derivs. with Ph_3P, and N-acylation with Ac_2O/py (cf. *71*, 160) s. Q. Li, Q.-H. Fan, L.-H. Zhang, X.-S. Ye, Synlett *2006* (15), 2464-8.

Palladium-carbon/ammonium formate *$Pd\text{-}C/HCOONH_4$*
Ar. amines from hydrazines s. *33*, 30s*72* $NHNR_2$ → NH_2

Palladium-carbon/triethylsilane *$Pd\text{-}C/Et_3SiH$*
Amines from azides under mild, neutral conditions s. *28*, 45s*72* N_3 → NH_2

Sulfur ↑ HN ↓↑ S

Barium carbonate/hydrogen chloride *$BaCO_3/HCl$*
1,2-Diamines from 1,2,5-thiadiazolidine 1,1-dioxides s. *72*, 173

1,1',3,3'-Bis(trimethylene)bis(imidazol-2-ylidene) ←
Organocatalyzed N-detosylation NTs → NH
with a super-electron donor under mild, neutral conditions s. *72*, 53

Carbon ↑ HN ↓↑ C

tert-*Butylamine or Methanolic ammonia* t-*$BuNH_2$ or NH_3*/MeOH
N-Decarbalkoxylation NCOOR → NH
with NH_3 cf. *26*, 24s*32*; with *t*-$BuNH_2$ in methanol s. O.R. Suárez-Castillo, L.A. Montiel-Ortega, M. Meléndez-Rodríguez, M. Sánchez-Zavala, Tetrahedron Lett. *2007*, *48* (1), 17-20; N-protective group removal with methanolic ammonia s. *72*, 513.

Ammonium cerium(IV) nitrate *CAN*
Periodic acid (or Trichloroisocyanuric acid)/sulfuric acid *H_5IO_6 (or TCCA)/H_2SO_4*
Oxidative removal of N-protective groups NR → NH
with CAN cf. *38*, 17(s*39*); N-de-*p*-methoxyphenylation with periodic acid (1 eq.) or trichloroisocyanuric acid (0.5 eq.) and 1 eq. H_2SO_4 in acetonitrile/water, also with simultaneous alcohol oxidation via addition of a little Na-dichromate, β-amino acids from Mannich adducts, s. J.M.M. Verkade, L.J.C. van Hemert, P.J.L.M. Quaedflieg, P.L. Alsters, F.L. van Delft, F.P.J.T. Rutjes, Tetrahedron Lett. *2006*, *47* (46), 8109-13; N-de(3,5-dimethyl-4-methoxy)phenylation with CAN s. under *46*, 47s*72*.

Rhodium trichloride/n-propanol *$RhCl_3$/PrOH*
N-Deallylation of acyclic N-subst. carboxylic acid N-allylamides $NCH_2CH{=}CH_2$ → NH

12.

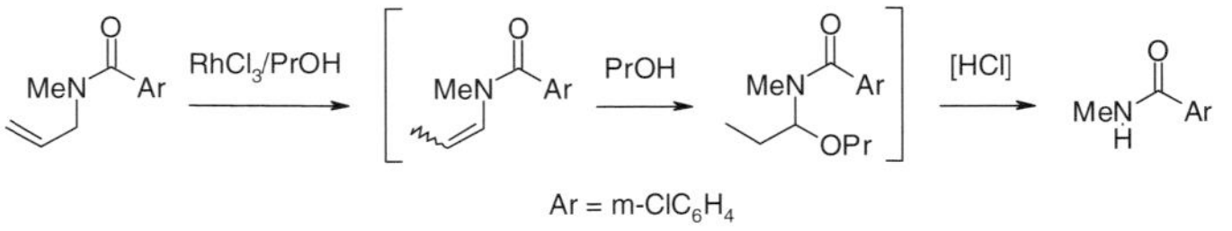

under mild conditions. A 0.5 *M* soln. of the startg. protected amide in anhydrous *n*-propanol degassed via N_2 sparging, 5 mol% $RhCl_3$-hydrate added under a positive pressure of N_2, heated at 97° for 4-6 h until TLC or HPLC indicated completion of reaction, cooled to room temp., diluted with water, and worked up with column chromatographic purification → product. Y 92%. This simple and convenient one-step procedure is generally applicable to aromatic or aliphatic tert. N-allylamides, and is tolerant of a wide range of steric and electronic environments, notably leaving ar. nitro compds. and bromides, as well as carbobenzoxy and piperonyloyl acetal groups unaffected. For large-scale conversions the catalyst loading can be reduced to 2 mol% over a longer period. Reaction is presumed to involve initial formation of a rhodium hydride species (from $RhCl_3$/PrOH) which catalyzes the isomerization to the enamide; this is followed by addition of a molecule of the alcohol to give the corresponding N,O-acetal prior to cleavage with HCl liberated during the conversion. F.e. and comparison of catalysts and alcoholic media s. M.J. Zacuto, F. Xu, J. Org. Chem. *2007, 72* (16), 6298-300.

Elimination ⇑

Oxygen ↑ HN ⇑ O

Lithium tetrahydridoaluminate or Palladium-carbon *$LiAlH_4$ or Pd-C*
Amines from hydroxylamines >NOH → >NH
1,2-Diamines with retention of diastereoselectivity s. *12*, 55s*72*

Formation of H-Rem Bond

Exchange ↓↑

Carbon ↑ HRem ↓↑ C

Diethyl phosphorazidate/triethylamine/potassium hydroxide/glycerol ←
Cleavage of α-(arylseleno)carboxylic acids s. *72*, 8 $ArSeC\text{-}CO_2H$ → ArSeH

Formation of H-C Bond

Uptake ⇓

Addition to Oxygen and Carbon HC ⇓ OC

Irradiation s. under (1-Benzyl-3-carbamoyl)pyridinium bromide ///
Potassium hydroxide s. under Iridium complexes *KOH*
Potassium tert-*butoxide s. under Ruthenium complexes* *KOBu*-t
Chiral calix[4]arene-based 2-aminoalcohols s. under Chlororuthenium complex ←

Chiral copper(II) bis(Δ^2-oxazoline) complexes/diisobutyl 2,6-dimethyl-1,4-dihydropyridine-3,5-dicarboxylate ←
α-Hydroxy- from α-keto-carboxylic acid esters CO → CHOH
Biomimetic asym. transfer-hydrogenation

13. 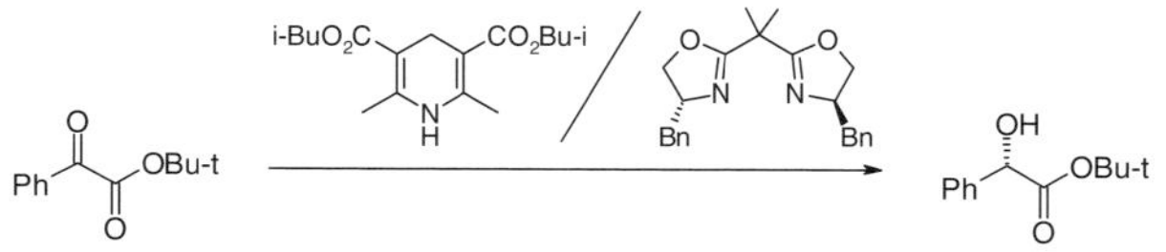

10 Mol% each of $Cu(OTf)_2$ and chiral bis(Δ^2-oxazoline) ligand dried *in vacuo* for 30 min in a Schlenk tube, distilled chloroform added, stirred for 1 h, *tert*-butyl benzoylformate added, cooled to -25°, treated with 1.4 eqs. diisobutyl 2,6-dimethyl-1,4-dihydropyridine-3,5-dicarboxylate, stirring continued for 48 h under argon, and worked up with column chromatographic purification → *tert*-butyl (S)-mandelate. Y 89% (e.r. >99:1). Reaction is applicable to aromatic and aliphatic substrates (the latter with slightly lower enantioselectivity), *tert*-butyl esters being more effective than methyl or benzyl. Of several Hantzsch esters tested, enantioselectivity was highest with the diisobutyl ester. The C_2-symmetric chiral copper(II) complex serves as an alcohol dehydrogenase mimic, while the Hantzsch ester serves as a synthetic NADH analog. F.e. and comparison of chiral ligands s. J.W. Yang, B. List, Org. Lett. *2007, 8* (24), 5653-5.

Borane-tetrahydrofuran s. under $CoCl_2$ *BH_3-THF*

Borane-dimethyl sulfide/(5S)-1-aza-2-imino-3-oxa-4,4-diphenylbicyclo[3.3.0]octane ←
Asym. reduction of ketones
with chiral 2-aminoalcohols as catalyst cf. *33*, 43s*69-71*; with (5*S*)-1-aza-2-imino-3-oxa-4,4-diphenylbicyclo[3.3.0]octane, incl. reduction of α-haloketones, s. D. Basavaiah, K.V. Rao, B.S. Reddy, Tetrahedron:Asym. *2007*, *18* (8), 963-7.

Decaborane *$B_{10}H_{14}$*
Prim. alcohols from aldehydes CHO → CH_2OH
with diborane cf. *20*, 51; with decaborane in aq. soln. s. S.H. Lee, M.H. Nam, M.Y. Cho, B.W. Yoo, H.J. Rhee, C.M. Yoon, Synth. Commun. *2006*, *36* (17), 2469-74.

Sodium tetrahydridoborate/(R,R)-2-phenyl-1,3,2-dioxaborolane-4,5-dicarboxylic acid ←
Sec. alcohols from ketones CO → CHOH
Asym. reduction under mild conditions

14. 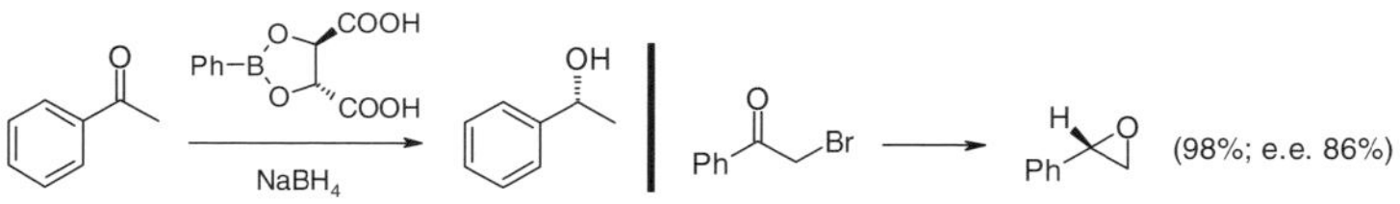

(R,R)-2-Phenyl-1,3,2-dioxaborolane-4,5-dicarboxylic acid (or the (S,S)-enantiomer), based on inexpensive tartaric acid, affords higher enantioselectivities than classical chiral auxiliaries for the asym. reduction of ketones with $NaBH_4$, proceeding smoothly *in air without requiring rigorous drying* of the components. **E:** A soln. of acetophenone and 2 eqs. 0.5 *M* (R,R)-2-phenyl-1,3,2-dioxaborolane-4,5-dicarboxylic acid in THF stirred for 15 min, 2 eqs. $NaBH_4$ added, stirring continued for 1 h, quenched by dropwise addition of 1 *M* HCl (*caution*: H_2 evolution), basified to pH 12 with 3 *M* NaOH, stirred for 30 min, and worked up → (R)-product. Y 83% (e.e. 99%). The procedure, which is adaptable on a large scale, is applicable to both aromatic and aliphatic ketones, *tert*-alkyl ketones giving excellent results and *sec*- and *prim*-alkyl ketones giving good to moderate results. Enantioselectivity was even higher with the *p*-nitrophenylboronate analog, but the phenyl deriv. is cheaper and can be easily recycled. F.e., **also epoxides from α-halogenoketones by asym. reduction,** s. S. Eagon, J. Kim, K. Yan, D. Haddenham, B. Singaram, Tetrahedron Lett. *2007, 48* (51), 9025-9.

Sodium tetrahydridoborate/acids *$NaBH_4$/acids*
Alcohols from oxo compds. CO → CHOH
cf. *5*, 40; with $NaBH_4$ activated by solid acids such as boric acid, benzoic acid or *p*-TsOH·H_2O in the absence of solvent at room temp. s. B.T. Cho, S.K. Kang, M.S. Kim, S.R. Ryu, D.K. An, Tetrahedron *2006*, *62* (34), 8164-8.

Aluminum oxyhydroxide s. under Rh and Pd-Fe-oxide *Al(O)OH*

Zwitterionic B,P-dihydrophosphonioborates ←
Prim. alcohols from aldehydes s. *72*, 24 CHO → CH_2OH

Isopropanol s. under Ni, Ruthenium complexes and Iridium complexes i-*PrOH*
1,4-Butanediol s. under Ruthenium complexes *$HO(CH_2)_4OH$*
Chiral N-formylpipecolinic acid amides s. under Trichlorosilane ←

Enzymes ←
Asym. reduction of ketones CO → CHOH
with ketoreductases cf. *41*, 38s*70*; of subst. benzophenones and benzoylpyridines s. M.D. Truppo, D. Pollard, P. Devine, Org. Lett. *2007*, *9* (2), 335-8; with W110A secondary alcohol dehydrogenase from *Thermoanaerobacter ethanolicus* (W110A TESADH) in Tris buffer/isopropanol, also (R)- from (S)-alcohols, s. M.M. Musa, K.I. Ziegelmann-Fjeld, C. Vieille, J.G. Zeikus, R.S. Phillips, J. Org. Chem. *2007*, *72* (1), 30-4; of hydrophobic ketones in organic solvents with the Xerogel-encapsulated enzyme s. idem., Angew. Chem. Int. Ed. Engl. *2007*, *46* (17), 3091-4; with 'designer cells' at high substrate concentrations, functionalized alcohols, s. H. Gröger, F. Chamouleau, N. Orologas, C. Rollmann, K. Drauz, W. Hummel, A. Weckbecker, O. May, ibid. *2006*, *45* (34), 5677-81; using whole cell recombinant *Escherichia coli* with intracellular cofactor regeneration in biphasic ionic liquid/water systems s. S. Bräutigam, S. Bringer-Meyer, D. Weuster-Botz, Tetrahedron:Asym. *2007*, *18* (16), 1883-7.

Alcohol dehydrogenase ←
2-Subst. 2-arylalcohols from α-subst. α-arylaldehydes CHO → CH_2OH
Asym. enzymatic reduction with dynamic kinetic resolution s. *72*, 21

One-pot Heck arylation-asym. enzymatic reduction s. *72*, 571 CO → CHOH

Reductase ←
β-Hydroxynitriles from α-cyanoketones
Asym. enzymatic reduction

15. 

α-Ethylation - an undesirable side reaction on asym. reduction of α-cyanoketones with whole cell biocatalysts - is completely eliminated by using an isolated carbonyl reductase instead. **E:** A soln. of 3-oxo-3-phenylpropanenitrile (170 mg) in DMSO added to a mixture of D-glucose (400 mg), D-glucose dehydrogenase (10 mg), NADPH (10 mg) and the recombinant carbonyl reductase from *Candida magnoliae* (10 mg) in a potassium phosphate buffer (pH 7), and stirred at room temp. for 60 h → (R)-3-hydroxy-3-phenylpropanenitrile. Y 85% (e.e. 98%). No ethanol is introduced or produced during the reaction so that α-ethylation is not possible. Reaction is applicable to aromatic derivs. possessing electron-withdrawing or -donating groups (on the 1 mmol scale) with high enantioselectivity (97-99%), and the products can be converted to the corresponding **chiral β-hydroxycarboxylic acids** by enzymatic hydrolysis with the nitrilase from *Synechocystis* sp. strain PCC 6803 (NIT6803) *at pH 7.2* or from a *Bradyrhizobium japonicum* strain. F.e.s. D. Zhu, H. Ankati, C. Mukherjee, Y. Yang, E.R. Biehl, L. Hua, Org. Lett. *2007, 9* (13), 2561-3.

*Chiral a-(carbo-*tert-*butoxyamino)-carboxylic acid thioamides or -hydroxamic acids* ←
s. under Chlororuthenium complex

Methyldiphenylsilyllithium *$MePh_2SiLi$*
β-Hydroxyketones from α,β-epoxyketones ▽O → C(OH)CH
Regiospecific reductive ring opening

16.

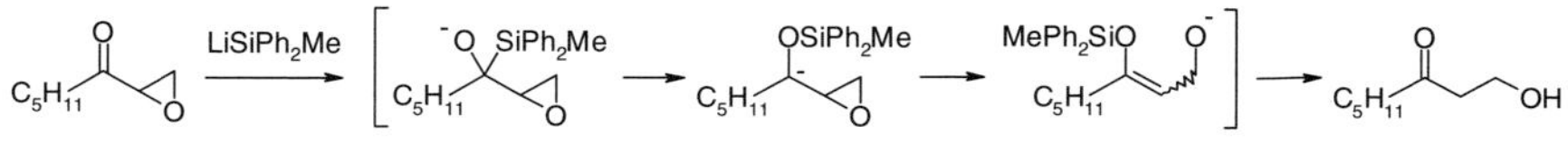

via Brook rearrangement. A soln. of 2.5-3 eqs. ca. 1 *M* methyldiphenylsilyllithium in dry THF added dropwise with stirring to a soln. of the startg. epoxyketone in the same solvent at -40° under argon, and the mixture quenched immediately with satd. aq. NH_4Cl → product. Y 89%. Reaction is presumed to take place by initial addition of the silyllithium reagent to the carbonyl group, followed by Brook rearrangement and conversion to the 3-oxidoenoxysilane; a second equivalent of the silyllithium reagent then desilylates the latter, and aqueous acidic work-up generates the hydroxyketone. The procedure is applicable to both aromatic and aliphatic substrates, but other silyllithium compds. ($Me_2PhSiLi$ and Ph_3SiLi) were less effective. F.e.s. S.C. Reynolds, S.E. Wengryniuk, A.M. Hartel, Tetrahedron Lett. *2007, 48* (38), 6751-3.

Diphenylsilane/chiral rhodium complexes *Ph_2SiH_2/[Rh]*
Asym. hydrosilylation of ketones CO → CHOH
cf. *49*, 42s*71*; using Ph_2SiH_2/[Rh(cod)Cl]$_2$/3-diphenylphosphino-2-(4(S)-*tert*-butyl-Δ^2-oxazolin-2-yl)thiophene [HETPHOX] s. A.G. Coyne, P.J. Guiry, Tetrahedron Lett. *2007, 48* (5), 747-50.

Trichlorosilane/chiral N-formylpipecolinic acid amides ←
Asym. reduction of ketones and ketimines
via organocatalyzed hydrosilylation

17.

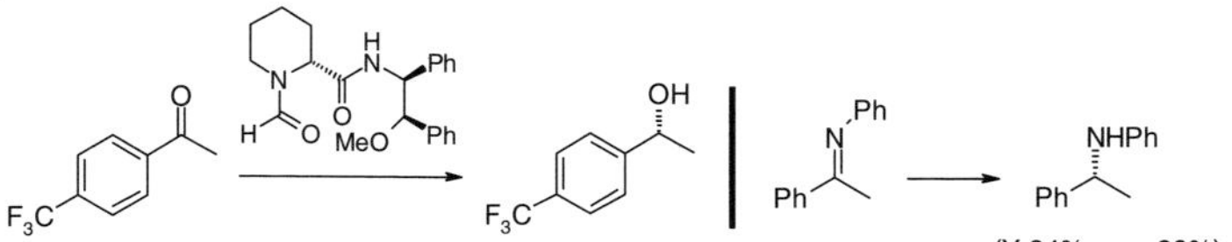

(Y 94%; e.e. 90%)

A chiral N-formylpipecolinic acid amide has evolved as organocatalyst with superior enantioselectivity for the asym. reduction of both aromatic and aliphatic ketones and ketimines via hydrosilylation. **E:** 2 Eqs. trichlorosilane added dropwise under argon to a stirred soln. of the startg. ketone and 10 mol% (2S,1′S,2′R)-1-formylpiperidine-2-carboxylic acid (2-methoxy-1,2-diphenylethyl)amide in dry toluene at -20°, stirring continued for 16 h at the same temp., and quenched with satd. aq. $NaHCO_3$ → (R)-product. Y 92%; e.e. 92% (after column chromatography). High enantioselectivities were recorded for the asym. reduction of aryl ketones possessing electron-donating or -withdrawing substituents, and enantioselectivity was moderate to good for the asym. reduction of aliphatic ketones. The procedure is also suitable for the asym. reduction of both aromatic and aliphatic ketimines, making this the most generally applicable system available for such functional group conversions. Mechanistically, the chiral formamide is thought to function as a tridentate activator through a heptacoordinate silicon transition state. F.e. and comparison of chiral formamides s. L. Zhou, Z. Wang, S. Wei, J. Sun, Chem. Commun. *2007* (28), 2977-9; asym. reduction of aryl ketones with N-formyl-α′-(2,4,6-triethylphenyl)-L-proline as organocatalyst cf. Y. Matsumura, K. Ogura, Y. Kouchi, F. Iwasaki, O. Onomura, Org. Lett. *2006, 8* (17), 3789-92.

Zwitterionic B,P-dihydrophosphonioborates s. under Borates ←
Triphenylphosphine s. under $RuCl_3$ Ph_3P

(1-Benzyl-3-carbamoyl)pyridinium bromide/sodium dithionite/sodium carbonate/irradiation ←
β-Hydroxyketones from α,β-epoxyketones → C(OH)CH
with Hantzsch ester under irradiation cf. *65*, 14; with (1-benzyl-3-carbamoyl)pyridinium bromide/Na-dithionite/Na_2CO_3 s. H.-J. Xu, Y.-C. Liu, Y. Fu, Y.-D. Wu, Org. Lett. *2006*, *8* (16), 3449-51.

Dicarbonyl(hydrido)iron oxycyclopentadienyl complex ←
Sec. alcohols from ketones CO → CHOH
cf. *23*, 51s*66*; with a dicarbonyl(hydrido)iron oxycyclopentadienyl complex under 3 atm. H_2 in toluene or ether at room temp., also transfer-hydrogenation in isopropanol, s. C.P. Casey, H. Guan, J. Am. Chem. Soc. *2007*, *129* (18), 5816-7.

Cobalt(II) chloride/borane-tetrahydrofuran $CoCl_2/BH_3$-*THF*
Reductive ring opening of cyclic acetals
cf. *15*, 38; of 4,6-*O*-benzylidene derivs. of hexopyranosides with $CoCl_2/BH_3$-THF, regiospecificity, s. S. Tani, S. Sawadi, M. Kojima, S. Akai, K. Sato, Tetrahedron Lett. *2007*, *48* (17), 3103-4.

Nickel/hydrogen chloride/isopropanol *Ni/HCl/i-PrOH*
Transfer-hydrogenation of oxo compds. CO → CHOH
with Raney Ni/NH_4-formate cf. *47*, 35; of ketones with Raney Ni in refluxing isopropanol containing a trace of HCl s. R.C. Mebane, K.L. Holte, B.H. Gross, Synth. Commun. *2007*, *37* (16), 2787-91.

Zwitterionic (η⁶-arene)(1-diisopropylphosphino-2-dimethylaminoindenide)-ruthenium(II) complex/isopropanol/potassium tert-*butoxide* ←
Transfer-hydrogenation of ketones with zwitterionic ruthenium(II) complexes

18. 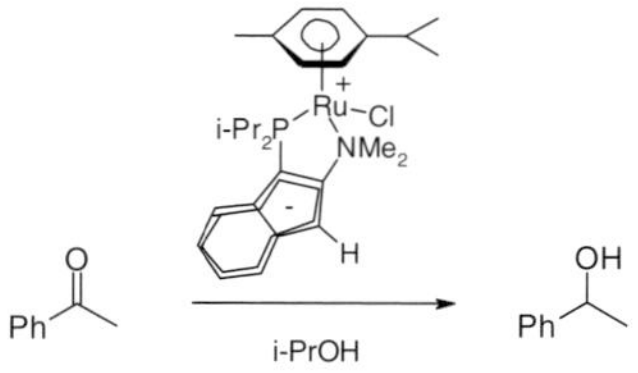

Zwitterionic (η^6-arene)(1-diisopropylphosphino-2-dimethylaminoindenide)ruthenium(II) complexes *lacking Noyori-type Ru-NH linkages* are among the most active and selective pre-catalysts for the transfer-hydrogenation of aromatic and aliphatic ketones, exhibiting consistently high turnover frequencies *up to 220,000* h^{-1} at catalyst loadings *as low as 0.05 mol%*. **E:** Benzophenone (5 mmol scale) and isopropanol added to a Schlenk flask containing a little zwitterionic (η^6-arene)(1-diisopropylphosphino-2-dimethylaminoindenide)ruthenium(II) complex, heated to reflux with stirring, 0.0025 *M* KOBu-*t* soln. (which had been heated to reflux) added [to give a catalyst/KOBu-*t*/ketone ratio of 1:50:10,000], and worked up after 3.5 h → product. Y 82%. The anionic nature of the P,N-ligand is thought responsible for the high activity of the catalyst, which in most instances leads to reduction of the ketone *in a matter of minutes*. F.e.s. R.J. Lundgren, M.A. Rankin, R. McDonald, G. Schatte, M. Stradiotto, Angew. Chem. Int. Ed. Engl. *2007, 46* (25), 4732-5.

*Carbonyl(dihydrido)tris(triphenylphosphine)ruthenium(II)/bis[*o*-(diphenylphosphino)phenyl]ether/1,4-butanediol/potassium* tert-*butoxide* ←

Alcohols from oxo compds. CO → CHOH

19. 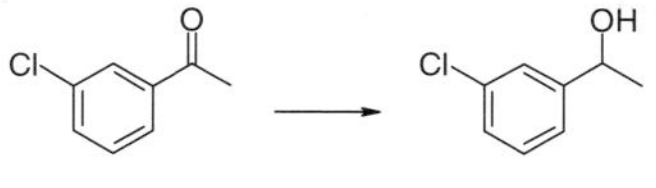

Improved transfer-hydrogenation. 3-Chloroacetophenone added to an oven-dried tube containing 2.5 mol% ruthenium(II) catalyst, 2.5 mol% di(phosphine) ligand, 5 mol% *t*-BuOK and *1 eq.* 1,4-butanediol in toluene, the mixture heated under reflux for 48 h, filtered through Celite and silica, washed with methylene chloride and concentrated *in vacuo* → 1-(3-chlorophenyl)ethanol. Y 79% (isolated yield). 1,4-Butanediol was the H-donor of choice, being significantly more effective than other 1,n-diols, and, since the equilibrium is driven by formation of the γ-butyrolactone, was not required in excess (unlike the familiar isopropanol). The method is successful for reduction of aralkyl and dialkyl ketones as well as ar. and aliphatic benzaldehydes in near quantitative yield. Isolated yields are somewhat reduced (61-92%) due to the volatility of products in some cases. However, the H-donor was not as effective as isopropanol for asym. transfer-hydrogenation using Noyori's catalyst. F.e.s. H.C. Maytum, B. Tavassoli, J.M.J. Williams, Org. Lett. *2007, 9* (21), 4387-9.

Ruthenium trichloride/triphenylphosphine $RuCl_3/Ph_3P$

1,3-Diols from β-hydroxyketones

Stereospecific reduction

with SmI_2/MeOH cf. *57*, 22(s*69*); with $RuCl_3/Ph_3P$ in methanol under 10 bar H_2 at 50°, chiral *anti*-1,3-diols, s. O. Labeeuw, C. Roche, P. Phansavath, J.-P. Genêt, Org. Lett. *2007*, *9* (1), 105-8.

*Chloro-ruthenium or -rhodium complex/chiral calix[4]arene-based 2-aminoalcohols or α-(carbo-*tert*-butoxyamino)-carboxylic acid thioamides or -hydroxamic acids* ←

Asym. transfer-hydrogenation of ketones

cf. *46*, 42s*49*; asym. transfer-hydrogenation of alkyl aryl ketones using $[Cp^*RhCl_2]_2$ or $[Ru(p\text{-cymene})Cl_2]_2$ and chiral α-Boc-aminocarboxylic acid thioamides in isopropanol with reversal of enantioselectivity compared to the corresponding amides s. A.B. Zaitsev, H. Adolfsson, Org. Lett. *2006*, *8* (22), 5129-32; with calix[4]arene-based 2-aminoalcohols/KOBu-*t* s. A. Quintard, U. Darbost, F. Vocanson, S. Pellet-Rostaing, M. Lemaire, Tetrahedron:Asym. *2007*, *18* (16), 1926-33; with $[Cp^*RhCl_2]_2$/chiral α-Boc-aminohydroxamic acids/NaOPr-*i*/LiCl s. K. Ahlford, A.B. Zaitsev, J. Ekström, H. Adolfsson, Synlett *2007* (16), 2541-4.

Dichloro[(S)-2,2′-bis(diphenylphosphino)-1,1′-binaphthyl]ruthenium(II)/triethylamine ←

Discrimination of β-ketocarboxylic acid esters by asym. homogeneous hydrogenation

20. 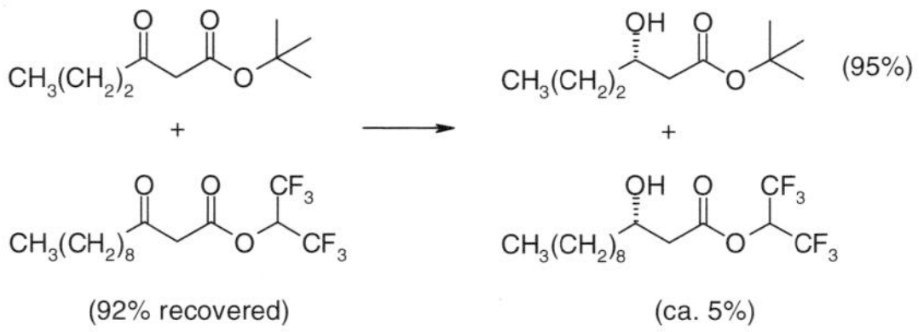

The classical Noyori asym. homogeneous hydrogenation of keto-esters has been applied to the discrimination of β-ketocarboxylic acid esters. **E:** A 1:1 mixture of *tert*-butyl 3-ketohexanoate and hexafluoroisopropyl 3-ketododecanoate in ethanol hydrogenated under 4 atm. H_2 at room temp. for 6 h in the presence of 0.25 mol% dichloro[(S)-2,2′-bis(diphenylphosphino)-1,1′-binaphthyl]ruthenium(II)-triethylamine complex → (S)-*tert*-butyl 3-hydroxyhexanoate (Y 95%; e.e. 97%) and (S)-hexafluoroisopropyl 3-hydroxydodecanoate (Y ca. 5% with 92% recovery of the corresponding ketoester). The rate of reduction depends on the electronic nature of the ester

O-alkyl group, being faster where the latter is more Lewis basic. F.e.s. R. Kramer, R. Brückner, Angew. Chem. Int. Ed. Engl. *2007, 46* (34), 6537-41.

Dichloro[2,2′-bis(di-3,5-xylylphosphino)-1,1′-binaphthyl]-((R,R)-1,2-diphenyl-1,2-ethylenediamine)ruthenium(II)/potassium tert-*butoxide* *[Ru(II)]/t-BuOK*

2-Subst. 2-arylalcohols from α-subst. α-arylaldehydes CHO → CH_2OH

Asym. homogeneous hydrogenation with dynamic kinetic resolution

21. 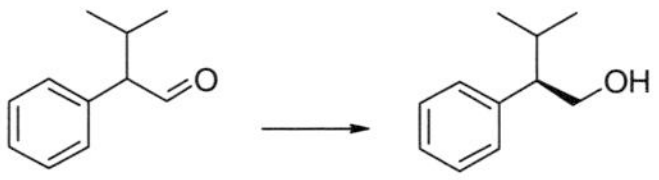

n-Hexanol (4.8 ml) added to *0.1 mol%* of the Noyori ruthenium catalyst derived from $RuCl_2$, (R)-xylyl-BINAP and (R,R)-DPEN in a glass vial purged with argon, stirred 5 min, then a soln. of KOBu-*t* (0.12 mmol) in *tert*-butanol added, followed by the startg. aldehyde (1 mmol), the vial transferred to a high-pressure autoclave, and the mixture stirred under H_2 (20 bar) at room temp. for 16-18 h → product. E.e. 98% (>99% conversion). Of a number of chiral di(phosphines) studied, (R)-xylyl-BINAP was found to give the highest enantioselectivity, which also improved with increasing steric bulk of the α-alkyl substituent. As anticipated, dynamic kinetic resolution is facilitated by rapid base-catalyzed racemization of the aldehyde prior to the asym. hydrogenation. F.e.s. X. Li, B. List, Chem. Commun. *2007* (17), 1739-41; with horse liver alcohol dehydrogenase in buffered aq. soln. s. D. Giacomini, P. Galletti, A. Quintavalla, G. Gucciardo, F. Paradisi, ibid. Chem. Commun. *2007* (39), 4038-40.

Chiral [7,7-bis(diarylphosphino)-1,1′-spirobiindane](1,2-diamine)(dichloro)-ruthenium(II)/potassium tert-*butoxide* ←

Cyclic *cis*-2-*tert*-aminoalcohols from α-*tert*-aminoketones CO → CHOH

Asym. homogeneous hydrogenation with dynamic kinetic resolution

22. 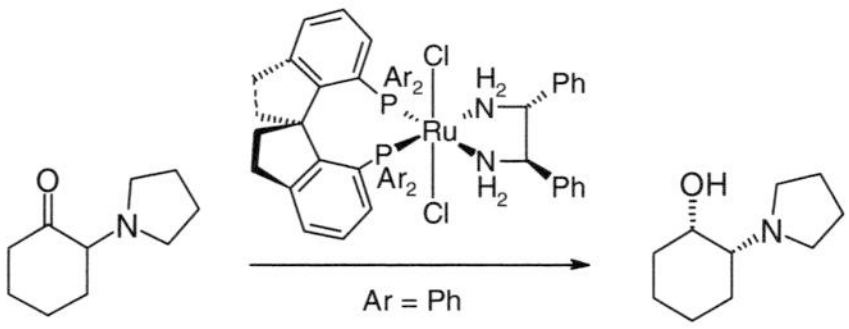

Chiral cyclic 2-*tert*-aminocyclohexanols. *0.1 Mol%* $RuCl_2$[(S)-SDP][(R,R)-DPEN] and isopropanol contained in a hydrogenation vessel purged three times with 10 atm. H_2 in an autoclave, the soln. stirred under 10 atm. H_2 for 5 min, the pressure released, startg. aminoketone and a soln. of KOBu-*t* in isopropanol added, the autoclave purged again with H_2 and pressurized to 10 atm., and stirring continued for 2 h → (1S,2R)-2-(pyrrolidin-1-yl)cyclohexanol. Y 90% (e.e. 99.8%; *cis/trans* >99:1). The procedure is highly enantioselective (e.e. 96-99.8%) in favour of the *cis*-product (>99:1) **at low catalyst loading**, the enantioselectivity and *cis*-selectivity in this example being exactly the same at a loading of *0.0033 mol%* (after 48 h). Reaction is applicable to 5- to 7-membered cycloalkanol and piperidol derivs., and is tolerant of various substitutents. F.e. and comparison of catalysts s. S. Liu, J.-H. Xie, L.-X. Wang, Q.-L. Zhou, Angew. Chem. Int. Ed. Engl. *2007, 46* (39), 7506-8.

Rhodium (or iridium) nanoparticles-in-aluminum oxyhydroxide *[Rh (or Ir)/AlO(OH)]*

Sec. alcohols from ketones

with benzene ring hydrogenation using Rh-Al_2O_3 cf. *26*, 52; hydrogenations of ketones or arenes using readily-prepared, recyclable nanoparticulate rhodium (or iridium) in aluminum oxyhydroxide [Rh/AlO(OH)] at room temp. under a hydrogen balloon s. I.S. Park, M.S. Kwon, K.Y. Kang, J.S. Lee, J. Park, Adv. Synth. Catal. *2007, 349* (11-12), 2039-47.

Rhodium complexes s. under Ph_2SiH_2 and Ruthenium complexes *[Rh]*

Palladium-ferric oxide nanoparticles-in-aluminum oxyhydroxide ←
Regiospecific hydrogenolysis of epoxides ▽O → C(OH)CH
over magnetically recoverable palladium nanoparticles

23. 

A mixture of epichlorhydrin (1 mmol scale) and palladium-ferric oxide nanoparticles-in-aluminum oxyhydroxide (2 mol% Pd; 2-3 nm in size) in ethyl acetate stirred for 4 h *at room temp.* under a balloon of H_2 *at atm. pressure*, the catalyst removed with an external magnet, washed five times with ethyl acetate, and the solution worked up → product. Conversion >99% (Y 99.8%). The readily recoverable catalyst (prepared by a sol-gel process) is highly active under ambient conditions, being a significant improvement on existing heterogeneous catalysts, and retains its activity *after 25 cycles*. Reaction is applicable to the hydrogenolysis of styrene oxides (with cleavage at the benzylic C-O bond), *trans*-stilbene oxide, cyclohexene oxide and alkene oxides (ring opening of the latter being dependent on steric factors). Phenoxy and hydroxyl groups were tolerated. F.e.s. M.S. Kwon, I.S. Park, J.S. Jang, J.S. Lee, J. Park, Org. Lett. *2007, 9* (17), 3417-9; cyclohexene hydrogenation over magnetically-removable palladium immobilized on silica coated magnetite nanoparticles cf. L.M. Rossi, F.P. Silva, L.L.R. Vono, P.K. Kiyohara, E.L. Duarte, R. Itri, R. Landers, G. Machado, Green Chem. *2007, 9* (4), 379-85.

(Acetoacetonato)bis[(2-carboxyphenyl)diphenylphosphine]palladium(II) fluoroborate ←
Prim. alcohols from aldehydes s. *23*, 51s*72* CHO → CH_2OH

Iridium s. under Rhodium *Ir*

Chiral iridium complex ←
Asym. homogeneous hydrogenation of ketones CO → CHOH
of benzo-fused ketones with a chiral cationic Ir-aminophosphine complex cf. *50*, 19; glycols from α-hydroxyketones with Cp*Ir(OTf)(MsDPEN) [MsDPEN = *N*-(methanesulfonyl)-1,2-diphenylethylenediamine] in methanol under 10 atm. H_2, chiral 1-aryl-1,2-ethanediols, s. T. Ohkuma, N. Utsumi, M. Watanabe, K. Tsutsumi, N. Arai, K. Murata, Org. Lett. *2007, 9* (13), 2565-7.

Iridium amidohydride complexes/isopropanol ←
Iridium-catalyzed transfer-hydrogenation of ketones
in water containing 5 eqs. formic acid cf. *65*, 21; in isopropanol with the air-sensitive Ir-amidodihydride complex, $IrH_2[(i\text{-}Pr_2PCH_2CH_2)_2N]$, or the moderately air-stable trihydride complex, $IrH_3[(i\text{-}Pr_2PCH_2CH_2)_2NH]$, s. Z.E. Clarke, P.T. Maragh, T.P. Dasgupta, D.G. Gusev, A.J. Lough, K. Abdur-Rashid, Organometallics *2006, 25* (17), 4113-7.

Iridium complex/chiral 1,2-bis(thien-2-ylmethyleneamino)- or 1,2-bis(thien-2-ylmethylamino)-cyclohexane/isopropanol/potassium hydroxide ←
Asym. transfer-hydrogenation of ketones
cf. *46*, 42s*70*; with chiral 1,2-bis(thien-2-ylmethyleneamino)- or 1,2-bis(thien-2-ylmethylamino)-cyclohexane/isopropanol/KOH s. X.-Q. Zhang, Y.-Y. Li, H. Zhang, J.-X. Gao, Tetrahedron:Asym. *2007, 18* (17), 2049-54.

Nanostructured, water-soluble, poly(diallyldimethylammonium chloride)-supported platinum carbonyl cluster complex ←
Prim. alcohols from aldehydes s. *23*, 51s*72* CHO → CH_2OH

Addition to Nitrogen and Carbon HC ⇓ NC

Potassium tert*-butoxide s. under Ruthenium complexes* *KOBu*-t

1,4-Dihydropyridines/chiral 1,1′-binaphthyl-2,2′-diyl hydrogen phosphates ←
Organocatalyzed asym. transfer-hydrogenation C=NR → CHNHR
of azomethines cf. *69*, 20s*70*; of benzoxazines, benzothiazines, and benzoxazinones at low catalyst loading s. M. Rueping, A.P. Antonchick, T. Theissmann, Angew. Chem. Int. Ed. Engl. *2006, 45* (40), 6751-5; of α-imino esters s. G. Li, Y. Liang, J.C. Antilla, J. Am. Chem. Soc. *2007, 129* (18), 5830-1.

Zinc/ammonium formate *Zn/HCOONH$_4$*
Amines from azomethines C=NR → CHNHR
with Zn in aq. NaOH cf. *56*, 33; by transfer-hydrogenation with Zn-dust/NH_4-formate s. K. Abiraj, B. Dinesh, G.R. Srinivasa, D.C. Gowda, J. Chem. Res. *2006* (8), 534-5.

Sodium tetrahydridoborate/methanol *NaBH$_4$/MeOH*
Lithium hydridotriethylborate *LiBHEt$_3$*
N-Sulfinylamines from N-sulfinylimines with asym. induction C=NS(O)R → CHNHS(O)R
cf. *66*, 33s*70,71*; with $LiBHEt_3$ or $NaBH_4$/MeOH, also conversion to chiral N-sulfinylaziridines, s. B. Denolf, E. Leemans, N. De Kimpe, J. Org. Chem. *2007*, *72* (9), 3211-7.

Zwitterionic B,P-dihydrophosphonioborates ←
Organocatalyzed hydrogenation ←

24. 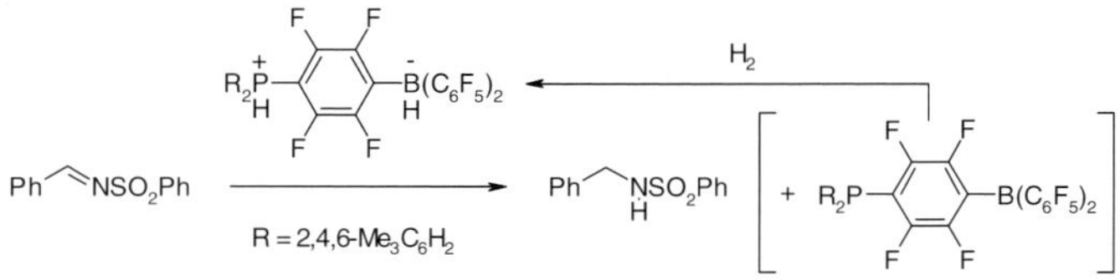

An inexpensive, *metal-free* catalytic hydrogenation has been achieved for the first time with air- and moisture-stable zwitterionic B,P-dihydrophosphonioborates which can be efficiently regenerated by hydrogen in the catalytic cycle. **E: Sec. amines from azomethines.** A glass bomb charged with the startg. imine, 5 mol% zwitterionic B,P-dihydrophosphonioborate and dry toluene (in a glove box), the vessel transferred to a vacuum/H_2 line, degassed three times with a freeze-pump-thaw cycle, cooled to -196°, 1 atm. H_2 introduced, the vessel sealed, warmed to room temp., stirred at 120° (now under 5 atm. H_2), and worked up after 10.5 h → product. Y 97%. Significantly, the catalyst was not deactivated by aminoborate formation, the high temperature (and/or incorporated bulky substituents) ensuring that the product amine remained dissociated. N-*tert*-Butylaldimines reacted more quickly, but N-benzylbenzaldimine deactivated the catalyst, permitting only stoichiometric reduction, as also occurred on reduction of aldehydes. Reaction is initiated by proton transfer from P to N, rather than by borohydride attack. F.e. incl. reduction of nitriles and reductive ring opening of aziridines, also activation of the substrates by complexation with $(C_6F_5)_3B$, s. P.A. Chase, G.C. Welch, T. Jurca, D.W. Stephan, Angew. Chem. Int. Ed. Engl. *2007, 46* (42), 8050-3.

Ammonium formate s. under Zn *HCOONH$_4$*
Triethylsilane s. under Pd-C *Et$_3$SiH*

Trichlorosilane/chiral fluorous-tagged α-(N-methylformamido)carboxylic acid amides ←
Sec. amines from azomethines via asym. hydrosilylation C=N → CHNH
cf. *36*, 670s*68,71*; with chiral fluorous-tagged α-(N-methylformamido)-amides s. A.V. Malkov, M. Figlus, S. Stoncius, P. Kocovský, J. Org. Chem. *2007*, *72* (4), 1315-25.

Trichlorosilane/chiral N-formylpipecolinic acid amides ←
Asym. reduction of ketimines via organocatalyzed hydrosilylation
s. *72*, 17

Chiral di(phosphines) s. under Pd(II)-trifluoroacetate ←
Chiral bis(phospholanes) s. under [Rh(cod)$_2$]BF$_4$ ←
Zwitterionic B,P-dihydrophosphonioborates s. under Borates ←

*(R)-1,1′-Binaphthyl-2,2′-diyl hydrogen phosphite/tris(2,4-di-*tert*-butylphenyl) phosphite s. under $[Ir(cod)Cl]_2$* ←

Chiral 1,1′-binaphthyl-2,2′-diyl hydrogen phosphates s. under 1,4-Dihydropyridines ←

Chiral chloro(η^5-pentamethylcyclopentadienyl)ruthenium(II) 2-aminophosphine complex/ potassium tert*-butoxide* ←

Hydroxycarboxylic acid amides from dicarboxylic acid imides C

Asym. homogeneous hydrogenation

25. 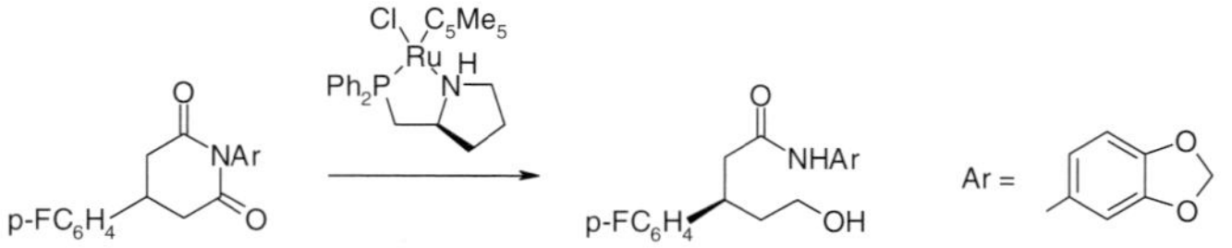

Chiral δ-hydroxycarboxylic acid amides. A degassed soln. of the startg. glutarimide in isopropanol added to a mixture of the chiral chloro(η^5-pentamethylcyclopentadienyl)ruthenium(II) 2-aminophosphine complex and KOBu-*t* in an autoclave, hydrogen introduced at 1-3 MPa, the mixture stirred at 80° for 24 h, the hydrogen vented, the mixture filtered through silica gel, and the soln. concentrated *in vacuo* → product. Y >99% (e.e. 98%). Five examples gave yields of 78-99% and e.e. of 64-98% with the highest enantioselectivities achieved with N-aryl derivs. In the illustrated example, the e.e. could be improved to 99% by conducting the reaction at 60°. A range of achiral derivs. (phthalimides, succinimides, glutarimides) generally gave near quantitative yields of products using the analogous achiral catalyst, chloro(η^5-pentamethylcyclopentadienyl)-ruthenium(II) 2-aminoethyl(diphenyl)phosphine complex. F.e. incl. use of the latter complex for the **cleavage of phthalimides** s. M. Ito, A. Sakaguchi, C. Kobayashi, T. Ikariya, J. Am. Chem. Soc. *2007, 129* (2), 290-1.

Bis(1,5-cyclooctadiene)rhodium(I) fluoroborate/chiral bis(phospholane) ←

Asym. homogeneous hydrogenation of ketimines C=N → CHNH

cf. *29*, 42s*60*; of α-imino esters with $[Rh(cod)_2]BF_4$/(S,S,R,R)-TangPhos under 50 atm. H_2 at 50° s. G. Shang, Q. Yang, X. Zhang, Angew. Chem. Int. Ed. Engl. *2006*, *45* (38), 6360-2.

Palladium-carbon/triethylsilane *Pd-C/Et_3SiH*

Amines from imines

under mild, neutral conditions s. *28*, 45s*72*

Palladium(II) trifluoroacetate/chiral di(phosphines) ←

Asym. hydrogenation of imines

chiral phosphinylamines with chiral di(phosphines) such as (S)-SegPhos in trifluoroethanol cf. *62*, 31s*71*; also chiral sulfonylamines or sultams s. Y.-Q. Wang, S.-M. Lu, Y.-G. Zhou, J. Org. Chem. *2007*, *72* (10), 3729-34.

Chiral cationic iridium P,N-ferrocenyl or spirobiindane-based phosphinooxazoline or tetrathiafulvalene-phosphinooxazoline complexes ←

Asym. homogeneous hydrogenation of ketimines

cf. *46*, 47s*71*; of N-(3,5-dimethyl-4-methoxy)phenylimines using chiral cationic P,N-ferrocenyl iridium complexes, also N-deprotection with $(NH_4)_2Ce(NO_3)_6$ (cf. *38*, 17), s. M.N. Cheemala, P. Knochel, Org. Lett. *2007*, *9* (16), 3089-92; with tetrathiafulvalene-phosphinooxazoline [TTF-PHOX] ligands (e.e. up to 68%) s. C. Réthoré, F. Riobé, M. Fourmigué, N. Avarvari, I. Suisse, F. Agbossou-Niedercorn, Tetrahedron:Asym. *2007*, *18* (15), 1877-82; with rigid, bulky spiro-biindane-based phosphinooxazoline [SIPHOX] ligands (e.e. up to 97%) s. S.-F. Zhu, J.-B. Xie, Y.-Z. Zhang, S. Li, Q.-L. Zhou, J. Am. Chem. Soc. *2006*, *128* (39), 12886-91.

*Chloro(1,5-cyclooctadiene)iridium(I) dimer/(R)-1,1′-binaphthyl-2,2′-diyl hydrogen phosphite/tris(2,4-di-*tert*-butylphenyl) phosphite* ←
Combinatorial asym. homogeneous hydrogenation of ketimines C=N → CHNH

26.

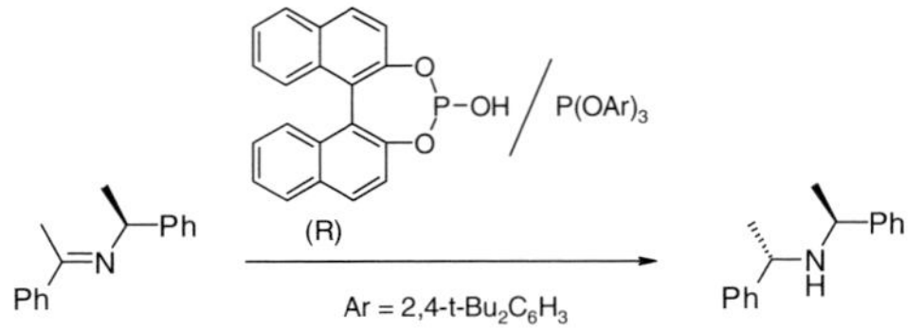

The mixing of a *chiral* phosphorous acid diester and a relatively simple *achiral* phosphine or phosphite can have a dramatic effect on the enantioselectivity of iridium-catalyzed asym. homogeneous hydrogenation of ketimines. **E:** 1 Mol% 0.005 *M* $[Ir(cod)Cl]_2$ in methylene chloride added to a 0.01 *M* soln. of 1 mol% (R)-1,1′-binaphthyl-2,2′-diyl hydrogen phosphite and a 0.01 *M* soln. of 2 mol% tris(2,4-di-*tert*-butylphenyl) phosphite in the same solvent under argon, stirred for 15 min at room temp., a 0.5 *M* soln. of the startg. (S)-imine in methylene chloride added, and hydrogenated in an autoclave for 48 h → product. Conversion 100% (d.r. 378:1). In the absence of the achiral ligand under the same conditions, the d.r. was 5.5:1. In certain instances, use of the chiral ligand alone can produce the racemic product, while in the presence of the achiral ligand enantioselectivities as high as 92% have been recorded. F.e. and parallel hydrogenations s. M.T. Reetz, O. Bondarev, Angew. Chem. Int. Ed. Engl. *2007, 46* (24), 4523-6.

Addition to Carbon-Carbon Bonds HC ⇓ CC

Lithium/polymer-based arene s. under $NiCl_2$ ←
Sodium hydroxide s. under CuF_2 *NaOH*
Lithium iodide s. under Bu_3SnH *LiI*

1,4-Dihydropyridines/chiral α-thioureidocarboxylic acid amides ←
Organocatalyzed asym. transfer-hydrogenation of 1,1-nitroethylene derivs. C=C → CHCH

27.

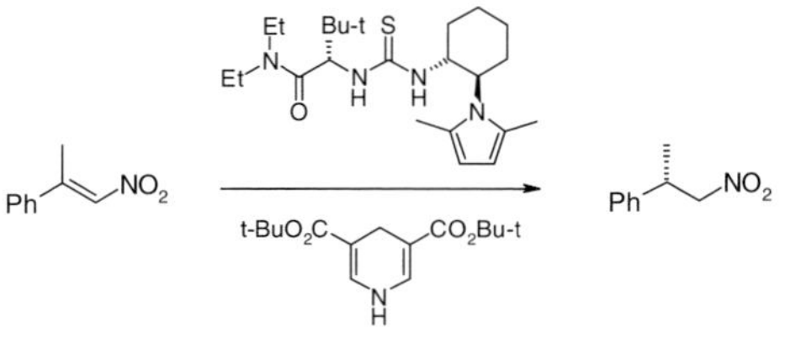

5 Mol% (S)-2-[3-[(1R,2R)-2-(2,5-dimethylpyrrol-1-yl)cyclohexyl]thioureido]-N,N-diethyl-3,3-dimethylbutanamide and 1.1 eqs. Hantzsch ester added to a 1.3 *M* soln. of the startg. nitroolefin in dibutyl ether, and stirred for 48 h at 40° under argon → (S)-2-phenyl-1-nitropropane. Y 97% (e.r. 97:3) after flash column chromatography. Reaction is inexpensive, complementary with transition metal-catalyzed procedures, and generally applicable in high yield and enantioselectivity (e.e up to 96%) to β-alkyl-subst. nitrostyrenes and aliphatic β-subst. 1,1-nitroethylene derivs. (notably β-*tert*-butyl derivs.), although, remarkably, the level of enantioselectivity is strongly dependent on the olefin configuration. Hydrogen-bonding between the chiral thiourea residue and the nitro group is presumed responsible for the asym. induction. Enantioselectivity was poor with chiral cyclic 1,1′-binaphthyl hydrogen phosphates. F.e. and comparison of chiral thiourea derivs. s. N.J.A. Martin, L. Ozores, B. List, J. Am. Chem. Soc. *2007, 129* (29), 8976-7.

1,4-Dihydropyridines/chiral 3,3′-diaryl-1,1′-binaphthyl-2,2′-diyl hydrogen phosphates ←
Organocatalyzed asym. transfer-hydrogenation of pyridines Ⓗ

28. 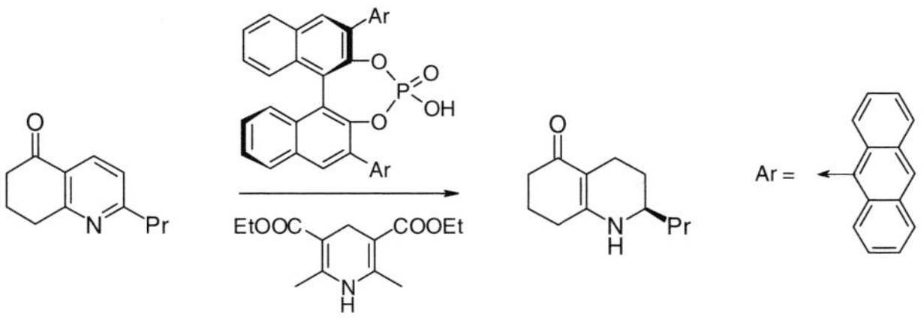

A suspension of the startg. pyridine, 5 mol% chiral 3,3′-bis(9-anthracenyl)-1,1′-binaphthyl-2,2′-diyl hydrogen phosphate and 4 eqs. Hantzsch ester in benzene stirred under argon in a screw-capped vial at 50° for 30-36 h → (2R)-2-propyl-2,3,4,6,7,8-hexahydroquinolin-5(1*H*)-one. Y 69% (e.e. 89%). cf. *71,* 20. This is the first example of an asym. reduction of pyridines catalyzed by a Brønsted acid, both yields and enantioselectivities being high (up to 92%) for the formation of chiral hexahydro- and tetrahydro-pyridines. Reaction is presumed to involve initial activation of the pyridine ring by catalytic protonation, followed by hydride transfer to the resulting chiral ion pair; acid-catalyzed isomerization of the formed 1,4-dihydropyridine to the corresponding iminium ion ensues, followed by a second hydride transfer with liberation of the chiral reagent. F.e. and comparison of chiral Brønsted acids and Hantzsch esters s. M. Rueping, A.P. Antonchick, Angew. Chem. Int. Ed. Engl. *2007, 46* (24), 4562-5.

*Cupric acetate/(R)-1-diphenylphosphino-2-[1-(dicyclohexylphosphino)ethyl]ferrocene/ polymethylhydrosiloxane/*tert-*butanol* ←
Asym. reduction of 3,3-diarylacrylonitriles C=C → CHCH

29. 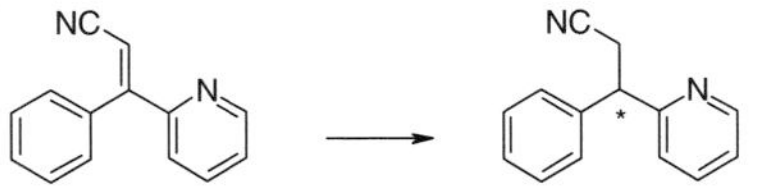

Asym. reduction of electron-deficient ethylene derivs. with chiral copper hydride complexes has been extended to 3-aryl-3-pyridylacrylonitriles with high enantioselectivity by using Josiphos as chiral ligand (cf. *58,* 32s*59*). **E:** A soln. of 3 mol% $Cu(OAc)_2$ and 3 mol% Josiphos ligand in toluene stirred under N_2 in a dry Schlenk tube for 10 min at room temp., treated with 4 eqs. polymethylhydrosiloxane, stirred for 5 min to activate the catalyst, cooled to 0°, the startg. nitrile and toluene added, followed by 4 eqs. *tert*-butanol, the tube sealed, allowed to react for 17 h, quenched with water, and worked up → 3-phenyl-3-(2-pyridyl)propionitrile. Y 90% (e.e. 96%). The procedure is applicable to pyrid-2-, 3- and 4-yl derivs. F.e. and inverse temperature effect on enantioselectivity, also comparison of chiral ligands, s. D. Lee, Y. Yang, J. Yun, Org. Lett. *2007, 9* (14), 2749-51.

Cupric fluoride/1,2-[(2R,5R)-2,5-dimethylphospholano]benzene monoxide/phenylsilane/ sodium hydroxide ←
Asym. reduction of β-subst. α,β-ethylenesulfones

30. 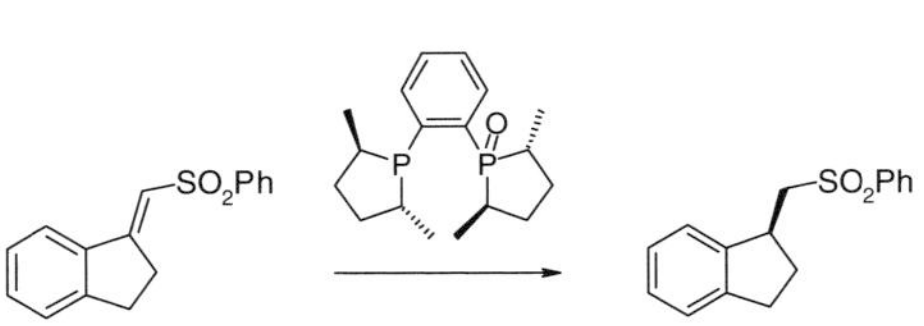

via hydrosilylation. A suspension of 5 mol% CuF_2-hydrate and 5.5 mol% 1,2-[(2R,5R)-2,5-dimethylphospholano]benzene monoxide in benzene stirred under argon *at room temp.* for 1 h (in a glove box), 1.5 eqs. phenylsilane added, stirred for a further 1 min, treated with 5.5 *M* aq. *NaOH* (20 mol% NaOH equal to 2 eqs. water), followed via syringe under argon by a soln. of the unsatd. sulfone in the minimum of benzene, and the heterogeneous mixture stirred for 12 h at room temp. → product. Y 90% (e.e. 99%). This is the first example of a general and convenient asym. reduction of this class of compd. The added base is necessary to eliminate competitive silylation of water. F.e. and comparison of solvents and chiral auxiliaries s. J.-N. Desrosiers, A.B. Charette, Angew. Chem. Int. Ed. Engl. *2007, 46* (31), 5955-7.

Magnesium bromide-etherate s. under Bu_3SnH — *$MgBr_2$-OEt_2*
Sodium tetrahydridoborate s. under $RuH_2(PPh_3)_4$ — *$NaBH_4$*
Alumina s. under Rh — *Al_2O_3*
Aluminum oxyhydroxide s. under Rh — *Al(O)OH*
Deuterium oxide s. under $[(allyl)PdCl]_2$ — *D_2O*
Alcohols s. under $Cu(OAc)_2$, $NiCl_2$, $[RuCl(\mu\text{-}Cl)(Me_6C_6)]_2$ and Pd(II)-di(phosphine) complex — *ROH*
Ammonium formate s. under $Pd(OH)_2$-C — *$HCOONH_4$*
Plant cell reductase — ←
γ-Lactones from 2(5*H*)-furanones — C=C → CHCH
Asym. enzymatic reduction

31. 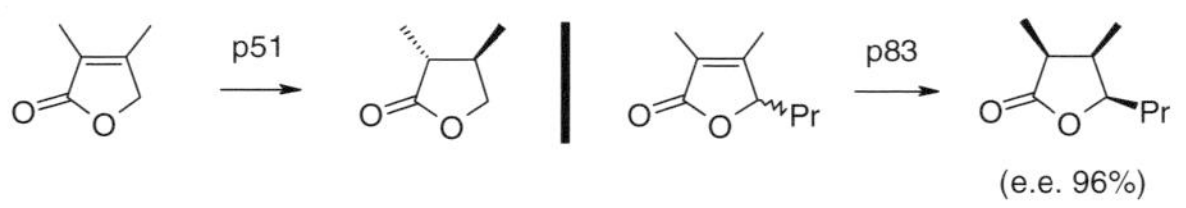

Chiral α,β-disubst. γ-lactones. A soln. of 2,3-dimethyl-2-butenolide (2 mmol) in Na-phosphate buffer (pH 7) treated with ca. 20 μg purified p51 reductase from the cultured plant cells of *Glycine max* at 36° for 12 h in the presence of 5 mmol NADPH → (2R,3R)-2,3-dimethylbutanolide. Conversion 82% (e.e. 99%). The (2S,3R)-isomer was obtained with the p83 reductase from the same source (conversion 90%; e.e. 96%). This is the first such highly enantioselective, high-yielding reduction of a 2,3-disubst. 2-butenolide. **(R)-γ-Subst. γ-lactones** were obtained similarly with resolution of the chiral centre using either enzyme. F.e.s. K. Shimoda, N. Kubota, T. Hirata, Y. Kondo, H. Hamada, Tetrahedron Lett. *2007, 48* (8), 1345-7.

Formic acid s. under $Pd(OAc)_2$ — *HCOOH*
Chiral α-thioureidocarboxylic acid amides s. under 1,4-Dihydropyridines — ←
Triethylsilane s. under Pd-C — *Et_3SiH*
Phenylsilane s. under CuF_2 — *$PhSiH_3$*
Polymethylhydrosiloxane s. under $Cu(OAc)_2$ and $Pd(OH)_2$ — *PMHS*
Hexamethyldisilane s. under $[(allyl)PdCl]_2$ — *$(Me_3Si)_2$*

Tributyltin hydride/magnesium bromide-etherate/lithium iodide — *$Bu_3SnH/MgBr_2$-OEt_2/LiI*
Carbonyl from α,β-ethylenecarbonyl compds.
oxo compds. with $Bu_3SnH/CuI/LiCl$ (and optionally Me_3SiCl) cf. *45*, 23; esters and amides with Bu_3SnH (2 eqs.)/$MgBr_2$-OEt_2 (3 eqs.)/LiI (0.1 eq.) in methylene chloride at 0° s. S. Hirasawa, Y. Tajima, Y. Kameda, H. Nagano, Tetrahedron *2007, 63* (45), 10930-8.

*Tri-*tert*-butylphosphine s. under $Pd(OAc)_2$* — t-*Bu_3P*
1,2-[(2R,5R)-2,5-Dimethylphospholano]benzene monoxide s. under CuF_2 — ←
Chiral phosphine-phosphoromonoamidites s. under $[Rh(cod)_2]BF_4$ — ←
Chiral cyclic phosphites s. under Rhodium(I) complex — ←
Chiral 3,3′-bis(triphenylsilyl)-5,5′,6,6′,7,7′,8,8′-octahydro-1,1′-binaphthyl-2,2′-diyl hydrogen phosphate s. under 1,4-Dihydropyridines — ←
p-*Toluenesulfonic or sulfuric acid s. under PtO_2* — *TsOH or H_2SO_4*
Chiral ferrocenyldiphosphines s. under $Cu(OAc)_2$ — ←

(Iminophosphoranyl)ferrocenes s. under Rhodium(I) complex ←

Nickel(II) chloride/lithium/polymer-based arene/ethanol ←

Carbonyl from α,β-ethylenecarbonyl compds. C=C → CHCH

Heterogeneous reduction under mild conditions with *in situ*-generated hydrogen

32. 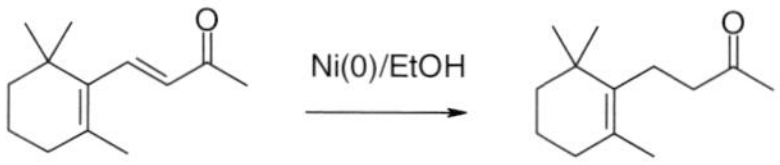

Dry $NiCl_2$ added to a suspension of 4.2 eqs. Li powder and 5 mol% vinylbiphenyl-divinylbenzene copolymer *as electron carrier* in THF under argon, stirred 10 min, the resulting black suspension of *nickel nanoparticles* diluted with more THF, cooled to 0°, 2.2 eqs. ethanol added (to generate hydrogen), followed by 1 eq. substrate, stirred for 12 h (GC-MS monitoring), filtered through Celite and silica gel, the filtrate dried over Na_2SO_4, and evaporated → product. Y 75% (after chromatography). Conditions are mild, work-up simple, and the method suitable for the conjugate reduction of α,β-unsaturated esters, ketones, carboxylic acids and amides, tolerating additional conjugate double bonds as illustrated here. However, the method is not yet suitable for reduction of conjugated nitriles or aldehydes. F.e.s. F. Alonso, I. Osante, M. Yus, Synlett *2006* (18), 3017-9.

Dihydridotetrakis(triphenylphosphine)ruthenium(II) (or Ruthenium trichloride)/ sodium tetrahydridoborate/water ←

Reduction of carbon-carbon multiple bonds with *in situ*-generated hydrogen ←

33. 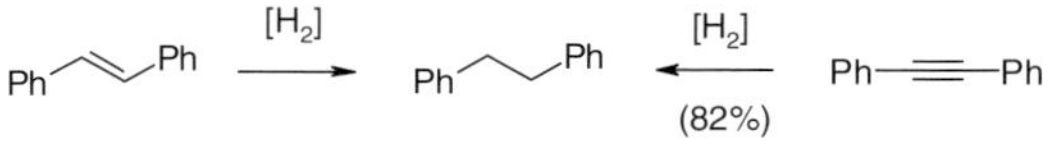

A soln. of (E)-stilbene, 5 eqs. $NaBH_4$, 2 eqs. water and 0.5 mol% $RuH_2(PPh_3)_4$ in toluene heated at 100° for 2.5 h in a pressure tube (*caution:* hydrogen evolution) → 1,2-diphenylethane. Y 95%. The procedure is applicable to the reduction of a range of ethylene derivs., incl. electron-deficient derivs. (without catalyst!), as well as acetylene derivs. (to the hydrocarbon), leaving amide and ester groups unaffected. N-Allyl and O-allyl groups were reduced similarly, in preference to isolated unsaturations, but tri- and tetra-subst. ethylene derivs. were left intact. F.e. and comparison of ruthenium catalysts s. G.R.A. Adair, K.K. Kapoor, A.L.B. Scolan, J.M.J. Williams, Tetrahedron Lett. *2006, 47* (50), 8943-4; reduction of mono- and di-subst. olefins with $RuCl_3$/ $NaBH_4$/water cf. P.K. Sharma, S. Kumar, P. Kumar, P. Nielsen, ibid. *2007, 48* (49), 8704-8.

Chloro(μ-chloro)(η⁶-hexamethylbenzene)ruthenium(II) dimer/ cesium carbonate/isopropanol *[Ru(II)]/Cs_2CO_3/i-PrOH*

Alcohols from 2-ethylenealcohols CH(OH)C=C → CH(OH)CHCH

via Ru-catalyzed isomerization-transfer-hydrogenation

34. 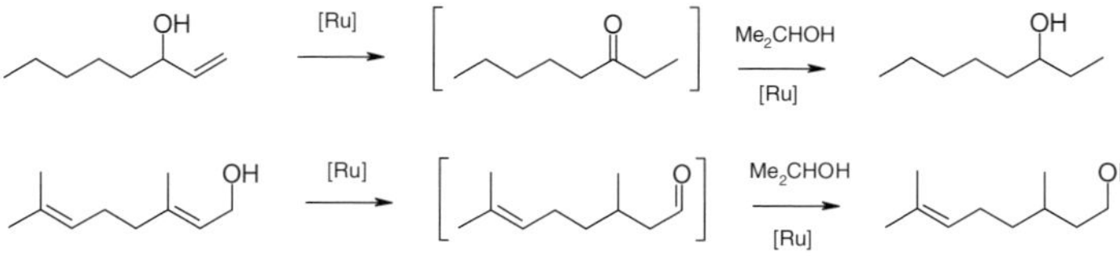

1-Octen-3-ol (2 mmol; 0.1 *M*), $[RuCl(\mu\text{-}Cl)(\eta^6\text{-}Me_6C_6)]_2$ (1 mol% Ru), and 2 mol% Cs_2CO_3 dissolved in isopropanol in a Schlenk flask under N_2, and the mixture stirred at 82° until reaction complete by GC (3.5 h) → octan-3-ol. Y 99% by GC (conversion >99%). The method is simple and efficient, proceeds under relatively mild conditions, and uses a readily-available catalyst and isopropanol as both solvent and hydrogen source. It is applicable to a variety of monosubst. allylic alcohols, and to di- and tri-subst. ones at higher ruthenium loadings (3-5 mol%). Selective reduction of geraniol to 3,7-dimethyl-6-octen-1-ol may also be achieved (5 mol% Ru; 24 h; Y 97%). F.e., Ru catalysts, and preparative (20 mmol) scale reaction s. V. Cadierno, J. Francos, J. Gimeno, N. Nebra, Chem. Commun. *2007* (23), 2536-8.

Rhodium-alumina *Rh-Al_2O_3*
Pyrrolidines from pyrroles by stereoselective hydrogenation Ⓗ

35. 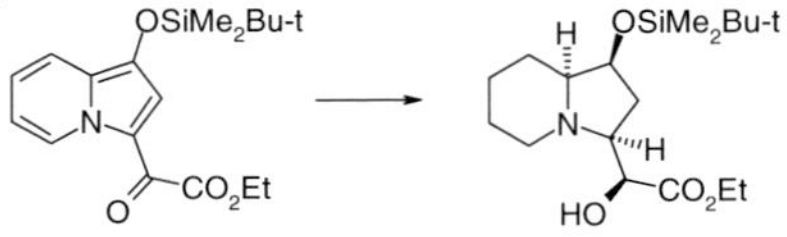

Polysubst. 2-α-hydroxypyrrolidines. A stirred mixture of the startg. pyrrole and 2.5 mol% rhodium-alumina in methanol hydrogenated in an open vial at 10 atm. for 24 h in a Parr apparatus, the mixture filtered through Celite, and the solvent removed under reduced pressure → perhydro-product. Y 91% (d.r. >20:1). Bicyclic and monocyclic pyrroles are hydrogenated under these conditions in the presence of ester, silyl ether and ether groups with yields generally >90% and diastereoselectivity >20:1. External double bonds are also reduced, as are imino groups in the side chain. It is thought that initial hydrogenation of the keto function assists in directing subsequent diastereoselective reduction of the pyrrole. Stereochemistry was assigned by analogy and is tentative. F.e. and 3-step synthesis of the pyrroles s. C. Jiang, A.J. Frontier, Org. Lett. *2007*, *9* (24), 4939-42.

Nanoparticulate rhodium (or iridium) in aluminum oxyhydroxide *[Rh (or Ir)/AlO(OH)]*
Benzene ring hydrogenation s. *26*, 52s*72*

Rhodium(I) complex/chiral cyclic phosphites or (iminophosphoranyl)ferrocenes ←
Asym. homogeneous hydrogenation C=C → CHCH
update s. *27*, 57s*71*; of dimethyl itaconate with deoxycholic acid-derived *tropos* biphenyl phosphites as ligands (e.e. up to 91%) s. A. Iuliano, D. Losi, S. Facchetti, J. Org. Chem. *2007*, *72* (22), 8472-7; with binaphthyl *ortho*- and *meta-closo*-dodecarboranyl phosphites s. S.E. Lyubimov, A.A. Tyutyunov, V.N. Kalinin, E.E. Said-Galiev, A.R. Khokhlov, P.V. Petrovskii, V.A. Davankov, Tetrahedron Lett. *2007*, *48* (46) 8217-9; of unfunctionalized olefins with chiral (iminophosphoranyl)ferrocenes s. T.T. Co, T.-J. Kim, Chem. Commun. *2006* (33), 3537-9.

Bis(1,5-cyclooctadiene)rhodium(I) fluoroborate/chiral phosphine-phosphoromonoamidites ←
Asym. homogeneous hydrogenation of α,β-ethylenephosphonic acid esters

36. 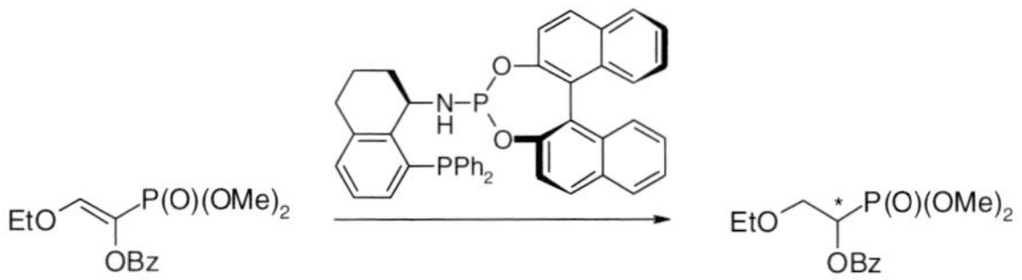

under mild conditions. 1 Mol% $[Rh(cod)_2]BF_4$ and 1.1 mol% chiral phosphine phosphoromonoamidite in methylene chloride degassed in a vial (contained in a N_2-filled glove box), stirred at room temp. for 15 min, a soln. of the startg. phosphonate in the same solvent added, the mixture transferred to an autoclave, charged with 10 atm. H_2, and worked up after 12 h → 1-benzoyloxy-1-dimethylphosphonyl-2-ethoxyethane. Conversion 100% (e.e. 99.9%). Enantioselectivities greater than 99% were recorded for the asym. hydrogenation of a series of β-aryl-, β-alkyl- and β-alkoxy-α-benzoyloxy-α,β-ethylenephosphonic acid esters at catalyst loadings as low as *0.05 mol%,* and a range of aromatic substituents was tolerated. The chiral ligand is easy to prepare and stable in moist air. F.e. and comparison of solvents s. D.-Y. Wang, X.-P. Hu, J.D. Huang, J. Deng, S.-B. Yu, Z.-C. Duan, X.-F. Xu, Z. Zheng. Angew. Chem. Int. Ed. Engl. *2007, 46* (41), 7810-3; with (R,R)-1,2-bis(2,5-dimethylphospholano)benzene as ligand (e.e. 91-5%) cf. H. Liu, Y.G. Zhou, Z.K. Yu, W.J. Xiao, S.H. Liu, H.W. He, Tetrahedron *2006, 62* (48), 11207-17.

Bis(1,5-cyclooctadiene)rhodium(I) fluoroborate/chiral phosphine-aminophosphines ←
N-Protected β-aryl-γ-aminocarboxylic C=C → CHCH
from β-aryl-γ-amino-α,β-ethylenecarboxylic acid esters
Asym. homogeneous hydrogenation

37. 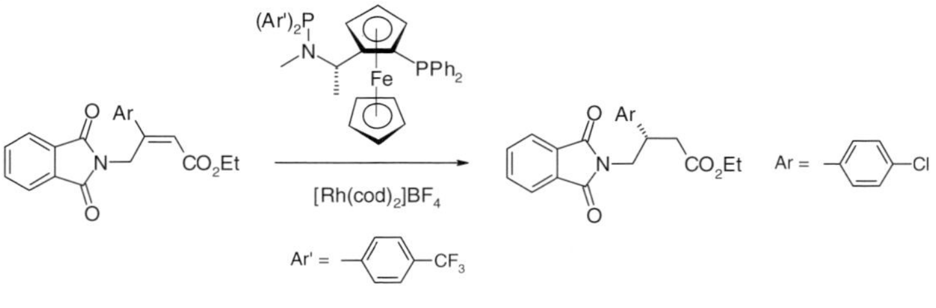

1.1 Mol% chiral phosphine-aminophosphine ligand and 1 mol% [Rh(cod)$_2$]BF$_4$ in dry degassed methylene chloride stirred for 30 min, a soln. of the startg. substrate in methylene chloride added, the mixture transferred to a Parr stainless steel autoclave, hydrogenated at 60 atm. for 24 h, and the solvent evaporated → product. Y 95% (e.e. 95%). Established catalysts for asym. hydrogenation of functionalized alkenes were ineffective in this case, leading to the development of novel *phosphine-aminophosphine ligands*. A series of β-aryl derivs. were hydrogenated in essentially quantitative yields and high enantioselectivity (93-97% e.e.). F.e. and details for preparation of the chiral ligand s. J. Deng, Z.-C. Duan, J.-D. Huang, X.-P. Hu, D.-Y. Wang, S.-B. Yu, X.-F. Xu, Z. Zheng, Org. Lett. *2007*, *9* (23), 4825-8.

P-Chiral rhodium P,P'-di(alk-1-ynyl)-1,2-di(phosphine) complexes ←
Asym. homogeneous hydrogenation s. *72*, 376

Chiral rhodium(I) bis(phospholane) complex ←
Asym. homogeneous hydrogenation Ⓗ
cf. *47*, 48; of N-protected 1,4,5,6-tetrahydropyridine-3-carbonyl compds. (prepared via partial hydrogenation of the corresponding pyridines over Pd-C, cf. *14*, 982) using the chiral Rh(I) complex, [Rh(TangPhos)nbd]SbF$_6$, chiral nipecotic acid derivs., s. A. Lei, M. Chen, M. He, X. Zhang, Eur. J. Org. Chem. *2006* (19), 4343-7.

Chiral rhodium(I) C_2-symmetric tetra(phosphines) or cyrhetrenyldi(phosphines) ←
Rh-Catalyzed asym. hydrogenation of enacylamines C(N-Ac)=C → CH(N-Ac)CH
cf. *71*, 26; of 1-arylenacylamines using Rh(cod)$_2$BF$_4$/chiral C_2-symmetric 1,2-bis(diarylphosphinomethyl)-3,4-bis(*o*-diphenylphosphinophenyl)cyclobutanes (e.e. up to 85%) or coumarin dimers, s. D. Zhao, Z. Wang, K. Ding, Tetrahedron Lett. *2007*, *48* (29), 5095-8; with chiral cyrhetrenyldi(phosphines) (e.e. up to 93%) s. R.T. Stemmler, C. Bolm, ibid. *48* (35), 6189-91.

Ruthenium nanoparticles-on-carbon nanofibers *Ru-C_{NF}*
Benzene ring hydrogenation Ⓗ
with Ru-C cf. *23*, 66s*67*; with Ru-nanoparticles on platelet-type carbon nanofibers [prepared by thermal decomposition of $Ru_3(CO)_{12}$] at 40-100° under 30 atm. H_2 s. Y. Motoyama, M. Takasaki, K. Higashi, S.-H. Yoon, I. Mochida, H. Nagashima, Chem. Lett. *2006*, *35* (8), 876-7.

Palladium nanoparticles *Pd*
Selective hydrogenation of carbon-carbon double bonds in water C=C → CHCH

38. 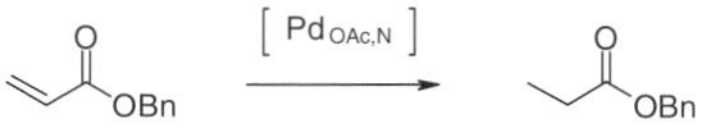

Palladium nanoparticles [prepared from a mixture of Pd(OAc)$_2$, Bu$_4$NBr and Bu$_3$N] are highly reactive and significantly more selective than Pd-C or existing stabilized or supported forms of nanoparticulate palladium for the hydrogenation of ethylene derivs. in water. **E:** A mixture of benzyl acrylate (1 mmol) and 2 ml water containing 1 mol% palladium nanoparticles ($Pd_{OAc,N}$) hydrogenated over a gas bag at room temp. for 4 h → benzyl propionate. Y 98%. Significantly, benzyl esters and ethers, ketones, aldehydes, nitriles, epoxides and aromatic chloro-, bromo- and nitro- groups remained unaffected, although the latter were reduced after 24 h. The reaction rate was lower in toluene, as also was selectivity. F.e.s. N.M. Callis, E. Thiery, J. Le Bras, J. Muzart, Tetrahedron Lett. *2007, 48* (46), 8128-31; 2-arylalcohols from styrene epoxides by hydrogenation over recyclable Pd nanoparticles in water cf. E. Thiery, J. Le Bras, J. Muzart, Green Chem. *2007, 9* (4), 326-7.

Palladium-carbon/ethylenediamine *Pd-C(en)*
Selective hydrogenation of carbon-carbon multiple bonds ←
of double bonds cf. *51*, 32s*56,60,62*; also of triple bonds with retention of ar. nitriles s. T. Maegawa, Y. Fujita, A. Sakurai, A. Akashi, M. Sato, K. Oono, H. Sajiki, Chem. Pharm. Bull. *2007, 55* (5), 837-9.

Palladium-carbon/triethylsilane *Pd-C/Et_3SiH*
Palladium(II) hydroxide-carbon/polymethylhydrosiloxane *$Pd(OH)_2$-C/PMHS*
Pd-Catalyzed reductions with organosilicon hydrides ←
ketones from enones with Pd-C/PMHS cf. *28*, 45; N-carbo-*tert*-butoxy-indolines from -indoles (cf. *21*, 78s*52*) with $Pd(OH)_2$-C/PMHS in ethanol, rapid procedure at room temp., s. S. Chandrasekhar, D. Basu, Ch.R. Reddy, Synthesis *2007* (10), 1509-12; Pd-C-catalyzed transfer-hydrogenation with triethylsilane under mild, neutral conditions of multiple bonds, azides, imines, and nitro groups, as well as cleavage of benzyl and allyl groups, s. P.K. Mandal, J.S. McMurray, J. Org. Chem. *2007, 72* (17), 6599-601.

Palladium(II) hydroxide-carbon/ammonium formate *$Pd(OH)_2$-C/$HCOONH_4$*
4,5,6,7-Tetrahydroisoindoles from isoindolines by transfer-hydrogenation ←

39.

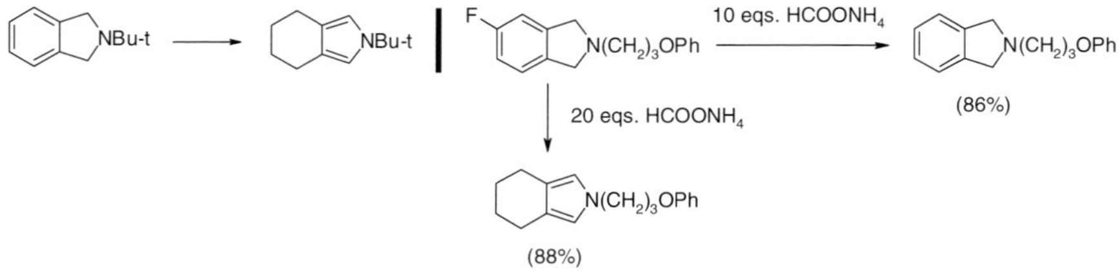

A mixture of 1,3-dihydro-2-*tert*-butylisoindole, 10 mol% $Pd(OH)_2$-C and 10 eqs. $HCOONH_4$ heated under reflux in methanol for 14 h, the mixture cooled, diluted with ethyl acetate, filtered, and concentrated → 4,5,6,7-tetrahydro-2-*tert*-butylisoindole. Y 95%. The method will tolerate alkyl groups in both rings affording products in 53-97% yields, but ether, ester, carboxamido and amine derivs. generally do not react (a 5-methoxy deriv. affording a 24% yield with 20 eqs. $HCOONH_4$). Halogeno (F or Cl) derivs. undergo clean dehalogenation under the experimental conditions in 86-88% yield, and use of 20 eqs. $HCOONH_4$ affords the dehalogenated 4,6,7,8-tetrahydroisoindoles. Surprisingly, 2,3-dihydroindoles undergo *oxidation* to **indoles** under these conditions (2 examples; 53 and 71%). F.e., also preparation of starting materials, and aromaticity calculations, s. D.-R. Hou, M.-S. Wang, M.-W. Chung, Y.-D. Hsieh, H.-H.G. Tsai, J. Org. Chem. *2007, 72* (24), 9231-9.

*Palladium(II) acetate/tri-*tert*-butylphosphine/formic acid* *$Pd(OAc)_2$/t-Bu_3P/HCOOH*
Mild and selective transfer-hydrogenation of carbon-carbon double bonds C=C → CHCH

40.

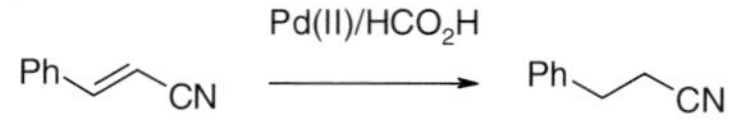

0.1 Eqs. *tri-*tert*-butylphosphine* added to 2.4 mol% $Pd(OAc)_2$ in THF under argon, the mixture refluxed for 10 min, 1 eq. cinnamonitrile and 5 eqs. formic acid added, heated at reflux with stirring for 15 min, cooled, stirring continued at room temp. for 12 h, filtered through Celite, and evaporated → 2-phenylpropionitrile. Y 86%. This mild and efficient method was used for selective hydrogenation of (Z)- and (E)-stilbene and styrene derivs., allylarenes, electron-deficient ethylene derivs., and unsatd. alcohols without loss of catalytic activity and in high yields (65-98%). α,β-Disubstituted alkenes such as limonene or pinene, and functionalized substrates such as cholesterol, however, did not give the required products. THF was the solvent of choice and other phosphine ligands were not effective. The catalytic mechanism is thought to involve formation of an active *formato(hydrido)palladium complex*. F.e.s. J.M. Brunel, Synlett *2007* (2), 330-2.

Chiral binuclear palladium(II) di(phosphine) complex/ethanol *[Pd]/EtOH*
Transfer-hydrogenation C=C → CHCH
with Pd-C/benzyl alcohol cf. *29*, 53; ketones from enones by **asym. transfer-hydrogenation** with $[((S)\text{-BINAP})Pd(OH)]_2(OTf)_2$ (0.25 mol%) in ethanol (as H-donor) at room temp. (e.e. up to 96%) s. Y. Tsuchiya, Y. Hamashima, M. Sodeoka, Org. Lett. *2006*, *8* (21), 4851-4.

(Acetoacetonato)bis[(2-carboxyphenyl)diphenylphosphine]palladium(II) fluoroborate ←
Hydrogenation of carbon-carbon multiple bonds ←
s. *11*, 67s*72*

Bis(η^3-allylpalladium chloride)/triphenylphosphine/hexamethyldisilane/deuterium oxide ←
α,β-Dideuterioketones from α,β-ethyleneketones C=C → CDCD

41. 

Startg. enone (0.4 mmol), hexamethyldisilane (0.6 mmol) and D_2O (4 mmol) added to a soln. of $[(\eta^3\text{-allyl})PdCl]_2$ (0.01 mmol) and triphenylphosphine (0.04 mmol) in dimethylacetamide (0.5 ml), and the mixture stirred at 60° for 24 h → product. Y 95%. The reaction is compatible with a variety of alkyl or aryl β-substituents and is not inhibited by electron-donating or -withdrawing substituents on either ring, except for a reduced yield when a *p*-trifluoromethyl group is present on the 1-phenyl group. Possible mechanisms are discussed for both the desired reaction and for over-deuteration at both the α- and β-positions; over-deuteration at the α-position is most likely caused by acid-catalyzed H-D exchange via an enol form of the product, whereas that at the β-position is caused by H-D exchange prior to reduction of the enone. F.e.s. H. Otsuka, E. Shirakawa, T. Hayashi, Chem. Commun. *2007* (18), 1819-21.

Nanoparticulate iridium in aluminum oxyhydroxide s. under Rh *[Ir/AlO(OH)]*

Chiral cationic iridium(I) phosphine complex *[Ir]*
Asym. homogeneous hydrogenation C=C → CHCH
with chiral cationic iridium(I) 4-phosphinomethyl-4,5,6,7-tetrahydro-1,3-benzothiazole complexes cf. *56*, 39s*68*; of enesilanes (e.e. up to 98%) s. K. Källström, I.J. Munslow, C. Hedberg, P.G. Andersson, Adv. Synth. Catal. *2006*, *348* (18), 2575-8.

Chiral cyclooctadiene(2-phosphinomethyl-Δ^2-oxazoline)iridium(I) tetrakis[3,5-bis-(trifluoromethyl)phenyl]borate ←
Asym. homogeneous hydrogenation of unfunctionalized tetrasubst. ethylene derivs.

42. 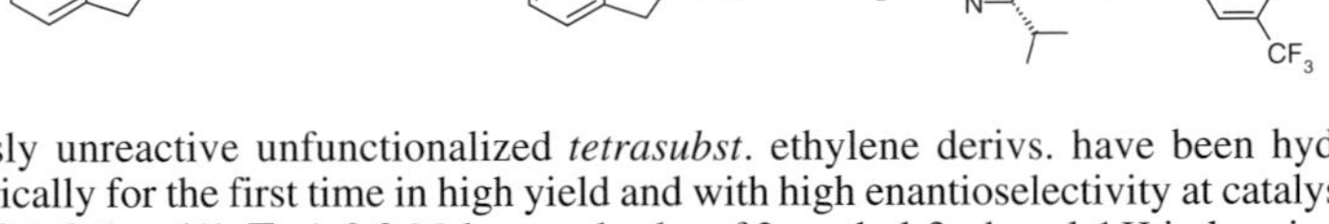

Notoriously unreactive unfunctionalized *tetrasubst.* ethylene derivs. have been hydrogenated asymmetrically for the first time in high yield and with high enantioselectivity at catalyst loadings as low as 0.1-0.5 mol%. **E:** A 0.2 *M* degassed soln. of 2-methyl-3-phenyl-1*H*-indene in methylene chloride containing 2 mol% chiral cyclooctadiene(2-phosphinomethyl-Δ^2-oxazoline)iridium(I) tetrakis[3,5-bis(trifluoromethyl)phenyl]borate introduced into a high-pressure steel autoclave (in a glove box), the vessel sealed, removed from the glove box, purged with H_2, sealed under 50 atm. H_2, and stirred for 4 h at room temp. → product. Conversion 99% (e.e. 95%). Reaction is generally applicable, with enantioselectivities of 75->99%, to tetrasubst. styrene derivs. and 2,3-disubst. indenes, but enantioselectivity was lower (58-77%) with 1,2-dihydronaphthalenes. The catalysts are also readily accessible. F.e. and comparison of ligands s. M.G. Schrems, E. Neumann, A. Pfaltz, Angew. Chem. Int. Ed. Engl. *2007, 46* (43), 8274-6.

Chiral iridium(I) di(phosphine) complexes ←
1,2,3,4-Tetrahydroquinolines from quinolines Ⓗ
Asym. homogeneous hydrogenation
with chiral 2,2′,6,6′-tetramethoxy-2,2′-bis(diphenylphosphino)biphenyl cf. *66*, 42s*68*; with SynPhos or DifluorPhos s. C. Deport, M. Buchotte, K. Abecassis, H. Tadaoka, T. Ayad, T. Ohshima, J.-P. Genet, K. Mashima, V. Ratovelomanana-Vidal, Synlett *2007* (17), 2743-7.

Chiral iridium(I) 4-α-phosphinoxy-Δ²-oxazoline complexes ←
Propylene carbonate as an environmentally friendly solvent ←
for asym. homogeneous hydrogenation

43.

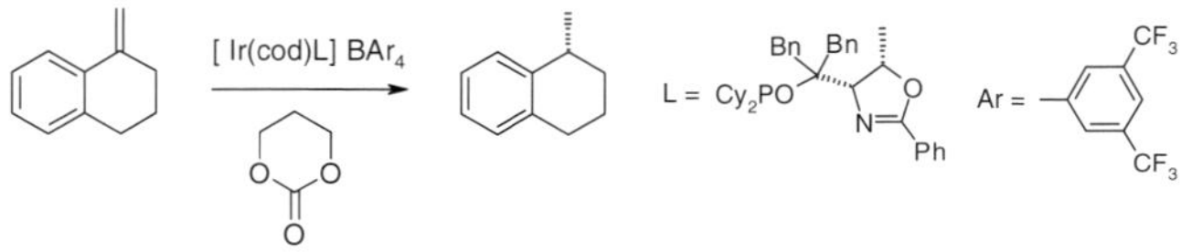

Dipolar propylene carbonate is a non-corrosive, non-toxic, and bio-degradable alternative to ‘unfriendly’ methylene chloride and other conventional organic solvents for asym. hydrogenation. Furthermore, it solubilizes transition metal complexes while being immiscible with non-polar solvents, so that, for asym. hydrogenation of non-functionized alkenes, extraction of the product can be simply effected in hexane. **E:** A soln. of 1-methylenetetralin in propylene carbonate hydrogenated at 25° with 1 mol% chiral iridium(I) 4-α-phosphinoxy-Δ²-oxazoline complex under 85 atm. H_2 for 4 h, hexane added to the mixture, shaken to extract the product, and the upper hexane layer removed for work-up → (R)-product. Conversion 100% (97% purity; e.e. 83.1%). The catalyst remaining in the propylene carbonate layer was recycled a further seven times with gradual erosion of enantioselectivity after the fifth cycle. With functionalized olefins, enantioselectivity was similar or better than that in methylene chloride. F.e.s. J. Bayardon, J. Holz, B. Schäffner, V. Andrushko, S. Verevkin, A. Preetz, A. Börner, Angew. Chem. Int. Ed. Engl. *2007, 46* (31), 5971-4.

*Platinum oxide/*p*-toluenesulfonic or sulfuric acid* *PtO_2/TsOH or H_2SO_4*
Piperidine from pyridine ring Ⓗ
with PtO_2/HCl cf. *4*, 86s*67* and with Pd-C/H_2SO_4 cf. *69*, 31; with PtO_2/*p*-TsOH or H_2SO_4 in ethanol under a H_2 balloon at room temp., part of a 2-step prepn. of mono-N-protected piperidinyloxy-N-heterocyclics from hydroxypyridines and N-protected hydroxy-N-heterocyclics via Mitsunobu etherification (cf. *31*, 170), s. J. Chao, M. Israiel, J. Zheng, C. Aki, Tetrahedron Lett. *2007*, *48* (5), 791-4; corrigendum s. ibid. *48* (13), 2435.

Nanostructured, water-soluble, poly(diallyldimethylammonium chloride)-supported platinum carbonyl cluster complex ←
Hydrogenation of carbon-carbon multiple bonds s. *11*, 67s*72* ←

Chloroplatinic acid *H_2PtCl_6*
Hydrogenation of N-heterocyclics Ⓗ
cf. *4*, 86s*67*; 1,2,3,4-tetrahydroquinolin-8-ols from quinolin-8-ols with retention of halogen using 1.28 eqs. H_2PtCl_6-$6H_2O$ in 1:75 water/isopropanol under 20 atm. H_2 at 90-95° s. V.V. Zhandarev, M.E. Goshin, V.N. Kazin, L.M. Ramenskaya, G.S. Mironov, A.L. Shishkina, Russ. J. Org. Chem. *2006*, *42* (7), 1093-4.

Exchange ↓↑

Oxygen ↑ HC ↓↑ O

*Electrolysis/biphenyl/*tert-*butanol* *ϟ/Ph-Ph/t-BuOH*
Hydrocarbon groups from mesylates $OSO_2Me \rightarrow H$
by electrolysis in the presence of diphenyl diselenide and diethyl malonate cf. *23*, 81s*48*; of alkyl mesylates in an undivided cell fitted with a Pt-cathode and an Mg-anode in the presence of biphenyl and *t*-BuOH, retention of epoxy, olefin, acetal, hydroxy, or cyano groups, s. H. Senboku, M. Takahashi, T. Fukuhara, S. Hara, Chem. Lett. *2007*, *36* (2), 228-9.

Sodium methoxide s. under Bis[2-aminoethyl(diphenyl)phosphine]dichloro-ruthenium(II) *NaOMe*
Diethylamine s. under Pd-C *Et_2NH*
Magnesium chloride s. under KBH_4 *$MgCl_2$*

Sodium tetrahydridoborate [s.a. under $NiCl_2(PPh_3)_2$] *$NaBH_4$*
Prim. alcohols from carboxylic acid pentafluorophenyl esters $COOC_6F_5 \rightarrow CH_2OH$

44. 

Startg. pentafluorophenyl ester added to a suspension of 1.2-1.5 eqs. $NaBH_4$ in dry THF, stirred *at room temp. for 30 min,* cooled to 0°, acidified with 1 *N* HCl, and worked up → product. Y 92%. Both simple prim. alcohols and **chiral N-protected 2-amino-*prim*-alcohols** can be obtained simply and in high yield under mild conditions, leaving a number of functional groups (halide, nitro, benzyl esters and ethers, *tert*-butyl esters and thioethers), as well as N-protective groups (Z, Boc, Fmoc) unaffected. F.e.s. E. Papavassilopoulou, P. Christofis, D. Terzoglou, P. Moutevelis-Minakakis, Tetrahedron Lett. *2007, 48* (47), 8323-5.

Sodium tetrahydridoborate/Amberlyst 15 ←
Lactones from ketocarboxylic acid esters ○
with Bu_4NBH_4 cf. *31*, 57 and with $NaBH_4$/HCl cf. *14*, 99s*69*; with $NaBH_4$/Amberlyst 15, 3-aryl-phthalides from *o*-aroyl esters, s. S.T. Patil, A.V. Karnik, Indian J. Chem. *2007*, *46B* (4), 710-12.

Potassium tetrahydridoborate/magnesium chloride *$KBH_4/MgCl_2$*
Prim. alcohols from carboxylic acids or esters $COOH,R \rightarrow CH_2OH$
with $NaBH_4/ZrCl_4$ cf. *44*, 33; with 1:1 $KBH_4/MgCl_2$ in refluxing THF s. Y.-C. Qiu, F.-L. Zhang, C.-N. Zhang, Tetrahedron Lett. *2007*, *48* (43), 7595-8.

*Tri-*n-*butylborane/water/air* *Bu_3B/H_2O/air*
Barton-type deoxygenation of sec. alcohols via xanthates OC(S)SR → H
with retention of chlorides and bromides s. *72*, 48

Indium(III) bromide s. under Et_3SiH *$InBr_3$*
*Biphenyl/*tert-*butanol s. under ϟ* *Ph-Ph/t-BuOH*
*Tri-*n-*butylamine-formic acid s. under $Pd(OAc)_2(PPh_3)_2$* *Bu_3N/HCOOH*
Amberlyst 15 s. under $NaBH_4$ ←

Triethylsilane/indium(III) bromide *$Et_3SiH/InBr_3$*
Ethers from carboxylic acid esters $COOR \rightarrow CH_2OR$

45. 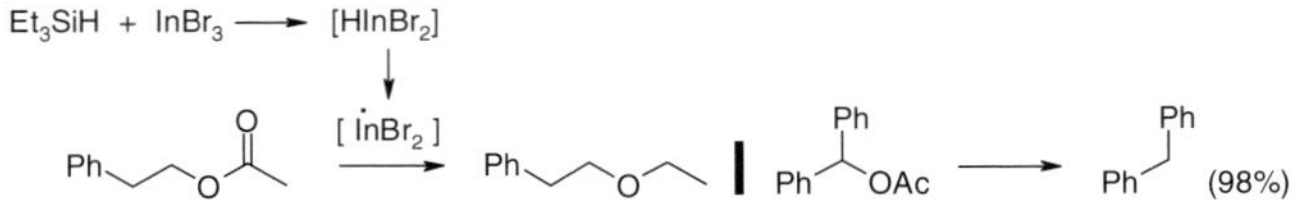

2-Phenethyl acetate, 5 mol% $InBr_3$ and 4 eqs. triethylsilane introduced in this order into a screw-capped vial containing chloroform under N_2, the sealed vial stirred at 60° for <1 h, water added, the orange suspension stirred vigorously until decolorized, and worked up with flash column chromatographic purification → ethyl 2-phenethyl ether. Y 86%. The procedure is mild, rapid and selective for the reduction of alkyl and arylalkyl carboxylates, while arylcarboxylic acid esters gave low yields, and benzyl or allyl acetates underwent predominant deacetoxylation. Nitro and bromine remained unaffected. A radical mechanism is invoked following transmetalation between the two reagents, the key indium radical being regenerated in the concluding step of the cycle. F.e. incl. cyclic ethers from lactones, and solvent effect s. N. Sakai, T. Moriya, T. Konakahara, J. Org. Chem. *2007, 72* (15), 5920-2.

Tetramethyldisiloxane/(μ_3,η^2:η^3:η^5-acenaphthylene)heptacarbonyltriruthenium ←
Amines from carboxylic acid amides CON< → CH_2N<
tert. amines from N,N-disubst. amides using a self-encapsulated ruthenium catalyst/polymethylhydrosiloxane cf. *70*, 37; sec. amines from N-subst. amides with tetramethyldisiloxane, also tert. amines via reductive N-alkylation with polymethylhydrosiloxane, s. S. Hanada, T. Ishida, Y. Motoyama, H. Nagashima, J. Org. Chem. *2007*, *72* (20), 7551-9.

Air s. under Bu_3B *O_2*

Dichlorobis(triphenylphosphine)nickel(II)/tricyclohexylphosphine/ sodium tetrahydridoborate ←
Replacement of sulfonyloxy groups by hydrogen OSO_2R → H
with $Pd(PPh_3)_4/NaBH_4$ cf. *42*, 58s*43*; with $NiCl_2(PPh_3)_2/Cy_3P/NaBH_4$, general procedure for ar. tosylates incl. hindered, deactivated, heterocyclic, or bifunctional ones, s. V. Kogan, Tetrahedron Lett. *2006*, *47* (43), 7515-8.

(Acenaphthylene)heptacarbonyltriruthenium s. under Tetramethyldisiloxane ←

Bis[2-aminoethyl(diphenyl)phosphine]dichlororuthenium(II)/sodium methoxide ←
Prim. alcohols from carboxylic acid esters COOR → CH_2OH
Homogeneous ruthenium-catalyzed hydrogenation under mild conditions

46. 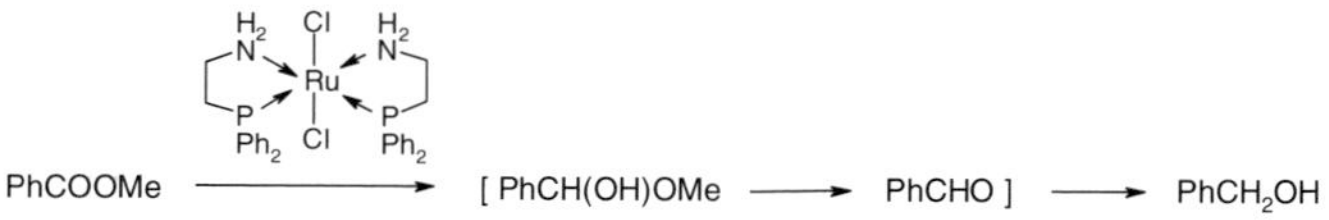

A soln. of methyl benzoate in THF added to a mixture of 0.05 mol% bis[2-aminoethyl(diphenyl)-phosphine]dichlororuthenium(II) and 5 mol% solid NaOMe in THF contained in a stainless-steel autoclave under argon, the vessel purged three times with H_2 (20 atm.), then pressurized with H_2 to 50 atm., and heated at 100° for 1 h → benzyl alcohol. Y 97%. The procedure is generally applicable to both aliphatic and aromatic esters and remarkably selective in that di- and tri-subst. ethylenic double bonds in the substrate are untouched. α,β-Ethylene and mono-subst. ethylene groups, however, underwent competitive hydrogenation of the unsaturation. There was no reaction with Noyori-type (diamine)ruthenium(II) bis(phosphine) complexes, nor with dimethylaminoethylphosphine analogs, indicating that the bis(*prim*-aminophosphine) residue is critical for reaction. The active catalyst is presumed to be a ruthenium *trans*-dihydride complex generated *in situ* by action of the base, hydrogenation possibly involving initial reduction to the corresponding hemiacetal, followed by formation and subsequent hydrogenation of the corresponding aldehyde. F.e. and comparison of catalysts s. L.A. Saudan, C.M. Saudan, C. Debieux, P. Wyss, Angew. Chem. Int. Ed. Engl. *2007, 46* (39), 7473-6.

Palladium-carbon/diethylamine *Pd-C/Et_2NH*
Replacement of ar. sulfonyloxy groups by hydrogen OSO_2R → H
with added Et_3N cf. *29*, 65: of triflyloxy or mesyloxy groups with added diethylamine (as promoter and acid scavenger) in methanol at room temp. s. A Mori, T. Mizusaki, T. Ikawa, T. Maegawa, Y. Monguchi, H. Sajiki, Tetrahedron *2007*, *63* (5), 1270-80.

*Diacetatobis(triphenylphosphine)palladium(II)/tri-*n*-butylamine/formic acid* ←
Methylene compds. from aldehydes CHCHO → C=CH(OTf) → C=CH_2
via enol triflates – One-pot procedure s. *72*, 630

Nitrogen ↑ HC ↓↑ N

Sodium methoxide/air *NaOMe/O_2*
Wolff-Kishner reduction ←
cf. *15*, 332; of electron-poor, tautomerizable, hetar. hydrazines with NaOMe (from NaH/methanol) at 60° *under air*, part of a prepn. of 2-alkoxy- from 2-chloro-3-hydrazino-pyrido[2,3-*b*]pyrazines, s. A. Unciti-Broceta, M.J. Pineda de las Infantas, M.A. Gallo, A. Espinosa, Chem. Eur. J. *2007*, *13* (6), 1754-62.

Copper(II) sulfate $CuSO_4$
β-Ketocarboxylic from α-diazo-β-hydroxycarboxylic acid esters $CH(OH)C{=}N_2 \rightarrow COCH_2$
in water s. *72*, 136

Tin/hydrogen chloride *Sn/HCl*
Replacement of nitro groups by hydrogen $NO_2 \rightarrow H$
in ar. nitro compds. cf. *24*, 85; in 2-subst. 4-nitroquinolines with Sn and concd. HCl in ethanol at reflux s. Y. Zhou, J. Li, H. Liu, L. Zhao, H. Jiang, Tetrahedron Lett. *2006*, *47* (48), 8511-4.

Palladium(II) hydroxide-carbon *Pd(OH)$_2$-C*
Reductive cleavage of benzylic nitro compds.

47. 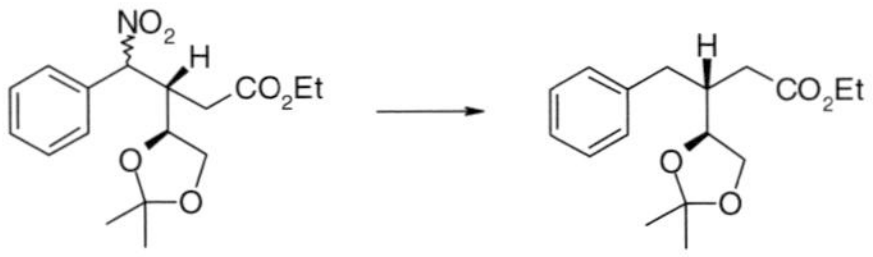

$Pd(OH)_2$/C (6 mol%) added to a suspension of the startg. nitroalkane *in isopropanol,* and the mixture stirred vigorously at 70° under H_2 until startg. material consumed (TLC) → product. Y 78%. A variety of functional groups are well-tolerated, including prim. benzyl ethers. A major advantage of the present method is that cleavage of the nitro group occurs without loss of optical activity at adjacent centres, affording new opportunities for asymmetric synthesis. Use of Pd-C in place of Pearlman's catalyst is also effective and mechanistic studies using this reagent system suggest that the solvent, rather than H_2, is the source of the benzylic hydrogen atom. F.e.s. T.C. Fessard, H. Motoyoshi, E.M. Carreira, Angew. Chem. Int. Ed. Engl. *2007*, *46* (12), 2078-81.

Halogen ↑ HC ↓↑ Hal

Indium/ammonium chloride *In/NH_4Cl*
2,2-Dichloroethyl from 2,2,2-trichloroethyl esters $COOCH_2CHCl_2$
s. *19*, 93s72

Sodium tetrahydridoborate/(R,R)-2-phenyl-1,3,2-dioxaborolane-4,5-dicarboxylic acid ←
Epoxides from α-halogenoketones C(O)C(Hal) → epoxide
by asym. reduction s. *72*, 14

*Tri-*n*-butylborane/water/air* *Bu_3B/H_2O/air*
Replacement of aliphatic iodine by hydrogen I → H

48. 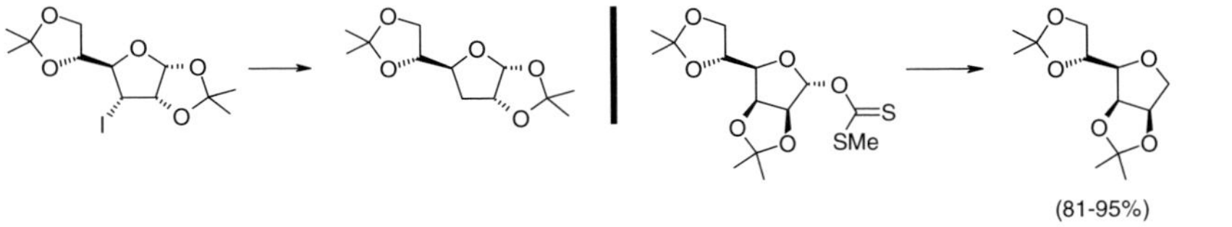

10 Eqs. water added to a 0.03 *M* soln. of the startg. iodide in benzene, degassed with N_2 while stirring for 1 h, treated with 4 eqs. tributylborane in one portion via syringe, the N_2 gas line

removed, air (0.8 eq. O_2) bubbled into the mixture via cannula beneath the surface, aq. NaOH (3 *M*) and 30% H_2O_2 added simultaneously (*exothermic*), stirred for 10 min, and worked up with chromatographic purification → product. Y 76%. Reaction is generally applicable to *prim*-, *sec*- and *tert*-alkyl iodides, leaving alkyl and aryl bromides and chlorides, epoxides, acetals and tosylamines unaffected. Aryl iodides were also generally inert, although deiodination of the hindered methyl 2-iodo-3-nitrobenzoate was effected in low yield. F.e., also Barton-type deoxygenation of sec. alcohols via xanthates (with retention of chlorides and bromides), s. M.R. Medeiros, L.N. Schacherer, D.A. Spiegel, J.L. Wood, Org. Lett. *2007, 9* (22), 4427-9.

Lithium tetrahydridoaluminate s, under $NbCl_5$ — $LiAlH_4$
Indium(III) acetate s. under $PhSiH_3$ — $In(OAc)_3$
Gallium chloride s. under Polymethylhydrosiloxane — $GaCl_3$

*Samarium/tri-*n*-butylphosphine/methanesulfonic acid or hydrogen chloride/ potassium carbonate* ←
Aldehydes from carboxylic acid chlorides under mild conditions COCl → CHO

49.
$$ArCOCl \xrightarrow{PBu_3} [\ ArCO\overset{+}{P}Bu_3 \xrightarrow{Sm/H^+} ArCH(OH)\overset{+}{P}Bu_3\ ] \longrightarrow ArCHO \qquad Ar = o\text{-}ClC_6H_4$$

1.2 Eqs. tri-*n*-butylphosphine added at -20° under N_2 to a stirred soln. of *o*-chlorobenzoyl chloride in *acetonitrile*, treated in succession after 15 min with 6 *M* HCl and 0.67 eq. samarium powder, stirred for 1 h, allowed to warm to room temp. gradually, poured into aq. 10% HCl, extracted with methylene chloride, and the organic phase washed with aq. 10% K_2CO_3 → *o*-chlorobenzaldehyde. Y 92%. The procedure is mild, inexpensive, manageable, and generally applicable to aromatic acyl chlorides, incl. *o*-subst. derivs., irrespective of the electronic nature of substituents. High yields were also obtained with aliphatic acyl chlorides, preferably with methanesulfonic acid as proton source and in THF. Significantly, there was no over-reduction of the products to the corresponding prim. alcohols. The key step in the conversion is the initial formation of an acylphosphonium ion, which is reduced to a 1,1-hydroxyphosphonium ion prior to cleavage during the weakly basic work-up. F.e.s. X. Jia, X. Liu, J. Li, P. Zhao, Y. Zhang, Tetrahedron Lett. *2007, 48* (6), 971-4.

Ammonium formate s. under $Pd(OH)_2$*-C* — $HCOONH_4$
Triethylsilane s. under Cyclometalated aryl(hydrido)iridium(III) bis(phosphinite) complex — Et_3SiH

Phenylsilane/indium(III) acetate — $PhSiH_3/In(OAc)_3$
Polymethylhydrosiloxane/gallium chloride — $PMHS/GaCl_3$
Radical replacement of aliphatic halogen by hydrogen Hal → H
cf. *45*, 35; of bromine or iodine with phenylsilane under $In(OAc)_3$ catalysis at 70° (or with added lutidine or Et_3B under air at room temp.) or with polymethylhydrosiloxane under $GaCl_3$ catalysis s. K. Miura, M. Tomita, Y. Yamada, A. Hosomi, J. Org. Chem. *2007, 72* (3), 787-92.

Phosphonio-supported tin hydrides/azodiisobutyronitrile ←
Radical replacement of bromine by hydrogen Br → H
cf. *17*, 117; with *in situ*-generated triarylphosphonio-subst. tin hydrides and added AIBN, improved removal, recovery and recycling of tin residues, also Stille coupling of triarylphosphonio-subst. enestannanes with added $Pd(PBu\text{-}t_3)_2$, s. J.-C. Poupon, D. Marcoux, J.-M. Cloarec, A.B. Charette, Org. Lett. *2007, 9* (18), 3591-4.

*Tri-*n*-butylphosphine s. under Sm* — Bu_3P

Niobium pentachloride/lithium tetrahydridoaluminate *$NbCl_5/LiAlH_4$*
Methyl from trifluoromethyl groups $CF_3 \rightarrow CH_3$
Selective reduction

50. 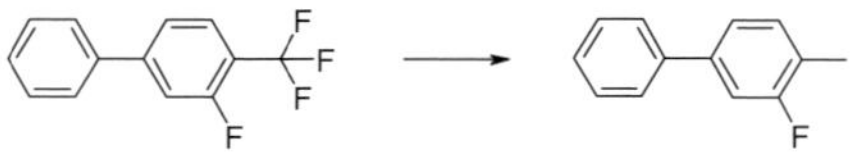

Solid $LiAlH_4$ (4 eqs.) added carefully to a soln. of startg. trifluoromethylarene (1.39 mmol) and $NbCl_5$ (5 mol%) in 1,2-dimethoxyethane, and the mixture refluxed for 4 h with stirring → product. Y 100%. Omission of the Nb(V) catalyst leads to mixtures of fully- and partially-reduced products, along with unreacted starting material. Unlike aryl fluorides, which are inert under the reaction conditions, aryl chlorides are readily and preferentially cleaved. Stepwise reduction of bis(trifluoromethyl)arenes to methyl(trifluoromethyl)arenes can be achieved by limiting the quantity of $LiAlH_4$ used to 3 eqs. More general reductive defluorination with $Et_3SiH/Tr^+B(C_6F_5)_4^-$ cf. *69,* 37. F.e.s. K. Fuchibe, Y. Ohshima, K. Mitomi, T. Akiyama, Org. Lett. *2007, 9* (8), 1497-9.

Air s. under Bu_3B *O_2*
Methanesulfonic acid or hydrogen chloride s. under Sm *$MeSO_3H$ or HCl*
Ammonium chloride s. under In *NH_4Cl*

Palladium nanoclusters-on-hydroxyapatite-encapsulated γ-ferric oxide nanocrystallites/sodium hydroxide ←
Heterogeneous replacement of halogen by hydrogen by catalytic hydrogenation over magnetically recoverable nanoclusters Hal → H

51. 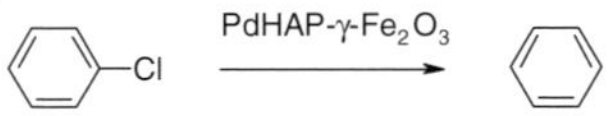

A mixture of chlorobenzene (2 mmol scale), 0.1 mol% palladium nanoclusters-on-hydroxyapatite-encapsulated γ-ferric oxide nanocrystallites, and 1.2 eqs. NaOH in isopropanol heated at 60° *under 1 atm. H_2* for 1 h → benzene. Y >99% (by GC). The catalyst was easily recovered, quickly and completely, with an external permanent magnet, and recycled at least three times without loss of activity. Turnover frequencies were very high (up to 2500 h^{-1}) for the reduction of aromatic bromides and chlorides, as well as benzyl and 4-pyridyl chloride, but ar. iodides were effectively unreactive. F.e.s. T. Hara, T. Kanets, K. Mori, T. Mitsudome, T. Mizugaki, K. Ebinati, K. Kaneda, Green Chem. *2007, 9* (11), 1246-51.

Palladium(II) hydroxide-carbon/ammonium formate *$Pd(OH)_2$-C/$HCOONH_4$*
4,5,6,7-Tetrahydroisoindoles from halogenoisoindolines by transfer-hydrogenation ←
s. *72*, 39

Cyclometalated aryl(hydrido)iridium(III) bis(phosphinite) complex/triethylsilane ←
Replacement of aliphatic halogen by hydrogen Hal → H

52. 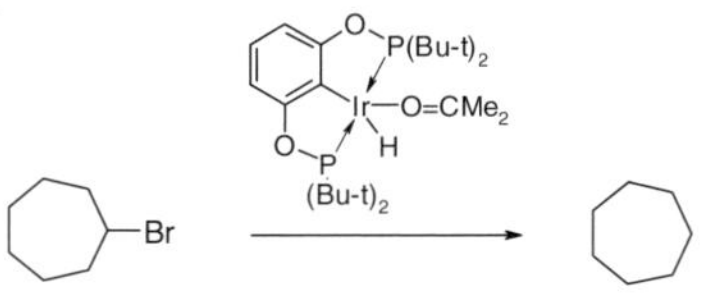

3 Eqs. triethylsilane added to a soln. of 0.5 mol% iridium(III) hydride complex in chlorobenzene-d_5 in a medium-walled J. Young NMR tube, the contents shaken, bromocycloheptane added, and

the soln. shaken at 23° for 20 min → cycloheptane. Y >99% (by NMR). The method is successful for prim. (incl. benzyl), sec. and tert. aliphatic halides giving the corresponding alkanes in near-quantitative conversions using 0.5 mol% (or less) of the iridium complex at 23 or 60°. In competition experiments the order of reactivity was determined to be RI > RBr > RCl, while 1-fluoropentane was reduced under more forcing conditions in 92% conversion (2 mol% catalyst/ 60°/50 h). The catalytic cycle is thought to involve initial formation of an equilibrated mixture of an iridium-alkyl halide complex and an electrophilic iridium silane complex; the latter then serves as a silylating agent to generate a silylhalonium ion prior to reduction with a liberated, nucleophilic dihydridoiridium complex. F.e. incl. synthesis and X-ray structure details of the iridium complex s. J. Yang, M. Brookhart, J. Am. Chem. Soc. *2007, 129* (42), 12656-7.

Sulfur ↑ HC ↓↑ S

Sodium amalgam/potassium dihydrogen phosphate *Na,Hg/KH_2PO_4*
Desulfonation SO_2R → H
syn-4-Ene-1,3-diols s. *72*, 96

1,4-Cyclohexadiene/Lewis acids ←
Thioethers from mercaptals $C(SR)_2$ → CHSR
with Ga_2Cl_4 cf. *70*, 44; with 1,4-cyclohexadiene/Lewis acids s. K. Ikeshita, N. Kihara, M. Sonoda, A. Ogawa, Tetrahedron Lett. *2007*, *48* (17), 3025-8.

1,1′,3,3′-Bis(trimethylene)bis(imidazol-2-ylidene) ←
Organocatalyzed desulfonylation with a super-electron donor under mild, neutral conditions SO_2R → H

53.

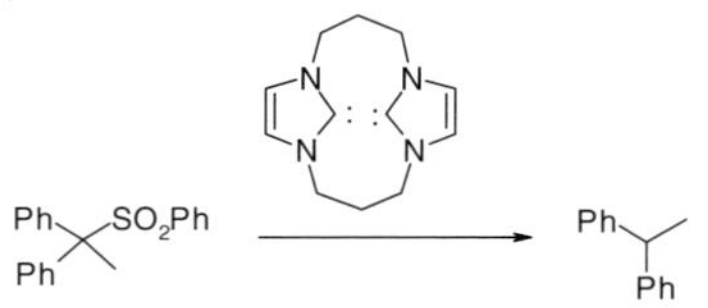

Replacement of sulfonyl groups by hydrogen. 3 Eqs. of the diimidazolium diiodide heated at 110° for 1 h *in vacuo*, cooled, 24 eqs. NaH added under argon, the mixture washed with hexane, dried under argon, dry deoxygenated DMF added dropwise to the resulting *super-electron donor*, stirred for 4 h, centrifuged, the supernatant liquid transferred via cannula to dry 1,1-diphenylethyl phenyl sulfone, the mixture heated at 110° for 18 h under argon, cooled, poured into water, extracted with ether, washed with water and brine, dried, and evaporated → 1,1-diphenylethane. Y 97%. The method is successful for the reduction of mildly activated sulfones, particularly *gem*-disulfones which undergo **monodesulfonation** in 94-98% yields. The method is also successful for **N-detosylation** of activated sulfonamides (e.g. N-tosylindole and N-tosyl-N-benzylaniline in 91 and 74% yields respectively). The generated electron-donor, which is a markedly stronger reducing agent than existing tetraazaalkenes, affords stabilized radical cations and dications on oxidation, which transfer an electron to the substrates; these fragments then take up a further electron to afford mixtures of anions, some of which have been trapped and isolated. F.e., also preparation of the catalyst, s. F. Schoenebeck, J.A. Murphy, S. Zhou, Y. Uenoyama, Y. Miclo, T. Tuttle, J. Am. Chem. Soc. *2007*, *129* (44), 13368-9.

Nickel *Ni*
2-Deoxy- from 2-arylthio-2-deoxy-furanosides s. *72*, 67 SR → H

Remaining Elements ↑ HC ↓↑ Rem

Potassium tert-*butoxide/tetra*-n-*butylammonium fluoride* *KOBu-t/Bu_4NF*
2-Ethyleneamines from 1,2-azasilacyclopent-3-enes s. *72*, 458 C

Carbon ↑ HC ↓↑ C

Trifluoromethanesulfonic acid or Hydrogen bromide *CF_3SO_3H or HBr*
Acid-catalyzed ar. C-deacylation C(O)R → H

54. 

with O-demethylation. 50% Trifluoromethanesulfonic acid in water added to a soln. of 1-(3-methoxyphenyl)ethanone in toluene (or xylene) as cosolvent, and refluxed for 48 h → phenol. Y 67%. The procedure is applicable to sterically non-hindered alkyl aryl ketones possessing strong electron-releasing groups (OMe, OH). Cleavage with aq. HBr in toluene, however, was limited to substrates possessing OH or OMe groups at the conjugating *o*- or *p*-site, but with either acid it was not possible to cleave substrates possessing more basic amino groups, as protonation blocks their activating effect. Cleavage with *anhydrous* acid in an aromatic hydrocarbon was facilitated by capture of the departing acylium ion by the solvent, but yields were lowish. F.e. incl. naphthalene derivs. s. M. Kozelj, A. Petric, Synlett *2007* (11), 1699-702.

Elimination ⇑

Oxygen ↑ HC ⇑ O

Carbon tetrabromide or Trifluoromethanesulfonic acid/isopropanol *CBr_4 or TfOH/i-PrOH*
Carbon tetrabromide or Trifluoromethanesulfonic acidisopropanol/microwaves ←
Diarylmethanes from benzhydrols OH → H
cf. *60*, 45; via disproportionation with 20 mol% CBr_4 or TfOH in isopropanol under thermal conditions or microwave irradiation s. N. L'Hermite, A. Giraud, O. Provot, J.-F. Peyrat, M. Alami, J.-D. Brion, Tetrahedron *2006*, *62* (51), 11994-2002.

N-*Isopropylidene*-N′-*2-nitrobenzenesulfonylhydrazine/triphenylphosphine/ diethyl azodicarboxylate* ←
Deoxygenation of activated alcohols via fragmentation of diazenes

55. 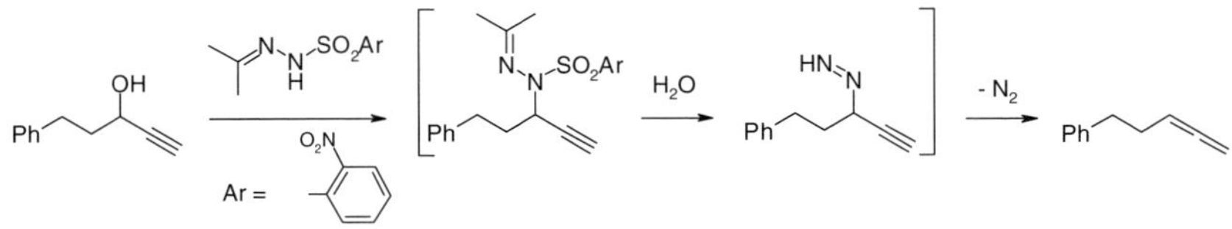

One-pot procedure under mild conditions. 1.2 Eqs. diethyl azodicarboxylate added dropwise to a soln. of 1.2 eqs. *N*-isopropylidene-*N′*-2-nitrobenzenesulfonylhydrazine, 1 eq. 5-phenylpent-1-yn-3-ol and 1.2 eqs. triphenylphosphine in dry THF at 0° under argon, the mixture warmed to 23° after 5 min, stirred 2 h, 50% aq. trifluoroethanol added, stirred for 2 h, partitioned between water/*n*-pentane, separated, dried, and evaporated → 5-phenylpenta-1,2-diene. Y 70%. The reagent has greater thermal stability than 2-nitrobenzenesulfonylhydrazine, offering greater flexibility in the initial **Mitsunobu reaction** and rapid thermal fragmentation of the intermediate diazene. The method is notably useful for the deoxygenation of allylic and propargylic prim. and sec. alcohols, the former undergoing double bond shift, and the latter affording allenes, on fragmentation of the diazene. Unactivated sec. alcohols, however, reacted sluggishly in low yield. The method is tolerant of ethers and carbonates with yields of 60-87%. F.e. incl. preparation of the reagent s. M. Movassaghi, O.K. Ahmad, J. Org. Chem. *2007*, *72* (5), 1838-41.

Via intermediates *v.i.*
Barton-type deoxygenation of sec. alcohols via xanthates
with retention of chlorides and bromides s. *72*, 48

Carbon ↑ **HC ⇑ C**

Thioureidoquinidines ←
α-Acylaminocarboxylic acid esters COOH → H
from α-acylaminomalonic acid monoesters
Organocatalyzed asym. decarboxylative protonation

56.

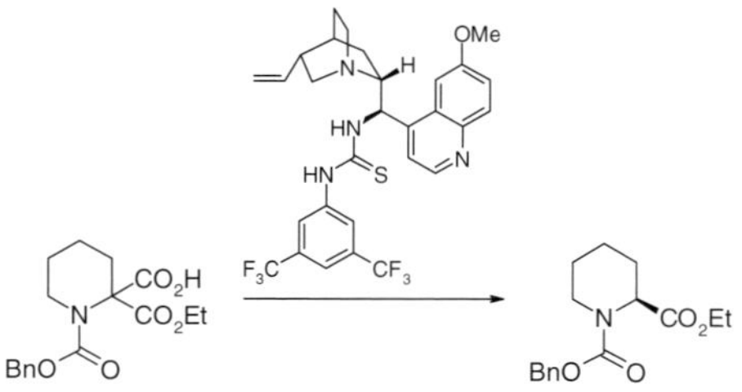

Equimolar amounts of the startg. malonic acid monoester (0.175 mmol scale) and the *bifunctional* thioureidoquinidine in THF stirred at 0° until TLC indicated completion of reaction (7 days) → (S)-1-benzyloxycarbonylpiperidine-2-carboxylic acid ethyl ester. Y 88% (e.e. 83%). Reaction is applicable to cyclic, acyclic or bicyclic substrates **under mild, metal-free conditions,** the (R)-products being obtained with the pseudoenantiomeric form of the catalyst. The amino function of the latter acts as a proton shuttle while the thiourea group serves as a hydrogen-bonding anchor to bring the chiral protonating agent into the vicinity of the enolate. Both steric and electronic properties of the N-acyl group dictate the enantioselectivity, which is highest in solvents of low polarity. The temperature also has a critical effect. F.e. and comparison of *cinchona*-based thioureas and ureas, also with N-acyl and N-carbomethoxy derivs., s. M. Amere, M.-C. Lasne, J. Rouden, Org. Lett. *2007, 9* (14), 2621-4.

Rhodium trichloride/1,3-bis(diphenylphosphino)propane *RhCl₃/dppp*
Decarbonylation of aldehydes CHO → H
cf. *21*, 121s*36*, *48*, 86; with $RhCl_3 \cdot 3H_2O$/dppp in diglyme s. M. Kreis, A. Palmelund, L. Bunch, R. Madsen, Adv. Synth. Catal. *2006*, *348* (15), 2148-54.

Palladous trifluoroacetate/trifluoroacetic acid *$Pd(OCOCF_3)_2/CF_3COOH$*
Catalytic decarboxylation of electron-rich arenes under mild conditions COOH → H

57.

0.2 Eqs. $Pd(OCOCF_3)_2$ added to a soln. of the startg. carboxylic acid in 5% DMSO/DMF, treated with 10 eqs. trifluoroacetic acid (as proton source), and heated at 70° for 24 h → product. Y 74%. Yields were high (70-100%) for the decarboxylation of *o,o'*-dimethoxyarenecarboxylic acids, but lower for *o*-methoxy-derivs. Reaction is thought to proceed through a 4-membered transition state, which is decarboxylated to give an arylpalladium(II) trifluoroacetate prior to proton transfer. No base must be present. F.e. and proton sources s. J.S. Dickstein, C.A. Mulrooney, E.M. O'Brien, B.J. Morgan, M.C. Kozlowski, Org. Lett. *2007, 9* (13), 2441-4.

Formation of O-O Bond

Uptake ⇓

Addition to Oxygen and Carbon OO ⇓ OC

Ammonium cerium (IV) nitrate or Sulfuric acid or Iodine *CAN or H_2SO_4 or I_2*
1,1-Bis(hydroperoxides) from oxo compds. CO → $C(OOH)_2$
with I_2 cf. *71*, 35; also 1,1-hydroxyhydroperoxides from aliphatic aldehydes, **1,1-alkoxyhydroperoxides** from ar. aldehydes in methanol, and **monoperoxyacetals** with *tert*-butyl hydroperoxide in methanol, s. K. Zmitek, M. Zupan, S. Stavber, J. Iskra, J. Org. Chem. *2007, 72* (17), 6534-40; with a little CAN in aq. H_2O_2 (50%)/acetonitrile s. B. Das, M. Krishnaiah, B. Veeranjaneyulu, B. Ravikanth, Tetrahedron Lett. *2007*, *48* (36), 6286-9; with 0.6 eq. concd. H_2SO_4 in aq. H_2O_2 (37%)/ THF s. A.O. Terent'ev, M.M. Platonov, Y.N. Ogibin, G.I. Nikishin, Synth. Commun. *2007*, *37* (8), 1281-7.

Formation of O-N Bond

Uptake ⇓

Addition to Hydrogen and Nitrogen ON ⇓ HN

Methyl(trifluoromethyl)dioxirane ←
Hydroxylamines from amines NH → NOH
N-subst. hydroxylamines with Na_2WO_4/urea-H_2O_2 cf. *55*, 49s*70*; N-hydroxylation of N-Boc-peptide methyl esters with methyl(trifluoromethyl)dioxirane, also side-chain hydroxylation of N-acetyl-derivs. (cf. *49*, 99), s. M.R. Rella, P.G. Williard, J. Org. Chem. *2007*, *72* (2), 525-31.

Addition to Nitrogen ON ⇓ N

Rhenium catalysts/sodium percarbonate ←
Methylrhenium trioxide/urea-hydrogen peroxide *$MeReO_3/(H_2N)_2CO\text{-}H_2O_2$*
N-Oxidation ←
cyclic N-oxides with $MeReO_3/H_2O_2$ cf. *51*, 45s*61*; with rhenium catalysts/Na-percarbonate s. S.L. Jain, J.K. Joseph, B. Sain, Synlett *2006* (18), 2661-3; nitrones from azomethines (cf. *44*, 83s*45*) with $MeReO_3$/urea-H_2O_2 s. G. Soldaini, F. Cardona, A. Goti, Org. Lett. *2007*, *9* (3), 473-6.

Exchange ↓↑

Hydrogen ↑ ON ↓↑ H

Polymer-based tungsten or molybdenum compds./hydrogen peroxide ←
Nitroso compds. from amines NH_2 → NO
with MoO_3/H_2O_2/KOH cf. *16*, 142s*67*; with recyclable, polymer-based tungsten or molybdenum compds./H_2O_2 s. A. Bordoloi, S.B. Halligudi, Adv. Synth. Catal. *2007*, *349* (13), 2085-8.

Hypofluorous acid-acetonitrile *HOF-MeCN*
Nitro compds. from amines NH_2 → NO_2
s. *46*, 90s*48,50*; α-alkylation of α-amino esters via oxidation to α-nitro esters with HOF-acetonitrile complex s. T. Harel, S. Rozen, J. Org. Chem. *2007*, *72* (17), 6500-3.

Formation of O-Hal Bond

Uptake ⇓

Addition to Halogen OHal ⇓ Hal

Ruthenium trichloride/peroxyacetic acid *$RuCl_3/MeCOO_2H$*
Iodoxy compds. from iodides $I \rightarrow IO_2$
with K-bromate cf. *19*, 125s*37*; with $RuCl_3$ (0.08 mol%) and peroxyacetic acid in acetic acid, iodoxyarenes, notably incl. trifluoromethylated derivs., s. A.Y. Koposov, R.R. Karimov, A.A. Pronin, T. Skrupskaya, M.S. Yusubov, V.V. Zhdankin, J. Org. Chem. *2006*, *71* (26), 9912-4.

Formation of O-S Bond

Uptake ⇓

Addition to Sulfur OS ⇓ S

Chiral copper Schiff base complexes/hydrogen peroxide *$[Cu]/H_2O_2$*
Chiral aluminum salalen complex/hydrogen peroxide *$[Al]/H_2O_2$*
Sulfoxides from thioethers by asym. oxidation $>S \rightarrow >SO$
cf. *46*, 106s*71*; chiral aryl benzyl sulfoxides (e.e. up to 81%) with chiral copper Schiff base complexes s. P. Kelly, S.E. Lawrence, A.R. Maguire, Synlett *2007* (10), 1501-6; with a chiral Al-salalen complex/H_2O_2 s. *72*, 59.

tert-*Butyl hydroperoxide s. under Tungsten oxide* t-*BuOOH*

Amberlyst 15 or Amberlite IR-400/hydrogen peroxide ←
L-*Proline/hydrogen peroxide* *$Pro\text{-}OH/H_2O_2$*
Sulfoxides from thioethers
cf. *5*, 101; with H_2O_2 and Amberlyst 15 or Amberlite IR-400 at room temp., dialkyl and diaryl sulfoxides, leaving functional groups such as hydroxyl, methoxy, amino, aldehyde and olefin intact, s. M.M. Lakouraj, M. Tajbakhsh, H. Tashakkorian, Monatsh. Chem. *2007*, *138* (1), 83-8; self-catalyzed chemoselective S-oxidation with H_2O_2 in the absence of solvent s. F. Shi, M.K. Tse, H.M. Kaiser, M. Beller, Adv. Synth. Catal. *2007*, *349* (16), 2425-30; chemoselective S-oxidation with L-proline/H_2O_2 under ambient conditions s. K. Rajender Reddy, C.V. Rajasekhar, A. Ravindra, Synth. Commun. *2006*, *36* (24), 3761-6.

Silica s. under Silica-sulfuric acid and Tungsten oxide *SiO_2*

Chiral azasteroidal-based oxaziridinium fluoroborate ←
Sulfoxides from thioethers by asym. oxidation
with chiral N-sulfonyloxaziridines cf. *35*, 57s*60*; with a chiral azasteroidal-based oxaziridinium fluoroborate (e.e. up to >99%) s. R.E. del Río, B. Wang, S. Achab, L. Bohé, Org. Lett. *2007*, *9* (12), 2265-8.

Vanadium trichloride/hydrogen peroxide *VCl_3/H_2O_2*
Sulfoxides from thioethers
with V_2O_5 cf. *25*, 80; with VCl_3 s. R. Trivedi, P. Lalitha, Synth. Commun. *2006*, *36* (24), 3777-82.

Hydrogen peroxide (s.a. under Al-salalen complex, Amberlyst 15, L-*Proline, VCl_3, Silica-sulfuric acid, MoO_3 and Hemoglobin)* *H_2O_2*
Sulfoxides from thioethers s. *5*, 101s*72*

Silica-sulfuric acid/hydrogen peroxide *SiO_2-HSO_4/H_2O_2*
Sulfoxides or sulfones from thioethers >S → >SO or >SO_2
with silica-supported tungstate cf. *37*, 81s*69*; with silica-sulfuric acid and aq. H_2O_2 at 25-90°, affording [alkyl or aryl] sulfoxides or sulfones depending on the amounts of oxidant and catalyst, s. A. Shaabani, A.H. Rezayan, Catal. Commun. *2007*, *8* (7), 1112-6.

*Tungsten oxide nanoclusters-on-mesoporous silica/*tert*-butyl hydroperoxide* ←
Sulfoxides from thioethers >S → >SO
Heterogeneous oxidation under mild conditions

58. 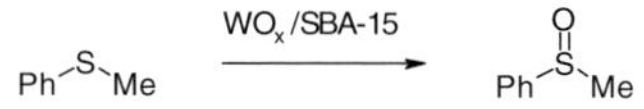

A mixture of 2 mmol methyl phenyl sulfide, 3 eqs. 70% *tert*-butyl hydroperoxide and 0.05 g tungsten oxide nanoclusters-on-mesoporous silica (SBA-15) in 1:1 methanol/methylene chloride stirred at room temp. until chromatography indicated completion of reaction, the catalyst filtered off, the solvent removed *in vacuo,* and the residue worked up with column chromatographic purification → methyl phenyl sulfoxide. Y 99%. The procedure is environmentally friendly and generally applicable to aryl alkyl, diaryl (incl. hindered substrates), dialkyl and dibenzyl thioethers in high yield and at high conversions, without over-oxidation to the sulfone or epoxidation of isolated double bonds, and permitting recycling of the catalyst without undue loss of activity. The high catalytic activity is attributed to highly dispersed tungsten oxide within the SBA-15 meso-channels, the unique pore size and low surface acidity facilitating discharge of the desired products from the channels at room temp. and under atmospheric pressure. F.e. and preparation of the catalyst s. A. Bordoloi, A. Vinu, S.B. Halligudi, Chem. Commun. *2007* (45), 4806-8.

Molybdenum trioxide/hydrogen peroxide *MoO_3/H_2O_2*
Sulfoxides or sulfones from thioethers >S → >SO or >SO_2
with MoO_2Cl_2/H_2O_2 cf. *17*, 153s*71*; with MoO_3/H_2O_2 in ethanol at 50°, selectivity, s. M.M. Khodaei, K. Bahrami, M. Khedri, Can. J. Chem. *2007*, *85* (1), 7-11.

N,N′-Dibenzyl-N,N,N′,N′-tetramethylethylenediammonium permanganate ←
Sulfones from thioethers >S → >SO_2
with $KMnO_4$ cf. *1*, 492(s*71*); with N,N′-dibenzyl-N,N,N′,N′-tetramethylethylenediammonium permanganate in acetic acid/acetonitrile at room temp. s. M.M. Lakouraj, M. Tajbakhsh, H. Tashakkorian, K. Ghodrati, Phosphorus, Sulfur Silicon Relat. Elem. *2007*, *182* (2), 485-90.

Hemoglobin/hydrogen peroxide ←
Sulfoxides from thioethers >S → >SO
biomimetic S-oxidation with N^5-ethylflavinium perchlorate cf. *45*, 255; with human hemoglobin and 30% H_2O_2 in phosphate buffer (pH 7.6) at room temp. s. A. Kumar, Akanksha, Tetrahedron Lett. *2007*, *48* (44), 7857-60.

Chiral chloroiron(III) salan complex/hydrogen peroxide ←
Sulfoxides from thioethers by asym. oxidation under mild conditions in water

59. 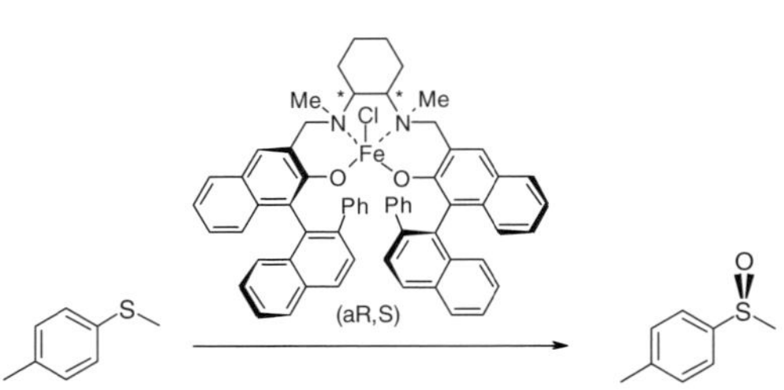

An *environmentally friendly* oxidation of thioethers is now at hand with water as solvent by using a novel, reduced salen-type, chiral iron(III) complex as catalyst and 'water-forming' *hydrogen peroxide* as oxidant, effecting high enantioselectivities *with high catalyst turnover* and *without* requiring a surfactant. **E:** The startg. thioether (0.2 mmol scale) added to 1 mol% chiral iron(III) salan complex in a Schlenk tube, diluted with water, stirred for 10 min at 20°, 1.5 eqs. H_2O_2 added to the suspension in one portion, stirring continued for 3 h *at room temp. in air,* and worked up with chromatographic purification on silica gel → (S)-product. Y 88% (e.e. 96%). The procedure is widely applicable to alkyl aryl and dialkyl thioethers (e.e. 81-96%), although the yield of sulfoxide from the latter is slightly lower due to a little over-oxidation to the sulfone. The yield from *o*-subst. aryl thioethers, however, was negligible, presumably for steric reasons. F.e.s. H. Egami, T. Katsuki, J. Am. Chem. Soc. *2007, 129* (29), 8940-1; with a chiral water-compatible aluminum salalen complex [where salalen is a hybrid salan/salen tetradentate ligand] and 30% aq. H_2O_2 in methanol containing a phosphate buffer (pH 7.4) for asym. oxidation of aryl alkyl thioethers and benzyl methyl thioethers with high enantioselectivity (e.e. 80-99%) cf. T. Yamaguchi, K. Matsumoto, B. Saito, T. Katsuki, Angew. Chem. Int. Ed. Engl. *2007, 46* (25), 4729-31.

Exchange ⇂↾

Halogen ↾ OS ⇂↾ Hal

Indium *In*

Sulfonic acid esters from alcohols OH → OSO_2R

cf. *4*, 212; with In (0.1 eq.) in acetonitrile at room temp., also sulfonamides (cf. *1*, 271), s. J.-G. Kim, D.O. Jang, Synlett *2007* (16), 2501-4.

Carbon ↾ OS ⇂↾ C

Peroxyformic acid *$HCOO_2H$*

Sulfonic acids from thiolacetic acid esters SAc → SO_3H

with $KHSO_5$/KOAc cf. *69*, 51; with formic acid/30% H_2O_2 on the multikilogram scale s. D.H. Brown Ripin, G.A. Weisenburger, D.J. am Ende, D.R. Bill, P.J. Clifford, C.N. Meltz, J.E. Phillips, Org. Proc. Res. Dev. *2007, 11* (4), 762-5.

Formation of O-Rem Bond

Uptake ⇓

Addition to Hydrogen and Oxygen ORem ⇓ HO

Oxorhenium(VII) complexes *[Re]*

Phosphoric acid monoesters from alcohols OH → OPO_3H

with trimethylsilyl polyphosphate cf. *41*, 101; under oxorhenium(VII) catalysis s. A. Sakakura, M. Katsukawa, K. Ishihara, Angew. Chem. Int. Ed. Engl. *2007, 46* (9), 1423-6.

Palladium/carbon *Pd/C*

Phosphorous acid [H-phosphonate] monoesters from alcohols P-H → P-OR

via palladium-catalyzed transfer-hydrogenation s. *72*, 62

Addition to Oxygen and Carbon ORem ⇓ OC

Silver triflate/triethylphosphine *AgOTf/Et$_3$P*
Alkoxysilanes from aldehydes and organosilicon hydrides CHO → $CH_2OSi\leqslant$
Silver-catalyzed hydrosilylation

60. 

20 Mol% triethylphosphine added to a soln. of 3 mol% AgOTf in THF, the colorless mixture stirred for 5 min, 4-acetylbenzaldehyde added, stirred for a further 5 min, 2 eqs. dimethylphenylsilane added, heated at 70° for 24 h, and the mixture filtered through Al_2O_3 → silyl ether product. Y >99%. This is the first application of silver catalysis for hydrosilylation of unsatd. compds. In initial tests $AgBF_4$, AgOTf and $AgSbF_6$ were identified as useful, but AgCl was less effective. The presence of a sterically accessible phosphine reduces the reaction rate but improves selectivity, notably for aliphatic aldehydes (*n*-BuCHO: 86%, cf. 9% without phosphine) while sterically hindered aldehydes give reduced yields even in the presence of phosphines (*t*-BuCHO: 48%; 2,6-dimethylbenzaldehyde: 76%). Toluene and DMSO were also useful solvents in this reaction but 1,2-dichloroethane was not. F.e.s. B.M. Wile, M. Stradiotto, Chem. Commun. *2006* (39), 4104-6.

Gold(I) phosphine complexes *[Au]*
Perrhenic acid *HReO$_4$*
Rhodium catalyst/silica-based phosphine ←
Alkoxysilanes from oxo compds. and organosilicon hydrides CO → CH($OSi\leqslant$)
with Cu(I)-imidazol-2-ylidene complexes cf. *66*, 57s*67,71*; from aldehydes with 3 mol% Me_2S-AuCl and 20 mol% indenyl- or trialkyl-phosphine ligand s. B.M. Wile, R. McDonald, M.J. Ferguson, M. Stradiotto, Organometallics *2007*, *26* (4), 1069-76; also from ketones with dimethylphenylsilane using perrhenic acid, and dehydrogenative silylation of alcohols in the absence of solvent at 25° (cf. *48*, 97), s. P.M. Reis, B. Royo, Catal. Commun. *2007*, *8* (7), 1057-9; from hindered ketones under rhodium catalysis with a silica-immobilized phosphine s. G. Hamasaka, A. Ochida, K. Hara, M. Sawamura, Angew. Chem. Int. Ed. Engl. *2007*, *46* (28), 5381-3.

Dirhodium(II) tetrakis(perfluorobutyrate) *Rh$_2$(pfb)$_4$*
1,4-Hydrosilylation of α,β-ethyleneoxo compds. CHC=C($OSi\leqslant$)
of α,β-ethyleneketones with Rh(I)-complexes cf. *49*, 91; also of α,β-ethylenealdehydes with dirhodium(II) complexes, especially dirhodium(II) tetrakis(perfluorobutyrate), s. M. Anada, M. Tanaka, K. Suzuki, H. Nambu, S. Hashimoto, Chem. Pharm. Bull. *2006*, *54* (11), 1622-3.

Addition to Remaining Elements ORem ⇓ Rem

Hexachloroacetone/(-)-menthol ←
Phosphine oxides from phosphines $\geqslant$P → $\geqslant$PO
Asym. oxidation with dynamic kinetic resolution

61. 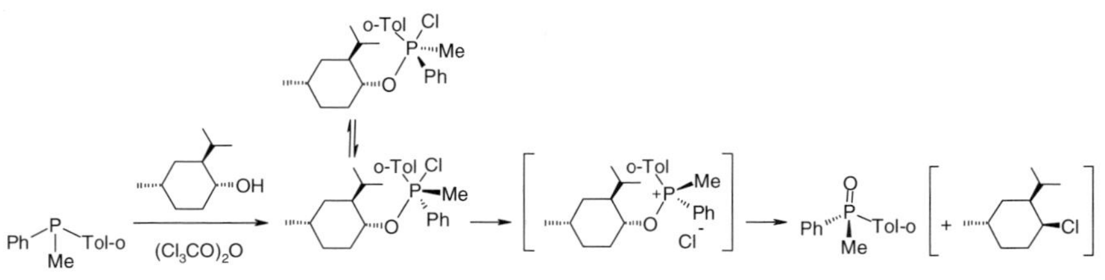

Anhydrous toluene, and stock solns. of the startg. phosphine and 1.2 eqs. (-)-menthol (each in toluene) added to a dry flask containing pre-dried 4 Å molecular sieves under N_2, cooled to -78°, stirred for 10 min, 1 eq. hexachloroacetone in dry toluene added steadily over 2 min, kept at -78° for 1 h, allowed to warm to room temp., stirring continued for 24 h, and worked up → (R)-product. Y >95% (e.e. 80%). Enantioselectivity is the highest on record for asym. oxidation of phosphines. Resolution is thought to be facilitated via low-temperature pseudorotation of pentavalent alkoxy(chloro)phosphoranes, which fragment via the corresponding chiral alkoxyphosphonium chlorides, although a thermodynamic resolution via the equilibrated phosphoranes cannot be ruled out. A number of cyclic chiral sec. alcohols were tested, but (-)-menthol proved to be the least expensive, comparing favourably with the cost of typical loadings of chiral metal-based catalysts. F.e. and oxidative homocoupling to **chiral 1,2-bis(phosphine oxides)** s. E. Bergin, C.T. O'Connor, S.B. Robinson, E.M. McGarrigle, C.P. O'Mahony, D.G. Gilheany, J. Am. Chem. Soc. *2007, 129* (31), 9566-7.

Exchange ↓↑

Hydrogen ↑ — ORem ↓↑ H

Tris(triphenylphosphine)copper(I) chloride — *$(Ph_3P)_3CuCl$*
Triphenylphosphine — *Ph_3P*
Acoxysilanes from organosilicon hydrides — ⩾SiH → ⩾SiOAc
with 15 mol% $ZnCl_2$ cf. *11*, 171s*71*; with 4 mol% $(Ph_3P)_3CuCl$ in acetonitrile at 80°, also with Ph_3P, s. G.-B. Liu, H.-Y. Zhao, T. Thiemann, Synth. Commun. *2007*, *37* (16), 2717-27; with the latter in DMF at 120° s. idem. Adv. Synth. Catal. *2007*, *349* (6), 807-11.

Perrhenic acid — *$HReO_4$*
O-Silylation with organosilicon hydrides s. *48*, 97s*72* — OH → OSi⩽

Cationic dirhodium(II) complexes — *[Rh]*
O-Silylation with organosilicon hydrides
with $RhCl(PPh_3)_3$ cf. *41*, 104; with cationic dirhodium(II) complexes in the presence or absence of solvent or in an ionic liquid 2-phase system s. A. Biffis, M. Basato, M. Brichese, L. Ronconi, C. Tubaro, A. Zanella, C. Graiff, A. Tiripicchio, Adv. Synth. Catal. *2007*, *349* (16), 2485-92.

Palladium/carbon — *Pd/C*
Phosphorous acid [*H*-phosphonate] monoesters from alcohols via palladium-catalyzed transfer-hydrogenation — P-H → P-OR

62.

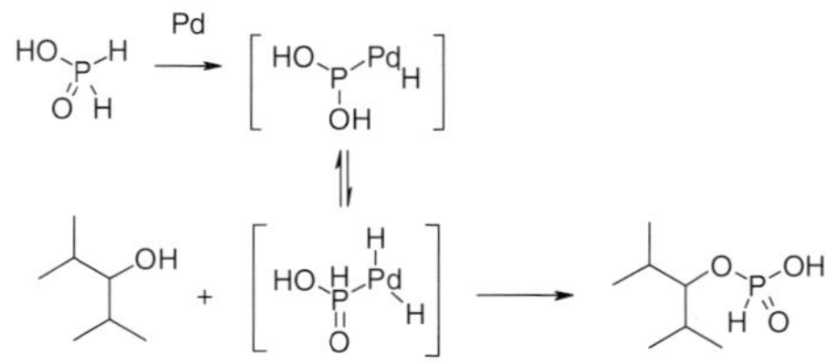

A mixture of the startg. alcohol, 1.5 eqs. H_3PO_2 (previously concentrated under 0.5 mm Hg at room temp.) and 10 wt% Pd/C (5 mol% Pd) heated at 85° in toluene/acetonitrile (1:1) under N_2 for 24 h, the mixture cooled, filtered through Celite, concentrated, dissolved in ethyl acetate, washed with brine, extracted with aq. $NaHCO_3$, the aq. extracts acidified to pH 1 (10% HCl), and extracted with ethyl acetate → 2,4-dimethylpent-3-yl *H*-phosphonate. Y 78%. The reaction conditions are optimized in terms of temperature, solvent and catalyst loading and afford *H*-phosphonate monesters of prim. and sec. alcohols under mild conditions in an atom-economical and environmentally friendly manner. Products are generally isolated as the free acid in 67-97% yields, however water-soluble products require precipitation as ammonium salts and give lower yields (37-56%). The method is tolerant of halide, alkene, alkyne, ketone and ether functionality. The reaction is thought to involve intermediate formation of a **phosphinidene oxide-dihydrido-**

palladium(II) complex, which undergoes transfer-hydrogenation with the alcohol to give the product. F.e.s. L. Coudray, I. Abrunhosa-Thomas, J.-L. Montchamp, Tetrahedron Lett. *2007*, *48* (37), 6505-8.

Oxygen ↑ ORem ↓↑ O

p-*Toluenesulfonic acid-Celite* *TsOH-Celite*
Phosphonic acid esters from phosphonic acids P-OH → P-OR
with silica chloride cf. *48*, 104s*71*; with *p*-TsOH-Celite, rapid procedure in the absence of solvent, alkanephosphonates, s. A.K. Gupta, R. Kumar, D.K. Dubey, M.P. Kaushik, J. Chem. Res. *2007* (6), 328-31.

Nitrogen ↑ ORem ↓↑ N

Indium tribromide or Ammonium thiocyanate or Trichloroisocyanuric acid or N-Bromosuccinimide or Sodium hydrogen sulfate-silica or Silica-sulfuric acid *$InBr_3$ or NH_4SCN or TCCA or NBS or $NaHSO_4$-SiO_2 or SiO_2-OSO_3H*
Catalytic O-trimethylsilylation with hexamethyldisilazane OH → $OSiMe_3$
of alcohols with I_2 cf. *60*, 55; of a variety of hydroxyl compds. with $InBr_3$ in methylene chloride at room temp. s. J.S. Yadav, B.V.S. Reddy, A.K. Basak, G. Baishya, A. Venkat Narsaiah, Synthesis *2006* (22), 3881-4; with trichloroisocyanuric acid under heterogeneous conditions, incl. phenols, s. A. Khazaei, M.A. Zolfigol, A. Rostami, A.G. Choghamarani, Catal. Commun. *2007*, *8* (3), 543-7; with NBS, also in the absence of solvent, and O-detrimethylsilylation in methanol (cf. *29*, 415s*64*) s. A. Khazaei, A. Rostami, A. Raiatzadeh, M. Mahboubifar, Can. J. Chem. *2007*, *85* (5), 336-40; with silica-sulfuric acid cf. H. Ghafuri, B. Eftekhari-Sis, M.M. Hashemi, Phosphorus, Sulfur, Silicon Relat. Elem. *2007*, *182* (7), 1645-51; of phenols with $NaHSO_4$-silica, leaving amino and carboxy groups unaffected, s. M.S. Khalili, H. Ghafuri, S. Mojahedi-Jahromi, M.M. Hashemi, ibid. *182* (1), 175-9; of prim., sec., tert., allylic, and sugar-derived alcohols using NH_4-thiocyanate under neutral conditions s. V.H. Jadhav, K.S.A. Kumar, V.D. Chaudhari, D.D. Dhavale, Synth. Commun. *2007*, *37* (8), 1363-70.

Halogen ↑ ORem ↓↑ Hal

Lithium diisopropylamide i-*Pr_2NLi*
O-Silyl O-alkyl keteneacetals from carboxylic acid esters CHCOOR → C=C(OSi⩽)OR
with LDA/DMPU cf. *39*, 97s*49*; (E)-isomers from *tert*-butyl esters with LDA in cyclopentyl methyl ether at 0-5°, also (1E,3Z)-1-alkoxy-1,3-disiloxy-1,3-dienes from β-keto esters, s. T. Okabayashi, A. Iida, K. Takai, Y. Nawate, T. Misaki, Y. Tanabe, J. Org. Chem. *2007*, *72* (21), 8142-5.

Chiral fluorous lithium dibenzylamides/lithium chloride *⋝NLi/LiCl*
Enoxysilanes from ketones with desymmetrization CHCO → C=C(OSi⩽)
with chiral Mg-bis(dialkylamides) cf. *41*, 166s*69*; with chiral Li-dibenzylamides having a fluorous ponytail s. H. Matsubara, L. Maeda, H. Sugiyama, I. Ryu, Synthesis *2007* (18), 2901-3.

Carbon ↑ ORem ↓↑ C

Chlorobis(cyclooctene)rhodium(I) dimer/hydrogen chloride *$[(coe)_2RhCl]_2$/HCl*
O-Silylation of alcohols with vinylsilanes OH → OSi⩽

63.

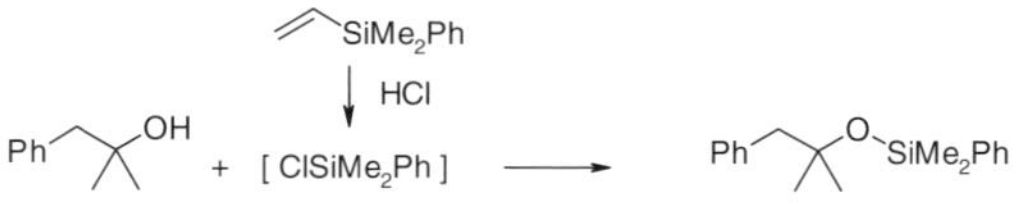

2-Methyl-1-phenyl-2-propanol and 3 eqs. dimethyl(phenyl)vinylsilane added to 2.5 mol% bis(cyclooctene)rhodium(I) chloride in chloroform in a pressure-capped vial, 15 mol% 4 *M* HCl in dioxane added, stirred for 4 h, and the crude mixture purified by chromatography →

dimethylphenyl(2-methyl-1-phenylprop-2-yloxy)silane. Y 88% (GC yield 100%). The procedure is successful with prim., sec. and tert. aliphatic alcohols, and gives high yields even with bulky vinylsilanes. The catalyst is the most efficient of several tested, and even at loadings as low as *0.01 mol%* (with 0.06 mol% HCl) silylation is complete within 24 h. The reaction is thought to involve initial ligand exchange to form the chlorobis(vinylsilane)rhodium(I) dimer, followed by addition of HCl and ultimately reductive elimination of reactive dimethylphenylsilyl chloride which, after silylation, is regenerated in the catalytic cycle. F.e. incl. use of the protocol to prepare di- and tri-alkoxysilanes s. J.-W. Park, C.-H. Jun, Org. Lett. *2007*, *9* (20), 4073–6.

Formation of O-C Bond

Uptake ⇓

Addition to Hydrogen and Carbon OC ⇓ HC

Copper piperazine hybrid polymer/lithium hydroxide *[Cu]/LiOH*
α-Hydroxylation H → OH
of lactones with $CuCl_2/Me_2NH$ under O_2 cf. *30*, 63; of cyclic or acyclic ketones with a Cu-piperazine hybrid polymer/LiOH s. T. Arai, T. Sato, H. Noguchi, H. Kanoh, K. Kaneko, A. Yanagisawa, Chem. Lett. *2006*, *35* (10), 1094-5.

Hydrogen peroxide s. under NMP·HBr$_3$ and Iron(II) complexes *H_2O_2*

2-(Arylsulfonyl)oxaziridines/chiral bis(Δ^2-oxazolines)/zinc acetate or nickel perchlorate ←
Asym. α-hydroxylation

64. 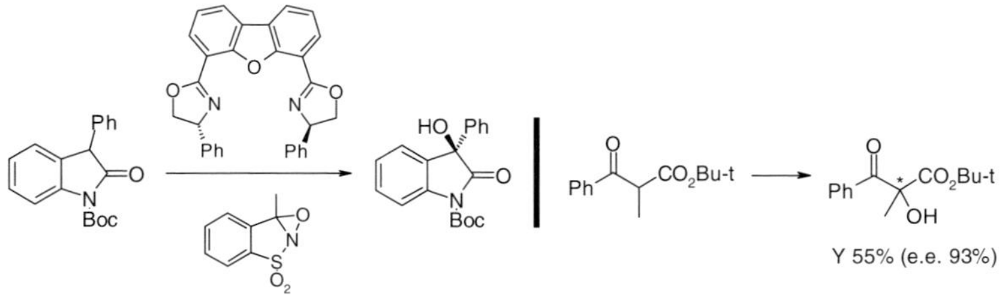

of N-protected oxindoles. 10 Mol% $Zn(OAc)_2$ and 11 mol% (R,R)-4,6-dibenzofurandiyl-2,2′-bis(4-phenyloxazoline) stirred *in vacuo* for 2 h at room temp., dry methylene chloride and 4 Å molecular sieves added under N_2, stirred for 1 h, a soln. of the startg. oxindole in the same solvent added, stirring continued for 30 min, treated directly with 1.2 eqs. of the tricyclic 2-(arylsulfonyl)-oxaziridine, stirred for a further 6 h, quenched with water, and worked up with column chromatographic purification → (R)-N-Boc-3-hydroxy-3-phenyl-2-oxindole. Y 82% (e.e. 93%). This is the first example of a catalytic asym. hydroxylation of oxindoles, enantioselectivities being very high for hydroxylation of 3-aryl derivs. (90-7%), and slightly lower (83-6%) for 3-alkyl- and 3-benzyl-derivs. F.e. and comparison of zinc salts, **also asym. α-alkylation of β-ketocarboxylic acid esters** (e.e. 91-7%) with $Ni(ClO_4)_2$ as catalyst, s. T. Ishimaru, N. Shibata, J. Nagai, S. Nakamura, T. Toru, S. Kanemasa, J. Am. Chem. Soc. *2006*, *128* (51), 16488-9.

Sodium hypochlorite s. under Ni(O)OH *NaOCl*
Sodium periodate or bromate s. under RuCl$_3$ *$NaIO_4$ or $NaBrO_3$*

N-Methyl-2-pyrrolidone hydrotribromide/hydrogen peroxide *$NMP·HBr_3/H_2O_2$*
Carboxylic acids from aldehydes CHO → COOH
with I_2/KI cf. *10*, 220; with N-methyl-2-pyrrolidone hydrotribromide/H_2O_2, arylcarboxylic acids, s. J.K. Joseph, S.L. Jain, B. Sain, Catal. Commun. *2007*, *8* (1), 83-7.

Bis(acetonitrile)[2-[[(S)-2-[(S)-1-(pyridin-2-ylmethyl)pyrrolidin-2-yl]pyrrolidin-1-yl]-methyl]pyridine]iron(II) bis(hexafluoroantimonate)/hydrogen peroxide/acetic acid ←
Remote hydroxylation of unactivated hydrocarbon groups H → OH

65.

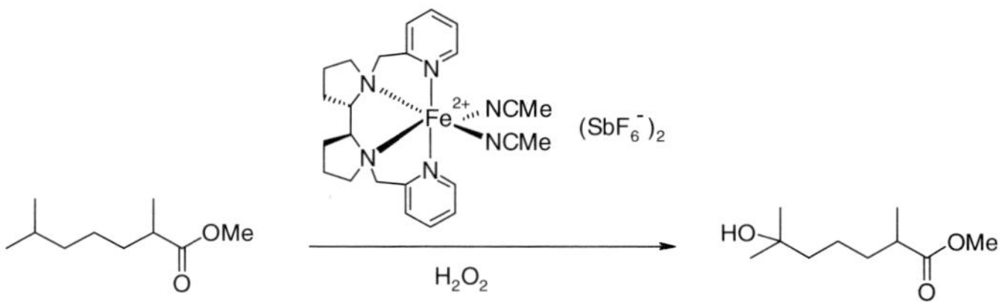

1.2 Eqs. 50% H_2O_2 in acetonitrile added by syringe over ca. 45-75 s to a vigorously stirred soln. of the startg. ester, 5 mol% iron(II) complex and 50 mol% acetic acid in the same solvent at room temp., after ca. 10 min further amounts of the catalyst and acetic acid in acetonitrile added via pipette, followed dropwise by a further amount of H_2O_2 (over the same period), the process repeated a third time after a further ca. 10 min, and worked up with column chromatographic purification → product. Y 56% (ε/α hydroxylation >99:1). Unusually high regioselectivity is achieved *at remote sites,* even in functionalized substrates possessing two tertiary sites. F.e. incl. remote hydroxylation of acoxy compds., ketones, acylamines and halides s. M.S. Chen, M.C. White, Science *2007*, *318* (5851), 783-7.

Nickel oxide hydroxide nanoparticles/sodium hypochlorite *Ni(O)OH/NaOCl*
Carboxylic acids from aldehydes CHO → COOH
Catalytic heterogeneous oxidation in water s. *72*, 95

Nickel perchlorate s. under 2-(Arylsulfonyl)oxaziridines *$Ni(ClO_4)_2$*

Ruthenium trichloride/sodium periodate or bromate *$RuCl_3/NaIO_4$ or $NaBrO_3$*
Steroidal γ-lactolides from tetrahydrofurans H → OH

66.

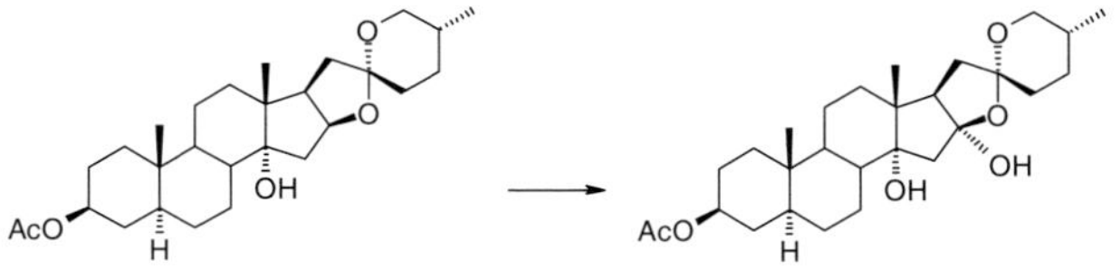

under very mild, neutral conditions. 2.5 Eqs. $NaIO_4$ added to a soln. of C_{14}-hydroxyhecogenin acetate in 1:1:2 ethyl acetate/acetonitrile/phosphate buffer (pH 7.5), the resulting slurry stirred for 5 min at room temp., 5 mol% $RuCl_3·xH_2O$ added in one portion, stirred for 10 h at room temp., diluted with ether to quench the remaining ruthenium peroxo species, stirring continued for 10 min, and filtered → product. Y 85%. The procedure is notably applicable to acid-sensitive subtrates, incl. steroidal [5.5]-spiro systems, and can be scaled up to the 100 g level. F.e. and with less expensive $NaBrO_3$ as the terminal oxidant s. J.S. Lee, H. Cao, P.L. Fuchs, J. Org. Chem. *2007, 72* (15), 5820-3.

Addition to Oxygen and Nitrogen OC ⇓ ON

Polymer-based 4-oxy-L-proline ←
Organocatalyzed asym. α-aminooxylation of oxo compds. H → ONHR
cf. *67*, 57s*69*; with a recyclable polymer-supported *trans*-4-hydroxyproline deriv. (e.e. 96-9%) s. D. Font, A. Bastero, S. Sayalero, C. Jimeno, M.A. Pericàs, Org. Lett. *2007*, *9* (10), 1943-6.

Addition to Oxygen and Carbon OC ⇓ OC

Microwaves s. under Lewis acids [\\\\\\\\]

(2S,6S)-2,6-Bis(o-hydroxyphenyl)-1-(2-pyridylmethyl)piperidine ←
Dicarboxylic acid monoesters from *meso*-anhydrides by asym. desymmetrization
with cinchonine cf. *41*, 118; with the chiral tripodal amine, (2*S*,6*S*)-2,6-bis(*o*-hydroxyphenyl)-1-(2-pyridylmethyl)piperidine (e.e. up to 81%), s. T. Okamatsu, R. Irie, T. Katsuki, Synlett *2007* (10), 1569-72.

Cupric nitrate *Cu(NO$_3$)$_2$*
Acylals from aldehydes CH(OAc)$_2$
cf. *51*, 55s*71*; *56*, 63s*71*; in the absence of solvent with $Cu(NO_3)_2 \cdot 3H_2O$ as catalyst, retention of ketones, s. M.M. Heravi, K. Bakhtiari, M.H. Tehrani, H.A. Oskooie, Synth. Commun. *2007*, *37* (7), 1121-6; from ar. aldehydes with a little recyclable dodecatungstophosphoric acid ($H_3PW_{12}O_{40}$), also deprotection in acetone, s. M.M. Heravi, F. Derikvand, F.F. Bamoharram, ibid. *2006*, *36* (21), 3109-15; with PEG-supported sulfonic acid s. Q.-Y. Wang, S.-R. Sheng, M.-H. Wei, Z.-L. Xie, X.-L. Liu, ibid. *37* (6), 1019-26; with a little *p*-TsOH, also deprotection in water under microwave irradiation, s. K. Manjula, M.A. Pasha, ibid. *37* (9), 1563-9; with HBF_4-SiO_2 (0.1 mol%), compatibility with other protective groups, s. V.T. Kamble, B.P. Bandgar, N.S. Joshi, V.S. Jamode, Synlett *2006* (17), 2719-22; with $Al(HSO_4)_3$ s. B.F. Mirjalili, M.A. Zolfigol, A. Bamoniri, M.A. Amrollahi, N. Sheikhan, Russ. J. Org. Chem. *2007*, *43* (6), 852-4; with FeF_3 (0.1 mol%) at room temp., chemoselectivity and large-scale procedure, s. V.T. Kamble, R.A. Tayade, B.S. Davane, K.R. Kadam, Australian J. Chem. *2007*, *60* (8), 590-4; with P_2O_5/Al_2O_3 s. A.R. Hajipour, A. Zarei, A.E. Ruoho, Tetrahedron Lett. *2007*, *48* (16), 2881-4; with Yb(III)-perfluorooctanesulfonate in a fluorous 2-phase system (toluene/perfluorodecalin) s. W.-B. Yi, Y.-Q. Yin, C. Cai, Can. J. Chem. *2006*, *84* (11), 1563-6.

Cupric triflate *Cu(OTf)$_2$*
Silver triflate/N-iodosuccinimide *AgOTf/NIS*
2-Tolylthio-2-deoxyfuranosides from tolyl 2,3-anhydrothiofuranosides ←
Stereospecific conversion via 1,2-tolylthio group migration

67.
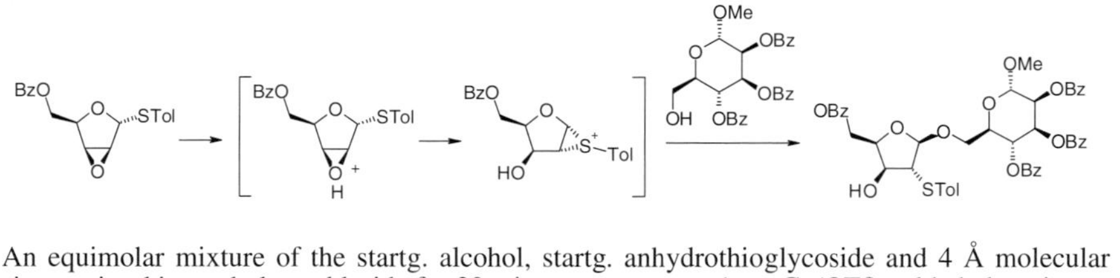

An equimolar mixture of the startg. alcohol, startg. anhydrothioglycoside and 4 Å molecular sieves stirred in methylene chloride for 20 min at room temp., 1 eq. $Cu(OTf)_2$ added, the mixture stirred overnight, neutralized with triethylamine, and concentrated *in vacuo* → the disaccharide. Y 77%. The reaction is clean and rapid with molecular sieves alone under reflux for simple prim., sec. and tert. alcohols, but requires treatment with added $Cu(OTf)_2$ overnight at room temp. for less active carbohydrate donors. The method is applicable to both 2β,3β-anhydro-α- and 2α,3α-anhydro-β-thioglycoside derivs. affording the corresponding 1,2-*trans*-products in 73-91% yields. It is suggested that reaction involves initial protonation of the epoxide (from triflic acid generated during the reaction) and subsequent tolylthio migration. The products are cleanly desulfurized using H_2/Raney nickel to give the corresponding **2-deoxyfuranosides** in moderate yields. F.e. and with AgOTf/NIS as activator s. D. Hou, T.L. Lowary, Org. Lett. *2007*, *9* (22), 4487-90.

Calcium, strontium or barium bis(trimethylsilyl)amide-bis(tetrahydrofuran) ←
Tishchenko reaction 2 RCHO → $RCOOCH_2R$
with Na cf. *4*, 154s*42*; with Ca-, Sr- or Ba-$[N(SiMe_3)_2]_2(THF)_2$ (0.1-10 mol%) in C_6D_6 at 25° s. M.R. Crimmin, A.G.M. Barrett, M.S. Hill, P.A. Procopiou, Org. Lett. *2007*, *9* (2), 331-3.

Alumina s. under P_2O_5 *Al_2O_3*

Mesoporous aluminosilicates ←
Glycol monoethers from epoxides → C(OR)C(OH)
cf. *3*, 115; heterogeneous catalysis with mesoporous aluminosilicates s. M.W.C. Robinson, R. Buckle, I. Mabbett, G.M. Grant, A.E. Graham, Tetrahedron Lett. *2007*, *48* (27), 4723-5; with poly(chloromethylstyrene-co-divinylbenzene)-supported Fe- or Ru-complexes under mild, neutral conditions, also glycols (cf. *2*, 151), s. S.H. Lee, E.Y. Lee, D.-W. Yoo, S.J. Hong, J.H. Lee, H. Kwak, Y.M. Lee, J. Kim, C. Kim, J.-K. Lee, New J. Chem. *2007*, *31* (9), 1579-82.

Lewis acid/microwaves ←
1,3-Dioxolanes from epoxides
cf. *30*, 72s*66*; *24*, 128s*70*; with Lewis acids under microwave irradiation, also Δ^2-oxazolines (cf. *32*, 268), β-lactam-containing derivs., s. F. Benfatti, G. Cardillo, L. Gentilucci, A. Tolomelli, M. Monari, F. Piccinelli, Adv. Synth. Catal. *2007*, *349* (7), 1256-64.

Fluoroboric acid-silica *HBF_4-SiO_2*
Acylals from aldehydes CHO → $CH(OAc)_2$
in the absence of solvent s. *51*, 55s*72*; *56*, 63s*72*

Ytterbium(III) perfluorooctanesulfonate *$Yb(OSO_2R_f)_3$*
Acylals from aldehydes
in a fluorous 2-phase medium s. *51*, 55s*72*; *56*, 63s*72*

β-Cyclodextrin/sodium salt *cyclodextrin/Na^+*
Glycol monoaryl ethers from epoxides → C(OAr)C(OH)
and Na-phenoxides/β-cyclodextrin in aq. acetone cf. *64*, 63s*66*; ring opening of *trans*-3-phenyloxirane-2-methanol in water at 50°, also with Na-thiophenoxides, s. M. Narender, M.S. Reddy, Y.V.D. Nageswar, K.R. Rao, Helv. Chim. Acta *2007*, *90* (6), 1107-11.

Polyethylene glycol s. under Sulfonic acids *PEG*

Dimethylformamide *DMF*
α-Aminocarboxylic acids ←
Polymer-based phenol/base ←
Silica-supported ionic liquid ←
1,3-Dioxolan-2-ones from epoxides
and CO_2 cf. *23*, 139s*70*; with DMF as catalyst in the absence of solvent, optionally with a little added water, s. J.-L. Jiang, R. Hua, Synth. Commun. *2006*, *36* (21), 3141-8; with a silica-supported ionic liquid as recyclable heterogeneous catalyst under supercritical conditions s. J.-Q. Wang, X.-D. Yue, F. Cai, L.-N. He, Catal. Commun. *2007*, *8* (2), 167-72; with naturally-occurring α-amino acids s. C. Qi, H. Jiang, Z. Wang, B. Zou, S. Yang, Synlett *2007* (2), 255-8; with recyclable polystyrene-supported phenol and an organic base (such as DBU, triethylamine, diethylamine, pyridine or, especially, DMAP), influence of reaction temperature, carbon dioxide pressure and reaction time on yield, s. C.-R. Qi, H.-F. Jiang, Z.-Y. Wang, B. Zou, Chin. J. Chem. *2007*, *25* (7), 1051-4; with a rhenium(I) tricarbonyl complex bearing a pyrrolidinium moiety in an ionic liquid medium s. W.-L. Wong, K.-C. Cheung, P.-H. Chan, Z.-Y. Zhou, K.-H. Lee, K.-Y. Wong, Chem. Commun. *2007* (21), 2175-7.

Ionic liquids s.a. under Re-carbonyl complexes ←
N-Iodosuccinimide s. under $Cu(OTf)_2$ or AgOTf *NIS*
Silica s. under HBF_4 and Ionic liquids *SiO_2*

Ammonium (or potassium) nitrate *NH_4NO_3 or KNO_3*
2-Ene-1,4-diones from furans
with CAN cf. *5*, 122s*39*; triarylated *cis*-but-2-ene-1,4-diones (as precursors of 3(2*H*)-furanones) from triphenylfurans with NH_4NO_3 or KNO_3 s. J.P. Rappai, S. Prathapan, M.V. Vishnu Unni, P.A. Unnikrishnan, Synth. Commun. *2007*, *37* (4), 569-72.

Phosphorus pentoxide/alumina *P_2O_5/Al_2O_3*
Acylals from aldehydes CHO → $CH(OAc)_2$
in the absence of solvent s. *51*, 55s*72*; *56*, 63s*72*

Bismuth(III) nitrate *$Bi(OTf)_3$*
Glycol mononitrates from epoxides s. *72*, 69 → $C(ONO_2)C(OH)$

p-*Toluenesulfonic acid* *TsOH*
Polyethylene glycol-supported sulfonic acid *$PEG\text{-}SO_3H$*
Aluminum hydrogen sulfate *$Al(HSO_4)_3$*
Dodecatungstophosphoric acid *$H_3PW_{12}O_{40}$*
Acylals from aldehydes CHO → $CH(OAc)_2$
in the absence of solvent s. *51*, 55s*72*; *56*, 63s*72*

Sodium persulfate s. under $KBrO_3$ *$Na_2S_2O_4$*

Chiral chromium(III) salen complexes *[Cr]*
Glycols from epoxides with kinetic resolution → C(OH)C(OH)
with chiral dimeric cobalt(III) Schiff base-aluminum chloride complexes cf. *53*, 59s*70*; with chiral chromium(III)-salen catalysts based on DIANANE (*endo,endo*-2,5-diaminonorbornane) s. A. Berkessel, E. Ertürk, Adv. Synth. Catal. *2006*, *348* (18), 2619-25.

Potassium bromate/sodium persulfate *$KBrO_3/Na_2S_2O_4$*
Diol monoesters from cyclic acetals
with O_3 or H_2O_2 cf. *39*, 107s*46*; regioselective oxidative cleavage of benzylidene acetals with $KBrO_3/Na_2S_2O_4$ in a 2-phase ($EtOAc/H_2O$) system at room temp. s. P.M. Senthilkumar, A. Aravind, S. Baskaran, Tetrahedron Lett. *2007*, *48* (7), 1175-8.

Pyrrolidinium-bearing rhenium(I) tricarbonyl complex/ionic liquid ←
1,3-Dioxolan-2-ones from epoxides s. *23*, 139s*72*

Iron(III) polyoxometalates ←
Glycol esters from epoxides → C(OAc)C(OAc)
with Ac_2O/H_2SO_4 cf. *13*, 192; with a little $(Bu_4N)_4[PFeW_{11}O_{39}]\cdot 3H_2O$ in Ac_2O s. B. Yadollahi, F.K. Esfahani, Chem. Lett. *2007*, *35* (5), 676-7.

Polymer-based iron complexes *[Fe]*
Glycol monoethers or glycols from epoxides → C(O-)C(OH)
s. *3*, 115s*72*; *2*, 151s*72*

Iron(III) fluoride *FeF_3*
Acylals from aldehydes s. *51*, 55s*72*; *56*, 63s*72* CHO → $CH(OAc)_2$

Polymer-based ruthenium complexes *[Ru]*
Glycol monoethers or glycols from epoxides → C(O-)C(OH)
s. *3*, 115s72; *2*, 151s*72*

Addition to Nitrogen and Carbon OC ⇓ NC

Microwaves s. under $ZnCl_2$ [\\\\]

1,8-Diazabicyclo[5.4.0]undec-7-ene *DBU*
Iminoesters from nitriles CN → C(OR)=NH
glycosyl trichloroacetimidates with KOH/Bu_4NHSO_4 cf. *14*, 181s*51*; glycosyl dichloro(cyano)-acetimidates from dichloro(cyano)acetonitrile with DBU, also conversion to glycosides in the presence of trimethylsilyl triflate (cf. *44*, 211s*71*), s. U. Schmelzer, Z. Zhang, R.R. Schmidt, J. Carbohyd. Chem. *2007*, *26* (4), 223-38.

Copper(II) triflate *Cu(OTf)$_2$*

Tosylaminoethers from small-ring N-heterocyclics and alcohols

Lewis acid-catalyzed regiospecific ring opening with inversion of configuration

68. 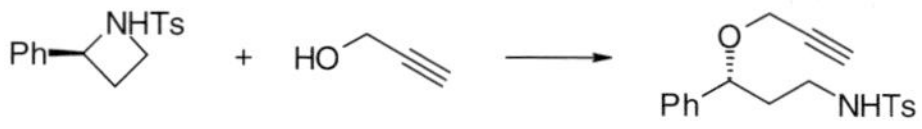

Chiral 1-aryl-3-tosylaminoethers from 2-aryl-1-tosylazetidines. A soln. of (S)-2-phenyl-1-tosylazetidine in propargyl alcohol added to 1 eq. dry $Cu(OTf)_2$ at -25° under argon, the mixture stirred for 1.5 h, quenched with satd. aq. $NaHCO_3$, the aq. layer extracted with methylene chloride, and the organic layer dried (Na_2SO_4) and evaporated → product. Y 91% (e.e. 86%). **Chiral 1-aryl-2-tosylaminoethers** were obtained similarly **from 2-aryl-1-tosylaziridines.** Single regioisomers were formed efficiently, highest yields (>70%) being obtained with prim. aliphatic alcohols (incl. ethylenic and acetylenic), while BnOH and *i*-PrOH gave lower yields (55-76%) and enantioselectivities. Sterically crowded alcohols, such as *t*-BuOH, gave no ring-opened products, while *o*-cresol gave a 1:1 mixture of the regioisomers in 60% yield. The use of methylene chloride as solvent gave a faster reaction but lower enantioselectivity (this example was complete at 0° in 10 min for a 76% e.e.). F.e.s. M.K. Ghorai, K. Das, D. Shukla, J. Org. Chem. *2007*, *72* (15), 5859-62.

Copper(II) triflate/chiral bis(Δ²-oxazolines) *Cu(OTf)$_2$/box*

Urethans from alcohols and isocyanates OH → OC(O)N<

chiral urethans with distannoxanes cf. *49*, 105s*51*; **asym. desymmetrization** of 5- to 8-membered cyclic *meso-vic*-diols (e.e. up to 93%) with $Cu(OTf)_2$/(S,S)-Ph-BOX (0.1 eq. each) s. K. Matsumoto, M. Mitsuda, N. Ushijima, Y. Demizu, O. Onomura, Y. Matsumura, Tetrahedron Lett. *2006*, *47* (48), 8453-6.

Zinc chloride/acetamide/microwaves *ZnCl$_2$/MeCONH$_2$/[\\\\]*

Carboxylic acid amides from nitriles s. *70*, 208s*72* CN → $CONH_2$

Alumina s. under MoO$_3$ *Al$_2$O$_3$*

N-Heterocyclic carbene ←

Nucleophilic ring opening of aziridines with anhydrides → C(OAc)C(N-Ac)

with Ac_2O in the presence of Bu_3P cf. *64*, 66s*66*; with acid anhydrides in the presence of a N-heterocyclic carbene s. X. Sun, S. Ye, J. Wu, Eur. J. Org. Chem. *2006* (21), 4787-90.

Immobilized lipase ←

Enzymatic kinetic resolution of 2-azetidinones

via hydrolysis to β-aminocarboxylic acids cf. *68*, 64; via alcoholysis in the presence of immobilized lipase PS from *Burkholderia cepacia* in *tert*-butyl methyl ether, especially for 3-fluoro-derivs., s. X.-G. Li, M. Lähitie, M. Päiviö, L.T. Kanerva, Tetrahedron:Asym. *2007*, *18* (13), 1567-73.

tert-*Butyl hydroperoxide s. under MoO$_3$* t-*BuOOH*

Bismuth(III) nitrate *Bi(NO$_3$)$_3$*

2-Tosylaminonitrates from N-tosylaziridines → C(ONO_2)C(NH)

Regio- and stereo-specific ring opening

69. 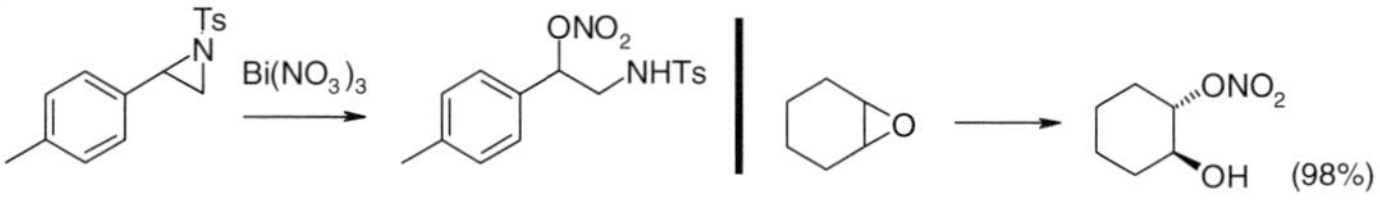

1.2 Eqs. $Bi(NO_3)_3$ added to a soln. of the startg. N-tosylaziridine in acetonitrile, stirred at room temp. for 15 min, quenched with water, and worked up with column chromatographic purification → product. Y 95%. 2-Aryl-N-tosylaziridines underwent ring opening at the benzylic site, while [less reactive] 2-alkyl derivs. underwent ring opening at the unsubstituted site. Condensed N-tosylaziridines were converted to *trans*-products by *anti*-addition of the nucleophile. There was no formation of 2-tosylaminoalcohols. F.e. and comparison of catalysts, **also glycol mononitrates**

from epoxides, s. B. Das, M. Krishnaiah, K. Venkateswarlu, V.S. Reddy, Helv. Chim. Acta *2007, 90* (1), 110-3.

*Molybdenum trioxide-alumina/*tert-*butyl hydroperoxide* *MoO_3-Al_2O_3/t-BuOOH*
Oxaziridines from azomethines ○
with Na_2WO_4/H_2O_2 cf. *50*, 56s*69*; with alumina-supported MoO_3/anhydrous *t*-BuOOH s. S. Singhal, S.L. Jain, V.V.D.N. Prasad, B. Sain, Eur. J. Org. Chem. *2007* (13), 2051-4.

Addition to Carbon-Carbon Bonds OC ⇓ CC

Electrolysis/cationic tetrakis(acetonitrile)palladium(II) complexes *↯/[Pd(II)]*
Methyl ketones from terminal ethylene derivs. $CH{=}CH_2 \rightarrow COCH_3$
Electrochemical Wacker-type reaction
cf. *49*, 107; in the presence of cationic palladium complexes, $[Pd(MeCN)_4][X]_2$ (X = BF_4, PF_6 or ClO_4), generated electrochemically from $Pd(OAc)_2$ and the appropriate tetra-*n*-butylammonium salt (as electrolyte) in acetonitrile s. K. Mitsudo, T. Kaide, E. Nakamoto, K. Yoshida, H. Tanaka, J. Am. Chem. Soc. *2007, 129* (8), 2246-7.

Microwaves s. under Lipase [\\\\]
Potassium hydroxide s. under 1,1-Bis(hydroperoxy)cyclohexane *KOH*
Chiral pyrrolidines s.a. under Chiral pyrrolidine/benzoic acid ←

2(S)-[Diphenyl(trimethylsiloxy)methyl]pyrrolidine/3-benzyl-4,5-dimethylthiazolium chloride/triethylamine/ethyldiisopropylamine/hydrogen peroxide ←
Asym. synthesis of β-hydroxycarboxylic acid esters from α,β-ethylenealdehydes ←
Sequential catalysis in one pot with two different organocatalysts s. *72*, 413

Silver acetate/1,8-diazabicyclo[5.4.0]undec-7-ene *AgOAc/DBU*
4-Alkylidene-1,3-dioxolan-2-ones from 2-acetylenealcohols ○
with CuCl in ionic liquids cf. *42*, 127s*67*; with a little AgOAc and a stoichiometric amount of DBU, stereospecificity, s. W. Yamada, Y. Sugawara, H.M. Cheng, T. Ikeno, T. Yamada, Eur. J. Org. Chem. *2007* (16), 2604-7; with a little *n*-Bu_3P in supercritical CO_2 s. Y. Kayaki, M. Yamamoto, T. Ikariya, J. Org. Chem. *2007, 72* (2), 647-9.

(Triphenylphosphine)gold(I) triflate *(Ph_3P)AuOTf*
3-β-Functionalized furans from 1-(alk-1-ynyl)cyclopropyl ketones ○

70.
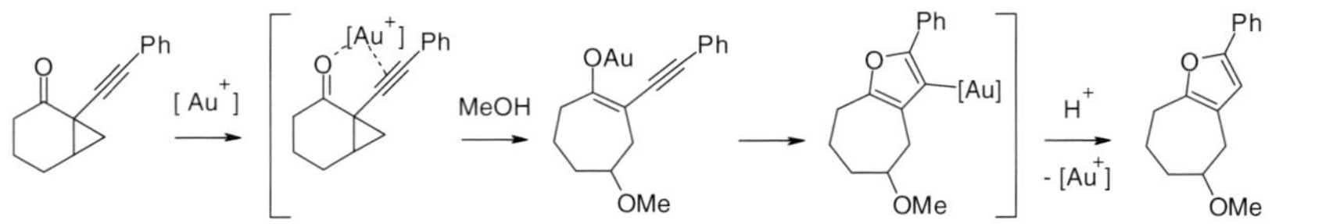

A soln. of the startg. cyclopropyl ketone and 2 eqs. methanol in dry methylene chloride added to a soln. of 1 mol% (Ph_3P)AuOTf [prepared *in situ* by mixing 1:1 (Ph_3P)AuCl and AgOTf in the same solvent, and filtering off the precipitated AgCl], stirred for 15 min at room temp., and worked up with column chromatographic purification → product. Y 91%. The procedure is mild and generally applicable to the reaction of condensed or isolated cyclopropyl ketones with a variety of O- and N-nucleophiles (alcohols, phenols, acetic acid, indole, 2-pyrrolidone), **regiospecific ring opening** taking place with 2-subst. 1-(alk-1-ynyl)cyclopropyl ketones by nucleophilic addition at the *more* substituted position of the ring. The alkyne group can be substituted by phenyl, alkyl, cyclopropyl or alkenyl groups, but yields were low with silylethynyl derivs. It is possible that reaction involves initial formation of a chelate complex which is attacked by the nucleophile in a regiospecific homo-Michael-type addition to form the gold(I) enolate prior to cycloisomerization; however, a cationic pathway cannot be ruled out. F.e. and comparison of catalysts s. J. Zhang, H.-G. Schmalz, Angew. Chem. Int. Ed. Engl. *2006, 45* (40), 6704-7.

Water-soluble phosphine-gold(I) complexes/sulfuric acid *[Au(I)]/H_2SO_4*
(N,N′-Disubst. imidazol-2-ylidene)gold(III) bromides *(NHC)$AuBr_3$*
Methyl ketones from terminal acetylene derivs. C≡CH → $COCH_3$
in aq. methanol with $MeAu(PPh_3)/CF_3SO_3H$ cf. *65*, 66; with 0.1 mol% water-soluble phosphine-containing gold(I) complexes, especially gold(I) alkynyl complexes, AuC≡CR(TPPTS) where R = *t*-Bu or 3-thienyl, and 10 mol% H_2SO_4 s. S. Sanz, L.A. Jones, F. Mohr, M. Laguna, Organometallics *2007, 26* (4), 952-7; with (N,N′-disubst. imidazol-2-ylidene)$AuBr_3$ s. P. de Frémont, R. Singh, E.D. Stevens, J.L. Petersen, S.P. Nolan, ibid. *26* (6), 1376-85.

Gold(III) chloride/copper(II) chloride *$AuCl_3/CuCl_2$*
Gold-catalyzed addition of alcohols to ethylene derivs. C=C → C(OR)CH

71. 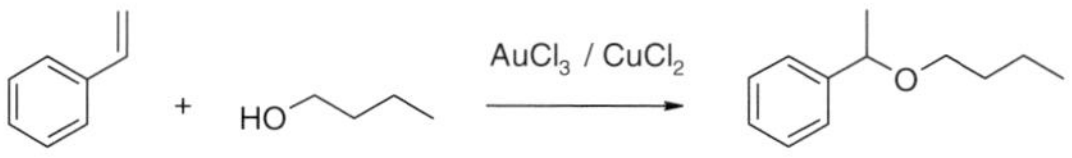

An equimolar mixture of styrene and 1-butanol heated with 8 mol% $AuCl_3$ and 16 mol% $CuCl_2$ in a sealed glass reactor at 120° for 4 h → butyl 1-phenylethyl ether. Y 86.7% (alkene conversion 95%). The presence of $CuCl_2$ retards reduction of the gold(III) catalyst leading to greatly improved conversions. The method is most effective for addition of prim. alcohols, with sec. alcohols reacting more slowly (2-butanol and styrene giving 68.5% conversion after 4 h) and tert. alcohols hardly at all (*t*-BuOH: 2.6% after 4 h). Reaction is also much slower with aliphatic and cycloaliphatic alkenes, 1-methylcyclohexene with methanol showing 60% conversion after 12 h (93% yield) to the 1-methoxy adduct. F.e.s. X. Zhang, A. Corma, Chem. Commun. *2007* (29), 3080-2.

Fluoroboric acid s. under PhIO *HBF_4*
Cerium(III) chloride s. under o*-Iodoxybenzoic acid* *$CeCl_3$*
Chiral cyclic ketones s. under H_2O_2 and K-monoperoxysulfate ←

2(S)-[Bis[3,5-bis(trifluoromethyl)phenyl]trimethylsiloxymethyl]pyrrolidine/benzoic acid ←
Organocatalyzed asym. Michael addition C=C → CHC(ON=CHPh)
of (E)-benzaldoxime as hydroxyl group equivalent
to α,β-ethylenealdehydes

72. 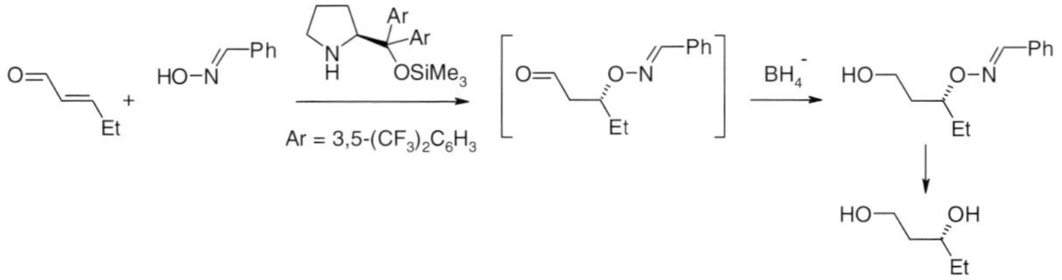

3 Eqs. (E)-benzaldoxime added at 4° to a stirred soln. of 10 mol% 2(S)-[bis[3,5-bis(trifluoromethyl)phenyl]trimethylsiloxymethyl]pyrrolidine, 10 mol% benzoic acid and the startg. enal in toluene, allowed to react for 1 h, and the product isolated after *in situ*-reduction on instantaneous treatment with methanol and 1.5 eqs. $NaBH_4$, followed by quenching with satd. aq. NH_4Cl → (R)-product. Y 72% (e.e. 96%). With aldoximes as nucleophile, complications due to simultaneous addition to the formyl group were minimized. Reaction is applicable to aliphatic enals [possessing *prim*-alkyl, *sec*-alkyl, alkenyl or carbalkoxy groups at the 3-position]. F.e. and comparison of aldoximes and chiral amines, also *gram-scale* conversion to the corresponding **chiral 1,3-diols** (by subsequent hydrogenation with $Pd(OH)_2/C$ in methanol), s. S. Bertelsen, P. Diner, R.L. Johansen, K.A. Jørgensen, J. Am. Chem. Soc. *2007, 129* (6), 1536-7; addition of 3,4-dimethoxybenzaldoxime with 9-amino(9-deoxy)epi-hydroquinine/D-*N*-Boc-phenylglycine cf. A. Carlone, G. Bartoli, M. Bosco, F. Pesciaioli, P. Ricci, L. Sambri, P. Melchiorre, Eur. J. Org. Chem. *2007* (33), 5492-5.

1-Methyl-5-chloroimidazole s. under $FeCl_3$ ←
Chiral 4-imidazolidones s.a. under PhIO ←

*(S)-3-Methyl-2-*tert*-butyl-4-imidazolidone* ←
2-Hydroxy-1,4-benzodioxanes from aldehydes and *o*-quinones via asym. organocatalyzed hetero-Diels-Alder reaction ○

73. 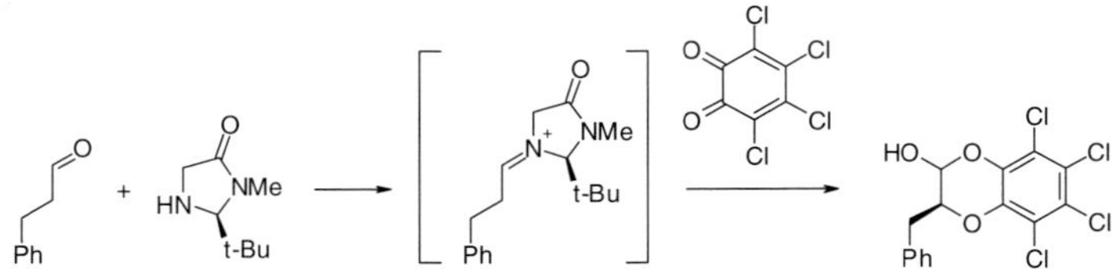

A mixture of 3 eqs. 3-phenylpropionaldehyde, 1 eq. of *o*-chloranil and 10 mol% (S)-3-methyl-2-*tert*-butyl-4-imidazolidone *in wet acetonitrile* stirred at 26° for 48 h → product. Y 68% (e.e. 81%). The adducts are viewed as **chiral cyclic α-aryloxyaldehyde derivs.** Seven examples gave yields of 68-75% and enantiomeric excesses of 75-80%. This mild inverse electron-demand cycloaddition is self-indicating in wet acetonitrile (the colour changing from deep-red to yellow) and complete within 72 h at room temp. The stereochemical outcome is consistent with attack of the *o*-quinone on the less-hindered side of the *in situ*-generated enamine. Reductive ring opening (with $NaBH_4$) afforded the corresponding **chiral glycol monocatechol monoethers**, which were converted to chiral 1,4-benzodioxanes via recyclization under Mitsunobu conditions. F.e. and comparison of chiral organocatalysts s. F.A. Hernandez-Juan, D.M. Cockfield, D.J. Dixon, Tetrahedron Lett. *2007*, *48* (9), 1605-8.

Peptides s. under H_2O_2 ←

Immobilized lipase/hydrogen peroxide/microwaves ←
Glycols from ethylene derivs. via epoxidation C=C → C(OH)C(OH)
Enzymatic conversion under mild conditions

74. 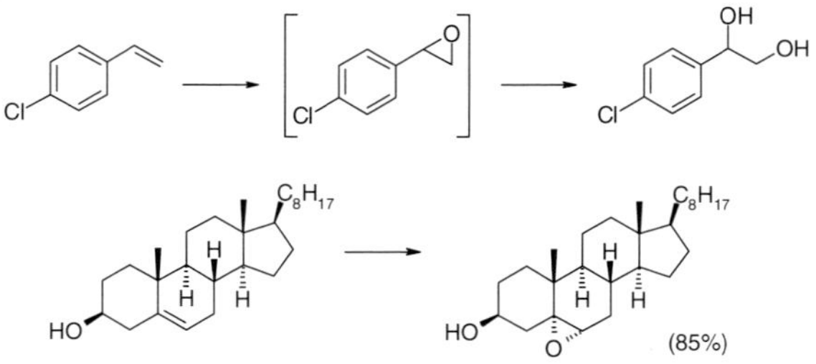

A mixture of 4-chlorostyrene, 3.6 eqs. of H_2O_2 (aq. 50%) and Lipase PSLG6 irradiated in *ethyl acetate* in a microwave oven (150 W) for 5 min, the reaction purged with air to ensure thorough mixing of reactants, the mixture cooled, diluted with ethyl acetate, the lipase filtered off and dried (for recycling), the organic layer washed sequentially with aq. Na_2SO_3, aq. $NaHCO_3$ and water, dried, and evaporated *in vacuo* → 1,2-diol product. Y 90%. Reaction is simple, safe, environmentally friendly and high yielding (70-90%) with styrenes, cyclohexenes and hexenes, but cholesterol and stigmasterol formed stable 5,6-epoxides under these conditions (83-85% yields). Formation of the glycols is rapid under microwave irradiation (5-10 min compared with 20 h under reflux) and does not denature the lipase which can be reused up to four times. *in situ*-Generated peroxyacetic acid is the effective oxidant. F.e.s. K. Sarma, N. Borthakur, A. Goswami, Tetrahedron Lett. *2007*, *48* (38), 6776-8.

Monooxygenases ←

Enzymatic asym. Baeyer-Villiger oxidation COR → COOR

s. *44*, 114; kinetic resolution of cyclic and linear ketones using phenylacetone monooxygenase and 4-hydroxyacetophenone monooxygenase s. C. Rodríguez, G. de Gonzalo, M.W. Fraaije, V. Gotor, Tetrahedron:Asym. *2007*, *18* (11), 1338-44; of α-subst. cyanocyclohexanones with cells of *Escherichia coli* expressing cyclohexanone monooxygenase s. N. Berezina, E. Kozma, R. Furstoss, V. Alphand, Adv. Synth. Catal. *2007*, *349* (11-12), 2049-53.

tert-*Butyl hydroperoxide s. under Polymer-based molybdenum carbonyl* t-*BuOOH*

Lithium tert-*butylperoxide* *LiOOBu*-t

Stereospecific nucleophilic epoxidation C=C →

cf. *14*, 194s*48,49*; of α-methylated α,β-ethylene-γ-hydroxy esters, and conversion to a β-hydroxy-α-methylene-γ-lactone s. I. López, S. Rodríguez, J. Izquierdo, F.V. González, J. Org. Chem. *2007*, *72* (17), 6614-7.

1,1-Bis(hydroperoxy)cyclohexane/potassium hydroxide ←

Epoxides from ethylene derivs.

with Triton B/*t*-BuOOH cf. *24*, 149; Weitz–Scheffer epoxidation of α,β-ethyleneketones with 1,1-bis(hydroperoxy)cyclohexane in 1 *M* KOH/dioxane at room temp. s. K. Jakka, J. Liu, C.-G. Zhao, Tetrahedron Lett. *2007*, *48* (8), 1395-8.

m-*Chloroperoxybenzoic acid* *$ArCOO_2H$*

2-Carbalkoxy-O-heterocyclics from cyclic enolethers via stereospecific oxidative rearrangement ←

75.

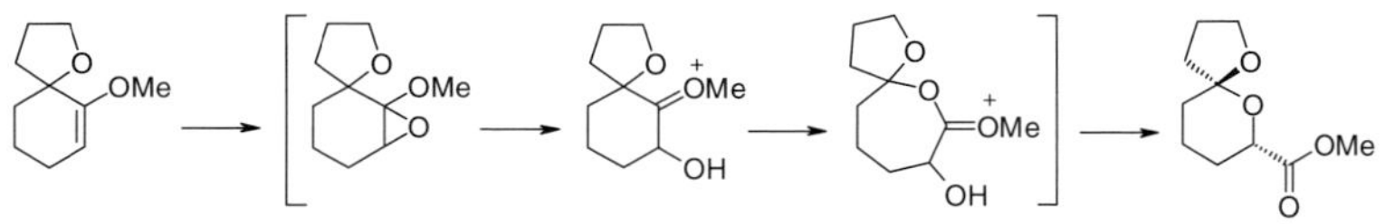

Spiroketals. 1.79 mmol ca. 70% *m*-chloroperoxybenzoic acid and 1.79 mmol Na_2HPO_4 added to a soln. of 0.595 mmol startg. cyclic enolether in dry methylene chloride, stirred for 12 h, treated with 2 *M* 2-methyl-2-butene in THF, stirring continued for 2 h, and worked up with column chromatographic purification → methyl 1,6-dioxaspiro[4.5]decane-7-carboxylate. Y 52% (single stereoisomer). Reaction is presumed to involve initial epoxidation, followed by ring opening to give a cyclic alkoxycarbenium ion prior to nucleophilic addition of the peroxy- carboxylate ion, then ring expansion and ring opening with intramolecular ketalization. Cyclic α-alkoxy-α,β-ethyleneketones reacted similarly to give the corresponding **1-carbalkoxylactones** (35%, 75%). F.e.s. D.L. Waller, C.R.J. Stephenson, P. Wipf, Org. Biomol. Chem. *2007*, *5* (1), 58-60.

3-Benzyl-4,5-dimethylthiazolium chloride s. under 2(S)-[Diphenyl(trimethylsiloxy)-methyl]pyrrolidine ←

Iodosobenzene/fluoroboric acid *$PhIO/HBF_4$*

Oxidative ring opening

of cyclic glycols with IOAc or $I(OAc)_3$ cf. *35*, 74; of cyclic ethylene derivs. (presumably via *vic*-dihydroxylation) with iodosobenzene/HBF_4 in the presence of water, also oxidative cleavage of acyclic ethylene derivs., and ar. aldehydes from styrene epoxides, s. K. Miyamoto, N. Tada, M. Ochiai, J. Am. Chem. Soc. *2007*, *129* (10), 2772-3.

in situ-*Generated iodosobenzene/chiral 4-imidazolidone* ←

α,β-Epoxyoxo compds. from α,β-ethyleneoxo compds.

Asym. epoxidation C=C →

of α,β-ethylenealdehydes with H_2O_2/chiral 2-(siloxymethyl)pyrrolidine cf. *70*, 63; with iodoso-benzene generated slowly *in situ* from an iminoiodinane and a chiral 2,3,5-trisubst. 4-imidazolidone (e.e. up to 93%) s. S. Lee, D.W.C. MacMillan, Tetrahedron *2006*, *62* (49), 11413-24; of α,β-ethyleneketones with *t*-BuOOH/chiral diaryl(2-pyrrolidinyl)carbinols (10 mol%) in hexane at room temp. (e.e. up to 90%) s. A. Lattanzi, A. Russo, ibid. *62* (52), 12264-9.

o-Iodoxybenzoic acid *$ArIO_2$* ←
o-Hemiquinols from phenols under mild conditions

76. 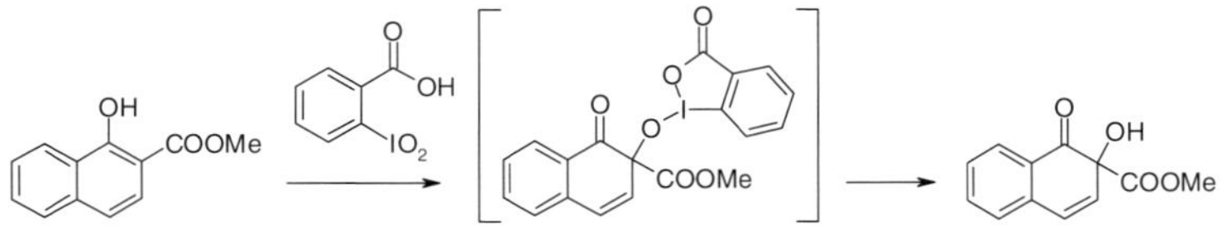

2 Eqs. 'non-explosive *o*-iodoxybenzoic acid' (a mixture of 49% *o*-iodoxybenzoic acid, 22% benzoic acid and 29% isophthalic acid) added in one portion to a soln. of the startg. naphthol in ethyl acetate, the resulting suspension stirred at 50° for 7 days, and worked up with column chromatographic purification → methyl 1,2-dihydro-2-hydroxy-1-oxonaphthalene-2-carboxylate. Y 74%. The reagent facilitates a direct, clean and safe conversion of 2-alkylated phenols and naphthols (incl. substrates with electron-withdrawing groups on the ring) into *o*-hemiquinols (or the corresponding [4+2]-cyclodimers). F.e.s. N. Lebrasseur, J. Gagnepain, A. Ozanne-Beaudenon, J.-M. Leger, S. Quideau, J. Org. Chem. *2007, 72* (16), 6280-3.

o-Iodoxybenzoic acid/cerium(III) chloride *$ArIO_2/CeCl_3$*
3-Hydroxyoxindoles from indoles C=CH → C(OH)C(O)
s. *17*, 219s*72*; *49*, 117s*72*

Phenyl iodosoacetate *$PhI(OAc)_2$*
Oxidative ring closures of ethylenealcohols ○
2-α-hydroxy-O-heterocyclics with *m*-CPBA cf. *28*, 107s*62*; *34*, 114s*62*; 2-[acylamino-(hydroxy)methyl]tetrahydropyrans from 6-acylamino-5-ethylenealcohols with $PhI(OAc)_2$ as oxidant and water as nucleophile at 0°, also with alcohols or acetic acid as nucleophile, s. X. Huang, N. Shao, A. Palani, R. Aslanian, Tetrahedron Lett. *2007*, *48* (11), 1967-71.

N,N'-Dibromo-N,N'-1,2-ethanediylbis(benzenesulfonamide) ←
Protection of hydroxyl groups as tetrahydropyran-2-yl ethers OH → OThp
s. *48*, 120s*72*

Chiral di-μ-oxotitanium salalen complex/hydrogen peroxide ←
Asym. epoxidation of unactivated ethylene derivs.

77. 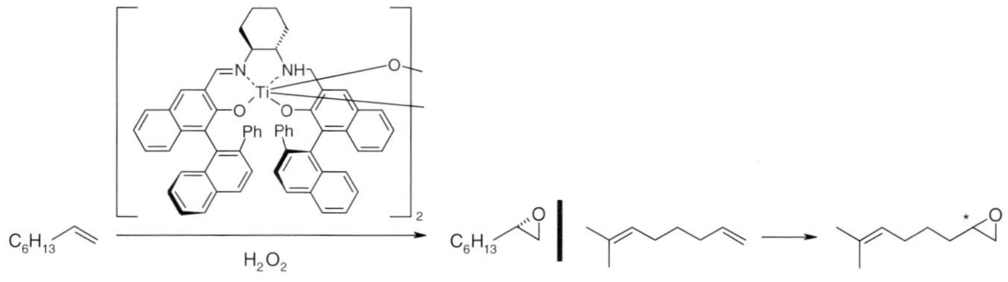

1-Octene and *2 mol%* chiral di-μ-oxotitanium salalen complex in methylene chloride treated with 1.5 eqs. 30% aq. H_2O_2, stirred at room temp. for 24 h, the solvent removed *in vacuo,* and the residue worked up with column chromatographic purification → product. Y 85% (e.e. 82%). The procedure is selective for the asym. epoxidation of *unactivated* terminal olefins and (Z)-olefins, enantioselectivities of 70-97% being the highest as yet recorded for the latter substrates. Interestingly, **regiospecific epoxidation** of *less rich* terminal alkene groups in the presence of internal alkene groups was also achieved. F.e.s. Y. Sawada, K. Matsumoto, T. Katsuki, Angew. Chem. Int. Ed. Engl. *2007, 46* (24), 4559-61; with readily modifiable, chiral, *in situ*-generated Ti-salalen complexes (e.e. up to 98%) s. A. Berkessel, M. Brandenburg, E. Leitterstorf, J. Frey, J. Lex, M. Schäfer, Adv. Synth. Catal. *2007*, *349* (14-15), 2385-91.

Titanium tetraisopropoxide s. under H_2O_2 — *Ti(OPr-*i*)$_4$*

*Titanium tetrachloride/*tert*-butylamine/aniline/hydrogen chloride* ←
Benzofurans from *o*-halogenoarylacetylenes via α-(*o*-halogenoaryl)ketones — C≡C → CH_2CO

78. 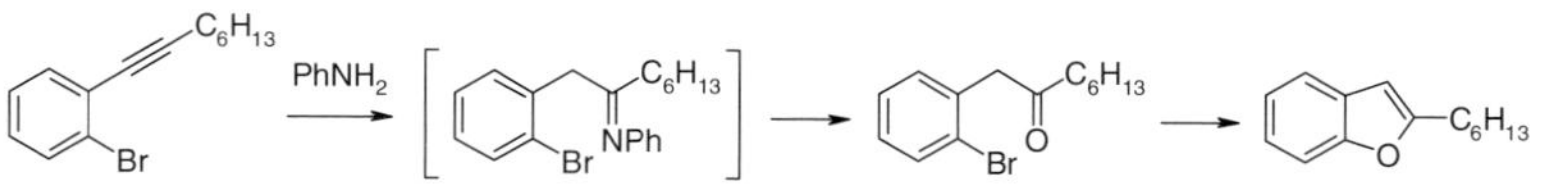

1.2 Eqs. *tert*-butylamine added to a soln. of 20 mol% $TiCl_4$ in toluene under N_2, followed by 1 eq. each of aniline and the startg. acetylene deriv., stirred at 105° for 22 h, treated with 2 *N* HCl at room temp., stirring continued at 20° for 24 h, quenched with water, and worked up with column chromatographic purification → intermediate ketone (Y 95%), in DMF under N_2 treated with 2 eqs. K_3PO_4 and 10 mol% CuI, stirred at 105° for 22 h, quenched with brine and ether at room temp., and worked up → 2-*n*-hexylbenzofuran (Y 82%). The initial step illustrates a general procedure for **anti-Markovnikov hydration of acetylene derivs. via hydroamination,** which is applicable to both internal and terminal alkynes. F.e. with anilines or alkylamines in the hydroamination phase s. L. Ackermann, L.T. Kaspar, J. Org. Chem. *2007, 72* (16), 6149-53.

*Tri-*n*-butylphosphine* — *Bu_3P*
4-Alkylidene-1,3-dioxolan-2-ones from 2-acetylenealcohols ○
in supercritical CO_2 s. *42*, 127s*72*

Vanadyl triisopropoxide/cumene hydroperoxide/chiral bis(hydroxamic acids) ←
Asym. epoxidation of 3-ethylenealcohols — C=C → epoxide
with chiral N-protected α-aminohydroxamic acids cf. *64*, 71; with chiral bis(hydroxamic acids) s. W. Zhang, H. Yamamoto, J. Am. Chem. Soc. *2007*, *129* (2), 286-7.

Hydrogen peroxide s.a. under Lipase, Chiral 2-a-siloxypyrrolidine, Chiral di-μ-oxo-titanium salalen complex, Chiral Mn(III)-salen complexes, $FeCl_3$ and Pt(II) complexes — *H_2O_2*

Hydrogen peroxide/chiral spirocyclic ketone/potassium carbonate ←
Asym. epoxidation
with chiral dioxiranes generated from chiral ketones/Oxone cf. *51*, 65s*60,62,63*; with an N-aryl-2-oxazolidone-containing spirocyclic ketone/H_2O_2/K_2CO_3 (e.e. up to 96%) s. C.P. Burke, L. Shu, Y. Shi, J. Org. Chem. *2007*, *72* (16), 6320-3.

Hydrogen peroxide/aspartyl-containing peptide/diisopropylcarbodiimide/4-dimethylaminopyridine ←
Aspartate-catalyzed asym. epoxidation

79. 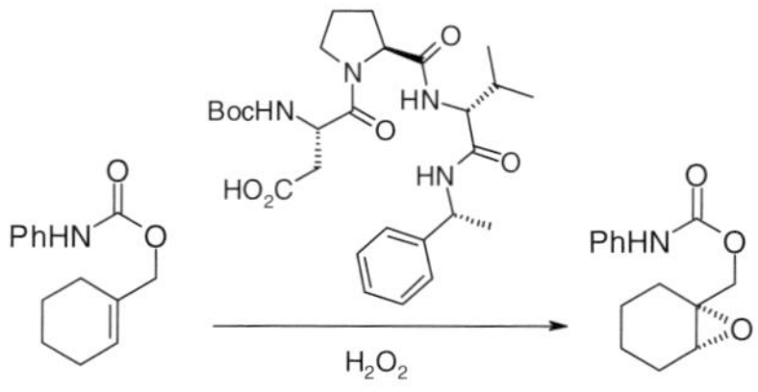

of 2-ethyleneurethans. 2.5 Eqs. 30% aq. H_2O_2 added to a soln. of the startg. alkene, 0.1 eq. aspartyl-containing peptide and 0.1 eq. 4-dimethylaminopyridine in methylene chloride, the 2-phase mixture cooled to -10°, treated with 2 eqs. diisopropylcarbodiimide, stirring continued for 78 h, diluted with ether, and directly chromatographed → product. Y 97% (e.e. 89%). The *electrophilic* catalytic cycle involves initial generation of the aspartyl peroxy acid, the effective oxidant, while the carbamate residue of the substrate complexes with the catalyst through hydrogen bonding to facilitate delivery of oxygen to the unsaturation with high enantioselectivity. Both electron-donating and -withdrawing groups were tolerated in the N-aryl residue, but enantioselectivity was low with the corresponding 3-ethyleneurethans. F.e. and proof of principle with benzyl N-Boc-aspartate as catalyst s. G. Peris, C.E. Jokobsche, S.J. Miller, J. Am. Chem. Soc. *2007, 129* (28), 8710-1.

Hydrogen peroxide/titanium tetraisopropoxide or peroxophosphotungstate complex ←
α-Hydroxyacetals from enolethers C=C(OR) → C(OH)C(OR)OR′
with *m*-CPBA cf. *33*, 130; α-hydroxy-γ- or -δ-lactolides from the corresponding cyclic enolethers with H_2O_2/Ti(OPr-*i*)$_4$ or Venturello′s peroxophosphotungstate compd., also α-hydroxy-δ-lactol esters from acids with the latter, s. P. Levecque, D. Gammon, H.H. Kinfe, P. Jacobs, D. De Vos, B. Sels, Org. Biomol. Chem. *2007, 5* (11), 1800-6.

Sulfuric acid s. under Phosphine-gold(I) complexes *H_2SO_4*

Potassium peroxymonosulfate/chiral cyclic ketones/potassium carbonate/acetic acid ←
γ-Aryl-γ-lactones from benzylidenecyclopropanes
One-pot conversion via asym. epoxidation, ring expansion to arylcyclobutanones and Baeyer-Villiger oxidation

80. 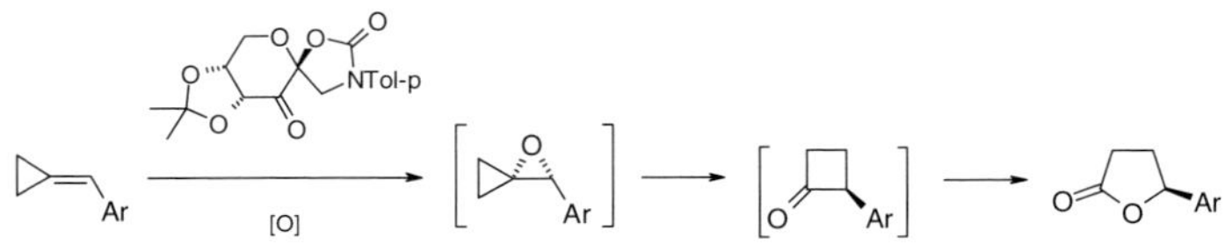

Ar = p-Tol

A mixture of the startg. benzylidenecyclopropane and 0.2 eqs. chiral 2-oxazolidone-containing ketone in DME/dimethoxymethane (3:1 v/v) buffered at pH 9.3 with 1 *M* K_2CO_3-AcOH in 0.1 *mM* EDTA, cooled to -10°, 3.2 eqs. Oxone in 0.1 *mM* EDTA and 13.4 eqs. K_2CO_3 in 0.1 *mM* EDTA added simultaneously via separate syringe pumps over 8 h, reaction quenched with ethyl acetate, washed with brine, and concentrated → (-)-5-*p*-tolyl-4,5-dihydro-2(3*H*)-furanone. Y 68% (e.e. 90%). Thirteen 5-aryl- or 5-aryl-5-methyl derivs. were prepared in 45-73% yields (e.e. 71-91%). The reaction tolerates methoxy-, *tert*-butyl- and halophenyl substitution and proceeds via initial asym. epoxidation, followed by ring expansion to the cyclobutanone, which undergoes further ring expansion via Baeyer-Villiger oxidation. By the use of less oxidant and more catalyst the intermediate chiral cyclobutanones can be isolated (2 examples; 51 and 68% yields, 85 and 90% e.e.). It is thought that the bulky cyclic ketone forms a favourable spirocyclic transition state leading to high enantioselectivity. F.e.s. B. Wang, Y.-M. Shen, Y. Shi, J. Org. Chem. *2006, 71* (25), 9519-21.

Peroxotungstophosphates s. under H_2O_2 ←

Heteropolyacids *$H_{14}[NaP_5W_{30}O_{110}]$*
Protection of hydroxyl groups as tetrahydropyran-2-yl ethers OH → OThp
of alcohols or phenols with $CeCl_3{\cdot}7H_2O$/NaI cf. *48,* 120s*71*; with a little $H_{14}[NaP_5W_{30}O_{110}]$, also deprotection, s. M.M. Heravi, M. Haghighi, F. Derikvand, F.F. Bamoharram, Synth. Commun. *2006, 36* (21), 3103-7; with a little N,N′-dibromo-N,N′-1,2-ethanediylbis(benzenesulfonamide) in methylene chloride (and deprotection in methanol) s. A. Khazaei, A. Rostami, M. Mahboubifar, Catal. Commun. *2007, 8* (3), 383-8; with 2 mol% Ru(acac)$_3$ in the absence of solvent, incl. selective monoprotection of diols, s. R. Varala, S.R. Adapa, Can. J. Chem. *2006, 84* (9), 1174-9.

*Polymer-based molybdenum carbonyl/*tert*-butyl hydroperoxide* ←
Epoxidation C=C →
with Mo(CO)$_6$ cf. *28*, 113s*44*; with Mo(CO)$_6$ immobilized using Merrifield resin-derived amines s. S. Tangestaninejad, V. Mirkhani, M. Moghadam, G. Grivani, Catal. Commun. *2007, 8* (5), 839-44.

Chiral pentadentate manganese(III) salen complexes/hydrogen peroxide ←
Asym. epoxidation s. *46*, 106s*72* C=C → ▽O

Cationic iron(III) salts *Fe^{3+}*
Acoxy compds. from ethylene derivs. C=C → CHC(OAc)
with $[Cp^*RuCl_2]_2$/AgOTf/dppb cf. *68*, 74; with cationic Fe(III)-salts s. K. Komeyama, Y. Mieno, S. Yukawa, T. Morimoto, K. Takaki, Chem. Lett. *2007*, *36* (6), 752-3.

Iron(III) chloride/hydrogen peroxide/5-chloro-1-methylimidazole ←
Iron(III)-catalyzed biomimetic epoxidation C=C → ▽O

81. 

3 Eqs. 30% aq. H_2O_2 added by syringe pump over 1 h to a stirred mixture of 4-bromostyrene, 5 mol% $FeCl_3·6H_2O$ and 15 mol% 5-chloro-1-methylimidazole in *tert*-amyl alcohol → epoxide. Y 82% (by GC; conversion 91%; selectivity 90%). The conditions were optimized with respect to choice of imidazole and stoichiometries of the reagents for an inexpensive, environmentally friendly and simple epoxidation of mono-, di- and tri-subst. ar. ethylene derivs. in 24-87% yields. Highest yields (generally >60%) were obtained with styrenes but *trans*-stilbene gave an 87% yield of *trans*-epoxide (the *cis*-isomer affording 24% of the *cis*-epoxide). The moderate yields for trisubst. derivs. still represent an improvement on previous iron-catalyzed methods. The reaction gave insignificant yields in the presence of TEMPO, suggesting a radical mechanism for the catalytic cycle. F.e.s. K. Schröder, X. Tong, B. Bitterlich, M.K. Tse, F.G. Gelalcha, A. Brückner, M. Beller, Tetrahedron Lett. *2007*, *48* (36), 6339-42.

Nickel oxide hydroxide nanoparticles/sodium hypochlorite ←
Glycidic from α,β-ethylenecarboxylic acids
Catalytic heterogeneous oxidation in water s. *72*, 95

Ruthenium tetroxide/sodium periodate *$RuO_4/NaIO_4$*
***cis*-2-α,5-α-Dihydroxytetrahydrofurans from 1,5-dienes** ○
with $RuO_4/NaIO_4$ cf. *61*, 71s*62*; regio- and diastereo-selective oxidative ring closure of 1,5,9-trienes and polyenes s. S. Göhler, C.B.W. Stark, Org. Biomol. Chem. *2007*, *5* (10), 1605-14.

Ruthenium(III) acetoacetonate *$Ru(acac)_3$*
Protection of hydroxyl groups as tetrahydropyran-2-yl ethers OH → OThp
in the absence of solvent s. *48*, 120s*72*

Cationic tetrakis(acetonitrile)palladium(II) complexes s. under ↯ *[Pd(II)]*

Chiral cationic (aquo)pentafluorophenylplatinum(II) di(phosphine) complexes/ hydrogen peroxide ←
Catalytic asym. epoxidation of unfunctionalized terminal ethylene derivs. C=C → ▽O

82. 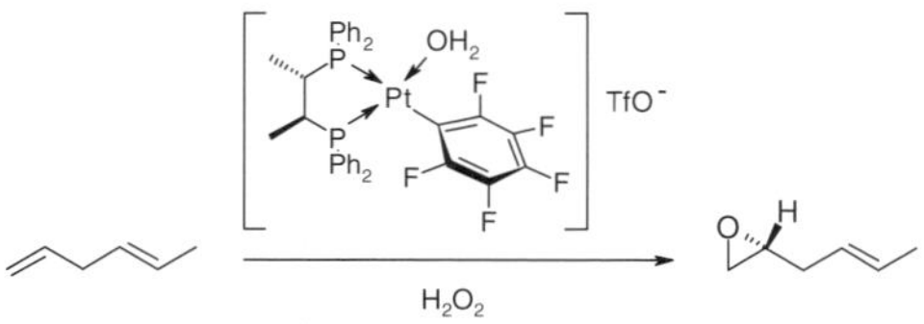

Regiospecific conversion. A soln. of 0.016 mmol chiral platinum(II) di(phosphine) complex in methylene chloride stirred for 10 min at -10° in a vial fitted with a screw-cap, 0.83 mmol startg. diene added, stirring continued for a further 10 min, treated in one portion with 0.83 mmol 35% H_2O_2, and worked up after 48 h → (R)-product. Y 93% (e.e. 63%; terminal/internal epoxidation 100:0). Reaction is highly enantioselective (up to 98% e.e.) and generally applicable to unfunctionalized *terminal* alkenes and allylarenes (not styrenes), catalysis arising, unusually, from an interaction between the metal centre and the *least* electron-rich unsaturation. The pentafluorophenyl residue imparts the requisite electronic properties to the metal, while at the same time increasing the rigidity and steric hindrance of the complex. F.e. and comparison of chiral di(phosphine) ligands s. M. Colladon, A. Scarso, P. Sgarbossa, R.A. Michelin, G. Strukul, J. Am. Chem. Soc. *2006, 128* (43), 14006-7.

Potassium osmate/chiral ligand/N-methylmorpholine N-oxide ←
Sharpless asym. dihydroxylation C=C → C(OH)C(OH)
cf. *47*, 114s*48*; application of nanofiltration with Starmem™120 to allow reuse of the catalytic system s. F.C. Ferreira, L.C. Branco, K.K. Verma, J.G. Crespo, C.A.M. Afonso, Tetrahedron:Asym. *2007, 18* (14), 1637-41.

Rearrangement ∩

Hydrogen/Oxygen Type OC ∩ HO

Silver mesylate/organic base/carbon dioxide ←
Cationic gold complexes *[Au]*
Meyer-Schuster-type reaction s. under CC∩HO C≡C-C(OH) → COCH=C

Acetate buffer/silica gel ←
6-α-Hydroxy-5,6-dihydro-4*H*-1,2-oxazines from γ,δ-epoxyketones
via (E)-γ,δ-epoxyhydrazones s. *72*, 208

Hydrogen/Carbon Type OC ∩ HC

Amines ←
Epimerization of hydroxy-O-heterocyclics ←
of carbohydrates with K_2CO_3 cf. *34*, 129s*37*; of 4-α-oxy-5-hydroxy-2(5*H*)-furanones with organic bases s. W.H. Miles, D.G. Duca, B.R. Selfridge, C.A. Palha De Sousa, K.B. Hamman, E.O. Goodzeit, J.T. Freedman, Tetrahedron Lett. *2007, 48* (44), 7809-12.

Copper(II) hexafluoroacetoacetonate *Cu(hfacac)₂*
2,5-Dihydrofurans from 3-ethyleneepoxides

83. 

0.5 Mol% Cu(II)-hexafluoroacetoacetonate in toluene added to a soln. of the startg. vinyloxirane in toluene contained in a threaded culture tube, the latter sealed, heated in an oil bath at 150° for 2 h, cooled to room temp., solvent removed *in vacuo,* the copper salt filtered off, and the mixture worked up → product. Y 94% (as a 17:1 mixture of 3-hexyl-2,5-dihydrofuran and isomeric 3-methylnonenals). The procedure is applicable to a range of mono- and di-subst. 1,3-butadiene monoxides, as well as cyclic 1,3-diene monoxides, the cleanest products (>20:1 chemoselectivity) being formed from 1,1-disubst., 4,4-disubst., and 1,3-disubst. 1,3-butadiene monoxides and from cycloocta-1,3-diene monoxide. Rearrangement is facilitated by simultaneous coordination of the metal with the olefin and epoxide residues, bringing the two ends into close proximity, the more electrophilic the catalyst the faster and more selective the conversion. Reaction can also be conducted **in the absence of solvent,** and hydroxyl and carbonyl groups were tolerated. F.e. and comparison of copper(II) catalysts s. L.A. Batory, C.E. McInnis, J.T. Njardarson, J. Am. Chem. Soc. *2006, 128* (50), 16054-5.

Chiral (η^6-arene)chloro(N-sulfonyl-1,2-ethylenediamine)ruthenium(II) complex/acetone/ potassium hydroxide/chiral ruthenium(II) Δ^2-oxazolin-2-ylferrocenylphosphine complex/ sodium isopropoxide/isopropanol ←
(Benzene)dichlororuthenium(II) dimer/(R)-2,2′-bis(diphenylphosphino)-1,1′-binaphthyl/ (R,R)-1,2-diphenylethylene-1,2-diamine/cyclohexanone/potassium tert-*butoxide* ←
Deracemization of sec. alcohols ←
via ruthenium-catalyzed oxidation-asym. hydrogenation

84. 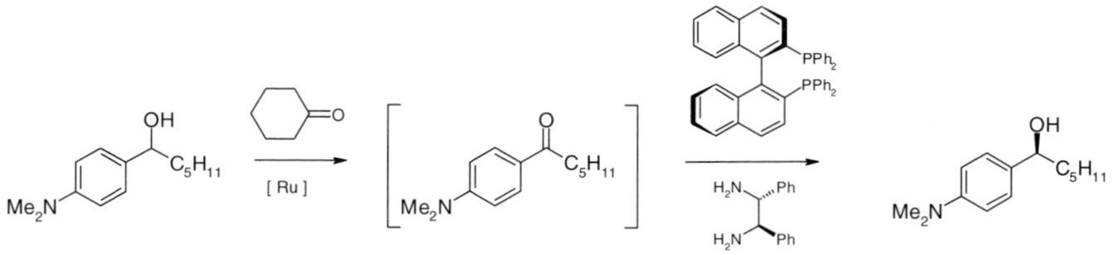

2.4 Eqs. cyclohexanone added to a mixture of 1 mol% $[RuCl_2(benzene)]_2$, 2 mol% (R)-BINAP and (R,R)-dpen in a flame-dried and argon-purged Schlenk tube, the mixture heated at 110° for 1 h (to effect complexation), allowed to cool to room temp., a further 0.6 eq. cyclohexanone, startg. *p*-dimethylaminobenzyl alcohol and THF added, followed by 8 mol% KOBu-*t*, the mixture heated at 60° for 20 h, allowed to cool, transferred via cannula under argon pressure to a Parr hydrogen bomb reactor (previously purged for 15 min by a rapid flow of argon), the bomb sealed and pressurized to 10 bar with H_2, stirred at 20° for 4 h, the pressure released, and the dark brown soln. filtered through silica and Celite, then purified → product. Y 96% (e.e. 92%). This non-biological method uses a single ruthenium catalyst for both the oxidation and reduction steps. Better results were obtained by avoiding achiral triphenylphosphine as ligand, and a large excess of the hydrogen acceptor was unnecessary if cyclohexanone was used in place of acetone. Higher enantioselectivities were achieved in two examples using (R)-xylyl-BINAP (Y 94%, e.e. 98% for the example above), although longer reaction times were necessary. This ligand was also used in two examples for the preparation of **(S)- from (R)-benzyl alcohols** (e.e. 80%, 87%). F.e.s. G.R.A. Adair, J.M.J. Williams, Chem. Commun. *2007* (25), 2608-9; deracemization of sec. benzyl alcohols in up to 95% e.e. with two different ruthenium complexes {1 mol% RuCl(S,S)-tsdpen(η^6-cymene) [or the mesitylene or msdpen analogs]/2 mol% KOH in acetone, then 0.5 mol% $RuCl_2(PPh_3)$(ip-FOXAP) [where ip-FOXAP = chiral 2-(4-isopropyl-Δ^2-oxazolin-2-yl)-ferrocenyl(diphenyl)phosphine]/2 mol% *i*-PrONa in isopropanol at room temp. and pressure} s. Y. Shimada, Y. Miyake, H. Matsuzawa, Y. Nishibayashi, Chem. Asian J. *2007*, *2* (3), 393-6.

Oxygen/Nitrogen Type OC ∩ ON

Sodium iodide/silica gel/microwaves s. under $CeCl_3$ *NaI/SiO_2/[\\\\]*
*Zinc chloride/*p*-toluenesulfonic acid* *$ZnCl_2$/TsOH*
Brønsted acid-mediated Lewis acid-catalyzed Beckmann rearrangement of ketoximes ←

85. 

A mixture of cyclododecanone oxime, 10 mol% TsOH and 12 mol% $ZnCl_2$ refluxed in acetonitrile for 5 h → cyclododecanamide. Y 93%. The procedure is inexpensive, mild and environmentally friendly. Similar treatment of cyclohexanone oxime gave only 40% of the corresponding lactam but a series of ar. ketoximes gave the corresponding amides in 90-97% yields while acetoxime did not react. The conditions were optimized with respect to stoichiometry, solvent, organic acid and Lewis acid ($ZnBr_2$ was also effective but $ZnCl_2$ is much cheaper). F.e.s. L.-F. Xiao, C.-G. Xia, J. Chen, Tetrahedron Lett. *2007, 48* (40), 7218-21.

Mercury(II) chloride *$HgCl_2$*
Cerium(III) chloride/sodium iodide/silica gel/microwaves *$CeCl_3/NaI/SiO_2$/[\\\\]*
Beckmann rearrangement C(R)=NOH → CONHR
with formic acid cf. *21*, 186; with a cerium catalyst (prepared from NaI, silica gel and $CeCl_3 \cdot 7H_2O$) under microwave irradiation in the absence of solvent s. Z. Li, Z. Lu, R.B. Ding, J.Y. Yang, J. Chem. Res. *2006* (10), 668-70; with a little $HgCl_2$ in refluxing acetonitrile s. C. Ramalingan, Y.T. Park, J. Org. Chem. *2007, 72* (12), 4536-8.

Alumina s. under Rh(OH)$_x$ *Al_2O_3*
Silica gel s. under $CeCl_3$ *SiO_2*
Hydroxylamine hydrosulfate s. under Rh(OH)$_x$ *$(NH_2OH)_2 \cdot H_2SO_4$*
p-*Toluenesulfonic acid s. under $ZnCl_2$ and Ru complexes* *TsOH*

*Carbonyl(dihydrido)tris(triphenylphosphine)ruthenium(II)/1,2-bis(diphenylphosphino)ethane/*p-*toluenesulfonic acid* ←
Rhodium hydroxide/alumina/hydroxylamine hydrosulfate *$Rh(OH)_x/Al_2O_3/(NH_2OH)_2 \cdot H_2SO_4$*
N-Unsubst. carboxylic acid amides from aldoximes CH=NOH → $CONH_2$

86.

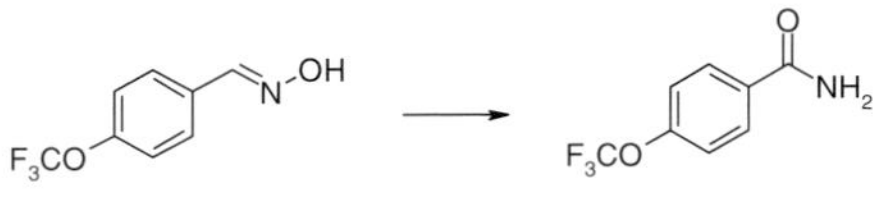

A N_2-purged mixture of the startg. benzaldoxime, *0.2 mol%* $Ru(CO)H_2(PPh_3)_3$, 0.2 mol% 1,2-bis-(diphenylphosphino)ethane and 0.8 mol% *p*-toluenesulfonic acid monohydrate in anhydrous toluene heated in a Schlenk tube at reflux for 6 h → product. Y 95%. Reaction is applicable at catalyst loadings as low as 0.04 mol% to aryl, aliphatic and α,β-ethylenic aldoximes (Z or E), and is compatible with electron-donating or -withdrawing groups in the aromatic ring. Ketoximes, however, were inert. F.e. and comparison of ligands and additives s. N.A. Owston, A.J. Parker, J.M.J. Williams, Org. Lett. *2007, 9* (18), 3599-601; under iridium catalysis at relatively higher catalyst loading and under forcing conditions cf. ibid. 73-5; from activated or unactivated aldoximes **under heterogeneous conditions in water** with $Rh(OH)_x/Al_2O_3$, **also** one-pot conversion **from aldehydes** with added $(NH_2OH)_2 \cdot H_2SO_4$ cf. H. Fujiwara, Y. Ogasawara, K. Yamaguchi, N. Mizuno, Angew. Chem. Int. Ed. Engl. *2007, 46* (27), 5202-5.

Oxygen/Sulfur Type OC ∩ OS

Trimethylsilyl triflate/amides *$Me_3SiOTf/RCON<$*
Pummerer rearrangement CH(SOR) → C(OAc)SR
with $MeSO_3H$ cf. *15*, 177; with Me_3SiOTf in amide solvents (especially DMAc or NMP) with asym. induction, chiral α-acoxy-α-(arylthio)acetic acid derivs., s. Y. Nagao, S. Miyamoto, M. Miyamoto, H. Takeshige, K. Hayashi, S. Sano, M. Shiro, K. Yamaguchi, Y. Sei, J. Am. Chem. Soc. *2006, 128* (30), 9722-9.

Nitrogen/Carbon Type OC ∩ NC

Scandium(III) triflate/trimethylsilyl chloride *$Sc(OTf)_3/Me_3SiCl$*
Ring expansion of N- to O,N-heterocyclics ○
1,3-O,N-heterocyclics with picric acid cf. *22*, 163; 5,6-dihydro-4*H*-1,2,4-oxadiazines from aziridin-1-yloximes (obtained from N-unsubst. aziridines and hydroximinochlorides) with $Sc(OTf)_3$ in the presence of Me_3SiCl s. S.Y. Cho, S.K. Kang, J.H. Ahn, J.D. Ha, J.-K. Choi, Tetrahedron Lett. *2006, 47* (51), 9029-33.

Carbon/Carbon Type OC ∩ CC

Irradiation ///
3-Alkylidenephthalides from indan-1,3-diones ○
cf. *21*, 192; 3-aroylenephthalides s. S. Mor, S.N. Dhawan, M. Kapoor, D. Kumar, Tetrahedron *2007*, *63* (3), 594-7.

Microwaves s. under [N,N′-Bis(2,6-diisopropylphenyl)imidazol-2-ylidene]gold(I) chloride [\\\\]

1,8-Diazabicyclo[5.4.0]undec-7-ene (s.a. under Au) *DBU*
Cycloisomerization of acetylenecarboxylic acids ○
3-alkylidenephthalides with CuI/Et_3N cf. *35*, 85s*51*; in high regioselectivity with a little DBU s. C. Kanazawa, M. Terada, Tetrahedron Lett. *2007*, *48* (6), 933-5.

Copper(II) triflate or Silver perchlorate *$Cu(OTf)_2$ or $AgClO_4$*
Cycloisomerization of *o*-ethylenephenols
of *o*-allylphenols with $ZrCl_4$ cf. *23*, 182s*68*; 2,3-dihydrobenzofuran, 3,4-dihydro-2*H*-1-benzopyran, or 3,4-dihydro-2*H*-1,5-benzodioxepin ring closures with $Cu(OTf)_2$ or $AgClO_4$, also ring closures of ethylenebenzoic acids and 2-allyl-*N*-tosylaniline, s. Y. Ito, R. Kato, K. Hamashima, Y. Kataoka, Y. Oe, T. Ohta, I. Furukawa, J. Organomet. Chem. *2007*, *692* (1-3), 691-7.

Silver triflate (s.a. under $FeCl_3$) *AgOTf*
Bicyclic ketals from acetylenediols
with AuCl or $AuCl_3$ cf. *69*, 80; with a little AgOTf s. C.H. Oh, H.J. Yi, J.H. Lee, New J. Chem. *2007*, *31* (6), 835-7.

*Silver tosylate s. under [(S)-2,2′-Bis[bis(3,5-di-*tert*-butyl-4-methoxyphenyl)-phosphino]-6,6′-dimethoxybiphenyl]bis(gold(I) chloride)* *AgOTs*

Polymer-based gold nanoclusters/1,8-diazabicyclo[5.4.0]undec-7-ene *Au/DBU*
Cyclic ethers from ethylenealcohols
tetrahydrofurans with AgOTf cf. *69*, 81s*70*; with 10 atm% poly(N-vinyl-2-pyrrolidone)-stabilized gold nanoclusters and 2 eqs. DBU in DMF/water under aerobic conditions (without Wacker-type oxidation) s. I. Kamiya, H. Tsunoyama, T. Tsukuda, H. Sakurai, Chem. Lett. *2007*, *36* (5), 646-7; with 10 mol% $FeCl_3$ and 30 mol% AgOTf in DCE at 80° s. K. Komeyama, T. Morimoto, Y. Nakayama, K. Takaki, Tetrahedron Lett. *2007*, *48* (18), 3259-61.

[N,N′-Bis(2,6-diisopropylphenyl)imidazol-2-ylidene]gold(I) chloride/silver fluoroborate ←
[N,N′-Bis(2,6-diisopropylphenyl)imidazol-2-ylidene]gold(I) chloride/silver fluoroborate/microwaves ←
Gold-catalyzed isomerization of acetoxy-2-ethylenes C(OAc)C=C → C=C-C(OAc)

87. 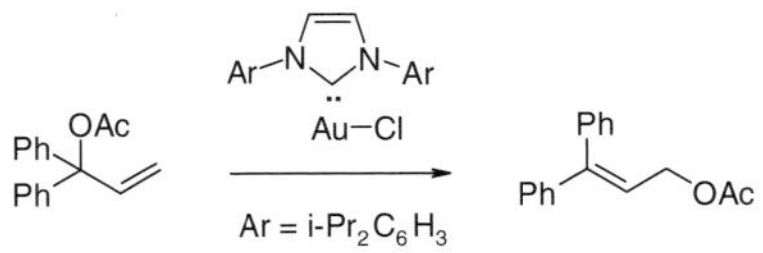

2 Mol% $AgBF_4$ added to a DCE soln. of 3 mol% [N,N′-bis(2,6-diisopropylphenyl)imidazol-2-ylidene]gold(I) chloride, the cloudy mixture stirred for 1 min, a soln. of the startg. allyl acetate in DCE added, the mixture heated in a microwave reactor at 80° for 12 min, diluted with pentane, filtered through Celite, and evaporated *in vacuo* → product. Y 88%. From a series of N-heterocyclic carbenes tested as potential ligands for this reaction, only the most sterically hindered afforded the isomerized product in high yield, others giving oligomerized products or being unreactive. 3-Aryl-3-acetoxyprop-1-enes were isomerized generally in >80% yield by refluxing (80°) for 12 h or under microwave irradiation for 12 min. Reaction failure where aryl is 2-nitrophenyl, 4-cyanophenyl or 2-pyridyl was attributed to coordination. The reaction conditions were optimized and it is worth noting that the reaction does not require strict anhydrous or anaerobic conditions. F.e.s. N. Marion, R. Gealageas, S.P. Nolan, Org. Lett. *2007*, *9* (14), 2653-6.

*[(S)-2,2′-Bis[bis(3,5-di-*tert*-butyl-4-methoxyphenyl)phosphino]-6,6′-dimethoxy-biphenyl]bis(gold(I) chloride)/silver tosylate* ←
2-Vinyl-O-heterocyclics from allenealcohols by asym. cycloisomerization ○

88. 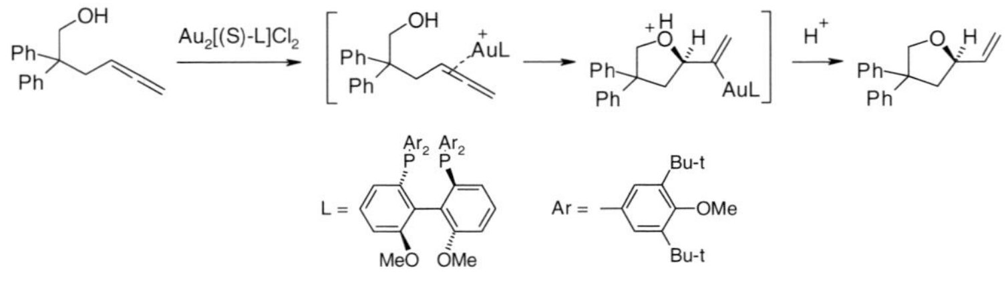

A mixture of 2.5 mol% [(S)-2,2′-bis[bis(3,5-di-*tert*-butyl-4-methoxyphenyl)phosphino]-6,6′-dimethoxybiphenyl]bis(gold(I) chloride) and 5 mol% Ag(I)-tosylate in *toluene* stirred at room temp. for 10 min, cooled to -20°, the startg. allenealcohol in toluene added, the resulting suspension stirred again for 18 h at the same temp., and worked up with column chromatographic purification → product. Y 67% (e.e. 93%). The procedure is applicable to the formation of both tetrahydrofuran and -pyran derivs. with high enantioselectivity, and tolerates *gem*-dialkyl or -diphenyl substitution between the hydroxyl and allene groups; however, enantioselectivity was low if there was no such substitution. Reaction involves initial coordination of Au(I) to the allenyl residue followed by asym. intramolecular oxyauration prior to protonolysis. F.e. and conversion of axially chiral allenealcohols s. Z. Zhang, R.A. Widenhoefer, Angew. Chem. Int. Ed. Engl. *2007, 46* (1-2), 283-5.

Gold(III) chloride or Gold(III) chloride/2,2'-bipyridyl *$AuCl_3$ or $AuCl_3$/bypy*
Cycloisomerization of 2-allenealcohols
2,3-dihydrofurans, stereoselectivity, cf. *36*, 148s*62*; 2-aryl-2,5-dihydrofurans (cf. *25*, 123s*37*), improved axis-to-center chirality transfer in the presence of an additive (such as 2,2′-bipyridyl), by using a moderately coordinating solvent (such as THF), or by lowering the reaction temp., s. C. Deutsch, B. Gockel, A. Hoffmann-Röder, N. Krause, Synlett *2007* (11), 1790-4.

2-Vinyltetrahydrofuran ring from 4-allenealcohols s. *72*, 148

Mercury(II) sulfate *$HgSO_4$*
Cyclic enolethers from acetylenealcohols
with $W(CO)_6$/Et_3N under irradiation cf. *31*, 135s*68*; 2-α-hydroxy-3(2*H*)-furanones from 1,4-enyn-3-ones via asym. dihydroxylation with AD-mix-α/$MeSO_2NH_2$ and cycloisomerization with aq. $HgSO_4$ in acetone (or one step process for electron-rich derivs.) s. C.M. Marson, E. Edaan, J.M. Morrell, S.J. Coles, M.B. Hursthouse, D.T. Davies, Chem. Commun. *2007* (24), 2494-6.

Zinc chloride *$ZnCl_2$*
Furans from β,γ-acetyleneketones
with $AuCl_3$ cf. *60*, 74; 2,5-di- and 2,3,5-tri-subst. furans with 10 mol% $ZnCl_2$ in methylene chloride at room temp. s. A. Sniady, A. Durham, M.S. Morreale, K.A. Wheeler, R. Dembinski, Org. Lett. *2007, 9* (7), 1175-8.

Silver fluoroborate s. under [N,N′-Bis(2,6-diisopropylphenyl)imidazol-2-ylidene]-gold(I) chloride *$AgBF_4$*

*1,3-Cyclohexadiene or triethylsilane or tri-*n*-butyltin hydride/iron(III) nitrate* ←
Ketones from cyclopropanols
with KOH cf. *19*, 211s*63*; with stoichiometric amounts of a H-donor, such as Bu_3SnH, 1,3-cyclohexadiene or Et_3SiH, and $Fe(NO_3)_3 \cdot 9H_2O$, incl. application to the total synthesis of (+)-spirolaxine methyl ether s. K.A. Keaton, A.J. Phillips, Org. Lett. *2007, 9* (14), 2717-9.

Silica *SiO_2*
2-α-Sulfonyloxy-1,3-dienes from 1-sulfonyloxy-1-vinylcyclopropanes
s. *72*, 225

Methanesulfonic acid *$MeSO_3H$*
Isomerization of 2-ethylenealcohols $C(OH)CH{=}CH_2 \rightarrow C{=}C\text{-}CH_2OH$

89.

Benzylic and alicyclic sec. and tert. vinylcarbinols isomerize to the corresponding **2-ethylene-*prim*-alcohols** in high yield and with high regioselectivity in the presence of methanesulfonic acid. **E:** 2 Eqs. methanesulfonic acid added dropwise over 5 min at room temp. to a soln. of the startg. vinylcarbinol in 4:1 THF/water, stirring continued for 12 h, quenched with satd. aq. $NaHCO_3$, and worked up with column chromatographic purification → product. Y 90%. The aromatic ring may possess electron-withdrawing or -donating substituents. Reaction did not go to completion wth triflic acid or TsOH, and there was no reaction with acetic or trifluoroacetic acid. F.e.s. R.R. Leleti, B. Hu, M. Prashad, O. Repic, Tetrahedron Lett. *2007, 48* (48), 8505-7.

Iron(III) nitrate s. under 1,3-Cyclohexadiene *$Fe(NO_3)_3$*

Iron(III) chloride/silver triflate *$FeCl_3/AgOTf$*
Cyclic ethers from ethylenealcohols s. *69*, 81s*72* ○

Bis(1,5-cyclooctadiene)rhodium(I) fluoroborate *$[Rh(cod)_2]BF_4$*
Bis(1,5-cyclooctadiene)iridium(I) fluoroborate *$[Ir(cod)_2]BF_4$*
***p*-Quinol ethers from *o*-quinone monoketals via 1,3-alkoxy group migration** ←

90.
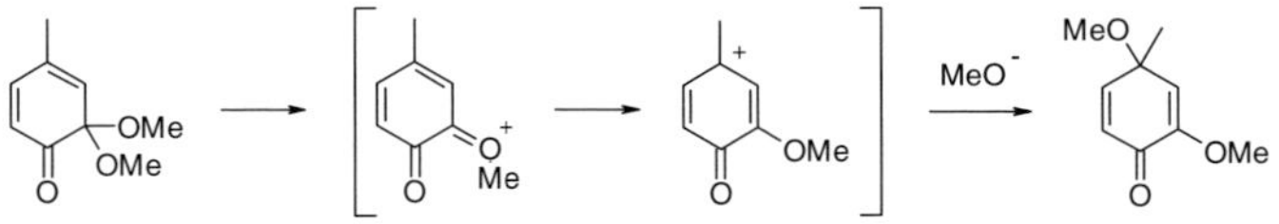

The startg. masked *o*-benzoquinone added to 5 mol% $[Rh(cod)_2]BF_4$ in methylene chloride, the mixture stirred 0.5 h, evaporated *in vacuo*, and purified by chromatography → 2,4-dimethoxy-4-methylcyclohexa-2,5-dienone. Y 80%. The reaction is general (Y 58-82%) for substrates possessing an electron-releasing group in the 4-position (4-H-5-Me, 4-TMS, 4-Br and 4-H-5-CO_2Me derivs. being recovered even at reflux). Yields were somewhat reduced, however, for bulky 4-alkyl derivs. (4-*t*-Bu: Y 64%; cf. 4-Me: Y 80%). Other Lewis acids gave only dimeric products, while rearrangement failed with neutral Rh complexes. However, the iridium analog $[Ir(cod)_2]BF_4$ was also effective at promoting migration. The proposed mechanism involves rearrangement of an intermediate oxonium species which would be stabilized by electron-releasing groups. F.e. and a one-pot procedure **from *o*-quinol monoethers** s. C.-S. Yang, C.-C. Liao, Org. Lett. *2007, 9* (23), 4809-12.

*Chloro(1,5-cyclooctadiene)rhodium(I) dimer/tris(*p*-fluorophenyl)phosphine* ←
Cycloisomerization of functionalized terminal acetylene derivs. ○
via intermolecular nucleophilic addition to rhodium vinylidene complexes – Benzofurans from terminal *o*-hydroxyarylacetylenes – 2(3*H*)-Furanones from α-ethynycarboxylic acids s. *72*, 190

Exchange ↓↑

Hydrogen ↑ OC ↓↑ H

Irradiation s. under Allyl bromide ⫽⫽

Microwaves s. under 2,3-Dichloro-5,6-dicyanoquinone [\\\\]

Sodium carbonate *Na_2CO_3*

Allyloxylamines $>N^{+}{=}O \rightarrow >N\text{-}O\text{-}C\text{-}C{=}C$

from trisubst. ethylene derivs. and oxoammonium salts

Ene-type reaction

91. 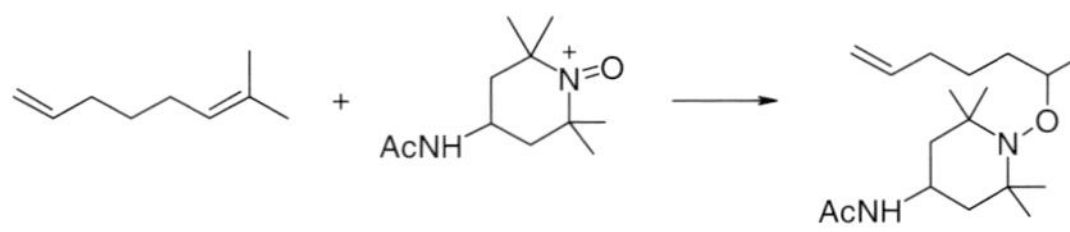

An equimolar mixture of 4-acetamido-2,2,6,6-tetramethylpiperidine-1-oxoammonium fluoroborate and 2-methylocta-2,7-diene in acetonitrile stirred at room temp. until complete consumption of the oxoammonium salt as indicated by starch-KI paper (2 h), the soln. concentrated *in vacuo*, the residue stirred in methylene chloride for 1 h, filtered, the filtrate washed sequentially with satd. aq. Na_2CO_3, water and brine, then dried, and concentrated *in vacuo* → product. Y 80%. The procedure is clean and rapid **under mild conditions,** and appears to be specific for *trisubst.* ethylene compds. giving yields of 60-95% (8 examples) in 1.5-4 h at room temp. (in each case reaction occurring at the least subst. carbon of the double bond). Mono- and di-subst. derivs. failed to react while tetra-subst. derivs. reacted very slowly. F.e.s. P.P. Pradhan, J.M. Bobbitt, W.F. Bailey, Org. Lett. *2006, 8* (24), 5485-7.

Potassium carbonate *K_2CO_3*

Base-catalyzed autoxidation of hydrocarbons ←

with KOBu-*t*/PEG cf. *20*, 119s*41*; with K_2CO_3, acyclic and cyclic aryl ketones from alkylarenes, incl. fluorene, xanthene, anthrone, 9,10-dihydroanthracene, and diphenylmethane derivs., leaving amino and aldehyde groups unaffected, s. K.K. Park, L.K. Tsou, A.D. Hamilton, Synthesis *2006* (21), 3617-20.

Cesium carbonate/tetra-n-*butylammonium bromide* *Cs_2CO_3/Bu_4NBr*

α-Ketocarboxylic from carboxylic acid amides $CH_2 \rightarrow CO$

by base-catalyzed aerobic oxidation under mild conditions

92. 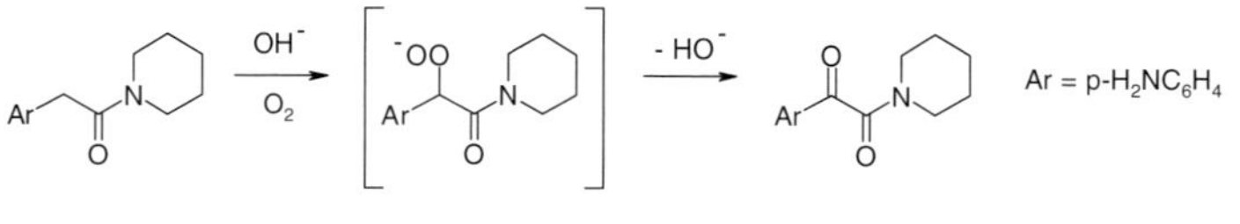

N,N-Disubst. arylglyoxylic acid amides. 1.1 Eqs. Cs_2CO_3 and 0.5 eqs. tetrabutylammonium bromide *as accelerator* added to a soln. of 1-(4-aminophenyl)-2-(piperidin-1-yl)ethanone in DMF, stirred for 9 h at 120° under dry air, and worked up with column chromatographic purification on silica gel → product. Y 75%. The procedure is remarkably clean, simple, avoids toxic chemicals and harsh conditions, and may be adapted on the industrial scale. Easily oxidizable *prim*-amino groups on the aromatic ring remained intact, whereas nitro derivs. presented complications. It is thought that benzylic deprotonation initiates the catalytic cycle, which then affords a peroxy anion on reaction with oxygen prior to formation of the product with elimination of hydroxide ion. F.e. and comparison of solvents and quaternary ammonium salts. s. B. Song, S. Wang, C. Sun, H. Deng, B. Xu, Tetrahedron Lett. *2007, 48* (51), 8982-6.

Sodium nitrite s. under (S)-5-Benzyl-2,2,3-trimethyl-4-imidazolidonium fluoroborate *$NaNO_2$*

2(S)-[Diphenyl(trimethylsiloxy)methyl]pyrrolidine/3-benzyl-4,5-dimethyl-thiazolium chloride/triethylamine/ethyldiisopropylamine ←

Asym. synthysis of β-functionalized carboxylic acid esters from α,β-ethylenealdehydes ←

Sequential catalysis in one pot with two different organocatalysts s. *72*, 413

1,8-Diazabicyclo[5.4.0]undec-7-ene s. under MnO_2 *DBU*

Copper(II) triflate/phenylhydrazine/chiral 2,6-bis(Δ²-oxazolin-2-yl)pyridines ←
Asym. Kharasch acoxylation of ethylene derivs. H → OAc
with $[Cu(MeCN)_4]PF_6$ cf. *51*, 72s*69*; with $Cu(OTf)_2$/phenylhydrazine and a variety of peresters (e.e. up to 98%) s. S.K. Ginotra, V.K. Singh, Org. Biomol. Chem. *2006*, *4* (23), 4370-4.

Indium(III) chloride s. under o-*Iodoxybenzoic acid* *$InCl_3$*
Imidazolium salts s. under ArI ←

(S)-5-Benzyl-2,2,3-trimethyl-4-imidazolidonium fluoroborate/iron(III) chloride/ oxygen/sodium nitrite ←
Organocatalyzed asym. radical α-aminooxylation of aldehydes H → ON<

93. 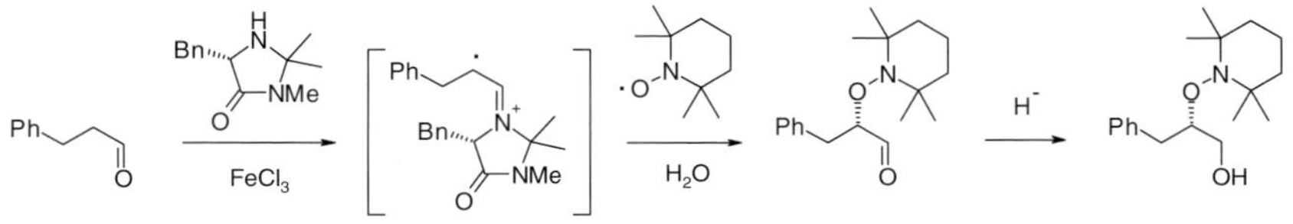

A mixture of the startg. aldehyde, 2 eqs. TEMPO, 30 mol% $NaNO_2$ and 20 mol% (S)-5-benzyl-2,2,3-trimethyl-4-imidazolidonium fluoroborate in DMF stirred for 5 min at room temp., 20 mol% $FeCl_3$ added and O_2 introduced via syringe, stirred at -10° for 24 h, quenched with aq. NH_4Cl, worked up, the resulting crude α-functionalized aldehyde dissolved in THF, cooled to 0°, 2 eqs. $NaBH_4$ added, the suspension warmed to room temp., stirred for 1 h, and worked up with chromatographic purification on silica gel → product. Y 68% (e.e. 82%). High enantioselectivity was recorded with aryl-subst. aldehydes (except phenylacetaldehyde) and 4-penten-1-al, but there was no selectivity with aliphatic aldehydes. Reaction is presumed to involve initial enamine formation, followed by SET with inexpensive $FeCl_3$ and quenching of the generated radical by TEMPO with asym. induction ($O_2/NaNO_2$ serving as reoxidant). F.e. and comparison of chiral amines, also conversion to **chiral glycols** (with Zn/AcOH), s. M.P. Sibi, M. Hasegawa, J. Am. Chem. Soc. *2007, 129* (14), 4124-5.

Chiral 2,6-bis(Δ²-oxazolin-2-yl)pyridines s. under $Cu(OTf)_2$ *Py-box*
1,2,4-Triazolium salts s. under MnO_2 ←
tert-*Butyl hydroperoxide (or N-hydroxyphthalimide) s. under $NaClO_2$* ←
Amberlyst s. under 2,3-Dichloro-5,6-dicyanoquinone ←
m-*Chloroperoxybenzoic acid s. under ArI* *$ArCOO_2H$*

o-*Iodoxybenzoic acid/indium(III) chloride* *$ArIO_2/InCl_3$*
Oxidation of indoles ←
oxindole- from indole-3-alkanoic acids with NBS/Pd-C cf. *17*, 219; isatins with *o*-iodoxybenzoic acid/$InCl_3$ in aq. acetonitrile at 80°, also aza-analogs, s. J.S. Yadav, B.V. Subba Reddy, Ch. Suresh Reddy, A.D. Krishna, Synthesis *2007* (5), 693-6; 3-hydroxyoxindoles from 3-alkylindoles (cf. *49*, 117s*71*) with *o*-iodoxybenzoic acid/$CeCl_3 \cdot 7H_2O$ at room temp. s. idem., Tetrahedron Lett. *2007*, *48* (11), 2029-32.

Phenyl iodosoacetate *$PhI(OAc)_2$*
Oxidative ring closures of ethylenealcohols ◯
with addition of nucleophiles – 2-[Acylamino-(acoxy)- or -(alkoxy)-methyl]tetrahydropyrans from 6-acylamino-5-ethylenealcohols s. *28*, 107s*72*; *34*, 114s*72*

Allyl bromide/irradiation *$CH_2{=}CHCH_2Br$/⇜*
Carboxylic acids from prim. alcohols $CH_2OH \rightarrow COOH$
halogen-mediated photooxidation with NBS cf. *70*, 82s*71*; arylcarboxylic acids from benzyl alcohols with a little allyl bromide, also from methylarenes (cf. *11*, 225s*69*), s. T. Sugai, A. Itoh, Tetrahedron Lett. *2007*, *48* (16), 2931-4.

*Chiral iodoarenes/*m*-chloroperoxybenzoic acid* *ArI/ArCOO$_2$H*

*Ionic liquid-based iodoarenes/*m*-chloroperoxybenzoic acid/1-ethyl-3-methyl-imidazolium tosylate* ←

α-Tosyloxylation of ketones H → OTs

with PhI(OH)OTs cf. *47*, 495 and with CuO cf. *54*, 62s*55*; with ionic liquid-supported iodobenzene/*m*-CPBA in [emim]OTs s. J. Akiike, Y. Yamamoto, H. Togo, Synlett *2007* (14), 2168-72; in moderate enantioselectivity (e.e. up to 28%) with iodine(III) reagents prepared *in situ* from chiral iodoarenes/*m*-CPBA (cf. *70*, 83) s. R.D. Richardson, T.K. Page, S. Altermann, S.M. Paradine, A.N. French, T. Wirth, Synlett *2007* (4), 538-42.

2,3-Dichloro-5,6-dicyanoquinone/Amberlyst/microwaves *DDQ/Amberlyst/*[\\\\]

Carboxylic acid esters from aldehydes CHO → COOR

with HCl under irradiation cf. *31*, 159; with DDQ/Amberlyst in alcohol/toluene under microwave irradiation, (E)-cinnamic acid esters and selectivity for oxidation of cinnamaldehydes over aliphatic and aromatic aldehydes, also from cinnamyl alcohols (cf. *66*, 99), s. A.K. Sinha, A. Sharma, A. Swaroop, V. Kumar, Tetrahedron *2007*, *63* (4), 1000-1007.

Phenylhydrazine s. under Cu(OTf)$_2$ *PhNHNH$_2$*

Oxygen s. under (S)-5-Benzyl-2,2,3-trimethyl-4-imidazolidonium fluoroborate and [Ir(coe)$_2$Cl]$_2$ *O$_2$*

Hydrogen peroxide/hydrogen bromide *H$_2$O$_2$/HBr*

Carbonyl from methylene groups CH$_2$ → CO

with DMSO/HBr cf. *25*, 129; with 30% H_2O_2/48% HBr in methylene chloride at room temp., benzylic oxidation, s. A.T. Khan, T. Parvin, L.H. Choudhury, S. Ghosh, Tetrahedron Lett. *2007*, *48* (13), 2271-4.

Sodium hypochlorite s. under Nickel oxide hydroxide *NaOCl*

*Sodium chlorite/*tert*-butyl hydroperoxide or N-hydroxyphthalimide* ←

Keto from methylene groups – Allylic and benzylic oxidation

with NaOCl/*t*-BuOOH cf. *68*, 90; with $NaClO_2$ and *t*-BuOOH (stoichiometric) or N-hydroxyphthalimide (catalytic) s. S.M. Silvestre, J.A.R. Salvador, Tetrahedron *2007*, *63* (11), 2439-45.

*Tetra-*n*-butylammonium bromide s. under Cs$_2$CO$_3$* *Bu$_4$NBr*

Hydrogen bromide s. under H$_2$O$_2$ *HBr*

Manganese dioxide/1,4-dimethyl-1,2,4-triazolium iodide/1,8-diazabicyclo-[5.4.0]undec-7-ene ←

Carboxylic acid esters from activated prim. alcohols and alcohols CH$_2$OH → COOR

N-Heterocyclic carbene-catalyzed oxidative esterification

94.

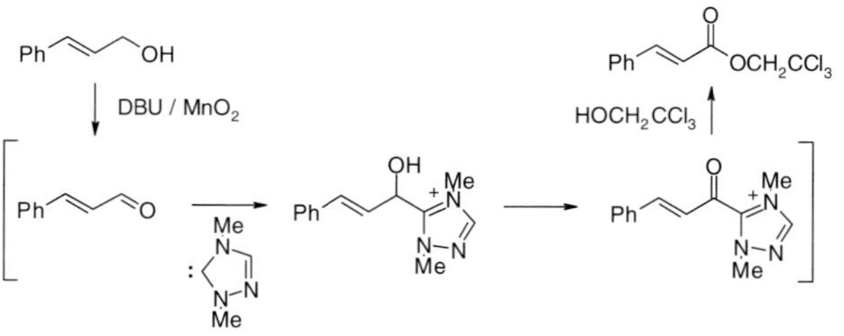

Cinnamyl alcohol, excess of 2,2,2-trichloroethanol and 15 eqs. MnO_2 added sequentially to 10 mol% 1,4-dimethyl-1,2,4-triazolium iodide under N_2, 0.1 eqs. DBU added by syringe, the mixture stirred to completion (with GC monitoring), filtered through Celite, and concentrated *in vacuo* → 2,2,2-trichloroethyl cinnamate. Y 82%. The reaction occurs by initial oxidation of the alcohol to the aldehyde, followed by coupling with the carbenoid catalyst and subsequent oxidation of the resulting 5-α-hydroxytriazolium ion to the ketone prior to reaction with the added alcohol and liberation of the carbene in a catalytic cycle. The procedure is successful (65-95%) for oxidation of allylic, propargylic and benzylic prim. alcohols and coupling with prim. or sec. alcohols to quench the intermediate (tert. butanol giving no product). 3-Phenylpropionaldehyde was also sufficiently activated by the triazolium salt to give methyl 3-phenylpropionate in 91% yield. The reaction is particularly useful for preparing protected carboxylates (e.g. 2,2,2-trichloroethyl, 2-methoxyethyl and 2-trimethylsilylethyl esters). F.e. and **desymmetrization of *meso*-glycols** (with a chiral N-condensed 1,2,4-triazolium salt) s. B.E. Maki, A. Chan, E.M. Phillips, K.A. Scheidt, Org. Lett. *2007, 9* (2), 371-4.

Iron(III) chloride s. under (S)-5-Benzyl-2,2,3-trimethyl-4-imidazolidonium fluoroborate $FeCl_3$

Nickel oxide hydroxide nanoparticles/sodium hypochlorite ←
Catalytic heterogeneous oxidations with nickel oxide hydroxide nanoparticles in water ←

95.

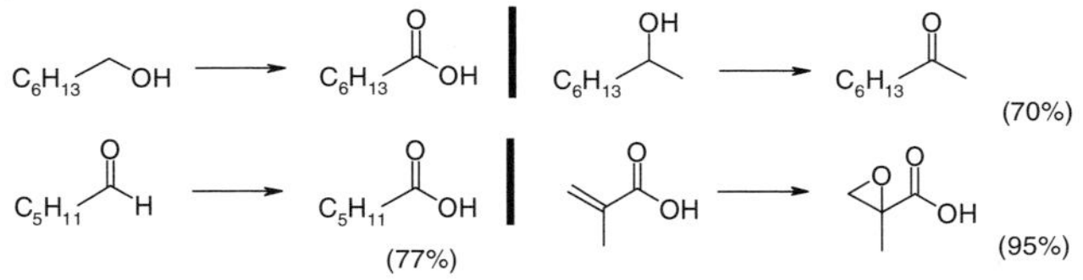

In the presence of commercial bleach (ca. 5% NaOCl) as reoxidant, catalytic amounts of $NiCl_2$ or $Ni(OAc)_2$ are converted to insoluble nickel oxide hydroxide nanoparticles of high surface area (ca. 4 nm), which serve as a useful, environmentally friendly heterogeneous catalyst for the controlled and selective oxidation of a variety of aliphatic and aromatic organic compds., in most instances **in the absence of organic solvent. E: Carboxylic acids from prim. alcohols.** 45 mmol *n*-Heptanol added to 1.14 mmol $NiCl_2 \cdot 6H_2O$ and 5 ml water, cooled in an ice-bath, 300 ml cold bleach added in a steady stream over 5 min, the resulting fine black slurry stirred for 2 h at 0° then for a further 2 h at room temp., acidified with 2 *M* HCl, and worked up → heptanoic acid. Y 86% (after distillation). A small amount of methylene chloride was added to the medium for oxidation of water-insoluble substrates. F.e., also ketones from sec. alcohols, dicarboxylic acids from cyclic glycols, carboxylic acids from aldehydes, and epoxidation of α,β-ethylenecarboxylic acids, s. J.M. Grill, J.W. Ogle, S.A. Miller, J. Org. Chem. *2006, 71* (25), 9291-6.

Chlorobis(cyclooctene)iridium dimer/air $[Ir(coe)_2Cl]_2/O_2$
Carboxylic acid esters from two prim. alcohol molecules $2\ RCH_2OH \rightarrow RCOOCH_2R$
with [Cp*(2,2-diphenylglycinolato-N,O)Ir(III)]/2-butanone/K_2CO_3 cf. *69,* 59; with $[Ir(coe)_2Cl]_2$ (3 mol%) at 95° under air in the absence of solvent s. A. Izumi, Y. Obora, S. Sakaguchi, Y. Ishii, Tetrahedron Lett. *2006, 47* (52), 9199-201.

Platinum nanoclusters/oxygen Pt/O_2
Carboxylic acids from prim. alcohols $CH_2OH \rightarrow COOH$
Base-free aerobic oxidation of alcohols in water s. *72,* 143

Via intermediates *v.i.*
Diaryl ethers from arenes and phenols H → OAr
via iridium-catalyzed borylation s. *61,* 100s*72*; *65,* 62s*72*

Oxygen ↑ OC ↓↑ O

Sodium hydroxide *NaOH*
α-Alkoxy-β,γ- or γ-alkoxy-α,β-ethylenecarboxylic acids ←
from 2′-ethylene-2,2,2-trichloroalcohols s. *72,* 326

Potassium tert-*butoxide* *KOBu*-t

O-Protected α-alkylidene-*syn*-β,δ-dihydroxysulfones from (E)-β′-acoxy-α,β-ethylene-δ-hydroxysulfones via stereospecific Michael addition-elimination ○

96.

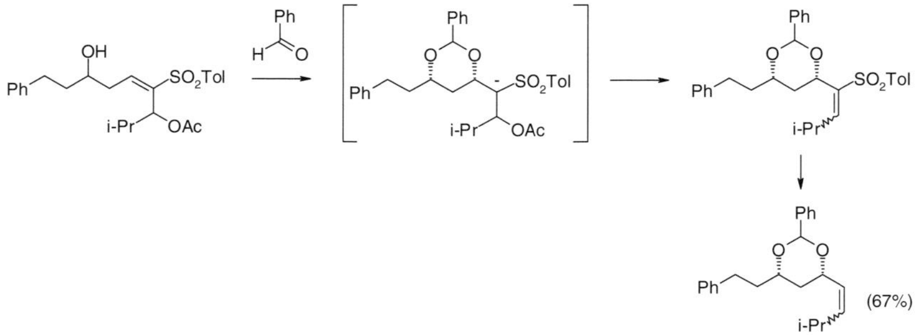

3.3 Eqs. benzaldehyde added to a ca. 0.1 *M* soln. of the startg. alkenyl sulfone in THF at 0°, treated with 1.1 eqs. *t*-BuOK, the mixture stirred at 0° for 45 min, quenched with satd. aq. NH_4Cl, diluted with ether, separated and dried → product. Y 83% (E/Z 1:1). When the base was added in portions (0.3 eq./15 min) the selectivity was improved (E/Z 3:1) but the yield was reduced to 65%. It is presumed that the (Z)-isomer (or its intermediate) was less stable than the (E)-isomer, and indeed the analogous sulfone with Ph replacing $PhCH_2CH_2$ is base-sensitive and gave a similar ratio of isomers at only 57% yield *even under optimized conditions*. Benzoate in place of acetate as leaving group gave similar results and its use is synthetically more efficient as it may be generated *in situ*. The sequence was applied to the synthesis of the C21-C25 fragment of Dolabelide C. F.e. and synthesis of the startg. m., also subsequent desulfonation (with Na/Hg/KH_2PO_4) to give ***syn*-4-ene-1,3-diols,** s. D. Rotulo-Sims, J. Prunet, Org. Lett. *2007*, *9* (21), 4147-50.

Triethylamine *Et_3N*

Carboxylic acid benzyl esters from acids COOH → COOBn

97.

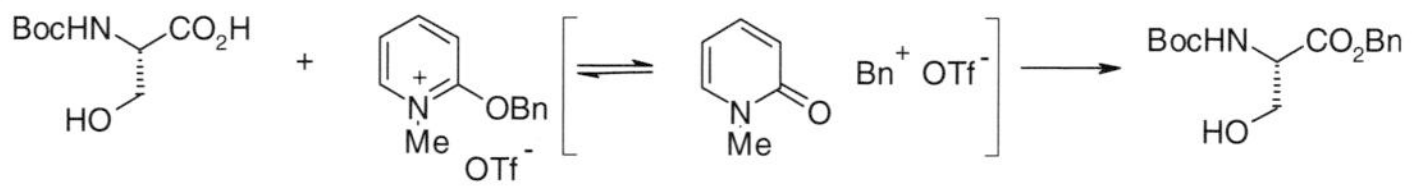

under nearly neutral conditions. A mixture of 1.9 eqs. 2-benzyloxy-1-methylpyridinium triflate, 1 eq. N-Boc-serine and 2 eqs. triethylamine in α,α,α-trifluorotoluene heated at 83° for 24 h, cooled, diluted with water, extracted with ethyl acetate, the organic layer washed with brine, dried, and evaporated → N-Boc-serine benzyl ester. Y 91%. Unlike the Mukaiyama agent (2-chloro-1-methylpyridinium iodide), the benzyloxypyridinium triflate plays no part in carboxylic acid activation and serves specifically as a **benzyl transfer agent**, benzylating aliphatic (incl. ethylenic and acetylenic) and aromatic carboxylic acids in 81-99% yields. The mild conditions are compatible with acetate groups, Boc-amines and acetonide groups, and leave alcohol, phenol and amide functionality unaffected. The reaction may be accomplished in the absence of base, but the use of triethylamine promotes the reaction and may also serve as a scavenger to effect complete suppression of dibenzyl ether formation. F.e.s. J. Tummatorn, P.A. Albiniak, G.B. Dudley, J. Org. Chem, *2007*, *72* (23), 8962-4.

4-Dimethylaminopyridine/1-methoxyethyl-3-methylimidazolium trifluoroacetate ←

O-Acylation OH → OAc

with 4-dialkylaminopyridines as catalysts cf. *29*, 184; O-benzoylation of ribo- and deoxyribonucleosides with DMAP in 1-methoxyethyl-3-methylimidazolium trifluoroacetate s. V. Kumar, V.S. Parmar, S.V. Malhotra, Tetrahedron Lett. *2007, 48* (5), 809-12.

Copper(II) triflate/phenylhydrazine/chiral 2,6-bis(Δ^2-oxazolin-2-yl)pyridines ←
Asym. Kharasch acoxylation of ethylene derivs. s. under OC↓↑H H → OAc

Copper(II) tosylate *$Cu(OTs)_2$*
Acetalation CO → $C(OR)_2$
with $Cu(OTf)_2$ cf. *58*, 78s*70*; 1,3-dioxolanes with $Cu(OTf)_2$ s. C.-H. Liu, X.-Y. Yu, X.-Z. Liang, W.-J. Wang, J.-G. Yang, M.-Y. He, Chinese J. Chem. *2007, 25* (1), 83-5.

Gold(III) chloride *$AuCl_3$*
2-Methylene-2-deoxy-α-glycosides from 2-propargyloxymethyl-2-deoxyglycals ←

98.
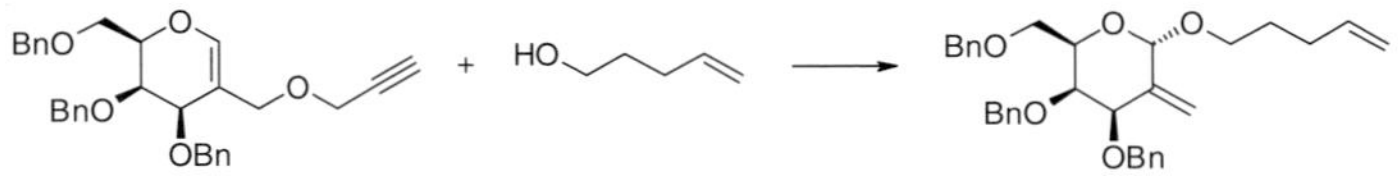

under mild conditions. 2 Eqs. 4-pentenol and *5 mol%* $AuCl_3$ in dry acetonitrile added at 0° to a soln. of the startg. 2-propargyloxymethyl-2-deoxyglycal in the same solvent, stirred at room temp. for 16 h, and worked up with column chromatographic purification → product. Y 68%. The procedure is fast and applicable to glucose, galactose and xylose derivs. on reaction with a range of aromatic, aliphatic, alicyclic and partially protected carbohydrate alcohols as nucleophile. Olefins, azides, O,O-isopropylidene derivs. and ethers remained unaffected. Reaction involves initial activation of the alkyne bond of the substrate with Au(III). F.e.s. S. Kashyap, S.R. Vidadala, S. Hotha, Tetrahedron Lett. *2007, 48* (51), 8960-2.

Aluminum silicate ←
Acetals from oxo compds. CO → $C(OR)_2$
with Kaolin cf. *28*, 141s*52,66*; with Al-silicate in (m)ethanol at room temp., also deprotection in aq. acetone at 55°, s. M.W.C. Robinson, A.E. Graham, Tetrahedron Lett. *2007, 48* (27), 4727-31; from aliphatic or aromatic aldehydes with polymer-based 1-ω-sulfoimidazolium triflates (cf. *63*, 102) s. R. Sugimura, K. Qiao, D. Tomida, C. Yokoyama, Catal. Commun. *2007, 8* (5), 770-2.

Potassium-exchanged molecular sieves ←
Aluminum triflate *$Al(OTf)_3$*
O-Acetylation OH → OAc
with molecular sieves or with $Al(HSO_4)_3$ cf. *54*, 79s*67*; of purine and pyrimidine nucleosides with K-exchanged molecular sieves (and acetic anhydride) s. M.M. Sá, L. Meier, Synlett *2006* (20), 3474-8; of alcohols, phenols, sugars and thiols with Al-triflate (0.01-0.1 mol%) in the absence of solvent at room temp. in short reaction times s. A. Kamal, M.N.A. Khan, K. Srinivasa Reddy, Y.V.V. Srikanth, T. Krishnaji, Tetrahedron Lett. *2007, 48* (22), 3813-8; O-acylation of alcohols, phenols, and amines with a variety of acid anhydrides and SO_4^{2-}/SnO_2 under solvent-free conditions at room temp., incl. chemoselective N-acetylation with retention of hydroxyl groups, s. J.R. Satam, M.B. Gawande, S.S. Deshpande, R.V. Jayaram, Synth. Commun. *2007, 37* (17), 3011-20.

Zeolites s. under Zirconium- or tin-containing zeolite *zeolites*

N-Ethylbenzimidazolium fluoroborate *$[Hebim]BF_4$*
Carboxylic acid esters from carboxylic acids COOH → COOR
in 1-methylimidazolium fluoroborate as Brønsted-acidic ionic liquid solvent cf. *65*, 93; in 3-ethyl-1*H*-benzimidazolium fluoroborate, improved solubility for aromatic alcohols and aromatic carboxylic acids, also from anhydrides, s. Y.-Y. Wang, W. Li, C.-D. Xu, L.-Y. Dai, Chinese J. Chem. *2007, 25* (1), 68-71.

Indium(III) chloride *$InCl_3$*
Cyclic acetals from oxo compds. ○
dibenzo[*d,f*][1,3]dioxepins s. *58*, 78s*69*; s.a. G. Tocco, M. Begala, G. Meli, G. Podda, Heterocycles *2007, 71* (3), 635-45.

Erbium(III) triflate *Er(OTf)₃*
Ferrier reaction under mild conditions ←

99. 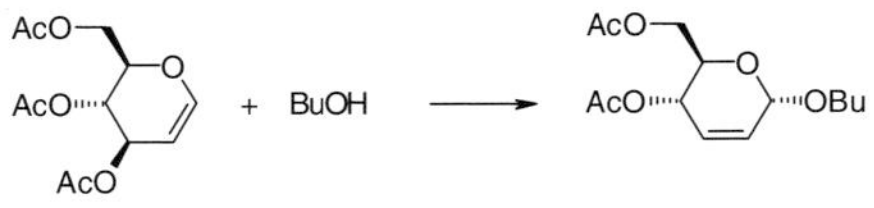

3,4,6-Tri-O-acetyl-D-glucal added to *5 mol%* $Er(OTf)_3$ *in nitromethane*, stirred *at room temp.* for 5 h, 1.5 eqs. *n*-butanol added, and stirring continued for 2 h → product. Y 95% (92% α). The procedure is a significant improvement on that with $Yb(OTf)_3$ (cf. *58, 76s61*), being widely applicable, notably to the reaction of prim., sec., benzyl, allyl and propargyl alcohols with both glucal and galactal derivs; it is also clean, highly stereoselective, and based on a commercially available reagent which may be recovered easily and retains its activity over several cycles (Y 80% after 4 cycles). Reaction with phenol, however, was more slow, requiring heating to 80° to force the conversion. F.e. and solvent effect, **also thio-Ferrier reaction** with mercaptans, s. A. Procopio, R. Dalpozzo, A. De Nino, L. Maiuolo, M. Nardi, M. Oliverio, B. Russo, Carbohyd. Res. *2007, 342* (14), 2125-31.

Ytterbium(III) triflate s. under N-Iodosuccinimide *$Yb(OTf)_3$*

Subcritical water *$_{sub}H_2O$*
Rapid and selective uncatalyzed O-acylation in subcritical water OH → OAc

100. 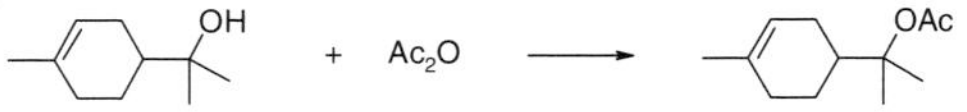

in a flow-type microreactor. The startg. alcohol and 4 eqs acetic anhydride placed across a high-speed flow of subcritical water, and the mixture introduced into a microreactor at 200° (5 MPa), where vigorous mixing was effected under a turbulent flow for 9.9 s → product. Y 90% (conversion 92%; chemoselectivity 97% over hydrolysis, hydration and pyrolysis products). The procedure is applicable in essentially quantitative yield to prim., sec. and tert. alcohols and phenols in *<10 s*, the speed of the reaction minimizing the possibility of any side-reactions, such as hydrolysis, hydration and pyrolysis, while the products accumulate at the bottom of the reactor or as water-immiscible oils which are easily separated. The use of batch conditions results in low yields, presumably due to hydrolysis of the products, and acylation of tert. alcohols requires excess of the anhydride (4-30 eqs.) to avoid formation of dehydration products. By adjusting the proportion of the anhydride, **selective O-acylation** of prim. alcohols may be effected in the presence of sec. Reaction benefits from the Lewis acidity of subcritical water so that addition of a base or acid catalyst is not required. F.e.s. M. Sato, K. Matsushima, H. Kawanami. Y. Ikuhsima, Angew. Chem. Int. Ed. Engl. *2007, 46* (33), 6284-8.

Imidazolium salts s. under 4-Dimethylaminopyridine, Fluoroborates and N-Iodosuccinimide ←
Azodicarboxylates s. under Ph_3P *ROOCN=NCOOR*

Dicyclohexylcarbodiimide/4-pyrrolidinopyridine ←
1-Ethyl-3-(3-dimethylaminopropyl)carbodiimide hydrochloride/4-pyrrolidinopyridine ←
Carboxylic acid esters from carboxylic acids COOH → COOR
cf. *16*, 236s*38,61,62*; α-cyclohexylideneacetic acid esters with DCC/4-pyrrolidinopyridine, also cyclohex-1-enylacetic acid esters by deconjugative esterification with EDC·HCl in place of DCC, s. S. Sano, E. Harada, T. Azetsu, T. Ichikawa, M. Nakao, Y. Nagao, Chem. Lett. *2006, 35* (11), 1286-7.

Benzoic anhydrides/(+)-benzotetramisole or tetramisole/ethyldiisopropylamine ←
Kinetic resolution of sec. benzyl alcohols by esterification OH → OAc

101. 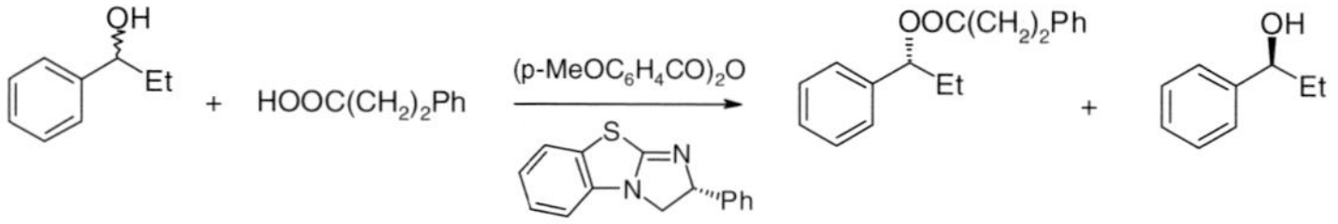

The first example of kinetic resolution of alcohols with *free* carboxylic acids is reported. **E:** 1.8 Eqs. ethyldiisopropylamine, 5 mol% (+)-benzotetramisole and racemic 1-phenyl-1-propanol added at room temp. to a soln. of 0.9 eq. *p*-methoxybenzoic anhydride and 0.75 eq. 3-phenylpropionic acid in methylene chloride, stirred for 12 h, quenched with satd. aq. $NaHCO_3$, and worked up with purification by TLC on silica → (R)-1-phenylpropyl 3-phenylpropanoate (Y 41%; e.e. 90%) and (S)-1-phenyl-1-propanol (Y 46%; e.e. 75%). Reaction proceeds with intermediate formation of reactive *p*-methoxybenzoic 3-phenylpropionic anhydride, the effective acylating agent. For isolation of the unreactive (S)-alcohols, benzoic anhydride was a superior auxiliary in terms of enantioselectivity. F.e. and comparison of benzoic anhydrides, also with tetramisole as chiral reagent, s. I. Shiina, K. Nakata, Tetrahedron Lett. *2007, 48* (47), 8314-7.

Amberlyst 15 ←
Ferrier rearrangement ←
with BF_3 cf. *24*, 240; with Amberlyst 15, incl. 2,3-unsatd. thioglycosides, s. Q. Tian, X.-M. Zhu, J.-S. Yang, Synth. Commun. *2007, 37* (5), 691-701.

(+)-Benzotetramisole or tetramisole s. under Benzoic anhydrides ←
Phenyl iodosoacetate s. under Ph_3P *$PhI(OAc)_2$*

Carbon tetrabromide *CBr_4*
Organocatalyzed acylation with carboxylic acid anhydrides in the absence of solvent OH → OAc

102. 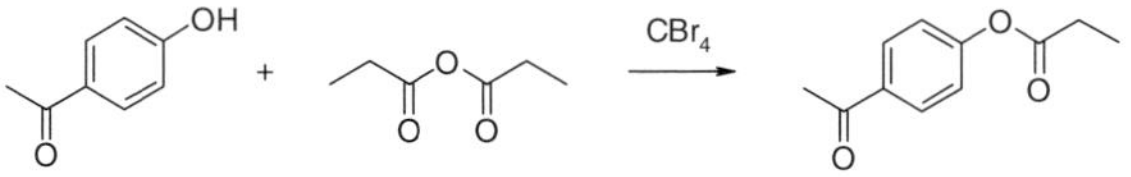

O-Acylation. Carbon tetrabromide (5 mol%) added to a mixture of 4-acetylphenol (0.5 mmol) and propionic anhydride (1.1 eqs.), and stirred at 60° for 3-6 h under air → 4-acetylphenyl propionate. Y 95%. The procedure is also applicable to the acylation of alcohols, although yields are poor with hindered substrates. The method avoids the use of solvents, hazardous acids, expensive Lewis acids and harsh reaction conditions, although it should be noted that carbon tetrabromide is both a poison and an irritant. F.e., **also S-acylation** of thiols and thiophenols, s. L. Zhang, Y. Luo, R. Fan, J. Wu, Green Chem. *2007, 9* (9), 1022-5.

N-Iodosuccinimide/ytterbium(III) triflate/1-butyl-3-methyl-imidazolium hexafluorophosphate *$NIS/Yb(OTf)_3/[bmim]PF_6$*
O²-Acylglycosides from carbohydrate pent-4-enyl 1,2-orthoesters s. *59*, 66s*72* C

Oxalyl chloride/triethylamine *$(COCl)_2/Et_3N$*
Protection of carboxyl groups as methylthiomethyl esters under mild conditions $COOH \rightarrow COOCH_2SMe$

103. 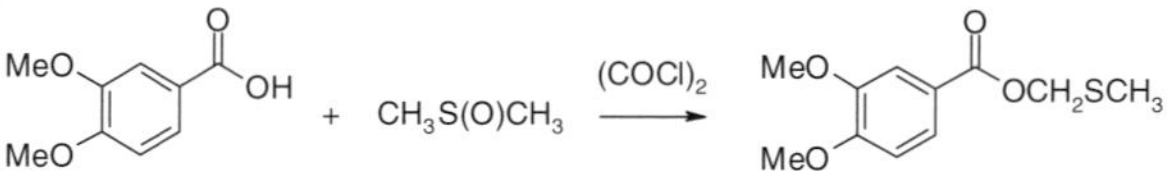

2.2 Eqs. DMSO in methylene chloride added dropwise to a stirred soln. of 1.1 eqs. oxalyl chloride in the same solvent at -60 to -65° under an inert atmosphere, the startg. carboxylic acid added after 5 min, stirring continued for 15 min, treated dropwise with 10 eqs. Et_3N, after a further 5 min the mixture allowed to warm to room temp., diluted with methylene chloride, quenched with water, and worked up with column chromatographic clean-up → product. Y 99%. The procedure is very mild and generally applicable to aliphatic, aromatic (incl. *p*-nitrobenzoic), α,β-ethylenic and N-protected α-amino-acids. F.e.s. S.B. Jadhav, U. Ghosh, Tetrahedron Lett. *2007, 48* (14), 2485-7.

Triethylsilane/iron(III) chloride *$Et_3SiH/FeCl_3$*
Ethers from oxo compds. and alcohols CO → CHOR
cf. *63*, 97; with $FeCl_3$ in place of $BiCl_3$ as catalyst s. K. Iwanami, K. Yano, T. Oriyama, Chem. Lett. *2007, 36* (1), 38-9.

Potassium triethylsilanolate s. under [Ir(cod)Cl]$_2$ *$KOSiEt_3$*
Tin(IV) oxide s. under Sulfate *SnO_2*

Zirconium- or tin-containing zeolite *Zr- or Sn-zeolite*
Etherification ROH + HOR′ → ROR′
benzyl ethers with zeolite/alumina cf. *54*, 83s*58*; with Sn- or Zr-containing zeolites as water-resistant, solid Lewis acids, also cyclic ethers from diols, s. A. Corma, M. Renz, Angew. Chem. Int. Ed. Engl. *2007*, *46* (1-2), 298-300.

Tri-n-butylphosphine *Bu_3P*
3-Ene-1,4-dioxan-2-ones from glycols ○
3-carbalkoxymethylene-1,4-dioxan-2-ones from acetylenedicarboxylates with dabco cf. *71*, 64; 3-benzylidene-1,4-dioxan-2-ones from arylpropiolates with Bu_3P, also 3-benzylidene-1,4-benzodioxin-2-ones from catechols, s. S. Gabillet, D. Lecerclé, O. Loreau, S. Dézard, J.-M. Gomis, F. Taran, Synthesis *2007* (4), 515-22.

Triphenylphosphine/diethyl azodicarboxylate *$Ph_3P/EtO_2CN{=}NCO_2Et$*
Carbonic acid esters from two alcohol molecules ROH + HOR′ → ROCOOR′
Mitsunobu-type reaction

104. 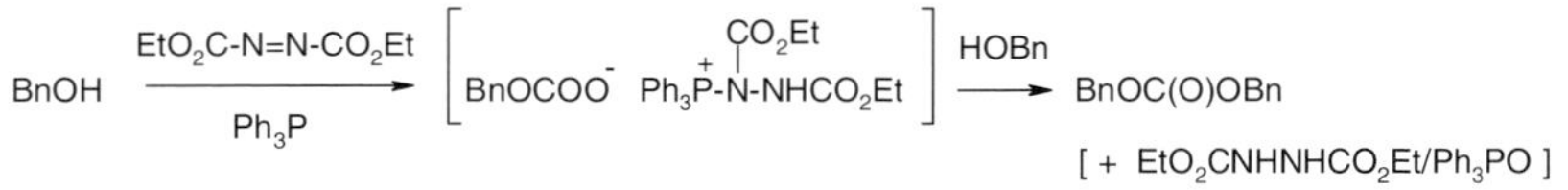

Sym. carbonic acid esters. Gaseous CO_2 bubbled through a soln. of benzyl alcohol in dry DMSO for 30 min, a mixture of equivalent amounts of triphenylphosphine and diethyl azodicarboxylate added slowly in 2-3 small portions, a further 1 eq. benzyl alcohol added, and worked up when TLC indicated completion of reaction (5 h) → dibenzyl carbonate. The procedure is mild, efficient, applicable to prim., sec and tert. alcohols, and suitable for preparing **unsym. carbonic acid esters** by interception of the intermediate hydrazophosphonium alkylcarbonate ion pair with a different alcohol molecule. F.e. and comparison of solvents s. D. Chaturvedi, N. Mishra, V. Mishra, Tetrahedron Lett. *2007, 48* (29), 5043-5.

Triphenylphosphine/diethyl azodicarboxylate/phenyl iodosoacetate ←
Organocatalyzed Mitsunobu reaction ←

105. 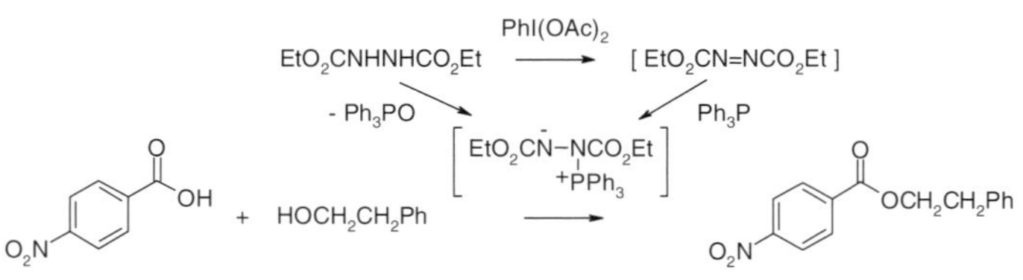

The Mitsunobu reaction can now be conducted with one component of the reagent combination, diethyl azodicarboxylate, *in catalytic amount*, with phenyl iodosoacetate as stoichiometric reoxidant. **E: Carboxylic acid esters from acids.** 3.6 Mmol triphenylphosphine, *0.18 mmol* diethylazodicarboxylate and 3.6 mmol phenyl iodosoacetate added to 1.98 mmol 4-nitrobenzoic acid and 1.8 mmol 2-phenylethanol in dry THF, stirred at room temp. for 16 h, diluted with ether, and worked up with column chromatographic purification → 2-phenylethyl 4-nitrobenzoate. Y 90%. Work-up is simplified as the by-product from azodicarboxylate is in small amount, and by-products from the reoxidant (iodobenzene and acetic acid) are easy to remove. The procedure is suitable for the coupling of a range of prim. alcohols (incl. benzyl alcohols) with aliphatic or aromatic carboxylic acids, arylacetic acids and Boc-protected glycine. For the coupling of sec. alcohols, however, a modified procedure was devised, whereby the phosphine was *added slowly* to a mixture of the alcohol, acid, reoxidant and azodicarboxylate. F.e. and with other azodicarboxylates s. T.Y.S. But, P.H. Toy, J. Am. Chem. Soc. *2006, 128* (30), 9636-7.

Triphenylphosphine/diisopropyl azodicarboxylate *$Ph_3P/ROOCN{=}NCOOR$*
Mitsunobu O-acylation OH → OAc
cf. *29*, 188; arylcarboxylic acid aryl esters (with DIAD) s. V.P. Fitzjarrald, R. Pongdee, Tetrahedron Lett. *2007*, *48* (20), 3553-7.

Triphenylphosphine/diisopropyl azodicarboxylate/triethylamine ←
Ethers from hydroxy compds. OH → OR
Mitsunobu etherification s. *31*, 170; aryl *sec*-alkyl ethers by sequential treatment with Et_3N, Ph_3P and DIAD s. P. Manivel, N.P. Rai, V.P. Jayashankara, P.N. Arunachalam, Tetrahedron Lett. *2007*, *48* (15), 2701-5.

*2-(Di-*tert*-butylphosphino)-2′,4′,6′-triisopropylbiphenyl s. under $PdCl_2(PhCN)_2$* ←
Chiral phosphoromonoamidites s. under $[Ir(cod)Cl]_2$ ←

Bismuth(III) triflate [or Ditriflimide] *$Bi(OTf)_3$ [or Tf_2NH]*
Lactolides from lactols OH → OR
with TsOH cf. *38*, 998s*65*; with 5 mol% $Bi(OTf)_3$ or Tf_2NH, non-reducing disaccharides (as trehalose mimics), s. T. Yamanoi, R. Inoue, S. Matsuda, K. Katsuraya, K. Hamasaki, Tetrahedron: Asym. *2006*, *17* (20), 2914-8.

Bismuth trichloride *$BiCl_3$*
Bismuth(III)-catalyzed nucleophilic substitution of 2-acetylenealcohols ←

106.

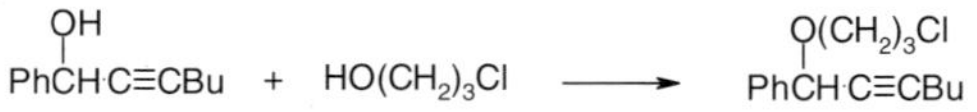

A variety of O-, N-, S- and C-nucleophiles can displace the hydroxyl group of 2-acetylenealcohols in the presence of a catalytic amount of $BiCl_3$. **E: Alkoxy-2-acetylenes from alcohols.** The startg. propargyl alcohol, 3 eqs. 3-chloropropanol, acetonitrile and *10 mol%* $BiCl_3$ added successively to a flask, the mixture stirred for 30 min at 35°, and worked up with column chromatographic purification → product. Y 89%. The procedure is generally applicable to the substitution of benzyl or *tert*-alcohols possessing internal or terminal propargyl groups; it is also inexpensive and a significant improvement on the classical [stoichiometric] Nicholas reaction and recent methods based on rhenium, gold or ruthenium catalysts. Nucleophiles include: alcohols, tosylamide, benzamide (not, alas, acetamide, aniline or piperidine), allylsilanes, electron-rich aromatics and heteroaromatics, and mercaptans (incl. hydroxymercaptans), reaction leaving a number of functional groups unaffected (chlorides, alkenyl groups, methoxy and cyano). F.e.s. Z.P. Zhan, W.Z. Yang, R.F. Yang, J.L. Yu, J.P. Li, H.J. Liu, Chem. Commun. *2006* (31), 3352-4.

Ditriflimide s. under $Bi(OTf)_3$ *Tf_2NH*

3,5-Bis(perfluorooctyl)phenylammonium triflate *$ArNH_2{\cdot}HOTf$*
Carboxylic acid esters from carboxylic acids COOH → COOR
under fluorous 2-phase conditions with $Ph_2NH{\cdot}HOTf$ cf. *59*, 92s*65*; with readily recoverable fluorous ammonium salts such as 3,5-bis(perfluorooctyl)phenylammonium triflate (1 mol%) at 80° s. L. Mercs, G. Pozzi, S. Quici, Tetrahedron Lett. *2007*, *48* (17), 3053-6.

p-Toluenesulfonic acid *TsOH*
Protection of carbonyl groups as cyclic acetals ○
as 5-(2-pyridyl)-1,3-dioxanes cf. *42*, 11; as 6-methoxy-4,4-diphenyl-1,3-benzodioxanes with (2-hydroxy-5-methoxyphenyl)diphenylcarbinol, also removal under UV irradiation (cf. *30*, 5s*68*), s. P. Wang, H. Hu, Y. Wang, Org. Lett. *2007*, *9* (8), 1533-5.

Polymer-based 1-ω-sulfoimidazolium triflates ←
Acetals from aldehydes s. *28*, 141s*72* CHO → $CH(OR)_2$

Sulfate/tin(IV) oxide SO_4^{2-}/SnO_2
O-Acetylation s. *54*, 79s*72* OH → OAc

Hydrogen chloride *HCl*
1,2,4-Trioxacycloalkanes from hydroxyhydroperoxides and ketones ○
1,2,4-trioxanes cf. *39*, 104s*40*; *48*, 170; *68*, 224s*71*; 7-vinyl-1,2,4-trioxepanes from 3-ethylene-alcohols via reaction with singlet oxygen and ring closure of 2-ethylene-3′-hydroxyhydroperoxides in the presence of HCl, also 7-ene-1,2,4-trioxocanes via 2-ene-4-hydroxyhydroperoxides, s. C. Singh, S. Pandey, G. Saxena, N. Srivastava, M. Sharma, J. Org. Chem. *2006*, *71* (24), 9057-61.

Rhenium(VII) oxide Re_2O_7
Mixed 5-oxa-1,9-diol esters by Lewis acid-catalyzed ring opening of tetrahydrofuran C

107.

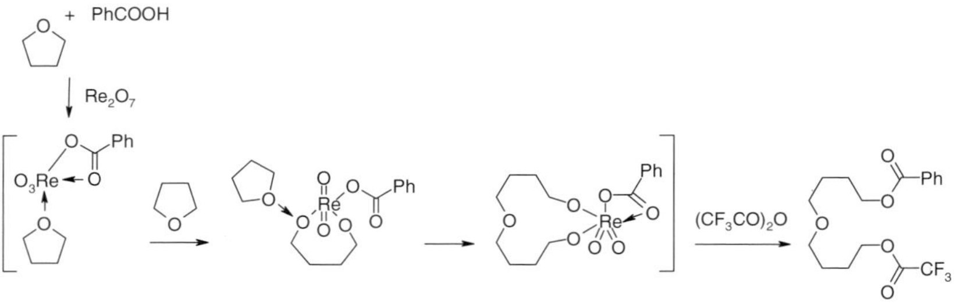

The unique *Lewis acidity* of rhenium(VII) oxide is exemplified in a novel heteroacylative ring opening-dimerization of tetrahydrofuran. **E:** Methylene chloride (8 ml), 4.035 mmol trifluoroacetic anhydride and 6.780 mmol tetrahydrofuran added sequentially via syringe to a Schlenk flask containing 0.268 mmol Re_2O_7 under argon, treated with 2.718 mmol benzoic acid under a positive argon pressure, stirred at room temp. for 24 h, and worked up with satd. aq. $NaHCO_3$ and 35 wt% H_2O_2 → product. Y 95% (after column chromatographic purification). The procedure is limited to tetrahydrofuran itself and trifluoracetic anhydride, while the carboxylic acid may be aromatic, α,β-ethylenic or aliphatic (except trifluoracetic acid). The catalytic cycle is thought to involve a multiple sequence of nucleophilic attacks, initiated by attack of the rhenium oxo ligand on a coordinated tetrahydrofuran molecule: two atoms of oxygen in the product arising from the cyclic ether, one from Re_2O_7, and the two carbonyl oxygens from the carboxylic acid and anhydride. Yields were low with tetrahydropyran and oxetane, while 2-methyl- and 2,5-dimethyl-tetrahydrofuran gave the corresponding mixed 1,4-diol esters. F.e.s. H.C. Lo, H. Han, L.J. D′Souza, S.C. Sinha, E. Keinan, J. Am. Chem. Soc. *2007*, *129* (5), 1246-53.

Magnetized 4-(dialkylamino)pyridine-supported iron oxide nanoparticles ←
Heterogeneous O-acylation ←
with a readily recoverable and recyclable nucleophilic catalyst s. *72*, 222

Iron(III) chloride s. under Et_3SiH $FeCl_3$

Tetrakis(triphenylphosphine)palladium(0) *$Pd(PPh_3)_4$*
Regiospecific O-allylation OH → O-C-C=C
of hydroxamic acids with 2-ethylenecarbonates cf. *65*, 99; of N-hydroxy-succinimide or -phthalimide with acoxy-2-ethylenes (derived from Baylis-Hillman adducts) in the presence of 10 mol% $Pd(PPh_3)_4$ s. Ch. Raji Reddy, N. Kiranmai, G.S. Kiran Babu, G. Dattatreya Sarma, B. Jagadeesh, S. Chandrasekhar, Tetrahedron Lett. *2007*, *48* (2), 215-8.

Palladium(II) chloride *$PdCl_2$*
Protection of alcohols as diphenylmethyl ethers OH → $OCHPh_2$

108. 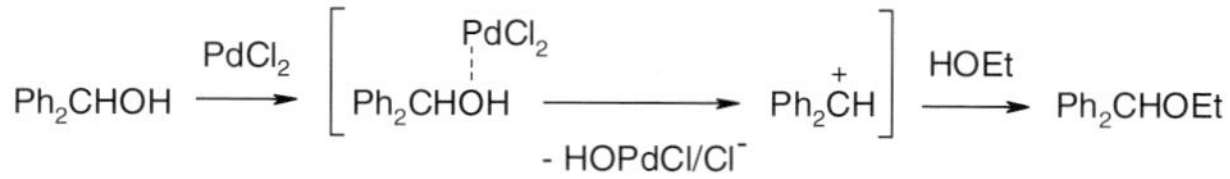

in the absence of solvent. 1 Eq. diphenylmethanol (100 mg) added to a soln. of 10 mol% $PdCl_2$ in ethanol (2.7 ml), heated at 80° until TLC indicated completion of reaction (4.3 h), excess of ethanol removed *in vacuo,* and the crude mixture subjected to column chromatography → ethyl diphenylmethyl ether. Y 97%. The procedure is simple, environmentally friendly, high-yielding, and generally applicable to the protection of prim., sec., benzyl and allyl alcohols, **selective protection** of prim. alcohols taking place in the presence of the slower reacting sec. alcohols. Tert. alcohols reacted sluggishly, while phenols gave a low yield and propargyl alcohols were unreactive. A number of protective groups (OBn, OTr, cyclic acetals, N-acetyl) and functional groups (incl. glycosidic linkages) remained unaffected. Reaction is thought to involve initial coordination of the catalyst to the hydroxyl group of diphenylmethanol, followed by generation of diphenylcarbenium ion and its quenching with the substrate alcohol. F.e. and *in dichloroethane*, **also cleavage of the protective group** (generally on heating with 10 mol% $PdCl_2$ in ethanol), and comparison of Lewis acids s. Y. Bikard, J.-M. Weibel, C. Sirlin, L. Dupuis, J.-P. Loeffler, P. Pale, Tetrahedron Lett. *2007, 48* (50), 8895-9.

*Bis(benzonitrile)dichloropalladium(II)/2-(di-*tert*-butylphosphino)-2′,4′,6′-triisopropylbiphenyl* ←
α-Glycosides from glycal O³-trichloroacetimidates by palladium-catalyzed Ferrier-type reaction ←

109. 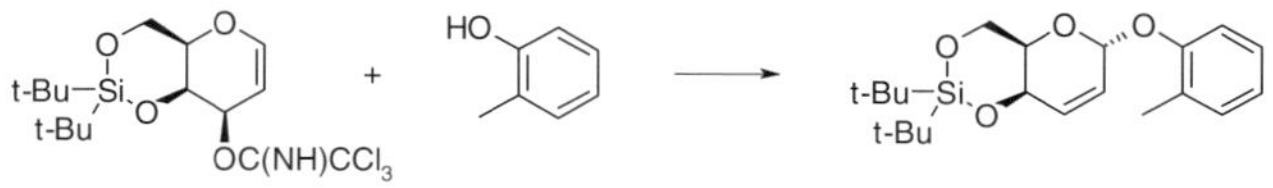

A mixture of 2.5 mol% $PdCl_2(PhCN)_2$ and 2.5 mol% 2-(di-*tert*-butylphosphino)-2′,4′,6′-triisopropylbiphenyl stirred in methylene chloride at room temp. for 3 h, 2 eqs. 2-methylphenol, 1 eq. startg. glycal trichloroacetimidate and methylene chloride added sequentially to the soln., stirred for 3 h, diluted with benzene, and purified by chromatography → product. Y 81% (predominantly α-anomer). The method relies on initial complexation of palladium with the glycal unsaturation before generation of an oxonium palladium-σ-complex stabilized by imidate nitrogen; the bulky ligand then controls the α-stereoselectivity (>20:1) at the anomeric carbon. The method is compatible with a wide range of protecting groups but, unlike most Ferrier reactions, does not rely on the latter to direct the incoming nucleophile. It is also efficient (61-97%) for both phenols and alkanols as acceptors, but the latter usually requires activation by treatment with Et_2Zn to generate the zinc alkoxide. F.e. incl α-disaccharides s. B.P. Schuff, G.J. Mercer, H.M. Nguyen, Org. Lett. *2007*, *9* (16), 3173-6.

*Chloro(cyclooctadiene)iridium(I) dimer/chiral phosphoromonoamidites/ potassium triethylsilanolate/*n*-propylamine* ←

Terminal 2-ethylenealcohols CH=CHCH$_2$OCOOR → CH(OH)CH=CH$_2$
from 2-ethylenecarbonic acid esters
Regiospecific asym. conversion with silanolates as hydroxyl ion equivalent

110.

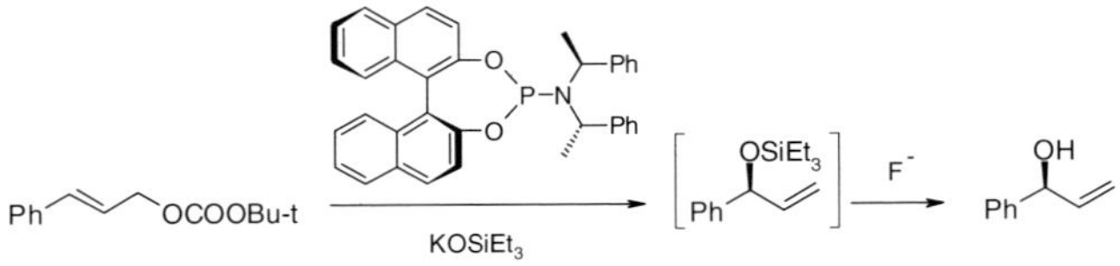

Chiral terminal siloxy-2-ethylenes as intermediates. 1:1 THF/*n*-propylamine added to a Schlenk flask containing 3 mol% [Ir(cod)Cl]$_2$ and 6 mol% (S)-(3,5-dioxa-4-phosphacyclohepta[2,1-*a*;3,4-*a'*]dinaphthalen-4-yl)bis[(1S)-1-phenylethyl]amine, stirred at 50° for 30 min, cooled to room temp., volatiles removed under high vacuum over 30 min, 2 eqs. K-triethylsilanolate in methylene chloride added, followed by 1 eq. of the startg. carbonate in the same solvent, stirring continued for 14 h, worked up, the crude siloxy-2-ethylene in THF treated with 2 eqs. 1 *M* Bu$_4$NF in THF at 0°, stirred for a further 2 h, and worked up with flash column chromatographic purification on silica gel → product. Y 88% (e.e. 97%). The procedure is applicable with high enantioselectivity (92-9%) to a series of cinnamyl carbonates (possessing electron-withdrawing or -donating groups in the ring), as well as to furyl and thienyl analogs, and to 2,4-dienyl carbonates using K-triethylsilanolate, -*tert*-butyldimethylsilanolate or -triisopropylsilanolate as water surrogate. Poor results were recorded, however, with cinnamyl acetate and cinnamyl carbonates possessing small alkyl groups. F.e. and with retention of acetal groups, also desilylation with 30% NaOH in methanol, and with isolation of the intermediate siloxy-2-ethylenes, s. I. Lyothier, C. Defieber, E.K. Carreira, Angew. Chem. Int. Ed. Engl. *2006, 45* (37), 6204-7.

Platinum-carbon *Pt-C*
Alkoxy-2-ethylenes from 2-ethylenealcohols and alcohols C(OH)C=C → C(OR)C=C

111.

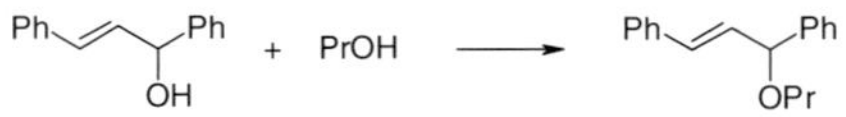

1,3-Diaryl-2-ethyleneethers. A mixture of (E)-1,3-diphenylprop-2-en-1-ol, 10 eqs. of propanol and a little 5 wt% Pt-C heated at 100° for 30 min → (E)-1,3-diphenylprop-2-enyl propyl ether. Y 95% (after column chromatography). Reaction is applicable to the coupling of (E)-1,3-diaryl- and (E)-1,3-dihetaryl-prop-2-en-1-ols with prim. or sec. alcohols, incl. allyl and benzyl alcohols, there being no homocoupling of the latter, and no reaction with tert. alcohols. Glycidol also took part in the conversion without epoxide ring opening. F.e.s. T. Thiemann, J. Chem. Res. *2007* (9), 528-34.

Nitrogen ↑ OC ↓↑ N

Microwaves s. under Silica chromate and Pd-C [\\\\]

Lithium hydroxide/hydrogen peroxide *LiOH/H$_2$O$_2$*
Carboxylic acids from N-acyl-2-oxazolidones or -2-thiazolidones CON< → COOH
Chiral aryl(fluoro)acetic acids s. *72*, 291

Sodium bis(trimethylsilyl)amide/boron fluoride *$NaN(SiMe_3)_2/BF_3$*

Allyl enol carbonates CHCO → C=C(OCOOC-C=C)

from ketones and N-carballyloxyimidazoles

112.

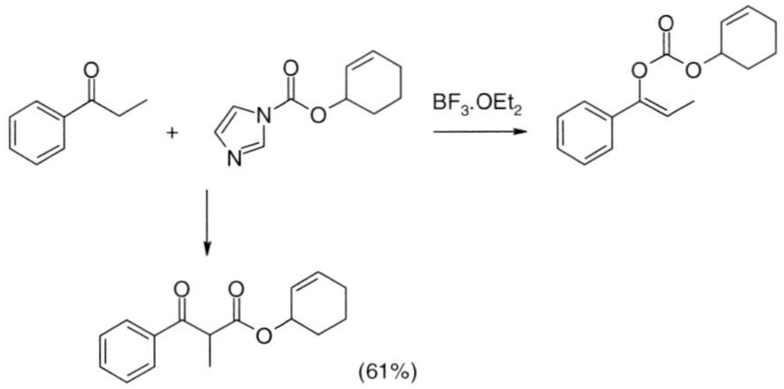

1.25 Eqs. of the startg. ketone stirred with $NaN(SiMe_3)_2$ in DME at -78° for 30 min, the soln. mixed under N_2 with 3 eqs. $BF_3{\cdot}OEt_2$ and 1 eq. 2-cyclohexenyl 1*H*-imidazole-1-carboxylate in DME (preliminarily stirred at -78° for 15 min), stirred for 30 min, quenched with aq. NH_4Cl at -78°, extracted with ether, dried, and evaporated → enol carbonate. Y 99%. It was reasoned that the complex formed between $BF_3{\cdot}OEt_2$ and a series of N-carballyloxyimidazoles should be a 'harder' electrophile than the imidazolide alone, thereby facilitating regiospecific O-acylation of a series of alkali metal enolates, generally in high yields. In the absence of the Lewis acid, however, C-acylated products were formed exclusively (61% in the illustrated example). In some examples, steric effects also disfavored O-acylation, with C-acylation predominating even in the presence of the Lewis acid; here, the use of potassium enolates was required to maximize O-acylation. F.e.s. B.M. Trost, J. Xu, J. Org. Chem. *2007*, *72* (24), 9372-5.

Copper(I) chloride/chiral spirocyclic bis(Δ^2-oxazolines)/sodium tetrakis[3,5-bis-(trifluoromethyl)phenyl]borate ←

α-Aryloxy- from α-diazo-carboxylic acid esters C=N_2 → CH(OAr)

Asym. carbene insertion

113.

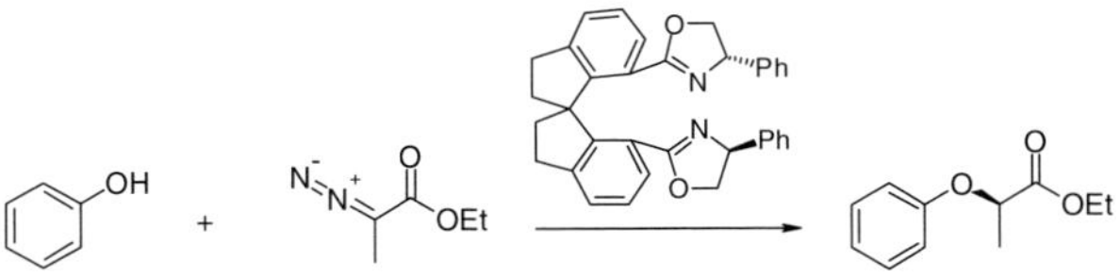

Methylene chloride injected into a mixture of 1 mol% CuCl, 1.2 mol% spirocyclic (S_a,S,S)-bis(Δ^2-oxazoline) ligand and 1.2 mol% Na-tetrakis[3,5-bis(trifluoromethyl)phenyl]borate and 5 Å molecular sieves in a dry Schlenk tube, stirred at room temp. under argon for 2 h, phenol and 0.2 eq. ethyl α-diazopropionate added, stirred at 25° for 3 h, and worked up with flash chromatographic clean-up → (R)-ethyl 2-phenoxypropionate. Y 87% (e.e. 99%). This is the first highly enantioselective insertion of carbenoids into the H-O bond of phenols (incl. naphthols), reaction being applicable to a number of diazo-esters, although benzyl α-diazobutyrate underwent β-elimination instead to give benzyl 2-butenoate. Enantioselectivity was much lower with aliphatic and allyl alcohols. By comparing a variety of chiral bis(Δ^2-oxazolines), it appears that both the backbone and oxazoline chirality is critical for high enantioselectivity, and there was no reaction in the absence of the borate salt. F.e. and with other copper(I) salts s. C. Chen, S.-F. Zhu, B. Liu, L.-X. Wang, Q.-L. Zhou, J. Am. Chem. Soc. *2007, 129* (42), 12616-7.

Sodium tetrakis[3,5-bis(trifluoromethyl)phenyl]borate s. under CuCl *$NaBAr_4$*
Boron fluoride s. under $NaN(SiMe_3)_2$ *BF_3*

Zinc fluoroborate *$Zn(BF_4)_2$*
Oxo compds. from their N-derivs. C=N- → CO
with $CuCl_2$ cf. *47*, 146s*68*; from oximes, hydrazones and semicarbazones with 2 eqs. 40% aq. $Zn(BF_4)_2$/methylene chloride at reflux s. A. Majee, S.K. Kundu, S. Islam, J. Indian Chem. Soc. *2007*, *84* (5), 496-7.

Chiral spirocyclic bis(Δ^2-oxazolines) s. under CuCl ←

Nitrilases ←
Carboxylic acids from nitriles CN → $CONH_2$
asym. hydrolysis with microorganisms cf. *46*, 172s*66,69*; chiral cyclic γ-amino acids with nitrilases in pH 8 buffer/DMSO s. M. Winkler, A.C. Knall, M.R. Kulterer, N. Klempier, J. Org. Chem. *2007*, *72* (19), 7423-6; hydrolysis of aromatic and aliphatic nitriles, asym. hydrolysis of phenyl-substituted β-hydroxynitriles, and partial hydrolysis of dinitriles having five or less methylene groups using the nitrilase from *Synechocystis* sp. PCC 6803, s. C. Mukherjee, D. Zhu, E.R. Biehl, L. Hua, Eur. J. Org. Chem. *2006* (23), 5238-42.

1,4-Dichloro-1,4-diazoniabicyclo[2.2.2]octane bis(chloride) ←
N,N′-Dichlorobis(2,4,6-trichlorophenyl)urea *$(ArNCl)_2CO$*
Oxo compds. from their N-derivs. C=N- → CO
from oximes with N-bromosaccharin cf. *31*, 185s*69*; with N,N′-dichlorobis(2,4,6-trichlorophenyl)urea at room temp. s. A.K. Gupta, J. Acharya, D. Pardasani, D.K. Dubey, J. Chem. Res. *2007* (7), 411-4; with 1,4-dichloro-1,4-diazoniabicyclo[2.2.2]octane bis(chloride) in water at 50°, also from phenylhydrazones and semicarbazones, and sym. disulfides from mercaptans (cf. *47*, 468s*71*), s. M. Tajbakhsh, S. Habibzadeh, ibid. *2007* (8), 486-9.

Silica s. under Silica chromate *SiO_2*
Hydrogen peroxide s. under LiOH and Polymer-based tungstate *H_2O_2*

Thionyl chloride *$SOCl_2$*
Carboxylic acid methyl esters from N-unsubst. amides $CONH_2$ → COOMe

114.
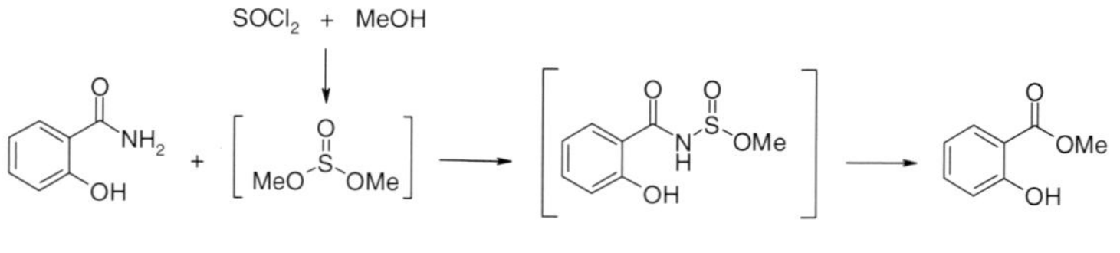

under mild conditions. Methanol and $SOCl_2$ stirred at -5° for 30 min, the soln. warmed to 0°, 2-hydroxybenzamide added, the mixture warmed at 55° for 6 h, methanol removed *in vacuo*, 80% aq. methanol added to the residue, the mixture extracted with ethyl acetate, the organic layer dried over Na_2SO_4, the solvent removed *in vacuo*, and the residue purified by chromatography → methyl 2-hydroxybenzoate. Y 90%. The method was effective in high yield (70-99%) with two aliphatic and seven aromatic amides, leaving hydroxyl, amino and ethylene groups unaffected. The mechanism is not thought to involve catalysis by HCl (since dry HCl/MeOH produced <5% conversion to the ester after 48 h), but rather the generation of dimethyl sulfite *in situ* as the effective methoxylating agent. The reaction has been studied using computational methods and several potential cyclic transitions states have been identified. F.e.s. L.-C. Li, J. Ren, T.-G. Liao, J.-X. Jiang, H.-J. Zhu, Eur. J. Org. Chem. *2007* (6), 1026-30.

Silica chromate/wet silica/microwaves ←
Oxo compds. from their N-derivs. C=N- → CO
from oximes with silica-supported chromic acid cf. *34*, 172s*71*; with silica chromate and wet SiO_2 (w/w 50%) under microwave irradiation in the absence of solvent s. M. Zolfigol, A. Khazaei, A. Ghorbani-Choghamarani, A. Rostami, Phosphorus, Sulfur Silicon Relat. Elem. *2006*, *181* (11), 2453-8.

Polymer-based tungstate/hydrogen peroxide *WO_4^{2-}/H_2O_2*
Oxo compds. from prim. amines $CHNH_2 \rightarrow CO$
with NaOCl on montmorillonite cf. *10*, 257s*61*; with polyacrylamide-bound tungstate/H_2O_2 in toluene at 60° s. H. Hamamoto, Y. Suzuki, H. Takahashi, S. Ikegami, Tetrahedron Lett. *2007*, *48* (24), 4239-42.

Triethylenediamine-bromine complex *$DABCO\text{-}Br_2$*
Oxo compds. from semicarbazones s. *19*, 253s*72* $C{=}NNCONH_2 \rightarrow CO$

Palladium-carbon/microwaves *Pd-C/*[\\\\]
Ar. aldehydes from benzylamines CHN< → CO
and one-pot conversion to β-arylcarboxylic acid esters via Wittig synthesis and hydrogenation s. *72*, 608

Halogen ↑ OC ↓↑ Hal

Without additional reagents *w.a.r.*
Diaryl ethers from ar. fluorides F → OAr
with KOBu-*t*/18-crown-6 cf. *33*, 200s*70*; from activated ar. fluorides at 58° under sonication in the absence of catalyst s. H. Xu, W.-M. Liao, H.-F. Li, Ultrasonics Sonochem. *2007*, *14* (6), 779-82.

Microwaves s. under Cu-C [\\\\]

Sodium hydride/2-nitropropane *NaH/Me_2CHNO_2*
Aldehydes from prim. halides $CH_2Hal \rightarrow CHO$
modified Hass procedure, 1-sulfonylindolecarboxaldehydes, s. *5*, 202s*70*; s.a. R. Balamurugan, A.K. Mohanakrishnan, Tetrahedron *2007*, *63* (45), 11078-85.

Sodium hydroxide *NaOH*
α-Alkoxy-β,γ- or γ-alkoxy-α,β-ethylenecarboxylic acids ←
from 2′-ethylene-2,2,2-trichloroalcohols s. *72*, 326

Lithium bromide s. under H_2O_2 *LiBr*

Ethyldiisopropylamine/trihexyl(tetradecyl)phosphonium triflimide ←
Carboxylic acid esters from carboxylic acids and halides in ionic liquids COOH → COOR
cf. *65*, 106s*70*; also further reactions such as Kornblum substitution of bromides, mesylates or tosylates s. J. McNulty, J.J. Nair, S. Cheekoori, V. Larichev, A. Capretta, A.J. Robertson, Chem. Eur. J. *2006*, *12* (36), 9314-22.

2,6-Lutidine/1-butyl-3-methylimidazolium hexafluorophosphate ←
Carbohydrate pent-4-enyl 1,2-orthoesters from O²-acylglycosyl bromides ○
s. *46*, 192s*72*

Triethylenediamine *dabco*
Regiospecific O-acylation with O,O-isopropylidene group migration ←

115. 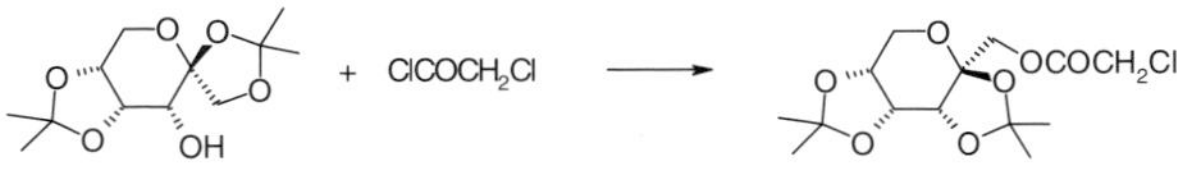

Protected fructose derivs. A mixture of 1,2:4,5-di-O-isopropylidene-D-fructose and 1.5 eqs. DABCO in dry acetonitrile treated dropwise with 1.5 eqs. chloroacetyl chloride, stirred at room temp. until TLC indicated completion of reaction (2-24 h), and quenched with methanol → 1-O-chloroacetyl-2,3:4,5-di-O-isopropylidene-β-D-fructopyranose. Y 100%. With less bulky pyridine as base, the free hydroxyl group is acylated without the acetal group migration. Reaction is thought to involve initial formation of an acylium ion, which preferentially attacks the most reactive O¹ prior to the migration. F.e.s. X.-B. Meng, Y.-F. Li, Z.-J. Li, Carbohyd. Res. *2007, 342* (8), 1101-4.

Copper nanoparticles/cesium carbonate *Cu/Cs_2CO_3*
Copper-charcoal/1,10-phenanthroline/cesium carbonate/microwaves ←
Silica-supported copper/potassium fluoride *$Cu\text{-}SiO_2/KF$*
Copper(I) iodide/(E)-N-phenylpyridine-2-carboxaldimine/potassium phosphate ←
Ullmann diaryl ether synthesis ArOH → ArOAr′
update s. *54*, 90s*71*; with recyclable Cu nanoparticles/Cs_2CO_3 in acetonitrile at 50-60° s. M. Kidwai, N.K. Mishra, V. Bansal, A. Kumar, S. Mozumdar, Tetrahedron Lett. *2007*, *48* (50), 8883-7; with Cu supported on 3-(2-aminoethylamino)propyl-functionalized silica gel (5 mol%)/KF in DMSO at 130° s. T. Miao, L. Wang, ibid. *48* (1), 95-9; with 5 mol% Cu (15 wt% on charcoal), 0.5 eqs. 1,10-phenanthroline and 2 eqs. Cs_2CO_3 in dioxane under microwave irradiation s. B.H. Lipshutz, J.B Unger, B.R. Taft, Org. Lett. *2007*, *9* (6), 1089-92; with CuI/(E)-N-phenylpyridine-2-carboxaldimine/K_3PO_4 s. A. Ouali, J.-F. Spindler, A. Jutand, M. Taillefer, Adv. Synth. Catal. *2007*, *349* (11-12), 1906-16; O-aryloximes with CuI/1,10-phenanthroline/Cs_2CO_3/Na,K-tartrate in toluene at 110° or DMSO at 30° s. P. De, Nonappa, K. Pandurangan, U. Maitra, S. Wailes, Org. Lett. *2007*, *9* (15), 2767-70.

Copper(I) chloride/chiral N,N-bis(1-sulfonyl-Δ^2-imidazolin-2-ylmethyl)benzylamines/ ethyldiisopropylamine ←
Asym. mono-O-benzoylation of glycols OH → OCOPh
s. *64*, 107; with CuCl/chiral N,N-bis(1-sulfonyl-Δ^2-imidazolin-2-ylmethyl)benzylamines s. T. Arai, T. Mizukami, A. Yanagisawa, Org. Lett. *2007*, *9* (6), 1145-7.

Copper(I) bromide/cesium carbonate *$CuBr/Cs_2CO_3$*
Ligand-free copper(I)-catalyzed arylation with ar. halides ArHal → ArOAr′

116.

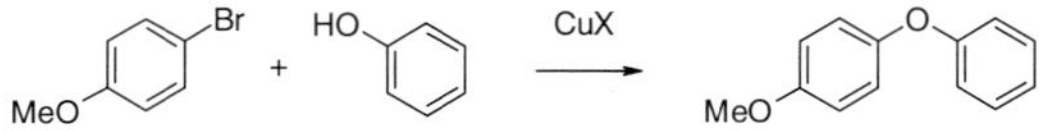

Diaryl ethers by Ullmann reaction. *p*-Methoxyphenyl bromide and 1.3 eqs. phenol in N-methyl-2-pyrrolidone added to 1.1 eqs. solid Cs_2CO_3, followed by *2.5 mol%* CuBr, the mixture heated under N_2 at 160° for 16 h, cooled to room temp, quenched with acetonitrile, filtered through Celite, and the solvent removed *in vacuo* → *p*-methoxyphenyl phenyl ether. Y 78% (isolated yield). The reaction is based on the use of *catalytic* amounts of simple, inexpensive and industrially viable copper halides *without* additives to enhance solubility. In initial tests, bromobenzene was found to be more effective than iodo- or chloro-benzene, but there was little difference between CuBr, CuCl or CuI as catalyst (although reaction with an aryl bromide appears to be preferred using CuBr). Six examples with different bromobenzenes gave yields of 68-89% but the reaction does not appear to be optimized. N-Arylation of imidazole was also effective (*72*, 238), but low yields were recorded for the N-arylation of benzylamine. F.e.s. E. Sperotto, J.G. de Vries, G.P.M. van Klink, G. van Hoten, Tetrahedron Lett. *2007*, *48* (41), 7366-70.

Lanthanum(III) nitrate *$La(NO_3)_3$*
O-Pivaloylation OH → OCOBu-*t*
regiospecific O-pivaloylation of carbohydrates in the presence of pyridine cf. *56*, 97; preferential O-pivaloylation of prim. alcohols in the presence of a little $La(NO_3)_3 \cdot 6H_2O$ in the absence of solvent s. P. Prabhakar, N. Suryakiran, Y. Venkateswarlu, Chem. Lett. *2007*, *36* (6), 732-3.

Chiral N,N-bis(1-sulfonyl-Δ^2-imidazolin-2-ylmethyl)benzylamines s. under CuCl ←
Imidazolium salts s.a. under 2,6-Lutidine ←

1-Butyl-3-methylimidazolium hydroxide *[bmim]OH*
Phenolethers from phenols and halides OH → OR
with NaOH/tridodecyl(methyl)ammonium chloride cf. *7*, 281s*71*; in 1-butyl-3-methylimidazolium hydroxide as [recyclable] ionic liquid solvent and reagent s. F. Mohanazadeh, M. Aghvami, Monatsh. Chem. *2007*, *138* (1), 47-9.

Dimethyl sulfoxide *DMSO*
Oxo compds from halides CH(Hal) → CO
cf. *13*, 287; 3-aroyl-2(1*H*)-quinoxalones, also prepn. of the startg. 3-(α-bromobenzyl)-2(1*H*)-

quinoxalones from ethyl 3-aryl-3-bromo-2-oxopropanoates and *o*-phenylenediamine in acetic acid, s. E.A. Gorbunova, V.A. Mamedov, Russ. J. Org. Chem. *2006*, *42* (10), 1528-31.

Bis(pentamethylcyclopentadienyl)oxozirconium(IV)-4-(3-phenylpropyl)pyridine ←
Siloxy-2-ethylenes from β,γ-ethylenechlorides C=C-C(Cl) → C(OSi≤)C=C
Regio- and stereo-specific conversion with double bond shift

117. 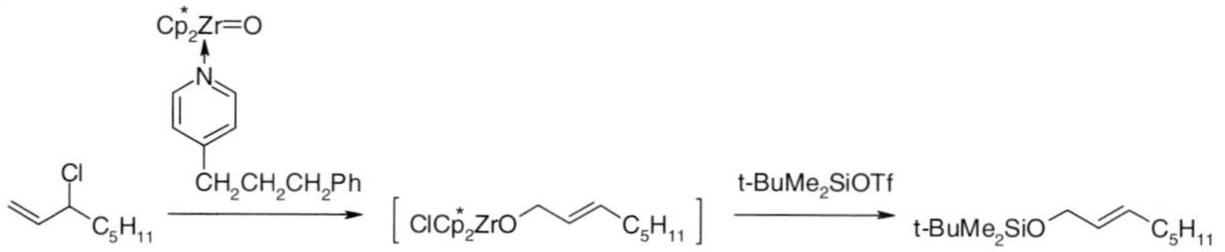

under mild conditions. A bomb charged with 3-chloro-1-octene and 1.3 eqs. bis(pentamethylcyclopentadienyl)oxozirconium(IV)-4-(3-phenylpropyl)pyridine complex in benzene (in a glove box), the vessel sealed and removed from the glove box, the resulting orange soln. warmed to 45°, cooled to room temp. after 8 h, returned to the glove box, treated with 2 eqs. *tert*-butyldimethylsilyl triflate, the bomb re-sealed, warmed to 80°, cooled to room temp. after 14 h, quenched with 50% satd. aq. $NaHCO_3$, pentane and brine, and worked up with flash chromatographic purification → product. Y 92%. Internal (E)-allyl chlorides reacted similarly in good yield, but the corresponding (Z)-isomers gave a mixture of products. Excellent functional group tolerance was demonstrated, with alkenes, alkyl chlorides, allylic *tert*-butyldimethylsilyl ethers, dithianes and dimethyl acetals being left intact. Reaction is presumed to involve silylation of an intermediate allyloxyzirconium(IV) species. F.e., also regiospecific conversion of chiral allyl chlorides with complete *syn*-selectivity, s. R.J. Fox, G. Lalic, R.G. Bergman, J. Am. Chem. Soc. *2007, 129* (46), 14144-5.

Basic lead carbonate *$2PbCO_3Pb(OH)_2$*
***tert*-Alkylation** OH → OCR_3
with *tert*-alkyl chlorides and ZnO cf. *36*, 198s*40*; *tert*-butyl ethers with *tert*-butyl bromide and basic lead carbonate s. N.P. Rai, P.N. Arunachalam, Synth. Commun. *2007*, *37* (17), 2891-6.

*Tetra-*n*-butylphosphonium hydroxide* *Bu_4POH*
Selective O-alkylation of phenols with retention of carboxyl groups ArOH → ArOR

118. 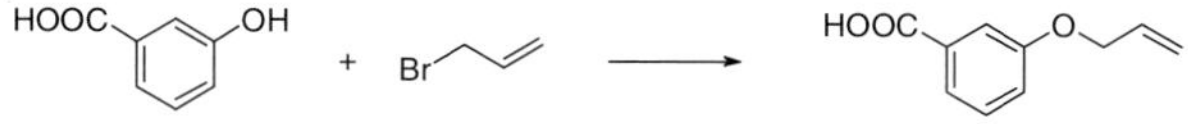

Allyl bromide added to a soln. of 1 eq. 3-hydroxybenzoic acid and 2 eqs. of 40% aq. tetra-*n*-butylammonium hydroxide in THF at 0°, the mixture stirred at room temp. until reaction complete by HPLC (1-3 h), THF removed *in vacuo*, the aq. residue acidified with 2 *N* aq. HCl, and the precipitate isolated by filtration → 3-allyloxybenzoic acid. Y 84%. The phosphonium base is able to exploit the difference in acidity between phenolic and carboxylic acid functions, giving ethers (Y 82-99%) as sole products on alkylation with ethyl, benzyl or allyl bromides over 1-24 h. Allyl and benzyl choride, however, required a longer reaction period. F.e. with retention of both aliphatic and aromatic carboxyl groups s. P. Liu, L. Huang, M.M. Faul, Tetrahedron Lett. *2007*, *48* (41), 7380-2.

Trihexyl(tetradecyl)phosphonium triflimide s. under i-*Pr_2NEt* *$[R_3PR']NTf_2$*

Hydrogen peroxide or Hydrogen peroxide/lithium bromide *H_2O_2 or H_2O_2/LiBr*
Oxo compds. from halides CH(Hal) → CO
with $NaIO_4$/DMF cf. *65*, 110; aryloxo compds. from benzyl bromides with H_2O_2 in refluxing ethanol, or from benzyl chlorides with added LiBr (10 mol%) s. J. Tang, J. Zhu, Z. Shen, Y. Zhang, Tetrahedron Lett. *2007*, *48* (11), 1919-21.

*Tris(dibenzylideneacetone)dipalladium/tri-*tert*-butylphosphine/potassium phosphate* ←
Phenols or phenolethers from ar. halides Hal → OH or OR
with KOH/$Pd_2(dba)_3$/*o*-phosphinobiphenyls/$C_{16}H_{33}NMe_3Br$ cf. *71*, 72; with $Pd_2(dba)_3$/*t*-Bu_3P/K_3PO_4 s. G. Chen, A.S.C. Chan, F.Y. Kwong, Tetrahedron Lett. *2007*, *48* (3), 473-6.

Via intermediates *v.i.*
Glycosides from glycosyl bromides Br → OR
cf. *5*, 192; O^2-acyl-glycosides from -glycosyl bromides via carbohydrate pent-4-enyl 1,2-orthoesters with 2 eqs. 2,6-lutidine in [bmim]PF_6 for the first step (cf. *46*, 192s*69*) and NIS (2 eqs.)/$Yb(OTf)_3$ (2 mol%) in [bmim]PF_6 for the second step (cf. *59*, 66s*65*) s. S. Anas, V.S. Sajisha, R. Rajan, R.T. Kumaran, K.V. Radhakrishnan, Bull. Chem. Soc. Japan *2007*, *80* (3), 553-60.

Sulfur ↑ OC ↓↑ S

Electrolysis/sodium triflate ot lithium perchlorate *ϟ/NaOTf or $LiClO_4$*
Glycosides from thioglycosides SR → OR
Anodic oxidation with a supporting electrolyte as catalytic activator

119. 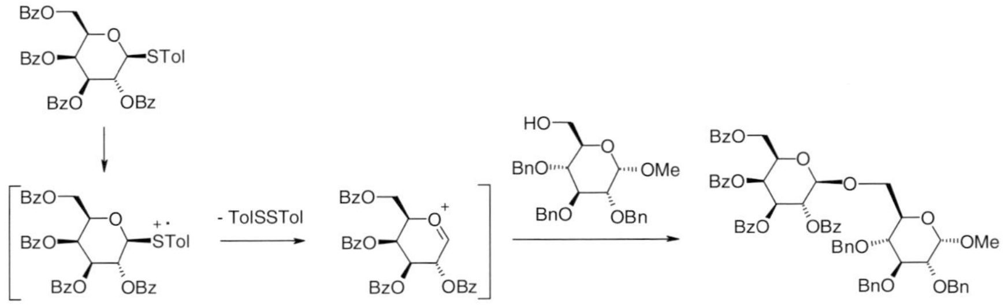

β-Glycosides. A soln. of 1.5 eqs. startg. thioglycoside, 1 eq. startg. acceptor alcohol and *12.5* mol% Na-triflate in acetonitrile electrolyzed in an undivided cell between Pt electrodes at 15° and 7-12 V with a constant current density of 5 mA/cm^2 until 1 F/mol had been supplied, the mixture evaporated *in vacuo*, and the residue purified by chromatography → 1,6-β-disaccharide. Y 80%. $LiClO_4$ was also an effective catalytic activator for coupling prim. (76-88%) and sec. alcohols (35-54%), but 10 mol% NBS and Br_2 were not. The method is compatible with standard protecting groups (benzyl ethers, acetate/benzoate esters, benzylidene acetals), and **selective coupling** at the 6-position of glycosyl donors unprotected at the 2- or 4-positions was demonstrated in high yield. The reaction is presumed to involve initial anodic oxidation to give a radical cation, followed by elimination of disulfide and quenching of the liberated oxonium species. F.e.s. N. Tanaka, F. Ohnishi, D. Uchihata, S. Torii, J. Nokami, Tetrahedron Lett. *2007*, *48* (41), 7383-7.

Copper(I) chloride *CuCl*
Carbonyl from thiocarbonyl compds. CS → CO
with Ag_2CO_3/Celite cf. *16*, 285s*66*; with 20 mol% CuCl in DMF at 80° under O_2, incl. ^{18}O-labelled compds., s. F. Shibahara, A. Suenami, A. Yoshida, T. Murai, Chem. Commun. *2007* (23), 2354-6.

Copper(I) bromide/potassium carbonate *CuBr/K_2CO_3*
Protection of carboxyl groups COOH → $COOC(Me)_2CH{=}CH_2$
as 1,1-dimethylprop-2-enyl esters

120. 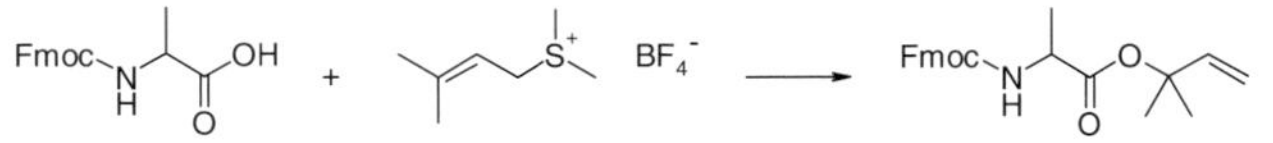

A soln. of dimethyl(3,3-dimethylprop-2-enyl)sulfonium fluoroborate added to a stirred suspension of 1 eq. of the startg. carboxylic acid, 1 eq. K_2CO_3 and 10 mol% CuBr in dry methylene chloride under N_2, stirred at room temp. until reaction complete by TLC (8-17 h), the mixture filtered, and evaporated → product. Y 88% (after flash chromatographic purification). 1,1-Dimethylprop-2-enyl esters are useful palladium-labile alternatives to *tert*-butyl and allyl esters (cf. *34,* 55), notably for the clean protection of α-amino acids in good-to-high yield (58-98%) regardless of side-chain substitution. The sulfonium salt is readily prepared from prenyl alcohol and may be stored for some time. In the absence of the copper catalyst, however, the reaction proceeds with opposite regioselectivity to afford the 3,3-dimethylprop-2-enyl ester. F.e.s. M. Sedighi, S. Çalimsiz, M.A. Lipton, J. Org. Chem. *2006, 71* (25), 9517-8.

Silver triflate s. under Methyl triflate *AgOTf*
1-Butyl-3-methylimidazolium bromide s. under $K_2S_2O_8$ *[bmim]Br*

Dimethyl disulfide/trifluoromethanesulfonic anhydride *$(MeS)_2/Tf_2O$*
Glycosides from thioglycosides at low temperature SR → OR′

121.
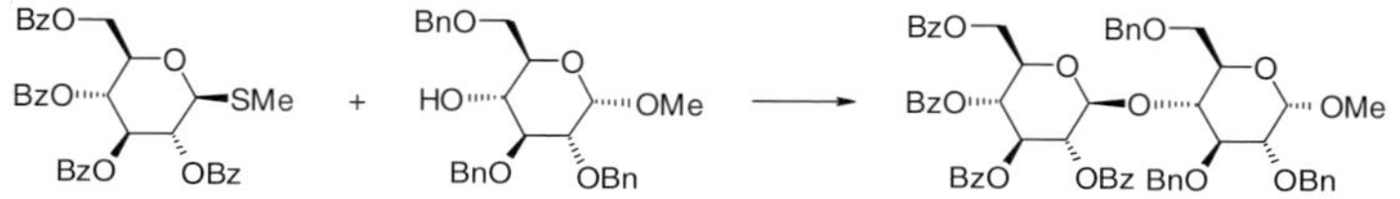

1,2-*trans*-β-Disaccharides. A mixture of triflic anhydride and dimethyl disulfide (in 1:1.1 proportion) stirred *at 0°* in methylene chloride for 0.5 h, 1.8 eqs. of this added to a mixture of 1.2 eqs. of the startg. thioglycoside, 1 eq. of the glycosyl acceptor and powdered 4 Å molecular sieves in methylene chloride, stirred *for 10 min*, quenched with an excess of Et_3N, diluted with methylene chloride, washed sequentially with 2 *M* aq. HCl, satd. aq. $NaHCO_3$ and water, the organic layer dried over $MgSO_4$ and concentrated *in vacuo* → β-disaccharide. Y 67%. The method is rapid for glycosylation of prim. or sec. hydroxyls (11 examples; 67-93%) at low temperature and is compatible with commonly used protecting groups (substrates with acid-sensitive acetal and *tert*-butyl groups requiring buffering with 2,6-di-*tert*-butyl-4-methylpyridine). Furthermore, it is suitable for the coupling not only of common neutral thioglycosides, but also aminothioglycosides and uronic acid thioglycosides, disarmed thioglycosides invariably affording *trans*-glycosides as a result of neighboring group participation. The new promoter is more reactive than dimethyl(methylthio)sulfonium triflate, and, once prepared, may be stored at 0° for a few days without loss of activity. F.e.s. J. Tatai, P. Fugedi, Org. Lett. *2007, 9* (22), 4647-50.

N-Iodosuccinimide or 1,3-Dibromo-5,5-dimethylhydantoin ←
Cyclic acetals from mercaptals ○
with NBS cf. *63,* 114; with *N*-iodosuccinimide or 1,3-dibromo-5,5-dimethylhydantoin, prepn. of pyruvate ketals of carbohydrates from methyl 2,2-di(ethylthio)propionate, s. G. Agnihotri, A.K. Misra, Tetrahedron Lett. *2006, 47* (48), 8493-7.

N-Iodosuccinimide or N-Iodosuccinimide/sulfuric acid-silica ←
Aldoses from thioglycosides SR → OH
with NIS/CF_3COOH cf. *71,* 77; with NIS/sulfuric acid-silica in 10:1 methylene chloride/water at 0° s. S. Dasgupta, B. Roy, B. Mukhopadhyay, Carbohydr. Res. *2006, 341* (16), 2708-13; with N-iodosaccharin without co-activator in 10:1 acetonitrile/water at room temp. s. P.K. Mandal, A.K. Misra, Synlett *2007* (8), 1207-10.

N-Iodosuccinimide/trifluoromethanesulfonic acid *NIS/CF_3SO_3H*
Glycosides from thioglycosides OR → SR′
s. *39,* 189s*70*; stereocontrolled synthesis via neighboring group participation with O^2-[3-(2′-benzyl-oxyphenyl)-3,3-dimethylpropanoyl]- and O^2-[3-(2′-benzyloxy-4′,6′-dimethylphenyl)-3,3-di-methylpropanoyl]-derivs., and their cleavage by hydrogenolysis (Pd-C/3 atm. H_2), also acyl glycosides from carboxylic acids, s. D. Crich, F. Cai, Org. Lett. *2007, 9* (8), 1613-5; highly β-selective glucosidation of prim., sec., and tert. alcohols via conformational control of the pyranose ring using O^2-triisopropylsilyl-derivs. in the presence of methyl triflate (cf. *39,* 189s*66*) s. Y. Okada, T. Mukae, K. Okajima, M. Taira, M. Fujita, H. Yamada, ibid. *9* (8), 1573-6; high β-selectivity with O^2-(glycosyl)thioglucosides as donors and methyl triflate/2,6-lutidine s. Y. Okada, O. Nagata, M. Taira, H. Yamada, ibid. *9* (15), 2755-8.

Silica s. under N-Iodosuccinimide and BuN_3 *SiO_2*

n-*Butyl azide/silica gel* *BuN_3/SiO_2*
4-Methylenecyclobutenones from alk-1-ynyl alk-2-ynyl thioethers □

122. 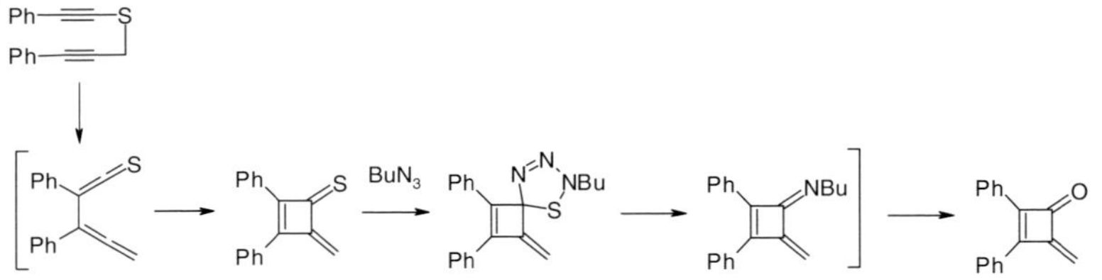

One-pot conversion. A soln. of the startg. thioether and 3 eqs. *n*-butyl azide in benzene heated to reflux for 14 h, and subjected to column chromatography on silica gel → product. Y 72%. Reaction takes place *without isolation of intermediates* by initial [3.3]-sigmatropic rearrangement to give the corresponding 4-methylenecyclobutenethione, which is then converted to the imine by the azide (via [3+2]-cycloaddition) prior to hydrolysis on the moist silica gel column. The product may be substituted by an alkyl, aryl or silyl group. F.e.s. S. Aoyagi, K. Kikuchi, K. Shimada, Y. Takikawa, Synlett *2007* (16), 2553-6.

Trifluoromethanesulfonic anhydride s.a. under $(MeS)_2$ *Tf_2O*

*Trifluoromethanesulfonic anhydride/N-(benzenesulfinyl)piperidine/2,6-di-*tert-*butyl-4-methylpyridine* ←
Trifluoromethanesulfonic anhydride/N-(benzenesulfinyl)morpholine ←
Methyl triflate or Methyl triflate/2,6-lutidine or silver triflate ←
Glycosides from thioglycosides SR → OR′
one-pot iterative oligosaccharide synthesis cf. *62*, 117s*70*; with triflic anhydride/N-(benzenesulfinyl)morpholine s. C. Wang, H. Wang, X. Huang, L.-H. Zhang, X.-S. Ye, Synlett *2006* (17), 2846-50; stereospecific disaccharide synthesis with O^2-(2,2-dimethyltrimethylene phosphoryl) as directing group using triflic anhydride/N-(benzenesulfinyl)piperidine/2,6-di-*tert*-butyl-4-methylpyridine s. T. Yamada, K. Takemura, J.-i. Yoshida, S. Yamago, Angew. Chem. Int. Ed. Engl. *2006*, *45* (45), 7575-8; chemoselective synthesis of 1,2-*trans*-linked oligosaccharides containing multiple 2-amino-2-deoxyglycose residues by 2-step activation with methyl triflate then Ag-triflate s. A.F.G. Bongat, M.N. Kamat, A.V. Demchenko, J. Org. Chem. *2007*, *72* (4), 1480-3; stereospecific glycosidation with methyl triflate or methyl triflate/lutidine s. *39*, 189s*72*.

Sodium triflate s. under ↯ *NaOTf*
Trifluoromethanesulfonic acid s. under N-Iodosuccinimide *CF_3SO_3H*
Sulfuric acid-silica s. under N-Iodosuccinimide *H_2SO_4-SiO_2*

Potassium persulfate/1-butyl-3-methylimidazolium bromide *$K_2S_2O_8$/[bmim]Br*
Oxo compds. from cyclic mercaptals $C(SR)_2$ → CO
with $(NH_4)_2S_2O_8$/montmorillonite K10 under microwave irradiation cf. *52*, 108s*67*; with $K_2S_2O_8$ in the ionic liquid solvent, [bmim]Br, s. A.R. Hajipour, M. Mostafavi, A.E. Ruoho, Monatsh. Chem. *2007*, *138* (6), 569-72.

Periodic acid *HIO_4*
Carbonyl from thiocarbonyl compds. CS → CO
with $[Bu_4N]IO_4$ cf. *59*, 103s*69*; N,N′-diaroyl-carbohydrazides from -thiocarbohydrazides by grinding with $HIO_4 \cdot 2H_2O$ in the absence of solvent s. Z. Li, Y. Zhao, J. Yang, Phosphorus, Sulfur Silicon Relat. Elem. *2007*, *182* (1), 79-84.

Remaining Elements ↑ OC ↓↑ Rem

Cesium fluoride *CsF*
O-Arylation of carboxylic acids with benzynes COOH → COOAr
with *o*-silyltriflates cf. *67,* 101; with (phenyl)[2-(trimethylsilyl)phenyl]iodonium triflate s. J. Xue, X. Huang, Synth. Commun. *2007, 37* (13), 2179-85.

Copper(II) acetate/potassium fluoride/oxygen *$Cu(OAc)_2/KF/O_2$*
Diaryl ethers from phenols OH → OAr
and arenes via iridium-catalyzed borylation s. *61,* 100s*72*; *65,* 62s*72*

Copper(II) acetate/dicyclohexyl(methyl)amine *$Cu(OAc)_2/Cy_2NMe$*
O-Arylation with arylbismuth(V) compds.
with triphenylbismuth dicarboxylates cf. *37,* 195s*41,42,44*; with tetraphenylbismuth fluoride in the presence of 15 or 20 mol% $Cu(OAc)_2$ and 2 eqs. dicyclohexyl(methyl)amine in toluene at room temp., notably for reaction of tert. alcohols, s. T. Mukaiyama, N. Sakurai, K. Ikegai, Chem. Lett. *2006, 35* (10), 1140-1.

Copper(II) acetate/4-dimethylaminopyridine *$Cu(OAc)_2/DMAP$*
Diaryl ethers from phenols
copper-mediated O-arylation with arylboronic acids cf. *55,* 166; complementary procedure with arylstannanes in the presence of $Cu(OAc)_2$/DMAP (stoichiometric) in acetonitrile at 80° under air s. A. Vakalopoulos, X. Kavazoudi, J. Schoof, Tetrahedron Lett. *2006, 47* (48), 8607-10.

Copper(II) triflate/triethylsilane *$Cu(OTf)_2/Et_3SiH$*
Regiospecific protection of glycosides ←
One-pot procedure under sequential catalysis with the same catalyst

123.

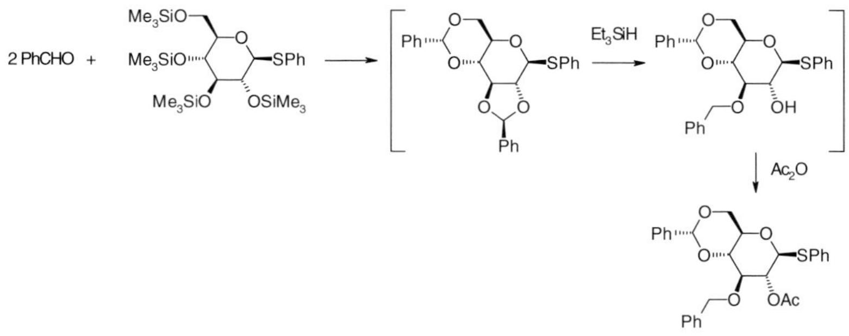

1 Mol% freshly dried $Cu(OTf)_2$ in acetonitrile added to an ice-cold soln. of the startg. persilylated thioglycoside and 3 eqs. benzaldehyde in methylene chloride, treated with 1.1 eqs. triethylsilane, concentrated to dryness under an argon flow over 30 min, the resulting yellow solid dissolved in methylene chloride, 3 eqs. acetic anhydride added, stirred for 1 h at room temp., quenched with satd. aq. $NaHCO_3$, and worked up → product. Y 79%. The same catalyst effects the initial bis(O,O-benzylidenation) at the 2,3- and 4,6-sites, as well as the subsequent regiospecific reductive ring opening of the 2,3-O,O-benzylidene deriv., and the concluding O-acylation at the 2-position. By adjusting the conditions the 3,4-di-O-benzyl deriv. and the 3,6-di-O-benzyl deriv. can also be secured, as may partially protected derivs. lacking the O-acyl group at the 2-position, while the use of $Ba(OTf)_2$, $Yb(OTf)_3$ or $Sm(OTf)_3$ afforded only the 4,6-O,O-benzylidene deriv. Significantly, the procedure is clean and does not require column chromatographic purification. F.e. and regiospecific protection of the corresponding methyl glycosides, also isolation of the intermediate bis(O,O-benzylidene derivs.) in the absence of triethylsilane, and application to an efficient iterative construction of a protected trisaccharide s. A. Francais, D. Urban, J.-M. Beau, Angew. Chem. Int. Ed. Engl. *2007, 46* (45), 8662-5.

Boron fluoride *BF_3*
2′-O-(2-Cyanoethyl)nucleosides from 2,2′-cyclonucleosides

124. 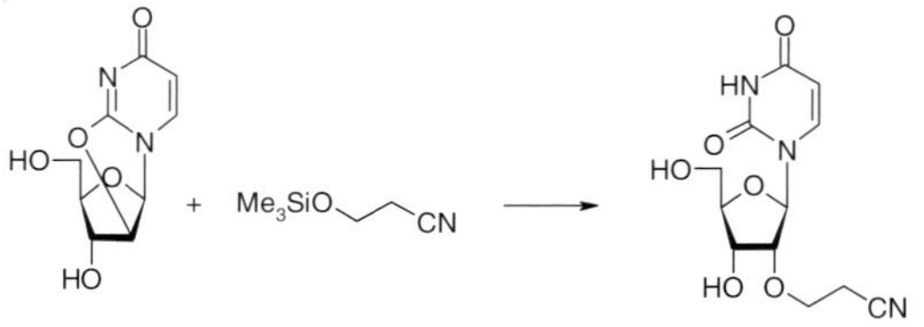

2,2′-Anhydrouridine coevaporated three times with pyridine and toluene, then dissolved in DMA under argon, 5 eqs. 2-cyanoethyl trimethylsilyl ether and 2.5 eqs. BF_3-etherate added, stirred at 120° for 15 h, methanol added, mixed with NH-silica gel, evaporated *in vacuo,* and the powder cleaned up on a silica gel column → 2′-O-(2-cyanoethyl)uridine. Y 65%. This is a significant improvement on base-catalyzed ring opening with free 2-cyanoethanol. With $AlCl_3$ or BCl_3 the corresponding 2′-chloro-2′-deoxynucleoside was obtained in high yield (94-5%). F.e. and conversion of the product to a protected uridine 3′-phosphoromonoamidite building block s. H. Saneyoshi, I. Okamoto, Y. Masaki, A. Ohkubo, K. Seio, M. Sekine, Tetrahedron Lett. *2007, 48* (48), 8554-7.

Triethylsilane s. under $Cu(OTf)_2$ *Et_3SiH*

Hydroxylamine hydrochloride/sodium hydroxide *$NH_2OH{\cdot}HCl/NaOH$*
Phenols from arylboronic acids or esters $B(OH,R)_2$ → OH
with Oxone/$NaHCO_3$ cf. *51*, 98; with 1.5 eqs. $NH_2OH{\cdot}HCl$ and 2 eqs. NaOH s. E. Kianmehr, M. Yahyaee, K. Tabatabai, Tetrahedron Lett. *2007*, *48* (15), 2713-5.

Carbon ↑ OC ↓↑ C

Microwaves s. under $Sc(OTf)_3$ [\\\\]
Potassium iodide s. under I_2 *KI*

1,8-Diazabicyclo[5.4.0]undec-7-ene *DBU*
Methyl carbonates from alcohols OH → OCOOMe
Selective O-carbomethoxylation in the presence of phenols

125. 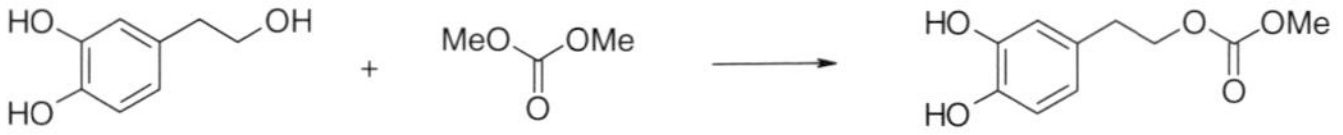

A mixture of 2-(3′,4′-dihydroxyphenyl)ethanol, an excess of dimethyl carbonate and 1.2 eqs. DBU heated under reflux (90°), the mixture cooled, excess of dimethyl carbonate removed as an azeotrope with methanol, the residue dissolved in ethyl acetate, washed with aq. 1 *N* HCl, separated, and solvent removed *in vacuo* → 3,4-dihydroxybenzyl methyl carbonate. Y 98%. Five substrates all gave >90% yield of the methyl carbonate in the presence of phenolic hydroxyl groups by using this cheap and green carbomethoxylating agent. The reactions were also successful using 20% sulfuric acid in place of DBU, and in one example the use of catalytic DBU (10 mol%) gave a similar high yield, albeit in 12 h compared to 4.5 h. F.e.s. R. Bernini, E. Mincione, F. Crisante, M. Barontini, G. Fabrizi, P. Gentili, Tetrahedron Lett. *2007*, *48* (39), 7000-3.

Silver nitrate/ammonium persulfate *$AgNO_3/(NH_4)_2S_2O_8$*
N-Acylcarboxylic acid amides CH(N-Ac)COOH → C(O)N-Ac
from α-(acylamino)carboxylic acids
oxidative cleavage via treatment of Δ^2-5-oxazolones with KOBu-*t*/O_2 cf. *40*, 127; directly with a little $AgNO_3$ and 2 eqs. $(NH_4)_2S_2O_8$ in chloroform/water s. W. Huang, L. Zhang, J. Chem. Res. *2006* (11), 738-9.

(Triphenylphosphine)gold(I) chloride/silver triflate *(Ph_3P)AuCl/AgOTf*
Alkylation with *o*-(alk-1-ynyl)benzoic acid esters OH → OR

126. 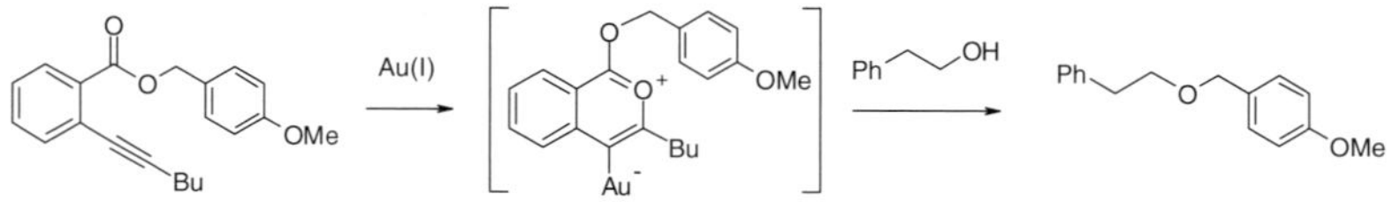

Ethers from alcohols. 3 Eqs. 2-phenylethanol and 1 eq. 4-methoxybenzyl 2-hex-1-ynylbenzoate added successively to a mixture of 5 mol% (Ph_3P)AuCl and 5 mol % AgOTf in chlorobenzene under argon, the mixture stirred for 1 h, satd. aq. $NaHCO_3$ added, extracted with ether, the extracts washed with brine, dried, and evaporated → 2-phenylethyl 4-methoxybenzyl ether. Y 92%. After indifferent results with $AuBr_3$, yields were significantly improved by use of the cationic gold(I) complex as catalyst. The reaction was successful with aryl- and alkyl-terminated acetylenes for the preparation of unsym. dialkyl ethers **under mild conditions** in >70% yields. Alkylation of prim., sec. and tert. alcohols was reported, and **cyclic ethers** were obtained from ω-hydroxyalkyl *o*-alkynylbenzoates in good yields. The method tolerates ester, ketone and nitrile functions without apparent loss in yields. The reaction is assumed to involve a 4-metalloisobenzopyrylium species, in keeping with the observation that there was no reaction in the absence of the acetylene moiety. F.e. and **Friedel-Crafts alkylation**, incl. the intramolecular variant (*72*, 628), s. N. Asao, H. Aikawa, S. Tago, K. Umetsu, Org. Lett. *2007*, *9* (21), 4299-302.

2(5*H*)-Furanones from (Z)-2,4-enynols ○
One-pot conversion by sequential catalysis with the same catalyst

127. 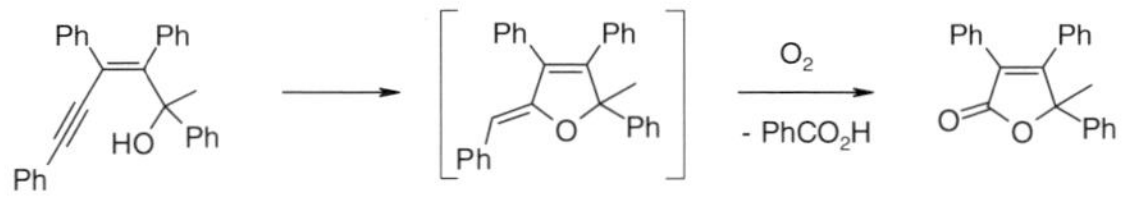

under mild conditions. 2 Mol% (Ph_3P)AuCl in THF (0.05 *M*) and 2 mol% AgOTf in THF (0.05 *M*) added sequentially to a soln. of the startg. (Z)-enynol in THF, O_2 gently bubbled through the mixture at 50° for 17 h (TLC), and worked up with chromatographic purification on silica gel → 5-methyl-3,4,5-triphenyl-5*H*-furan-2-one. Y 97%. The gold(I) complex catalyzes both the initial cyclization and the subsequent cleavage of the intermediate 2-benzylidene-2,5-dihydrofuran with oxygen (in fact, the first example of a gold-catalyzed C-C cleavage with molecular oxygen). The alkyne residue may be substituted by an alkyl or aryl group (electron-donating *para*-groups accelerating the process), and the α-carbon (relative to the OH group) may be substituted by alkyl, aryl or hetaryl (not silyl). A radical mechanism is presumed. F.e. incl. 5-spirofuranones, also with [($Ph_3PAu)_3O$]BF_4, s. Y. Liu, F. Song, S. Guo, J. Am. Chem. Soc. *2006, 128* (35), 11332-3.

Gold(III) chloride *$AuCl_3$*
Glycosides from propargyl glycosides $OCH_2C{\equiv}CH$ → OR

128. 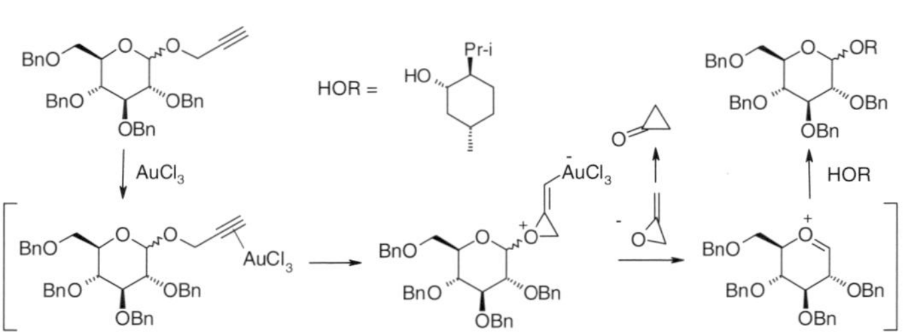

Per-O-benzylated propargyl glycosides serve as useful glycosyl donors in a new Au(III)-catalyzed glycosidation. **E:** 3 Mol% $AuCl_3$ in acetonitrile added to a soln. of the startg. propargyl glycoside (readily prepared from D-glucose and propargyl alcohol in dry dioxane/HCl, followed by per-O-benzylation) and 1.2 eqs. of the startg. aglycone in the same solvent, heated to 60° for 6 h, concentrated *in vacuo,* and purified by column chromatography → product. Y 68% (as a mixture of α/β-anomers). There was no reaction with the corresponding per-O-benzoylated or per-O-acetylated propargyl glycosides, and use of other alkyne activators [$PtCl_2$, $Co_2(CO)_8$ or $RuCl_3$] resulted in decomposition or recovery of the donor. Reaction is presumed to involve initial activation of the triple bond, followed by generation of a glycosyl oxocarbenium ion prior to quenching with the aglycone. The propargyl residue cyclizes on cleavage to 2-methyleneoxirane, which isomerizes to cyclopropanone. F.e. incl. mannose and galactose derivs. s. S. Hotha, S. Kashyap, J. Am. Chem. Soc. *2006, 128* (30), 9620-1.

Indium(III) triflate s. under MnO_2 — *In(OTf)*$_3$
Fluoroboric acid s. under PhIO — *HBF*$_4$

Scandium(III) triflate or Scandium(III) triflate/microwaves — *Sc(OTf)*$_3$ *or Sc(OTf)*$_3$/[\\\\\\\\]
Transesterification — COOR → COOR′
with CAN cf. *61*, 79s*65*; with $Sc(OTf)_3$ (10 mol%) in boiling alcoholic solvent (or under microwave irradiation with shorter reaction times) s. N. Remme, K. Koschek, C. Schneider, Synlett *2007* (3), 491-3.

Ytterbium(III) triflate — *Yb(OTf)*$_3$
Unsym. from sym. carbonic acid esters — ROH + $CO(OR')_2$ → ROCOOR′
with Sc_2O_3 cf. *57*, 105; with $Yb(OTf)_3$ s. C. Yu, B. Zhou, W. Su, Z. Xu, Synth. Commun. *2007*, *37* (4), 645-51.

Lipase ←
Kinetic resolution of alcohols by asym. enzymatic O-acylation — OH → OAc
with enolesters cf. *44*, 214s*70*; of *anti*-1,3-diols or -1,3-diol monoesters (obtained via Zr-catalyzed aldol-Tishchenko reaction) using *Candida antarctica* Lipase B (CALB) s. F. Jakob, C. Schneider, Eur. J. Org. Chem. *2007* (18), 2958-63.

Amberlyst 15 ←
Glycosides from glycosyl trichloroacetimidates — $OC(=NH)CCl_3$ → OR
on-column conversion with $HClO_4$-silica cf. *44*, 211s*70*; gluco-, galacto-, rhamno-, or lactosylation of a range of alcohols with 20 mol% Amberlyst 15 and 100 mol% 4 Å molecular sieves in ether at room temp., incl. oligosaccharides and glycosylated amino acids, also thioglycosides, s. Q. Tian, S. Zhang, Q. Yu, M.-B. He, J.-S. Yang, Tetrahedron *2007*, *63* (10), 2142-7; α-glycosylation with 2-azido-2-deoxy-glucosyl trichloroacetimidates with trimethylsilyl triflate in the presence of PhSEt or thiophene s. J. Park, S. Kawatkar, J.-H. Kim, G.-J. Boons, Org. Lett. *2007*, *9* (10), 1959-62; with trimethylsilyl triflate, 2-(dimethylphosphorylamino)-2-deoxyglycosides, s. Y. Yang, B. Yu, Tetrahedron Lett. *2007*, *48* (26), 4557-60; with 1.5 eqs. iodine and 10 mol% Et_3SiH, β-linked 2,6-dideoxyoligosaccharides, s. H. Tanaka, A. Yoshizawa, T. Takahashi, Angew. Chem. Int. Ed. Engl. *2007*, *46* (14), 2505-7.

PhSEt or thiophene s. under Trimethylsilyl triflate ←

*Phenylselenyl triflate/2,4,6-tri-*tert-*butylpyrimidine* ←
Glycosides from pent-4-enoyl glycosides — OCOR → OR′
with 1,3-dithian-2-ylium fluoroborate/Drierite cf. *46*, 228; with PhSeOTf (from PhSeBr/AgOTf)/ 2,4,6-tri-*tert*-butylpyrimidine at -78 to 0° s. T.J. Choi, J.Y. Baek, H.B. Jeon, K.S. Kim, Tetrahedron Lett. *2006*, *47* (52), 9191-4; β-mannopyranosylation s. J.Y. Baek, T.J. Choi, H.B. Jeon, K.S. Kim, Angew. Chem. Int. Ed. Engl. *2006*, *45* (44), 7436-40.

Iodosobenzene/fluoroboric acid — *PhIO/HBF*$_4$
Oxidative C-cleavage ←
of ethylene derivs. – Ar. aldehydes from styrene epoxides s. *35*, 74s*72*

Triethylsilane s. under I_2 — Et_3SiH
*2-(Di-*tert-*butylphosphino)-2′,4′,6′-triisopropylbiphenyl s. under* $PdCl_2(PhCN)_2$ ←
(S,S)-N,N′-Bis[2-(diphenylphosphino)-1-naphthoyl]-1,2-diaminocyclohexane s. under $Pd_2(dba)_3$ ←

Ozone *O_3*

Oxo compds. from ethylene derivs. by ozonolysis C=C → CO

cf. *5*, 216; oxo-subst. organo(trifluoro)borates, cleavage of the intermediate ozonides with Zn/AcOH, N-oxides or water, s. G.A. Molander, D.J. Cooper, J. Org. Chem. *2007*, *72* (9), 3558-60.

Trimethylsilyl triflate *Me_3SiOTf*

O-Benzylation with benzyl N-phenyl-2,2,2-trifluoroacetimidate OH → OBn

129.

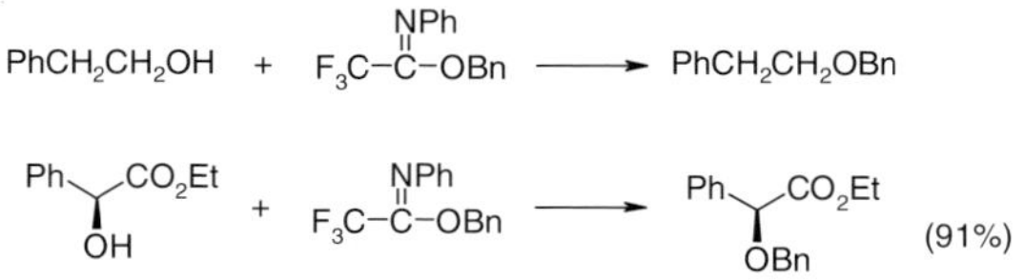

0.1 Eq. trimethylsilyl triflate added under N_2 to a stirred mixture of 2-phenylethanol and 1.1 eqs. benzyl N-phenyl-2,2,2-trifluoroacetamidate in dioxane containing 5 Å molecular sieves, and quenched with triethylamine after 30 min → benzyl 2-phenylethyl ether. Y 94%. The new benzylating agent is more stable (in air at room temp. over several months) than benzyl 2,2,2-trichloroacetimidate, and is effective for the benzylation of prim., sec. and tert. alcohols, incl. hindered substrates, and to base-sensitive α-hydroxy esters (without racemization). Phenols, however, were unaffected. F.e. and comparison of Brønsted and Lewis acid activators s. Y. Okada, M. Ohtsu, M. Bando, H. Yamada, Chem. Lett. *2007, 36* (8), 992-3.

Glycosides from glycosyl dichloro(cyano)acetimidates OC(=NH)CCl_2CN → OR

s. *44*, 211s*72*

Glycosyl carbamates OC(=NH)CCl_3 → OC(O)NHR

from trichloroacetimidoyl glycosides and carbo-*tert*-butoxyamines

130.

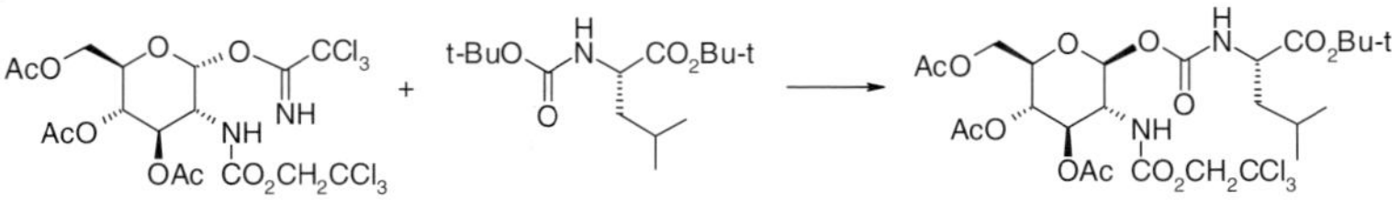

N-Protected 2-amino-2-deoxyglycosyl carbamates. The startg. Boc-amino ester, 1.5 eqs. of the glycosyl trichloroacetimidate, 10 mol% trimethylsilyl triflate and 4 Å molecular sieves stirred in methylene chloride at -35° until reaction complete → glycosyl carbamate. Y 87%. Glycosylation of *tert*-butyl carbamates derived from amino acid esters or electron-rich anilines takes place **under mild conditions** in high yields. The reaction is selective for *tert*-butyl carbamates (neither benzyl nor fluorenylmethyl carbamates reacting) and is compatible with common O-protecting groups (TIPS ethers, acetate, benzyl, *tert*-butyl and pentafluorophenyl esters) and N-protecting groups (Troc, Cbz, Fmoc). The mechanism is presumed to involve electrophilic attack of the glycosyl cation on the carbonyl moiety of the carbamate and subsequent loss of the more stable *tert*-butyl cation. F.e.s. K.J. Henry Jr, J.P. Lineswala, Tetrahedron Lett. *2007*, *48* (10), 1791-4.

Trimethylsilyl triflate *Me_3SiOTf*

Trimethylsilyl triflate/PhSEt or thiophene ←

Glycosides from glycosyl trichloroacetimidates OC(=NH)CCl_3 → OR

2-Phosphorylamino-2-deoxyglycosides – 2-Azido-2-deoxy-α-glucosides in the presence of PhSEt or thiophene s. *44*, 211s*72*

Tetra-n-*butylammonium triflate s. under $Pd_2(dba)_3$* *Bu_4NOTf*

Sulfuric acid *H_2SO_4*

Methyl carbonates from alcohols OH → OCOOMe

Selective O-carbomethoxylation s. *72*, 125

Potassium peroxymonosulfate *$KHSO_5$*

Furans from 1,4-dihydropyridines by oxidative ring contraction

131.

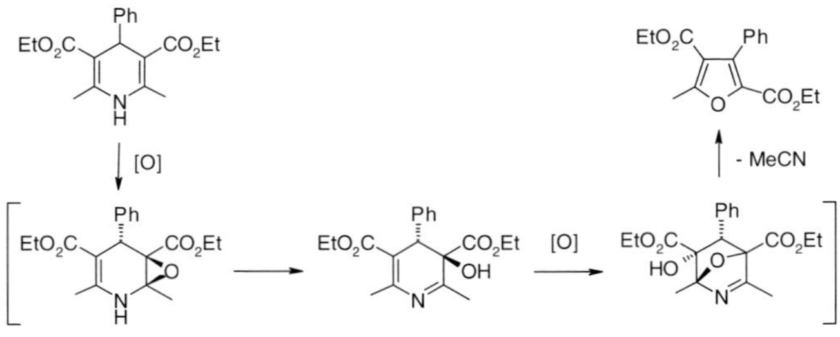

3,5-Disubst. furan-2,4-dicarboxylic acid esters. 1.5 Eqs. aq. Oxone added dropwise to a stirred soln. of the startg. Hantzsch ester *in acetonitrile* at reflux temp. over 5 min, stirring continued for a further 5 min at reflux, cooled to room temp., quenched with water, and treated with satd. aq. $NaHCO_3$ → product. Y 44% (after silica gel column chromatography). The anticipated pyridines were only formed in small amount in acetonitrile. The moderate product yields (20-40%) should be balanced against the difficulty of obtaining such polysubst. furans by alternative means, and by the ready availability of Hantzsch esters. The 4-unsubst. Hantzsch ester, however, underwent aromatization exclusively. Mechanistically, it is thought that reaction involves initial epoxidation, followed by ring-opening isomerization to the 3-hydroxy-3,4-dihydropyridine and a second epoxidation; intramolecular ring opening of the generated epoxide with the hydroxyl group at C_3 then provides a 3,4-oxido-1,4-dihydropyridine which undergoes retro-Diels-Alder reaction with elimination of acetonitrile to give the product. F.e. and solvent/pH effects s. Z. Liu, W. Yu, L. Yang, Z.-L. Liu, Tetrahedron Lett. *2007, 48* (30), 5321-4.

Ammonium persulfate s. under $AgNO_3$ *$(NH_4)_2S_2O_8$*

Iodine *I_2*

Regiospecific O,O-alkylidenation-O-acetylation of carbohydrates with enolesters ←

132.

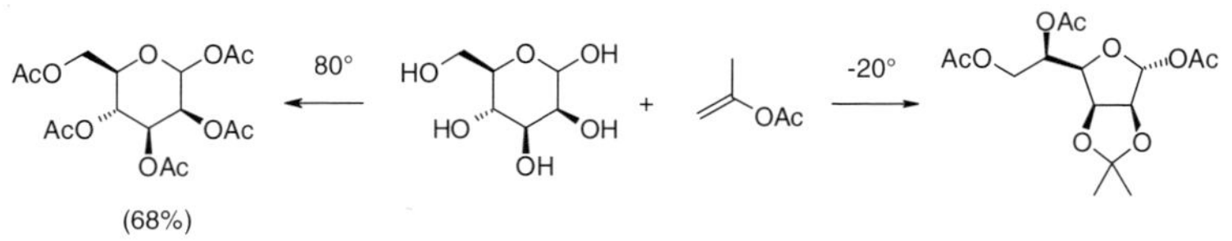

in the absence of solvent. 0.1 Eqs. Iodine added to a mixture of 1 eq. D-mannose and 4 eqs. isopropenyl acetate *at -20°* under N_2, stirred at room temp. for 1 h (with TLC monitoring), extracted with methylene chloride, washed with satd. $Na_2S_2O_3$, the organic layer dried, and evaporated *in vacuo* → 1,5,6-triacetyl-2,3-isopropylidene-α-D-mannofuranose. Y 81%. The regioselectivity is markedly affected by the choice of substrate with, for example, D-glucose affording the 3,5,6-triacetyl-1,2-isopropylidene-α-D-glucofuranose (69%) and D-galactose the 6-acetyl-1,2:3,4-di-isopropylidene-α-D-galacto*pyranose* (81%). In several examples, per-acetylated glycopyranoses are also obtained in low yield and become the major product by conducting the reaction *at 80°* instead of -20° (5 examples; 65-84%). The use of solvents such as acetonitrile, THF or DMF reduce the rate of reaction while in ether, toluene or methylene chloride no reaction is observed. The method is attractive due to the ease of preparation of orthogonally protected sugar derivs. previously only available from multi-step synthesis. F.e. and with vinyl acetate s. D. Mukherjee, B.A. Shah, P. Gupta, S.C. Taneja, J. Org. Chem. *2007, 72* (23), 8965-8.

Iodine/potassium iodide *I_2/KI*

Arylcarboxylic acid esters from arylacetonitriles $CH_2CN \rightarrow COOR$

133.

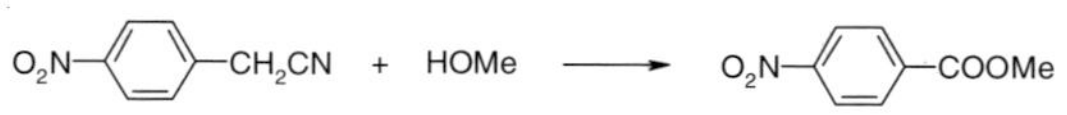

under mild conditions. A mixture of 4-nitrophenylacetonitrile, 1.1 eqs. KI and 10 mol% I_2 in methanol stirred for 3 h at room temp. → methyl 4-nitrobenzoate. Y 78%. Reaction is applicable to the cleavage of nitro- and chloro-phenylacetonitrile, as well as phenylacetonitrile, with aliphatic, benzyl and cycloalkyl alcohols, but substrates with electron-donating groups were unreactive. F.e.s. Z. Wang, J. Kang, M. Yu, J. Chem. Res. *2007* (6), 323-4.

Iodine/triethylsilane *I_2/Et_3SiH*

Glycosides from glycosyl trichloroacetimidates $OC(=NH)CCl_3 \rightarrow OR$

β-Linked 2,6-dideoxyoligosaccharides s. *44*, 211s*72*

Sodium periodate s. under $RuCl_3$ *$NaIO_4$*

Manganese dioxide/indium(III) triflate *MnO_2/In(OTf)$_3$*

Acetals from prim. alcohols via aldehydes in one pot $CH_2OH \rightarrow CH(OR)_2$

134.

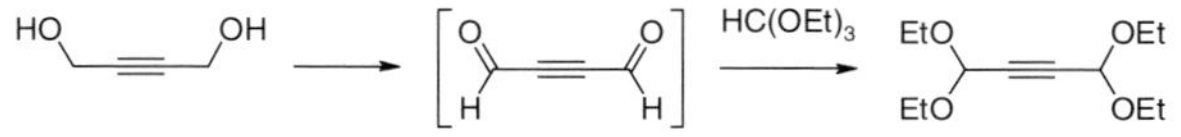

1 Mol% In(OTf)$_3$ added to a mixture of 2-butyne-1,4-diol, 2 eqs. ethyl orthoformate and 10 eqs. MnO_2 in chloroform, stirred for 6 h at reflux, a further 5 eqs. MnO_2, 2 eqs. ethyl orthoformate and 1 mol% In(OTf)$_3$ added, heating continued at reflux for 24 h, cooled to room temp., filtered, and the filtrate worked up → 1,1,4,4-tetraethoxybut-2-yne. Y 66%. The procedure is inexpensive, may be performed on the gram scale, is generally applicable to activated alcohols (benzylic, propargylic), and especially useful where the intermediate aldehydes are highly unstable. Alternative oxidants failed. F.e. incl. cyclic acetals from diols s. B.M. Smith, A.E. Graham, Tetrahedron Lett. *2007, 48* (28), 4891-4.

Ruthenium trichloride/sodium periodate *$RuCl_3$/$NaIO_4$*

Carboxylic acids from ethylene derivs. by C-cleavage CH=C → COOH

with RuO_4/$NaIO_4$/acetonitrile cf. *37*, 220; with $RuCl_3 \cdot xH_2O$/$NaIO_4$ in acetonitrile/ethyl acetate/ water, hydroxydicarboxylic acids, part of a multi-step synthesis via addition of unsatd. Grignard compds. to ethylene-aldehydes or -epoxides, s. C. Chiruta, S. Jachak, D.L.J. Clive, Tetrahedron Lett. *2007*, *48* (18), 3141-3.

*Tris(dibenzylideneacetone)dipalladium/(S,S)-N,N¢-bis[2-(diphenylphosphino)-1-naphthoyl]-1,2-diaminocyclohexane/tetra-*n*-butylammonium triflate* ←

Regiospecific asym. allylation with 1,4-dien-3-ol carbonates ←

135.

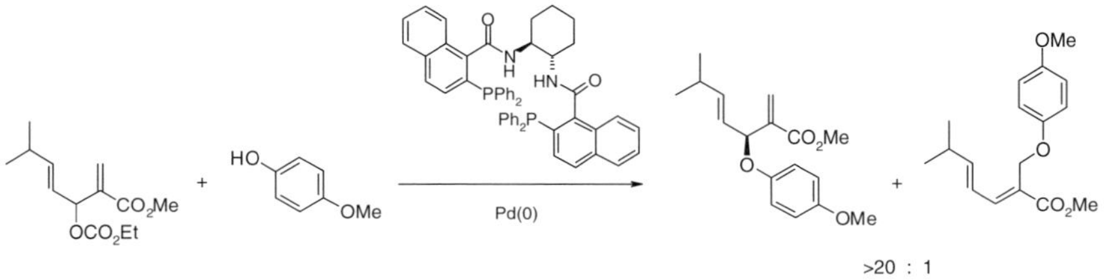

Chiral 3-aryloxy-1,4-dienes. 4-Methoxyphenol, 1 mol% tris(dibenzylideneacetone)dipalladium-chloroform, 3 mol% (S,S)-N,N′-bis[2-(diphenylphosphino)-1-naphthoyl]-1,2-diaminocyclohexane and 6 mol% tetrabutylammonium triflate stirred in degassed dimethoxyethane under N_2

for 5 min (color change from purple to orange), a soln. of the Baylis-Hillman dienol carbonate in dimethoxyethane added, and the mixture stirred 24-30 h → methyl 3-(4-methoxyphenoxy)-6-methyl-2-methylenehept-4-enoate. Y 83% (e.e. 89%). The conditions were optimized with respect to choice of solvent and ligand. Variations in the substituents on the carbonate substrate had a significant effect on yields and selectivity as did variations in substituents on the phenolic nucleophile. In most examples minor regioisomers were also formed in low yield by allylic substitution at the unsubstituted methylene group. Meldrum's acid was the only C-nucleophile that produced acceptable results. F.e. incl. detailed comparison of ligands and substituent effects s. B.M. Trost, M.K. Brennan, Org. Lett. *2007*, *9* (20), 3961-4.

*Bis(benzonitrile)dichloropalladium(II)/2-(di-*tert*-butylphosphino)-2′,4′,6′-triisopropylbiphenyl* ←

Ferrier-type reaction s. under OC⇃↾O

←

Elimination ⥮

Hydrogen ↾ OC ⥮ H

Irradiation s. under I_2 ///

Microwaves s. under $KMnO_4$ [\\\\]

Sodium bromide s. under $MeReO_3$ *NaBr*

Potassium bromide s. under $PhI(OAc)_2$ *KBr*

*Copper(II) acetoacetonate/*tert*-butyl hydroperoxide/ 1-butyl-3-methylimidazolium hexafluorophosphate* *$Cu(acac)_2$/t-BuOOH/[bmim]PF_6*

Ketones from sec. alcohols CHOH → CO

with $MoCl_5$/hydroperoxides cf. *27*, 184; with [reusable] $Cu(acac)_2$/*t*-BuOOH in [bmim]PF_6 (cf. *52*, 119) s. C. Liu, J. Han, J. Wang, Synlett *2007* (4), 643-5.

Copper(II) sulfate *$CuSO_4$*

β-Ketocarboxylic from α-diazo-β-hydroxycarboxylic acid esters CH(OH)C=N_2 → $COCH_2$

Copper(II)-catalyzed conversion in water

136. 

The startg. α-diazo-β-hydroxy-ester stirred in water, 5 mol% $CuSO_4$ added, stirring continued for 4 h, the mixture extracted with methylene chloride, dried over Na_2SO_4, and concentrated *in vacuo* → β-keto-ester. Y 84%. This method is mild, environmentally friendly, based on an inexpensive catalyst at low catalyst loading (1 or 5 mol%), and produces nitrogen as the only by-product. The method is successful with a range of substrates possessing β-aryl or β-alkyl substituents in 72-92% yields, and also with β-2-furyl, β-ethenyl and β-alk-1-ynyl derivs., albeit in moderate yields (41-69%). Reaction times are longer (>24 h) for less soluble substrates and the reaction fails altogether for solid substrates, such as β-4-methoxyphenyl-, β-2,6-dichlorophenyl- and β-2,4-dichlorophenyl-derivs., and also for β,β–disubst. derivs. Under these conditions no 1,2-alkyl/aryl migration was observed, and there were no products resulting from O-H bond insertion. F.e. and comparison of catalysts s. M. Liao, J. Wang, Tetrahedron Lett. *2006*, *47* (50), 8859-61.

Copper(II) triflate/chiral bis(Δ²-oxazolines)/N-bromosuccinimide or N-bromophthalimide/potassium carbonate ←

α-Hydroxyketones from glycols by asym. oxidation CHOH → CO

137. 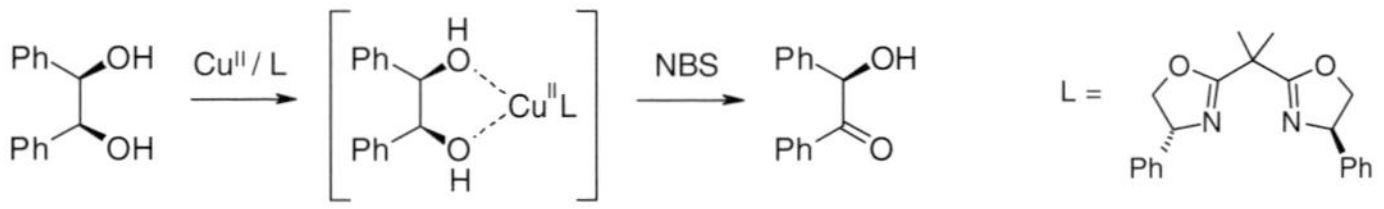

with desymmetrization. A soln. of 10 mol% $Cu(OTf)_2$ and 10 mol% chiral bis(Δ²-oxazoline) ((R,R)-Ph-BOX) in chloroform stirred for 10 min under an aerobic atmosphere, *meso*-hydrobenzoin, 2 eqs. K_2CO_3 and 2 eqs. NBS added, stirring continued for 3 h at room temp., the soln. poured into 10% aq. $Na_2S_2O_3$, and worked up with column chromatographic purification on silica gel → (R)-product. Y 83% (e.e. 72%). Reaction is presumed to involve intermediate formation of a chiral copper(II) glycolate, followed by asym. oxidation with NBS. **Kinetic resolution** of *dl*-hydrobenzoin and *rac*-cycloalkane-*cis*-1,2-diols was also effected under the same conditions. F.e. and comparison of bases and solvents, also with N-bromophthalimide as oxidant s. O. Onomura, H. Arimoto, Y. Matsumura, Y. Demizu, Tetrahedron Lett. *2007, 48* (49), 8668-72

Triethylenediamine-copper(I) chloride/2,2,6,6-tetramethylpiperidine nitroxyl/oxygen ←

Oxo compds. from alcohols

with CuCl/TEMPO in [bmim]PF_6 cf. *52*, 119s*63*; with DABCO-CuCl/TEMPO (5 mol% each) in nitromethane, allylic and benzylic oxidation, s. S. Mannam, S.K. Alamsetti, G. Sekar, Adv. Synth. Catal. *2007*, *349* (14-15), 2253-8.

Polystyrene-based gold nanoclusters/oxygen ←

Oxo compds. from alcohols

with nanocrystalline Pd-Au alloy on TiO_2 cf. *70*, 119; with polystyrene-supported gold nanoclusters s. H. Miyamura, R. Matsubara, Y. Miyazaki, S. Kobayashi, Angew. Chem. Int. Ed. Engl. *2007*, *46* (22), 4151-4.

Isopropylmagnesium chloride-lithium chloride/benzaldehyde i-*PrMgCl-LiCl/PhCHO*

Oxo compds. from alcohols

catalytic Oppenauer oxidation with a cyclic methylaluminum sulfonylaminophenoxide complex/pivaldehyde cf. *56*, 28s*66*; with *i*-PrMgCl-LiCl/PhCHO, ketones, incl. polyfunctional biaryl ketones and ferrocenyl ketones, s. R.J. Kloetzing, A. Krasovskiy, P. Knochel, Chem. Eur. J. *2007*, *13* (1), 215-27.

Diethylaluminum ethoxide/1,1,1-trifluoroacetone *Et_2AlOEt/CF_3COMe*

Ketones from sec. alcohols via modified Oppenauer oxidation

138. 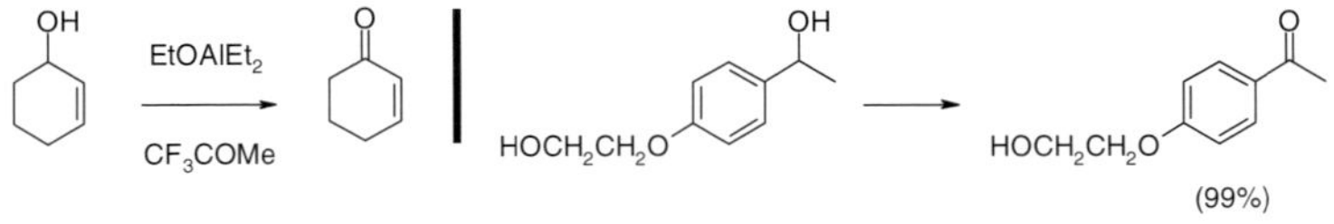

0.1 Eqs. diethylaluminum ethoxide (1.6 *M* in toluene) added to cyclohex-2-enol in methylene chloride at room temp., stirred for 5 min, 2 eqs. 1,1,1-trifluoroacetone (as hydride acceptor) added, the flask tightly closed, the mixture stirred for 7 h, treated with water and $MgSO_4$, filtered, and evaporated *in vacuo* → 2-cyclohexenone. Y >99%. With a hydride acceptor of enhanced activity and reduced Lewis basicity (relative to traditionally used carbonyl compds.), oxidation is efficient and selective for sec. alcohols with yields generally >99%, notably proceeding **with retention of prim. alcohols** (incl. allyl alcohols). The work-up is also simple due to the volatility of the ketone and its reduction product. Conducting the reaction in a sealed tube at 40° gave lower conversions, attributed to the high vapour pressure of 1,1,1-trifluoroacetone which reduces its availability in solution. Use of acetone as hydride acceptor gave low conversions as did the use of magnesium alkoxides in place of the aluminum alkoxide. F.e.s. R. Mello, J. Martinez-Ferrer, G. Asensio, M.E. Gonzalez-Nunez, J. Org. Chem. *2007*, *72* (24), 9376-8.

Benzaldehyde s. under i-PrMgCl-LiCl *PhCHO*
p-*Benzoquinone s. under Pd(OCOCF$_3$)$_2$* ←
1-Butyl-3-methylimidazolium salts s. under Cu(acac)$_2$ and MeReO$_3$ *[bmim]$^+$*
Chiral bis(Δ^2-oxazolines) s. under Cu(OTf)$_2$ and Pd(OCOCF$_3$)$_2$ ←
N,N,N′,N′-Tetramethylguanidine s. under ArPh$_3$Bi$^+$BF$_4^-$ *TMG*
2,2,6,6-Tetramethylpiperidine nitroxyl s.a. under DABCO-CuCl *TEMPO*
*2,2,6,6-Tetramethylpiperidine nitroxyl/hydrogen bromide/*tert-*butyl nitrite/oxygen* ←

Oxo compds. from alcohols by metal-free aerobic oxidation
with TEMPO/Br_2/$NaNO_2$ cf. *67*, 113; with TEMPO/HBr/*tert*-butyl nitrite (TON up to 16,000) s. Y. Xie, W. Mo, D. Xu, Z. Shen, N. Sun, B. Hu, X. Hu, J. Org. Chem. *2007*, *72* (11), 4288-91.

2,2,6,6-Tetramethylpiperidine nitroxyl/iodobenzene dichloride/pyridine ←
Polyethylene glycol-based or fluorous-tagged 2,2,6,6-tetramethylpiperidine nitroxyl/ sodium hypochlorite ←
Silica-supported 4-amino-2,2,6,6-tetramethylpiperidine nitroxyl/potassium iodide/ sodium hypochlorite ←

Oxo compds. from alcohols CHOH → CO
with a 4-triazinyloxy-TEMPO/PhI(OAc)$_2$ cf. *45*, 120s*68*; with TEMPO/PhICl$_2$/pyridine, incl. α-hydroxyketones or α-diketones from glycols (depending on the amount of PhICl$_2$), also convenient prepn. of ar. iododichlorides from ar. iodides with 5 eqs. NaOCl or $NaClO_2$ in HCl (cf. *35*, 182s*67*), s. X.-F. Zhao, C. Zhang, Synthesis *2007* (4), 551-7; with a silica-supported 4-amino-TEMPO/KBr/NaOCl (cf. *58*, 103) s. A. Michaud, G. Gingras, M. Morin, F. Béland, R. Ciriminna, D. Avnir, M. Pagliaro, Org. Proc. Res. Dev. *2007*, *11* (4), 766-8; with recyclable 4-[1-[(2-perfluorooctyl)ethyl]-1,2,3-triazol-4-ylmethoxy]-TEMPO (prepared via ‘click’ chemistry)/NaOCl s. A. Gheorghe, E. Cuevas-Yañez, J. Horn, W. Bannwarth, B. Narsaiah, O. Reiser, Synlett *2006* (17), 2767-70; with poly(vinyl alcohol)-grafted-poly(ethylene glycol) supported 1,2,3-triazol-4-ylmethoxy-TEMPO/NaOCl s. J. Luo, C. Pardin, W.D. Lubell, X.X. Zhu, Chem. Commun. *2007* (21), 2136-8.

tert-*Butyl hydroperoxide s. under Cu(acac)$_2$* t-*BuOOH*
Diphenyl diselenide s. under PhI(OCOCF$_3$)$_2$ *(PhSe)$_2$*

Phenyl iodosoacetate/potassium bromide *PhI(OAc)$_2$/KBr*

1-Aryllactones from ω-arylcarboxylic acids ○

139.
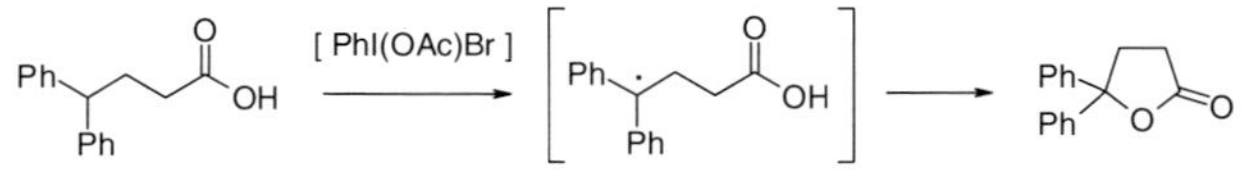

under mild conditions. 1.2 Eqs. phenyl iodosoacetate added to a stirred suspension of 4,4-diphenylbutyric acid and 1 eq. KBr in dry methylene chloride, the mixture vigorously stirred at 30° for 8 h (TLC monitoring), satd. aq. $NaHCO_3$ added to the mixture, stirred for 5 min, the organic layer separated, washed with satd. aq. $NaHCO_3$ and dil. aq. $Na_2S_2O_3$, dried (Na_2SO_4), and evaporated *in vacuo* → dihydro-5,5-diphenyl-2(3*H*)-furanone. Y 85%. The method affords 5- and 6-membered aryl-subst. lactones in 50-80% yield, but a 6-aryl-subst. hexanoic acid gave only 21% of the lactone. The reaction is assisted by electron-releasing substituents in the aryl ring but reactions are slow with electron-withdrawing substituents (4-nitrophenylpropanoic acid giving 47% yield after 18 h). The method was also suitable for the conversion of 2-methyl-, 2-ethyl- and 2-cyclohexyl-benzoic acids to the corresponding **phthalides** (81-91%), and the *o*-(2-phenylethyl) deriv. to the corresponding isocoumarin (81%) by cyclization at the more reactive benzylic site. Lactonization does not occur with hypervalent iodine(III) agents alone or with Lewis acid activation, but finely powdered KBr has a dramatic catalytic effect on the reaction in organic solvents. The authors suggest that the active agent is *in situ*-generated PhI(OAc)Br which delivers a benzyl radical prior to ring closure. F.e.s. T. Dohi, N. Akihiro, A. Goto, A. Maruyama, Y. Kita, Org. Lett. *2007*, *9* (16), 3129-32.

Phenyl iodosotrifluoroacetate/diphenyl diselenide $PhI(OCOCF_3)_2/(PhSe)_2$
2(5*H*)-Furanones from (E)-β,γ-ethylenecarboxylic acids ○
via catalytic intramolecular oxyselenation

140. 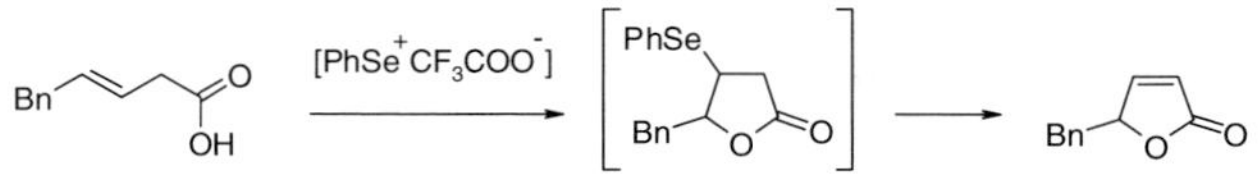

5-Phenylpent-3-enoic acid and 1.05 eqs. phenyl iodosotrifluoroacetate added sequentially to a soln. of *5 mol%* diphenyl diselenide in acetonitrile, the mixture stirred under argon at room temp. until reaction complete (TLC monitoring), the mixture evaporated *in vacuo*, and the residue purified by chromatography → 5-benzylfuran-2(5*H*)-one. Y 59%. The lactonization is successful with 4-aryl-, 4-benzyl- and 4-alkyl-3-butenoic acids (10 examples; 49-96%) and complete at room temp. in 3 h with acetonitrile, toluene or methylene chloride as solvent. The use of enantiomerically pure diselenides achieved moderate enantioselectivity (best enantiomeric ratio of 86:14). Catalytic phenylselenenyl trifluoroacetate is thought to be the active selenating species, the intermediate β-phenylseleno-γ-lactones undergoing oxidative elimination with regeneration of the catalytic electrophile. F.e. and comparison of high-valent iodine oxidants s. D.M. Browne, O. Niyomura, T. Wirth, Org. Lett. *2007*, *9* (16), 3169-71.

o-*Iodoxybenzoic acid* $ArIO_2$
in situ-*Generated* o-*iodoxybenzoic acid/tetraphenylphosphonium peroxymonosulfate* ←
Oxo compds. from alcohols CHOH → CO
with *o*-iodoxybenzoic acid cf. *50*, 123; catalytic procedure with *o*-iodoxybenzoic acid generated *in situ* from *o*-iodobenzoic acid with tetraphenylphosphonium peroxymonosulfate as re-oxidant for the prepn. of aldehydes without over-oxidation to acids s. P.C.B. Page, L.F. Appleby, B.R. Buckley, S.M. Allin, M.J. McKenzie, Synlett *2007* (10), 1565-8; *in situ*-oxidation of β-hydroxy-α-methylenecarboxylic acid esters with *o*-iodoxybenzoic acid, followed by Sc(III)-catalyzed 1,4-addition of allyltrimethylsilane s. *72*, 592.

1,1,1-Trifluoroacetone s. under Et_2AlOEt CF_3COMe

4,4′-Bis(dichloroiodo)biphenyl/tetraethylammonium bromide ←
Oxo compds. from alcohols
with PhIO/KBr cf. *59*, 84(s*63*); with [recyclable] 4,4′-bis(dichloroiodo)biphenyl/tetraethylammonium bromide at room temp. s. V.N. Telvekar, O.P. Herlekar, Synth. Commun. *2007*, *37* (5), 859-63.

N-Bromosuccinimide [s.a. under $Cu(OTf)_2$] NBS
Aldehydes from prim. alcohols CH_2OH → CHO
cf. *12*, 329; ar. aldehydes s. S. Adimurthy, P.U. Patoliya, Synth. Commun. *2007*, *37* (9), 1571-7.

Silica s. under Phosphotungstic acid sodium salt SiO_2
tert-*Butyl nitrite* s. *under Tetramethylpiperidine nitroxyl* t-*BuONO*

Mesityltriphenylbismuthonium fluoroborate/N,N,N′,N′-tetramethylguanidine ←
Oxo compds. from alcohols CHOH → CO
with *o*-Tol_3BiCl_2/DBU cf. *64*, 127; with mesityltriphenylbismuthonium fluoroborate/N,N,N′,N′-tetramethylguanidine s. Y. Matano, T. Suzuki, T. Shinokura, H. Imahori, Tetrahedron Lett. *2007*, *48* (16), 2885-8.

Vanadium ion s.a. Iron-exchanged vanadium-incorporated 12-molybdophosphoric acid ←
Oxygen s. under Triethylenediamine-copper(I) chloride, Polymer-based Au, 2,2,6,6-Tetramethylpiperidine nitroxyl, Phosphotungstic acid sodium salt, I_2, Perruthenate and Pt O_2
Hydrogen peroxide s. under $MeReO_3$ and Pt H_2O_2

Tetraphenylphosphonium peroxymonosulfate s. under in situ-*Generated o-iodoxybenzoic acid* *[Ph_4P]HSO_5*

Phosphotungstic acid sodium salt/silica/oxygen ←
Iron-exchanged vanadium-incorporated 12-molybdophosphoric acid/oxygen ←
Oxo compds. from alcohols CHOH → CO
with activated carbon-supported heteropolyacid sodium salt cf. *47*, 192; heterogeneous oxidation of allylic, benzylic or aliphatic alcohols using the silica-supported Preyssler type heteropolyacid, $H_{14}[NaP_5W_{30}O_{110}]/SiO_2$, s. M.M. Heravi, V. Zadsirjan, K. Bakhtiari, H.A. Oskooie, F.F. Bamoharram, Catal. Commun. *2007*, *8* (3), 315-8; with an iron-exchanged vanadium-incorporated 12-molybdophosphoric acid catalyst in refluxing acetonitrile s. P. Nagaraju, N. Pasha, P.S.S. Prasad, N. Lingaiah, Green Chem. *2007*, *9* (10), 1126-9.

Hexadecyltrimethylammonium bromochromate *[Me_3RN][CrO_3Br]*
Oxo compds. from alcohols
with dodecyltrimethylammonium bromochromate cf. *42*, 235s*71*; with the hexadecyl analog s. S. Ghammamy, H. Eimanieh, M.K. Mohammady, Synth. Commun. *2007*, *37* (4), 599-605.

Iodine/oxygen/irradiation *I_2/O_2/↯*
Oxo compds. from alcohols
allylic and benzylic photooxidation with 2 eqs. I_2 in diisopropyl ether cf. *62*, 135; with 25 mol% I_2 in acetonitrile under O_2 s. S. Farhadi, A. Zabardasti, Z. Babazadeh, Tetrahedron Lett. *2006*, *47* (50), 8953-7; with 10 mol% I_2 in ethyl acetate s. H. Nakayama, A. Itoh, Chem. Pharm. Bull. *2006*, *54* (11), 1620-21.

Manganese dioxide *MnO_2*
Ring closure to O-heterocyclics ○
with PbO cf. *18*, 307; isoxazoles from α,β-ethyleneoximes with 2 eqs. MnO_2 in refluxing chloroform, 3,5-disubst. derivs., s. R.F. Kurangi, R. Kawthankar, S. Sawal, V.G. Desai, S.G. Tilve, Synth. Commun. *2007*, *37* (4), 585-7.

Potassium permanganate/microwaves *$KMnO_4$/*[\\\\]
1,3,4-Oxadiazoles from acylhydrazones
under microwave irradiation with $PhI(OAc)_2$/alumina in the absence of solvent cf. *44*, 301s*71*; with $KMnO_4$ in DMF, 2-aryl-5-(9′-ethylcarbazol-3′-yl)-1,3,4-oxadiazoles, s. Z. Li, Y. Xing, Xuelin, S. Xiao, Z. Lu, Synth. Commun. *2006*, *36* (22), 3287-95.

Tetraethylammonium bromide s. under 4,4′-Bis(dichloroiodo)biphenyl *Et_4NBr*
Hydrogen bromide s. under Tetramethylpiperidine nitroxyl *HBr*

Methylrhenium oxide/hydrogen peroxide/sodium bromide/1-butyl-3-methylimidazolium fluoroborate ←
Oxo compds. from alcohols CHOH → CHOH
with $MeReO_3/H_2O_2$ and $MgSO_4$ (to trap water) in acetonitrile, α-diketones, cf. *67*, 116; in the presence of NaBr in [bmim]BF_4 as ionic liquid solvent s. S.L. Jain, V.B. Sharma, B. Sain, Bull. Chem. Soc. Jpn. *2006*, *79* (10), 1601-3.

Iron(III) nitrate *$Fe(NO_3)_3$*
Oxo compds. from alcohols in the absence of solvent

141. $PhCH_2OH \longrightarrow PhCHO$

Benzyl alcohol admixed with 0.67 eqs. $Fe(NO_3)_3{\cdot}9H_2O$ in a test-tube, warmed at 80° with stirring for 15 min on a water bath, cooled to room temp., water added, and worked up → benzaldehyde. Y 99%. The procedure is environmentally friendly, based on a cheap and relatively non-toxic reagent, produces the minimum of waste, and is generally applicable to prim. and sec. benzyl alcohols and aliphatic and alicyclic sec. alcohols. Halogeno-subst. benzyl alcohols, however, were less reactive, while aliphatic prim. alcohols gave a mixture of products. It is thought that reaction proceeds through initial dehydration to give the sym. ether, which is then oxidized. F.e. and comparison of inorganic nitrates s. V.V. Namboodiri, V. Polshettiwar, R.S. Varma, Tetrahedron Lett. *2007, 48* (50), 8839-42.

Iron salts s. Iron-exchanged vanadium-incorporated 12-molybdophosphoric acid ←

Nickel oxide hydroxide nanoparticles/sodium hypochlorite *Ni(O)OH/NaOCl*
Ketones from sec. alcohols CHOH → CO
Catalytic heterogeneous oxidation in water s. *72*, 95

Polymer-based tetraalkylammonium perruthenate/oxygen ←
Oxo compds. from alcohols
with sol-gel-entrapped tetrapropylammonium perruthenate cf. *54*, 111s*62*; with poly(ethylene glycol-co-N-isopropylacrylamide-co-(3-acryloylaminopropyl)dodecyldimethylammonium perruthenate) s. H. Hamamoto, M. Kudoh, H. Takahashi, H. Natsugari, S. Ikegami, Chem. Lett. *2007*, *36* (5), 632-3.

Chloro(1,5-cyclooctadiene)(η^5-pentamethylcyclopentadienyl)ruthenium(II)/ [2-((di)methylamino)ethyl]diphenylphosphine/potassium tert-*butoxide* ←
Lactones from diols ○
with $Ru(acac)_3/(C_8H_{17})_3P$ cf. *37*, 236s*66*; γ-lactones with Ru(Cl)cod(Cp*)/[2-(methyl- or dimethyl-amino)ethyl]diphenylphosphine/KOBu-*t* in acetone (>1,000 TOF/h at 30°), incl. flavor lactones hinokinin and muricatacin, s. M. Ito, A. Osaku, A. Shiibashi, T. Ikariya, Org. Lett. *2007*, *9* (9), 1821-4.

*Bis(dibenzylideneacetone)palladium(0)/1,3-bis(2,5-di-*tert*-butylphenyl)imidazolidine-2-thione/sodium acetate* ←
Oxo compds. from alcohols – Aerobic oxidation CHOH → CO
with a diacetoxopalladium(II) imidazol-2-ylidene complex under air cf. *67*, 119; with $Pd(dba)_2$/ 1,3-bis(2,5-di-*tert*-butylphenyl)imidazolidine-2-thione/NaOAc under O_2 s. M. Yang, K.-T. Yip, J.-H. Pan, Y.-C. Chen, N.-Y. Zhu, D. Yang, Synlett *2006* (18), 3057-60.

*Palladium(II) trifluoroacetate/chiral bis(Δ^2-oxazolines)/*p*-benzoquinone* ←
Asym. Wacker-type ring closure of *o*-allylphenols ○
cf. *53*, 109; *67*, 118; with Pd(II)-trifluoroacetate and a series of chiral bis(Δ^2-oxazoline) ligands having axis-unfixed biphenyl backbones (e.g. 3,3′-dimethoxy-6,6′-bis((4S)-4-phenyl-Δ^2-oxazolin-2-yl)biphenyl]/*p*-benzoquinone s. F. Wang, Y.J. Zhang, G. Yang, W. Zhang, Tetrahedron Lett. *2007*, *48* (24), 4179-82.

Platinum black/hydrogen peroxide *Pt/H_2O_2*
α,β-Ethyleneoxo compds. from 2-ethylenealcohols in the absence of organic solvent CHOH → CO

142. 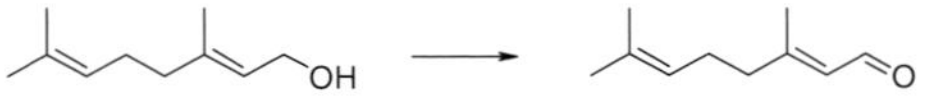

1.1 Eqs. 30% aq. H_2O_2 added dropwise over 1 h to a vigorously stirred mixture of the startg. allyl alcohol and 1 mol% Pt black at 90°, heating continued for a further 2 h, cooled to room temp., the organic phase separated, washed with satd. aq. $Na_2S_2O_3$, dried, and evaporated → product. Y 97%. The method is straightforward, effective and environmentally friendly for the oxidation of prim. and sec. allyl alcohols in high yield within 3 h (Y 81-99%). It can also be effected with 1 mol% Pt and on a 100 *g* scale, and the catalyst may be reused for up to seven cycles. No epoxidation of internal double bonds was observed but allyl alcohols with terminal unsaturation failed to react, presumably because of a poisoning effect by strong complexation with the metal. F.e. and benzaldehyde from benzyl alcohol s. Y. Kon, Y. Usui, K. Sato, Chem. Comm. *2007* (42), 4399-400.

Platinum nanoclusters/oxygen *Pt/O_2*
Base-free aerobic oxidation of alcohols in water ←

143. 

Soluble metallic platinum nanoclusters [of 1.5 nm mean diameter, prepared in deionized water via reduction of $H_2PtCl_6 \cdot 6H_2O$ with glycol and stabilization with poly(N-vinyl-2-pyrrolidone)] are highly efficient for the base-free, environmentally friendly oxidation of a range of activated *and unactivated* alcohols in water. **E: Ketones from sec. alcohols.** A mixture of platinum nanoclusters in deionized water (containing 0.0290 mmol Pt) and 0.579 mmol octan-2-ol stirred at 80° for 24 h under a balloon of O_2 (1 atm.), cooled to room temp., and extracted with ether → octan-2-one. Y 94.2% (conversion 95.2%). **Carboxylic acids** were obtained similarly **from prim. alcohols**, reaction being applicable to a wide range of benzyl alcohols, allyl alcohols, alicyclic alcohols and aliphatic alcohols, the catalyst being simply recyclable without significant loss of activity. F.e.s. T. Wang, C.-X. Xiao, L. Yan, L. Xu, J. Luo, H. Shou, Y. Kou, H. Liu, Chem. Commun. *2007* (42), 4375-7.

Oxygen ↑ OC ⥣ O

Microwaves s. under $CF_3COOH/AcOH$ [\\\\\]

Trifluoroacetic anhydride *$(CF_3CO)_2O$*
Enollactones from ketocarboxylic acids ○
2(3*H*)-furanones with H_2SO_4/Ac_2O cf. *16*, 331s*36*; 3,5,6-trisubst. 3,4-dihydro-2-pyrones with $(CF_3CO)_2O$, part of a synthesis from Baylis-Hillman adducts and ketones, s. S.J. Kim, H.S. Lee, J.N. Kim, Tetrahedron Lett. *2007*, *48* (6), 1069-72.

Trifluoroacetic acid/acetic acid/microwaves *$CF_3COOH/AcOH$/*[\\\\\]
Benzoxazoles from *o*-(acylamino)phenols
cf. *16*, 326s*54*; *21*, 543s*68*; with 1:1 CF_3COOH/AcOH under microwave irradiation, benzoxazole-7-carboxylic acid esters, s. A. Huxley, Synlett *2006* (16), 2658-60.

Benzoic acid s. under $[Ru(CO)H_2(PPh_3)_3]$ *PhCOOH*
9,9-Dimethyl-4,5-bis(diphenylphosphino)xanthene s. under $[Ru(CO)H_2(PPh_3)_3]$ *Xantphos*

Thionyl chloride *$SOCl_2$*
1,3,4-Oxadiazoles from N,N′-diacylhydrazines
with $ZrCl_4$ cf. *19*, 563s*68*; in refluxing $SOCl_2$, also 2-mercapto-1,3,4-oxadiazoles from acylhydrazines with 1.2 eqs. CS_2/1 eq. aq. KOH in refluxing ethanol (cf. *23*, 383), s. M. Zareef, R. Iqbal, N.A. Al-Masoudi, J.H. Zaidi, M. Arfan, S.A. Shahzad, Phosphorus, Sulfur Silicon Relat. Elem. *2007*, *182* (2), 281-98.

Carbonyl(dihydrido)tris(triphenylphosphine)ruthenium(II)/9,9-dimethyl-4.5-bis(diphenylphosphino)xanthene/benzoic acid ←
Furans from 2-acetylene-1,4-diols via γ-diketones in one pot

144. 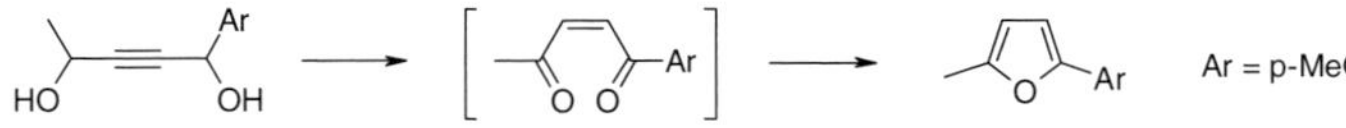

The startg. alkynediol, 5 mol% benzoic acid and 1 mol% each of $[Ru(CO)H_2(PPh_3)_3]$ and Xantphos in toluene (contained in a dry, clean Radley's carousel tube) heated under N_2 at 110° for 24 h, cooled, and quenched with base → product. Y 100% (conversion 100%). Reaction involves initial isomerization of the substrate to the corresponding γ-diketone, followed by acid-catalyzed cyclodehydration, and is a useful alternative to the classical Paal-Knorr route in those instances where γ-diketones are unavailable or unstable. Other carboxylic acids may be used as co-catalyst, but stronger acids, e.g. TsOH, completely stopped the reaction. F.e. incl. 2,5-dialkylfurans, also comparison of catalysts and ligands, s. S.J. Pridmore, P.A. Slatford, J.M.J. Williams, Tetrahedron Lett. *2007, 48* (29), 5111-4.

Nitrogen ↑ OC ⥣ N

Copper(II) sulfate *$CuSO_4$*
β-Ketocarboxylic from α-diazo-β-hydroxycarboxylic acid esters $CH(OH)CN_2 \rightarrow COCH_2$
Copper-catalyzed conversion in water s. *72*, 136

Halogen ↑ OC ⇑ Hal

Copper(I) iodide/potassium phosphate ←
Benzofurans from α-(*o*-halogenoaryl)ketones s. *72*, 78 ○

Copper(I) iodide/1,10-phenanthroline/cesium carbonate ←
2-Alkylideneoxetanes from 3-ethylene-3,1-bromhydrins by copper-catalyzed intramolecular O-vinylation □o

145. 

Relatively inaccessible 2-alkylideneoxetanes can now be obtained in high yield by 4-*exo*-ring closure under copper catalysis. **E:** The startg. γ-bromohomoallyl alcohol in dry acetonitrile added to a mixture of 10 mol% CuI, 20 mol% 1,10-phenanthroline and 2 eqs. Cs_2CO_3 at room temp. under N_2, stirred at reflux temp. for 8 h, cooled to room temp., and worked up with column chromatographic purification → product. Y 98%. With substrates possessing both unsaturated 1,3- and 1,4-bromhydrin residues, **selective ring closure** in favour of the 4-membered ring takes place, while under palladium catalysis 5-*exo* cyclization is favoured. Nevertheless, 5-*exo* and 6-*exo* cyclization are still possible under copper catalysis with certain unsatd. 1,4- and 1,5-bromhydrins, although clearly with less facility than 4-*exo* ring closure. Reaction is applicable to substrates having prim., sec. or tert. hydroxyl groups, leaving isolated hydroxyl and OBz groups unaffected and proceeding with retention of the olefinic configuration. There was no reaction, however, with the corresponding chlorhydrins, while substrates substituted at the terminal position of the vinyl residue required a much longer reaction time. F.e. and comparison of ligands and bases s. Y. Fang, C. Li, J. Am. Chem. Soc. *2007, 129* (26), 8092-3.

Sulfur ↑ OC ⇑ S

Silver nitrate *$AgNO_3$*
p-*Toluenesulfonyl chloride/pyridine* *$TsCl/C_5H_5N$*
Ring closure of N[1]-(acyl)thiosemicarbazides ○
cf. *6*, 311s*71*; 5-phosphorylimino-Δ^2-1,3,4-oxadiazolines with 2 eqs. $AgNO_3$ in ethanol s. D. Feng, Y. Huang, R. Chen,Y. Yu, H. Song, Synthesis *2007* (12), 1779-84; 5-aryl- or 5-alkyl-2-amino-1,3,4-oxadiazoles with tosyl chloride/pyridine s. S.J. Dolman, F. Gosselin, P.D. O'Shea, I.W. Davies, J. Org. Chem. *2006*, *71* (25), 9548-51.

Remaining Elements ↑ OC ⇑ Rem

1,4-Dichloro-1,4-diazoniabicyclo[2.2.2]octane dichloride ←
Oxo compds. from trimethylsilyl ethers CH(OSi≤) → CO
in water under neutral conditions s. *72*, 149

Tetraethylammonium superoxide *$[Et_4N]O_2$*
Triethylenediamine-bromine complex *DABCO-Br_2*
Oxo compds. from alkoxysilanes
cf. *37*, 264s*71*; cleavage of trimethylsilyl ethers with $[Et_4N]O_2$ s. R.S. Raghuvanshi, K.N. Singh, Synth. Commun. *2007*, *37* (8), 1371-4; with DABCO-bromine complex, also oxidative cleavage of tetrahydropyran-2-yl ethers (cf. *34*, 249s*70*) or semicarbazones (cf. *19*, 253s*44*), s. M. Tajbakhsh, M.M. Heravi, S. Habibzadeh, ibid. *37* (17), 2967-73.

Carbon ↑ OC ⇑ C

Microwaves s. under $MeSO_2Cl$ [\\\\\]

Potassium hydroxide *KOH*

Spiroketals ○

from α,β-acetylene-ω-benzyloxy-ω′-hydroxyketones with Pd-C cf. *68*, 138; from ω,ω′-diacoxy-ketones with KOH s. M. de Greef, S.Z. Zard, Org. Lett. *2007, 9* (9), 1773-6.

*Silver hexafluoroantimonate s. under [(Ar)*t*-Bu*$_2$*P]AuCl and RhCl(PPh*$_3$*)*$_3$ $AgSbF_6$

(Triphenylphosphine)gold(I) triflimide $(Ph_3P)AuNTf_2$

α,β-Ethyleneketones from acoxy-2-acetylenes via [3.3]-sigmatropic rearrangement ←

146. 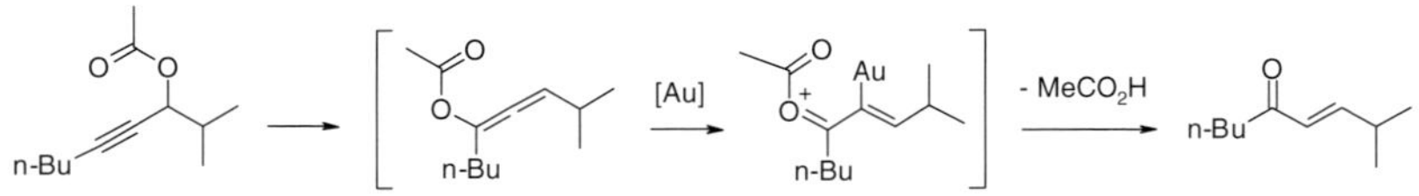

under mild conditions. 2 Mol% $(Ph_3P)AuNTf_2$ added to a soln. of startg. propargyl acetate *in 2-butanone*, the mixture stirred at room temp. for 16 h, quenched with Et_3N, concentrated, and the residue purified by chromatography → product. Y 92% (single (E)-isomer). The gold catalyst was used in place of the more reactive (triphenylphosphine)gold(I) chloride and found to be generally useful for reaction of propargyl acetates derived from both α,β-acetylene-aldehydes and -ketones to give the corresponding β-monosubst. and β,β-disubst. enones in 78-98% yields (13 examples). (E)-Isomers were formed exclusively except for a cyclohexylethynyl analog which afforded an 8:1 mixture of E/Z isomers, while the ethynyl analog, 1-phenylpropargyl acetate, gave 1-acetoxy-1-phenylpropan-2-one via gold-catalyzed Markovnikov hydration of the triple bond. Regular 2-butanone as solvent and a mixture of dry 2-butanone and water (160:1) gave similar yields, whereas 80:1 acetonitrile/water was the solvent of choice for the preparation of β,β-disubst. enones. F.e.s. M. Yu, G. Li, S. Wang, L. Zhang, Adv. Synth. Catal. *2007*, *349* (6), 871-5.

Cyclic enol carbamates from acetylene(N-carbo-*tert*-butoxy)amines ○

with $(Ph_3P)AuCl/AgSbF_6$ cf. *71*, 99; with $(Ph_3P)AuNTf_2$, 5-methylene-2-oxazolidones, s. A. Buzas, F. Gagosz, Synlett *2006* (17), 2727-30.

(Triphenylphosphine)gold(I) chloride/silver triflate $(Ph_3P)AuCl/AgOTf$

Cyclic ethers from ω-hydroxyalkyl *o*-alkynylbenzoates s. *72*, 126

*[o-Biphenylyl(di-*tert*-butyl)phosphine]gold(I) chloride/silver hexafluoroantimonate* ←

Cyclohex-4-ene-1,2-diol cyclic carbonates from 1,5-enynol 3-carbonates
Stereospecific gold(I)-catalyzed double ring closure

147. 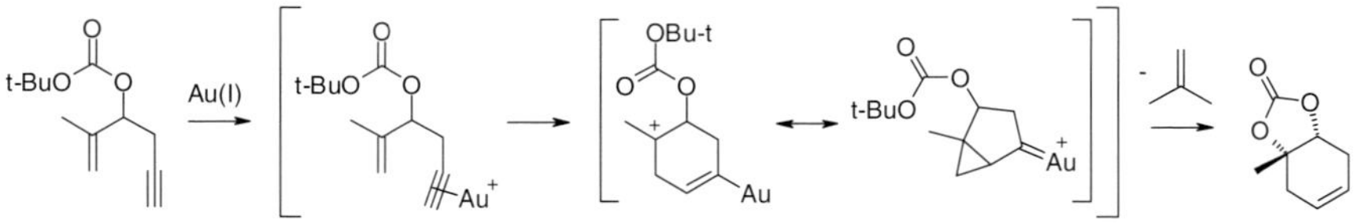

A soln. of the startg. enyne in methylene chloride added to a soln. of 2.5 mol% [*o*-biphenylyl(di-*tert*-butyl)phosphine]gold(I) chloride and 2.5 mol% $AgSbF_6$ in the same solvent at room temp., stirred for 20 min, quenched with triethylamine, and worked up with flash chromatographic purification → product. Y 78% (with 10% minor diastereoisomer). Substrates possessing a trisubst. ethylene residue yielded products with three contiguous stereocentres with high diastereoselectivity in accord with a 6-*endo-dig* transition state. Reaction is presumed to involve initial activation of the alkyne group by Au(I), followed by nucleophilic attack of the ethylenic function prior to quenching of the generated carbocation with the carbonato residue on elimination of isopropene. Cyclic substrates and those possessing an internal alkyne group also participated in the ring closure. F.e. and catalysts s. C. Lim, J.-E. Lee, S. Shin, Org. Lett. *2007, 9* (18), 3539-42.

Gold(III) chloride *$AuCl_3$*
Regiospecific ring closures of protected *versus* unprotected 4-allenealcohols ○

148.

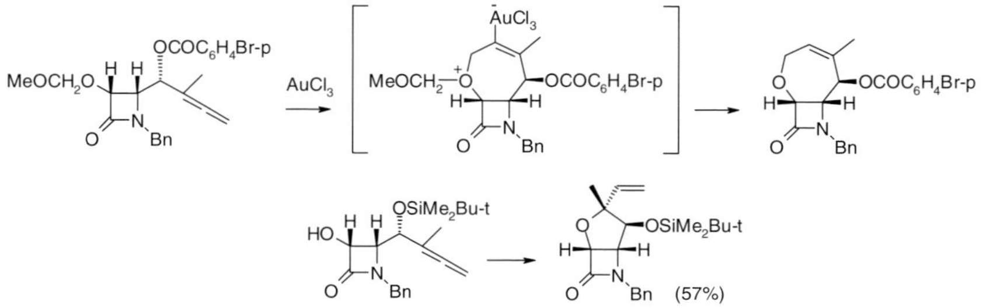

2,5,6,7-Tetrahydrooxepin ring from methoxymethoxy-4-allenes. 5 Mol% $AuCl_3$ added to a stirred soln. of the startg. protected allenol in methylene chloride under argon, stirring continued at room temp. for 72 h, and quenched with brine → product. Y 58%. Reaction involves initial activation of the allene residue with Au(III), followed by exclusive 7-*endo*-cyclization with participation of the methoxymethoxy group prior to protonolysis (possibly by traces of water in the medium). Substrates with an unprotected hydroxyl group, however, underwent 5-*exo*-cyclization to give the corresponding **2-vinyltetrahydrofuran ring**. F.e.s. B. Alcaide, P. Almendros, T. Martínez del Campo, Angew. Chem. Int. Ed. Engl. *2007, 46* (35), 6684-7.

β-*Cyclodextrin s. under* o-*Iodoxybenzoic acid* *cyclodextrin*

Polymer-based phenyl iodosoacetate *Ⓟ-ArI(OAc)$_2$*
Aldehydes from glycols – Oxidative C-cleavage CH(OH)CHOH → CHO
with $NaIO_4/SiO_2$ cf. *44*, 249s*55*; with polymer-based phenyl iodosoacetate s. F.-E. Chen, B. Xie, P. Zhang, J.-F. Zhao, H. Wang, L. Zhao, Synlett *2007* (4), 619-22.

o-*Iodoxybenzoic acid/β-cyclodextrin* ←
Oxo compds. from 1,3-oxathiolanes C
with NBS cf. *23*, 234s*63*; with *o*-iodoxybenzoic acid/β-cyclodextrin in water s. M. Somi Reddy, M. Narender, A. Mahesh, Y.V.D. Nageswar, K. Rama Rao, Synth. Commun. *2006, 36* (24), 3771-5.

p-*Iodophenoxyacetic acid/potassium peroxymonosulfate* p-*$IC_6H_4OCH_2COOH/KHSO_5$*
***p*-Quinones from *p*-quinol monoethers** ←
with $PhI(OCOCF_3)_2$ cf. *44*, 245; with *p*-iodophenoxyacetic acid/Oxone in 2:1 acetonitrile/water s. T. Yakura, T. Konishi, Synlett *2007* (5), 765-8.

1,4-Dichloro-1,4-diazoniabicyclo[2.2.2]octane dichloride ←
Oxo compds. from tetrahydropyran-2-yl ethers in water under neutral conditions CH(OThp) → CO

149.

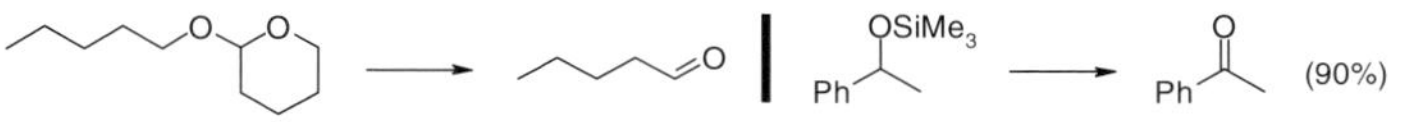

A mixture of the startg. tetrahydropyran-2-yl ether and 0.6 eqs. 1,4-dichloro-1,4-diazonia-bicyclo[2.2.2]octane dichloride in water warmed to 50°, and stirred for 20 min → pentanal. Y 90%. The procedure is inexpensive, environmentally friendly, applicable on the 60 mmol scale, and the work-up is straightforward, the reagent being simply recyclable from the isolable by-product (DABCO). Reaction is applicable to aliphatic, aromatic and allylic tetrahydropyran-2-yl ethers, and is not complicated by over-oxidation or chlorination of isolated ethylenic bonds. Yields were lower, however, with nitroaryl derivs. F.e. **and from trimethylsilyl ethers** s. M. Tajbakhsh, S. Habibzadeh, J. Chem. Res. *2007* (7), 377-80.

Phenylhydrazine hydrochloride *$PhNHNH_2{\cdot}HCl$*
Tetrahydrofurans from benzyl 4,5-epoxyethers ○
Regio- and stereo-specific debenzylative ring closure

150. 

Chiral 2,3,4-trisubst. polyoxytetrahydrofurans. *6.9 Mol%* phenylhydrazine hydrochloride added to a soln. of the startg. chiral benzyl epoxyether (readily prepared via epoxidation of the corresponding glucal) in abs. ethanol, stirred at reflux for 2 h, and worked up with column chromatographic purification → (2S,3S,4S,5R)-3,6-anhydro-4-O-benzyl-D-galactitol. Y 82%. Reaction is presumed to involve protonation of the epoxide oxygen, followed by intramolecular nucleophilic attack by the benzyloxy group prior to elimination. The hydroxyl group at the 5-position may be unprotected or protected (as the acetate or benzyl ether). F.e. and comparison of hydrazinium and amine hydrochlorides s. L.V.R. Reddy, A.D. Roy, R. Roy, A.K. Shaw, Chem. Commun. *2006* (32), 3444-6.

Tetraethylammonium superoxide *$[Et_4N]O_2$*
Cleavage of glycols under mild conditions ←

151. 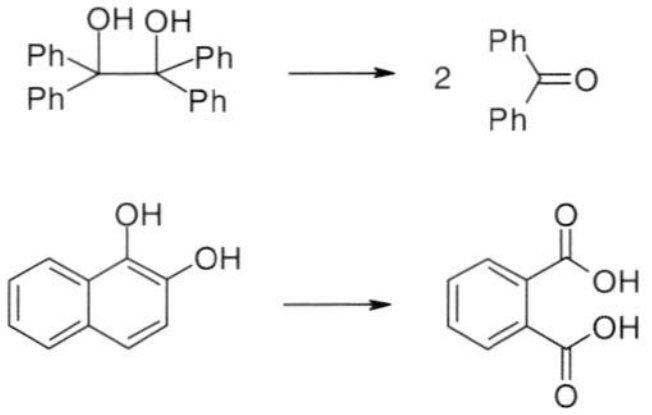

2 Eqs. tetraethylammonium bromide added to a mixture of 4 eqs. KO_2 in dry DMF under N_2, stirred for 15 min until homogeneous, the startg. diol added, stirring continued *at room temp.* under N_2 for 15-20 h (TLC monitoring), and quenched with cold brine and $NaHCO_3$ → benzophenone. Y 73%. F.e., **also *o*-dicarboxylic acids from 1,2-naphthoquinols**, s. K.N. Singh, R. Kumar, A.K. Shukla, Indian J. Chem. *2007, 46B* (8), 1347-51; oxidative cleavage of 3-hydroxyflavones s. ibid. 1554-7.

Methanesulfonyl chloride/ethyldiisopropylamine/microwaves ←
Cyclic ethers from diol monobenzyl ethers ○
chiral polyoxytetrahydrofurans with Et_2NSF_3 cf. *53*, 116; C-furanosyl glycines with $MeSO_2Cl$/*i*-Pr_2NEt under microwave irradiation, part of a thiazole-based synthesis from hexopyranoses, s. A. Dondoni, A. Massi, A. Nuzzi, Synlett *2007* (2), 303-7.

Triethylenediamine-bromine complex *$DABCO$-Br_2*
Oxo compds. from tetrahydropyran-2-yl ethers s. *34*, 249s*72* CH(OThp) → CO

Chlorotris(triphenylphosphine)rhodium(I)/silver hexafluoroantimonate *$RhCl(PPh_3)_3/AgSbF_6$*
Benzofurans from *o*-alkoxyacetylenes ○
with LiI/collidine cf. *41*, 263; with $RhCl(PPh_3)_3/AgSbF_6$, N-protected 2-amino-derivs., s. J. Oppenheimer, W.L. Johnson, M.R. Tracey, R.P. Hsung, P.-Y. Yao, R. Liu, K. Zhao, Org. Lett. *2007, 9* (12), 2361-4.

Formation of N-N Bond

Exchange ↓↑

Hydrogen ↑ NN ↓↑ H

Copper(II) sulfate/imidazole-1-sulfonyl azide hydrochloride/potassium carbonate ←
Azides from prim. amines $NH_2 \rightarrow N_3$

152.
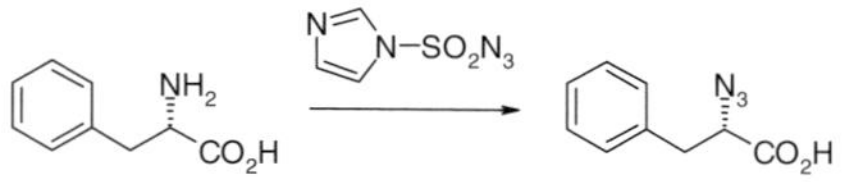

Imidazole-1-sulfonyl azide hydrochloride (1.2 eqs.) added to phenylalanine, K_2CO_3 (1.5 eqs.) and $CuSO_4 \cdot 5H_2O$ (10 mol%) in methanol, the mixture stirred at room temp. for 12 h, concentrated, diluted with water, and extracted with ethyl acetate → (S)-2-azido-3-phenylpropanoic acid. Y 75%. The diazo transfer reagent is prepared in a one-pot reaction from imidazole and $ClSO_2N_3$ (generated *in situ* from NaN_3 and SO_2Cl_2) and is shelf-stable, crystalline and inexpensive. Primary aromatic and aliphatic amines are converted by the reagent to the corresponding azides in comparable yields to those using triflyl azide. Nickel(II) and zinc(II) salts can replace copper(II).
F.e.s. E.D. Goddard-Borger, R.V. Stick, Org. Lett. *2007*, *9* (19), 3797-800.

Aroyloxylamines/potassium tert-*butoxide* *$ArCOONH_2$/KOBu*-t
N-*prim*-Amination of N-heterocyclics NH → $N\text{-}NH_2$
with O-diphenylphosphinylhydroxylamine cf. *38*, 252; with *p*-subst. O-benzoylhydroxylamines/KOBu-*t* s. L. Parlanti, R.P. Discordia, J. Hynes, Jr., M.M. Miller, H.R. O'Grady, Z. Shi, Org. Lett. *2007*, *9* (19), 3821-4.

Ammonium nitrate/sulfuric acid *NH_4NO_3/H_2SO_4*
N-Nitration NH → $N\text{-}NO_2$
with NH_4NO_3/trifluoroacetic anhydride cf. *22*, 286s*44*; selective mono-N-nitration of the sulfonamide in *prim*-aminoarenesulfonic acid amides with NH_4NO_3 and concd. H_2SO_4 at 0-5°, incl. application to the preparation of dabsyl nitroamide sodium salt, s. K. Minksztym, Synthesis *2007* (12), 1819-22.

Elimination ⇑

Hydrogen ↑ NN ⇑ H

Potassium fluoride-alumina s. under Air *$KF\text{-}Al_2O_3$*

Dimethyl sulfoxide/oxalyl chloride/triethylamine *$DMSO/(COCl)_2/Et_3N$*
Diazo compds. from hydrazones $C{=}NNH_2 \rightarrow C{=}N_2$
cf. *19*, 354s*49*; with DMSO/oxalyl chloride/Et_3N at -78° followed by removal of Et_3NHCl by vacuum filtration s. M.I. Javed, M. Brewer, Org. Lett. *2007*, *9* (9), 1789-92.

Titanium trichloride/hydrogen bromide/hydrogen peroxide *$TiCl_3/HBr/H_2O_2$*
Azo from hydrazo compds. by catalytic oxidation NHNH → N=N

153. PhNHNHPh ⟶ PhN=NPh

0.2 Mol% $TiCl_3$ (15 wt% in 25 wt% aq. HCl) and 1.05 mol% 33% HBr in acetic acid added at room temp. to a soln. of 1,2-diphenylhydrazine in methanol, 1.34 eqs. aq. H_2O_2 (33%) added dropwise (with a colour change from pale yellow to deep red), volatiles removed *in vacuo*, and the residue worked up with column chromatographic purification → 1,2-diphenyldiazene. Y

95%. The procedure is rapid, applicable to substrates possessing electron-donating or -withdrawing groups, and generally more efficient than with NH_4VO_3 in place of $TiCl_3$/HBr. A dual catalytic cycle is thought implicated, hydroxyl radical being initially generated from water with Ti(IV), and subsequently oxidizes HBr to HOBr; the latter then dehydrogenates the hydrazo compd. There was no reaction using CuI or $VO(acac)_2$. F.e.s. E. Drug, M. Gozin, J. Am. Chem. Soc. *2007, 129* (45), 13784-5.

Air/alumina or potassium fluoride-alumina — *O_2/Al_2O_3 or O_2/KF-Al_2O_3*

Azo from hydrazo compds. — NHNH → N=N

with $NaNO_2/NaHSO_4/SiO_2$ cf. *16*, 58s*65*; with air/basic alumina or air/KF-alumina in the absence of solvent, incl. N-carbamyl derivs., s. M. Mihara, T. Nakai, T. Iwai, T. Ito, T. Mizuno, Synlett *2007* (13), 2124-6.

Oxygen ↑ — NN ⇑ O

Sodium hydroxide/indium — *NaOH/In*

Reductive cinnoline ring closure ○

benzo[*c*]cinnolines from *o,o′*-dinitrodiaryls with PhCOMe in ethanolic Na-ethoxide (or the N-oxide in aq. NaOH) cf. *14*, 366s*68*; from *o*-amino-*o′*-nitrodiaryls with NaOH followed by reduction of the N-oxide with In, also pyrido[3,2-*c*]cinnolines, quinoxalino[6,7-*c*]cinnolines and [with KOBu-*t*] thieno[3,2-*c*]cinnolines, s. Å. Slevin, T. Koolmeister, M. Scobie, Chem. Commun. *2007* (24), 2506-8.

Sulfur ↑ — NN ⇑ S

Irradiation — ///

Ring closures with sulfilimines ○

2*H*-1,2,3-triazole ring cf. *67*, 127; 2-alkylthio-1,2,4-triazolo[3,2-*b*]benzothiazoles under irradiation s. M. Takahashi, S. Satoh, Heterocycles *2007, 71* (6), 1407-11.

Formation of N-S Bond

Uptake ⇓

Addition to Sulfur — NS ⇓ S

N-Bromosuccinimide or Iodine — *NBS or I_2*

N-Cyanosulfilimines from thioethers — >S → >S=NCN

with $PhI(OAc)_2$ cf. *36*, 270; with NBS or I_2 in *tert*-butyl methyl ether, also oxidation to N-cyano-sulfoximines and N-decyanation to N-unsubst. sulfoximines, s. O. García Mancheño, O. Bistri, C. Bolm, Org. Lett. *2007, 9* (19), 3809-11.

Exchange ⇓↑

Nitrogen ↑ — NS ⇓↑ N

Irradiation s. under [Ru(CO)(salalen)] — ///

Trimethylphosphine *Me_3P*
N-Sulfenylimines from oximes under mild conditions C=NOH → C=NSR

154. 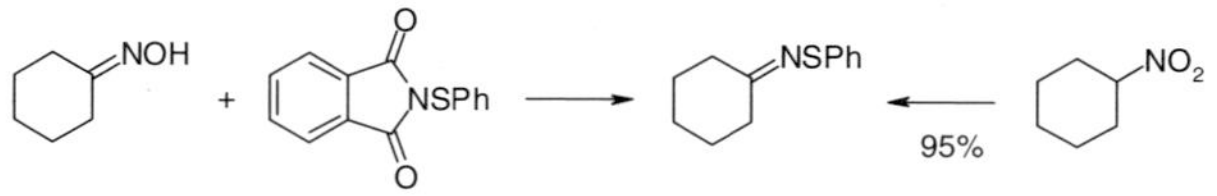

N-Sulfenylketimines. 1.1 Eqs. 1 *M* trimethylphosphine in THF added to a mixture of the startg. ketoxime and 1.2 eqs. N-(phenylsulfenyl)phthalimide at 0° under N_2, the ice-bath removed, the mixture stirred at room temp. for 30 min, water added, stirring continued for 5 min, and worked up with flash chromatographic purification on basic alumina → product. Y 96%. The same products were also obtained **from sec. nitro compds.** (with 2.2 eqs. trimethylphosphine), both reactions proceeding with retention of α-chirality. Other phosphines were ineffective or gave low yields, and there was no reaction with N-(*tert*-butylsulfenyl)phthalimide. F.e.s. J. Burés, C. Isart, J. Vilarrasa, Org. Lett. *2007*, *9* (22), 4635-8.

Chiral carbonylruthenium salalen complexes/irradiation *[Ru(CO)(salalen)]/⫻*
N-Functionalized sulfilimines from thioethers >S → >S=N-
Asym. conversion
with chiral carbonylruthenium salen complexes cf. *62*, 146s*67*; with *cis*-β-Ru(salalen)$(CO)_2$ complexes under irradiation (e.e. up to 90%) s. H. Fujita, T. Uchida, R. Irie, T. Katsuki, Chem. Lett. *2007*, *36* (9), 1092-3.

Halogen ↑ NS ↓↑ Hal

Indium *In*
N-Sulfonylation s. *1*, 271s*72* NH → NSO_2R

β-Cyclodextrin *cyclodextrin*
N-Sulfonylation NH → NSO_2R
under microwave irradiation cf. *2*, 312s*68*; of amines or amino acids with β-cyclodextrin in water at room temp. s. R. Sridhar, B. Srinivas, V. Pavan Kumar, M. Narender, K. Rama Rao, Adv. Synth. Catal. *2007*, *349* (11-12), 1873–6.

Sulfur ↑ NS ↓↑ S

Copper(I) iodide/N,N,N′,N′-tetramethylethylenediamine/air *CuI/TMEDA/air*
Sulfenamides from disulfides and amines ArSSAr → ArSN<

155. 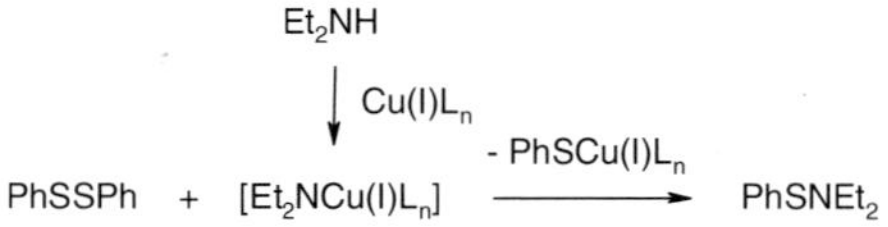

Diphenyl disulfide (0.5 mmol) and diethylamine (1.05 mmol) added to CuI (0.05 mmol), TMEDA (0.05 mmol) and DMSO (0.3 ml), and stirred at 65° for 18 h *in air* → N-phenylsulfenyldiethylamine. Y 74%. The procedure is generally applicable to the condensation of diaryl disulfides with prim. and sec. alkylamines (the former with added NH_4PF_6 in DMSO), but there was no reaction with aniline. The copper(I) salt serves either as a Lewis acid or co-oxidant with air. Both phenylthio groups of the disulfide are incorporated, one being initially liberated as a copper(I) thiophenoxide and subsequently re-oxidized to the startg. disulfide. Dibutyl disulfide, however, was relatively unreactive. F.e. and copper catalysts s. N. Taniguchi, Synlett *2007* (12), 1917-20.

Carbon ↾ NS ⇃↾ C

Without additional reagents *w.a.r.*
N-Subst. sulfonic acid amides from acids and isonitriles $SO_3H \rightarrow SO_2NHR$
Uncatalyzed procedure under neutral conditions

156.

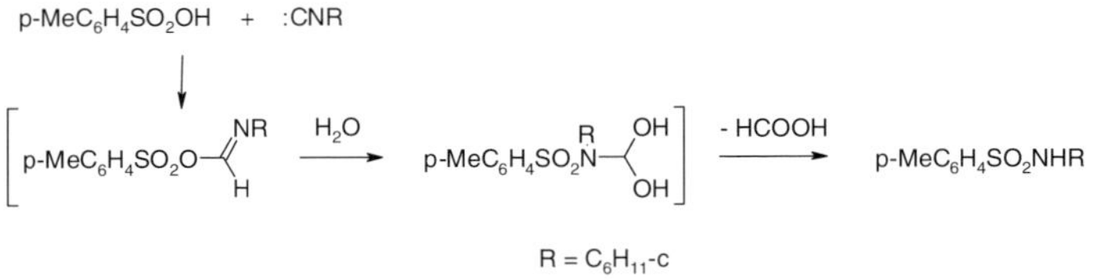

1 eq. of the startg. isonitrile added to a stirred mixture of *p*-toluenesulfonic acid and water (1 ml) in methylene chloride (10 ml), and stirred for 20 min at room temp. → product. Y 90%. Reaction is generally applicable in high yield to the coupling of alkyl or aryl isonitriles with alkane- or arene-sulfonic acids, and is thought to involve intermediate formation of an O-sulfonyl-formiminoester after the initial protonation of the isonitrile. F.e.s. A. Shaabani, E. Soleimani, A.H. Rezayan, Tetrahedron Lett. *2007*, *48* (12), 2185-8.

Elimination ⥮

Hydrogen ↾ NS ⥮ H

Chromium trioxide/silica *CrO_3/SiO_2*
Chromium trioxide/silica/microwaves *CrO_3/SiO_2/[\\\\]*
Isothiazoles from β-amino-α,β-ethylenethiooxo compds. ○
cf. *45*, 150s*68*; with CrO_3/silica in methylene chloride at room temp. or under microwave irradiation in the absence of solvent s. M. Mishra, K.K. Mahalanabis, Indian J. Chem. *2007*, *46B* (1), 204-6.

Carbon ↾ NS ⥮ C

Phenyl iodosotrifluoroacetate/trifluoroacetic acid *$PhI(OCOCF_3)_2/CF_3COOH$*
5-(Alkylthio)isothiazol-3(2*H*)-ones from carbamylketene mercaptals ○

157.

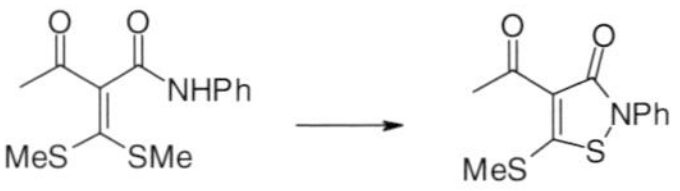

Phenyl iodosotrifluoroacetate (1.2 eqs.) and trifluoroacetic acid (3 eqs.) *in methylene chloride* added dropwise at room temp. to a stirred soln. of the startg. ketene mercaptal in the same solvent, and quenched with satd. aq. $NaHCO_3$ after 3 h → 4-acetyl-5-(methylthio)-2-phenylisothiazol-3(2*H*)-one. Y 65%. The procedure is simple, and based on an innocuous oxidant and readily available substrates. It is limited, however, to N-arylamides, proceeds with difficulty without trifluoroacetic acid, is sluggish in other solvents, and subject to steric inhibition. The activating acetyl group in the substrate also appears important. Reaction is thought to involve initial oxidation to an N-acylnitrenium ion, followed by intramolecular amidation. F.e.s. J. Huang, Y. Lu, B. Qiu, Y. Liang, N. Li, D. Dong, Synthesis *2007* (18), 2791-6.

Formation of N-Rem Bond

Exchange ↓↑

Hydrogen ↑ NRem ↓↑ H

Cyanuric chloride/potassium fluoride-Celite ←
Phosphoromonoamidates from phosphorous acid diesters and amines P-OH → P(O)N<
cf. *2*, 314s*70*; by treatment of dialkyl phosphites with cyanuric chloride in acetonitrile then reaction with amines in the presence of KF-Celite, dialkyl N-cyclooctylphosphoramidates [Ptychodiscus brevis toxin and derivs.], s. A.K. Gupta, J. Acharya, D.K. Dubey, M.P. Kaushik, J. Chem. Res. *2007* (1), 29-33

Formation of N-C Bond

Uptake ⇓

Addition to Oxygen and Nitrogen NC ⇓ ON

(2S)-[Diphenyl(trimethylsiloxy)methyl]pyrrolidine/sodium tetrahydridoborate ←
Improved organocatalyzed asym. α-hydroxylamination of aldehydes H → N(R)OH

158.
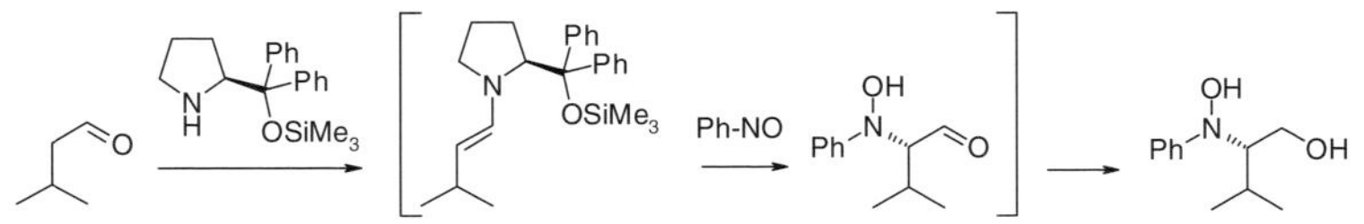

Chiral 2-hydroxylamino-*prim*-alcohols. 20 Mol% (2S)-[diphenyl(trimethylsiloxy)methyl]-pyrrolidine added to a greenish soln. of 3 eqs. 3-methylbutanal and 1 eq. nitrosobenzene in methylene chloride at 0°, the colorless soln. stirred at 0° for 30 min, $NaBH_4$ and ethanol added, the mixture stirred at 0° for 30 min, quenched with satd. NaCl, warmed to room temp, extracted with methylene chloride, the organic layer dried over $MgSO_4$, concentrated under reduced pressure, and purified by chromatography → product. Y 60% (e.e. 99%). Almost perfect regioselectivity (>99:1) for hydroxylamination over aminooxylation is achieved with this catalyst (cf. *71*, 105) with 91-9% enantioselectivity arising from approach of nitrosobenzene from the *si* face of the intermediate enamine. The reaction is rapid (5-30 min) at 0° or room temp. for ethanal to butanals, but longer-chain aldehydes required adduct formation at -20° overnight for optimum conversion. It is thought that the product hydroxylamines or water from the initial enamine formation are responsible for activation via hydrogen-bonding to the nitroso oxygen. F.e.s. C. Palomo, S. Vera, I. Velilla, A. Mielgo, E. Gómez-Bengoa, Angew. Chem. Int. Ed. Engl. *2007*, *46* (42), 8054-6.

Addition to Oxygen and Carbon NC ⇓ OC

Without additional reagents *w.a.r.*
N-β-Hydroxyalkylation of N-heterocyclics NH → N-C-C(OH)
of hydantoins with KCl cf. *28*, 267; of imidazoles or benzimidazole at 60° in the absence of solvent s. R. Torregrosa, I.M. Pastor, M. Yus, Tetrahedron *2007*, *63* (2), 469-73.

Sodium azide s.a. under Dehalogenase *NaN_3*

Sodium azide/polyethylene glycol *NaN_3/PEG*

2-Azidoalcohols from epoxides ▽O → $C(N_3)C(OH)$

in acetonitrile containing molecular sieves cf. *40*, 176s*68*; in polyethylene glycol-400 at room temp., regio- and stereo-selectivity, also ring opening of N-tosylaziridines (cf. *60*, 123s*67*), s. B. Das, V.S. Reddy, F. Tehseen, M. Krishnaiah, Synthesis *2007* (5), 666-8.

Sodium sulfite s. under $ClSO_2NCO$ *Na_2SO_3*

(S,S)-6,6′-Bis(1-hydroxy-2,2-dimethylpropyl)-2,2-bipyridyl s. under $In(OTf)_3$ ←

Zinc perchlorate *$Zn(ClO_4)_2$*

2-Aminoalcohols from epoxides in the absence of solvent ▽O → C(N<)C(OH)

regio- and stereo-specific ring opening with $Sc(OTf)_3$ cf. *49*, 254s*70*; with $Zn(ClO_4)_2 \cdot 6H_2O$ at room temp. or 80°, also comparison with other perchlorate and zinc salts, s. Shivani, B. Pujala, A.K. Chakraborti, J. Org. Chem. *2007*, *72* (10), 3713-22.

Dicyclohexylboron chloride/triethylamine *Cy_2BCl/Et_3N*

1,1-Acylaminoalcohols from aldehydes and acylamines CHO → CH(OH)NHAc

159. 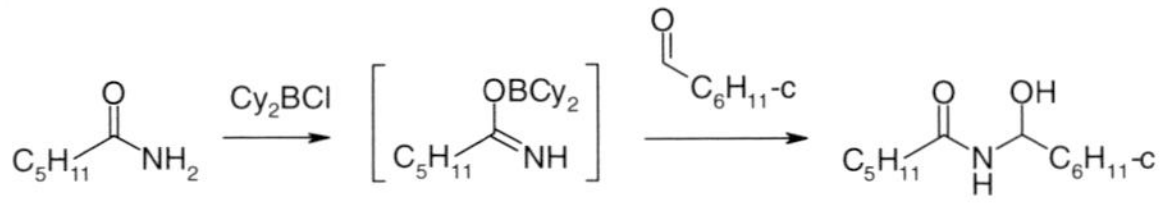

2 Eqs. triethylamine added to 1.2 eqs. 0.2 *M* hexanoic acid amide in ether, cooled to 0°, treated dropwise with 1.3 eqs. 1 *M* dicyclohexylboron chloride in hexane, the heterogeneous mixture stirred for 15 min, cyclohexanecarboxaldehyde added, and quenched after 30 min with a mixture of methanol/phosphate buffer (pH 7.4)/30% H_2O_2 → product. Y 88%. Even products without hydrogen-bonding heteroatomic substituents were found to be surprisingly stable to routine handling and storage. Reaction is presumed to involve initial formation of a boro imidate which then adds to the aldehyde. The same reaction via aluminum imidates, however, was capricious. F.e. incl. reaction with O-protected α-hydroxy-β-methoxycarboxylic acid amides s. S. Kiren, N. Shangguan, L.J. Williams, Tetrahedron Lett. *2007*, *48* (42), 7456-9.

Mesoporous aluminosilicates ←

2-Aminoalcohols from epoxides ▽O → C(N<)C(OH)

with silica at room temp. in the absence of solvent cf. *13*, 361s*68*; with mesoporous aluminosilicates s. M.W.C. Robinson, D.A. Timms, S.M. Williams, A.E. Graham, Tetrahedron Lett. *2007*, *48* (36), 6249-51.

Indium(III) triflate/(S,S)-6,6′-bis(1-hydroxy-2,2-dimethylpropyl)-2,2-bipyridyl ←

2-Aminoalcohols from epoxides with desymmetrization

cf. *68*, 58s*70*; with $In(OTf)_3$ in place of Sc(III)-dodecyl sulfate in methylene chloride (e.e. up to 98%) s. E. Mai, C. Schneider, Synlett *2007* (13), 2136-9.

Polyethylene glycol s. under NaN_3 *PEG*

Dehalogenase/sodium azide ←

2-Azidoalcohols from epoxides ▽O → $C(N_3)C(OH)$

Regiospecific ring opening with kinetic resolution

cf. *61*, 134; notably incl. chiral 2-azido-*tert*-alcohols, also β-hydroxynitriles with cyanide, s. M.M. Elenkov, H.W. Hoeffken, L. Tang, B. Hauer, D.B. Janssen, Adv. Synth. Catal. *2007*, *349* (14-15), 2279-85.

Chiral tetradentate 1,1′-bi-2-naphthol s. under $Nb(OMe)_5$ ←

Zirconium(IV) dodecyl sulfate *$Zr(OSO_2OR)_4$*

2-Aminoalcohols from epoxides in water ▽O → C(N<)C(OH)

2-(arylamino)alcohols with β-cyclodextrin cf. *59*, 134s*70*; with Zr(IV)-dodecyl sulfate [prepared from $ZrCl_4$ (5 mol%)/Na-dodecyl sulfate (10 mol%)] at pH 5, regioselectivity, s. S. Bonollo, F.

Fringuelli, F. Pizzo, L. Vaccaro, Synlett *2007* (17), 2683-6.

1-(1-Carboxyethyl)-3-methylimidazolium azide ←
2-Azidoalcohols from epoxides s. *72*, 279 ⊽o → $C(N_3)C(OH)$

Niobium pentamethoxide/chiral tetradentate 1,1′-bi-2-naphthol ←
2-(Arylamino)alcohols from *meso*-epoxides ⊂
Niobium(V)-catalyzed desymmetrization with a high degree of molecular recognition

160.

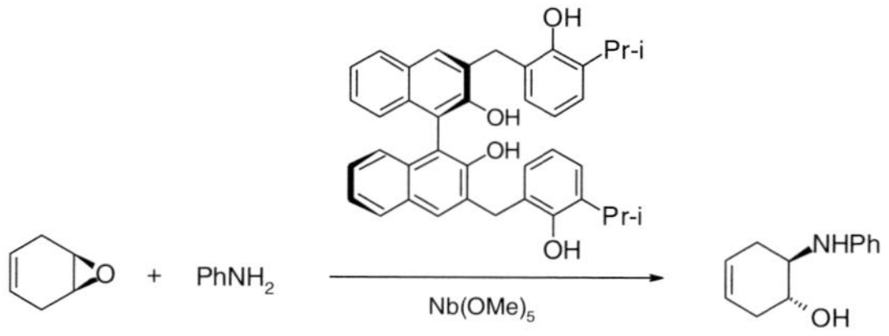

10 Mol% $Nb(OMe)_5$ added to a soln. of 11 mol% chiral 1,1′-bi-2-naphthol ligand in toluene, the mixture stirred at 60° for 3 h, added to 4 Å molecular sieves in toluene under argon, stirred for 30 min, the mixture cooled to -25°, 1.2 eqs. *cis*-cyclohexa-1,4-diene monoepoxide and 1 eq. aniline in methylene chloride added sequentially, the mixture stirred at -25° for 48 h, quenched with satd. aq. $NaHCO_3$, the aq. layer extracted with methylene chloride, the organic layer separated, washed with brine, dried over anhydrous Na_2SO_4, the solvent removed *in vacuo*, and the residue purified by preparative TLC → 6-phenylaminocyclohex-3-en-1-ol. Y 100% (e.e. 86%). The novel Lewis acidic *tetradentate* niobium catalyst showed both high enantioselectivity (82-96%) and particular substrate selectivity, affording 54-100% yields with *cis*-but-2-ene oxide and epoxides derived from cyclohexenes, cyclopentenes and 2,3-dihydrofuran, but failing to react appreciably with *higher* alkene oxides, e.g. *cis*-hex-3-ene oxide and *cis*-oct-4-ene oxide. A variety of subst. anilines (Me, CF_3, OMe, Br) were accommodated, and the catalyst maintained its activity at loadings as low as *0.25 mol%*. F.e.s. K. Arai, M.M. Salter, Y. Yamashita, S. Kobayashi, Angew. Chem. Int. Ed. Engl. *2007*, *46* (6), 955-7.

Chlorosulfonyl isocyanate/sodium carbonate/sodium sulfite $ClSO_2NCO/Na_2CO_3/Na_2SO_3$
Carbobenzoxyamines from benzyl ethers OBn → NHCOOBn
with retention of configuration

161.

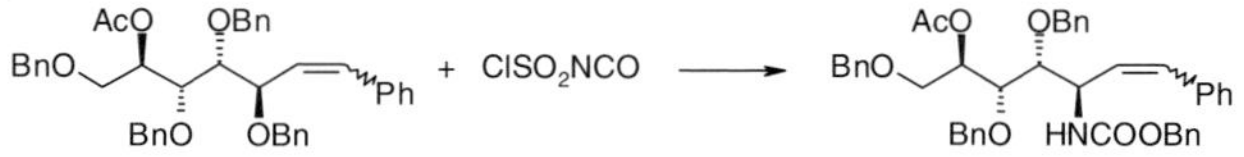

0.047 Mol Na_2CO_3 and 0.031 mol chlorosulfonyl isocyanate added at 0° under N_2 to a stirred soln. of 5.18 mmol (2R,3R,4R,5R)-7-phenyl-1,3,4,5-tetrabenzyloxyhept-6-en-2-yl acetate in dry toluene, stirred for 24 h at 0°, quenched with water, the aq. layer extracted with ethyl acetate, the organic layer added to a soln. of 25% aq. Na_2SO_3, and stirred for a further 24 h at room temp. → product. Y 76% (*cis/trans* 3.7:1; *anti/syn* 16:1). The retention of stereochemistry is the result of a double inversion of configuration. F.e.s. I.S. Kim, S.J. Kim, J.K. Lee, Q.R. Li, Y.H. Jung, Carbohydr. Res. *2007*, *342* (11), 1502-9.

Dodecyl sulfate ion s. under Zirconium(IV) dodecyl sulfate ←
Perchlorate ion s. under $Zn(ClO_4)_2$ ClO_4^-

Bis(η³-allylpalladium chloride)/(S,S)-N,N′-bis(2-diphenylphosphino-1-naphthoyl)-cyclohexane-1,2-diamine ←
N-Carbalkoxy-3-ethylene-2-aminoalcohols from 3-ethyleneepoxides ⊂
and N-carbalkoxyaroylamines – Dynamic kinetic asym. transformation s. *72*, 165

Addition to Nitrogen and Carbon NC ⇓ NC

Microwaves s. under NaN_3, CuOTf and Hexamethyldisilazane [\\\\]

Sodium azide/zinc chloride/microwaves *$NaN_3/ZnCl_2$/*[\\\\]
Sodium azide/zinc bromide *$NaN_3/ZnBr_2$*
Tetrazoles from nitriles ○
with $NaN_3/ZnBr_2$ in aq. isopropanol cf. *13*, 371s*64*; with $NaN_3/ZnCl_2$ in water under microwave irradiation s. L.V. Myznikov, J. Roh, T.V. Artamonova, A. Hrabalek, G.I. Koldobskii, Russ. J. Org. Chem. *2007*, *43* (5), 765-7; N-tetrazol-5-yl- from N-cyano-sulfoximines with $NaN_3/ZnBr_2$ in aq. methanol s. O. García Mancheño, C. Bolm, Org. Lett. *2007*, *9* (15), 2951-4.

Sodium azide/polyethylene glycol *NaN_3/PEG*
2-(Tosylamino)azides from N-tosylaziridines → C(NH-)C(N_3)
Regio- and stereo-selective conversion s. *60*, 123s*72*

Copper(I) triflate/microwaves *CuOTf/*[\\\\]
5-Acyltetrazoles from acylcyanides and azides ○
cycloaddition under high pressure cf. *43*, 249; with $(CuOTf)_2$-benzene in methylene chloride under microwave irradiation, also 5-sulfonyl derivs. (cf. *64*, 137), and use of *p*-methoxybenzyl azide as a HN_3 equivalent, s. L. Bosch, J. Vilarrasa, Angew. Chem. Int. Ed. Engl. *2007*, *46* (21), 3926-30.

Zinc chloride or bromide s. under NaN_3 *$ZnCl_2$ or $ZnBr_2$*

Dialkylaluminum azides *R_2AlN_3*
Tetrazoles from nitriles under mild conditions

162. 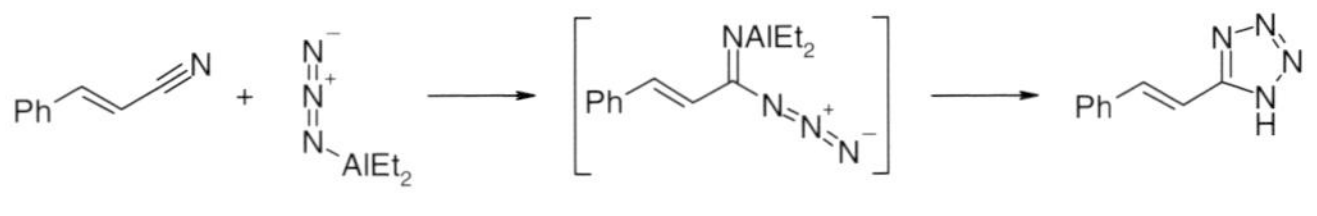

Cinnamonitrile added in two portions, initially and after 5 min, to a suspension of 1.4 eqs. Et_2AlN_3 *in toluene* under argon at room temp., the mixture heated with stirring at 50-70° for 18 h until conversion >98% complete (HPLC/TLC), the mixture cooled to 0°, quenched with aq. $NaOH/NaNO_2$ (to destroy excess of azide), the pH adjusted to 1.5 with 6 *M* HCl (to destroy hydrazoic acid), the mixture extracted with ethyl acetate, washed with water, concentrated *in vacuo*, the crude product dissolved in ethyl acetate, extracted with aq. K_2CO_3 soln., precipitated by acidification to pH 2.5 with 6 *M* HCl, the product extracted with ethyl acetate, washed with water, and evaporated *in vacuo* → 5-β-styryltetrazole. Y 97%. The procedure is mild, inexpensive, safe, rapid, clean (no by-product formation!), potentially applicable on the industrial scale, and does *not* rely on a toxic metal or dangerous hydrazoic acid (or salts thereof). Work-up is also straightforward. The reaction is successful for a range of nitriles (yields generally >60% for 30 substrates), with sterically hindered or deactivated nitriles requiring higher temperatures (up to 100°) and reaction times up to 72 h. Alkyl (linear and branched), aryl (incl. *o*-subst. derivs.) and various hetaryl substituents are accepted, and the reaction is compatible with ester groups and *tert*-butoxycarbonyl or benzyloxycarbonyl protection of amines if conducted under mild conditions. Free hydroxyl groups require prior treatment with Et_3Al with the protecting group being removed during standard work-up. It is presumed that the aluminum centre acts as a Lewis acid, activating the cyano group for the azide addition. F.e.s. V. Aureggi, G. Sedelmeier, Angew. Chem. Int. Ed. Engl. *2007*, *46* (44), 8440-4.

Scandium(III) triflate *$Sc(OTf)_3$*
1,2-Diamines from aziridines → C(NH-)C(N<)
with $BiCl_3$ cf. *50*, 151s*65*; from *meso*-*N*-phenylaziridines with 1 mol% $Sc(OTf)_3$ in methylene chloride at room temp. s. S. Peruncheralathan, M. Henze, C. Schneider, Synlett *2007* (14), 2289-91.

Ytterbium(III) triflate *Yb(OTf)$_3$*
Oxazolidines from epoxides and azomethines ○
cf. *18*, 373; with 5 mol% Yb(OTf)$_3$ at 40° in the absence of solvent s. C. Yu, X. Dai, W. Su, Synlett *2007* (4), 646-8.

[Ethylenebis(η^5-indenyl)][bis(trimethylsilyl)amido]lanthanide(III) complexes *[Ln(III)]*
Guanidines from carbodiimides RN=C=NR′→RN=C(NHR′)N<
cf. *11*, 536(s*65*); with 2 mol% [ethylenebis(η^5-indenyl)][bis(trimethylsilyl)amido]lanthanide(III) complex (Ln = Y, Sm or Yb) in toluene at room temp. or 110°, also α,β-acetyleneamidines from terminal acetylene derivs. in THF at 80° (cf. *28*, 617), s. S. Zhou, S. Wang, G. Yang, Q. Li, L. Zhang, Z. Yao, Z. Zhou, H. Song, Organometallics *2007*, *26* (15), 3755-61.

Polyethylene glycol s. under NaN$_3$ *PEG*

*Hexamethyldisilazane/*p*-toluenesulfonic acid/metal salt/microwaves* ←
Metalloporphyrazines from maleonitriles ○
cf. *34*, 285s*71*; under microwave irradiation s. M. Chandrasekharam, Ch. Srinivasa Rao, S.P. Singh, M. Lakshmi Kantam, M. Ramesh Reddy, P. Yella Reddy, T. Toru, Tetrahedron Lett. *2007*, *48* (14), 2627-30.

Trimethylsilyl azide/chiral cyclic phosphoric acid diester ←
2-Acylaminoazides from N-acylaziridines ▽N → C(NH-)C(N$_3$)
Catalytic desymmetrization
with Me$_3$SiN$_3$/Y(OPr-*i*)$_3$/(2S,3S,4R)-3-hydroxy-4-(2-hydroxy-4,5-difluorophenoxy)-2-(diphenylphosphinylmethyl)tetrahydropyran cf. *71*, 109; with Me$_3$SiN$_3$/chiral VAPOL-phosphoric acid (e.e. up to 95%) s. E.B. Rowland, G.B. Rowland, E. Rivera-Otero, J.C. Antilla, J. Am. Chem. Soc. *2007*, *129* (40), 12084-5.

1-(1-Carboxyethyl)-3-methylimidazolium azide ←
2-Tosylaminoazides from N-tosylaziridines s. *72*, 279

(S,S)-N,N′-Bis(2-diphenylphosphino-1-naphthoyl)cyclohexane-1,2-diamine ←
s. under [(η^3-allyl)PdCl]$_2$

Chiral cyclic phosphoric acid diesters (s.a. under Me$_3$SiN$_3$) ←
N-α-(Carbalkoxyamino)dicarboxylic acid imides ←
from N-carbalkoxyimines and dicarboxylic acid imides
Organocatalyzed asym. 1,2-addition

163. 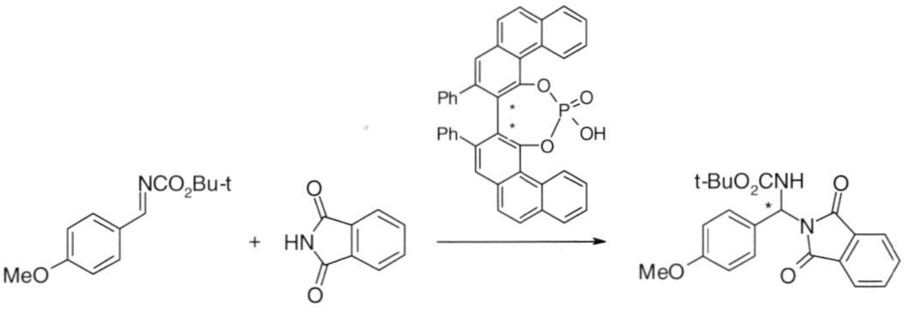

Chiral differentially N,N′-protected 1,1-diamines. Dry ether added via syringe to a mixture of the startg. N-Boc-protected imine, 0.5 eq. phthalimide and 5 mol% chiral VAPOL-phosphoric acid diester in a test-tube, and stirred at room temp. overnight → (-)-*tert*-butyl (1,3-dioxoisoindolin-2-yl)(4-methoxyphenyl)methylcarbamate. Y 92% (e.e. 96%). This is the first example of a highly enantioselective 1,2-addition of an imide to an imine, reaction being generally applicable to the asym. addition of phthalimides and maleimide to ar. aldimine derivs. possessing an electron-donating or -withdrawing group in the ring. Enantioselectivity was low, however, with carboxamides. F.e. incl. hindered naphthyl derivs., also with chiral 1,1′-binaphthyl-2,2′-diyl diesters (with lower enantioselectivity) s. Y. Liang, E.B. Rowland, G.B. Rowland, J.A. Perman,

J.C. Antilla, Chem. Commun. *2007* (43), 4477-9.

p-*Toluenesulfonic acid s. under Hexamethyldisilazane* *TsOH*

2,4-Dinitrobenzenesulfonic acid *$ArSO_3H$*
Brønsted acid-catalyzed Ritter reaction OH → NHAc

164.

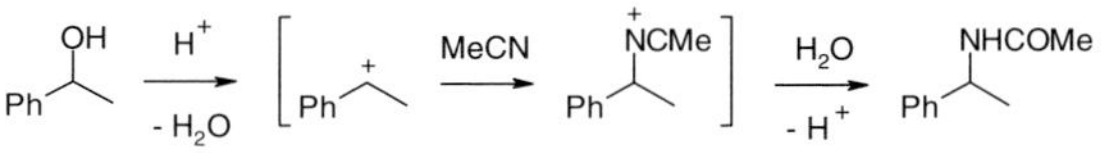

with sec. benzyl alcohols. *10 Mol%* 2,4-dinitrobenzenesulfonic acid added to a soln. of the startg. sec. benzyl alcohol in acetonitrile, stirred at reflux for 15 h, solvent removed, and the mixture worked up with column chromatographic purification → product. Y 82%. The procedure is clean, non-corrosive, environmentally friendly, metal-free, tolerant of air and moisture, adaptable on a large scale, and generally applicable to alkyl(aryl)carbinols and benzhydrols (even those possessing electron-withdrawing groups). It is also the first example of a *catalytic* Ritter reaction with a Brønsted acid, being thus a significant improvement on classical approaches with stoichiometric amounts of acid or with metal complexes. Reaction with tert. benzyl alcohols, however, gave a mixture of dimeric products. F.e.s. R. Sanz, A. Martinez, V. Guilarte, J.M. Alvarez-Gutierrez, F. Rodriguez, Eur. J. Org. Chem. *2007* (28), 4642-5.

Bis(η^3-allylpalladium chloride)/(S,S)-N,N′-bis(2-diphenylphosphino-1-naphthoyl)-cyclohexane-1,2-diamine ←
N^1-Aroyl-N^2-carbalkoxy-3-ethylene-1,2-diamines ←
from 2-vinylaziridines and N-carbalkoxyaroylamines
Dynamic kinetic asym. transformation

165.

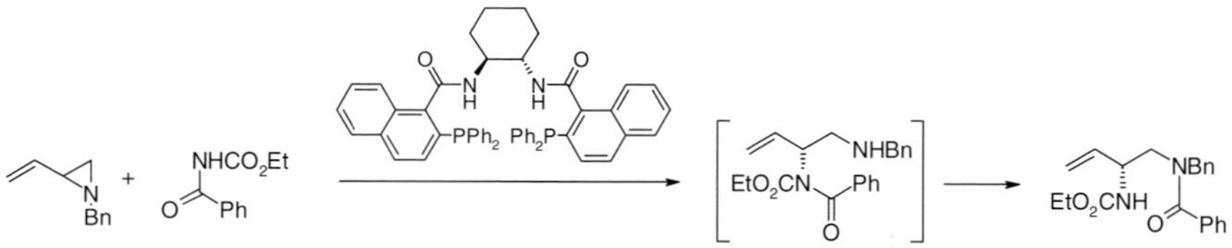

A soln. of 2 mol% bis(η^3-allylpalladium chloride) and 6 mol% (S,S)-N,N′-bis(2-diphenylphosphino-1-naphthoyl)cyclohexane-1,2-diamine in dry, degassed methylene chloride stirred under argon for 15 min at room temp., the startg. 2-vinylaziridine added via syringe, the resulting soln. transferred by cannula into a vessel containing 1.05 eqs of the startg. N-carbalkoxyaroylamine under argon, stirred at 35° for 18 h, and worked up with column chromatographic purification → product. Y 90% (e.e. 89%). Reaction involves initial nucleophilic attack of the N-carbalkoxyaroylamine at the 2-position of the 2-vinylaziridine ring, followed by **1,4-N-aroyl migration,** the dynamic kinetic transformation benefiting from preferential amination of one diastereoisomeric η^3-allylpalladium complex from a pair which rapidly interconverts through a π-σ-π mechanism. **Chiral mixed N^1,N^2-dicarbalkoxy-3-ethylene-1,2-diamines** were obtained similarly from unsym. dicarbalkoxyamines, the smaller carbalkoxy group migrating in preference to the larger. Enantioselectivity was poor, however, with aliphatic N-carbalkoxyacylamines. F.e. and **chiral N-carbalkoxy-3-ethylene-2-aminoalcohols from 3-ethyleneepoxides** s. B.M. Trost, D.R. Fandrick, T. Brodmann, D.T. Stiles, Angew. Chem. Int. Ed. Engl. *2007*, *46* (32), 6123-5.

Addition to Carbon-Carbon Bonds NC ⇓ CC

Without additional reagents *w.a.r.*

α-Carbamyloxyketones from 2-acetylenealcohols and sec. amines ←

Uncatalyzed conversion in compressed carbon dioxide without added solvent

166. 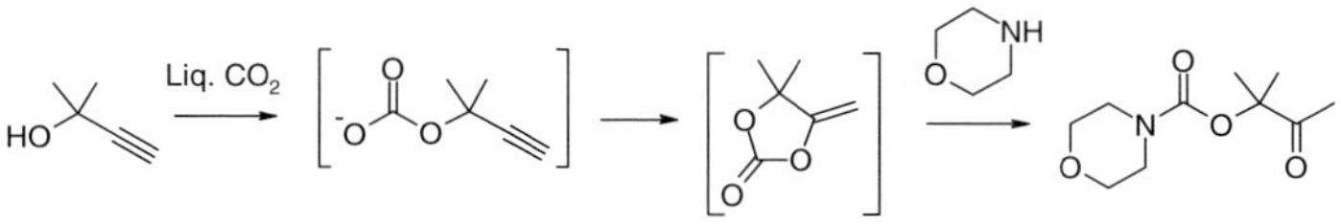

2 Eqs. morpholine and 1 eq. 2-methyl-3-butyn-2-ol pressurized (14 MPa) with CO_2 in a stainless steel autoclave, heated at 130° for 48 h, cooled to 0°, vented, the residue extracted with ether, evaporated *in vacuo*, and purified by chromatography → product. Y 78%. The formation of the illustrated carbamate is notably due to the low basicity of morpholine, but yields were low with more sterically-demanding alcohols which gave ketones as the predominant product via retro-Favorskii reaction. Reactions with diethylamine and pyrrolidine were more general in scope, affording carbamates in 35-88% yields (8 examples). The reaction conditions were optimized with respect to temp. and pressure, and the postulated mechanism, involving intermediate formation of a cyclic enol carbonate, is based on some experimental investigation. F.e.s. C.-R. Qi, H.-F. Jiang, Green Chem. *2007*, *9* (12), 1284-6.

N-Triflylaziridines from ethylene derivs. 

via stereospecific metal-free addition of N-triflylnitrene under mild conditions

167. 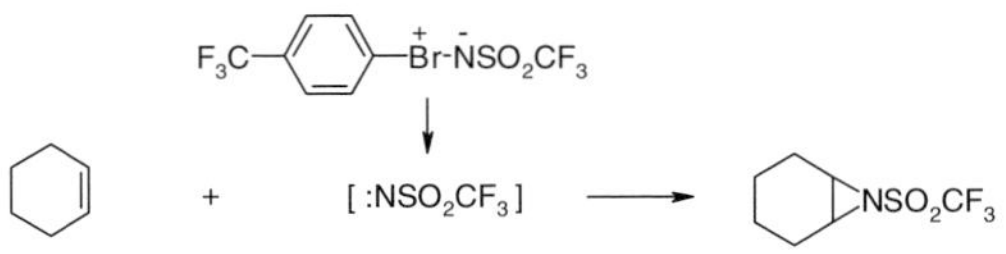

Cyclohexene (0.07 mmol) added *at room temp.* under argon to a stirred soln. of N-triflylimino[4-(trifluoromethyl)phenyl]-λ^3-bromane (0.084 mmol) in acetonitrile (readily prepared from trifluoromethanesulfonamide and difluoro[4-(trifluoromethyl)phenyl]-λ^3-bromane in acetonitrile), stirred for 5 h, concentrated, and worked up with column chromatographic purification on silica gel → product. Y 75%. Reaction is applicable to cyclic and acyclic (incl. terminal) ethylene derivs., *cis*-olefins affording *cis*-products, but *trans*-olefins reacting sluggishly. The nitrene precursor is fairly stable in the solid state, and may be stored for ca. 2 months at 4° under argon without decomposition. F.e.s. M. Ochiai, T. Kaneaki, N. Tada, K. Miyamoto, H. Chuman, M. Shiro, S. Hayashi, W. Nakanishi, J. Am. Chem. Soc. *2007*, *129* (43), 12938-9.

Microwaves s. under Polystyrenesulfonic acid and [(Cp)RuCl]$_4$* [\\\\]

Lithium amides *LiNR$_2$*

Hydroamination of ethylene derivs. C=C → CHC(N<)

of styrenes with added TMEDA cf. *6*, 367s*67*; 2-amino- from 2-ethylene-alcohols (cinnamyl alcohols) s. C.S. Barry, N.S. Simpkins, Tetrahedron Lett. *2007*, *48* (46), 8192-5; 2-ethyleneamines from allenes using an *aliphatic* amine (morpholine) in the presence of $(Ar_3P)AuCl/AgOTs$ s. N. Nishina, Y. Yamamoto, Synlett *2007* (11), 1767-70.

Sodium azide s. under $CuSO_4$ or CuI *NaN_3*

(2S)-[Diphenyl(trimethylsiloxy)methyl]pyrrolidine ←

N-Protected 5-hydroxyisoxazolidines from α,β-ethylenealdehydes ○
via organocatalyzed asym. Michael addition of N-protected hydroxylamines

168. 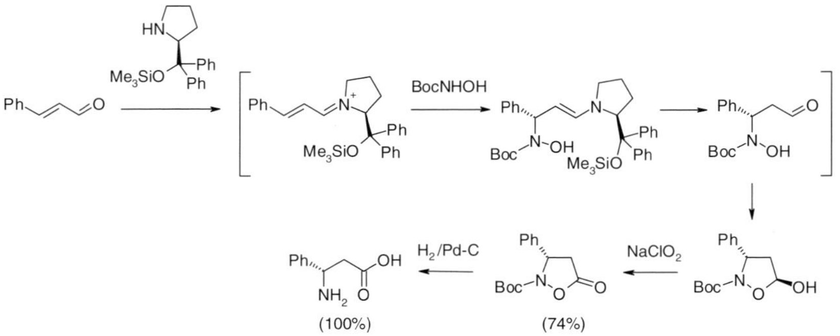

3-Phenylpropenal and 1.2 eqs. N-Boc-hydroxylamine added to 20 mol% (2S)-[diphenyl-(trimethylsiloxy)methyl]pyrrolidine in chloroform at 4°, the mixture vigorously stirred for 3 h, worked up, and purified directly by chromatography → (3S,5S)-2-*tert*-butoxycarbonyl-3-phenyl-5-hydroxyisoxazolidine.Y 80% (e.e. 99%). The method is highly chemo- and enantio-selective, affording products in 75-94% yield (e.e. 90-99%) from N-Boc- or -Cbz-protected hydroxylamines and 3-aryl- or 3-alkyl-subst. acroleins. The reaction is scalable and compatible with nitro, cyano, ester and halo functionality. *in situ*-Oxidation of the products ($NaClO_2$) affords the corresponding **chiral N-protected 5-isoxazolidones** (3 examples; Y 74-81%), which undergo *quantitative* hydrogenolysis (H_2-Pd/C) to afford **chiral β-aminocarboxylic acids**. The bulky catalyst promotes selective *re*-face attack of the hydroxylamine nitrogen on the β-carbon of the intermediate enamine, and the released β-(hydroxylamino)aldehyde undergoes favoured hemiacetal formation to give the *trans*-product. F.e. and comparison of chiral amines, also conversion to **chiral 3-amino-*prim*-alcohols,** s. I. Ibrahem, R. Rios, J. Vesely, G.-L. Zhao, A. Cordova, Chem. Commun. *2007* (8), 849-51.

(2S)-[Bis[3,5-bis(trifluoromethyl)phenyl](trimethylsiloxy)methyl]pyrrolidine ←
s. under Pyrrolidines/benzoic acid

Planar-chiral ferrocene-condensed 4-methylpyridine/hydrazoic acid ←

Urethans from ketenes C=C=O → CHNHCOOR
via asym. addition of hydrazoic acid and Curtius rearrangement

169. 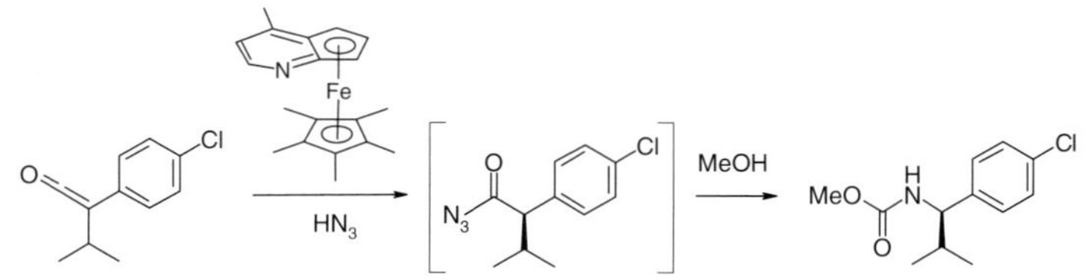

A soln. of 4-chlorophenyl(isopropyl)ketene in toluene/hexanes prepared under argon in a glove box, the reaction vessel capped with a septum, the mixture cooled to -78°, treated with a soln. of 10 mol% planar-chiral ferrocene-condensed *4-methylpyridine* in toluene, 1.1 eqs. of hydrazoic acid in toluene/methylene chloride (prepared in a glove box) added over 1 h via pump to the purple soln., stirred an additional 5 h at -78°, warmed to room temp. over ca. 30 min, benzene added, the mixture heated at 90° for 1 h, methanol added, the reaction mixture heated at 75° for 12 h, cooled to room temp., the soln. concentrated *in vacuo*, and the residue purified by chromatography → methyl N-[1-(4-chlorophenyl)-2-methylpropyl]carbamate. Y 90% (e.e. 92%). The reaction is thought to occur by initial protonation of the chiral base, followed by Brønsted acid-catalyzed addition of hydrazoic acid to the ketene and subsequent Curtius rearrangement of the intermediate acid azide. Significantly, the conversion was less effective with the corresponding (and more familiar) 4-pyrrolidinopyridine catalyst due to its higher basicity. Twelve examples with aryl(alkyl)ketenes gave 89-94% yields with enantiomeric excesses generally >90%, the highest selectivity being achieved with ketenes possessing electron-rich aryl substituents and sec. alkyl groups. An example with hindered *tert*-butyl gave 76% e.e. while phenyl(ethyl)ketene gave only 4% e.e. due to rapid uncatalyzed addition of hydrazoic acid. The presence of an *o*-substituent, as in 2-methoxyphenyl(methyl)ketene, slowed the uncatalyzed reaction affording a 70% e.e. F.e. incl. preparation of ketenes and the chiral ferrocene s. X. Dai, T. Nakai, J.A.C. Romero, G.C. Fu, Angew. Chem. Int. Ed. Engl. *2007*, *46* (23), 4367-9.

Copper(II) trifluoroacetate/hexamethylphosphoramide *$Cu(OCOCF_3)_2$/HMPA*
3-Sulfonyloxazolidines from ethylene derivs.

170.
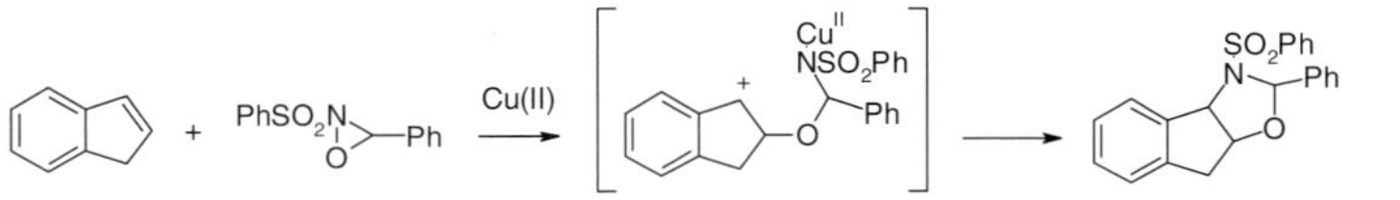

Regio- and stereo-specific conversion. A soln. of indene and 10 mol% HMPA in freshly distilled methylene chloride added via syringe to a soln. of 2 mol% $Cu(OCOCF_3)_2$-hydrate and *trans*-2-(phenylsulfonyl)-3-phenyloxaziridine in the same solvent, the reaction vial sealed with a septum, stirred under N_2 until reaction complete (1.5 h), and worked up with column chromatographic purification → product. Y 81% (*cis/trans* 3.5:1). Reaction is applicable to the oxyamination of mono-, di-, tri- and tetra-subst. styrenes (possessing electron-withdrawing or -donating groups) as well as enolethers and enesilanes, in keeping with a mechanism involving Lewis acid-catalyzed activation of the oxaziridine ring and initial generation of a carbocation. The corresponding **2-aminoalcohols** were obtained by subsequent ring opening with trifluoroacetic acid in dioxane/water or perchloric acid in DME/water, followed by desulfonylation with Red-Al in toluene. F.e.s. D.J. Michaelis, C.J. Shaffer, T.P. Yoon, J. Am. Chem. Soc. *2007*, *129* (7), 1866-7.

Copper(II) oxide-copper(I) oxide-carbon *$CuO/Cu_2O/C$*
Copper(II) sulfate/sodium ascorbate ←
1,2,3-Triazoles from terminal acetylene derivs. and azides
with $CuSO_4$/Na-ascorbate cf. *64*, 141; 4-[2,2-bis(phosphonyl)ethyl]-1,2,3-triazoles s. H. Skarpos, S.N. Osipov, D.V. Vorob′eva, I.L. Odinets, E. Lork, G.-V. Röschenthaler, Org. Biomol. Chem. *2007*, *5* (15), 2361-7; with a $CuO-Cu_2O$-carbon catalyst in dioxane under air s. B.H. Lipshutz, B.R. Taft, Angew. Chem. Int. Ed. Engl. *2006*, *45* (48), 8235-8.

Copper(II) sulfate/sodium azide/sodium ascorbate ←
Copper(I) iodide/sodium azide *CuI/NaN_3*
1-β-Hydroxy-1,2,3-triazoles from acetylene derivs. and epoxides
One-pot regiospecific conversion via 2-azidoalcohols

171.
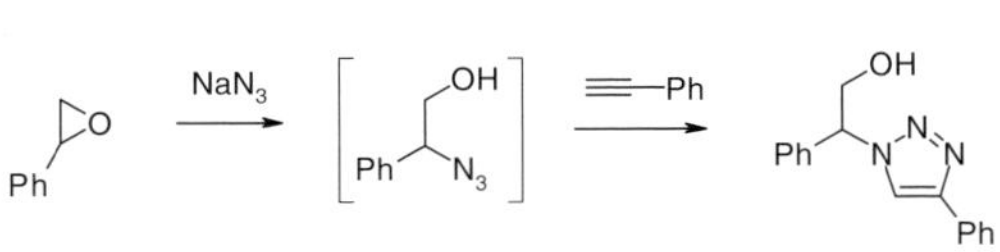

in water. 10 Mol% $CuSO_4 \cdot 5H_2O$ and 20 mol% Na-ascorbate added to a suspension of styrene oxide and 1.2 eqs. NaN_3 in water, stirred for 2 h at room temp., 1 eq. phenylacetylene added, stirring continued for a further 2 h, and worked up with column chromatographic purification →

product. Y 92%. Reaction is mild, simple, and highly regiospecific, the initial ring opening of the epoxide with azide taking place at the benzylic site (with styrene oxides) and at the terminal position with 2-alkyloxiranes. With cyclohexene oxide, the expected *trans*-products were obtained. Interestingly, yields were significantly lower in organic solvents, and the reaction time was longer. F.e. with terminal acetylene derivs. (incl. ethynyltrimethylsilane) s. J.S. Yadav, B.V. Subba Reddy, G.M. Reddy, D.N. Chary, Tetrahedron Lett. *2007*, *48* (49), 8773-6; with 5 mol% CuI *in polyethylene glycol-400*, **also 1-β-tosylamino-1,2,3-triazoles** from N-tosylaziridines, incl. chiral compds., s. G. Kumaraswamy, K. Ankamma, A. Pitchaiah, J. Org. Chem. *2007*, *72* (25), 9822-5.

Copper(I) iodide/2,6-lutidine ←
1-Sulfonyl-1,2,3-triazoles from terminal acetylene derivs. and sulfonic acid azides ○

172. 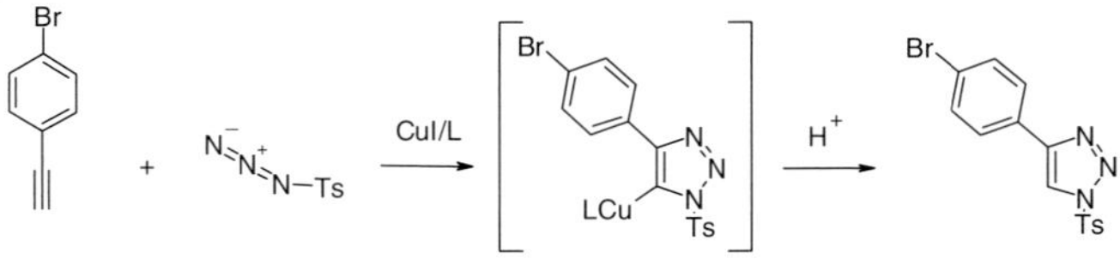

via 5-cupro-1-sulfonyl-1,2,3-triazoles. 1.2 Eqs. 2,6-lutidine added slowly to a stirred mixture of 1.2 eqs. 4-bromophenylacetylene, *p*-toluenesulfonyl azide, and 10 mol% CuI *in chloroform at 0°*, the mixture stirred at room temp. for 12 h, diluted with methylene chloride and aq. NH_4Cl, stirred for an additional 30 min at room temp., the aq. layer extracted with methylene chloride, dried over $MgSO_4$, filtered, concentrated *in vacuo*, and the crude residue purified by flash column chromatography → 1-(4-toluenesulfonyl)-4-(4-bromophenyl)-1,2,3-triazole. Y 95%. The procedure is generally applicable to reaction of terminal acetylene derivs. bearing aryl groups (irrespective of electronic variation), thiazol-3-yl, cyclohex-1-enyl, carbomethoxyalkyl, chloroalkyl, Boc-aminomethyl or 2-hydroxyprop-2-yl groups with arene- or alkane-sulfonyl azides, incl. silylalkyl derivs. Yields were in the range 56-95% (13 examples) and the reaction may be accomplished on a gram scale without problems. Under standard aq. conditions (Na-ascorbate, aq. *t*-BuOH) only a low yield of the triazole is obtained with the [presumed] intermediate cuprotriazole undergoing predominant loss of nitrogen on ring opening. F.e.s. E.J. Yoo, M. Ahlquist, S.H. Kim, I. Bae, V.V. Fokin, K.B. Sharpless, S. Chang, Angew. Chem. Int. Ed. Engl. *2007*, *46* (10), 1730-3.

*(Tri-*n*-butylphosphine)copper(I) chloride* *(Bu_3P)CuCl*
1,2,5-Thiadiazolidine 1,1-dioxides from activated terminal ethylene derivs. ○

173. 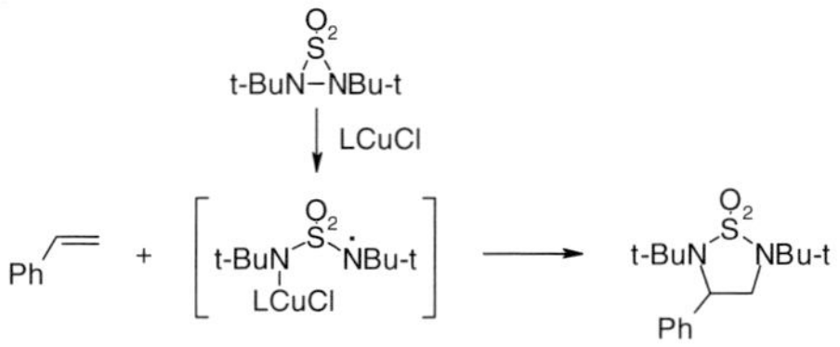

under mild conditions. Deuteriochloroform and 0.08 mmol tri-*n*-butylphosphine added under argon to 0.08 mmol CuCl, stirred at room temp. for 10 min, 0.4 mmol styrene and 0.6 mmol di-*tert*-butylthiadiaziridine 1,1-dioxide added, stirred at 50° for 24 h, and worked up with flash chromatographic purification on silica gel → product. Y 65%. Reaction is generally applicable to styrenes (especially those possessing electron-donating groups), enynes and enolethers, but there was no diamination of 1-octene or ethyl acrylate. An intermediate aminyl radical is invoked. The corresponding **1,2-diamines** were obtained by subsequent treatment with $BaCO_3$/concd. HCl. F.e.s. B. Zhao, W. Yuan, H. Du, Y. Shig, Org. Lett. *2007*, *9* (24), 4943-5.

Silver salts s. under Tungstophosphoric acid *Ag^+*

(Triphenylphosphine)gold(I) triflate *$(Ph_3P)AuOTf$*
3-β-Functionalized furans from 1-(alk-1-ynyl)cyclopropyl ketones s. *72*, 70

(Triarylphosphine)gold(I) chloride/silver tosylate *$(Ar_3P)AuCl/AgOTs$*
2-Ethyleneamines from allenes s. *6*, 367s*72* C=C=C → C(N<)CH=C

Cadmium chloride *$CdCl_2$*
Cadmium-catalyzed Michael addition of amines C=C → CHC(N<)

174. 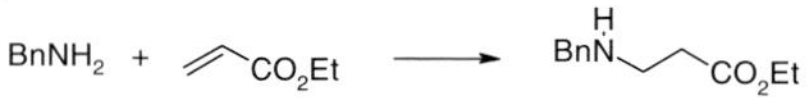

0.4 Mol% $CdCl_2$ added to a soln. of benzylamine and 1.1 eqs. ethyl acrylate in methylene chloride, stirred at room temp. for 3 h, and worked up with column chromatographic purification → product. Y 93%. Reaction is inexpensive and rapid with a moisture-resistant Lewis acid for the Michael addition of prim. and sec. alkylamines to nitroolefins and α,β-ethylenic esters or nitriles (aniline reacting more slowly). F.e.s. M. Vijender, P. Kishore, B. Satyanarayana, Synth. Commun. *2007*, *37* (4), 589-92.

(2S)-[Bis[3,5-bis(trifluoromethyl)phenyl](trimethylsiloxy)methyl]pyrrolidine/benzoic acid ←
Regiospecific organocatalyzed asym. Michael addition of N-heteroarenes to α,β-ethylenealdehydes ←

175. 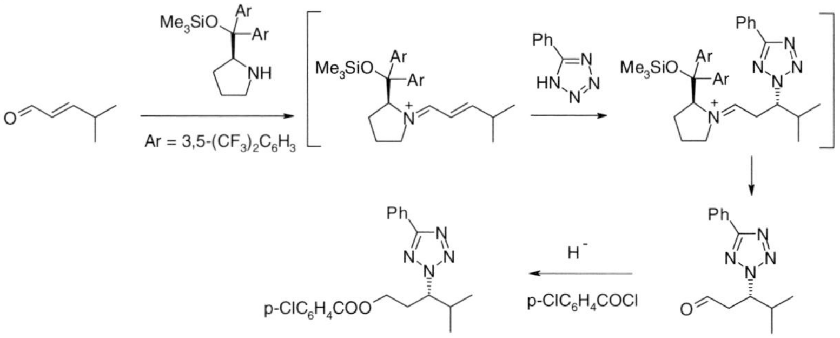

1.5 Eqs. 5-phenyltetrazole, 10 mol% (2S)-[bis[3,5-bis(trifluoromethyl)phenyl](trimethylsiloxy)-methyl]pyrrolidine and 10 mol% benzoic acid stirred in toluene *at -20°* for 30 min, 1 eq. 4-methyl-2-pentenal added, the mixture stirred for 51 h (^1H NMR monitoring), and the adduct isolated after reduction to the prim. alcohol (with 1.5 eqs. $NaBH_4$) and acylation with 2 eqs. 4-chlorobenzoyl chloride → product. Y 66% (e.e. 91%). The reaction is successful with a number of β-alkyl-α,β-ethylenealdehydes affording adducts in 45-87% yields and high enantioselectivities. Substrates with bulky alkyl groups reacted more slowly than those with linear alkyl groups, but gave slightly higher enantioselectivities, while reaction with the β-phenyl analog (cinnamaldehyde) gave only a 20% yield. 1,2,4-Triazole also reacted regioselectively (at the N^1 position) whereas benzotriazole and 1,2,3-triazole gave mixtures of regioisomers via reaction at both possible ring nitrogens. Addition takes place below the unshielded *re* face of the intermediate azomethinium ion. F.e.s. P. Dinér, M. Nielsen, M. Marigo, K.A. Jørgensen, Angew. Chem. Int. Ed. Engl. *2007*, *46* (12), 1983-7.

Acylase ←
Enzymatic Michael addition of N-nucleophiles C=C → C(N<)CH
with lipase B cf. *67*, 144s*69*; of aromatic N-heterocyclics with a promiscuous zinc-active-site acylase in DMSO at 50° s. C. Qian, J.-M. Xu, Q. Wu, D.-S. Lv, X.F. Lin, Tetrahedron Lett. *2007*, *48* (35), 6100-4.

Amberlyst 15 ←
Michael addition with amines C=C → CHC(N<)
heterogeneous conversion with SiO_2 in the absence of solvent cf. *63*, 147s*68*; to α,β-ethylenesulfones with Amberlyst 15 at room temp. s. A.P. Esteves, M.E. Silva, L.M. Rodrigues, A.M.F. Oliveira-Campos, R. Hrdina, Tetrahedron Lett. *2007*, *48* (51), 9040-43.

Sodium ascorbate s. under $CuSO_4$ ←

Chiral hydroxythioureas ←
Organocatalyzed asym. Michael addition of alkoxylamines

176. 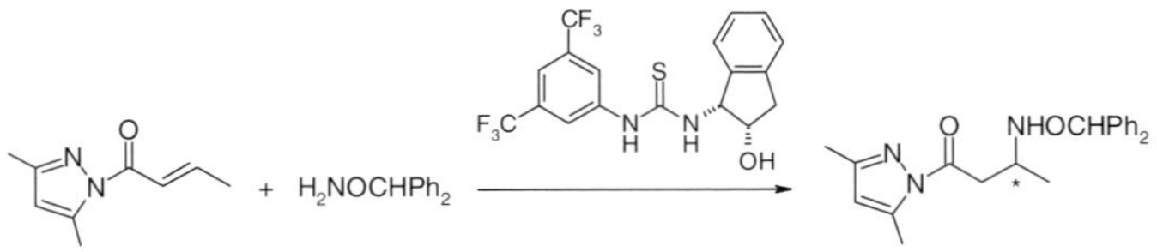

to 1-(α,β-ethyleneacyl)pyrazoles. 1.2 Eqs. of the startg. alkoxylamine in *dry α,α,α-trifluorotoluene* added dropwise at 0° under N_2 to a soln. of the startg. Michael acceptor, 1 eq. of the chiral hydroxythiourea and 4 Å molecular sieves in the same solvent, stirred for several days, poured into satd. aq. NH_4Cl, and worked up with flash chromatographic purification → 3,5-dimethyl-1-[3-(diphenylmethoxylamino)butanoyl]pyrazole. Y 86% (e.e. 89%). The high enantioselectivity (up to 98%) is the consequence of hydrogen-bonding between the *bifunctional* hydroxythiourea and the Michael acceptor (through N^2 and the carbonyl oxygen). The *cis*-configuration of the two substituents on the indan ring, however, appears critical to high asym. induction. Poor results were obtained with N-cinnamoyl derivs., and reaction was much longer with catalytic amounts of the organocatalyst. F.e. and with ring-subst. pyrazolyl derivs. s. M.P. Sibi, K. Itoh, J. Am. Chem. Soc. *2007*, *129* (26), 8064-5.

Bromodimethylsulfonium bromide *$BrMe_2S^+Br^-$*
Michael addition with amines C=C → CHC(N<)
in the absence of solvent with CAN as catalyst cf. *71*, 112; with bromodimethylsulfonium bromide s. A.T. Khan, T. Parvin, S. Gazi, L.H. Choudhury, Tetrahedron Lett. *2007*, *48* (22), 3805-8.

Hafnium(IV) chloride *$HfCl_4$*
Michael addition with N-heterocyclics
with phthalimide and NaOMe cf. *47*, 248; of pyrazole and imidazole to enones in the presence of a little $HfCl_4$ s. S. Aburatani, M. Kawatsura, J. Uenishi, Heterocycles *2007*, *71* (1), 189-96.

Hydrazoic acid s. under Planar-chiral ferrocene-condensed 4-methylpyridine *HN_3*

1,3-Bis(diphenylphosphino)propane *dppp*
Satd. 5-membered N-sulfonyl-N-heterocyclics from electron-deficient acetylene derivs. via di(phosphine)-catalyzed stereospecific double Michael addition of 2-functionalized sulfonylamines ○

177. 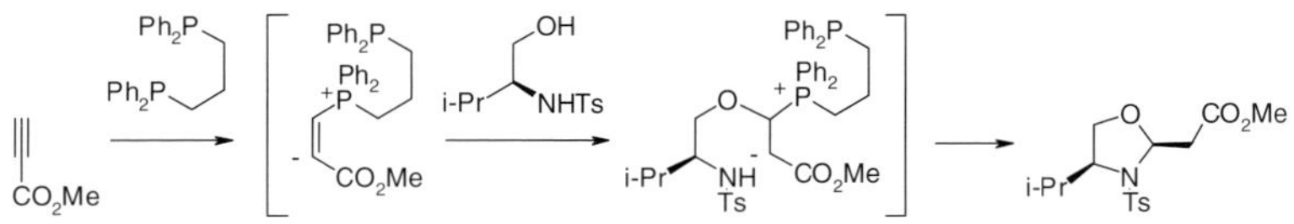

3-Sulfonyloxazolidines. 10 Mol% 1,3-bis(diphenylphosphino)propane and acetonitrile added sequentially under argon to the startg. N-tosylamino pronucleophile, heated in an oil bath at 80°, 1.1 eqs. of the acetylene deriv. in acetonitrile added over 2 h via syringe pump, stirred under argon at 80° for 9 h, concentrated, and worked up with flash column chromatographic purification on silica gel → product. Y 92% (*cis/trans* 96:4). **3-Sulfonylthiazolidines and 1-sulfonylpyrrolidines** were obtained similarly from the appropriate functionalized sulfonylamines. Reaction is triggered by 1,4-addition of the di(phosphine) to the electron-deficient alkyne, coordination of the free phosphino group [to the phosphonium atom] serving as a stabilizing handle [yields being low with monophosphines and other di(phosphines)]. F.e. and with α,β-acetylene-ketones and -sulfones s. V. Sriramurthy, G.A. Barcan, O. Kwon, J. Am. Chem. Soc. *2007*, *129* (43), 12928-9.

Diphenylphosphinyloxylamine/N-methylmorpholine/sodium hydroxide ←
Aziridines from ethylene derivs. 
with sulfonyloxylamines cf. *36*, 301; with 1.05 eqs. each of N-methylmorpholine and diphenylphosphinyloxylamine and 2 eqs. NaOH, **also asym. conversion** in the presence of (-)-quinine, s. A. Armstrong, C.A. Baxter, S.G. Lamont, A.R. Pape, R. Wincewicz, Org. Lett. *2007*, *9* (2), 351-3.

Chiral phosphoromonoamidites s. under $Pd_2(dba)_3$ ←
Hexamethylphosphoramide s. under $Cu(OCOCF_3)_2$ *HMPA*

Polystyrenesulfonic acid/microwaves ←
Michael addition with amines C=C → CHC(N<)
double Michael addition with a little $Bi(OTf)_3$ cf. *65*, 144; with polystyrenesulfonic acid in water under microwave irradiation s. V. Polshettiwar, R.S. Varma, Tetrahedron Lett. *2007*, *48* (49), 8735-8.

Tungstophosphoric acid ←
Hydroamination of styrenes
with NBS cf. *69*, 68; with tungstophosphoric acid at 80° in the absence of solvent s. N. Seshu Babu, K. Mohan Reddy, P.S. Sai Prasad, I. Suryanarayana, N. Lingaiah, Tetrahedron Lett. *2007*, *48* (43), 7642-5; addition of non-nucleophilic compds. such as *p*-toluenesulfonamide, *p*-nitroaniline or *p*-nitrobenzamide in the presence of $FeCl_3$ s. J. Michaux, V. Terrasson, S. Marque, J. Wehbe, D. Prim, J.-M. Campagne, Eur. J. Org. Chem. *2007* (16), 2601-3.

Silver-exchanged tungstophosphoric acid ←
Ketimines from acetylene derivs. C≡C → C(=NR)CH_2
Heterogeneous silver-catalyzed hydroamination in the absence of solvent

178. 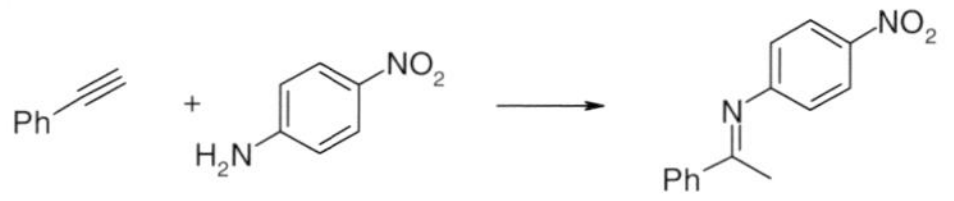

A mixture of phenylacetylene, 1.2 eqs. 4-nitroaniline and a little silver-exchanged tungstophosphoric acid stirred at 70° under N_2 for 2 h (TLC monitoring), the mixture cooled, hexane added, the solid filtered off, and the filtrate worked up → 4-nitro-N-(α-methylbenzylidene)aniline. Y 97%. The method is based on an inexpensive, environmentally friendly and recyclable [up to four times] catalyst, and is highly regiospecific and high-yielding (79-99%) for the Markovnikov addition of ar. or aliphatic amines to terminal acetylenes, accommodating electron-rich or -poor ar. amines and bulky substituents. Internal acetylenes also reacted, but yields were somewhat reduced (2 examples; 52 and 56%). F.e. incl. preparation of the catalyst by treatment of 12-tungstophosphoric acid with $AgNO_3$ s. N. Lingaiah, N.S. Babu, K.M. Reddy, P.S.S. Sai, I. Suryanarayana, Chem. Commun. *2007* (3), 278-9.

Rhenium carbonyl *$Re_2(CO)_{10}$*
N-Vinylation of amides NH → NC=CH
with terminal acetylene derivs.
under Ru-catalysis cf. *69*, 149; with $Re_2(CO)_{10}$ (5 mol%) in refluxing toluene s. S. Yudha S., Y. Kuninobu, K. Takai, Org. Lett. *2007*, *9* (26), 5609-11.

Ferrocenes s. under Planar-chiral ferrocene-condensed 4-methylpyridine ←

Iron(III) chloride *$FeCl_3$*
Hydroamination of styrenes s. *69*, 68s*72* C=C → CHC(N<)

Chloro(η^5-pentamethylcyclopentadienyl)ruthenium(II) tetramer/microwaves [(Cp)RuCl]$_4$/*∿
1,2,3-Triazoles from acetylene derivs. and azides ○
cf. *8*, 404; 1-aryl-derivs. with $[(Cp^*)RuCl]_4$ in DMF under microwave irradiation s. L.K. Rasmussen, B.C. Boren, V.V. Fokin, Org. Lett. *2007*, *9* (26), 5337-9.

Tris(dibenzylideneacetone)dipalladium/chiral phosphoromonoamidites ←
Tetrakis(triphenylphosphine)palladium(0) *$Pd(PPh_3)_4$*
4-Vinyl-2-imidazolidones from 1,3-dienes and diaziridinones ◌

179. 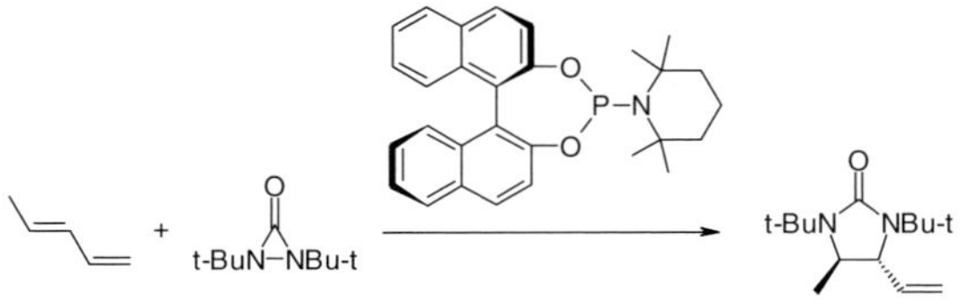

Regiospecific asym. conversion. A mixture of 5 mol% $Pd_2(dba)_3$ and 22 mol% chiral phosphoromonoamidite in benzene-d_6 (contained in an NMR tube under argon) heated with stirring at 65° (in an oil bath) for ca. 10 min, *trans*-1,3-pentadiene and 1.25 eqs. di-*tert*-butyldiaziridinone added sequentially to the homogeneous soln., heating continued at 65° for 1.5 h, and worked up with flash column chromatographic purification on silica gel → product. Y 91% (e.e. 91%). Reaction is highly diastereo- and enantio-selective with regiospecific diamination of *internal (E)-configured* double bonds. Enantioselectivity was low with chiral phosphines, di(phosphines) and bis(phosphinites). F.e. and comparison of chiral phosphoromonoamidites s. H. Du, W. Yuan, B. Zhao, Y. Shi, J. Am. Chem. Soc. *2007*, *129* (38), 11688-9; stereospecific conversion with $Pd(PPh_3)_4$ s. H. Du, B. Zhao, Y. Shi, ibid. *129* (4), 762-3.

Rearrangement ∩

Hydrogen/Oxygen Type NC ∩ HO

Dimethylsulfoxonium methylid *$Me_2S(O)=CH_2$*
2-Alkylidene-3,4-oxido-1-tosylpyrrolidines ←
from 1-(1-tosylaziridin-2-yl)-2-acetylenealcohols
via stereospecific aza-Payne rearrangement-intramolecular hydroamination

180. 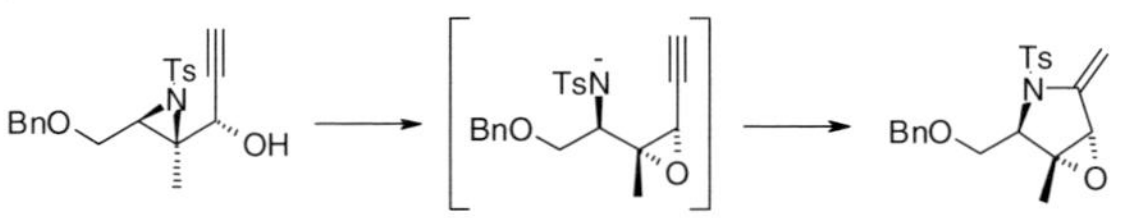

under mild conditions. 4 eqs. Trimethylsulfoxonium iodide added in portions to 4 eqs. NaH (60% dispersion in mineral oil) in dry DMSO, the milky-white suspension stirred for a further 30 min, 1 eq. of the startg. aziridine in the same solvent added, stirred overnight at room temp., quenched with water, and worked up with column chromatographic purification → product. Y 76%. The alkyne group may be terminal or substituted by phenyl, but yields were low with silylethynyl and alkynyl derivs. The initial aza-Payne rearrangement is followed by intramolecular hydroamination to produce **densely-subst. pyrrolidines**. However, only *syn*-diastereoisomers underwent the hydroamination, being aided by pre-orientation of the nitrogen atom and the alkyne residue on the same side of the epoxide ring. F.e.s. J.M. Schomaker, A.R. Geiser, R. Huang, B. Borhan, J. Am. Chem. Soc. *2007*, *129* (13), 3794-5.

Hydrogen/Carbon Type NC ꟗ HC

1,3-Dimesitylimidazolium chloride/1,8-diazabicyclo[5.4.0]undec-7-ene ←
Succinimides from 2-azetidinone-4-carboxaldehydes ◌
N-Heterocyclic carbene-catalyzed ring expansion

181. 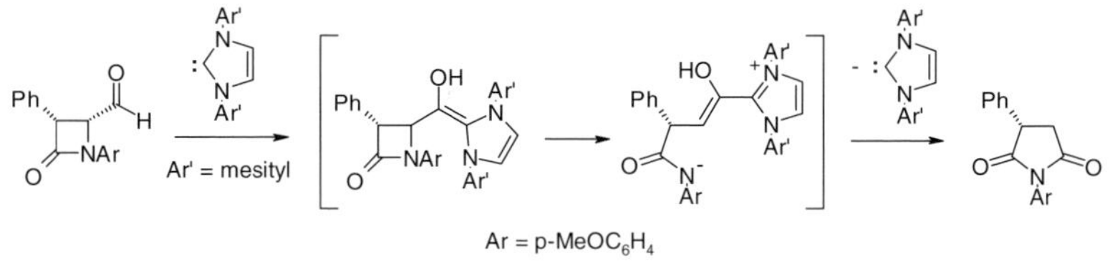

1 Mol% DBU added to a soln. of the startg. β-lactam and 1 mol% 1,3-dimesitylimidazolium chloride in methylene chloride, and stirred at reflux for 8 h → product. Y 99%. The procedure is mild and based on the inexpensive *in situ* generation of a carbenoid reagent at low catalyst loading. Alkyl or aryl substitution was tolerated at the α-position of the β-lactam to give succinimides possessing tertiary or quaternary carbon atoms; yields were lower, however, with mesityl as N-substituent and in THF. Reaction is presumed to involve classical activation of the aldehyde as a Breslow intermediate, followed by ring opening and intramolecular attack of amide anion with liberation of the carbene catalyst. F.e. and comparison of N-heterocyclic carbenes s. G.-Q. Li, Y. Li, L.-X. Dai, S.-L. You, Org. Lett. *2007*, *9* (18), 3519-21.

Rhodium(II) trifluoroacetate dimer *$Rh_2(OCOCF_3)_4$*
Indoles from 3-aryl-Δ^1-azirines
by heating in a sealed tube cf. *70*, 156; with a little $Rh_2(OCOCF_3)_4$ in 1,2-dichloroethane at room temp. to reflux s. S. Chiba, G. Hattori, K. Narasaka, Chem. Lett. *2007*, *36* (1), 52-3.

Oxygen/Carbon Type NC ꟗ OC

Microwaves (s.a. under H_2SO_4) [\\\\\]
2-Ethyleneacylamines from 2-ethyleneiminoesters ←
under Pd- or Pt-catalysis cf. *47*, 254s*70*; aza-Claisen rearrangement of 2-ethylenetri-chloro- or -fluoro-acetimidates in *o*-xylene under microwave irradiation, also 2-ethyleneisothiocyanates from 2-ethylenethiocyanates (cf. *8*, 411), s. J. Gonda, M. Martinková, A. Zadrošová, M. Šoteková, J. Raschmanová, P. Èonka, E. Gajdošíková, C.O. Kappe, Tetrahedron Lett. *2007*, *48* (39), 6912-5.

Sulfuric acid/microwaves *H_2SO_4/[////]*
Acid-catalyzed isomerization of 2-aminoalcohols under microwave irradiation C(N<)C(OH) → C(OH)C(N<)

182. 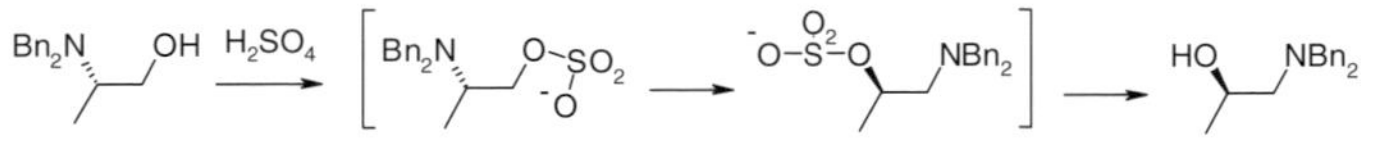

with inversion of configuration. 5 Mol% H_2SO_4 added to a soln. of the startg. chiral 2-aminoalcohol in THF at room temp., heated at 180° for 2 h under microwave irradiation in a sealed tube, quenched with satd. aq. Na_2CO_3, and worked up with flash chromatographic purification → product. Y 97% (e.e. 99%). The procedure is inexpensive and applicable to dibenzyl- and benzyl-amines, as well as benzylprolinol derivs., other acids (AcOH, HCl, TsOH) being generally ineffective. Reaction is thought to involve intermediate formation of a 2-aminosulfate monoester, followed by synchronous intramolecular nucleophilic displacements. F.e.s. T.-X. Metro, D. Gomez Pardo, J. Cossy, Synlett *2007* (18), 2888-90.

Chiral dimeric [ferrocenylbis(1-sulfonyl-Δ²-imidazoline)]bispalladacyclic complexes ←
Asym. [3.3]-sigmatropic [aza-Claisen] rearrangement ←
of (Z)-N-(*p*-methoxyphenyl)-2-ethylenetrifluoroacetimidates

183.

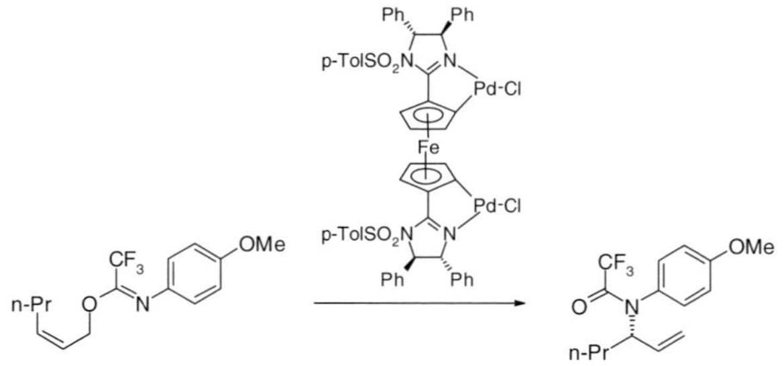

A soln. of 0.6 mol% AgOTs in acetonitrile evaporated to dryness, *0.1 mol%* chiral dimeric [ferrocenylbis(1-sulfonyl-Δ²-imidazoline)]bispalladacyclic complex in methylene chloride added to the residue under N_2 in the absence of light, after 46 h the mixture filtered over CaH_2/Celite, the filter cake washed with methylene chloride (until colourless), the solvent removed in a stream of N_2, the solid activated catalyst dissolved in dry chloroform, added to the startg. (Z)-2-ethylenetrifluoroacetimidate, the mixture heated at 55° for 3 days in the absence of light under N_2, the residue suspended in cyclohexane/ethyl acetate (9:1), and purified by chromatography → N-(2-ethylene)trifluoroacetamide. Y 94% (e.e. 97%). This is the first example of a highly enantioselective aza-Claisen (Overman) rearrangement of (Z)-configured substrates. The catalyst requires activation with 4-6 eqs. of AgOTs but is robust enough to be used at 55° with a catalyst loading of *0.1 mol%* in some cases without loss of yield or selectivity in gram-scale preparations. Eleven examples gave 86-99% yields (e.e. 93-9%), substrates having linear alkyl groups reacting to completion at room temp. with 1 mol% catalyst; furthermore, previously unreactive substrates having *α-branched* alkyl groups are also converted by this catalyst with high enantioselectivity (64% yield). The possibility of intermediate formation of ferrocenium species, although not detected, could not be excluded. F.e. incl. preparation and comparison of four analogous catalysts s. S. Jautze, P. Seiler, R. Peters, Angew. Chem. Int. Ed. Engl. *2007*, *46* (8), 1260-4.

Sulfur/Carbon Type NC ∩ SC

Microwaves [\\\\]
2-Ethyleneisothiocyanates from 2-ethylenethiocyanates s. *8*, 411s*72* ←

Carbon/Carbon Type NC ∩ CC

Without additional reagents *w.a.r.*
1,2,3-Triazole ring from acetyleneazides ○
cf. *31*, 325s*69*; regioselective synthesis of chiral N-Boc-4,5,6,7-tetrahydro[1,2,3]triazolo[1,5-*a*]-pyrazines (derived from α-amino acids) via intramolecular 'click' reaction in refluxing methylene chloride or chloroform s. D.K. Mohapatra, P.K. Maity, R.G. Gonnade, M.S. Chorghade, M.K. Gurjar, Synlett *2007* (12), 1893-6; on-resin macrocyclization with 1 eq. CuBr, 3 eqs. Na-ascorbate, 10 eqs. *i*-Pr_2NEt and 10 eqs. 2,6-lutidine, cyclic tetra-, penta-, hexa-, and hepta-peptides, s. R.A. Turner, A.G. Oliver, R.S. Lokey, Org. Lett. *2007*, *9* (24), 5011-4.

n-*Butyllithium/chiral bis(Δ²-oxazolines)/diisopropylamine* ←
Lithium-catalyzed regiospecific asym. intramolecular hydroamination ○

184. 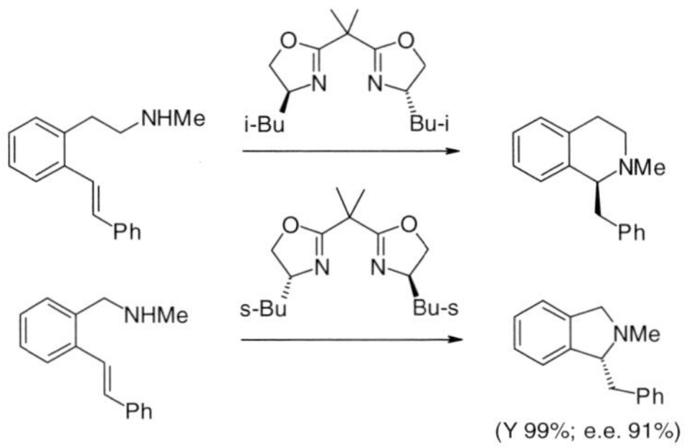

Chiral 1,2,3,4-tetrahydroisoquinolines. A mixture of 2-(2-methylaminoethyl)stilbene, 0.4 eq. chiral bis(Δ²-oxazoline), 0.4 eq. of *n*-butyllithium and 0.2 eq. of diisopropylamine (as proton source) *in toluene* stirred at -60° for 5 h → (S)-1-benzyl-2-methyl-1,2,3,4-tetrahydroisoquinoline. Y 97% (e.e. 84%). Enantioselectivity was high, and cyclization took place under kinetic control with no formation of the *endo*-product. A number of related chiral bis(Δ²-oxazolines) were tested giving (S)- or (R)-products, enantioselectivity apparently being disfavoured by bulky groups at the 4-position in the oxazoline ring. The analogous aminomethylstilbene gave the corresponding chiral isoindoline in 89-99% yields with e.e. up to 91%. F.e.s. T. Ogata, A. Ujihara, S. Tsuchida, T. Shimizu, A. Kaneshige, K. Tomioka, Tetrahedron Lett. *2007*, *48* (38), 6648-50.

Copper(I) bromide/sodium ascorbate/ethyldiisopropylamine/lutidine ←
1,2,3-Triazole ring from acetyleneazides
On-resin macrocyclization - Cyclic tetra-, penta-, hexa-, and hepta-peptides s. *31*, 325s*72*

Silver triflate *AgOTf*
Cycloisomerization of acetyleneamines
2,3-Disubst. 1,2-dihydroisoquinoline-1-phosphonic acid esters s. *48*, 294s*72*

Chiral [di(phosphine)]gold(I) p-*nitrobenzoate complexes* ←
N-Protected 2-vinyl-N-heterocyclics from alleneamines
Intramolecular asym. hydroamination

185. 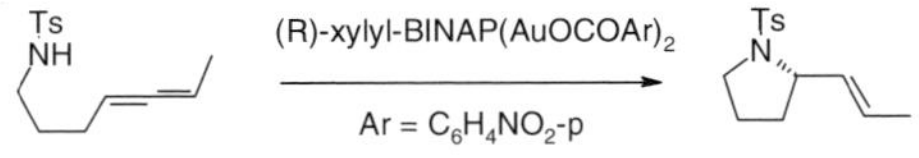

3 Mol% chiral [di(phosphine)]gold(I) *p*-nitrobenzoate complex added to a soln. of the startg. allenetosylamine in DCE or nitromethane, the homogeneous mixture protected from ambient light, stirred at 23° for 15 h, and worked up with flash chromatographic purification → product. Y 98% (e.e. 99%). There was an interesting counterion effect, the corresponding AuCl complex being less enantioselective. F.e. and with (R)-SegPhos as ligand, also piperidine ring closure [with (R)-Cl(MeO)BIPHEP as ligand], s. R.L. LaLonde, B.D. Sherry, E.J. Kang, F.D. Toste, J. Am. Chem. Soc. *2007*, *129* (9), 2452-3; intramolecular hydroamination of N-carbobenzoxyallene-amines **with dynamic kinetic resolution** using $AgClO_4$/(S)-(3,5-di-Bu^t-4-MeO)MeOBIPHEP cf. Z. Zhang, C.F. Bender, R.A. Widenhoefer, ibid. *129* (46), 14148-9; with the corresponding Au(I) complex cf. Org. Lett. *2007*, *9* (15), 2887-9.

(Triphenylphosphine)gold(I) triflate *$[(Ph_3P)Au]OTf$*

Δ³-Isoxazolines from propargyloxylamines

with K_2CO_3 in refluxing methanol cf. *70*, 157; with 5 mol% $[(Ph_3P)Au]OTf$ at room temp. s. H.-S. S. Yeom, E.-S. Lee, S. Shin, Synlett *2007* (14), 2292-4.

Sodium tetrachloroaurate/1-butyl-3-methylimidazolium fluoroborate *$NaAuCl_4/[bmim]BF_4$*
Mercury(II) triflate *$Hg(OTf)_2$*

Indoles from *o*-acetyleneamines

N-unsubst. indoles with $InBr_3$ cf. *56*, 136s*70*; with $NaAuCl_4{\cdot}H_2O$ in $[bmim]BF_4$ s. I. Ambrogio, A. Arcadi, S. Cacchi, G. Fabrizi, F. Marinelli, Synlett *2007* (11), 1775-9; N-tosyl-derivs. with a little $Hg(OTf)_2$ in methylene chloride at room temp. s. T. Kurisaki, T. Naniwa, H. Yamamoto, H. Imagawa, M. Nishizawa, Tetrahedron Lett. *2007*, *48* (10), 1871-4.

Magnesium halides/N,N,N′,N′-tetramethylethylenediamine *$MgHal_2/TMEDA$*

1-Tosyl-1,2-dihydropyridazines from 2-methylenecyclopropyl tosylhydrazones

186. 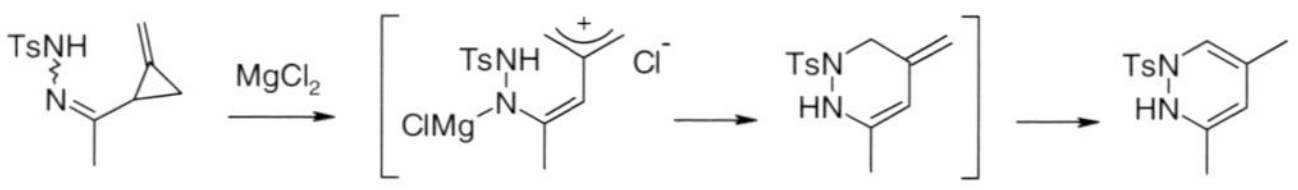

The startg. hydrazone, DME and 1 eq. TMEDA added sequentially with stirring to an oven-dried vial under argon, treated with 10 mol% $MgCl_2$, the vial sealed, and heated at 120° for 3.5 h (TLC monitoring) → 3,5-dimethyl-1-tosyl-1,2-dihydropyridazine. Y 85%. The **1,6-dihydro-deriv.** was favoured with MgI_2 in refluxing THF. It is thought that the hydrazone is initially activated by the Lewis acid prior to rearrangement. F.e. and with $MgBr_2$ s. M.E. Scott, Y. Bethuel, M. Lautens, J. Am. Chem. Soc. *2007*, *129* (6), 1482-3.

Ammonium trifluoroacetate *CF_3COONH_4*

Brønsted acid-catalyzed intramolecular hydroamination

187. 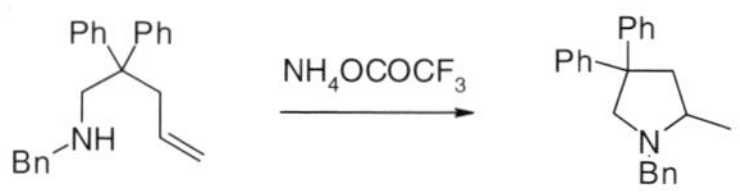

Pyrrolidines. A soln. of 5-benzylamino-4,4-diphenyl-1-pentene and 20 mol% ammonium trifluoroacetate in dry dioxane heated under N_2 in a sealed tube at 130° for 24 h, the mixture cooled, partitioned between ether and aq. $NaHCO_3$, the organic layers dried over $MgSO_4$, concentrated *in vacuo*, and the residue purified by chromatography → 1-benzyl-2-methyl-4,4-diphenylpyrrolidine. Y 74%. The method is applicable to both prim. and sec. 1-pentenyl-5-amines (11 examples; 64-93%) and *gem*-disubstitution at the 4-position is *not* a requirement for success. The method is compatible with ether, ester, free hydroxyl, nitrile, halo and nitro functionality. Other amine salts with weakly coordinating anions were also effective [the best being dimethylphenylammonium tetrakis(pentafluorophenyl)borate at 80°], but ammonium trifluoroacetate is easier to prepare and economically attractive. F.e. and with Group 4 metal halides as catalyst for the intramolecular hydroamination of prim. 1-pentenyl-5-amines (single example) s. L. Ackermann, L.T. Kaspar, A. Althammer, Org. Biomol. Chem. *2007*, *5* (12), 1975-8.

2(S)-[Bis[3,5-bis(trifluoromethyl)phenyl]trimethylsiloxymethyl]pyrrolidine/benzoic acid ←
N-Protected 2-β-hydroxy-N-heterocyclics
via organocatalyzed asym. intramolecular Michael addition
of N-protected amino-α,β-ethylenealdehydes

188.
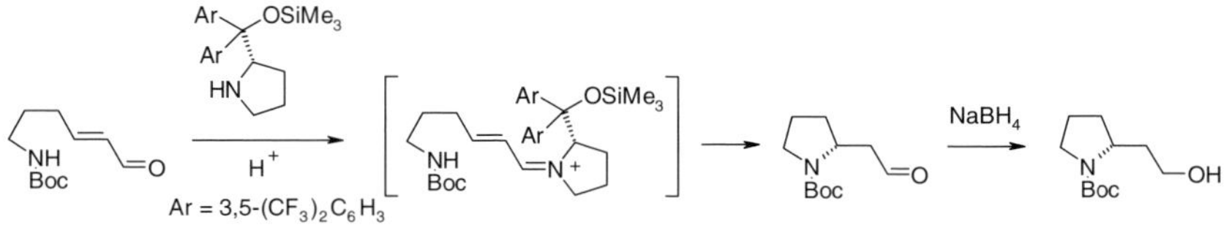

A mixture of 20 mol% 2(S)-[bis[3,5-bis(trifluoromethyl)phenyl]trimethylsiloxymethyl]-pyrrolidine and 20 mol% benzoic acid in dry chloroform added to a soln. of the startg. aldehyde in the same solvent at -50°, stirred for 22 h at the same temp., diluted with methanol, 3 eqs. $NaBH_4$ added in portions, allowed to warm to 0°, quenched with satd. aq. NH_4Cl after 30 min, and worked up with flash chromatographic purification over silica gel → (2R)-N-*tert*-butoxycarbonyl-2-hydroxyethylpyrrolidine. Y 71% (e.e. 93%). The high enantioselectivity is believed to be a consequence of shielding of the *si*-face of the intermediate (E)-iminium ion by the chiral diaryl(trimethylsiloxy)methyl group of the catalyst. F.e. and chiral 5- or 6-membered N,N-, N,O- and N,S-heterocyclic analogs, also from N-Cbz-derivs., s. S. Fustero, D. Jiménez, J. Moscardó, S. Catalán, C. del Pozo, Org. Lett. *2007*, *9* (25), 5283-6.

Imidazolium salts s. under $NaAuCl_4$ ←
Chiral bis(Δ^2-oxazolines) s. under BuLi *box*
Sodium ascorbate s. under CuBr ←

Chiral zirconium(IV) bis(amidate) complex *[Zr(IV)]*
Zirconium(IV)-catalyzed asym. intramolecular hydroamination ○

189.
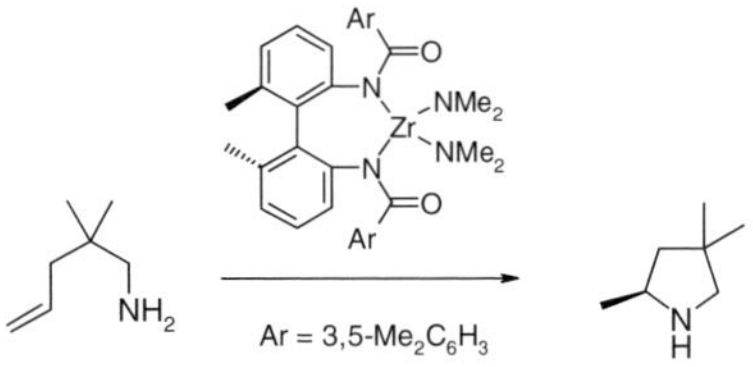

Chiral N-unsubst. pyrrolidines. A mixture of 2,2-dimethylpent-4-enylamine and 10 mol% (-)-Zr(IV)-bis(amidate) complex heated in perdeuteriotoluene at 110° for 3 h, the mixture filtered, evaporated and purified by chromatography → (S)-2,4,4-trimethylpyrrolidine. Y >98% (e.e. 93% as the benzamide deriv.). The (R)-enantiomer was similarly prepared in 80% yield (e.e. 93%) using the (+)-zirconium(IV) catalyst. A series of 2,2-disubst. pent-4-enylamines gave chiral pyrrolidines in 80-98% yield (e.e. 74-93%), unsym. derivs. affording mixtures of two diastereomers. The novel modular, easily prepared complex was the best of three neutral zirconium(IV) bis(amidates) tested (attributed to the bulky mesitylene group), and the catalyst loading may be reduced to 5 mol% over a longer period. The failure of an N-methyl derivative to undergo cyclization suggests that reaction involves an intermediate zirconium-imido species. F.e. and preparation of the catalyst s. M.C. Wood, D.C. Leitch, C.S. Yeung, J.A. Kozak, L.L. Schafer, Angew. Chem. Int. Ed. Engl. *2007*, *46* (3), 354-8.

*Chloro(1,5-cyclooctadiene)rhodium(I) dimer/tris(*p*-fluorophenyl)phosphine* ←
Cycloisomerization of functionalized terminal acetylene derivs.
via intramolecular nucleophilic addition to rhodium vinylidene complexes

190. 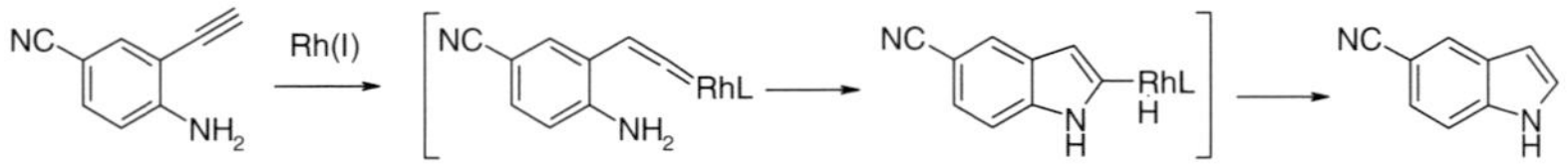

Indoles from *o*-ethynylamines. A mixture of 2-ethynyl-4-cyanoaniline, 5 mol% $[Rh(cod)Cl]_2$, and 0.6 eqs. tris(*p*-fluorophenyl)phosphine stirred under argon in degassed DMF at 85° for 2 h, the dark orange soln. cooled to 25°, quenched with satd. aq. $NaHCO_3$, extracted with ether, dried over $MgSO_4$, and evaporated *in vacuo* → 5-cyanoindole. Y 98%. This efficient and atom-economical preparation of indoles illustrates a novel ring closure via nucleophilic addition to *in situ*-generated rhodium vinylidene complexes, and is compatible with chloro, ester, nitrile, ketone, nitro and ether functionality giving yields of 79-98%. The procedure is also applicable to N-benzyl- and N-*p*-methoxybenzyl-derivs. (in 88 and 94% yield), but an N-allyl analog gave decomposition products. The reaction can also be effected with cheaper triphenylphosphine but yields are generally lower. F.e., **also benzofurans** from *o*-ethynylphenols, and **2(3*H*)-furanones** from α-ethynylcarboxylic acids, s. B.M. Trost, A. McClory, Angew. Chem. Int. Ed. Engl. *2007*, *46* (12), 2074-7.

Bis(acetonitrile)dichloropalladium(II) *$PdCl_2(MeCN)_2$*
Cycloisomerization of acetyleneamines
with $PdCl_2(MeCN)_2$ cf. *43*, 271s*47*; 2,3-disubst. 2*H*-isoindol-1-ylphosphonic acid esters from α-amino-α-(*o*-acetylenearyl)phosphonic acid esters, also 2,3-disubst. 1,2-dihydroisoquinoline-1-phosphonic acid esters with 5 mol% AgOTf in acetonitrile (cf. *48*, 294s*69*), s. Q. Ding, Y. Ye, R. Fan, J. Wu, J. Org. Chem. *2007*, *72* (14), 5439-42.

Exchange ⇂↿

Hydrogen ↿ NC ⇂↿ H

Potassium hydroxide *KOH*
2-Mercapto-1,3,4-oxadiazoles from acylhydrazines s. *23*, 383s*72* ○

*Sodium azide/*tert*-butyl hypochlorite* *NaN_3/t-BuOCl*
Carboxylic acid azides from aldehydes RCHO → $RCON_3$

191. 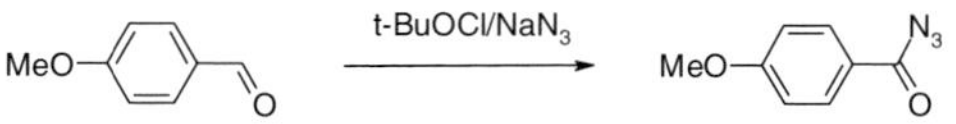

under mild conditions. 4-Methoxybenzaldehyde added to 2 eqs. *tert*-butyl hypochlorite and 2 eqs. NaN_3 in carbon tetrachloride, the mixture stirred at room temp. until reaction complete (TLC monitoring), the mixture diluted with methylene chloride, the organic layer washed sequentially with 5% aq. $Na_2S_2O_3$, 5% aq. $NaHCO_3$, water and brine, dried over anhydrous Na_2SO_4, concentrated *in vacuo*, and purified by chromatography → 4-methoxybenzoyl azide. Y 83%. The method is successful with ar., hetar. and aliphatic aldehydes (the latter requiring reaction at 5-10° to limit rearrangement to isocyanates) giving yields of 72-87% (12 examples) in the presence of methoxy, nitro, halo and amine functionalities. Carbon tetrachloride was the solvent of choice, but chloroform, methylene chloride and cyclohexane were also acceptable while acetonitrile gave slow reaction and *tert*-butanol produced anhydride by-products. F.e.s. N.D. Arote, K.G. Akamanchi, Tetrahedron Lett. *2007*, *48* (32), 5661-4.

Sodium nitrate/benzyltriphenylphosphonium persulfate *$NaNO_3/[BnPh_3P]_2S_2O_8$*
Ar. nitration H → NO_2
with KNO_3/triflic acid/triflic anhydride cf. *4*, 337s*33*; with $NaNO_3$/benzyltriphenylphosphonium

persulfate in acetonitrile s. H. Tajik, M.A. Zolfigol, J. Albadi, R. Eslami, Synth. Commun. *2007*, *37* (16), 2771-6.

Sodium iodide s. under t-*BuOCl and Cobalt(II) salen complexes* *NaI*

Copper(I) imidazol-2-ylidene complex ←
Copper-catalyzed oxidative carbonylation of amino derivs. under mild conditions ○

192.

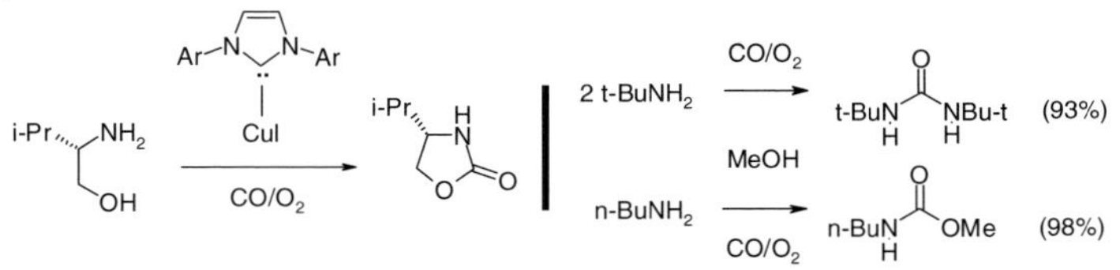

2-Oxazolidones from 2-aminoalcohols. A mixture of (S)-2-amino-3-methylbutanol and 10 mol% Cu(I)-imidazol-2-ylidene complex in dioxane heated under CO/O_2 (4.8/0.2 MPa respectively) at 100° for 3 h → (S)-4-isopropyl-2-oxazolidone. Y 93%. Six examples gave high yields (86-96%) of 2-oxazolidones, with diethanolamine giving the mono-carbonylated product, N-(2-hydroxyethyl)-2-oxazolidone, while 2-aminophenol underwent oxidative dimerization to 2-aminophenoxazin-3-one in 90% yield. The catalyst is air- and moisture-stable, inexpensive, eco-friendly, readily prepared from the corresponding imidazolium chloride, and also effective for the formation of **sym. ureas from prim. amines,** and **urethans from amines and alcohols.** However, there was no reaction with unfunctionalized ar. amines. F.e.s. S. Zheng, F. Li, J. Liu, C. Xia, Tetrahedron Lett. *2007*, *48* (33), 5883-6.

Copper(II) sulfate/imidazole-1-sulfonyl azide hydrochloride/potassium carbonate ←
Diazo compds. from active methylene groups $CH_2 \rightarrow C{=}N_2$

193.

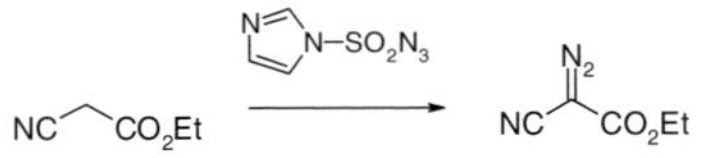

1.2 Eqs. imidazole-1-sulfonyl azide hydrochloride added to ethyl cyanoacetate and 5 eqs. K_2CO_3 in acetonitrile, the mixture stirred at 40° for 9 h, diluted with ethyl acetate, washed with 1 *M* HCl, and dried → ethyl 2-cyanodiazoacetate. Y 61%. The diazo transfer reagent is prepared in a one-pot reaction using imidazole and $ClSO_2N_3$ (generated *in situ* from NaN_3 and SO_2Cl_2) and is shelf-stable, crystalline and inexpensive. The yield is relatively low due to product volatility but is still comparable with results using triflyl azide. Diethyl malonate and ethyl 2-acetoacetate were similarly converted but no reaction occurred with benzenesulfonyl or diethoxyphosphinyl analogs. F.e.s. E.D. Goddard-Borger, R.V. Stick, Org. Lett. *2007*, *9* (19), 3797-800.

*Copper(I) bromide/*tert-*butyl hydroperoxide* *CuBr/t-BuOOH*
N′,N′-Disubst. N-monoacyl-1,1-diamines from tert. amines and acylamines H → NHAc
Oxidative radical acylamination

194.

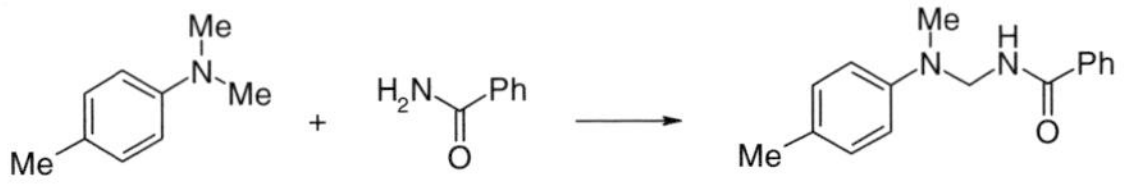

2 Eqs. N,N-dimethyl-*p*-toluidine, 1 eq. benzamide and 1.5 eqs. *tert*-butyl hydroperoxide stirred under N_2 in decane for 30 min, 5 mol% CuBr added, stirred for 15 min, heated at 80° for 6 h, cooled, ethyl acetate added, filtered, and the filtrate purified by column chromatography → product. Y 77%. Yields are variable with highest yields being obtained with electron-rich anilines and electron-deficient amides (prim. more reactive than sec.). Surprisingly, reaction of N-methyl-N-benzylanilines occurred at both methyl and benzyl groups affording dehydrogenative and dephenylation coupling products with the latter predominating in all cases (3 examples). CuBr

was the best copper catalyst of several tried and the reaction does not proceed in the absence of either CuBr or *t*-BuOOH or at room temperature. Decane was the best solvent with mixtures of decane/DMSO suitable for insoluble substrates, and yields are low in the absence of a nitrogen atmosphere. A radical mechanism is invoked with generation of an intermediate copper-complexed iminium ion, which then reacts with the acylamine. F.e. and intramolecular process s. Y. Zhang, H. Fu, Y. Jiang, Y. Zhao, Org. Lett. *2007*, *9* (19), 3813-6.

Silver fluoroborate s. under Pd(OAc)$_2$ *AgBF$_4$*
Imidazolium salts s. under PdCl$_2$(phen) ←

tert-*Butyl hydroperoxide (s.a. under CuBr)* t-*BuOOH*
N,N-Disubst. arylcarboxylic acid amides CHO → CON<
from ar. aldehydes and sec. amines
Metal-free oxidative amination

195. 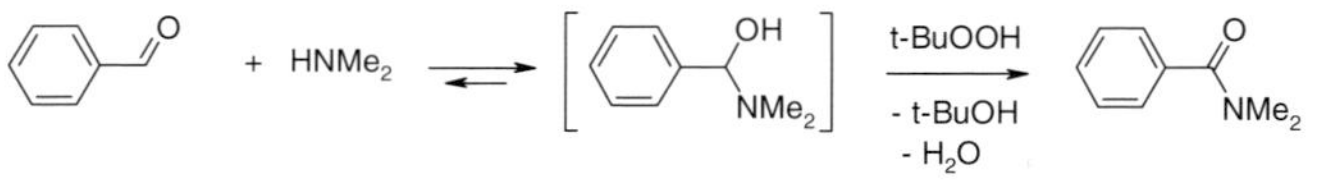

under mild conditions. A mixture of benzaldehyde, 1.3 eqs. dimethylamine and 1.3 eqs. *tert*-butyl hydroperoxide in dry acetonitrile stirred at 80° for 5 h, cooled to room temp., quenched with water, and worked up with flash chromatographic purification on silica gel → N,N-dimethylbenzamide. Y 99%. The procedure is high yielding (irrespective of the electronic nature of ring substituents), inexpensive, avoids transition metals, bases, and harsh conditions, and an excess of the amine is not required. However, yields were lowish with diethylamine. F.e.s. K. Ekoue-Kovi, C. Wolf, Org. Lett. *2007*, *9* (17), 3429-32.

Phenyl iodosocarboxylates s. under Rh(II)-α,α,α',α'-tetramethyl-1,3-benzenedipropionate and Pd(OAc)$_2$ *PhI(OCOR)$_2$*

Triphenylphosphine/diethyl azodicarboxylate *Ph$_3$P/ROOCN=NCOOR*
Urethans from alcohols and amines by Mitsunobu carboxylation NH → NCOOR
s. *66*, 169; s.a. D. Chaturvedi, N. Mishra, V. Mishra, Monatsh. Chem. *2007*, *138* (1), 57-60.

Oxygen s. under Cobalt(II) salen complexes and PdCl$_2$(phen) *O$_2$*
Imidazole-1-sulfonyl azide hydrochloride s. under CuSO$_4$ ←
tert-*Butyl hypochlorite s.a. under NaN$_3$* t-*BuOCl*

tert-*Butyl hypochlorite/sodium iodide* t-*BuOCl/NaI*
Δ^2-Oxazolines from ethylene derivs. and N-unsubst. carboxylic acid amides ○

196. 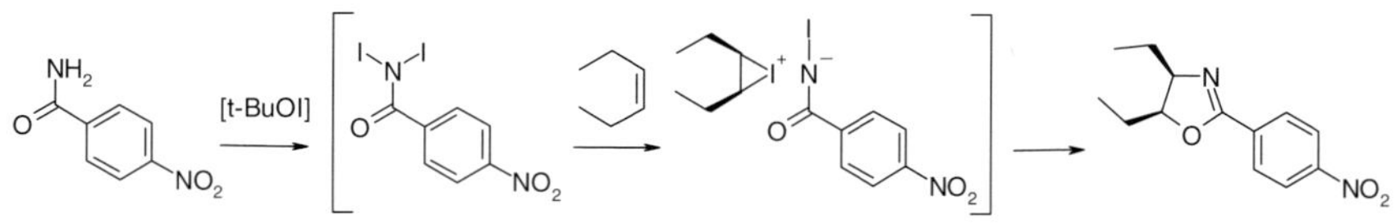

3 Eqs. *tert*-butyl hypochlorite added to a mixture of (Z)-3-hexene, 0.5 eq. 4-nitrobenzamide and 3 eqs. NaI in acetonitrile at room temp., the mixture stirred in the dark under N_2 for 96 h, quenched with aq. 0.3 *M* $Na_2S_2O_3$, extracted with methylene chloride, dried over $MgSO_4$, concentrated *in vacuo*, and purified by chromatography → *cis*-2-(4-nitrophenyl)-4,5-diethyl-2-oxazoline. Y 68%. The (E)-hexene gave the *trans*-isomer in 71% yield. Reaction is successful with a number of alkyl- and aryl-subst. alkenes, monosubst. derivs. giving 2,5-disubst. 2-oxazolines as the sole or major product with minimal formation of N-acylaziridines. Yields were in the range 50-79% for the electron-deficient *p*-nitrobenzamide but low for electron-rich *p*-methoxybenzamide (1 example; 37%) and *n*-butyramide (1 example; 28%). The reaction is presumed to involve an intermediate iodonium species, generated with the aid of *in situ*-generated tert-*butyl hypoiodite*. F.e. and comparison of iodinating agents s. S. Minakata, Y. Morino, T. Ide, Y.

Oderaotoshi, M. Komatsu, Chem. Commun. *2007* (31), 3279-81.

Cobalt(II) salen complexes/sodium iodide/oxygen *Co(salen)/NaI/O_2*
2-Oxazolidones from 2-aminoalcohols and carbon monoxide ○
s. *54*, 143s*72*

Rhodium hydroxide/alumina/hydroxylamine hydrosulfate *$Rh(OH)_3/Al_2O_3/(NH_2OH)_2$-$H_2SO_4$*
N-Unsubst. carboxylic acid amides from aldehydes s. *72*, 86 CHO → $CONH_2$

Rhodium(II) α,α,α′,α′-tetramethyl-1,3-benzenedipropionate/phenyl iodosopivalate ←
N-Acylsulfonic acid amides from aldehydes and sulfonic acid amides CHO → $CONHSO_2R$
Catalytic oxidative sulfonylamination under mild conditions

197. 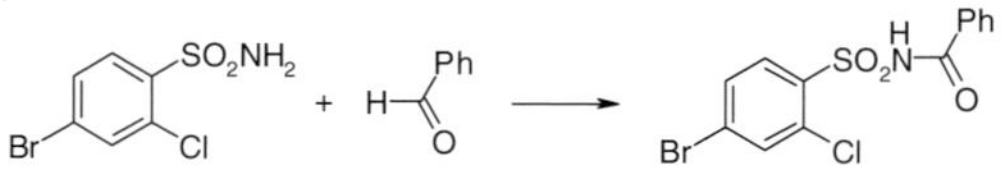

1.2 Eqs. benzaldehyde added to a mixture of 4-bromo-2-chlorophenylsulfonamide and 2 eqs. phenyl iodosopivalate in isopropyl acetate, stirred for 5 min until homogeneous, 2 mol% Rh(II)-α,α,α′,α′-tetramethyl-1,3-benzenedipropionate [$Rh_2(esp)_2$] added in one portion, stirred for 2 h, concentrated *in vacuo*, slurried in 4:1 hexane/methylene chloride for 15 min, filtered, and worked up → N-(4-bromo-2-chlorophenylsulfonyl)benzamide. Y 80%. The procedure is mild, free from by-products, generally applicable to the reaction of alkane- or arene-sulfonamides with aliphatic or aromatic aldehydes, and leaves a number of functional groups unaffected (NO_2, halogen, CF_3, activated olefins, N-heterocyclics) as well as enolizable functions. It is thought that reaction involves a concerted asynchronous nitrene insertion into the aldehydic C-H group. F.e.s. J. Chan, K.D. Baucom, J.A. Murry, J. Am. Chem. Soc. *2007*, *129* (46), 14106-7.

Palladium(II) acetate/phenyl iodosoacetate/silver fluoroborate ←
Stereospecific palladium(IV)-catalyzed intramolecular oxyamination ○

198. 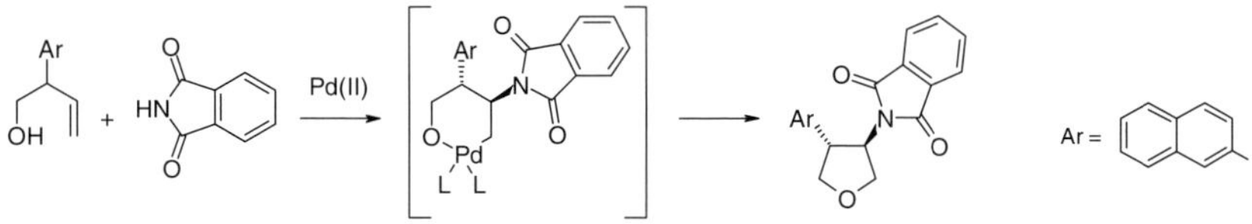

N-Protected 4-subst. *trans*-3-aminotetrahydrofurans from 3-ethylenealcohols. 1 Eq. phthalimide, 2 eqs. phenyl iodosoacetate, 10 mol% $AgBF_4$ and 5 mol% $Pd(OAc)_2$ in acetonitrile stirred in a vial, 2 eqs. 2-(2-naphthyl)-3-buten-1-ol added, the vial sealed with a Teflon-lined cap, the mixture heated at 60° for 1 h, cooled to room temp., an additional 1 eq. phenyl iodosoacetate, 5 mol% $Pd(OAc)_2$, 10 mol% $AgBF_4$ and 1 eq. of the alcohol added, the mixture stirred for a further 8 h at 60°, evaporated to dryness, the residue dissolved in ether, washed sequentially with 10% aq. triethylamine, water, satd. aq. $NaHCO_3$ and NaCl, dried over $MgSO_4$, and concentrated → *trans*-3-(2-naphthyl)-4-phthalimidotetrahydrofuran. Y 80% (d.r. 12:1). Reaction involves an unusual *Pd(II)/Pd(IV)-catalytic cycle*, facilitated by the hydroxyl group which, after the initial aminopalladation, coordinates to palladium(II) and prevents the familiar β-elimination; oxidation of palladium(II) to palladium(IV) by the iodine(III) reagent thereby becomes feasible to facilitate the ensuing O-C bond-formation. Interestingly, the concluding elimination takes place with *retention* of configuration, as demonstrated with a substrate possessing an internal ethylene group. A series of 2-subst. 3-butenols were cyclized by this method with variable *trans*-selectivity (1.4:1 to >20:1), lower selectivities being observed with substrates possessing *o*-subst. aryl groups. Yields were in the range 54-80% for aryl derivs. but alkyl (methyl, benzyl, isopropyl) analogs gave only 5-30% yields; in addition the 1,1-dimethyl deriv. gave a 47% yield of the 2,2-dimethyl-4-phthalimidotetrahydrofuran. F.e.s. L.V. Desai, M.S. Sanford, Angew. Chem. Int. Ed. Engl. *2007*, *46* (30), 5737-40.

Palladium N-heterocyclic carbene complexes ←
Sym. ureas from amines 2 >NH → $(>N)_2CO$
and CO with montmorillonite-supported $Pd(OAc)_2$(bipy)/$CuCl_2$ cf. *47*, 270; oxidative carbonylation with palladium carbene complexes containing both an aniline and an N-heterocyclic carbene ligand s. S.-Z. Zheng, X.-G. Peng, J.-M. Liu, W. Sun, C.-G. Xia, Chinese J. Chem. *2007*, *25* (8), 1065-8.

Tetrakis(triphenylphosphine)palladium(0) *$Pd(PPh_3)_4$*
4-Vinyl-2-imidazolidones from terminal ethylene derivs. ○
Stereospecific catalytic diamination in the absence of solvent

199. 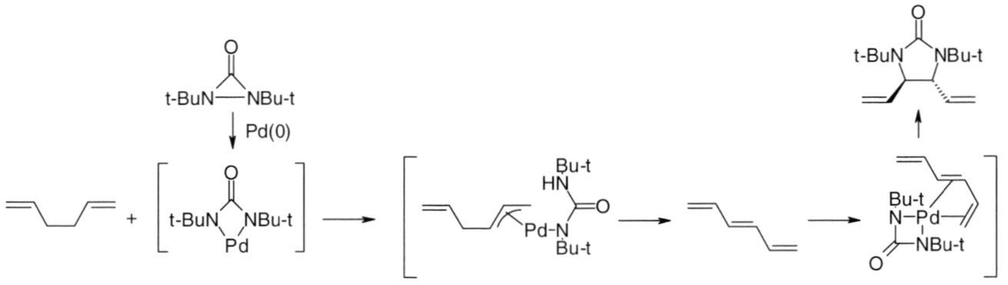

1,5-Hexadiene added to 5 mol% $Pd(PPh_3)_4$ contained in a vial under argon, stirred in an oil bath at 65°, 2.75 eqs. di-*tert*-butyldiaziridinone added *over 11 h* by syringe pump (at the rate of 0.4 mmol/h), stirring continued for a further 1 h, and worked up with flash chromatographic purification on silica gel → product. Y 76%. The procedure is highly stereoselective and generally applicable to mono- and 1,1-di-subst. olefins. Reaction is thought to involve initial insertion of palladium(0) into the N-N bond of the diaziridinone, followed by generation of a 1,3-diene via a palladium π-allyl complex prior to a second palladium complexation and reductive elimination on ring closure. F.e.s. H. Du, W. Yuan, B. Zhao, Y. Shi, J. Am. Chem. Soc. *2007*, *129* (24), 7496-7.

Dichloro(1,10-phenanthroline)palladium(II)/oxygen/ 1-butyl-3-methylimidazolium iodide *$PdCl_2$(phen)/O_2/[bmim]I*
2-Oxazolidones from 2-aminoalcohols and carbon monoxide
with PdI_2/KI cf. *54*, 143s*65*; with $PdCl_2$(1,10-phenanthroline) in a 2-phase medium comprising 1-butyl-3-methylimidazolium iodide (as stabilizer against deactivation and promoter as well as solvent) and glyme under CO/O_2, also one example of a 2-benzoxazolone, s. F. Li, C. Xia, Tetrahedron Lett. *2007*, *48* (28), 4845-8; with a little salen-Co(II)/NaI or salen-Co(III)-I s. J.-M. Liu, X.-G. Peng, J.-H. Liu, S.-Z. Zheng, W. Sun, C.-G. Xia, ibid. *48* (6), 929-32.

Via intermediates *v.i.*
N-Arylation NH → NAr
via iridium-catalyzed borylation of arenes s. *55*, 166s*72*; *65*, 62s*72*

Oxygen ↑ NC ↓↑ O

Without additional reagents *w.a.r.*
N-Acylation of amines NH → NAc
with Ac_2O in AcOH cf. *10*, 295; in the absence of solvent and additive, selectivity in the presence of thiols, alcohols, and phenols, also S-acylation, s. M.M. Mojtahedi, M.S. Abaee, M.M. Heravi, F.K. Behbahani, Monatsh. Chem. *2007*, *138* (1), 95-9.

Selective N-carbo-*tert*-butoxylation NH → NCOOBu-*t*
cf. *49*, 284; in the absence of solvent and catalyst, retention of hydroxyl groups, s. X.S. Jia, Q. Huang, J. Li, S.Y. Li, Q.S. Yang, Synlett *2007* (5), 806-8; with HCl in aq. methanol, preferential protection of prim. over sec. amino groups, s. D.W. Lee, H.J. Ha, Synth. Commun. *2007*, *37* (5), 737-42.

1,2,4-Oxadiazoles from amidoximes ○
and carboxylic acid esters in the absence of solvent cf. *50*, 183s*71*; s.a. W. Du, Q. Truong, H.B. Qi, Y. Guo, H.R. Chobanian, W.K. Hagmann, J.J. Hale, Tetrahedron Lett. *2007, 48* (13), 2231-5.

1-Acyl-1,2-dihydro-4*H*-3,1-benzoxazin-4-ones from *o*-aminocarboxylic acids and aldehydes via *o*-(alkylideneamino)carboxylic acids

200. 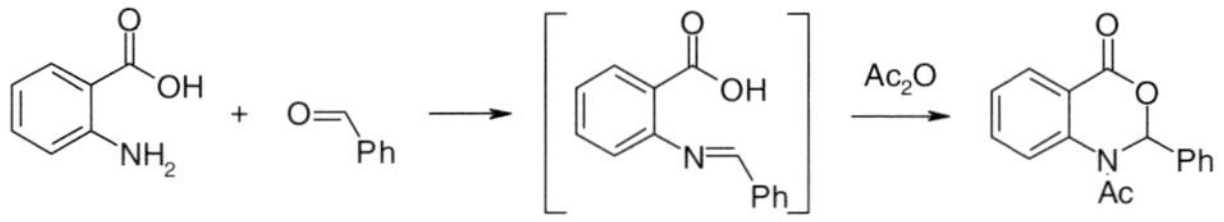

One-pot conversion. A mixture of anthranilic acid and 1.2 eqs. benzaldehyde stirred in methanol at room temp. for ca. 1 h, the solvent removed, and the formed imine refluxed in acetic anhydride for 2 h → product. Y 84%. No separation or purification of the intermediate was required. F.e. and with isolation of the intermediate s. F. Nikpour, D. Sheikh, M. Saraji, Chem. Lett. *2007, 36* (8), 1074-5.

Microwaves s. under Montmorillonite, Acetic acid, 2,4-Dimethoxy-1,3,2,4-dithiadiphosphetane 2,4-disulfide, Na-dithionite and Trimethylsilyl triflate [\\\\]

Sodium azide (s.a. under N-Tosylimidazole) *NaN_3*
4-α-Functionalized 2-oxazolidones from N-carbo-*tert*-butoxyaziridin-2-ylcarbinols Regio- and stereo-specific conversion under mild conditions

201. 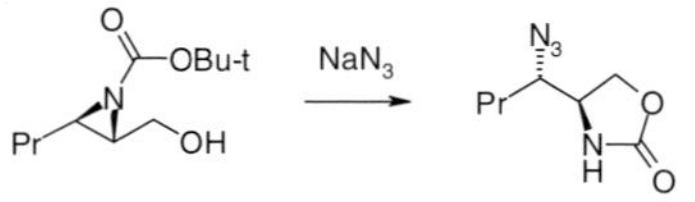

4-α-Azido-2-oxazolidones. 2.5 Eqs. NaN_3 added to a soln. of the startg. aziridine in dry DMF at room temp., stirred at 70° for 8-10 h (TLC monitoring), cooled, washed with satd. aq. NH_4Cl, and worked up with flash chromatographic purification → (1′S*,4S*)-4-(1′-azidobutyl)-2-oxazolidone. Y 82%. Reaction is both regio- and stereo-specific, involving attack of azide ion at C_3, followed by intramolecular cyclization with inversion at C_3 and retention of configuration at C_2. F.e. and 4-α-bromo-derivs. with LiBr, also in DMSO and with $NaN_3/LiClO_4$, s. G. Righi, S. Ciambrone, A. Pompili, F. Caruso, Tetrahedron Lett. *2007, 48* (43), 7713-6.

Sodium nitrite *$NaNO_2$*
1,2-Nitroethylene derivs. from 2-ethylenealcohols with double bond shift $C(NO_2)C{=}C$
Polymer-based conversion on a soluble support

202. 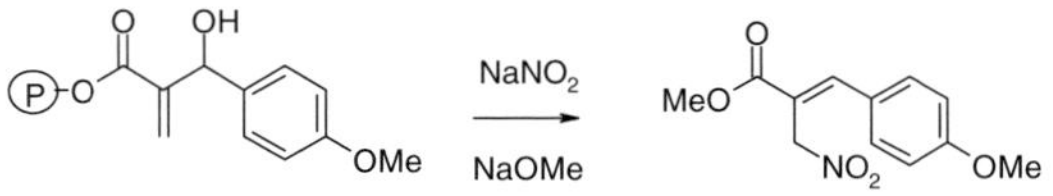

α,β-Ethylene-α-(nitromethyl)carboxylic acid esters. 4 Eqs. $NaNO_2$ added to the startg. PEG-supported Baylis-Hillman adduct in DMF, stirred for 5 h at 80°, ether added to precipitate the supported product, chilled at 0° for 30 min, filtered, and the support removed by stirring in 0.1 *M* methanolic NaOMe at room temp. for 8 h → product. Y 94%. F.e.s. S.R. Sheng, S.Y. Lin, Q. Wang, X.L. Liu, M. Lin, Synth. Commun. *2007, 37* (6), 1011-7.

Triethanolamine s. under Trimethylsilyl triflate *$(HOCH_2CH_2)_3N$*

2(S)-[Diphenyl(trimethylsiloxy)methyl]pyrrolidine ←
N-Protected aziridine-2-carboxaldehydes from α,β-ethylenealdehydes 
Organocatalyzed asym. aziridination with N-acoxyurethans

203.

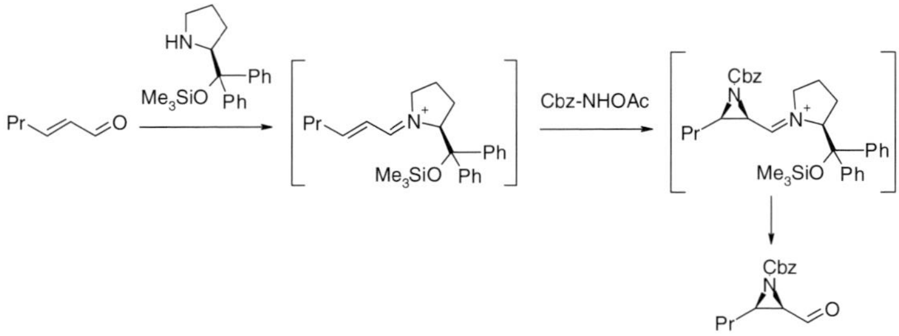

A mixture of 1.2 eqs. 2-hexenal, N-carbobenzoxy-O-acetylhydroxylamine and 20 mol% 2(S)-[diphenyl(trimethylsiloxy)methyl]pyrrolidine stirred in chloroform at 40° for 0.5 h, and the mixture purified directly by chromatography (to avoid decomposition) → product. Y 78% (e.e. 96%; d.r. 4:1). The yield and selectivity were reduced at 10 mol% catalyst loading while reaction at room temp. increased the selectivity (e.e. 99%; d.r. 10:1) at the expense of the reaction rate and yield (62%) - the latter due to instability of the aziridine ring. Ten examples gave 60-78% yields at 40° with e.e. 82-99% and d.r. 4-7:1, not only in the presence of Boc- and Cbz-protected amines but also benzyl and benzoate esters and terminal double bonds. Selectivity is attributed to efficient shielding of the *si*-face in the iminium intermediate leading to preferential *re*-face Michael addition of the N-acoxyurethan; mechanistically, the formed 1,4-adduct, now possessing an *electrophilic* nitrogen, then undergoes 3-*exo-tet*-cyclization to give the aziridine. F.e. incl. N-Boc derivs., also conversion to chiral β-aminocarboxylic acid esters (by treatment with a thiazolium chloride in the presence of ethanol and subsequent N-deprotection with H_2/Pd-C), s. J. Veseley, I. Ibrahem, G.L. Zhao, R. Rios, A. Cordova, Angew. Chem. Int. Ed. Engl. *2007*, *46* (5), 778-81.

*2-Hydroxypyridine s. under Zr(OBu-*t*)$_4$* *2-HOPy*
4-(Dialkylamino)pyridines s. under Pyridine-3-carboxylic anhydride and Iron oxide ←

Copper(II) or Silver triflate *Cu(OTf)$_2$ or AgOTf*
N-Benzylsulfonylamines from benzyl alcohols OH → $N(R)SO_2R'$

204.

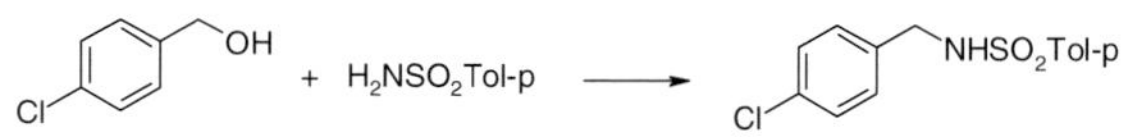

A mixture of 4-chlorobenzyl alcohol, 2 eqs. tosylamine and 5 mol% AgOTf *in nitromethane* stirred at 100° for 8 h, and worked up with column chromatographic purification on silica gel → N-(4-chlorobenzyl)-4-methylbenzenesulfonamide. Y 85%. The procedure is straightforward (not requiring activation of the hydroxyl group), environmentally friendly, and applicable to the reaction of prim. benzyl and 2-naphthylmethyl alcohols with N-unsubst. or N-monosubst. arene- or alkane-sulfonamides, with water as the only by-product and without requiring any other additives. Yields were low, however, from benzyl alcohols possessing strong electron-withdrawing groups (e.g. *p*-NO_2). The corresponding **sym. N,N-dibenzylsulfonylamines** were obtained with 2 eqs. of the benzyl alcohol. F.e and with Cu(OTf)$_2$, Sc(OTf)$_3$ or Bi(OTf)$_3$ as catalyst s. B. Sreedhar, P.S. Reddy, M.A. Reddy, B. Neelima, R. Arundhathi, Tetrahedron Lett. *2007*, *48* (46), 8174-7; N-sulfonyl-, N-carbobenzoxy- and N-carbo-*tert*-butoxy-benzylamines from sec. benzyl alcohols with 5 mol% $MoCl_5$ cf. C.R. Reddy, P.P. Madhavi, A.S. Reddy, ibid. *48* (40), 7169-72.

Zinc salts s. under Prolinates *Zn^{2+}*
Triphenylborane s. under $[Ir(cod)Cl]_2$ *Ph_3B*

Montmorillonite K10/microwaves *clay/[\\\\]*
Azomethines from oxo compds. and amines CO → C=NR
under microwave irradiation in the absence of solvent cf. *1*, 391s*71*; trifluoromethyl ketimines on montmorillonite K10 s. M. Abid, M. Savolainen, S. Landge, J.B. Hu, G.K.S. Prakash, G.A. Olah, B. Torok, J. Fluorine Chem. *2007*, *128* (6), 587-94; aldimines from aliphatic, α,β-unsatd. or (het)ar. aldehydes with silica in ethanol under ultrasonication s. K.P. Guzen, A.S. Guarezemini, A.T.G. Orfao, R. Cella, C.M.P. Pereira, H.A. Stefani, Tetrahedron Lett. *2007*, *48* (10), 1845-8.

1-Methylimidazolium fluoroborate *$[Hmim]BF_4$*
2-Aryl-1-benzylbenzimidazoles from *o*-diamines and two ar. aldehyde molecules ○
in ionic liquid solvent s. *71*, 146s*72*

Ammonium cerium(IV) nitrate *CAN*
β-Amino-α,β-ethylene- from β-keto-carbonyl compds. C(O)CH → C(N<)=C
in the ionic liquid $[EtNH_3]NO_3$ cf. *26*, 331s*71*; with CAN in ethanol s. V. Sridharan, C. Avendano, J.C. Menendez, Synlett *2007* (6), 881-4; with $ZrCl_4$ in the absence of solvent s. J. Lin, L.F. Zhang, Monatsh. Chem. *2007*, *138* (1), 77-81.

Scandium(III) triflate *$Sc(OTf)_3$*
N-Benzylsulfonylamines from benzyl alcohols s. *72*, 204 OH → $N(R)SO_2R'$

Scandium(III) or Dysprosium(III) triflate *$Sc(OTf)_3$ or $Dy(OTf)_3$*
***trans*-4,5-Diamino-2-cyclopentenones** ←
from furan-2-carboxaldehyde and two amine molecules

205.

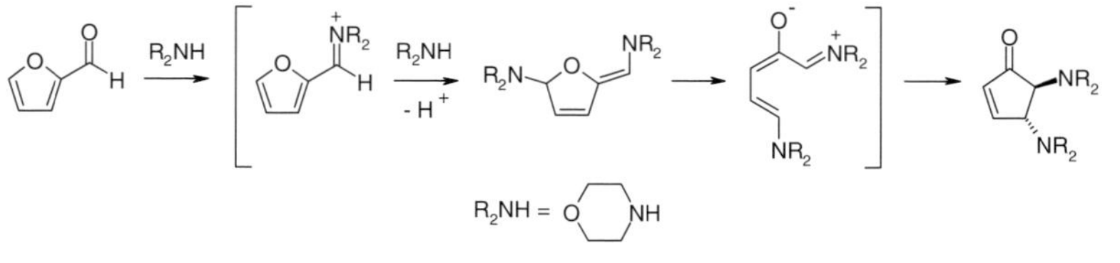

10 Mol% $Dy(OTf)_3$ added to a soln. of 2-furancarboxaldehyde and 2 eqs. morpholine in acetonitrile containing 4 Å molecular sieves, and stirred for 16 h at room temp. → *trans*-4,5-dimorpholinocyclopent-2-enone. Y 100%. Reaction is presumed to involve initial Lewis acid-catalyzed formation of a deprotonated Stenhouse salt, followed by a thermal conrotatory π4a electrocyclization (Nazarov-type ring closure) in keeping with the exclusive *trans*-stereochemistry. A variety of aliphatic sec. amines, sec. anilines and prim. anilines [preferably with $Sc(OTf)_3$] take part in the conversion, but aliphatic prim. amines failed to react. F.e. and comparison of Lewis and Brønsted acids s. S.-W. Li, R.A. Batey, Chem. Commun. *2007* (36), 3759-61.

Ytterbium(III) triflate s. under NH_4OAc *$Yb(OTf)_3$*
Ammonium carbonate s. under P_2I_4 *$(NH_4)_2CO_3$*

Ammonium acetate [s.a. under $PhI(OAc)_2$] *NH_4OAc*
Ammonium acetate/ytterbium(III) triflate *$NH_4OAc/Yb(OTf)_3$*
Imidazoles from aldehydes and α-dioxo compds. ○
from α-diketones under microwave irradiation or in ionic liquids cf. *23*, 423s*69*; in the presence of $Yb(OTf)_3$ s. L.M. Wang, Y.H. Wang, H. Tian, Y.F. Yao, J.H. Shao, B. Liu, J. Fluorine Chem. *2006*, *127* (12), 1570-3; from α-ketoaldehydes with NH_4OAc in methanol at room temp. s. V. Zuliani, G. Cocconcelli, M. Fantini, C. Ghiron, M. Rivara, J. Org. Chem. *2007*, *72* (12), 4551-3.

1-Butyl-3-methylimidazolium hexafluorophosphate *[bmim]PF_6*
2-Aryl-2,3-dihydro-4(1*H*)-quinazolones ○
from *o*-aminocarboxylic acid amides and ar. aldehydes
Eco-friendly synthesis in ionic liquids without additional catalysis

206. 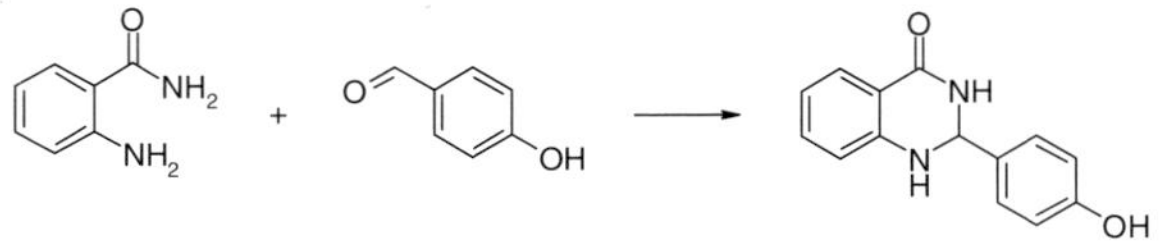

1-Butyl-3-methylimidazolium hexafluorophosphate (2 ml) added to a mixture of anthranilamide (5.5 mmol) and 4-hydroxybenzaldehyde (5 mmol), and the soln. stirred at 75° for 30 min → product. Y 92%. Use of ionic liquids affords higher yields and shorter reaction times than with organic solvents (cf. *19*, 481). The reaction is compatible with heteroaryl aldehydes and a variety of substituted benzaldehydes possessing electron-donating or -withdrawing groups. The reaction is also effected by 3-component condensation **from 3,1-benzoxazine-2,4-diones,** NH_4OAc and aryl aldehydes *in 3:2 ionic liquid-water*. F.e.s. J. Chen, W. Su, H. Wu, M. Liu, C. Jin, Green Chem. *2007, 9* (9), 972-5.

Dialkyl azodicarboxylates s. under Ph_3P and Polymer-based Ar_3P *ROOCN=NCOOR*
1,5,7-Triazabicyclo[4.4.0]dec-5-ene s. under $[IrCl(cod)]_2$ ←

Immobilized lipase ←
Enzymatic desymmetrization of *meso*-1,3-diamines in an organic solvent NH → NCOOR

207. 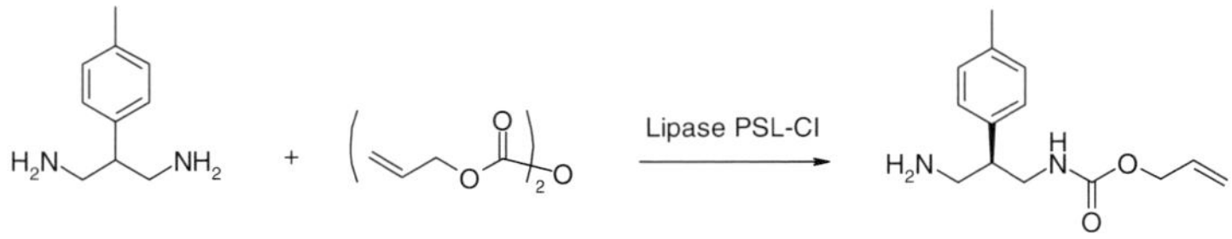

by N-carballyloxylation. 1 Eq. diallyl carbonate added to a suspension of the startg. *meso*-diamine and ceramic-supported lipase (from *Pseudomonas cepacia*) in dry *dioxane* under N_2, shaken at 30° and 250 rpm for 70 h (with TLC monitoring), the enzyme filtered off, washed with methanol, and solvent evaporated → product. Y 72% (e.e. 96%). The immobilized lipase was the most effective of several tested and accepted other *p*-subst. (F and MeO) 2-aryl-1,3-diamines in 68-72% yield (e.e. 86-96%). Dioxane was the solvent of choice due to the high solubility and stability of the startg. diamine. Other carbonates were unreactive whereas non-activated esters (e.g. ethyl acetate) were non-selective. F.e. incl. synthesis of the startg. diamines s. E. Busto, V. Gotor-Fernández, J. Montejo-Bernardo, S. García-Granda, V. Gotor, Org. Lett. *2007, 9* (21), 4203-6.

Novozyme-433/palladium nanoparticles ←
Carboxylic acid amides from carboxylic acid esters and amines COOR → CON<
with kinetic resolution
with lipase cf. *45*, 210s*67,70*; with Novozyme-433/Pd-nanoparticles with dynamic kinetic resolution for benzylic or aliphatic prim. amines s. M.J. Kim, W.H. Kim, K. Han, Y.K. Choi, J. Park, Org. Lett. *2007, 9* (6), 1157-9.

Pyridine-3-carboxylic anhydride/4-(dimethylamino)pyridine *$(3-PyCO)_2O$/DMAP*
N-Subst. carboxylic acid amides from acids and amines COOH → CON<
s. *72*, 219

Acetate buffer/silica gel ←
6-α-Hydroxy-5,6-dihydro-4*H*-1,2-oxazines from γ,δ-epoxyketones under mild conditions C○

208. 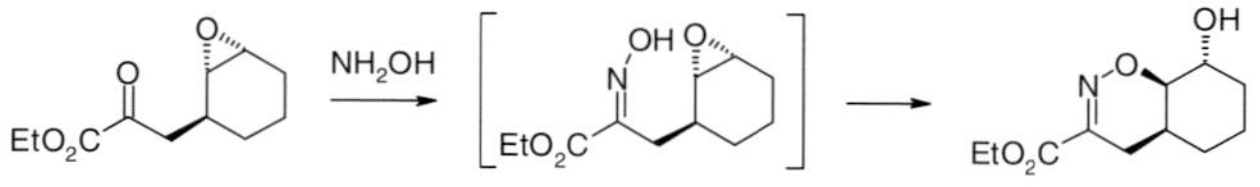

One-pot conversion via (E)-γ,δ-epoxyoximes. 1.25 Eqs. each of hydroxylamine hydrochloride and Na-acetate in water added via pipette to a stirred soln. of the startg. epoxyketone in acetonitrile at room temp., stirred for 12 h, carefully concentrated, the crude oxime extracted, the residue (not isolated) dissolved in ethyl acetate, silica gel added in one portion, the slurry stirred vigorously at room temp. for 4 h, a further amount of silica gel and solvent added, stirring continued for 8 h, filtered, and worked up with flash column chromatographic purification → product. Y 72%. F.e. and with K_2CO_3 in ethanol for the ring closure s. J.R. Donald, M.G. Edwards, R.J.K. Taylor, Tetrahedron Lett. *2007, 48* (30), 5201-4.

Acetic acid *AcOH*
Acetic acid/microwaves *AcOH/[\\\\]*
Paal-Knorr pyrrole synthesis ←
with acetic acid in ethanol under microwave irradiation cf. *1*, 335s*69*; s.a. T. Ullrich, M. Ghobrial, K. Weigand, Synth. Commun. *2007, 37* (7), 1109-19; under thermal conditions s. R.R. Sagyam, H. Vurimidi, P.R. Padi, M.R. Ghanta, J. Heterocycl. Chem. *2007, 44* (4), 923-6.

L-Proline *Pro-OH*
Zinc prolinate *$(Pro\text{-}O)_2Zn$*
2-Aryl-1-benzylbenzimidazoles from *o*-diamines and two ar. aldehyde molecules ○
cf. *71*, 146; with L-proline in chloroform s. R. Varala, A. Nasreen, R. Enugala, S.R. Adapa, Tetrahedron Lett. *2007, 48* (1), 69-72; with Zn-prolinate in water at room temp. s. V. Ravi, E. Ramu, K. Vijay, A.S. Rao, Chem. Pharm. Bull. *2007, 55* (8), 1254-7; in ionic liquid solvent (1-methylimidazolium fluoroborate) at 60° s. H.Q. Ma, Y.L. Wang, J.P. Li, J.Y. Wang, Heterocycles *2007, 71* (1), 135-40.

Trifluoroacetic acid s. under Polymer-based triphenylphosphine *CF_3COOH*

Phenyl iodosoacetate *$PhI(OAc)_2$*
Cyclic iminoesters from aminoalcohols and aldehydes

209. 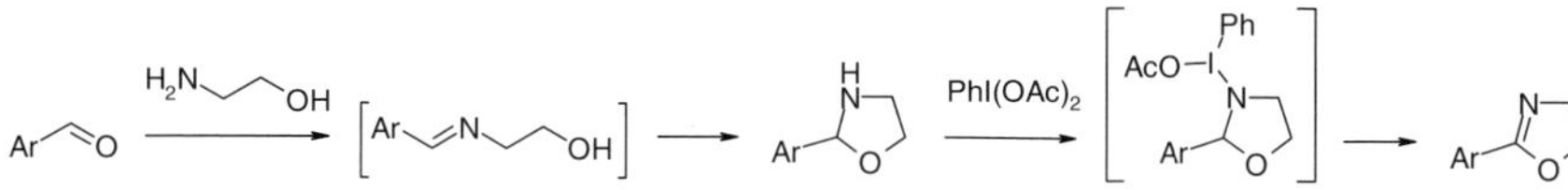

Ar = p-$MeOC_6H_4$

Δ^2-Oxazolines. Equimolar amounts of 4-methoxybenzaldehyde and 2-aminoethanol stirred in acetonitrile at room temp. for 4 h, 1.2 eqs. phenyl iodosoacetate added, the mixture stirred for 4 h (TLC monitoring), water added, the mixture extracted with methylene chloride, the extracts dried over dry Na_2SO_4, concentrated *in vacuo*, and purified by chromatography → 2-(4-methoxyphenyl)-2-oxazoline. Y 70%. The reaction is successful with aromatic (electron-rich or -poor) and aliphatic aldehydes affording good yields (52-71%) of both Δ^2-oxazolines and **5,6-dihydro-4*H*-1,3-oxazines** (from 3-aminoalcohols). The reaction involves formation of an intermediate satd. 1,3-O,N-heterocyclic which suffers mild dehydrogenation by the hypervalent iodine reagent (while other sources of electrophilic iodine were not effective). F.e. and chiral products s. N.K. Karade, G.B. Tiwari, S.V. Gampawar et al., Synlett *2007* (12), 1921-4.

Benzimidazoles from *o*-diamines and aldehydes ○
2-Arylbenzimidazoles s. *69*, 171s*72*; *46*, 321s*72*

Phenyl iodosoacetate/ammonium acetate *PhI(OAc)$_2$/NH$_4$OAc*
Imidazoles from α-hydroxyketones and aldehydes

210. 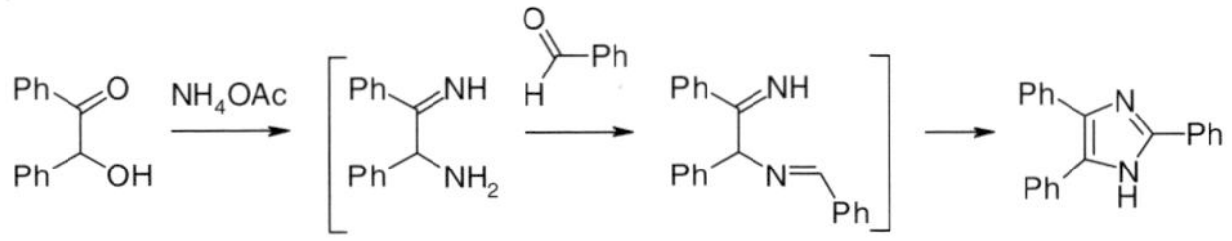

A mixture of benzaldehyde, 1 eq. benzoin, 1.5 eqs. NH_4OAc and 1 eq. phenyl iodosoacetate in ethanol refluxed for 30 min, cooled, and worked up with column chromatographic purification on silica gel → product. Y 98%. The procedure is rapid, and generally applicable to aliphatic and aromatic aldehydes, irrespective of the electronic character of aryl substituents. Reaction is thought to involve oxidative ring closure of an intermediate α-(alkylideneamino)imine. F.e.s. B. Das, Y. Srinivas, H. Holla, M. Krishnaiah, R. Narender, Chem. Lett. *2007, 36* (10), 1270-1.

o-*Iodoxybenzoic acid* *ArIO$_2$*
Nitriles from aldehydes CHO → CN
with NBS and 30% aq. NH_3 cf. *22*, 44s*71*; with *o*-iodoxybenzoic acid in aq. NH_3/acetonitrile s. N.D. Arote, D.S. Bhalerao, K.G. Akamanchi, Tetrahedron Lett. *2007*, *48* (21), 3651-3.

Carbon tetrachloride s. under Ph$_3$P *CCl$_4$*

N-Bromosuccinimide *NBS*
Δ²-Imidazolines from 1,2-diamines and aldehydes ○
with NBS s. *69*, 169; s.a. H. Fujioka, K. Murai, O. Kubo, Y. Ohba, Y. Kita, Tetrahedron *2007*, *63* (3), 638-43; with *t*-BuOCl in *tert*-butanol (and optionally KI, especially for electron-rich aldehydes) s. M. Ishihara, H. Togo, Synthesis *2007* (13), 1939-42.

Oxalyl chloride s. under NH$_2$OH·HCl *(COCl)$_2$*

Bromodimethylsulfonium bromide *BrMe$_2$S$^+$Br$^-$*
Benzimidazoles from *o*-diamines and aldehydes
cf. *69*, 171; *46*, 321s*71*; with bromodimethylsulfonium bromide in acetonitrile at room temp. s. B. Das, H. Holla, Y. Srinivas, Tetrahedron Lett. *2007*, *48* (1), 61-4; 2-aryl-derivs. with H_2O_2/HCl/LiCl in water at 100° s. K. Bahrami, M.M. Khodaei, I. Kavianinia, J. Chem. Res. *2006* (12), 783-4; with aq. 30% H_2O_2/37% HCl in acetonitrile at 25°, incl. large-scale synthesis, s. idem. Synthesis *2007* (4), 547-50; with PhI(OAc)$_2$ in dioxane s. L.-H. Du, Y.-G. Wang, ibid. *2007* (5), 675-8; 4,5-fused imidazoles from purine sulfates in DMSO at 90° s. Q. Dai, D.W. Xu, K. Lim, R.G. Harvey, J. Org. Chem. *2007*, *72* (13), 4856-63; from acetals with 1 *M* HCl/methanol at reflux or, in lower yield, with an acidic cation exchange resin, 2-C-glycosylated benzimidazoles, s. M. Vojtech, M. Petrusova, E. Slavikova, S. Bekesova, L. Petrus, Carbohydr. Res. *2007*, *342* (1), 119-23.

Silica (s.a. under Acetate buffer and Silica-sulfuric acid) *SiO$_2$*
Azomethines from aldehydes and amines CHO → CH=NR
under ultrasonication s. *1*, 391s*72*

N,O-Bis(trimethylsilyl)acetamide s. under Trimethylsilyl triflate ←

Zirconium tetra-tert-*butoxide/2-hydroxypyridine* *Zr(OBu*-t*)$_4$/2-HOPy*
Urethans from amines and carbonic acid esters NH → NCOOR
cf. *21*, 244; with 5 mol% Zr(OBu-*t*)$_4$ and 10 mol% 2-hydroxypyridine, also ureas, s. C. Han, J.A. Porco, Org. Lett. *2007*, *9* (8), 1517-20.

Trimethylsilyl chloride/pyridine/triethylamine *$Me_3SiCl/C_5H_5N/Et_3N$*

Benzimidazolines from N,N-disubst. *o*-diamines and oxo compds. ○

211. 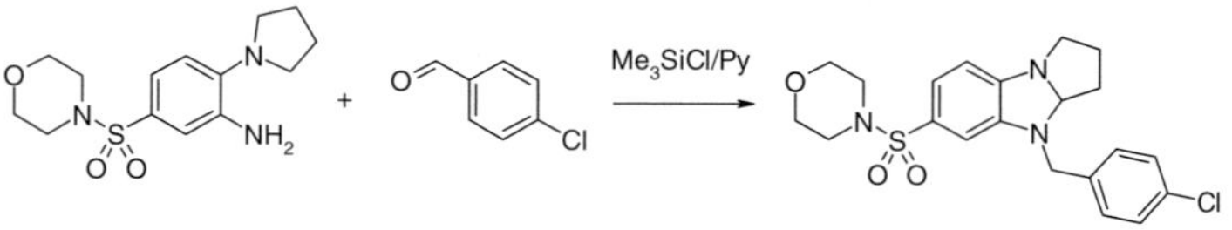

One-pot synthesis. Trimethylsilyl chloride (6 mmol) added dropwise to a soln. of startg. dialkylaminoaniline (2 mmol) and *p*-chlorobenzaldehyde (2 mmol) in a pressure tube, the tube sealed, heated at 100° for 2-8 h with magnetic stirring, cooled (*caution: high pressure*) then triethylamine (6.6 mmol) added dropwise, the soln. sonicated for 1 h at room temp., water added, and the resulting suspension sonicated for 1 h → product. Y 75%. The use of pyridine as solvent neutralises HCl as formed and prevents unwanted side-products, otherwise formed by disproportionation of the desired benzimidazolines. Ketones require longer reaction times than aldehydes and afford lower yields. F.e., discussion of the activating effect of trimethylsilyl chloride, and mechanism s. S.V. Ryabukhin, A.S. Plaskon, D.M. Volochnyuk, A.N. Shivanyuk, A.A. Tolmachev, J. Org. Chem. *2007, 72* (19), 7417-9.

Zirconium tetrachloride *$ZrCl_4$*

β-Amino-α,β-ethylene- from β-keto-carbonyl compds. C(O)CH → C(N<)=C

in the absence of solvent s. *26*, 331s*72*

Benzimidazoles from *o*-diamines and orthocarboxylic acid esters ○

s. *52*, 147s*72*

Tin(IV) oxide s. under SO_4^{2-} *SnO_2*

Lead(II) nitrate *$Pb(NO_3)_2$*

Pyrazines from α-hydroxyketones and 1,2-di-*prim*-amines

with activated MnO_2 then KOH cf. *66*, 182; with $Pb(NO_3)_2$ s. S.A. Kotharkar, D.B. Shinde, Chin. J. Chem. *2007, 25* (1), 105-7.

*Di-*tert*-butyl iminodicarboxylate s. under Polymer-based Ar_3P* *$HN(COOBu$-t$)_2$*

Hydroxylamine O-benzyl ether s. under $Pd(OAc)_2$ *H_2NOBn*

*Tri-*tert*-butylphosphonium fluoroborate s. under $Pd(OAc)_2$* t-*Bu_3PHBF_4*

Triphenylphosphine *Ph_3P*

Δ^3-Pyrazoline-1,2-dicarboxylic acid esters
from α,β-ethyleneketones and azodicarboxylic acid esters

212. 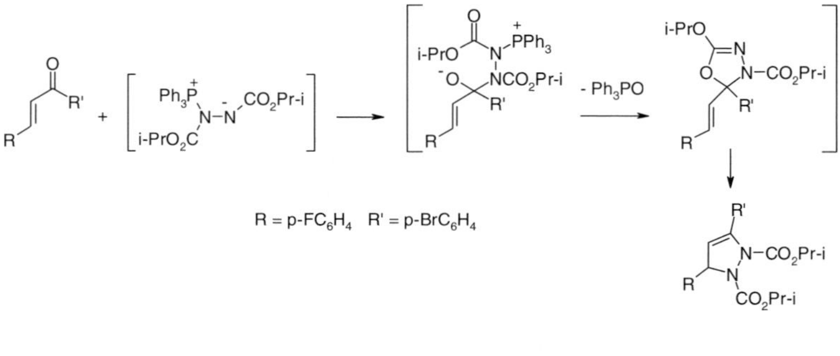

1.5 Eqs. diisopropyl azocarboxylate added to a stirred soln. of the startg. chalcone in toluene under argon, stirring continued at reflux, a soln. of 1.5 eqs. triphenylphosphine in the same solvent added dropwise, stirred at reflux for a further 3 h, cooled, solvent removed, and the residue worked up with column chromatographic purification on silica gel → product. Y 62%. The procedure is generally applicable to subst. chalcones and dialkyl azodicarboxylates, the expected 2-alkoxy-5-vinyl-Δ^2-1,3,4-oxadiazolines not being formed. Reaction is presumed to involve initial formation of the 'Huisgen zwitterion', followed by addition to the carbonyl group then formation of a 1,3,4-oxadiazoline (with elimination of Ph_3PO) prior to ring opening and recyclization of the product. Cross-conjugated dienones, however, gave the novel **2,3,3a,4,5,6-hexahydro-1*H*-pyrazolo[3,4-*c*]pyridazine-1,2,4,5-tetracarboxylates** via Diels-Alder reaction of the initially formed 3-vinylpyrazoline with a second molecule of the azodicarboxylate. F.e.s. V. Nair, S.C. Mathew, A.T. Biju, E. Suresh, Angew. Chem. Int. Ed. Engl. *2007, 46* (12), 2070-3.

Triphenylphosphine/dialkyl azodicarboxylates *Ph_3P/ROOCN=NCOOR*
N-Alkylation with alcohols NH → NR
cf. *28*, 753s*70*; 1-subst. pyrazolo[3,4-*d*]pyrimidines (with DIAD) s. M. Braendvang, L.L. Gundersen, Tetrahedron Lett. *2007*, *48* (17), 3057-9; β-amino-α-(carbalkoxyamino)- from α-(carbalkoxyamino)-β-hydroxy-carboxylic acids via *in situ* formation of the corresponding β-lactones (with DEAD) s. S. Moura, E. Pinto, ibid., *48* (13), 2325-7.

Triphenylphosphine/carbon tetrachloride/triethylamine *$Ph_3P/CCl_4/Et_3N$*
N-Acylindolines from 2-(*o*-aminoaryl)alcohols and carboxylic acids ○

213.

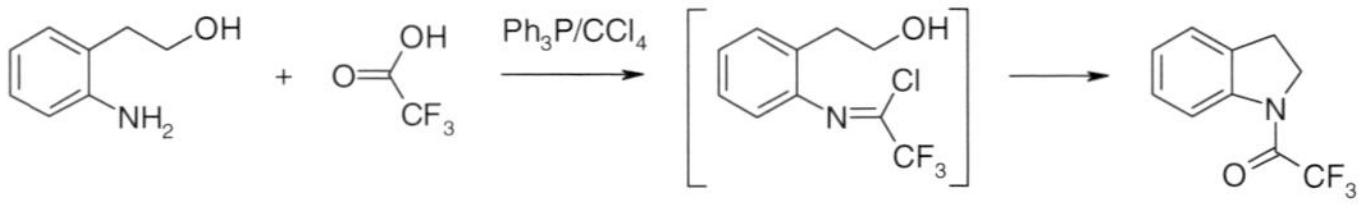

A soln. of trifluoroacetic acid, 3 eqs. triphenylphosphine and 3 eqs. triethylamine in 41.9 eqs. carbon tetrachloride stirred under N_2 at 0° for 10 min, a soln. of 1 eq. 2-aminophenethyl alcohol in 22 eqs. carbon tetrachloride added dropwise, refluxed until reaction complete (3-12 h), and worked up with column chromatographic purification on neutral alumina → N-trifluoroacetyl-indoline. Y 97%. The procedure is clean, efficient, low-cost, direct and scalable (to the 100 g level) with a range of carboxylic acids (of pK_a 0.5-4.87), and may be performed without rigorous exclusion of air or moisture. The yield was low, however, with trichloroacetic acid. Interestingly, a 1.1:1 mixture of *threo*- and *erythro*-2-(2-aminophenyl)-1-phenylpropan-1-ol gave uniquely the *threo*-product, indicative of the intermediate formation of a carbonium ion on cleavage of the hydroxyl group. F.e. and isolation of the intermediate iminochlorides s. Z. Wang, W. Wan, H. Jiang, J. Hao, J. Org. Chem. *2007, 72* (24), 9364-7.

Benzoxazoles from *o*-aminophenols and carboxylic acids
Also benzimidazoles s. *12*, 465s*72*

*Polymer-based triphenylphosphine/di-*tert-*butyl azodicarboxylate/trifluoroacetic acid/ di-*tert-*butyl iminodicarboxylate* ←
Prim. amines from prim. or sec. alcohols OH → NH_2
Modified Mitsunobu reaction with a simplified work-up

214.

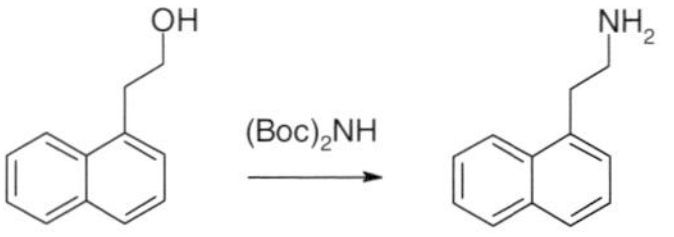

One-pot procedure. 2-(1-Naphthyl)ethanol, 4 eqs. polystyrene-based phosphine resin and 4 eqs. di-*tert*-butyl iminodicarboxylate in dry methylene chloride stirred at 0° under N_2 for 10 min, 3 eqs. di-*tert*-butyl azodicarboxylate added in a single portion, the mixture stirred for 30 min, treated with trifluoroacetic acid, stirring continued at room temp. for 1 h, the mixture filtered through Celite, the residue washed with methanol/methylene chloride (1:1), the filtrates treated with aq. Na_2CO_3, extracted with ethyl acetate, the extracts washed with brine, dried over $MgSO_4$, and evaporated *in vacuo* → 2-(1-naphthyl)ethylamine. Y 88%. The method is clean, simple and based on commercially available reagents to afford prim. amines from prim. alcohols (8 examples; 74-91%) and from sec. alcohols (one example; 55%). The key to the protocol is easy removal of reagents and their by-products by standard work-up techniques. The products are obtained >90% pure and can be purified to >99% by chromatography. F.e.s. W. Sun, J.C. Pelletier, Tetrahedron Lett. *2007, 48* (44), 7745-6.

Bis[o-(diphenylphosphino)phenyl] ether s. under Pt(cod)Cl_2 *DPEphos*
9,9-Dimethyl-4,5-bis(diphenylphosphino)xanthene s. under Ru(CO)H_2(PPh_3)$_3$ *Xantphos*

(N-Isocyanimino)triphenylphosphorane *$Ph_3P{=}N\text{-}NC$:*
2-Aryl-1,3,4-oxadiazoles from arylcarboxylic acids ○

215. 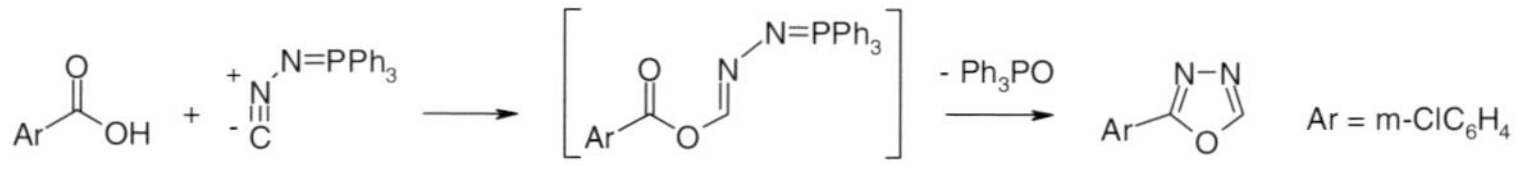

A soln. of 3-chlorobenzoic acid in methylene chloride added dropwise to 1 eq. (N-isocyanimino)-triphenylphosphorane over 15 min, and stirred at room temp. for 12 h → product. Y 86%. Reaction is thought to involve addition of the acid to the isocyano group, followed by intra-molecular aza-Wittig synthesis. Conditions are mild, and work-up easy. F.e.s. A. Souldozi, A. Ramazani, Tetrahedron Lett. *2007, 48* (9), 1549-51.

Diphenyl phosphorazidate/potassium carbonate *$(PhO)_2PON_3/K_2CO_3$*
Sulfonylureas from carboxylic acids and sulfonylamines $RCOOH \rightarrow RNHCONHSO_2R'$
One-pot conversion via isocyanates

216. 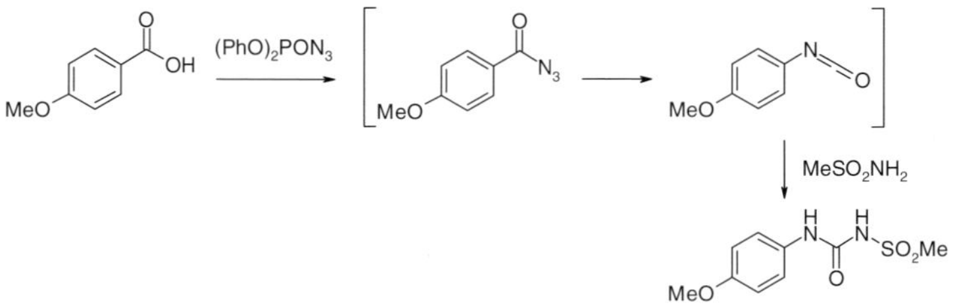

1.2 Eqs. diphenyl phosphorazidate and 3 eqs. K_2CO_3 added to an equimolar mixture of 4-methoxybenzoic acid and methanesulfonamide in dioxane under N_2, heated at 85° for 2 h, cooled, dissolved in water, adjusted to pH 4-5 by dropwise addition of concd. HCl, the mixture filtered, and the solid triturated with ether → product. Y 51%. Twenty-six examples are given using a range of aromatic, heteroaromatic and aliphatic carboxylic acids with K_2CO_3/dioxane or diisopropylamine/benzene as base/solvent. Yields are generally moderate (<50%; maximum 77%), attributed to formation of sym. ureas as by-products, but existing routes to these compds. are limited by the availability and stability of isocyanates, which here are generated *in situ* **via Curtius rearrangement**. No advantage was gained by conducting the reaction in a sealed tube or in the presence of molecular sieves. F.e.s. C.A. Luckhurst, I. Millichip, B. Parker, J. Reuberson, M. Furber, Tetrahedron Lett. *2007, 48* (50), 8878-82.

Chiral phosphoromonoamidites s. under [IrCl(cod)]$_2$ ←

2,4-Dimethoxy-1,3,2,4-dithiadiphosphetane 2,4-disulfide/microwaves ←
Benzoxazoles from *o*-aminophenols and carboxylic acids ○
under microwave irradiation with polymer-based $ArPPh_2/Cl_3CCN$ cf. *12*, 465s*71*; with Lawesson's reagent in the absence of solvent, also benzothiazoles (cf. *19*, 674s*71*), s. J.A. Seijas, A.P. Vazquez-Tato, A.R. Carballido-Reboredo, J. Crecente-Campo, L. Romar-Lopez, Synlett *2007*

(2), 313-7; with Ph_3P and excess Et_3N in CCl_4 under reflux, also benzimidazoles and benzothiazoles, s. F.L. Ge, Z.X. Wang, W. Wan, W.C. Lu, J. Hao, Tetrahedron Lett. *2007*, *48* (18), 3251-4; 2-subst. oxazolo[4,5-*b*]pyridines under microwave irradiation in the absence of solvent s. M.J. Myllymaki, A.M.P. Koskinen, ibid. *48* (13), 2295-8.

Ethyl phosphorodichloridate s. under $NH_2OH{\cdot}HCl$ — $EtOP(O)Cl_2$

Diphosphorus tetraiodide/ammonium carbonate — $P_2I_4/(NH_4)_2CO_3$

Nitriles from carboxylic acids — COOH → CN

217. 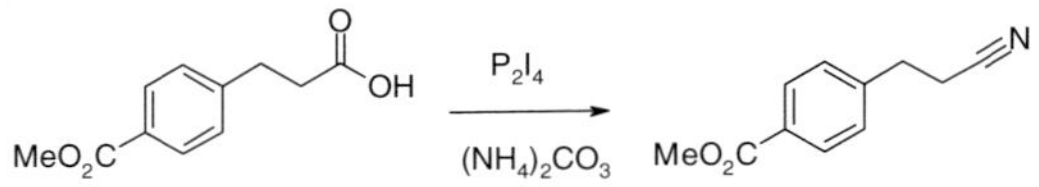

under mild conditions. 3-(4-Methoxycarbonylphenyl)propanoic acid added to 1.1 eqs. P_2I_4 in dry carbon disulfide, the mixture stirred for 2 min at room temp., 1.1 eqs. $(NH_4)_2CO_3$ added, stirring continued for 7 h (TLC monitoring), the mixture diluted with carbon disulfide, washed sequentially with 10% aq. $NaHCO_3$ and water, the organic layer dried over Na_2SO_4, concentrated *in vacuo*, and the residue purified by chromatography → 3-(4-methoxycarbonylphenyl)propanonitrile. Y 80%. This simple and high-yielding method is successful with both ar. and aliphatic acids (the latter requiring somewhat longer reaction times) affording high yields (17 examples; 80-92%) in the presence of ester, ether and alkene functionality. Carbon disulfide was the solvent of choice but initial reactions were successful in carbon tetrachloride and slow in chloroform or methylene chloride (no reaction in acetonitrile). F.e. and with retention of α-chirality s. V.N. Telvekar, R.A. Rane, Tetrahedron Lett. *2007*, *48* (34), 6051-3.

Bismuth(III) triflate — $Bi(OTf)_3$
Bismuth(III) triflate/potassium hexafluorophosphate — $Bi(OTf)_3/KPF_6$
Bismuth(III) trichloride — $BiCl_3$
Catalytic replacement of α,β-unsatd. hydroxyl groups by amido nucleophiles — OH → NH-

218. 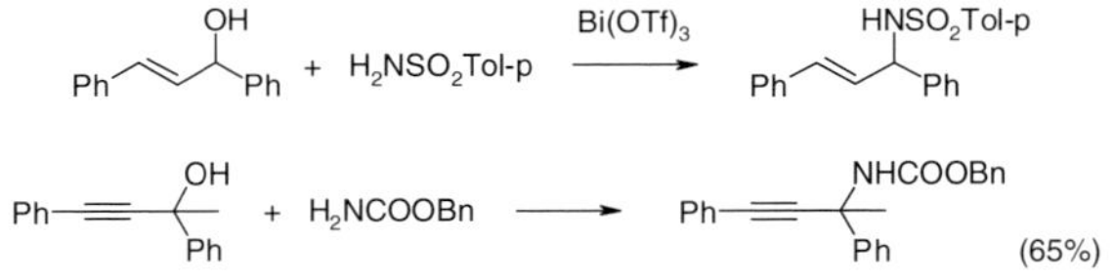

under mild conditions. Dioxane added to a mixture of *2 mol%* $Bi(OTf)_3$, *2 mol%* KPF_6 and Drierite contained in a test-tube, stirred for 10 min at room temp., 1.5 eqs. tosylamine and 1 eq. of the unsatd. alcohol added sequentially, stirred at 23-6° for 10 min, diluted with ether, treated with silica gel, filtered, and the filtrate worked up with column chromatographic purification on silica gel → product. Y 96%. The procedure is generally applicable to the nucleophilic substitution of 2-ethylene-*prim*- and 2-ethylene-*sec*-alcohols (without double bond shift), 2-acetylene-*sec*- and 2-acetylene-*tert*-alcohols, and sec. benzyl alcohols with various arenesulfonamides, urethans and acylamines (the latter with 5 mol% catalyst). With chiral 2-ethylenealcohols, however, racemic products were obtained, suggestive of intermediate formation of a carbenium ion. Reaction was slower with $BiCl_3$. F.e.s. H. Qin, N. Yamagiwa, S. Matsunaga, M. Shibasaki, Angew. Chem. Int. Ed. Engl. *2007, 46* (3), 409-13; N-benzylsulfonylamines with $Bi(OTf)_3$ s. *72*, 204; nucleophilic substitution of 2-acetylenealcohols by benzamide or tosylamide with $BiCl_3$ s. *72*, 106.

Niobium pentaethoxide s. under $[Ir(cod)Cl]_2$ — $Nb(OEt)_5$

Hydrogen peroxide/hydrogen chloride — H_2O_2/HCl
Hydrogen peroxide/hydrogen chloride/lithium chloride — $H_2O_2/HCl/LiCl$
Benzimidazoles from *o*-diamines and aldehydes ○
2-Arylbenzimidazoles s. *46*, 321s*72*; *69*, 171s*72*

Sodium dithionite/microwaves *$Na_2S_2O_4$/[\\\\]*
Benzimidazoles from *o*-nitramines and aldehydes ○
with $SnCl_2$, polymer-based synthesis, cf. *60*, 139; with Na-dithionite in DMF/water under microwave irradiation s. M.P. Surpur, P.R. Singh, S.B. Patil, S.D. Samant, Synth. Commun. *2007*, *37* (8), 1375-9.

N-Tosylimidazole/sodium azide/triethylamine/tetra-n-*butylammonium iodide* ←
Azides from alcohols OH → N_3
via benzenesulfonates cf. *27*, 401; one-pot procedure with N-tosylimidazole/NaN_3/Et_3N and a little Bu_4NI in DMF at reflux s. M.N.S. Rad, S. Behrouz, A. Khalafi-Nezhad, Tetrahedron Lett. *2007*, *48* (19), 3445-9.

Benzenesulfonic anhydride/4-(dimethylamino)pyridine *$(PhSO_2)_2O$/DMAP*
N-Subst. carboxylic acid amides from acids and amines COOH → CON<

219. $$PhCH_2CH_2COOH + H_2NCH_2CH_2CH_2Ph \xrightarrow{(PhSO_2)_2O} PhCH_2CH_2CONHCH_2CH_2CH_2Ph$$

Anhydrous benzenesulfonic anhydride (ca. 1.1 eqs.) and DMAP (ca. 2.2 eqs.) added successively at room temp. to a stirred soln. of 3-phenylpropionic acid in methylene chloride, stirred for 10 min, ca. 1 eq. 3-phenylpropylamine in the same solvent added, stirring continued for 1 h, and quenched with satd. aq. $NaHCO_3$ → 3-phenyl-N-(3-phenylpropyl)propanamide. Y 100%. The procedure is mild, simple, high-yielding, generally applicable to the condensation of both simple and hindered carboxylic acids with prim. or sec. amines [in stoichiometric amount!], and is the first such reaction with a sulfonic anhydride as dehydrating agent. F.e. and comparison of bases s. S. Funasaka, K. Kato, T. Mukaiyama, Chem. Lett. *2007*, *36* (12), 1456-7; with pyridine-3-carboxylic anhydride/DMAP cf. S. Funasaka, T. Mukaiyama, ibid. *36* (5), 658-9.

Trimethylsilyl triflate/N,O-bis(trimethylsilyl)acetamide/triethanolamine/microwaves ←
Nucleosides from acyl glycosides OAc → N<
with K-perfluorobutanesulfonate/$(Me_3Si)_2NH$/Me_3SiCl cf. *34*, 360; with N,O-bis(trimethylsilyl)-acetamide/Me_3SiOTf/triethanolamine under microwave irradiation s. B.C. Bookser, N.B. Raffaele, J. Org. Chem. *2007*, *72* (1), 173-9.

Sulfamic acid *H_2NSO_3H*
Benzimidazoles from *o*-diamines and orthocarboxylic acid esters ○
with $Yb(OTf)_3$ cf. *52*, 147s*69*; with 5 mol% sulfamic acid in methanol at room temp. s. Z.H. Zhang, T.S. Li, J.J. Li, Monatsh. Chem. *2007*, *138* (1), 89-94; with 10 mol% $ZrCl_4$ in ethanol s. Z.H. Zhang, L. Yin, Y.M. Wang, Catal. Commun. *2007*, *8* (7), 1126-31.

Sulfuric acid salts ←
Imidazole ring from *o*-diamines and aldehydes s. *46*, 321s*72*; *69*, 171s*72*

Sulfate/tin(IV) oxide *SnO_4^{2-}/SnO_2*
N-Acylation with anhydrides s. *54*, 79s*72* NH → NAc

Silica-sulfuric acid *SiO_2-OSO_3H*
1,3,4-Oxadiazoles from carboxylic acid hydrazides ○
and orthocarboxylic acid esters with BF_3 cf. *50*, 182; with silica-sulfuric acid s. M. Dabiri, P. Salehi, M. Baghbanzadeh, M.A. Zolfigol, M. Bahramnejad, Synth. Commun. *2007*, *37* (7), 1201-9.

Molybdenum pentachloride *$MoCl_5$*
N-Protected benzylamines from benzyl alcohols s. *72*, 204 OH → N<

tert-*Butyl hypochlorite* t-*BuOCl*
Δ^2-Imidazolines from aldehydes and 1,2-diamines s. *69*, 169s*72* ○

Hydrochlorides ←
1*H*-Indazoles from *o*-hydroxyoxo compds. and hydrazines ○

220. 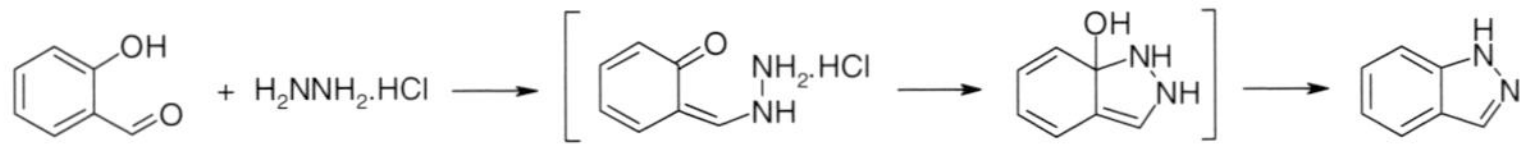

Excess of hydrazine hydrochloride added to a soln. of salicylaldehyde in ethanol, refluxed for 2-3 h, concentrated under reduced pressure, and filtered through a pad of silica → 1*H*-indazole. Y 70%. The procedure is simple, rapid, based on readily available substrates, and is generally applicable to the reaction of hydrazine or arylhydrazine hydrochlorides with salicylaldehydes or *o*-hydroxyacetophenones possessing electron-donating or -withdrawing groups. Yields were poor under basic conditions, and there was no reaction (other than hydrazone formation) with *o*-halogeno-, *o*-methoxy-, *o*-allyloxy- or *o*-benzyloxy-salicylaldehydes. A Fischer-indole-type mechanism is invoked, involving initial tautomerization to the *o*-quinone methid. F.e.s. P.D. Lokhande, A. Raheem, S.T. Sabale, A.R. Chabukswar, S.C. Jagdale, Tetrahedron Lett. *2007, 48* (39), 6890-2.

Hydroxylamine hydrochloride/triethylamine/oxalyl chloride *$NH_2OH{\cdot}HCl/Et_3N/(COCl)_2$*
Hydroxylamine hydrochloride/1,8-diazabicyclo[5.4.0]undec-7-ene/ ethyl phosphorodichloridate ←
Nitriles from aldehydes CHO → CN
with hydroxylamine hydrochloride in DMF cf. *31*, 373; with added Et_3N and oxalyl chloride in refluxing acetonitrile s. B. Movassagh, A. Fazeli, Synth. Commun. *2007*, *37* (4), 623-8; with added DBU then $EtOP(O)Cl_2$ in methylene chloride at room temp. s. J.L. Zhu, F.Y. Lee, J.D. Wu, C.W. Kuo, K.S. Shia, Synthesis *2007* (8), 1317-9.

Tetra-n-*butylammonium iodide s. under N-Tosylimidazole* *Bu_4NI*

Hydrogen chloride (s.a. under H_2O_2 and Fe) *HCl*
Selective N-carbo-*tert*-butoxylation NH → NCOOBu-*t*
of prim. amines s. *49*, 284s*72*

Benzimidazoles from *o*-diamines ○
and acetals – 2-C-Glycosylated benzimidazoles s. *69*, 171s*72*; *46*, 321s*72*

Iron/hydrogen chloride *Fe/HCl*
Ar. N-arylaldimines from ar. aldehydes and ar. nitro compds. CHO → CH=NAr

221. 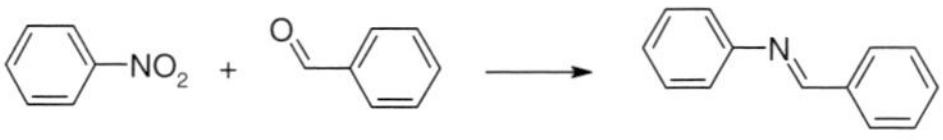

4.5 mmol HCl added to a mixture of 0.72 mmol nitrobenzene, 0.72 mmol benzaldehyde and *7.32 mmol* iron powder in 2:1 ethanol/water, and heated to 65° for 1.5 h → product Y 88%. The procedure is mild, tolerant of electron-withdrawing and -donating groups on the aromatic ring, and leaves esters and halogen unaffected. Reaction requires an optimum of 6 eqs. iron(0) for high yields, and the added acid is essential for the conversion. However, there was no reaction with $FeCl_2$ or $FeCl_3$. F.e. and comparison of acids s. A. Korich, T. Hughes, Synlett *2007* (16), 2602-4.

Magnetized 4-(dialkylamino)pyridine-supported iron oxide nanoparticles ←
Magnetized 4-(dialkylamino)pyridine-supported iron oxide nanoparticles as a readily recoverable and recyclable nucleophilic catalyst ←

222.

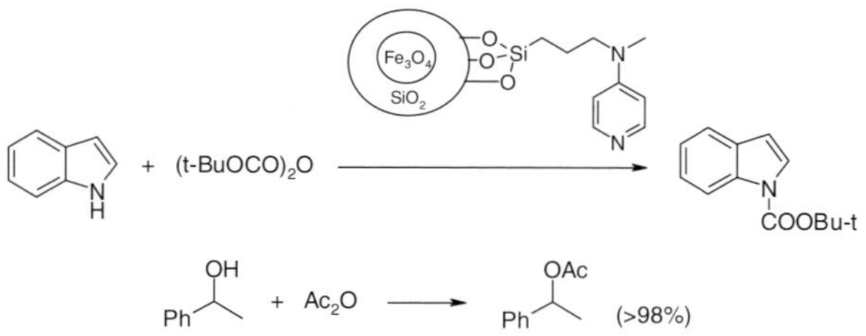

Heterogeneous N-acylation. 0.327 mmol *tert*-butoxyformic anhydride in THF added to a mixture of 0.164 mmol (10 mol%) magnetized 4-(dialkylamino)pyridine-supported iron oxide nanoparticles and 0.164 mmol indole under argon, the suspension shaken for 16 h at room temp., the reaction vessel placed in the vicinity of an external magnet to remove the supported reagent, and the soln. worked up with flash chromatographic purification → product. Y ca. 100%. The reagent is robust, possesses unprecedented activity at room temp. (at catalyst loadings as low as *0.2 mol%*), and is recyclable as many as *thirty times* without degradation or deactivation (in this example, the yield being 97-8% after the 24th cycle!). The catalyst is also simple to prepare by condensation of silica-coated magnetite nanoparticles (9.6 nm average diameter) with a triethoxysilyl-functionalized 4-dialkylaminopyridine deriv. F.e. and reactions (incl. O-acylation and classical azlactone formation by rearrangement of 5-acoxyoxazoles) s. C.Ó. Dálaigh, S.A. Corr, Y. Gun'ko, S.J. Connon, Angew. Chem. Int. Ed. Engl. *2007, 46* (23), 4329-32.

Carbonyl(dihydrido)tris(triphenylphosphine)ruthenium(II)/ 9,9-dimethyl-4,5-bis(diphenylphosphino)xanthene *$Ru(CO)H_2(PPh_3)_3$/Xantphos*
Pyrroles from 2-yne-1,4-diols and prim. amines via γ-diketones ○

223.

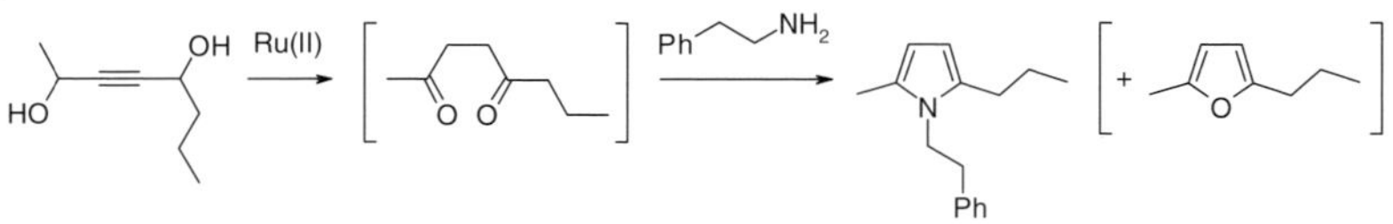

One-pot conversion via Paal-Knorr cyclization. 3-Octyne-2,5-diol, 2.5 mol% $Ru(CO)H_2(PPh_3)_3$ and 2.5 mol% Xantphos heated under reflux (110°) in dry toluene for 30 min, 2 eqs. 2-phenylethylamine added, the mixture refluxed for 24 h, cooled, diluted with methanol/toluene (1:1), concentrated *in vacuo*, and purified by chromatography → 1-(2-phenylethyl)-2-methyl-5-*n*-propylpyrrole. Y 86% (containing 14% 2-methyl-5-*n*-propylfuran; GC yields). The method is most selective for condensing unhindered 1,4-alkynediols with aliphatic amines, hindered substrates affording pyrroles as major products (with 63-100% chemoselectivity). It is notable that *addition of a base* to the reaction mixture *had no effect on product distribution*. Selectivity is also high for neutral electron-rich anilines (4-methoxyaniline and 3-hexyne-2,5-diol giving a 98:2 mixture of pyrrole:furan) but 4-nitroaniline and 2-aminopyridine gave mainly γ-diketones. The method is compatible with ester, halide, nitrile and furyl functionality. F.e.s. S.J. Pridmore, P.A. Slatford, A. Daniel, M.K. Whittlesey, J.M.J. Williams, Tetrahedron Lett. *2007, 48* (29), 5115-20.

Palladium s. under Novozyme-433 *Pd*

*Palladium(II) acetate/tri-*tert*-butylphosphonium fluoroborate/sodium* tert*-butoxide/ hydroxylamine O-benzyl ether* ←

Sym. triarylamines and sym. sec. diarylamines from ar. bromides NH → NAr
Hydroxylamine O-benzyl ether as ammonia equivalent

224. 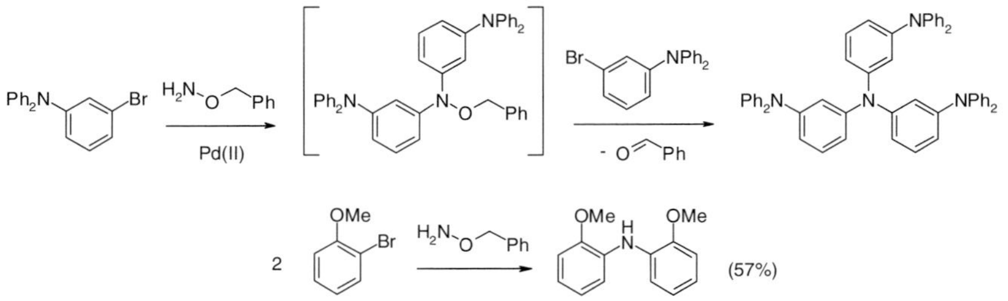

The Buchwald-Hartwig amination reaction is an effective method for the formation of N-C bonds, working efficiently with aliphatic and ar. amines. Recent work has extended the scope of the reaction using various ammonia equivalents but these typically provide prim. amines only. The illustrated method extends the scope of the reaction further to provide di- and triaryl-amines using a *commercially available ammonia equivalent.* **E: Sym. triarylamines.** Hydroxylamine-O-benzyl ether and 4 eqs. 3-diphenylaminophenyl bromide added to a suspension of 4 mol% $Pd(OAc)_2$, 5 mol% tri-*tert*-butylphosphonium fluoroborate and 5 eqs. *t*-BuONa in *toluene*, the mixture refluxed (110°) for 18 h, cooled, quenched with 2 *M* aq. HCl, the aq. layer extracted with methylene chloride, the organic layer dried over $MgSO_4$, filtered, concentrated *in vacuo*, and the residue purified by chromatography → tris(3-diphenylaminophenyl)amine. Y 66% (by NMR). Triarylamines were formed exclusively with *m*- and *p*-subst. aryl bromides (little or no reaction taking place with aryl chlorides or triflates), whereas *o*-subst. analogs afforded diarylamines exclusively or predominantly (97:3). NMR yields were in the range 42-86% with isolated yields (where carried out) some 10% lower. 2-Bromopyridine gave 2-benzyloxypyridine (Y 36%) under these conditions. The authors suggest that an O-benzyl-N-arylhydroxylamine is formed which undergoes palladium-induced fragmentation to amine products and benzaldehyde. F.e.s. R.B. Bedford, M. Betham, Tetrahedron Lett. *2007, 48* (51), 8947-50.

Tris(dibenzylideneacetone)dipalladium/triphenylphosphine *$Pd_2(dba)_3/Ph_3P$*
2-α-Amino-1,3-dienes from 1-sulfonyloxy-1-vinylcyclopropanes C

225. 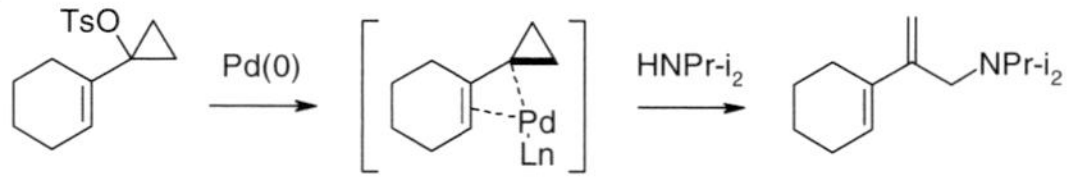

Cyclic 2-α-amino-1,3-dienes. A soln. of 5 mol% $Pd_2(dba)_3$ and 20 mol% triphenylphosphine in THF stirred for 10 min under argon, 2 eqs. diisopropylamine and the startg. tosylate added sequentially, heated at 65° overnight, cooled to room temp., and worked up with column chromatographic purification on basic alumina → product. Y 88%. The procedure, which presumably involves nucleophilic addition to a palladium π-allyl species, is applicable to prim. or sec. aliphatic or aromatic amines. The driving force for the reaction is attributed to minimalization of allylic strain on ring opening, which is energetically more favourable than the familiar cyclobutane ring expansion. With silica in 1:5 ethyl acetate/hexanes the startg. sulfonates isomerized to the corresponding **2-α-sulfonyloxy-1,3-dienes.** F.e.s. L.G. Quan, H.G. Lee, J.K. Cha, Org. Lett. *2007, 9* (22), 4439-42.

Chloro(cyclooctadiene)iridium(I) dimer/chiral cyclic phosphoromonoamidites/ n-propylamine/N-potassio compd. ←

Regiospecific asym. formation of N-protected 2-ethylene-*prim*-amines from 2-ethylenecarbonic acid esters C=C-C-OCO$_2$R → C(NH-)C=C

226.

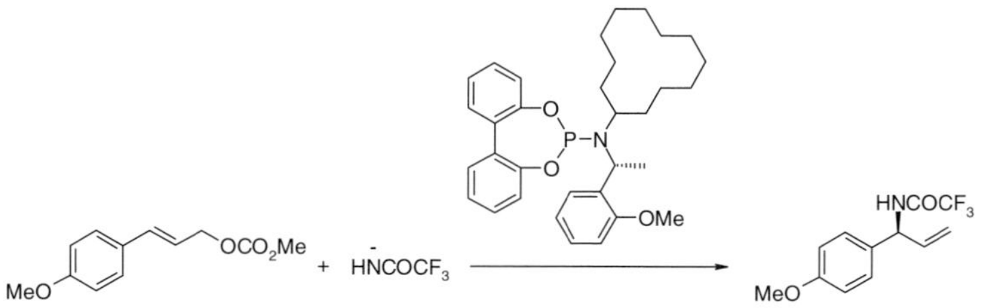

1 Mol% [Ir(cod)Cl]$_2$ and 2 mol% chiral phosphoromonoamidite ligand heated in THF containing *n*-propylamine (as catalyst activator) at 50° for 30 min, evaporated to dryness in a reaction vial, N-potassiotrifluoroacetamide **as protected ammonia equivalent** and the startg. allyl carbonate added, the vial sealed with a PTFE septum, stirred for 12 h, diluted with hexanes, and quenched with water → 3-(4-methoxyphenyl)-3-trifluoroacetylaminopropene. Y 93% (e.e. 98%). The illustrated ammonia equivalent is inexpensive and gave products cleanly (47-93% yield) with high regio- and enantio-selectivity (e.e. 85-96%) for electron-rich, -neutral and -poor cinnamyl carbonates as well as aliphatic allyl carbonates. The more reactive di-*tert*-butyl N-lithio-*tert*-butyliminodicarboxylate reacted with a similar range of substrates (Y 45-82%; e.e. 92-97%) generating bis-Boc-protected allylic amines which can be selectively mono-deprotected. With ammonia itself, however, the corresponding chiral diallylamines were formed. F.e. incl. catalyst and ligand comparison s. M.J. Pouy, A. Leitner, D.J. Weix, S. Ueno, J.F. Hartwig, Org. Lett. *2007*, *9* (20), 3949-52.

Chloro(cyclooctadiene)iridium(I) dimer/chiral cyclic phosphoromonoamidites/ propylamine/niobium pentaethoxide or triphenylborane ←

Regiospecific iridium-catalyzed asym. N-allylation with Lewis acid-activated terminal 2-ethylenealcohols C=C-C-OH → C(N<)C=C

227.

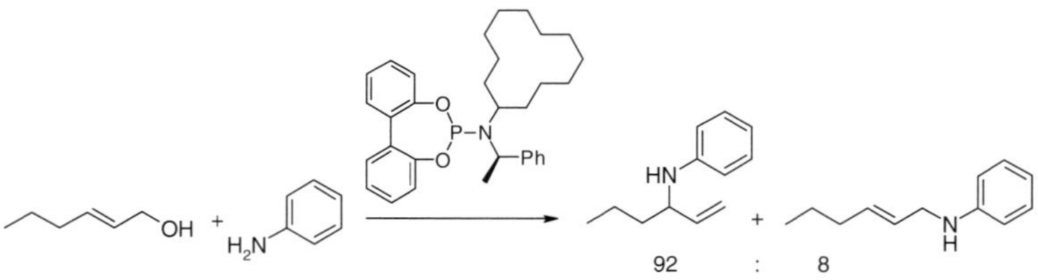

A mixture of [Ir(cod)Cl]$_2$ and 2 eqs. chiral phosphoromonoamidite ligand heated at 50° in propylamine/THF (1:1) for 1 h, 2 mol% of this stock catalyst soln. added to a mixture of 2-hexenol, 1.2 eqs. Nb(OEt)$_5$, *powdered 4 Å molecular sieves* and 1.5 eqs. aniline in THF, the mixture stirred at 50° for 24 h, cooled, added to satd. aq. NaHCO$_3$ and methylene chloride, stirred vigorously for 1 h, filtered through Celite, the residue and aq. filtrate washed well with methylene chloride, the organic phases washed with brine, dried over anhydrous Na$_2$SO$_4$, and concentrated *in vacuo* → product. Y 70% (e.e. 90%). Significantly, molecular sieves are essential for high yields as is activation of the allyl alcohol with a stoichiometric amount of the Lewis acid, but yields and selectivity are not affected by the choice of solvent (toluene, dioxane, dimethoxyethane, *tert*-butyl methyl ether all being acceptable). The method is highly regioselective (generally >90:10) for the reaction of aryl and aliphatic allyl alcohols with electron-rich or -poor anilines, as well as *benzylamines* and *morpholine*, giving e.e. of 81-94% and yields of 66-90% (bulky 4-methyl-2-pentenol affording a 45% yield and selectivity of 79:14). A second series of experiments effected similarly high regio- and enantio-selectivity using 8 mol% of triphenylboron as Lewis acid, but yields and the scope of the reaction were somewhat reduced. F.e.s. Y. Yamashita, A. Gopalarathnam, J.F. Hartwig, J. Am. Chem. Soc. *2007*, *129* (24), 7508-9.

Chloro(1,5-cyclooctadiene)iridium(I) dimer/chiral 1,1′-binaphthyl-2,2′-diyl phosphoromonoamidites/sodium hydride or 1,5,7-triazabicyclo[4.4.0]dec-5-ene ←

Regiospecific asym. iridium(I)-catalyzed N-allylation NH → N-C-C=C

of sulfonamides cf. *68*, 178s*69*; of amides with 1,5,7-triazabicyclo[4.4.0]dec-5-ene or NaH s. S. Spiess, C. Berthold, R. Weihofen, G. Helmchen, Org. Biomol. Chem. *2007*, *5* (15), 2357-60.

Dichloro(pentamethylcyclopentadienyl)iridium(III) dimer/ sodium hydrogen carbonate *[Cp*IrCl$_2$]$_2$/NaHCO$_3$*

Piperazines from 1,2-diamines and glycols in water ○

228. 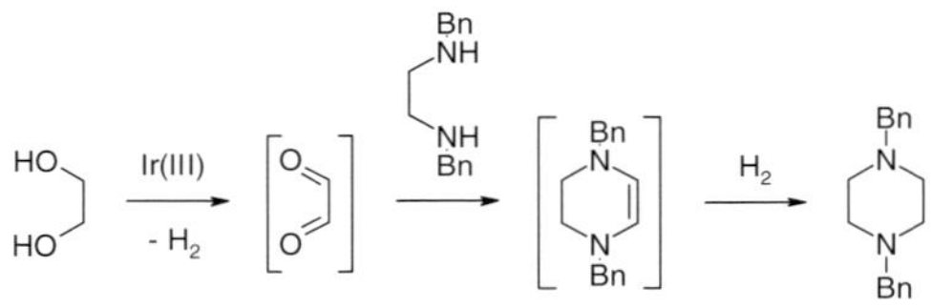

Equimolar amounts of 1,2-bis(benzylamino)ethane and ethylene glycol heated in a screw-top vial with 0.5 mol% [Cp*IrCl$_2$]$_2$ and 5 mol% NaHCO$_3$ in *water* under argon at 140° for 24 h, the mixture cooled, filtered, and washed with water and a little ether → 1,4-dibenzylpiperazine. Y 73%. The procedure is general (six examples giving 60-100% yields), atom-economical (water being the only by-product) and environmentally friendly. The use of *water* as solvent is also notable since the reaction involves intermediate formation of imines. It is also significant that the hydrogen liberated in the initial phase of the reaction is used in the ultimate hydrogenation. Reactions with *sec*-amino derivs. and alcohols were slower compared to those with *prim*-amino analogs, requiring a higher temp. (140° vs. 110°), and addition of trifluoroacetic acid to the mixture did not increase the reaction rate. The illustrated product was also prepared in 94% yield using a mixture of ethylene glycol and benzylamine, and **unsym. piperazines** were obtained in one-pot (2 examples; 63 and 90%) **from two different glycol molecules** (with benzylamine). F.e. and chiral products, also with toluene as solvent, s. L.U. Nordstrøm, R. Madsen, Chem. Commun. *2007* (47), 5034-6.

*Dichloro(cyclooctadiene)platinum(II)/bis[*o*-(diphenylphosphino)-phenyl] ether* *Pt(cod)Cl$_2$/DPEphos*

Platinum-catalyzed N-allylation with 2-ethylenealcohols NH → N-C-C=C

of ar. amines with Pt(acac)$_2$/Ph$_3$P/Ti(OPr-*i*)$_4$ cf. *62*, 168; with Pt(cod)Cl$_2$ (1 mol%)/DPEphos (2 mol%), also of aliphatic amines, s. M. Utsunomiya, Y. Miyamoto, J. Ipposhi, T. Ohshima, K. Mashima, Org. Lett. *2007*, *9* (17), 3371-4.

Nitrogen ↑ NC ↓↑ N

Microwaves [\\\\\]

Substitution of amidines C(=NR)N<

cf. *19*, 490; of cyclic amidines in THF under microwave irradiation s. M. de Fatima Pereira, V. Thiéry, T. Besson, Tetrahedron Lett. *2007*, *48* (43), 7657-9.

Copper(I) chloride/chiral spirocyclic bis(Δ²-oxazolines)/sodium tetrakis[3,5-bis-(trifluoromethyl)phenyl]borate ←

α-(Arylamino)- from α-diazo-carboxylic acid esters and ar. amines $C{=}N_2 \rightarrow CHN{<}$

Copper-catalyzed asym. nitrene insertion

229. 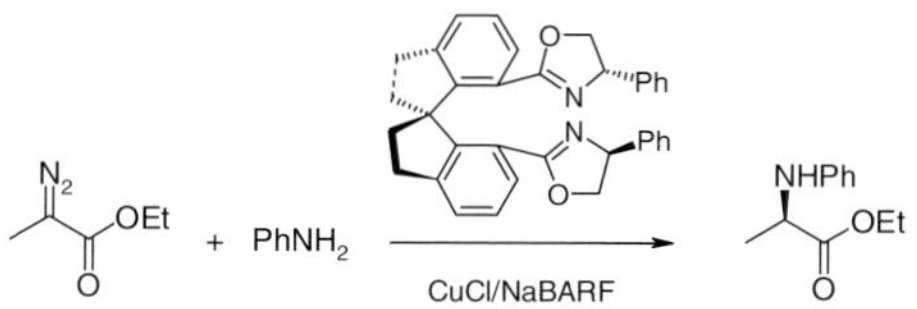

Enantioselectivities as high as 98% are now achievable for copper-catalyzed insertion of nitrenes into the N-H bond by using chiral *spirocyclic* bis(Δ²-oxazolines) as ligand. **E:** Methylene chloride injected into a Schlenk tube containing 5 mol% CuCl, 6 mol% chiral spirocyclic bis(Δ²-oxazoline) and 6 mol% Na-tetrakis[3,5-bis(trifluoromethyl)phenyl]borate [in an argon-filled glove box], stirred at room temp. for 2 h under argon, equivalent amounts of aniline and ethyl α-diazo-propionate added, stirring continued for 2 h, and worked up with flash chromatographic purification → (R)-ethyl 2-(phenylamino)propionate. Y 94% (e.e. 98%). High enantioselectivity (85-98%) was recorded for the reaction of aliphatic α-diazo-esters with a range of anilines (incl. *o*-subst. derivs.) and naphthylamines, but aliphatic amines and cyclohexylamine were inert. Reaction was slower with the corresponding α-diazobutyrate (Y 51%; e.e. 94%), while enantio-selectivity was very low (8%) with the phenylacetic analog. The nature of the counterion [tetrakis-[3,5-bis(trifluoromethyl)phenyl]borate being optimum] is critical, as is also the spirocyclic moiety of the ligand since Evans′ (S,S)-Ph-Box gave only 5% enantiomeric excess. F.e.s. B. Liu, S.-F. Zhu, W. Zhang, C. Chen, Q.-L. Zhou, J. Am. Chem. Soc. *2007*, *129* (18), 5834-5.

Copper(I) iodide *CuI*

N-Phosphorylamidines ←

from terminal acetylene derivs., phosphoromonoazidates and amines

230. 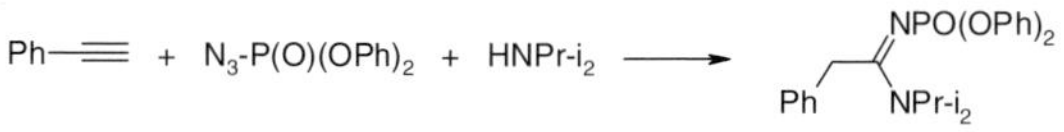

0.75 Eqs. diisopropylamine added slowly at room temp. under N_2 to a stirred soln. of phenyl-acetylene, 0.5 eqs. diphenyl phosphorazidate and 5 mol% CuI in THF, diluted after 12 h with methylene chloride and aq. NH_4Cl, stirred for a further 30 min, and worked up with flash column chromatographic purification → N¹,N¹-diisopropyl-N²-(diphenylphosphoryl)-2-phenylacet-amidine. Y 90%. The procedure is generally applicable to the coupling of a range of terminal acetylene derivs. (incl. alkynes, electron-rich or -poor arylacetylenes, 3-thienylacetylene, enynes and functionalized alkynes) with *diaryl* phosphorazidates and prim. or sec. amines. Dialkyl phos-phorazidates gave poor yields, but the desired N-dialkoxyphosphinylamidines can be simply obtained from the di(aryloxy)phosphinyl derivs. by transesterification. Chiral products were also obtained with (S)-1,1′-bi-2-naphthol as reactant, and the products readily underwent C-α-alkylation with asym. induction. F.e.s. S.H. Kim, D.Y. Jung, S. Chang, J. Org. Chem. *2007*, *72* (25), 9769-71; glycosyl-linked **N-sulfonylamidines from sulfonic acid azides** (in place of the phosphor-azidates) s. S. Mandal, H.M. Gauniyal, K. Pramanik, B. Mukhopadhyay, ibid. 9753-6.

Zinc/ammonium chloride *Zn/NH₄Cl*

β-Amino-α,β-ethylenecarboxylic $COCH \rightarrow C(NHR){=}C$

from β-ketocarboxylic acid esters and azides in aq. medium

231. 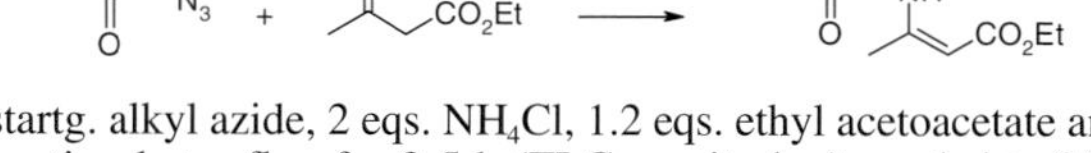

A mixture of the startg. alkyl azide, 2 eqs. NH_4Cl, 1.2 eqs. ethyl acetoacetate and 1 eq. Zn dust in 20:3 ethanol/water stirred at reflux for 3-5 h (TLC monitoring), cooled to 25-35°, filtered, and the filtrate worked up with column chromatographic purification on silica gel → product. Y 89%. The (Z)-isomer (depicted) predominated. The procedure is simple, cost-effective and environmentally friendly. Furthermore, there was no over-reduction or isolation of the intermediate amines, and there was no loss of O-benzyl groups. F.e.s. A.S. Prabhakar, S. Sashikanth, P.P. Reddy, P. Cherukupally, Tetrahedron Lett. *2007*, *48* (49), 8709-11.

Magnesium perchlorate/chiral bis(Δ^2-oxazolines) *$Mg(ClO_4)_2$/Box*

N^1-Subst. 3-pyrazolidones from N-acyl-α,β-ethylenecarboxylic acid amides via regiospecific asym. Michael addition of monosubst. hydrazines ○

232. 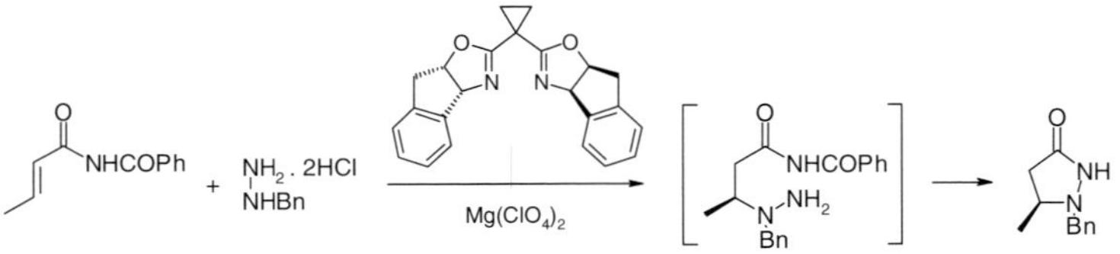

A mixture of 30 mol% $Mg(ClO_4)_2$, 33 mol% chiral bis(Δ^2-oxazoline) and 4 Å molecular sieves in methylene chloride stirred for 2 h at room temp., the startg. imide added at room temp., stirred for 30 min, cooled *to -50°*, stirring continued for 15 min, a soln. of 1.5 eqs. hydrazine hydrochloride and 1.5 eqs. triethylamine in the same solvent added, stirred for 24 h, and worked up with column chromatographic purification on silica gel → (S)-1-benzyl-5-methyl-3-pyrazolidone. Y 92% (e.e. 84%; N^1-benzyl/N^2-benzyl deriv. 98:2). Enantioselectivity was high for the addition of a series of monoalkylhydrazines to α,β-ethylenic N-benzoylamides possessing alkyl, functionalized alkyl or carbo-*tert*-butoxy groups at the β-position, but yields were low with cinnamic acid derivs. Regioselectivity was ca. 100%, the more nucleophilic and bulky nitrogen atom of the hydrazine preferentially attacking the β-carbon, the (R)-selectivity being consistent with hydrazine addition to an *s-cis* rotamer of the magnesium-complexed imide (with the metal in a *cis*-octahedral geometry). Other Lewis acids and other N-functions on the α,β-ethylenecarboxamide were less effective. F.e.s. M.P. Sibi, T. Soeta, J. Am. Chem. Soc. *2007*, *129* (15), 4522-3.

Sodium tetrakis[3,5-bis(trifluoromethyl)phenyl]borate s. under CuCl *$NaBAr_4$*

Boron fluoride *BF_3*

3,4-Dihydro-2-pyridones from cross-conjugated dienones and azides via Lewis acid-catalyzed Nazarov cyclization and Schmidt rearrangement Regiospecific and stereoselective conversion

233. 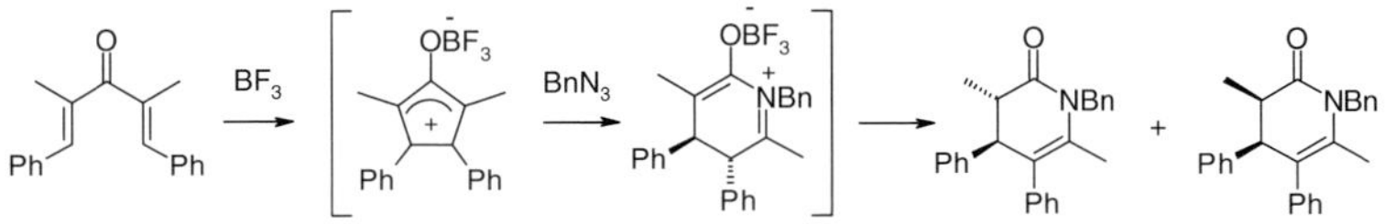

1.1 Eqs. BF_3-etherate added to a soln. of the startg. dienone and 2 eqs. benzyl azide in methylene chloride at -78°, stirred for 10 min, quenched with satd. aq. $NaHCO_3$, and worked up with column chromatographic clean-up of the product → *trans*-1-benzyl-3,6-dimethyl-4,5-diphenyl-2-pyridone (Y 55%) and *cis*-isomer (Y 27%). Reaction is presumed to involve initial Lewis acid-catalyzed Nazarov cyclization to give an intermediate 2-oxidocyclopentenyl ion, which is trapped by the azide prior to Schmidt rearrangement to give a transient 1,4-dipole; the latter then rearranges to the product via either proton transfer or 1,5-hydride shift. For unsymmetrical substrates, complete regioselective addition of the azide was observed in favour of attack on the less substituted side of the cyclopentenyl ion. The *trans*-product predominated or was formed exclusively. F.e. and stereochemical considerations s. D. Song, A. Rostami, F.G. West, J. Am. Chem. Soc. *2007*, *129* (39), 12019-22.

Boron fluoride/polymer-based triphenylphosphine BF_3/Ⓟ-$ArPPh_2$
Cyclic iminoesters from azidoalcohols and aldehydes ○
with BF_3 cf. *11*, 489s*51*,*64*; with BF_3 then polymer-based triphenylphosphine, Δ^2-oxazoline and 5,6-dihydro-4*H*-oxazine library synthesis, s. P. Chaudhry, F. Schoenen, B. Neuenswander, G.H. Lushington, J. Aubé, J. Comb. Chem. *2007*, *9* (3), 473-6.

Chiral bis(Δ^2-oxazolines) s. under CuCl and Mg(ClO_4)$_2$ *Box*

Polymethylhydrosiloxane/(μ_3,η^2:η^3:η^5-acenaphthylene)heptacarbonyltriruthenium ←
Tert. amines 2 RCONHR′ → $RCH_2N(R')CH_2R$
from two N-subst. carboxylic acid amide molecules via reductive N-alkylation s. *70*, 37s*72*

Triphenylphosphine Ph_3P
N-Acylamines from azides s. *71*, 160s*72* N_3 → NHAc

Polymer-based triphenylphosphine s. under BF_3 Ⓟ-$ArPPh_2$
Ammonium chloride s. under Zn NH_4Cl

Diruthenium [Shvo] hydride complex ←
N-Alkylation of prim. ar. amines with aliphatic prim. amines $ArNH_2$ → ArNHR

234.

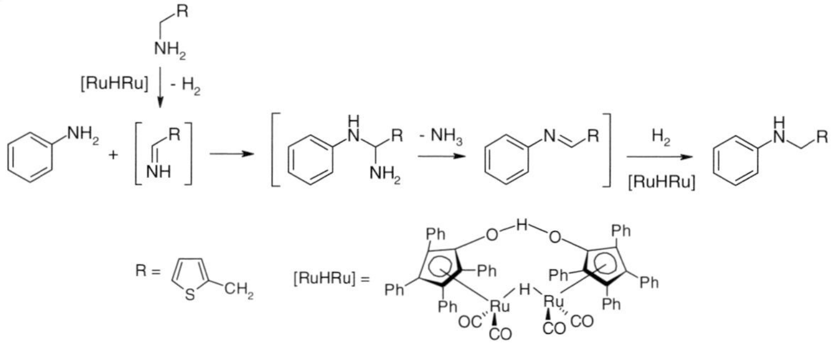

2-(Thiophen-2-yl)ethanamine, 2 eqs. aniline and 1 mol% Shvo complex dissolved in 2-methylbutan-2-ol under argon in an ACE-pressure tube, the tube fitted with a Teflon cap, heated at 150° for 24 h, solvent removed, and the residue worked up with column chromatographic purification → N-[2-(thiophen-2-yl)ethyl]aniline. Y 97%. The procedure is salt- and base-free, does not require high-pressure equipment, and is applicable to the N-alkylation of a range of activated and electron-rich anilines (incl. halogenated derivs. which cannot easily participate in the Pd-catalyzed Buchwald-Hartwig reaction) with both simple and branched aliphatic prim. amines. Reaction is also applicable to aminopyridines but hindered anilines gave a low yield. Mechanistically, a sequential dehydrogenation-hydrogenation ['hydrogen-borrowing'] is invoked with participation of the ruthenium catalyst in both phases: initial dehydrogenation of the alkylamine yields an imine, to which the aniline adds prior to elimination of NH_3 and hydrogenation of the resulting N-arylaldimine [with hydrogen evolved in the initial step!]. Nitro, nitrile and amide groups were tolerated. F.e.s. D. Hollmann, S. Bähn, A. Tillack, M. Beller, Angew. Chem. Int. Ed. Engl. *2007*, *46* (43), 8291-4.

Halogen ↑ NC ↓↑ Hal

Without additional reagents *w.a.r.*
N-Triflylaziridines from ethylene derivs. 
and N-triflylimino[4-(trifluoromethyl)phenyl]-λ^3-bromane s. *72*, 167

Electrolysis/1-butyl-3-methylimidazolium fluoroborate ⚡/[bmim]BF_4
Urethans from amines and halides >NH + RHal → >NCOOR
and CO_2 s. *50*, 214s*72*

Microwaves s. under Zn [\\\\\\\\]

Potassium fluoride-Celite *KF-Celite*
N-Alkylation NH → NR
with KF/alumina cf. *37*, 364s*59*; N-monoallylation of anilines with β,γ-ethylenehalides and KF-Celite in refluxing acetonitrile s. V. Pace, F. Martínez, M. Fernández, J.V. Sinisterra, A.R. Alcántara, Org. Lett. *2007*, *9* (14), 2661-4.

1,2-Diamines s. under CuI and FeCl$_3$ ←

1,8-Diazabicyclo[5.4.0]undec-7-ene/N,N′-ditosylhydrazine *DBU/TsNHNHTs*
Diazo compds. from bromides CHBr → C=N_2

235. 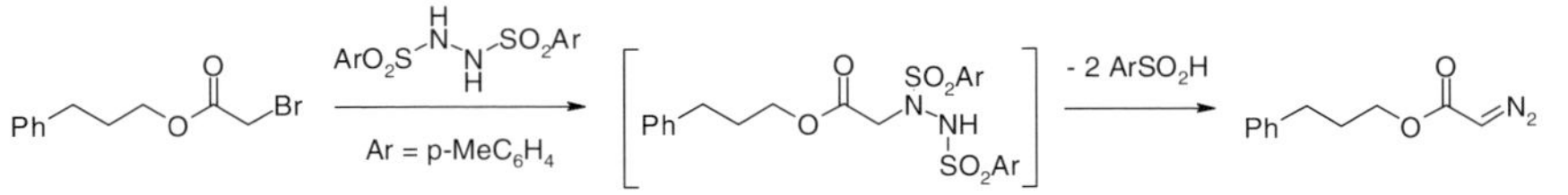

Diazoacetic acid esters. 3-Phenylpropyl bromoacetate (prepared crude from a mixture of 3-phenylpropanol and bromoacetyl bromide in acetonitrile containing $NaHCO_3$) and 2 eqs. N,N′-ditosylhydrazine (readily prepared from *p*-toluenesulfonic acid hydrazide and tosyl chloride in methylene chloride with pyridine as base) dissolved in THF, cooled to 0°, 5 eqs. DBU added dropwise, stirred for 10 min, quenched with satd. $NaHCO_3$ soln., and worked up with purification on neutral silica gel → 3-phenylpropyl diazoacetate. Y 88% (from bromoacetyl bromide). The procedure is simple and rapid, and the reagent (a stable crystalline solid) is easy to handle. F.e. and α-diazoketones s. T. Toma, J. Shimokawa, T. Fukuyama, Org. Lett. *2007*, *9* (16), 3195-7.

Copper(II) oxide nanoparticles/potassium hydroxide or carbonate *CuO/KOH or K_2CO_3*
Ligand-free copper-catalyzed N-arylation s. *72*, 238 NH → NAr

Copper(II) oxide/iron(III) acetoacetonate/cesium carbonate *CuO/Fe(acac)$_3$/Cs_2CO_3*
Iron/copper-catalyzed N-arylation of N-heterocyclics

236. 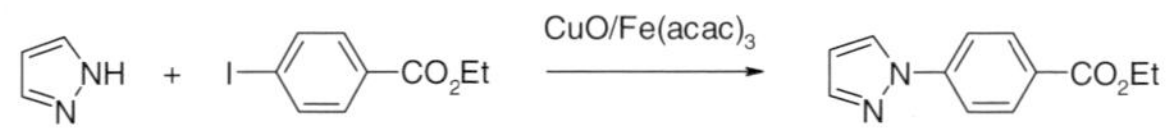

under mild conditions. Ethyl *p*-iodobenzoate and dry DMF added to a mixture of 1.5 eqs. 1*H*-pyrazole, 30 mol% Fe(acac)$_3$, 10 mol% CuO and 2 eqs. Cs_2CO_3 under N_2 in a Schlenk tube, the latter sealed, heated at 90° for 30 h, cooled to room temp., diluted with methylene chloride, filtered, the filtrate washed with water, the aq. phase extracted with methylene chloride, the organic phase dried over Na_2SO_4, and concentrated *in vacuo* → 1-(4-ethoxycarbonyl)pyrazole. Y 93%. This rare bimetal-catalyzed conversion is simple, mild (90°), environmentally friendly, inexpensive and clean (none of the usual Ullmann by-products being observed). The procedure is applicable to N-arylation of triazole, pyrazole, pyrrole, imidazole, indole and 2-pyrrolidone nucleophiles (Y 41-98%) with aryl bromides or iodides (irrespective of substitution), and one example is cited with an *aryl chloride* at 140° giving 40% yield. The method is compatible with ester, amine, ether, nitrile, nitro and keto functionality, with some electron-rich aryl halides requiring heating at a higher temperature (120°). No reaction occurs in the absence of either iron or copper pre-catalysts, and it appears that the acetoacetonate anion is critical for high yields. F.e. and comparison of copper and iron salts s. M. Taillefer, N. Xia, A. Ouali, Angew. Chem. Int. Ed. Engl. *2007*, *46* (6), 934-6.

Copper(II) 2,2,6,6-tetramethyl-3,5-heptanedionate/potassium tert-*butoxide* ←
Copper-mediated N-arylation
cf. *1*, 376; of aliphatic, aromatic and heteroaromatic amines with ar. iodides or bromides using

Cu(II)-2,2,6,6-tetramethyl-3,5-heptanedionate/KOBu-*t* at 120° s. N.S. Nandurkar, M.J. Bhanushali, M.D. Bhor, B.M. Bhanage, Tetrahedron Lett. *2007, 48* (37), 6573-6.

Tetrakis(acetonitrile)copper(I) hexafluorophosphate/Chloramine-T ←
Benzylic tosylamination H → NHTs

237. 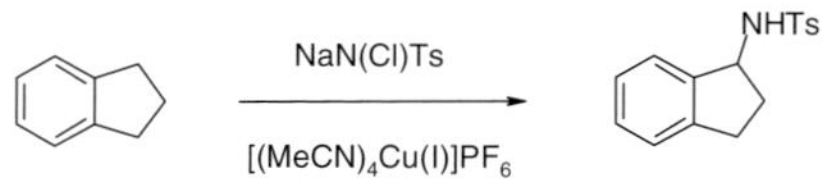

Indan stirred with a soln. of 1.3 eqs. *anhydrous* Chloramine-T, 10 mol% tetrakis(acetonitrile)-copper(I) hexafluorophosphate and 4 Å molecular sieves in acetonitrile at 65-70° for 10 h, the mixture cooled, and filtered → 1-(4-toluenesulfonamido)indan. Y 64%. The reaction is successful with prim., sec. and tert. benzylic substrates. However, the electron-poor substrate, 4-nitro-1-ethylbenzene, gave only a 15% yield after 48 h, while cumene gave a 2.5:1 mixture of the expected product and α-(4-toluenesulfonamidomethyl)styrene. The use of anhydrous conditions and reagents reduced the amount of by-products and increased the regioselectivity. F.e. and comparison of copper catalysts, also α-tosylamination of cyclic ethers (THF and dioxane), s. R. Bhuyan, K.M. Nicholas, Org. Lett. *2007, 9* (20), 3957-9.

Copper(I) bromide/potassium carbonate *CuBr/K_2CO_3*
Ligand-free copper-catalyzed arylation at low catalyst loading NH → NAr

238. 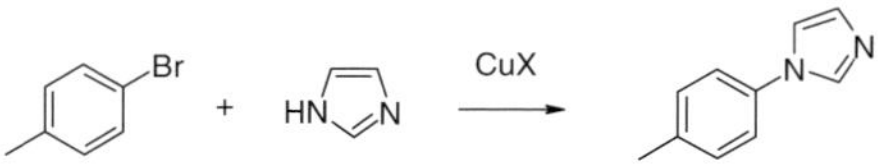

N-Arylation of imidazoles. *p*-Bromotoluene and 1.3 eqs. imidazole in N-methyl-2-pyrrolidone added to 1.1 eqs. solid K_2CO_3, followed by *2.5 mol%* CuBr, the mixture heated under N_2 at 160° for 16 h, cooled to room temp., quenched with acetonitrile, filtered through Celite, and the solvent removed *in vacuo* → 1-*p*-tolylimidazole. Y 89% (by GC). The procedure is simple, inexpensive and industrially viable. In initial tests, bromobenzene was found to be more effective than iodo- or chloro-benzene, but there was little difference between CuBr, CuCl or CuI as catalyst, although the presence of bromide ions in the medium is clearly favoured. Six conversions with different bromobenzenes, incl. *p*-subst. derivs., gave yields of 67-91%, but the reaction does not appear to be optimized. A second series with benzylamine as the amine component, however, was less successful giving yields in the range 9-50%. F.e. and **diaryl ethers from phenols** s. E. Sperotto, J.G. de Vries, G.P.M. van Klink, G. van Hoten, Tetrahedron Lett. *2007, 48* (41), 7366-70; N-phenylation of amines with, especially, iodobenzene using 1.26 mol% CuO nanoparticles and 1 eq. KOH in DMSO under air at 80-110° s. L. Rout, S. Jammi, T. Punniyamurthy, Org. Lett. *2007, 9* (17), 3397-9; N-arylation of heterocyclic nitrogen with activated chloro- or fluoro-arenes using CuO nanoparticles and K_2CO_3 in DMF at 120° s. M. Lakshmi Kantam, J. Yadav, S. Laha, B. Sreedhar, S. Jha, Adv. Synth. Catal. *2007, 349* (11-12), 1938-42; N-arylation of imidazole with CuBr/Cs_2CO_3 s. *72*, 116.

Copper(I) bromide/N-condensed 3,5-bis(2-hydroxyphenyl)-1,2,4-diazaphospholidine/ cesium carbonate ←
Copper(I) iodide/cesium carbonate *CuI/Cs_2CO_3*
Copper(I) iodide/N-hydroxydicarboxylic acid imides/sodium methoxide ←
Copper-catalyzed N-arylation
cf. *62*, 171s*71*; of N-heterocyclics, such as (benz)imidazoles, pyrroles, pyrazoles, indoles and triazoles, with (het)ar. iodides, bromides or chlorides using 20 mol% CuI/Cs_2CO_3 in DMF at 120°, tolerating functional groups, such as ester, cyano, nitro, keto, hydroxyl, or prim. amino, s. L. Zhu, P. Guo, G. Li, J. Lan, R. Xie, J. You, J. Org. Chem. *2007, 72* (22), 8535-8; of pyrroles, indoles, and imidazole using 2.5-10 mol% CuI/5-20 mol% N-hydroxy-succinimide, -maleimide, or -phthalimide and NaOMe in DMSO at 90-110° s. H.-C. Ma, X.-Z. Jiang, ibid. *72* (23), 8943-6; of alkyl or heteroaryl amines with ar. iodides using 10 mol% CuBr/12 mol% air-stable, N-

condensed 3,5-bis(2-hydroxyphenyl)-1,2,4-diazaphospholidine/Cs_2CO_3 in DMSO at 55-80° s. M. Yang, F. Liu, ibid. 8969-71; of acylamines with *o*-bromo- or *o*-iodo-ketones using CuI/ diamine ligand/base in toluene at 90-110°, followed by Camps ring closure to 4(1*H*)-quinolones with NaOH in 1,4-dioxane at 110° (cf. *28*, 903), 2-aryl- or 2-vinyl-derivs., s. C.P. Jones, K.W. Anderson, S.L. Buchwald, ibid. *72* (21), 7968-73.

Copper(I) iodide/N,N′-dimethylethylenediamine *CuI/DMEDA*

N-Vinylation NH → N-C=C

cf. *47*, 337s*71*; of α-aminocarboxylic acid amides with 5 mol% CuI/10 mol% DMEDA in THF at 70° s. R.R. Cesati III, G. Dwyer, R.C. Jones, M.P. Hayes, P. Yalamanchili, D.S. Casebier, Org. Lett. *2007*, *9* (26), 5617-20.

Copper(I) iodide/N,N′-dimethylethylenediamine/cesium carbonate *CuI/DMEDA/Cs_2CO_3*

N-Protected pyrroles from 1,4-dihalogeno-1,3-dienes ○

239. 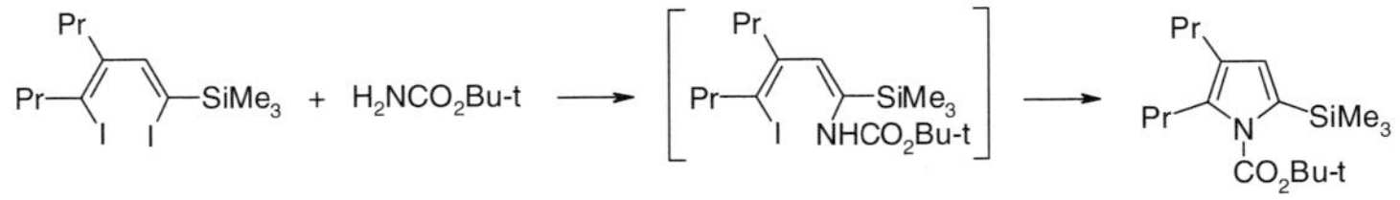

20 Mol% N,N′-dimethylethylenediamine [DMEDA] added via syringe under a positive argon pressure to an oven-dried, screw-capped Schlenk tube containing 5 mol% CuI, 1 eq. startg. 1,4-dihalogeno-1,3-diene, 1.2 eqs. *tert*-butyl carbamate and 3 eqs. Cs_2CO_3, diluted with THF, the tube sealed, stirred at 80° for 6 h, cooled to room temp., and worked up with column chromatographic purification on silica gel (2.5% Et_3N) → *tert*-butyl 2,3-dipropyl-5-trimethylsilyl-1*H*-pyrrole-1-carboxylate. Y 89%. The procedure is useful for preparing di-, tri- and tetra-subst. pyrroles, as well as 2,3-arene-condensed and 2,3:4,5-bis(arene)-condensed pyrroles, reaction tolerating a variety of functional groups (esters, ethers, alkyl halides, Boc-protected hydrazines, alkenes, and silyl groups) as well as electron-donating and -withdrawing groups in the aromatic ring. F.e. and comparison of bases and ligands s. R. Martín, C.H. Larsen, A. Cuenca, S.L. Buchwald, Org. Lett. *2007*, *9* (17), 3379-82.

*Copper(I) iodide/N,N′-dimethylethylenediamine/iodine/potassium or cesium carbonate/ 1,8-diazabicyclo[5.4.0]undec-7-ene/*p*-toluenesulfonic acid* ←

Oxazoles from α,β-ethylenehalides via enacylamines

240. 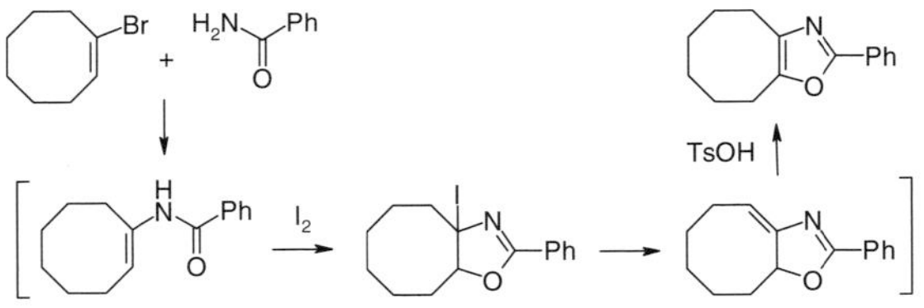

One-pot procedure. 20 Mol% N,N′-dimethylethylenediamine added via syringe under a positive argon pressure to a screw-capped, oven-dried Schlenk tube containing 5 mol% CuI, bromocyclooct-1-ene, 1.2 eqs. benzamide and 2 eqs. K_2CO_3, diluted with toluene, the tube sealed, stirred at 110° for 8 h, cooled to room temp., a further 2 eqs. K_2CO_3 added, cooled to 0°, 1.1 eqs. I_2 in THF added dropwise, stirred at 80° for 4 h, cooled to room temp., 2 eqs. 1,8-diazabicyclo[5.4.0]undec-7-ene added, stirred again for 4 h at 80°, cooled to room temp., worked up, and the crude mixture aromatized by stirring in toluene at 110° for 16 h with 15 mol% *p*-toluenesulfonic acid → 2-phenyl-4,5,6,7,8,9-hexahydrocycloocta[*d*]oxazole. Y 72% (after column chromatographic purification on silica gel). The procedure is based on readily available substrates (aliphatic or aromatic amides), and is equally facile with α,β-ethyleneiodides (with Cs_2CO_3 in THF). A number of functional groups remained unaffected (silyl, cyano, electron-deficient ethylenes, ar. halides and CF_3 groups). F.e. and comparison of iodinating agents s. R. Martín, A. Cuenca, S.L. Buchwald,

Org. Lett. *2007*, *9* (26), 5521-4.

*Copper(I) iodide/*trans-*N,N′-dimethylcyclohexane-1,2-diamine/potassium carbonate* ←
N-Acyl-Δ²-pyrrolines from (Z)-1,2-ethylene-1,4-diiodides and acylamines ○

241.

Toluene, 20 mol% *rac-trans*-N,N′-dimethylcyclohexane-1,2-diamine and (Z)-3-propyl-1,4-diiodohept-3-ene added sequentially to a Schlenk tube containing 1 eq. valeramide, 20 mol% CuI and 2 eqs. K_2CO_3, heated at 110° for 48 h, cooled, quenched with aq. $NaHCO_3$, and worked up with column chromatographic purification on silica gel → product. Y 69%. There was significant elimination of HI with the stronger Cs_2CO_3 as base, and there was no coupling or low yield with $NaHCO_3$ or Et_3N as base. Reaction is presumed to involve tandem N-vinylation-intramolecular N-alkylation. F.e. and comparison of ligands and solvents, also N-carbalkoxy-derivs. with N-unsubst. alkyl carbamates, and 3,4-condensed derivs. s. X. Zhou, H. Zhang, J. Yuan, L. Mai, Y. Li, Tetrahedron Lett. *2007*, *48* (40), 7236-9.

Copper(I) iodide/L-proline/potassium carbonate *CuI/Pro-OH/K_2CO_3*
1,2-Disubst. benzimidazoles from *o*-iodoacylamines and prim. amines

242.

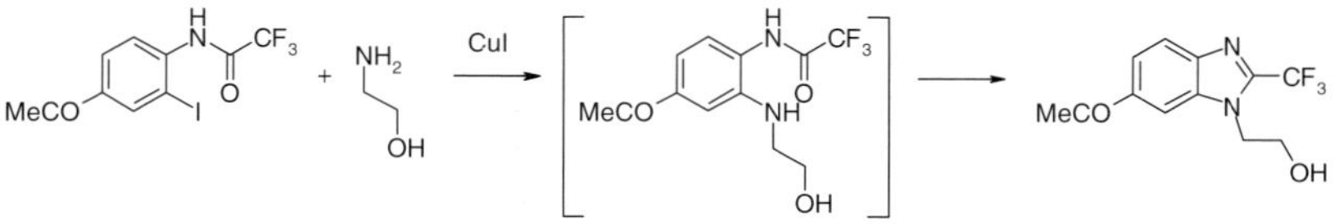

One-pot conversion under mild conditions. 1.5 Eqs. 2-aminoethanol and DMSO sequentially added to a mixture of the startg. aryl iodide, 10 mol% CuI, 20 mol% L-proline and 2 eqs. K_2CO_3 in a Schlenk tube under argon, stirred at room temp. for 24 h (TLC monitoring), the mixture partitioned between ethyl acetate and satd. aq. NH_4Cl, the organic layer washed with satd. aq. $NaHCO_3$ and brine, dried over Na_2SO_4, and concentrated *in vacuo* → product. Y 90%. Sterically accessible prim. amines couple with electron-rich and electron-poor 2-iodotrifluoroacetanilides *at room temp.* with spontaneous cyclization to the benzimidazole in 10-24 h (7 examples; 72-94% yields). For more sterically-demanding amines, the intermediates were cyclized by heating at 120-140° or by treatment with acetic acid at 40-50° (11 examples; 61-97%), with retention of ether, ester, keto, hydroxy and alkene functionality. It is notable that phenyl iodide and benzylamine do not couple under these conditions. Reaction is also applicable to bromoaryl and bromohetaryl analogs (12 examples; 62-90%), but requires more vigorous conditions. F.e.s. B. Zou, Q. Yuan, D. Ma, Angew. Chem. Int. Ed. Engl. *2007*, *46* (15), 2598-601.

Copper(I) iodide/L-proline or trans-*4-hydroxy-L-proline/potassium carbonate or phosphate* ←
N-Subst. 1,3-dihydrobenzimidazol-2-ones from *o*-halogenourethans via *o*-aminourethans

243.

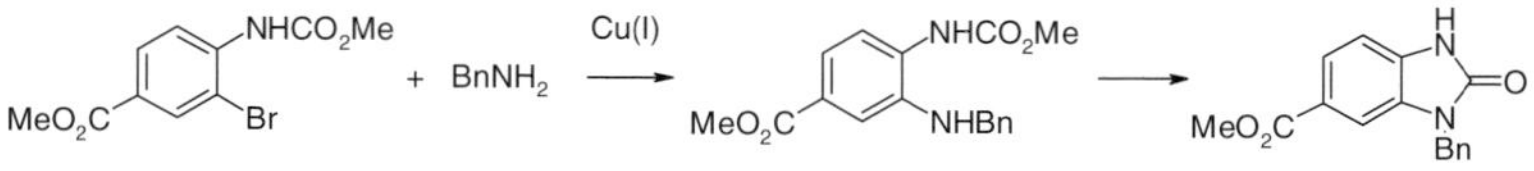

One-pot conversion. 1 Eq. benzylamine and DMSO added successively to a mixture of the startg. aryl bromide, 2 eqs. K_3PO_4, 0.2 eqs. CuI and 0.4 eqs. *trans*-4-hydroxy-L-proline in a Schlenk tube under argon, stirred at 70° for 12 h, then at 130° for 12 h (to effect ring closure), the mixture cooled, partitioned between ethyl acetate and satd. aq. NH_4Cl, the organic layer separated, washed with brine, dried, and concentrated *in vacuo* → 1-benzyl-6-methoxycarbonyl-1,3-dihydrobenzimidazol-2-one. Y 74%. The procedure is applicable to the coupling of *o*-bromo- and *o*-iodo-urethans (the latter at 50° with K_2CO_3/L-proline), irrespective of the electronic character of ring substituents, and the amine may be functionalized or hindered. The reaction conditions were optimized in terms of temperature, base and ligand, and were tolerant of ethenyl, nitro, carboxylate, carboxamide, ester, ketone and silyl ether groups. F.e.s. B. Zou, Q. Yuan, D. Ma, Org. Lett. *2007*, *9* (21), 4291-4.

Zinc/microwaves *Zn/[\\\\]*
Carboxylic acid amides from acid chlorides COCl → CON<
peptides with In cf. *67*, 185; from [hindered] N-benzylaminoisobutyric acid with commercial zinc dust under microwave irradiation in methylene chloride s. J. Cianci, J.B. Baell, A.J. Harvey, Tetrahedron Lett. *2007*, *48* (34), 5973-5.

1-Butyl-3-methylimidazolium fluoroborate s. under Electrolysis *[bmim]BF_4*
L-*Proline(s) s. under CuI* *Pro-OH*
Chloramine-T s. under [Cu(MeCN)$_4$]PF_6 *TsN(Na)Cl*
Triethylsilane/hexamethyldisilazane s. under $FeSO_4$ *$Et_3SiH/(Me_3Si)_2NH$*
Silica s. under Silica-sulfuric acid *SiO_2*
N,N′-Ditosylhydrazine s. under 1,8-Diazabicyclo[5.4.0]undec-7-ene *TsNHNHTs*
Tert. phosphines s. under $Pd_2(dba)_3$ *⩾P*

Tetraethylammonium superoxide *$[Et_4N]O_2$*
Urethans from amines and halides >NH + RHal → >NCOOR
and CO_2 with Triton B cf. *50*, 214s*70*; with $[Et_4N]O_2$ (generated *in situ* from KO_2/Et_4NBr) s. K.N. Singh, Synth. Commun. *2007*, *37* (16), 2651-4; electrochemically in [bmim]BF_4 as ionic liquid solvent (without the need for a supporting electrolyte) s. M. Feroci, M. Orsini, L. Rossi, G. Sotgiu, A. Inesi, J. Org. Chem. *2007*, *72* (1), 200-3.

p-*Toluenesulfonic acid s. under CuI* *TsOH*

Silica-sulfuric acid *SiO_2-HSO_4*
N-Carbalkoxylation NH → NCOOR
with sulfated yttria-zirconia cf. *41*, 377; chemoselective N-carbobenzoxylation of a variety of amines (incl. aminoalcohols and heterocyclic amines) using silica-sulfuric acid under solvent-free conditions at room temp. s. M.B. Gawande, V. Polshettiwar, R.S. Varma, R.V. Jayaram, Tetrahedron Lett. *2007*, *48* (46), 8170-3.

Iodine s. under CuI *I_2*
Iron(III) acetoacetonate s. under CuO *$Fe(acac)_3$*

Iron(II) sulfate/triethylsilane/hexamethyldisilazane *$FeSO_4/Et_3SiH/(Me_3Si)_2NH$*
N-Protected prim. amines from oxo compds. via hydrosilylation of N-protected imines CO → CHNHCOOBu-*t*

244.

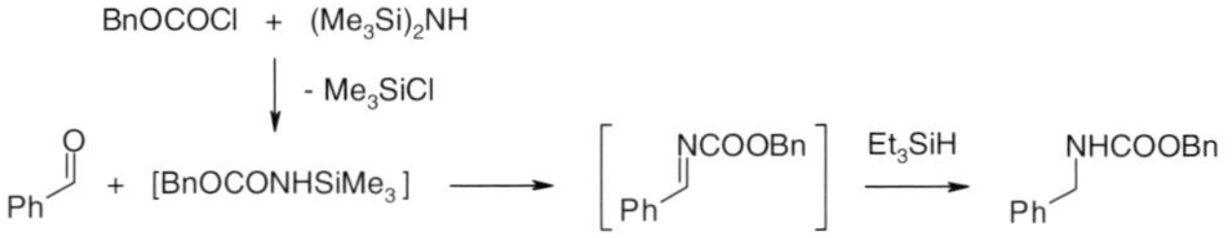

N-Carbobenzoxy-*prim*-amines. 1.2 Eqs. each of benzyl chloroformate, hexamethyldisilazane and triethylsilane, and 5 mol% $FeSO_4 \cdot 7H_2O$ added sequentially at room temp. to a stirred soln. of benzaldehyde in dry methylene chloride, allowed to react for 24 h, and worked up with column chromatographic purification on silica gel → product. Y 95%. The procedure is mild, clean [no formation of sec. amines!], and generally applicable to aliphatic and aromatic aldehydes (incl. hindered compds.), cyclic and acyclic ketones (incl. α-keto-esters), as well as to their surrogates (acetals, ketals and enolethers), leaving nitro groups, olefins, TBS ethers and Boc groups unaffected. Yields were optimum with inexpensive and environmentally friendly iron sulfate, which can be simply recovered. F.e., incl. N-carbophenoxy- and N-benzoyl-derivs., and comparison of Lewis acids, solvents and reducing agents s. B.-L. Yang, S.-K. Tian, Eur. J. Org. Chem. *2007* (28), 4646-50.

Iron(III) chloride/N,N′-dimethylethylenediamine/potassium phosphate *$FeCl_3$/DMEDA/K_3PO_4*
Iron-catalyzed N-arylation NH → NAr

245. 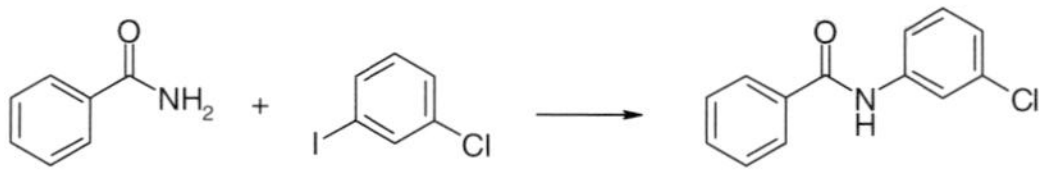

1.5 Eqs. 3-chlorophenyl iodide, 20 mol% DMEDA and dry toluene added to a mixture of benzamide, *10 mol%* $FeCl_3$ and 2 eqs. K_3PO_4 under argon, the tube sealed, the mixture heated at 135° for 24 h, cooled, diluted with methylene chloride, filtered through silica gel, and concentrated *in vacuo* → N-(3-chlorophenyl)benzamide. Y 97%. The procedure is notably cheaper and friendlier than traditional copper- or palladium-catalyzed conversions, clean, and applicable to the N-arylation of pyrazole, triazole, indole, 7-azaindole, 2-pyrrolidone and benzamide with aryl bromides and iodides (the former reacting more slowly) in the presence of halo, ether, and ester functionality. Yields are in the range 46-97% for aryl iodides and 34-64% for aryl bromides but are low for *o*-subst. aryl iodides (2-chloro- and 2-methoxy-phenyl iodide giving 18 and 41% yields, respectively), and only traces of product were detected on attempted arylation of aromatic and aliphatic amines. $Fe(ClO_4)_2$ was also an effective catalyst, but $FeCl_3$ is less expensive and environmentally benign. F.e.s. A. Correa, C. Bolm. Angew. Chem. Int. Ed. Engl. *2007*, *46* (46), 8862-5.

*Tris(dibenzylideneacetone)dipalladium/tri-*tert*-butylphosphine or 9,9-dimethyl-4,6-bis(diphenylphosphino)xanthene/cesium carbonate* ←
N-Arylation
of a fluorous benzophenone imine *en route* to prim. ar. amines (using BINAP as ligand) cf. *59*, 173s*68*; of fluorous carbo-*tert*-butoxyamine(s) *en route* to prim. or sec. anilines (using Xantphos as ligand) s. A.A. Trabanco, J.A. Vega, M.A. Fernández, J. Org. Chem. *2007*, *72* (21), 8146-8; of N,N-dialkylhydrazines with ar. bromides (using *t*-Bu_3P as ligand in toluene at 100°) and reductive cleavage to prim. anilines with Pd-C/$HCOONH_4$ in refluxing methanol (cf. *33*, 30), compatibility with, e.g., cyano, ester, keto and Boc-amino groups, s. X. Liu, M. Barry, H.-R. Tsou, Tetrahedron Lett. *2007*, *48* (48), 8409-12.

Tris(dibenzylideneacetone)dipalladium/2-(dicyclohexylphosphino)-2′,4′,6′-triisopropylbiphenyl/potassium phosphate ←
1-Arylbenzimidazoles from *o*-halogenacylamines and prim. ar. amines ○

246. 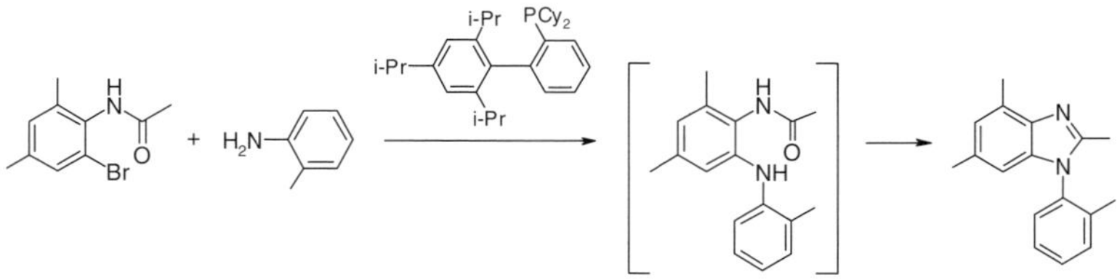

tert-Butanol added under a positive argon flow to a degassed Schlenk tube containing the startg. *o*-halogenacylamine, 1.5 eqs. of the startg. aniline, 2.5 eqs. K_3PO_4, 2 mol% $Pd_2(dba)_3$ and 8 mol% 2-(dicyclohexylphosphino)-2′,4′,6′-triisopropylbiphenyl, the tube sealed, and heated at 110° for 18 h → product. Y 91%. Otherwise elusive 1-arylbenzimidazoles can be thus prepared in regioisomerically pure form from either *o*-chloro- or *o*-bromo-acylamines, permitting a wide variety of substitution at the 2-position of the ring (e.g. alkyl, phenyl, furyl, cyclopropyl, benzyloxymethyl); electronically diverse anilines and *o*-subst. anilines were also tolerated, but reaction was slow with *o,o′*-disubst. substrates. 2-(Dicyclohexylphosphino)-2′,6′-diisopropoxybiphenyl, however, was the ligand of choice for reaction with hindered anilines. F.e. incl. condensation with heteroaryl halides s. N. Zheng, K.W. Anderson, X. Huang, H.N. Nguyen, S.L. Buchwald, Angew. Chem. Int. Ed. Engl. *2007*, *46* (39), 7509-12.

Palladium phosphine complexes/base ←

N-Arylation NH → NAr

update s. *51*, 171s*71* and *52*, 171s*71*; with ar. chlorides or bromides using tert. 2,2-diarylene-phosphines as ligands cf. K. Suzuki, Y. Hori, T. Nishikawa, T. Kobayashi, Adv. Synth. Catal. *2007*, *349* (13), 2089-91; of aziridines with BINAP as ligand s. B. Witulski, S. Senft, J. Bonet, O. Jost, Synthesis *2007* (2), 243-50; 2,3-diamino- from 2-amino-3-chloro-*p*-quinones with 1,1′-bis(diphenylphosphino)ferrocene s. X.L. Wang, X.F. Zheng, L. Wang, J. Reiner, W.L. Xie, J.B. Chang, ibid. *2007* (7), 989-98; N-pyridylation of deactivated aminothiophenes with 2,6-dihalogenopyridines using Xantphos s. A. Begouin, S. Hesse, M.J.R.P. Queiroz, G. Kirsch, Eur. J. Org. Chem. *2007* (10), 1678-82; of amines with 5-bromo-2-chloro-3-fluoropyridine, selectivity depending on the reaction conditions, s. B.W. Stroup, P.V. Szklennik, C.J. Forster, M.H. Serrano-Wu, Org. Lett. *2007*, *9* (10), 2039-42; hetarylation of amides and lactams with added Et_3B s. Q.L. Shen, J.F. Hartwig, J. Am. Chem. Soc. *2007*, *129* (25), 7734-5.

Sulfur ↾ NC ⇃↾ S

Microwaves s. under Diisopropylcarbodiimide [\\\\]

Cesium carbonate Cs_2CO_3

N-Subst. carboxylic acid amides COSH → CONHR

from thiolic acids and 2,4-dinitrophenylsulfonylamines

247. 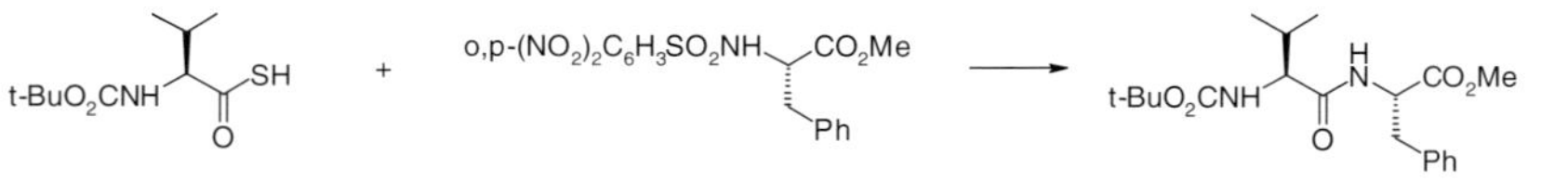

Peptide synthesis. 1.5 Eqs. Cs_2CO_3 added under N_2 to a stirred soln. of the startg. N-*tert*-butoxycarbonyl-α-aminothiolic acid (prepared by cleavage of the 9-fluorenylmethyl thioester with piperidine) in DMF, stirred for 10 min at room temp., a soln. of the startg. N-[(2,4-dinitrophenyl)sulfonyl]-α-amino acid methyl ester in DMF added, stirring continued for 1 h, and worked up with chromatographic purification → N-*tert*-butoxycarbonyl-L-valinyl-L-phenylalanine methyl ester. Y 81% (from the 9-fluorenylmethyl thioester). The procedure is mild, compatible with typical amino acid protective groups, as well as hydroxyl groups, and is applicable to hindered amino acid derivs. F.e. incl. tetrapeptides, **also amide-linked neoglycoconjugates** from 2-(sulfonylamino)ethyl glycosides, s. D. Crich, K. Sana, S. Guo, Org. Lett. *2007*, *9* (22), 4423-6.

Copper(I) chloride/potassium carbonate $CuCl/K_2CO_3$

N,N″-Dicarbalkoxyguanidines from amines ROOCN=C(NHCOOR)N<

and N,N′-dicarbalkoxyisothioureas with DMAP cf. *5*, 346s*68*; N,N″-di(carbomethoxy)-derivs. with $CuCl/K_2CO_3$ in THF at 40°, also N,N″-di(carbo-*tert*-butoxy)-derivs. from N,N′-di(carbo-*tert*-butoxy)thiourea (cf. *48*, 367s*68*), s. H. Ube, D. Uraguchi, M. Terada, J. Organomet. Chem. *2007*, *692* (1-3), 545-9.

Diisopropylcarbodiimide i-*PrN=C=NPr*-i

Diisopropylcarbodiimide/barium hydroxide/microwaves ←

Cyclic guanidines from isothiocyanates and N,N-dinucleophiles ○

2-aminobenzimidazoles from *o*-diamines with diisopropylcarbodiimide [DIC] cf. *34*, 417s*71*; 2-(carboxyarylamino)benzimidazoles (and subsequent amide formation), also polymer-based method, s. R.D. Carpenter, P.B. DeBerdt, K.S. Lam, M.J. Kurth, J. Comb. Chem. *2006*, *8* (6), 907-14; benzimidazo[2,1-*b*]quinazolin-12(5*H*)-ones from *o*-isothiocyanatocarboxylic acid esters and *o*-diamines by reaction with DIC in methylene chloride at room temp., then completion of the second ring closure with $Ba(OH)_2 \cdot 8H_2O$ in dioxane under microwave irradiation, also polymer-based variant, s. R.D. Carpenter, K.S. Lam, M.J. Kurth, J. Org. Chem. *2007*, *72* (1), 284-7.

Phosphorus pentoxide *P_4O_{10}*
Oxazolidines from 2-aminoalcohols under mild conditions
Dimethyl sulfoxide as formaldehyde equivalent

248. 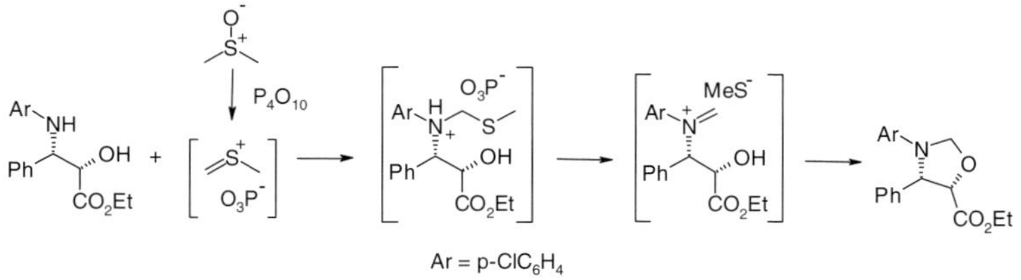

A mixture of 2 eqs. P_4O_{10} and dimethyl sulfoxide ultrasonicated for 10 min, ethyl *anti*-2-hydroxy-3-phenyl-3-(4-chlorophenylamino)propionate added, stirring continued for 24 h (TLC monitoring), the mixture quenched with satd. aq. $NaHCO_3$ and water, extracted with ethyl acetate, the organic layer washed with water (to remove unreacted dimethyl sulfoxide) and brine, dried over $MgSO_4$, concentrated *in vacuo*, and purified by chromatography → *cis*-3-(4-chlorophenyl)-4-phenyl-5-carbethoxyoxazolidine. Y 90%. The reaction came to light while attempting to oxidize the alcohol to the ketone. Cyclization takes place in high yield with retention of configuration for electron-rich arylamino derivs. (4-Cl or 4-OMe subst. derivs. giving 85-95% yields, but the anilino deriv. gave a complex mixture of products), and was used for the preparation of both racemic *cis*- and enantiopure *trans*-products. The mechanism is thought to involve competitive reaction of the amine with a generated sulfonium intermediate, followed by Pummerer rearrangement/cyclization of the resulting adduct. F.e. incl. preparation of the startg. m. via aminolysis of *cis/trans*-phenyl glycidates s. J. Wang, F.D. Rochon, Y. Yang, L. Hua, M.M. Kayser, Tetrahedron:Asym. *2007*, *18* (9), 1115-23.

Remaining Elements ↑ NC ↓↑ Rem

Potassium or cesium fluoride s. under $Cu(OAc)_2$ *KF or CsF*

*Copper/iron(III) chloride/tetra-*n*-butylammonium fluoride* *$Cu/FeCl_3/Bu_4NF$*
Co-catalyzed N-arylation of N-heterocyclics with aryl(trimethoxy)silanes at low catalyst loading under solvent-free conditions with exposure to air NH → NAr

249. 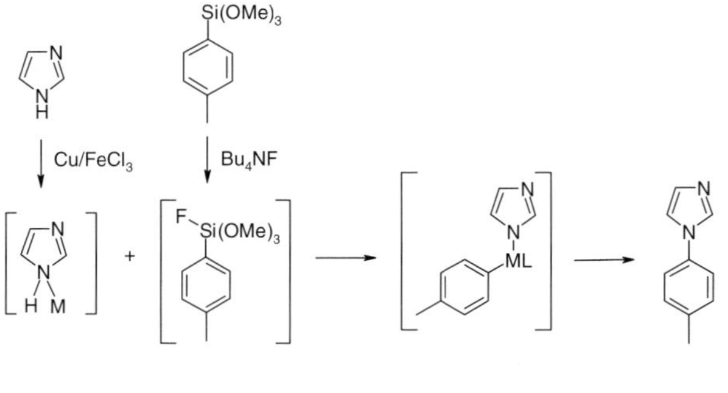

1-Arylimidazoles. A mixture of imidazole, 2 eqs. 4-methylphenyl(trimethoxy)silane, *3 mol%* Cu, *3 mol%* $FeCl_3$ and 3 eqs. Bu4NF·$3H_2O$ stirred in air at 50° for 18 h (TLC monitoring), the mixture diluted with ethyl acetate, washed with brine, the aq. layer extracted with ethyl acetate, the organic layers dried over Na_2SO_4, evaporated *in vacuo*, and the residue purified by chromatography → 1-(4-methylphenyl)imidazole. Y 58%. N-Arylation **and N-vinylation** were successful with benzimidazole, imidazole and 1,2,3- or 1,2,4-triazoles using the respective aryl- or vinyl-(trimethoxy)silanes (12 examples; 33-99% yields), while only traces of products were obtained from benzotriazole or hindered 2-phenylimidazole or by using allyl(trimethoxy)silane or phenyl(trimethyl)silane. Use of either metal catalyst alone gave reduced yields as did reaction at room temp. or under an argon or oxygen atmosphere. The mechanism is not established but is believed to involve coordination of a metal with the N-heterocycle which undergoes transmetalation with a pentavalent silicate, and subsequent formation of the products by air oxidation, disproportionation or reductive elimination. It is thought that a pure oxygen atmosphere may lead to unwanted side-reactions. F.e.s. R.-J. Song, C.-L. Deng, Y.-X. Xie, J.-H. Li, Tetrahedron Lett. *2007*, *48* (44), 7845-8.

Copper(II) acetate/potassium fluoride/oxygen *$Cu(OAc)_2/KF/O_2$*
Copper(II) acetate/cesium fluoride/potassium tert-*butoxide* *$Cu(OAc)_2/CsF/KOBu$-t*
N-Arylation with arylboronic acids or esters NH → NAr
of pyrazinones with arylboronic acids using $Cu(OAc)_2/Et_3N$/pyridine under microwave irradiation cf. *55*, 166s*71*; of alkyl or aryl amines with pinacol arylboronates using $Cu(OAc)_2$/KF and powdered molecular sieves in acetonitrile at 80° under O_2 as part of a 2-step conversion from arenes via Ir-catalyzed borylation (cf. *65*, 62), also diaryl ethers from phenols (cf. *61*, 100), s. C.C. Tzschucke, J.M. Murphy, J.F. Hartwig, Org. Lett. *2007*, *9* (5), 761-4; **N-(1,3-dienyl)ation** of azoles with 1,3-dienylboronic acid esters using $Cu(OAc)_2$/CsF/KOBu-*t* in methylene chloride under air s. A. Deagostino, C. Prandi, C. Zavattaro, P. Venturello, Eur. J. Org. Chem. *2007* (8), 1318-23.

Copper(II) acetate/triethylamine or pyridine *$Cu(OAc)_2/Et_3N$ or C_5H_5N*
Copper-catalyzed N-substitution with bismuth(III) compds. NH → NR
N-arylation of sec. amines with triarylbismuthines using $Cu(OAc)_2$ cf. *50*, 230; of N-carbalkoxy-hydrazones with added Et_3N s. P. Starkov, I. Zemskov, R. Sillard, O. Tsubrik, U. Maeorg, Tetrahedron Lett. *2007*, *48* (7), 1155-7; N-cyclopropylation of azoles, lactams and cyclic urethans with tricyclopropylbismuthine or dicyclopropylbismuth(III) chloride and added pyridine s. A. Gagnon, M. St-Onge, K. Little, M. Duplessis, F. Barabé, J. Am. Chem. Soc. *2007*, *129* (1), 44-5.

Copper(II) sulfate/sodium ascorbate ←
1-Aryl-1,2,3-triazoles from terminal acetylene derivs. and arylboronic acids ←
One-pot procedure via ar. azides

250.

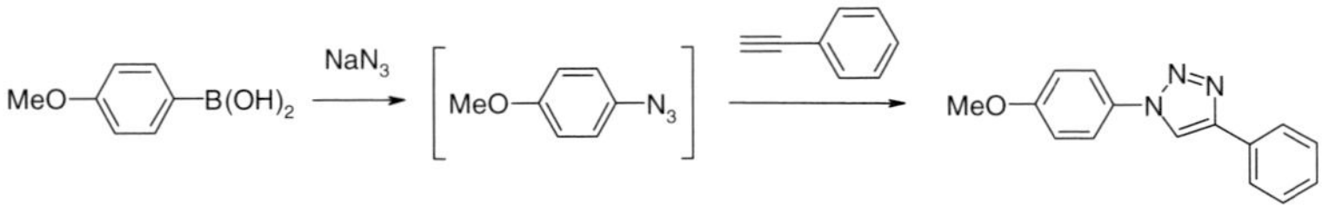

under mild conditions. A mixture of 1.2 eqs. NaN_3, 4-methoxyphenylboronic acid and 10 mol% $CuSO_4$ stirred vigorously in methanol at room temp. until azidonation complete (TLC monitoring), 1.1 eqs. phenylacetylene, 0.5 eqs. Na-ascorbate and water added, stirring continued for 12 h (TLC monitoring), the mixture filtered, the precipitate washed with water, and dried → 1-methoxyphenyl-3-phenyl-1,2,3-triazole. Y 95%. The method is applicable to the coupling of alkynes or arylacetylenes with various arylboronic acids, incl. o-*subst.* derivs., and avoids the need to isolate unstable ar. azides which were prepared in initial experiments in 70-98% yields (8 examples). The conversion was also accomplished in 63-96% yield at *room temp. in air* with *methanol* as solvent (water giving a lower yield while less polar solvents such as methylene chloride are unsuitable). Halides and ether functionality were unaffected. F.e. and with other copper(II) or copper(I) catalysts [$Cu(OAc)_2$, CuI and CuCl] s. C.-Z. Tao, X. Cui, J. Li, A.-X. Liu, L. Liu, Q.-X. Guo, Tetrahedron Lett. *2007*, *48* (20), 3525-9.

Trifluoroacetic acid s. under $[Rh(OCOR)_2]_2$ *CF_3COOH*

m-Chloroperoxybenzoic acid/potassium hydrogen phosphate/hydrogen chloride/ benzoyl isocyanate ←
2-Oxazolidones from 2-hydroxyselenides ○

251. 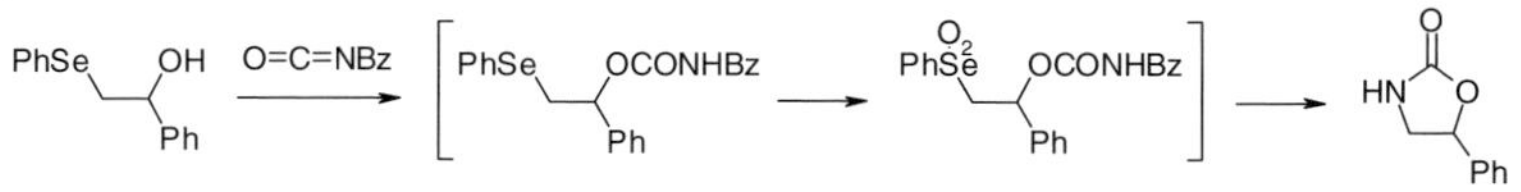

One-pot conversion. 1.1 Eqs. benzoyl isocyanate added to a soln. of the startg. 2-hydroxyselenide in dry THF, stirred at room temp. for 15-17 h (TLC monitoring), treated with excess of *m*-chloroperoxybenzoic acid and 5 eqs. K_2HPO_4, stirred for 6-8 h, 4 *N* HCl added, stirring continued at 65° for 4 h, cooled, and quenched by slow addition of satd. aq. $KHCO_3$ → product. Y 85%. F.e.s. S.-R. Sheng, H.-R. Luo, Z.-Z. Huang, W.-K. Sun, X.-L. Liu, Synth. Commun. *2007*, *37* (16), 2693-9.

Trimethylsilyl azide s. under $FeCl_3$ *Me_3SiN_3*
Tetra-n-butylammonium fluoride s. under Cu *Bu_4NF*
Iron(III) chloride s.a. under Cu *$FeCl_3$*

Iron(III) chloride/trimethylsilyl azide *$FeCl_3/Me_3SiN_3$*
1-Alkoxy-1-azides from oxo compds. and alkoxysilanes CO → $C(OR)N_3$

252. 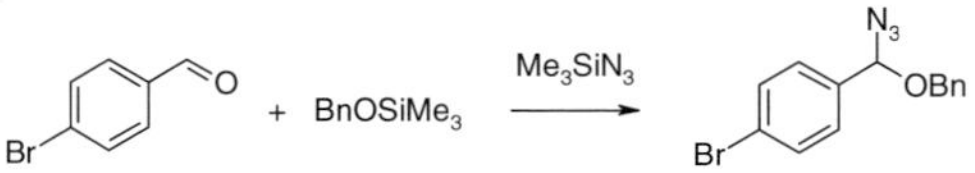

A suspension of 2 mol% $FeCl_3$ (anhydrous) and the startg. aldehyde treated with 2.4 eqs. benzyloxytrimethylsilane, stirred at 0° under argon for 2 h, 1.5 eqs. trimethylsilyl azide in propionitrile added, stirring continued at -78° for 1 h, and quenched with pH 7 phosphate buffer → product. Y 96%. The procedure is mild and generally applicable to ar. aldehydes (possessing electron-withdrawing or -donating groups), α,β-ethylenic aldehydes and a number of ketones. F.e.s. M. Omura, K. Iwanami, T. Oriyama, Chem. Lett. *2007*, *36* (4), 532-3.

Chiral dirhodium(II) tetracarboxylates/trifluoroacetic acid *$[Rh(OCOR)_2]_2/CF_3COOH$*
α-(Sulfonylamino)carbonyl compds. from enoxysilanes C(OSi⩽)=C → $COC(NHSO_2R)$
Catalytic asym. conversion

253. 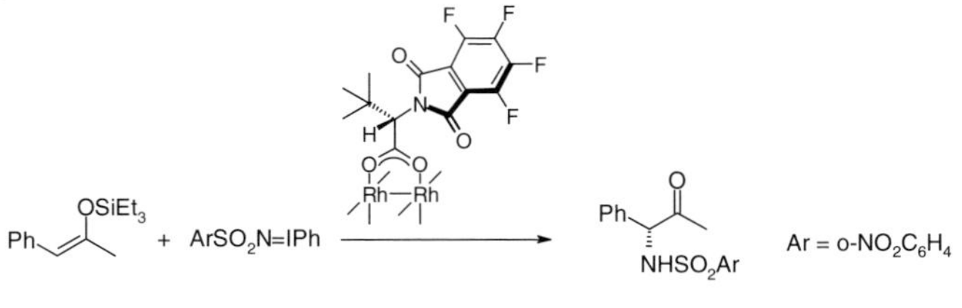

Chiral α-(sulfonylamino)ketones. 1.05 Eqs. *o*-nitrobenzenesulfonylimino(phenyl)iodinane added in one portion to a soln. of the startg. enoxysilane (Z/E 96:4) and 3 mol% chiral dirhodium(II) tetrakis[N-tetrafluorophthaloyl-(S)-*tert*-leucinate] in methylene chloride at -40°, stirred for 5 h, treated with 90% aq. trifluoroacetic acid, stirring continued for 1 h at room temp., and worked up with column chromatographic purification on silica gel → (R)-1-[(2-nitrophenylsulfonyl)amino]-1-phenyl-2-propanone. Y 94% (e.e. 95%). This is the first highly enantioselective, generally applicable, catalytic asym. amination of enoxysilanes. Aryl propyl ketone silyl enolates also participated in the reaction (irrespective of the nature of ring substituents), but enantioselectivity was low with the corresponding aryl butyl ketone silyl enolates. F.e. and comparison of catalysts, **also from α,β-ethyleneketones** in one-pot via 1,4-hydrosilylation-asym. amination, s. M. Anada,

M. Tanaka, T. Washio, M. Yamawaki, T. Abe, S. Hashimoto, Org. Lett. *2007*, *9* (22), 4559-62; **chiral α-(sulfonylamino)carboxylic acid esters** with the chiral tetrachlorophthalimido complex as catalyst s. M. Tanaka, Y. Kurosaki, T. Washio, M. Anada, S. Hashimoto, Tetrahedron Lett. *2007*, *48* (50), 8799-802.

Palladadiaziridine phenanthroline complexes ←

N-Arylhydrazodicarbonyl N═N → NHN(Ar)

from azodicarbonyl compds. and arylboronic acids

254. 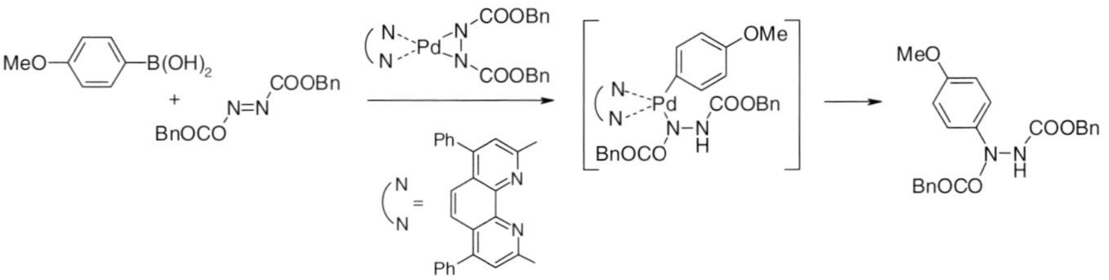

The startg. azodicarboxylate and 5 mol% palladadiaziridine phenanthroline complex added to a Schlenk tube containing 1.5 eqs. *p*-methoxyboronic acid in dichloroethane, diluted with more solvent, stirred at 60° overnight, quenched with satd. aq. Na_2SO_3, and worked up with flash chromatographic purification → product. Y 99%. The procedure is completely selective, base-free, waste-free, mild and applicable to a range of substituted arylboronic [and hetarylboronic] acids, incl. *o*-subst. derivs., *m*- and *p*-halogeno-derivs., and, notably, electron-rich derivs. with which the rate is fastest. The key step of the reaction is the first general reductive C-N bond formation from a palladium amidate complex, the catalytic cycle being initiated by aryl transfer to the palladium centre with concomitant opening of the palladadiaziridine ring, which is reformed in the closing phase. However, it does appear that the stabilizing phenanthroline ligand is essential for high yields. F.e. and with the *in situ*-generated complex s. K. Muniz, A. Iglesias, Angew. Chem. Int. Ed. Engl. *2007*, *46* (33), 6350-3.

Carbon ↑ NC ↓↑ C

Without additional reagents *w.a.r*

N-Subst. arylcarboxylic acid amides from acids and isonitriles COOH → CONHR

255. 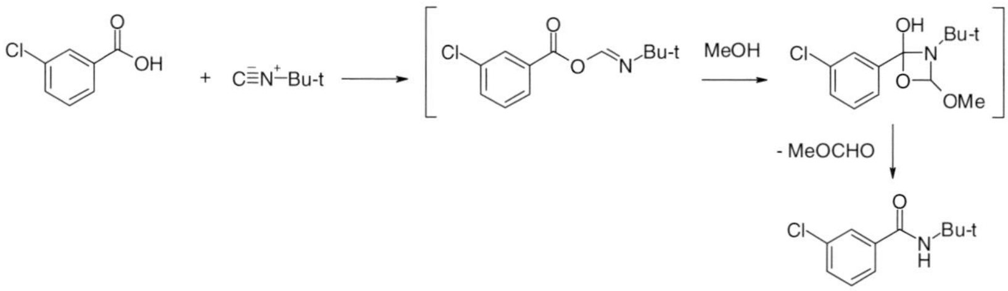

under mild conditions. 1 Eq. *tert*-butyl isocyanide added to 3-chlorobenzoic acid in methanol, the mixture stirred at room temp. for 24 h, the solvent removed *in vacuo*, and the residue washed with ether → N-*tert*-butyl-3-chlorobenzamide. Y 90%. The method is mild, efficient, *requires no catalyst or prior activation* or dry solvents with *methanol* being the solvent of choice (yields being lower in ethanol, with no reaction in methylene chloride). Ten subst. benzoic acids gave 65-92% yields of the corresponding amides but no products were obtained from aliphatic carboxylic acids. The mechanism is thought to involve initial formation of an aroyl formimidate (in a reaction conducted in water such a product was obtained in 95% yield), which affords amides via an intermediate 1,3-oxazetidine. F.e.s. A. Shaabani, E. Soleimani, A.H. Rezaya, Tetrahedron Lett. *2007*, *48* (35), 6137-41.

β-Ketocarboxylic acid thioamides from 3-(acylene)-1,3-dithiolanes C

256. 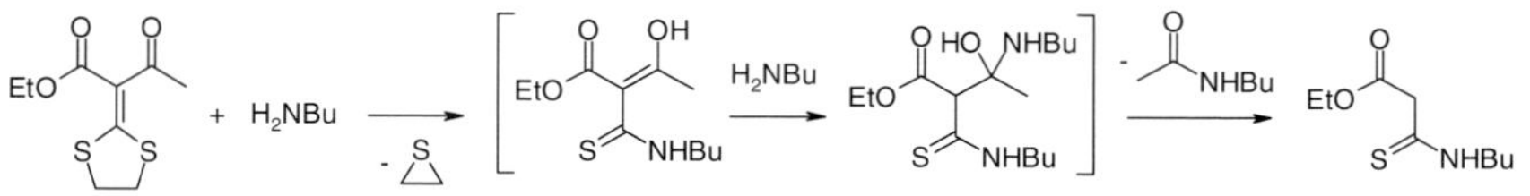

1.5 Eqs. *n*-butylamine added to a stirred soln. of the startg. 1,3-dithiolane *in DMSO*, and heated to 120° for 1.5 h → product. Y 87%. The procedure is catalyst-free, high-yielding, and generally applicable to aliphatic prim. amines (reaction with sec., tert. and ar. amines being inefficient). The presence of two acyl groups is essential, and there was no conversion with the corresponding 1,3-dithianes. Reaction is thought to involve an initial nucleophilic vinylic substitution associated with fragmentation of the dithiolane ring, followed by intramolecular elimination of thiirane prior to addition of the amine and C-cleavage. F.e. incl. 1-thionomalonic acid ester amides s. F. Liang, Y. Li, D. Li, X. Cheng, Q. Liu, Tetrahedron Lett. *2007*, *48* (45), 7938-41.

1-Butyl-3-methylimidazolium hexafluorophosphate/ammonium acetate *[bmim]PF_6/NH_4OAc*
2-Aryl-2,3-dihydro-4(1*H*)-quinazolones ←
from 3,1-benzoxazine-2,4-diones and ar. aldehydes in ionic liquid/water s. *72*, 206

Low valent transition metal carbonyl complexes ←
Sym. carbodiimides from isocyanates 2 RN=C=O → RN=C=NR
with polymer-based triarylarsine oxide cf. *39*, 406; with low valent transition metal carbonyl complexes, especially $[(\eta^5\text{-}C_5H_5)Fe(CO)_2]_2$, $[(\eta^5\text{-}C_5Me_5)Fe(CO)_2]_2$, or $Me(\eta^5\text{-}C_5H_5)Mn(CO)_3$, s. A.K.F. Rahman, K.M. Nicholas, Tetrahedron Lett. *2007*, *48* (34), 6002-4.

Elimination ⇈

Hydrogen ↿ NC ⇈ H

Sodium nitrite s. under Phosphomolybdic acid *$NaNO_2$*
Copper(II) acetate/silver trifluoroacetate s. under Pd(OAc)$_2$ *$Cu(OAc)_2/AgOCOCF_3$*

*Copper(I) bromide/*tert-*butyl hydroperoxide* *CuBr/t-BuOOH*
Intramolecular oxidative radical acylamination s. *72*, 194 ○

Magnesium oxide s. under Rh(OCOR)$_2$ *MgO*
tert-*Butyl hydroperoxide s. under CuBr and Dirhodium(II) tetrakis[caprolactamate]* t-*BuOOH*
Dimethyl sulfoxide s. under O$_2$ *DMSO*
Phenyl iodosoacetate s.a. under Rh(OCOR)$_2$ and Pd(OAc)$_2$ *PhI(OAc)$_2$*

Phenyl iodosoacetate/iodine *PhI(OAc)$_2$/I$_2$*
Metal-free radical oxidation of sulfonic acid amides under mild conditions ←

257. 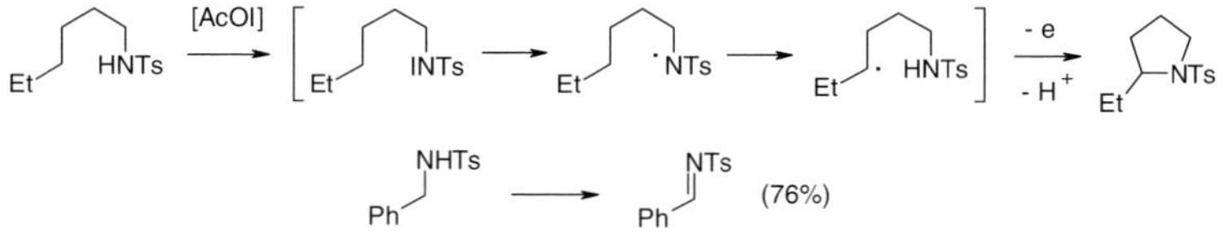

N-Sulfonylpyrrolidines via intramolecular sulfonylamination. 1 Eq. iodine and dichloroethane added sequentially to a Schlenk tube containing the startg. sulfonamide and 3 eqs. phenyl iodosoacetate under argon, stirred at room temp. until TLC indicated completion of reaction, quenched with $Na_2S_2O_3$, extracted with ethyl acetate, and worked up with column chromatographic purification on silica gel → 2-ethyl-1-tosylpyrrolidine. Y 82%. Activation of the remote C-H site was effected by a radical mechanism via initial N-iodination with *in situ*-generated AcOI, followed by formation of a sulfonylaminyl radical and its quenching by 1,5-hydrogen shift. **N-Sulfonylimines** were obtained under the same conditions **from sulfonylamines**, and can be trapped with suitable nucleophiles in a one-pot process. There was no reaction, however, with the corresponding acylamines. F.e.s. R. Fan, D. Pu, F. Wen, J. Wu, J. Org. Chem. *2007*, *72* (23), 8994-7; **N-protected pyrrolidine-fused carbohydrates from C-(2-aminoethyl)glycosides** s. C.G. Francisco, A.J. Herrera, A. Martin, I. Perez-Martin, E. Suarez, Tetrahedron Lett. *2007*, *48* (36), 6384-8.

Silica s. under Phosphomolybdic acid *SiO_2*

Oxygen/dimethyl sulfoxide *O_2/DMSO*

Pyridines from 1,4-dihydropyridines ←

with urea-H_2O_2/maleic anhydride cf. *25*, 649s*71*; under air in solvent-grade DMSO as solvent and catalyst, also pyrido[2,3-*d*]pyrimidines from 6-aminouracils via *in situ* oxidation of dihydropyrido[2,3-*d*]pyrimidines, s. A. Saini, S. Kumar, J.S. Sandhu, Synth. Commun. *2007*, *37* (14), 2317-24; with $[Bu_4N]IO_4$ and the Mn-amino acid Schiff base complex derived from $Mn(OAc)_2 \cdot 4H_2O$, salicylaldehyde and L-tryptophan s. G.R. Karimipour, M. Nasr-Esfahani, G. Valipour, J. Chem. Res. *2007* (7), 415-7.

Phosphomolybdic acid/sodium nitrite/silica *$H_3PMo_{12}O_{40}/NaNO_2/SiO_2$*

Cyclic azomethines from cyclic sec. amines CHNH → C=N

with NaOCl cf. *20*, 373; quinolines from 1,2-dihydroquinolines (cf. *25*, 649s*64*) with phosphomolybdic acid/$NaNO_2$/silica s. K. Niknam, B. Karami, S. Molazadeh, A. Pourkhosro, Synth. Commun. *2007*, *37* (7), 1091-6.

Iodine s. under $PhI(OAc)_2$ *I_2*

Tetrabutylammonium periodate/manganese(II) Schiff base complex *$[Bu_4N]IO_4/[Mn]$*

Pyridines from 1,4-dihydropyridines s. *25*, 649s72 ←

Tetramethylammonium chloride s. under $Pd(OAc)_2$ *Me_4NCl*

*Rhodium(II) acetate or dirhodium(II) tetrakis(N-tetrafluorophthaloyl-*L-tert-*leucinate) bis(ethyl acetate)/phenyl iodosoacetate/magnesium oxide* ←

Cyclic sulfamic acid esters by intramolecular oxidative amination ○

with $Rh(OAc)_2/PhI(OAc)_2/MgO$ cf. *62*, 181; 1,2,3-oxathiazolidine 2,2-dioxides with dirhodium(II) tetrakis(N-tetrafluorophthaloyl-L-*tert*-leucinate) bis(ethyl acetate) s. M. Yamawaki, S. Kitagaki, M. Anada, S. Hashimoto, Heterocycles *2006*, *69* (1), 527-37; 1,2,3,5-oxathiadiazacycloheptane S,S-dioxide ring from cyclic N-(sulfamyloxyacetyl)amines with $Rh(OAc)_2$ s. M.S.T. Morin, S. Toumieux, P. Compain, S. Peyrat, J. Kalinowska-Tluscik, Tetrahedron Lett. *2007*, *48* (48), 8531-5.

*Dirhodium(II) tetrakis[caprolactamate]/*tert-*butyl hydroperoxide* *$Rh_2(cap)_4$/t-BuOOH*

Azomethines from sec. amines CHNHR → C=NR

258. 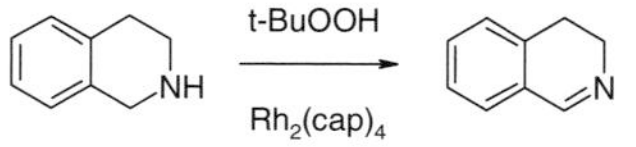

under mild conditions. 2 Eqs. *tert*-butyl hydroperoxide added to a soln. of 1,2,3,4-tetrahydroisoquinoline and 1 mol% dirhodium(II) tetrakis[caprolactamate] in acetonitrile with exclusion of air, the deep red soln. stirred for 16 h, diluted with hexane and ether, filtered through a short plug of Florisil wet with hexane, washed down with ether, and the filtrates evaporated *in vacuo* → 3,4-dihydroisoquinoline. Y 85%. Indoline was similarly oxidized (87%) but tetrahydroquinoline suffered overoxidation to quinoline (82%), while a series of sec. N-benzyl-, N-furylmethyl- and N-cinnamyl-amines gave the corresponding imines (11 examples; 74-95%) using *acetonitrile* as solvent. The method is also successful for electronically-diverse benzylamines, occurs in the presence of free hydroxyl, ether and nitro functionality, and chemoselectively affords the aldimine rather than the ketimine with benzyl(α-methylbenzyl)amine. F.e. incl. use of the technique for N-deprotection of methyl N-benzyl- L-phenylalaninate without racemization s. H. Choi, M.P. Doyle, Chem. Commun. *2007* (7), 745-7.

Palladium(II) acetate/sodium acetate/tetramethylammonium chloride/phenyl iodosoacetate ←
Double ring closure ○
via stereospecific palladium-catalyzed intramolecular 1,2-di(tosylamination) of ethylene derivs.

259.

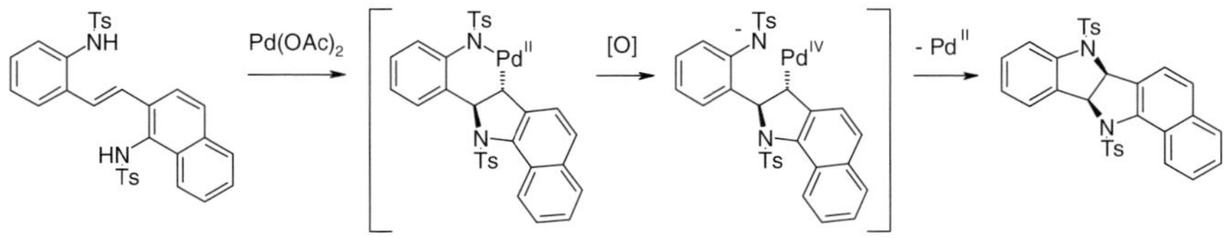

N,N′-Ditosyl-*cis*-bis(indolines). Reagent grade *DMF* added to a mixture of the startg. bis(tosylamine), 10 mol% $Pd(OAc)_2$, 1.1 eqs. each of NaOAc and tetramethylammonium chloride, and 1.6 eqs. phenyl iodosoacetate, the soln. sealed in a Pyrex tube, and stirred at room temp. for 16 h → product. Y 89% (after column chromatography on silica gel). Both ring closures are facilitated by the same catalyst in a catalytic sequence: initially, after pre-coordination of the catalyst, regiospecific, *endo*-selective *anti*-aminopalladation takes place, followed by oxidation of Pd(II) to Pd(IV) prior to amide dissociation and selective *anti*-amination/depalladation. Reaction is highly stereoselective from (E)-alkenes, is not influenced by steric or electronic factors, and is equally facile with aliphatic bis(tosylamines). F.e. incl. N,N′-ditosylpyrrolidino[2,3-*b*]pyrrolidines, and other annelated indolines s. K. Muniz, J. Am. Chem. Soc. *2007*, *129* (47), 14542-3.

Palladium(II) acetate/copper(II) acetate/silver trifluoroacetate ←
1-Tosylindazoles from ar. tosylhydrazones
Oxidative ring closure

260.

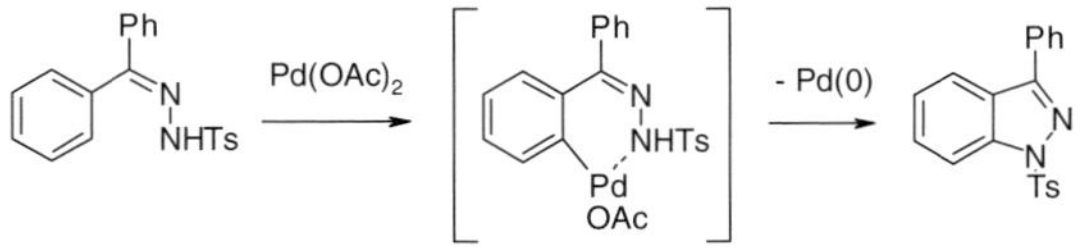

A mixture of benzophenone tosylhydrazone, 10 mol% $Pd(OAc)_2$, 1 eq. $Cu(OAc)_2$ and ca. 2 eqs. $AgOCOCF_3$ in DMSO stirred at 50° for 12 h, and worked up with column chromatographic purification on silica gel → 1-(4-methylphenylsulfonyl)-3-phenylindazole. Y 90%. This relatively mild procedure was optimized by assessing various palladium catalysts, copper(II) oxidants, solvents, reaction conditions and additives. With ring-substituted ar. tosylhydrazones, the regioselectivity of the ring closure is determined by the character of the substituent. Reaction is thought to involve initial association of $Pd(OAc)_2$ with the NHTs group, followed by activation of the *ortho*-position via a 6-membered palladacycle prior to reductive elimination. Functional groups such as ester, cyano, and halogen were unaffected. F.e. incl. 3-alkyl derivs. s. K. Inamoto, T. Saito, M. Katsuno, T. Sakamoto, K. Hiroya, Org. Lett. *2007*, *9* (15), 2931-4.

Oxygen ↑ NC ⇑ O

Microwaves s. under $ZnCl_2$, 1-Ethyl-3-methylimidazolium hexafluorophosphate and $(EtO)_3P$ [\\\\]

Copper(I) iodide/L-proline/potassium carbonate *$CuI/Pro\text{-}OH/K_2CO_3$*
1,2-Disubst. benzimidazoles from N-monoacyl-*o*-diamines s. *72*, 242 ○

Copper(I) iodide/L-proline or trans-*4-hydroxy-L-proline/potassium carbonate or phosphate* ←
N-Subst. 1,3-dihydrobenzimidazol-2-ones from *o*-aminourethans s. *72*, 243 ○

Zinc/ammonium chloride *Zn/NH_4Cl*
Quinolines from 1-(*o*-nitroaryl)-2-acetylenealcohols
Reductive ring closure s. *72*, 266

Zinc chloride/acetonitrile/microwaves *$ZnCl_2$/MeCN/*[\\\\]
Nitriles from carboxylic acid amides in aq. media $CONH_2$ → CN
with $PdCl_2$, also reverse reaction, cf. *70*, 208; with $ZnCl_2$ in 1:1 acetonitrile/water under microwave irradiation, also reverse reaction in THF/water with added acetamide, s. K. Manjula, M.A. Pasha, Synth. Commun. *2007*, *37* (9), 1545-50.

1-Ethyl-3-methylimidazolium hexafluorophosphate/toluene/microwaves ←
Microwave-assisted ring closures via iminyl radicals ○

261.

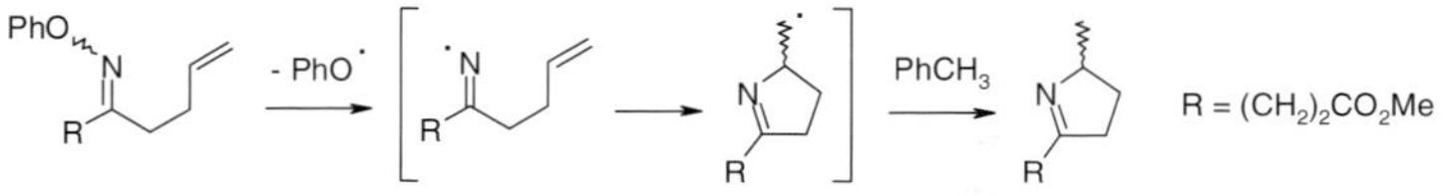

By comparision with thermal processes, microwave-assisted ring closures of ethylenic and aromatic O-phenyloximes via iminyl radicals are very clean, affording high yields of N-heterocyclics relatively quickly. **E: Cyclic azomethines.** 1-Ethyl-3-methylimidazolium hexafluorophosphate (2 mmol) added to a soln. of 1.91 mmol of the startg. iminyl radical precursor in toluene *as H-donor*, and the sealed vessel subjected to microwave irradiation for 15 min at 160° in a 300 MHz Biotage Initiator microwave reactor → product. Y 78%. **Benzo-condensed N-heterocyclics** were obtained similarly from aromatic substrates, the arene ring serving as radical acceptor to give a cyclohexadienyl-type radical prior to aromatization by transfer of hydrogen. Attempted intermolecular coupling failed, and there was no 6-*exo* ring closure of 1-phenylhex-5-en-1-one. F.e.s. F. Portela-Cubillo, J.S. Scott, J.C. Walton, Chem. Commun. *2007* (39), 4041-3.

Dimethylformamide *DMF*
Nitriles from aldoximes by thermal dehydration CH=NOH → CN

262.

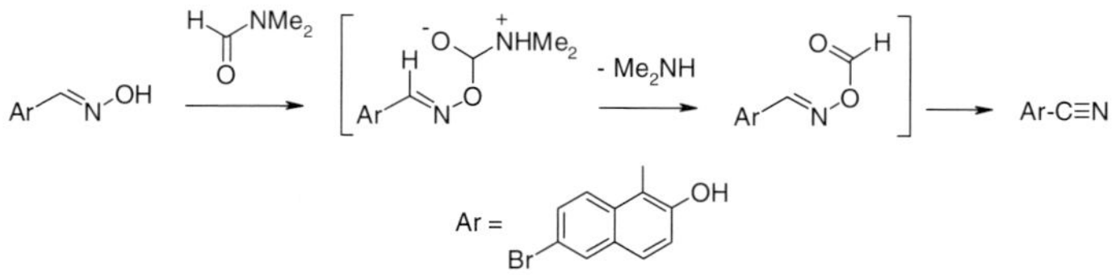

3 Mmol startg. aldoxime (as the (E)-isomer) in 10 ml DMF heated at 135° for 48 h, cooled, diluted with water, extracted with ethyl acetate, and worked up with column chromatographic purification → 6-bromo-2-hydroxynaphthalene-1-carbonitrile. Y 68%. DMF serves as both solvent [just sufficient to dissolve the substrate] and catalyst, reaction being generally applicable to both aliphatic and aromatic (E)- or (Z)-aldoximes, leaving a number of functional groups unaffected, notably hydroxyl groups. Significantly, no expensive or corrosive catalyst or electrophile is required, nor a solid support. Reaction is presumed to involve intermediate formation of the O-formylaldoxime, which eliminates formic acid to give the product. F.e.s. P. Supsana, T. Liaskopoulos, P.G. Tsoungas, G. Varvounis, Synlett *2007* (17), 2671-4.

Formic acid *HCOOH*
Parallel polymer-based synthesis of 1,2,3,4-tetrahydro-1,4-benzodiazepin-5-ones via Leuckart-Wallach reaction ○

263. 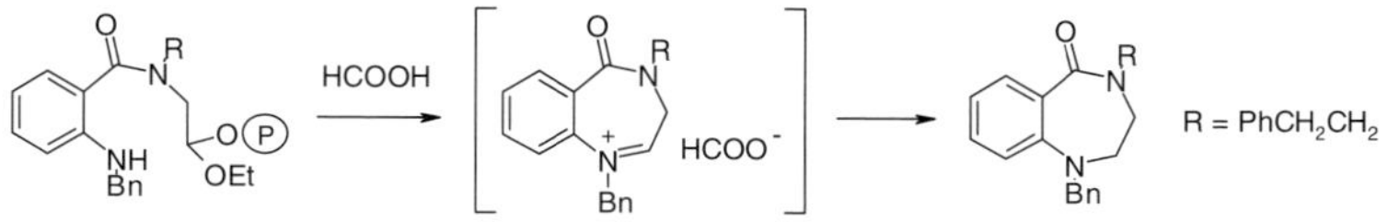

Startg. Wang-based resin dried *in vacuo* in a cell of a Robbins 48-well filtration block (along with 47 other supported substrates), treated with 100% formic acid for 3 h at 60°, filtered, and the filtrate worked up → product. Y 88%. The other 47 substrates were treated in the same way simultaneously. This is the last step of a 4-step parallel polymer-based synthesis from a bromoacetal resin affording products with three points of molecular diversity without generation of any non-volatile by-products. It was not possible, however, to prepare such benzodiazepinones substituted at N[1] by an electron-withdrawing group. F.e.s. S.-C. Lee, S.B. Park, Chem. Commun. *2007* (36), 3714-6.

2,2′-Dithiobis(benzothiazole)-triphenylphosphine ←
2-Azetidinones from β-aminocarboxylic acids s. *42*, 420s*72*

2,3-Dichloro-5,6-dicyanoquinone s. under Ph_3P *DDQ*

Hexamethyldisilane/iodine *$Me_3SiSiMe_3/I_2$*
4(3*H*)-Pyrimidinone ring from *o*-(acylamino)carboxylic acid amides ○
with NH_3 cf. *28*, 437; 4(3*H*)-quinazolones with hexamethyldisilane/iodine in methylene chloride at room temp. s. U.A. Kshirsagar, S.B. Mhaske, N.P. Argade, Tetrahedron Lett. *2007*, *48* (18), 3243-6.

Tin(II) chloride/hydrogen chloride *$SnCl_2/HCl$*
Quinolines from 1-(*o*-nitroaryl)-2-acetylenealcohols
Reductive ring closure s. *72*, 266

Triphenylphosphine s.a. under 2,2′-Dithiobis(benzothiazole) *Ph_3P*

Triphenylphosphine/diethyl azodicarboxylate/triethylamine hydrochloride ←
Intramolecular Mitsunobu N-alkylation
cf. *35*, 312s*70*; 1-sulfonylpiperazines in the presence of $Et_3N{\cdot}HCl$ [or 1-β-(sulfonylamino)-aziridines in its absence] s. J.C. Anderson, H.A. Chapman, Org. Biol. Chem. *2007*, *5* (15), 2413-22.

Triphenylphosphine/2,3-dichloro-5,6-dicyanoquinone *Ph_3P/DDQ*
Nitriles from aldoximes CH=NOH → CN
with Ph_3P/I_2 cf. *25*, 346s*70*; under neutral conditions with Ph_3P/DDQ in methylene chloride at room temp. s. G. Aghapour, M. Arnirabadi, Indian J. Chem. *2007*, *46B* (4), 649-52.

9,9-Dimethyl-4,5-bis(diphenylphosphino)xanthene s. under $RuH_2(CO)(PPh_3)_3$ *Xantphos*

Triethyl phosphite/microwaves *$(EtO)_3P$/[\\\\]*
N-Unsubst. benzo-condensed *o*-vinyl-N-heterocyclics from (nitroaryl)ethylene derivs. Regiospecific reductive ring closure ○

264. 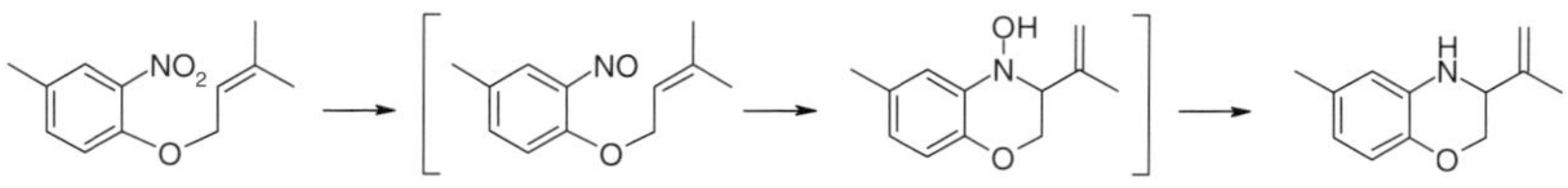

N-Unsubst. 3-vinyl-3,4-dihydro-2*H*-1,4-benzoxazines. A soln. of the nitroarene and 6 eqs. triethyl phosphite in toluene sealed in a septum reaction vial, subjected to microwave irradiation

(Discover, CEM; 2450 MHz; 300 W; 200°) for 20 min, excess of triethyl phosphite and generated triethyl phosphate removed, and worked up with flash chromatographic purification on silica gel → product. Y 57%. The solvent is critical in avoiding the formation of undesirable N-ethyl derivs. Reaction is thought to involve initial reduction of the nitro group to nitroso, followed by intramolecular nitroso-ene reaction, and finally reduction of the generated N-hydroxy deriv. Halides and esters remained unaffected. F.e. incl. N-unsubst. 2-vinyl-1,2,3,4-tetrahydroquinoxalines and -quinolines, also thermal method, s. E. Merisor, J. Conrad, I. Klaiber, S. Mika, U. Beifuss, Angew. Chem. Int. Ed. Engl. *2007*, *46* (18), 3353-5.

*2,4-Bis(*p*-methoxyphenyl)-1,3,2,4-dithiadiphosphetane 2,4-disulfide/triethylamine* ←
2-Azetidinones from β-aminocarboxylic acids
with $POCl_3/Et_3N$ cf. *42*, 420s*68*; with 0.5 eq. Lawesson's reagent and excess Et_3N in methylene chloride, also with 2,2′-dithiobis(benzothiazole)-triphenylphosphine, s. S. Kanwar, S.D. Sharma, Bull. Chem. Soc. Jpn. *2006*, *79* (11), 1748-52.

Ethyl or phenyl phosphorodichloridate/1,8-diazabicyclo[5.4.0]undec-7-ene *$ROP(O)Cl_2$/DBU*
Nitriles from carboxylic acid amides under mild conditions $CONH_2 \rightarrow CN$

265. 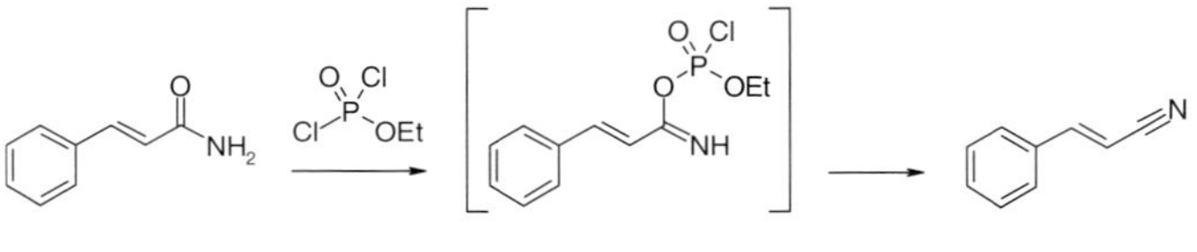

A mixture of cinnamic acid amide and 3 eqs. DBU in methylene chloride stirred at room temp. for 10 min, 2 eqs. ethyl phosphorodichloridate added, stirring continued for 3 h, quenched with aq. NH_4Cl, extracted with methylene chloride, and purified by chromatography → cinnamonitrile. Y 92%. Seventeen ar., heteroar. and aliphatic prim. amides gave yields of 73-95% in the presence of ether, halo, nitro and Boc-protected amine functionality. The reaction is rapid at room temp. (1-3 h), notably applicable to thermodynamically labile nitriles, occurs without racemization of chiral centres alpha to the amide, and work-up is clean and simple. Phenyl phosphorodichloridate was also an effective dehydrating agent but dimethylaminophosphoryl dichloride and the P(III)-analog were not. An intermediate O-phosphoryliminoester is possibly involved. F.e.s. C.-W. Kuo, J.-L. Zhu, J.-D. Wu, C.-M. Chu, C.-F. Yao, K.-S. Shia, Chem. Commun. *2007* (3), 301-3.

Iodine s. under $Me_3SiSiMe_3$ *I_2*
Ammonium chloride s. under Zn *NH_4Cl*
Hydrogen chloride s. under $SnCl_2$ and Fe *HCl*
Triethylamine hydrochloride s. under Ph_3P *Et_3NHCl*

Iron/hydrogen chloride *Fe/HCl*
Quinolines from 1-(*o*-nitroaryl)-2-acetylenealcohols ○
Reductive ring closure

266. 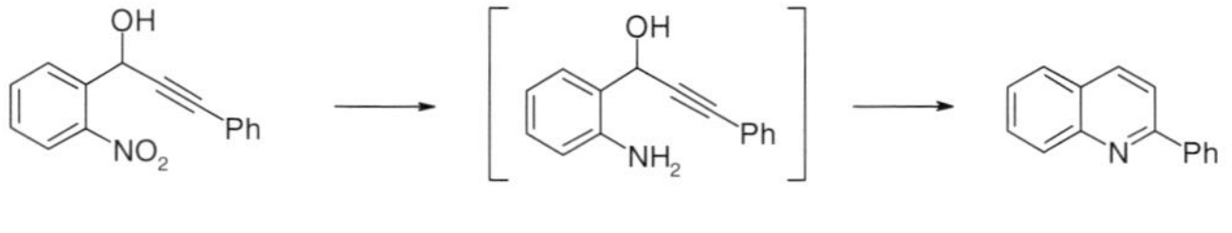

A soln. of 3.49 mmol iron metal in ethanol heated to 80°, 2 drops of concd. HCl added, stirred for 1 h, 0.727 mmol of 1-(2-nitrophenyl)-3-phenylprop-2-yn-1-ol in the same solvent added, stirring continued at 80° until the startg. m. was consumed (2 h), treated with 10% aq. HCl, stirred for 25 h (adding additional aq. HCl if the pH exceeded 4), quenched with satd. aq. $NaHCO_3$, and worked up → 2-phenylquinoline. Y 64%. Reaction tolerates electron-donating and -withdrawing groups, and the hydroxyl group may be secondary or tertiary. It is thought that, after the initial reduction of the nitro group, an acid-catalyzed Meyer-Schuster-type rearrangement takes place prior to heteroannulation. F.e. incl. 2-alkylquinolines, also with Zn-dust/NH_4Cl in aq. ethanol or with $SnCl_2$/HCl, s. M.J. Sandelier, P. DeShong, Org. Lett. *2007*, *9* (17), 3209-12.

Nickel *Ni*

Quinolines from β-(*o*-nitroaryl)-α,β-ethylenealdehydes [*o*-nitrocinnamaldehydes] ○

with $FeSO_4/NH_3$ cf. *10*, 394; by hydrogenation over Raney Ni in ethanol, and comparison with use of In/NH_4Cl s. S. Some, J.K. Ray, M.G. Banwell, M.T. Jones, Tetrahedron Lett. *2007*, *48* (20), 3609-12.

Carbonyl(dihydrido)tris(triphenylphosphine)ruthenium(II)/9,9-dimethyl-4,5-bis-(diphenylphosphino)xanthene ←

Nitriles from O-alkylaldoximes under neutral conditions CH═NOR → CN

267.

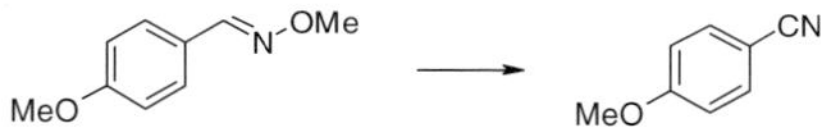

3.5 Mol% $RuH_2(CO)(PPh_3)_3$ and 3.6 mol% Xantphos heated at reflux in dry toluene under N_2 for 10-15 min, the startg. alkoxime in the same solvent added, heated at reflux for 24 h, cooled, evaporated, and worked up with column chromatographic purification on silica → product. Y 81%. Reaction is clean, high-yielding, and applicable to aliphatic or aromatic O-methyl-, O-benzyl- or O-*tert*-butyl-aldoximes, although yields were low with O-methyl-*p*-nitrobenzaldoxime (28%) and slightly lower with O-methyl-4-pyridylaldoxime (73%). No acidic or basic co-catalyst is required. It is believed that elimination of alcohol is facilitated by N-coordination of the substrate to the metal to give a nitrile complex which then dissociates. F.e. and comparison of ligands, also solvent effect, s. N. Anand, N.A. Owston, A.J. Parker, P.A. Slatford, J.M.J. Williams, Tetrahedron Lett. *2007*, *48* (44), 7761-3.

Nitrogen ↑ NC ⇑ N

Without additional reagents *w.a.r.*

Imidazoles from α-(aziridin-1-yl)-N-carbalkoxyhydrazones ○
via 1,5-electrocyclization of N-β-(N-carbalkoxyhydrazono)azomethinium ylids

268.

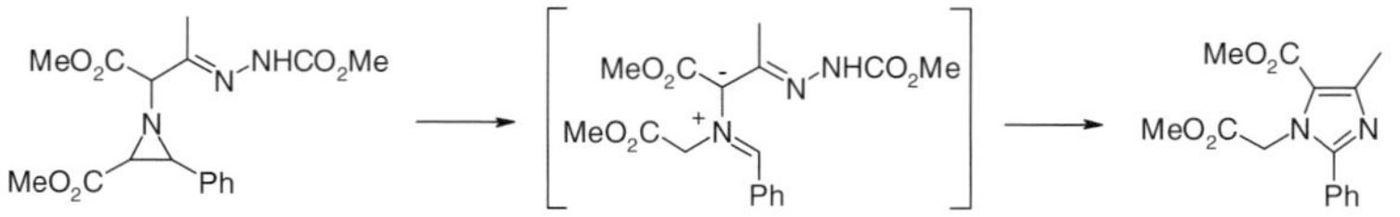

A soln. of the startg. hydrazone (readily prepared by 1,4-addition of the N-unsubst. aziridine with the corresponding N-carbalkoxyenazo compd. in the absence of solvent) in toluene heated at reflux for 6.5 h (TLC monitoring), concentrated, and worked up with column chromatographic purification on silica gel → 3-methoxycarbonylmethyl-5-methyl-2-phenyl-3*H*-imidazole-4-carboxylic acid methyl ester. Y 60%. Reaction is thought to involve initial generation of a stabilized azomethinium ylid which undergoes 1,5-electrocyclization prior to aromatization (unusually, without C-C bond formation!). F.e.s. O.A. Attanasi, P. Davoli, G. Favi, P. Filippone, A. Forni, G. Moscatelli, F. Prati, Org. Lett. *2007*, 9 (17), 3461-4; **2-arylimidazoles** from N-carbalkoxyenazo compds. and ar. amidines cf. O.A. Attanasi et al., Synlett *2007* (11), 1691-4.

Copper(II) triflate or Zinc iodide *$Cu(OTf)_2$ or ZnI_2*

Pyrrole-2-carbonyl from 2-azido-2,4-dienecarbonyl compds. ○

269. 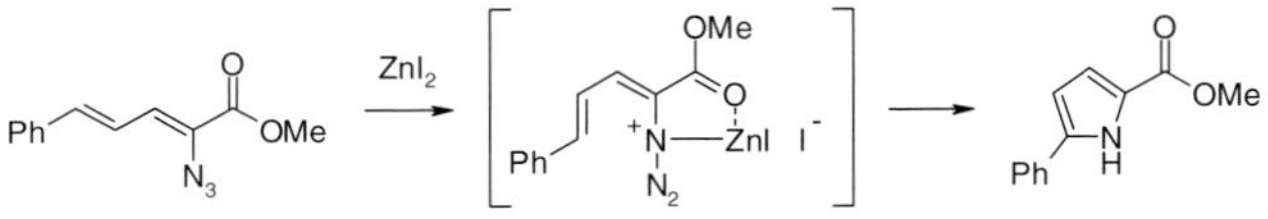

Pyrrole-2-carboxylic acid esters. Methylene chloride added to a mixture of the startg. dienyl azide and 5 mol% ZnI_2, stirred *at room temp.* for 15 h, filtered through SiO_2, concentrated *in vacuo*, and purified chromatographically → methyl 5-phenylpyrrole-2-carboxylate. Y 93%. For electron-deficient substrates the catalyst loading was increased to 10 mol%. The substituent at the 5-position of the substrate may be aryl (irrespective of electronic character), hetaryl or alkyl. Reaction is thought to involve a Schmidt-like Lewis acid activation, rather than a nitrenoid mechanism. F.e. **and 2-acylpyrroles** [under more stringent conditions], also with $Cu(OTf)_2$ or $Rh_2(OCOC_3F_7)_4$, s. H. Dong, M. Shen, J.E. Redford, B.J. Stokes, A.L. Pumphrey, T.G. Driver, Org. Lett. *2007*, *9* (25), 5191-4.

Triphenylphosphine *Ph_3P*

Peptidyl Δ^2-thiazolines or -oxazolines by intramolecular aza-Wittig synthesis

270. 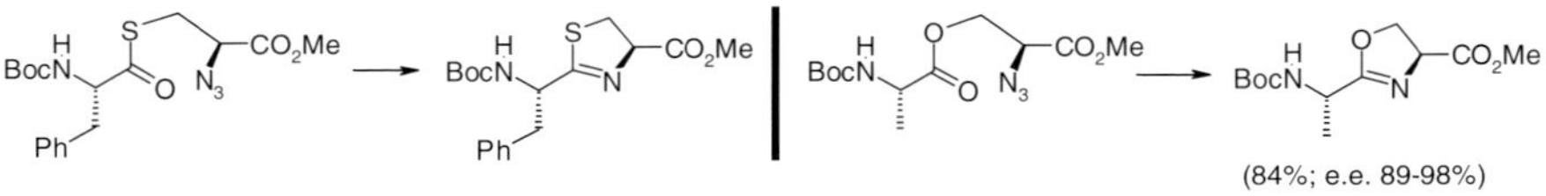

under mild conditions. Triphenylphosphine (ca. 2 eqs.) in dry THF added to a soln. of the startg. azide in the same solvent, and stirred for 2 h at -20 to 20° → product. Y 82% (e.e. 94%). The procedure is clean, tolerates a range of functionality, is water insensitive, and leaves chiral centres intact. The corresponding peptidyl Δ^2-oxazolines were obtained similarly (65-87% yield), although reaction was retarded with electron-withdrawing side-chains and/or β-branching of the side-chain. F.e. incl. bis(azolines) by **double ring closure**, and oxidation to the heteroarenes (with bromotrichloromethane/DBU), s. M. Riedrich, S. Harkal, H.-D. Arndt, Angew. Chem. Int. Ed. Engl. *2007*, *46* (15), 2701-3.

p-*Toluenesulfonic acid* *TsOH*

Dihydro-1,4-benzodiazepinone ring closure with azides

1,3-dihydro-2*H*-1,4-benzodiazepin-2-ones from *o*-(α-azidoacylamino)ketones with 1,3-propane-dithiol/Et_3N cf. *47*, 390; 4,5-dihydro-3*H*-1,4-benzodiazepin-3-ones from N-(α,β-ethyleneacyl)-*o*-azidobenzylamines via intramolecular 1,3-dipolar cycloaddition-cycloreversion in the presence of a little *p*-TsOH in refluxing toluene, chiral compds., s. G. Molteni, P. Del Buttero, Tetrahedron: Asym. *2007*, *18* (10), 1197-201.

(5,10,15,20-Tetraphenylporphyrinato)cobalt(II) *(TPP)Co(II)*

Regiospecific cobalt(II)-catalyzed intramolecular sulfonylamination with arenesulfonic acid azides

271. 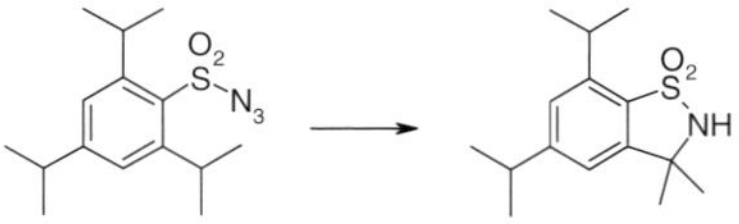

2,3-Dihydro-1,2-benzisothiazole 1,1-dioxides. Chlorobenzene added under N_2 through a rubber septum to a mixture of the startg. azide, *2 mol%* (5,10,15,20-tetraphenylporphyrinato)cobalt(II) and 5 Å molecular sieves contained in an oven-dried Schlenk tube, purged again with N_2 for 2 min, the septum replaced by a Teflon screw cap, heated at 80° for 18 h, concentrated, and worked up with flash chromatographic purification → product. Y 96%. Conditions are mild and neutral, the procedure clean (N_2 being the only by-product), and the catalyst is commercially available and effective at low loadings without requiring any additives. Insertion can take place at prim., sec. or tert. hydrocarbon groups, and there was no evidence of six-membered ring closure. F.e. and comparison of catalysts s. J.V. Ruppel, R.M. Kamble, X.P. Zhang, Org. Lett. *2007*, *9* (23), 4889-92.

Rhodium(II) perfluorobutyrate *$Rh_2(OCOC_3F_7)_4$*
Pyrrole-2-carbonyl from 2-azido-2,4-dienecarbonyl compds. s. *72*, 269 ○
Indole-2-carboxylic from α-azidocinnamic acid esters via generation of rhodium(II) nitrenoids under mild conditions

272.

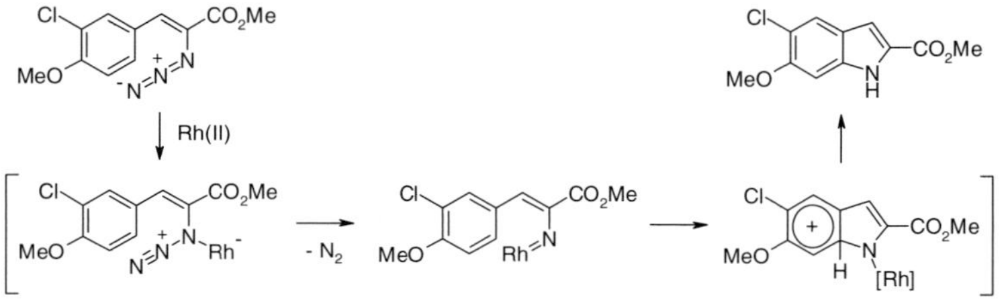

A mixture of startg. azidocinnamate and 3 mol% $Rh_2(OCOC_3F_7)_4$ heated in *toluene* at 40° for 16 h, cooled to room temp., filtered through silica, the filtrate concentrated *in vacuo*, and the residue purified by flash chromatography → 2-methoxycarbonyl-5-chloro-6-methoxyindole. Y 92% (as a 92:8 mixture with the 7-chloro-6-methoxy-isomer). Sixteen examples gave 71-98% yields of indoles and benzindoles with a predominance of the least sterically-encumbered regioisomer from *m*- and *m,p*-subst. derivs. The reaction tolerates both electron-rich and -poor aryl substituents with optimal reaction temps. of 40-60° (lower temps. requiring higher catalyst loadings) and reaction times of 3-5 h. β-Hetaryl (2-furyl, 2-pyrrolyl and 2-thienyl) derivs. gave the corresponding pyrrolo-fused heterocycles (5 examples; 79-94%). Deuterium-labelling experiments suggest that C-N bond formation occurs after loss of N_2 via a rhodium(II) nitrenoid and stepwise electrophilic substitution of the aryl ring. F.e. and preparation of startg. azidoacrylates by condensation of aryl/hetaryl aldehydes with methyl azidoacetate s. B.J. Stokes, H. Dong, B.E. Leslie, A.L. Pumphrey, T.G. Driver, J. Am. Chem. Soc. *2007*, *129* (24), 7500-1.

Halogen ↑ NC ⇑ Hal

Copper(I) chloride/acetylacetone/potassium carbonate *$CuCl/(MeCO)_2CH_2/K_2CO_3$*
Copper(I) iodide/ligand/potassium carbonate ←
Intramolecular N-arylation ○
with CuI/proline/K_2CO_3 cf. *63*, 191s*70*; oxindoles from *o*-bromoarylacetic acid amides with CuCl/acetylacetone/K_2CO_3 s. A. van den Hoogenband, J.H.M. Lange, J.A.J. den Hartog, R. Henzen, J.W. Terpstra, Tetrahedron Lett. *2007*, *48* (26), 4461-5; pyrazolo[1,5-*a*]indoles from 3(5)-(*o*-bromobenzyl)pyrazoles with CuI/ligand/K_2CO_3 s. Y.-M. Zhu, L.-N Qin, R. Liu, S.-J. Ji, H. Katayama, ibid. *48* (36), 6262-6.

Sulfur ↑ NC ⇑ S

Dichlorodimethylsilane/triethylamine *Me_2SiCl_2/Et_3N*
Isothiocyanates from dithiocarbamic acids NHCSSH → N=C=S
with 2-chloropyridine methiodide/Et_3N cf. *30*, 340s*32*; with Me_2SiCl_2/Et_3N s. G.F. Bian, H.Y. Qiu, J.X. Jiang, J.R. Wu, G.Q. Lai, Phosphorus, Sulfur Silicon Relat. Elem. *2007*, *182* (3), 503-8.

Remaining Elements ↑ NC ⇑ Rem

Sodium tetrahydridoborate/sodium hydride *$NaBH_4/NaH$*
Azetidines from β-(phosphorylamino)ketones 

273. 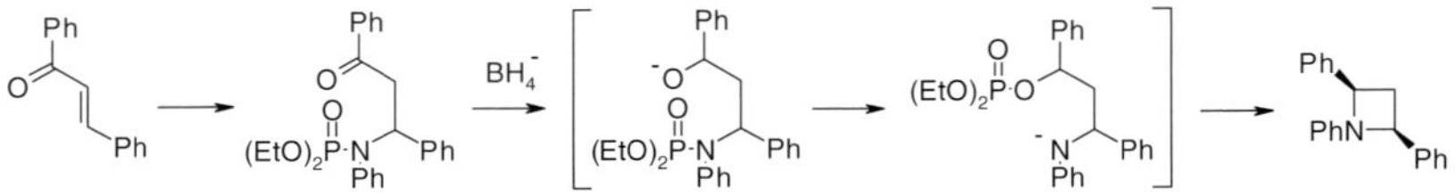

***cis*-1,2,4-Triarylazetidines.** Equivalent amounts of the startg. phosphorylaminoketone and $NaBH_4$ in dry *tert*-butanol stirred at room temp. for 2 h, heated at 60° for 2 h, 1 eq. NaH added at room temp., stirring continued for 3 h, then at 60° for a further 2 h, and worked up → product. Y 80%. This is part of a **two-step synthesis from α,β-ethyleneketones** via Michael addition of phosphoromonoamidates. The procedure is highly diastereoselective (with 93-6% of the *cis*-isomer being formed), and is thought to involve intramolecular attack of alkoxide ion on phosphorus prior to an N→O-phosphoryl group migration and elimination of phosphate. F.e.s. L. Dhar, S. Yadav, C. Awasthi, V.K. Rai, A. Rai, Tetrahedron Lett. *2007*, *48* (45), 8037-9.

Carbon ↑ NC ⇑ C

Without additional reagents *w.a.r.*
N-Condensed 3,4-dihydro-1,2,4-triazine-5(2*H*)-thiones from 2-allylthio-2-aminoenazo compds. 

274. 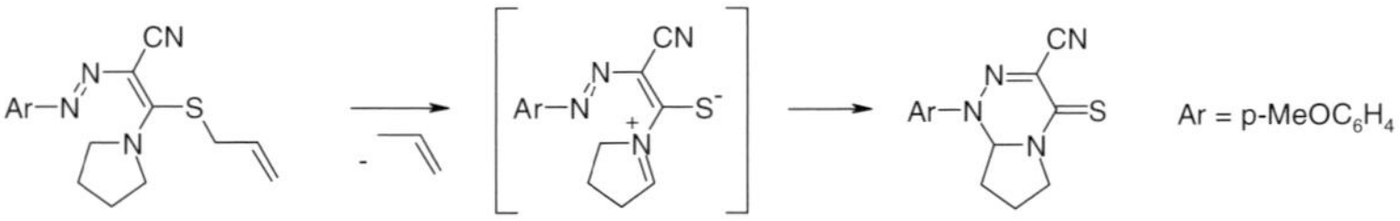

A soln. of the startg. 2-aminoenazo compd. in acetonitrile kept under argon at 40° for 40 h (TLC monitoring), solvent removed, and the residue worked up with column chromatographic purification → product. Y 70%. Although the mechanism of the reaction was not established, it is possible that it involves initial elimination of propene, followed by a 6π-electrocyclization. F.e. and from the corresponding 2-amino-2-(propargylthio)enazo compds. (with elimination of propyne) s. N.P. Belskaia, T.G. Deryabina, A.V. Koksharov, M.I. Kodess, W. Dehaen, A.T. Lebedev, V.A. Bakulev, Tetrahedron Lett. *2007*, *48* (52), 9128-31.

Formic acid/triethylamine s. under $Pd(PPh_3)_4$ ←

o-*Iodoxybenzoic acid/tetraethylammonium bromide* *$ArIO_2/Et_4NBr$*
N,N-Disubst. cyanamides from glycinamides $>NCH_2CONH_2 \rightarrow >NCN$

275. 

under mild conditions. 3 Eqs. tetraethylammonium bromide added to a stirred suspension of 3 eqs. *o*-iodoxybenzoic acid in acetonitrile, stirred for 5 min, the startg. glycinamide added in one portion to the resulting yellow suspension, heated for 20 min at 60°, cooled, and worked up with column chromatographic purification on silica gel → product. Y 85%. Dess-Martin periodinane was equally effective in place of *o*-iodoxybenzoic acid. The procedure is rapid, non-toxic, and relatively inexpensive. F.e. and with retention of N-benzyl groups s. K.H. Chaudhari, U.S. Mahajan, D.S. Bhalerao, K.G. Akamanchi, Synlett *2007* (18), 2815-8.

Chiral phosphoromonoamidites s. under $[Ir(cod)Cl]_2$ ←

Tetraethylammonium bromide s. under o-*Iodoxybenzoic acid* *Et_4NBr*

Tetrakis(triphenylphosphine)palladium(0)/formic acid/triethylamine ←

N-Unsubst. indoles from 3-(*o*-trifluoroacetylamino)aryl-2-acetylenecarbonic acid esters ○

Palladium-catalyzed reductive ring closure

276.

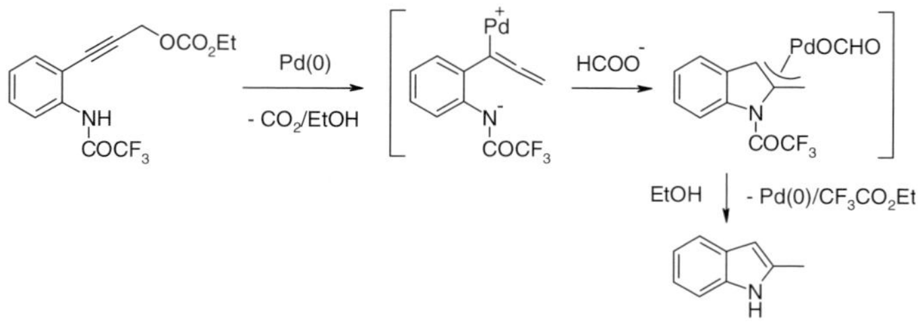

N-Unsubst. 2-alkylindoles. A soln. of the startg. 2-acetylenecarbonate, 5 mol% $Pd(PPh_3)_4$, 3 eqs. triethylamine and 2 eqs. formic acid in acetonitrile stirred at 80° under argon for 1 h in a Carousel Tube Reaction (Radley Discovery), cooled, concentrated, and worked up with column chromatographic purification on silica gel → product. Y 91%. The procedure is simple, high-yielding and applicable to a wide range of substrates, which are readily obtainable from ethynylmagnesium chloride and aldehydes, followed by Sonogashira coupling. Reaction may involve initial generation of a σ-allenylpalladium complex (in equilibrium with a π-propargylpalladium species), followed by intramolecular attack by nitrogen on the central carbon atom, conversion of the resulting carbene to a π-allylpalladium complex, and finally a concerted decarboxylation with hydride transfer to one of the allylic termini. F.e. and comparison of catalysts, additives, solvents and reaction temperatures s. I. Ambrogio, S. Cacchi, G. Fabrizi, Tetrahedron Lett. *2007*, *48* (43), 7721-5.

Chloro(cyclooctadiene)iridium(I) dimer/chiral cyclic phosphoromonoamidite/ 1,8-diazabicyclo[5.4.0]undec-7-ene/1,8-bis(dimethylamino)naphthalene ←

N-Protected 2-ethyleneamines from 2-ethyleneurethans ←

Decarboxylative asym. intramolecular N-allylation with double bond shift

277.

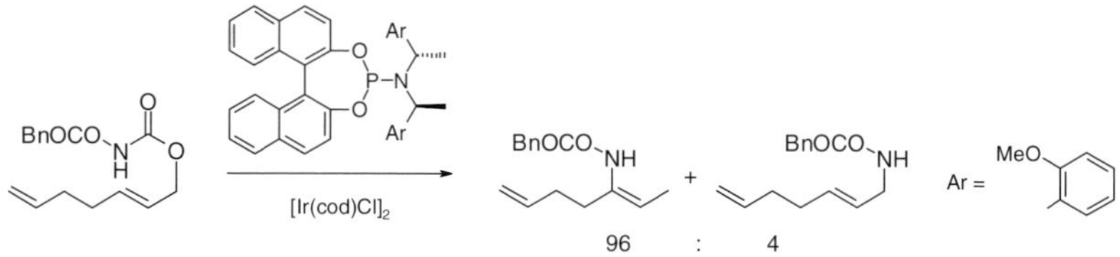

under mild conditions. A mixture of 2 mol% $[Ir(cod)Cl]_2$ and 4 mol% chiral phosphoromonoamidite stirred in THF at room temp. under N_2, 1 eq. DBU, 1 eq. startg. ethyleneurethan and 1 eq. Proton Sponge added (colour change from orange to yellow), the homogeneous soln. stirred for 3 h, diluted with chloroform, washed with 1 *M* HCl, water and satd. aq. $NaHCO_3$, the organic layer dried over Na_2SO_4, evaporated *in vacuo*, and the residue purified by chromatography → product. Y 86% (e.e. 96%; selectivity 96%). Eleven examples gave 78-92% yields with e.e. >91% and selectivity (for branched vs. linear regioisomers) >89% at room temp. (generally in 1-3 h), additional conjugated and unconjugated double bonds, furyl groups and *tert*-butyldiphenylsilyl ethers being unaffected. The reaction does not proceed to completion in the absence of either DBU or Proton Sponge. Initial experiments suggest the intermediacy of an iridium π-allyl complex rather than a [3.3]-sigmatropic rearrangement, and that decarboxylation occurs after N-C bond formation. F.e. incl. preparation of the substrates s. O.V. Singh, H. Han, J. Am. Chem. Soc. *2007*, *129* (4), 774-5.

Formation of Hal-Hal Bond

Uptake ⇓

Addition to Halogen HalHal ⇓ Hal

Sodium hypochlorite or chlorite/hydrogen chloride *NaOCl or $NaClO_2$/HCl*
Iododichlorides from iodides I → ICl_2
Ar. iododichlorides s. *35*, 182s*72*

Formation of Hal-S Bond

Exchange ↓↑

Hydrogen ↑ HalS ↓↑ H

Trimethylsilyl chloride/potassium nitrate *Me_3SiCl/KNO_3*
Sulfonic acid chlorides from mercaptans SH → SO_2Cl
with KNO_3 cf. *48*, 417; with Me_3SiCl/KNO_3 in methylene chloride at 50°, also from disulfides (cf. *18*, 561), s. G.K.S. Prakash, T. Mathew, C. Panja, G.A. Olah, J. Org. Chem. *2007*, *72* (15), 5847-50.

Sulfur ↑ HalS ↓↑ S

Trimethylsilyl chloride/potassium nitrate *Me_3SiCl/KNO_3*
Sulfonic acid chlorides from disulfides $(RS)_2$ → RSO_2Cl
s. *18*, 561s*72*

Formation of Hal-Rem Bond

Exchange ↓↑

Hydrogen ↑ HalRem ↓↑ H

Trichloroisocyanuric acid *TCCA*
Replacement of Si-hydrogen by chlorine ⩾SiH → ⩾SiCl
cf. *6*, 540s*70*; in silanes or siloxanes with trichloroisocyanuric acid s. S. Varaprath, D.H. Stutts, J. Organometal. Chem. *2007*, *692* (10), 1892-7.

Formation of Hal-C Bond

Uptake ⇓

Addition to Oxygen and Carbon HalC ⇓ OC

Monobromoborane-dimethyl sulfide *$BrBH_2$-SMe_2*
1,2-Bromohydrins from epoxides → C(Br)C(OH)
with $MgBr_2$ cf. *49*, 398s*68*; with $BrBH_2$-SMe_2, regio- and chemo-selectivity, s. C.D. Roy, H.C. Brown, Aust. J. Chem. *2007*, *60* (2), 139-45.

*Trifluoroacetic acid/tetra-*n*-butylammonium halides* *CF_3COOH/Bu_4NHal*
1,2-Halogenhydrins from epoxides → C(Hal)C(OH)
with Ph_3P/DDQ/Bu_4NHal cf. *38*, 461s*69*; with 1.5 eqs. CF_3COOH and 3 eqs. Bu_4NHal in alcohol-free chloroform, regio- and stereo-specific ring opening of glycidyl derivs., s. S.D. Stamatov, J. Stawinski, Tetrahedron Lett. *2007*, *48* (11), 1887-9; with Silphos [[$PCl_{3-n}(SiO_2)_n$]] in the presence of Br_2, I_2, or N-halogenosuccinimides in acetonitrile or in the absence of solvent s. N. Iranpoor, H. Firouzabadi, A. Jamalian, Phosphorus, Sulfur, Silicon Relat. Elem. *2006*, *181* (11), 2615-21.

Cyanuric chloride/morpholine ←
Zirconyl chloride *$ZrOCl_2$*
1,2-Chlorohydrins from epoxides → C(Cl)C(OH)
with $ZrCl_4$ cf. *41*, 457s*68*; with $ZrOCl_2$ in acetonitrile, also aziridine ring opening (cf. *61*, 191s*69*), s. B. Das, M. Krishnaiah, K. Venkateswarlu, Chem. Lett. *2007*, *36* (1), 82-3; with cyanuric chloride/morpholine in water, regio- and stereo-selectivity, s. B. Das, K. Venkateswarlu, M. Krishnaiah, Helv. Chim. Acta *2007*, *90* (1), 149-52.

Silica-based phosphorus(III) chloride/N-halogenosuccinimides ←
1,2-Halogenhydrins from epoxides s. *38*, 461s*72* → C(Hal)C(OH)

Iodine *I_2*
Acoxyhalides from cyclic ethers and carboxylic acid halides
4-acoxyhalides with a little $BiCl_3$ cf. *43*, 402s*69*; with 2.5 mol% I_2 s. M.A. Pasha, K. Manjula, Synth. Commun. *2007*, *37* (6), 927-32.

*Tetra-*n*-butylammonium halides s. under CF_3COOH* *Bu_4NHal*

Addition to Nitrogen and Carbon HalC ⇓ NC

Without additional reagents *w.a.r.*
2-Bromo-N-benzylamines from aziridines and benzyl bromides via regiospecific ring opening of N-benzylaziridinium ions

278. 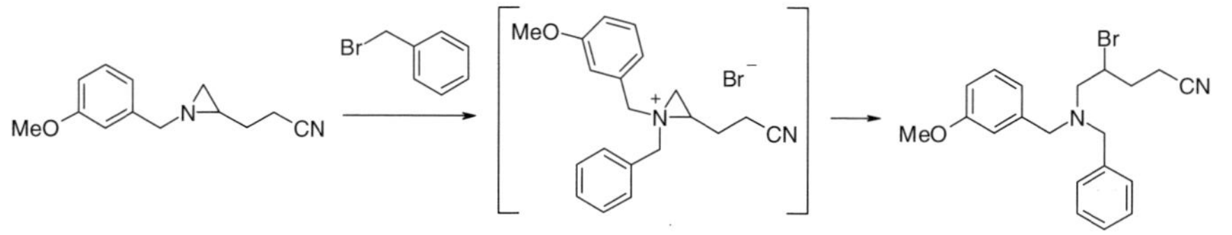

γ-Bromo-δ-(dibenzylamino)nitriles. Benzyl bromide (1 eq.) added to a soln. of 1-(3-methoxyphenylmethyl)-2-(2-cyanoethyl)aziridine (5 mmol) in acetonitrile (10 ml), and the mixture heated under reflux for 5 h → product. Y 98%. This novel approach offers a convenient and direct method for the preparation of functionalized γ-bromonitriles, which are otherwise difficult to prepare. Reaction is presumed to involve initial N-benzylation, followed by regiospecific ring opening of the intermediate N-benzylaziridinium ring with halide ion. F.e.s. M. D'hooghe, K. Vervisch, N. De Kimpe, J. Org. Chem. *2007, 72* (19), 7329-32.

Zirconyl chloride *$ZrOCl_2$*
1,2-Chloramines from aziridines s. *61*, 191s*72* ▽N → C(Cl)C(NH)

1-(1-Carboxyethyl)-3-methylimidazolium halides ←
Imidazolium [pseudo]halides as both ionic liquids and sources of reactive anions ←

279. 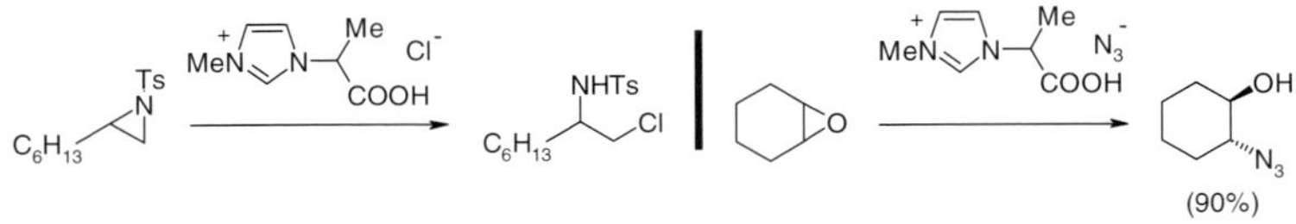

2-Halogenotosylamines from N-tosylaziridines. A mixture of 2-hexyl-1-tosylaziridine and 1.3 eqs. 1-(1-carboxyethyl)-3-methylimidazolium chloride heated at 60° with stirring for 50 min (TLC monitoring), quenched with satd. brine, extracted with ether, evaporated, and worked up with column chromatographic purification on silica gel → N-(1-chloromethylheptyl)-4-methylbenzenesulfonamide. Y 90%. The procedure is very simple, clean, high-yielding, highly regio- and stereo-specific, and equally facile with imidazolium bromides, iodides, azides and thiocyanates to give the corresponding 2-functionalized tosylamines, notably with retention of chloro, methoxy and phenoxy groups. F.e., **also regio- and stereo-specific ring opening of epoxides** with the corresponding imidazolium azide or thiocyanate, and the conversion of a cyclic α,β-epoxyketone to the corresponding cyclic α-azido-α,β-ethyleneketone, s. B.C. Ranu, L. Adak, S. Banerjee, Can. J. Chem. *2007, 85* (5), 366-71.

Addition to Carbon-Carbon Bonds HalC ⇓ CC

Sodium azide/potassium iodide/potassium peroxymonosulfate/alumina ←
2-Iodoazides from ethylene derivs. C=C → C(I)C(N_3)
s. *43*, 456s*64*; 2-iodo-2-deoxyglycosyl azides from glycals, stereoselectivity, also with KI/N-iodosuccinimide, s. G.K. Rawal, S. Rani, K.P. Madhusudanan, Y.D. Vankar, Synthesis *2007* (2), 294-8.

Lithium bromide/sodium periodate/hydrogen chloride *LiBr/$NaIO_4$/HCl*
α-Bromo-β-hydroxy- from α,β-ethylene-carbonyl compds. C=C → C(Br)C(OH)
regio- and stereo-specific conversion with LiBr/$NaIO_4$/H_2SO_4 cf. *66*, 224; asym. conversion of chiral N-(α,β-ethyleneacyl)-2-oxazolidones with HCl as the acid, incl. applications to (-)-cytoxazone and (-)-droxidopa synthesis, s. S. George, S.V. Narina, A. Sudalai, Tetrahedron Lett. *2007*, *48* (8), 1375-8.

Chiral ferrocenyl-fused 4-pyrrolidinopyridine ←
α-*tert*-Chlorocarboxylic acid esters from ketenes ←
Catalytic asym. conversion

280. 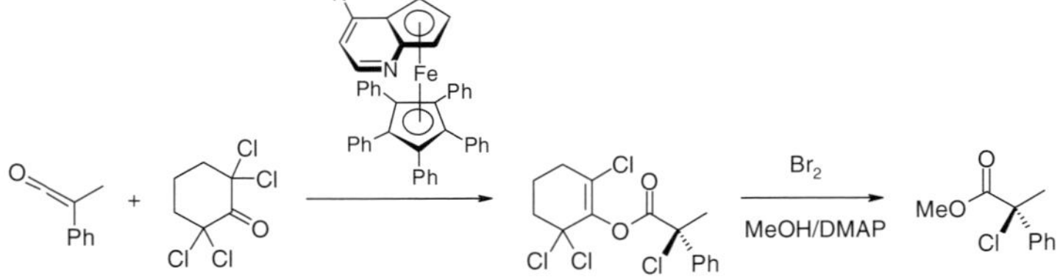

via 2,6,6-trichlorocyclohex-1-enyl esters. A soln. of ca. 1.2 eqs. 2,2,6,6-tetrachlorocyclohexanone in ether added in one portion to a soln. of methyl(phenyl)ketene in the same solvent at -78°, treated dropwise over ca. 15 s with a soln. of 3 mol% chiral ferrocenyl-fused 4-pyrrolidinopyridine in 1:1 ether/toluene, stirred at -78° for 45 min, solvent removed, and the residue worked up with flash chromatographic purification → product. Y 76% (e.e. 90%). Conversion to the corresponding methyl ester was simply effected (without loss of chirality) by subsequent activation with Br_2 in methylene chloride at 0°, followed by treatment with methanol at room temp. in the presence of DMAP. Enantioselectivity was noticeably lower with less sterically demanding hexachloroacetone as chlorinating agent. The procedure is generally applicable to a range of alkyl(aryl)ketenes (e.e. 82-95%), although enantioselectivity was slightly lower (65-67%) with substrates possessing branched alkyl groups or *o*-subst. aryl groups. The mechanism of the reaction has not been deduced, but appears to involve either addition of the N-chlorinated reagent to an *achiral* enolate, or addition of an achiral chlorinating agent to a *chiral* enolate. Other reagents were less effective, as were other solvents or variations in the temperature. F.e.s. E.C. Lee, K.M. McCauley, G.C. Fu, Angew. Chem. Int. Ed. Engl. *2007, 46* (6), 977-9.

Copper(I) chloride *CuCl*

Regio- and stereo-selective copper(I)-catalyzed aminochlorination of α,β-ethylenenitriles under mild conditions C=C → C(Cl)C(NH-)

281.

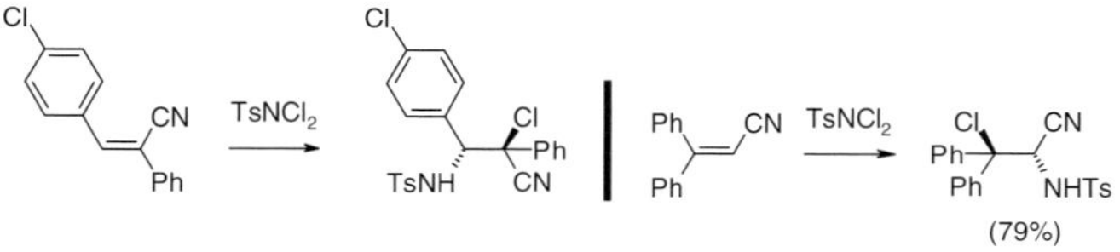

***anti*-α-Chloro-β-(tosylamino)nitriles.** A soln. of 2 eqs. N,N-dichloro-*p*-toluenesulfonamide in acetonitrile added to a mixture of startg. ethylenenitrile, 10 mol% CuCl and 4 Å molecular sieves in *acetonitrile* under argon during 5 min, the mixture stirred at room temp. *without argon protection* for 24 h, quenched with satd. aq. Na_2SO_3, the aq. layer separated, extracted with ethyl acetate, the combined organic layers washed with brine, dried over anhydrous Na_2SO_4, and purified by TLC → α-chloro-β-(tosylamino)nitrile. Y 89% (*anti/syn* 10:1). The reaction is tolerant of temperature variation (5-50°) but is conveniently completed within 2 h at room temperature for subst. cinnamonitriles with *2 eqs.* of the dichlorosulfonamide in yields of 55-89% and *anti/syn* selectivities at least 7:1. CuCl was the catalyst of choice and low yields were obtained in the absence of molecular sieves or with other solvents in place of acetonitrile. Ethylenenitriles lacking an α-substituent gave the β-chloro regioisomer whereas the opposite regioselectivity was observed for α,β-disubst. derivs. These observations are consistent with the proposed S_N2-type cleavage of a stabilized aziridinium intermediate. F.e.s. J.-L. Han, S.-J. Zhi, L.-W. Wang, Y. Pan, G. Li, Eur. J. Org. Chem. *2007* (8), 1332-7.

Silver nitrate/oxalyl chloride *$AgNO_3/(COCl)_2$*

2-Chloroacylamines from ethylene derivs. Regio- and stereo-specific conversion C=C → C(Cl)C(NHAc)

282.

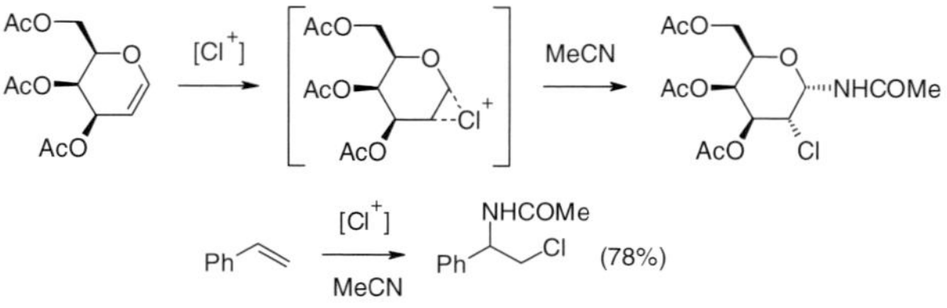

N-Acyl-2-chloro-2-deoxyglycosylamines. 1.5 Eqs. oxalyl chloride added slowly to a soln. of 1.5 eqs. $AgNO_3$ in acetonitrile at -20°, stirred for 5 min, cooled to -30°, the startg. glycal in acetonitrile added at once, stirring continued for 45 min (TLC monitoring), the mixture carefully decanted into satd. aq. $NaHCO_3$, the precipitate (AgCl) removed and washed, and the mixture worked up with column chromatographic purification → product. Y 80%. The reagent combination provides a mild and useful source of chloronium ion. F.e. and conversion to **glycopeptides**, also reaction with styrene and 1,2-disubst. olefins, and preparation of **β-acylamino-α-chlorocarboxylic acid esters** from α,β-ethylenecarboxylic acid esters, s. G.K. Rawal, A. Kumar, U. Tawar, Y.D. Vankar, Org. Lett. *2007, 9* (25), 5171-4.

1,3-Bis(2,6-diisopropylphenyl)imidazolidin-2-ylidenegold(I) tert-*butoxide/ triethylamine hydrofluoride/N,N-dimethylanilinium triflate/potassium hydrogen sulfate* ←
Regiospecific gold-catalyzed *trans*-hydrofluorination C≡C → C(F)=CH
of acetylene derivs. with reversible carbon-fluorine bond formation

283. 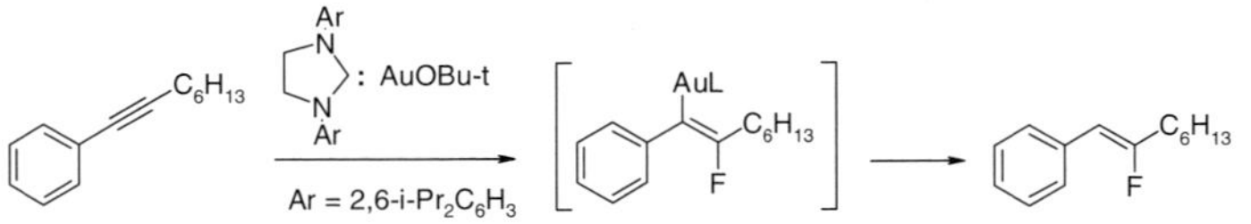

under mild conditions. In a glove box, methylene chloride added to a mixture of 1-phenyloct-1-yne, 2.5 mol% gold(I) carbene complex and 10 mol% N,N-dimethylanilinium triflate in a vial, 2 eqs. triethylamine hydrofluoride and 1 eq. $KHSO_4$ added without special precaution, the vial capped, the heterogeneous mixture stirred at room temp. in the absence of light for 22 h, the mixture quenched with satd. aq. $NaHCO_3$, extracted with methylene chloride, the extract dried over $MgSO_4$, concentrated *in vacuo*, and the residue purified by chromatography on silica gel → *trans*-2-fluoro-1-phenyloct-1-ene. Y 78% (*trans/cis* 13:1; single regioisomer). The method is successful with both diaryl- and aryl(alkyl)-subst. acetylenes giving yields of 63-86% (6 examples). The presence of an electron-releasing 4-MeO group on the aryl ring reduced regioselectivity to 5:1 while electron-withdrawing 4-CO_2Me gave the β-fluorostyrene deriv. as sole product. In all cases, only *trans*-hydrofluorination was observed and no *gem*-difluoroalkanes were detected. The results are consistent with formation of an intermediate, *reversibly formed* 2-fluorovinylgold(I) complex which was isolated in one case for X-ray structural analysis. F.e.s. J.A. Akana, K.X. Bhattacharyya, P. Müller, J.P. Sadighi, J. Am. Chem. Soc. *2007, 129* (25), 7736-7.

Alumina s. under NaN_3 *Al_2O_3*
Phenyl iodosoacetate s. under TsN(Cl)Na *$PhI(OAc)_2$*
1,2-Dichloroethane s. under Bu_4NI *$ClCH_2CH_2Cl$*
N-Bromosuccinimide s. under $Cp_2Zr(H)Cl$ *NBS*

Chloramine-T/phenyl iodosoacetate *TsN(Cl)Na/$PhI(OAc)_2$*
2-Chlorotosylamines from electron-deficient ethylene derivs. C=C → C(Cl)C(NHTs)
s. *24*, 555s*72*

Chloramine-T/sulfuric acid/benzyltriethylammonium chloride *TsN(Cl)Na/H_2SO_4/$BnEt_3NCl$*
β-Chloro-α-tosylamino- from α,β-ethylene-carbonyl compds.
Regio- and stereo-specific conversion under Brønsted acid catalysis

284. 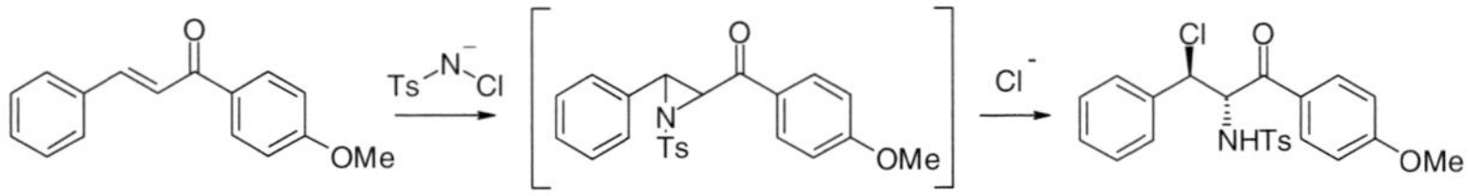

in water under mild conditions. *1.2 Eqs. H_2SO_4* added to the startg. enone, 2 eqs. Chloramine-T and 10 mol% benzyltriethylammonium chloride in water, the mixture stirred vigorously at room temp. for 8 h, extracted with ethyl acetate, the organic phase dried over $MgSO_4$, evaporated *in vacuo*, and the residue purified by chromatography → *trans*-3-chloro-1-(4-methoxyphenyl)-3-phenyl-2-(tosylamino)propan-1-one. Y 73% (*anti/syn* 93:7). The method is ecofriendly, efficient and mild, and is *not* promoted by any other common phase-transfer agent. The reaction is successful with chalcones and cinnamyl esters/amides giving 65-83% yields of predominantly *anti* isomers (92-99%). Low yields were obtained, however, with methyl styryl ketone (41%) and 4-nitrostyryl phenyl ketone (40%), while 1-butenyl phenyl ketone did not react. *p*-Toluenesulfonic (but not methanesulfonic) acid can replace sulfuric acid. F.e. and optimization of the reaction conditions s. X.-L. Wu, G.-W. Wang, J. Org. Chem. *2007*, *72* (24), 9398-40.

Bromodichloroisocyanuric acid ←
1,2-Oxybromination of ethylene derivs. C=C → C(O-)C(Br)
also ar. bromination cf. *71*, 196; with bromodichloro- in place of tribromo-isocyanuric acid, also prepn. of the reagent, s. L.S. de Almeida, P.M. Esteves, M.C.S. de Mattos, Synlett *2007* (11), 1687-90.

Triiodoisocyanuric acid *TICA*
1,2-Oxyiodination of ethylene derivs. C=C → C(I)C(O-)

285. 

Regio- and stereo-specific conversion to 1,2-iodohydrins. Triiodoisocyanuric acid (TICA) (0.67 mmol) added to a soln. of 1-methylcyclohexene (2 mmol) in 2:1 acetone/water (12 ml), the mixture stirred at room temp. for 10 min, methylene chloride added, cyanuric acid filtered off, the resulting soln. treated with 10% aq. $NaHSO_3$, and extracted with methylene chloride → *trans*-iodohydrin. Y 93%. The reaction is also highly regioselective for styrene derivs. (Markovnikov addition) but gives a mixture of regioisomers for 1-octene, affording predominantly the prim. iodide (75:25). A high-yielding, multigram procedure for preparing TICA from trichloroisocyanuric acid and I_2 is also described. F.e., also **α-iodoacetals** from enolethers, and **1,2-alkoxyiodides** and **1,2-acetoxyiodides** from alkenes using TICA with alcohols and acetic acid, respectively, s. R. da S. Ribeiro, P.M. Esteves, M.C.S. de Mattos, Tetrahedron Lett. *2007*, *48* (49), 8747-51.

Oxalyl chloride s. under $AgNO_3$ *$(COCl)_2$*

Polymethylhydrosiloxane *PMHS*
Regiospecific hydroiodination under mild conditions ←

286. 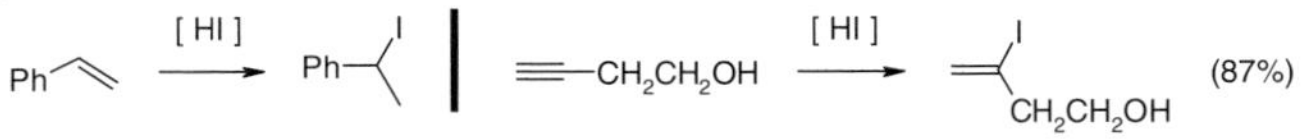

of ethylene derivs. A mixture of styrene and 1.5 eqs. polymethylhydrosiloxane in chloroform treated with 1 eq. iodine, stirred at room temp. for 30 min, solvent removed, and the residue worked up with column chromatographic purification on silica gel → product. Y 96%. The procedure is simple, mild, rapid and highly regioselective in favour of the Markovnikov adduct, except with methyl acrylate and acrylonitrile which afforded the 3-chloro derivs. Iodine is thought to react initially with polymethylhydrosiloxane to give a hydrosilyl hypoiodite (with elimination of trimethylsilyl iodide), which then eliminates *hydrogen iodide* as the reactive species. However, yields were poor or zero with other halogenating agents (e.g. NBS, NCS, NIS, I_2). The stereochemistry is consistent with approach of halogen from the *si*-face. **α,β-Ethyleneiodides** were obtained similarly (although more slowly) **from acetylene derivs.** (the Markovnikov product predominating as with ethylene derivs.). F.e. incl. functionalized 5- to 8-membered and 10-membered O-heterocyclics, also comparison of other chiral ligands, s. B. Das, Y. Srinivas, H. Holla, R. Narender, Chem. Lett. *2007, 36* (6), 800-1.

Chlorobis(cyclopentadienyl)hydridozirconium *$Cp_2Zr(H)Cl$*
Chlorobis(cyclopentadienyl)hydridozirconium/N-bromosuccinimide *$Cp_2Zr(H)Cl/NBS$*
α,β-Ethylenehalides from acetylene derivs. via hydrozirconation C≡C → CH=C(Hal)
1-halogenene- from acetylene-stannanes with I_2 or NBS cf. *43*, 625s*68*; (1E,3Z)-1-halogeno-3-silyl-1,3-dienes from (Z)-3-silyl-3-en-1-ynes (prepared via Sonogashira coupling of (E)-α-iodoenesilanes with (trimethylsilyl)acetylene followed by monodesilylation) s. X. Ye, P. Wang, M.Z. Cai, J. Chem. Res. *2007* (6), 319-22.

Antimony pentafluoride *SbF_5*
2-Fluoroamines from 2-ethyleneamines C=C → C(F)CH

287. 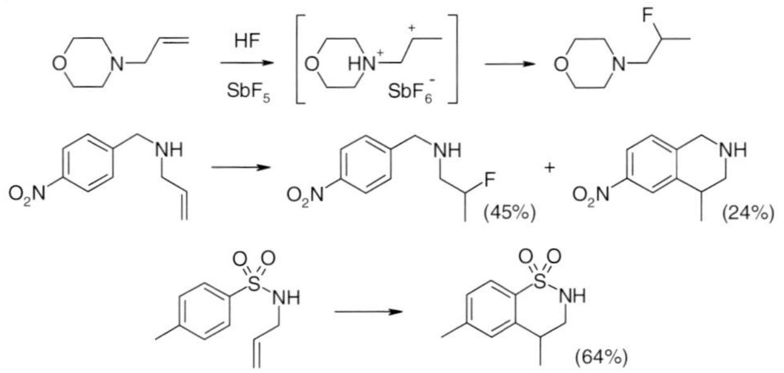

N-Allylmorpholine added to a mixture of SbF_5/HF (*1:7*) at -20°, the mixture stirred for 1 h, neutralized with ice/Na_2CO_3, extracted with methylene chloride, the organic phases dried over $MgSO_4$, concentrated *in vacuo*, and the residue purified by column chromatography → N-(2-fluoropropyl)morpholine. Y 61%. Using optimized conditions, hydrofluorination was successful with a number of sec. and tert. 2-ethyleneamines and N-allylindole giving 40-85% yields of the β-fluoro derivs. N-Allylbenzylamine gave a complex mixture of products while N-allyl-4-nitrobenzylamine gave 45% of the fluorinated product and 24% of a tetrahydroisoquinoline formed by **intramolecular Friedel-Crafts** reaction. Similar treatment of amide and sulfonamide analogs was largely unsuccessful affording hydration products only, but N-allyl-4-nitrobenzenesulfonamide gave the fluoride in 74% yield while N-allyl-4-methylbenzenesulfonamide underwent cyclization to a 1,2,3,4-tetrahydro-1,2-benzothiazine 1,1-dioxide in 64% yield. The observed products were largely explained by the formation of dicationic intermediates of differing reactivities. F.e.s. S. Thibaudeau, A. Martin-Mingot, M.-P. Jouannetaud, O. Karam, F. Zunino, Chem. Commun. *2007* (30), 3198-200.

Hydrogen peroxide *H_2O_2*
1,2-Ethylene-1,2-diiodides from acetylene derivs. C≡C → C(I)=C(I)
with I_2 in water containing CO_2, *trans*-isomers, cf. *13*, 561s*67*; with H_2O_2 in THF, (E)-isomers, s. A.O. Terent'ev, D.A. Borisov, I.B. Krylov, G.I. Nikishin, Synth. Commun. *2007*, *37* (18), 3151-64.

Hydrogen peroxide/ammonium iodide *H_2O_2/NH_4I*
Acyl 2-iodo-2-deoxyglycosides from glycals C=C → C(OAc)C(I)
with $PhI(OAc)_2$/quaternary ammonium iodides cf. *55*, 180; with H_2O_2/NH_4I and acetic anhydride in 1:1 acetic acid/acetonitrile s. D.W. Gammon, H.H. Kinfe, J. Carbohyd. Chem. *2007*, *26* (3), 141-57.

1-Butyl-3-methylimidazolium triflimide *[bmim]NTf_2*
2-Chlorosulfonylamines from ethylene derivs. C=C → C(Cl)C(N-SO_2R)
with N,N-dichloro-*o*-nitrobenzenesulfonamide and CuOTf cf. *24*, 555s*61*; α-chloro-β-(*o*-nitrophenylsulfonylamino)ketones from enones with high regio- and stereo-selectivity by heating in [bmim]NTf_2 (as ionic liquid solvent) without the need for a transition metal catalyst s. Y.N. Wang, B.K. Ni, A.D. Headley, G.G. Li, Adv. Synth. Catal. *2007*, *349* (3), 319-22; 2-chlorotosylamines from electron-deficient ethylene derivs. with Chloramine-T and $PhI(OAc)_2$ in a ball mill (30 Hz) s. G.-W. Wang, X.-L. Wu, Adv. Synth. Catal. *2007*, *349* (11-12), 1977-82.

N,N-Dimethylanilinium triflate/potassium hydrogen sulfate *$Me_2(Ar)NOTf/KHSO_4$*
s. under Gold(I) complexes
Sulfuric acid s. under TsN(Cl)Na *H_2SO_4*
Potassium peroxymonosulfate s. under NaN_3 *$KHSO_5$*

Sodium periodate s. under LiBr *$NaIO_4$*
Ammonium iodide s. under H_2O_2 *NH_4I*
Triethylamine hydrofluoride s. under Gold(I) complexes *Et_3NHF*

*Tetra-*n*-butylammonium iodide/1,2-dichloroethane* *$Bu_4NI/ClCH_2CH_2Cl$*
1,2-Ethylene-1,2-chloroiodides from acetylene derivs. C≡C → C(Cl)=C(I)
by addition of ICl cf. *37*, 446s*68*; with 3 eqs. Bu_4NI in refluxing dichloroethane (probably via *in situ*-generation of ICl) s. M.L. Ho, A.B. Flynn, W.W. Ogilvie, J. Org. Chem. *2007*, *72* (3), 977-83.

Benzyltriethylammonium chloride s. under TsN(Cl)Na *$BnEt_3NCl$*
Hydrogen chloride s. under LiBr *HCl*
Ferrocenes s.a. Chiral ferrocenyl-fused 4-pyrrolidinopyridine ←

Palladium(II) acetate/1-butyl-3-methylimidazolium triflimide *$Pd(OAc)_2/[bmim]NTf_2$*
2-Chlorosulfonylamines from ethylene derivs. C=C → C(Cl)C(N-SO_2R)
with asym. induction
chiral N-[β-chloro-α-(tosylamino)acyl]-2-oxazolidones with $TsNCl_2$ using a little CuOTf in $[bmim]BF_4$ cf. *68*, 204; under chelation control using 10 mol% $Pd(OAc)_2$ in $[bmim]NTf_2$/acetonitrile s. Y.-N. Wang, A. Kattuboina, T. Ai, D. Banerjee, G. Li, Tetrahedron Lett. *2007*, *48* (44), 7894-8.

Exchange ↓↑

Hydrogen ↑ HalC ↓↑ H

Irradiation s. under $PhI(OAc)_2$ and Mesoporous silica *///*

Potassium carbonate *K_2CO_3*
Intramolecular iodoamidation of unsatd. sulfonic acid amides ○
with *t*-BuOCl/NaI cf. *71*, 201; with K_2CO_3/I_2, 4-iodo-2*H*-1,2-benzothiazine 1,1-dioxides from *o*-acetylenesulfonic acid amides, s. D.K. Barange, V.R. Batchu, D. Gorja, V.R. Pattabiraman, L.K. Tatini, J.M. Babu, M. Pal, Tetrahedron *2007*, *63* (8), 1775-89.

Potassium bromide s. under $(BnPh_3P)_2S_2O_8$ *KBr*
*4,4′-Di-*tert*-butylbipyridyl s. under $[Ir(cod)OMe]_2$* ←
Copper(II) halides s.a. under $PdHal_2$ and $[Ir(cod)OMe]_2$ *$CuHal_2$*

Copper(II) bromide *$CuBr_2$*
Ar. bromination s. *20*, 417s*72*; *17*, 598s*72* H → Br

Silver nitrate/acetic acid *$AgNO_3/AcOH$*
Silver sulfate *Ag_2SO_4*
Ar. iodination H → I
cf. *17*, 598s*47*; *7*, 564s*46*,*48*; of ar. amines with 1 eq. I_2 and 1 eq. Ag_2SO_4 in ethylene glycol s. Y. Zhang, T.R. Ren, W.W. Zhu, Y.H. Xie, Org. Prep. Proced. Int. *2007*, *39* (1), 90-3; of activated or deactivated arenes by grinding with 1 eq. I_2 and 1 eq. $AgNO_3$ and a little acetic acid in a pestle and mortar s. M.S. Yusubov, E.N. Tveryakova, E.A. Krasnokutskaya, I.A. Perederyna, V.V. Zhdankin, Synth. Commun. *2007*, *37* (8), 1259-65.

Bis(pinacolato)diboron s. under $[Ir(cod)OMe]_2$ ←
Ammonium cerium(IV) nitrate s. under 1-Methyl-3-carboxymethylimidazolium halides *CAN*

Cerium(III) chloride/sodium hypochlorite *$CeCl_3/NaOCl$*
Halogenolactonization ○
cf. *33*, 477s*60*; chloro-β- or γ-lactones with $CeCl_3$/NaOCl s. J.A. Lopez-Lopez, F.M. Gurrea, J. Moreno-Dorado, Z.D. Jorge, G.M. Massanet, Tetrahedron Lett. *2007*, *48* (10), 1749-52.

Imidazolium salts s. under Halides ←
Benzoyl peroxide s. under Cl_3CCN or $Br_3CCOOEt$ *$[PhC(O)O]_2$*

m-*Chloroperoxybenzoic acid* *$ArCOO_2H$*
Diaryliodonium salts from ar. iodides and arenes ArH → ArI^+Ar'
with $K_2S_2O_8/CF_3COOH$ then ion exchange with Na-triflate cf. *51*, 204s*71*, with *m*-chloroperoxybenzoic acid in methylene chloride at room temp., followed by heating in a sealed tube with 2 eqs. triflic acid, also from two arene molecules and I_2, s. M. Bielawski, B. Olofsson, Chem. Commun. *2007* (24), 2521-3.

Dimethyl sulfoxide/oxalyl halides/triethylamine *$DMSO/(COHal)_2/Et_3N$*
α,β-Ethylene-α-halogenoketones from 2-ethylenealcohols CH(OH)CH=C → COC(Hal)=C
One-pot conversion via Swern oxidation

288. 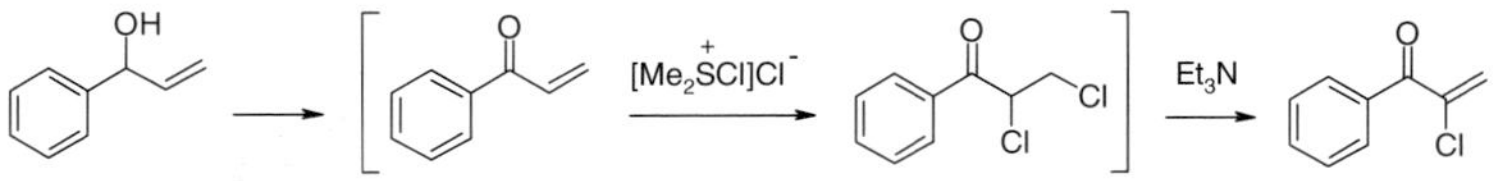

Aryl 1-halogenovinyl ketones. 3 Eqs. oxalyl chloride and 4 eqs. dimethyl sulfoxide added sequentially to dry methylene chloride at -78°, after 15 min the startg. α-vinylbenzyl alcohol in the same solvent added, kept at -78° for a further 1 h, treated with 6 eqs. Et_3N, allowed to warm to room temp. over 30 min, and poured onto ice-cold 1 *M* HCl → 2-chloro-1-phenyl-2-propen-1-one. Y 81%. The procedure is mild and tolerates a variety of functional groups. Substrates with an electron-donating group at the *o*- or *p*-position, however, gave the corresponding γ-aryl-β,γ-ethylenechlorides, while there was no reaction with higher alkenylarenes. A chloronium ion is thought to be implicated after the initial Swern oxidation. F.e. and aryl 1-bromovinyl ketones with oxalyl bromide s. J. Yin, C.E. Gallis, J.D. Chisholm, J. Org. Chem. *2007, 72* (18), 7054-7.

Phenyl iodosoacetate/irradiation *$PhI(OAc)_2$/*
Halogeno-N-heterocyclics by intramolecular halogenamination ○
bromo-derivs. with NBS cf. *48*, 434s*66*; iodo-derivs. with $PhI(OAc)_2/I_2$ under irradiation, mixed 3,3-dihalogenopiperidines from 4-ethylene-4-halogenamines and 2-(dihalogenomethyl)-pyrrolidines from 4-ethylene-5-halogenamines, s. H.J. Lu, Q. Chen, C.Z. Li, J. Org. Chem. *2007, 72* (7), 2564-9.

Trichloroacetonitrile or ethyl tribromoacetate/benzoyl peroxide ←
Carboxylic acid halides from aldehydes RCHO → RC(O)Hal

289. 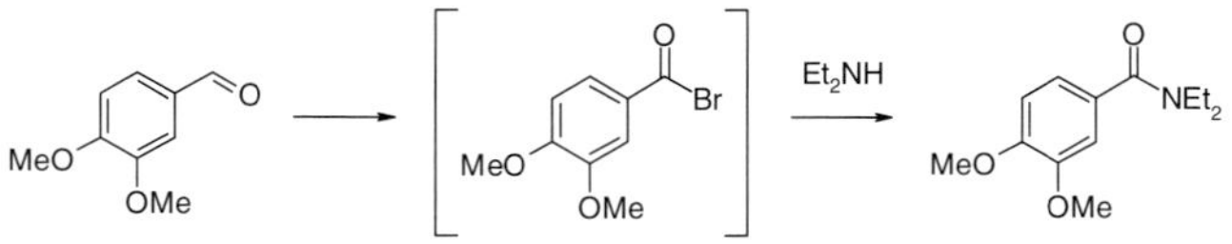

Carboxylic acid amides via acid bromides. A mixture of 3,4-dimethoxybenzaldehyde (0.5 mmol) and ethyl tribromoacetate (1.5 mmol) in chlorobenzene (4 ml) refluxed under argon for 24 h with addition of benzoyl peroxide (0.5 mmol) in four portions, the mixture cooled to room temp., Et_3N (1.5 mmol) added followed by diethylamine (0.75 mmol), then stirred for 1 h → product. Y 73%. The reaction affords moderate to good yields [of amides] with both aromatic and aliphatic aldehydes, although yields are best for ar. aldehydes bearing electron-donating substituents, consistent with a radical mechanism. Due to their moisture sensitivity, the intermediate acid bromides were further reacted without isolation or purification. F.e., **also carboxylic acid chlorides** using Cl_3CCN as the chlorinating agent, s. D.H. Kang, T.Y. Joo, W. Chavasiri, D.O. Jang, Tetrahedron Lett. *2007, 48* (2), 285-7.

N,N′-Dichloro-N,N′-bis(2,4,6-trichlorophenyl)urea *[Ar(Cl)N]$_2$CO*
1,1-Nitrosochlorides from oximes C=NOH → C(Cl)NO
with *t*-BuN(Cl)CN cf. *15,* 432s*70*; with N,N′-dichloro-N,N′-bis(2,4,6-trichlorophenyl)urea s. A.K. Gupta, J. Acharya, D. Pardasani, D.K. Dubey, Tetrahedron Lett. *2007, 48* (5), 767-70.

N-Chlorosuccinimide/triphenylphosphine *NCS/Ph$_3$P*
1,1,1-Trichlorides from methyl groups CH_3 → CCl_3
under photoirradiation cf. *10,* 451s*33*; with N-chlorosuccinimide/Ph$_3$P, trichloromethyl-heterocyclics, s. S. Goswami, A.C. Maity, H.K. Fun, Chem. Lett. *2007, 36* (4), 552-3.

N-Iodosuccinimide/chiral phosphoromonoamidites/chlorosulfonic acid ←
Asym. halogenative ring closure of polyprenoids ○

290.

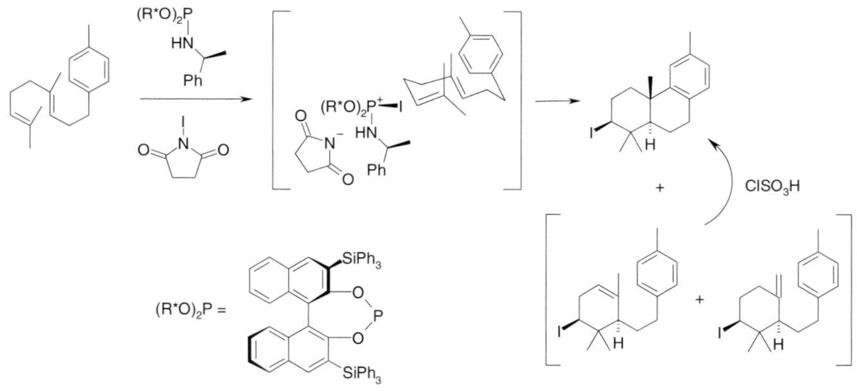

The first successful examples of enantioselective halogenopolycyclization of polyprenoids are reported by using a *nucleophilic* chiral promoter, thus placing the activated halogen atom closer to the chiral environment of the promoter than would be the case under Lewis acid-mediated conditions. **E: Double ring closure**. 4-(Homogeranyl)toluene and 1.1 eqs. N-iodosuccinimide added successively at -78° to a soln. of 1 eq. of the (S)-N-(1-phenethyl)-(R)-binaphthyl phosphoromonoamidite in toluene, the mixture stirred at -78° for 0.5 h, then at -40° for 24 h, quenched with 20% aq. $Na_2S_2O_3$, extracted with hexane, dried, filtered, concentrated, then purified by column chromatography on silica, the obtained ca. 6:4 mixture of the desired polycyclic product and (2-arylethyl)cyclohexene/(2-arylethyl)methylenecyclohexane taken up in 2-nitropropane, treated dropwise at -78° with $ClSO_3H$, stirred for 4 h, then quenched with satd. aq. $NaHCO_3$, extracted with hexane, dried, filtered, concentrated, and purified by column chromatography on silica → (3S,5R,10S)-(+)-product. Y 57% (e.e. 95%). The racemic product was obtained in methylene chloride. Reaction with N-bromosuccinimide proceeded in higher yield (75%) but lower enantioselectivity (36%), while asym. halogenocyclization with N-chlorosuccinimide failed. Transhalogenation of the product with retention of configuration was achieved, however (*t*-BuLi/ BrF_2CCF_2Br or ClF_2CCF_2Cl; Y 85% or 70%, respectively). It is believed in toluene the phosphonium salt forms a tight ion pair in which the succinimide anion may form a hydrogen bond with the NH group to fix the rotation of the N-(1-phenethyl)amino group; approach at the *si*-face of the terminal isobutenyl group of the substrate is then preferred for steric reasons. F.e. incl. asym. triple ring closure and diastereoselective ring closures with Bu$_3$P/N-halogenosuccinimides s. A. Sakakura, A. Ukai, K. Ishihara, Nature *2007, 445,* 900-3.

N-Fluorobenzenesulfonimide/chiral dichloro[2,2′-bis(diphenylphosphino)-1,1′-binaphthyl]nickel(II)/triethylsilyl triflate/2,6-lutidine ←
Asym. α-fluorination of arylacetic acid derivs. H → F

291. 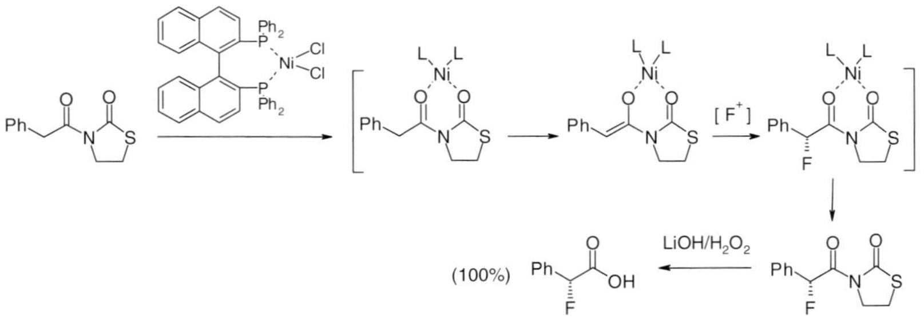

A mixture of N-phenylacetyl-2-thiazolidone, 1.5 eqs. N-fluorobenzenesulfonimide, 0.75 eqs. triethylsilyl triflate and 5 mol% chiral nickel(II) complex stirred in toluene at -20° for 10 min, 2,6-lutidine added, and stirring continued for 24 h → product. Y 99% (e.e. 88%). The reaction is successful with subst. phenylacetyl derivs. of 2-thiazolidone or 2-oxazolidone and is largely unaffected by the electronic nature of the aryl ring for *p*-subst. derivs. (Y 90-99%; e.e. 81-88%), while *o*- and *m*-subst. derivs required 10 mol% of catalyst for satisfactory results (Y 56-95%; e.e. 61-82%) and aliphatic derivs. were of low reactivity. *Toluene* appears to be the solvent of choice since, unlike methylene chloride, it does not appear to promote the non-catalyzed racemic reaction at low temperatures. No difluorination or racemization of the monofluorinated products was observed. The exact role of the triflate is not certain but it is presumed to increase the reactivity of the sulfonimide as a secondary Lewis acid. The absolute stereochemistry of the product was determined by conversion to the Weinreb amide. F.e. incl. quantitative cleavage of the products (with $LiOH/H_2O_2$) to the corresponding **chiral aryl(fluoro)acetic acids,** s. T. Suzuki, Y. Hamashima, M. Sodeoka, Angew. Chem. Int. Ed. Engl. *2007*, *46* (28), 5345-9.

*N-Fluorobenzenesulfonimide/chiral [2,2′-bis[di(3,5-xylyl)phosphino]-1,1′-binaphthyl]-(acetonitrile)(aqua)palladium(II) triflate/2,6-di-*tert*-butyl-4-methylpyridine* ←
Palladium-catalyzed asym. α-fluorination
of β-ketophosphonic acid esters cf. *65*, 199s*69*; of α-cyanophosphonic acid esters in the presence of chiral [2,2′-bis[di(3,5-xylyl)phosphino]-1,1′-binaphthyl](acetonitrile)(aqua)palladium(II) triflate/2,6-di-*tert*-butyl-4-methylpyridine (e.e. up to 91%) s. Y.K. Kang, M.J. Cho, S.M. Kim, D.Y. Kim, Synlett *2007* (7), 1135-8.

Bromodichloroisocyanuric acid ←
Ar. bromination s. *71*, 196s*72* H → Br

N-Chloromethyl-N′-fluoro-1,4-diazoniabicyclo[2.2.2]octane bis(fluoroborate) *Selectfluor*
Regiospecific α-fluorination H → F
cf. *39*, 458s*69*; of β-dicarbonyl compds. by heating in the presence of 1 drop of water in the absence of solvent s. G. Stavber, M. Zupan, S. Stavber, Tetrahedron Lett. *2007*, *48* (15), 2671-3.

Oxalyl halides s. under Dimethyl sulfoxide *(COHal)$_2$*
Silica s.a. under KHSO$_5$ *SiO_2*

Mesoporous silica/irradiation *SiO_2////*
α-Iodoketones from ethylene derivs. C=CH → C(I)CO
with Ag_2CrO_4/pyridine cf. *33*, 487(s*54*); aryl iodomethyl ketones from styrenes by irradiation with 1 eq. I_2 and FSM mesoporous silica in dry ethyl acetate under O_2 s. H. Nakayama, A. Itoh, Tetrahedron Lett. *2007*, *48* (7), 1131-3.

Triphenylphosphine s. under N-Chlorosuccinimide *Ph_3P*
Chiral phosphoramidates s. under N-Iodosuccinimide ←
Oxygen s. under HI *O_2*
Triethylsilyl triflate s. under N-Fluorobenzenesulfonimide *Et_3SiOTf*

Potassium peroxymonosulfate/hydrogen chloride/silica gel *$KHSO_5/HCl/SiO_2$*
Hydroximinochlorides from aldoximes CH=NOH → C(Cl)=NOH
with HCl cf. *48*, 442; by grinding with HCl/silica gel in the absence of solvent, also conversion to isoxazolines via 1,3-dipolar cycloaddition, s. M.A. Bigdeli, G.H. Mahdavinia, S. Jafari, J. Chem. Res. *2007* (1), 26-8.

Benzyltriphenylphosphonium persulfate/potassium bromide *$(BnPh_3P)_2S_2O_8/KBr$*
1,4-Dioxane dibromide ←
Ar. bromination H → Br
with Cl_2/NaBr cf. *20*, 417; regiospecific monobromination with 1.5 eqs. 1,4-dioxane dibromide (cf. *23*, 553s*70*) or dibromination with 2.5-4 eqs. of the reagent in the absence of solvent s. S.K. Chaudhuri, S. Roy, M. Saha, S. Bhar, Synth. Commun. *2007*, *37* (4), 579-83; regiospecific monobromination of electron-rich arenes (incl. phenols, phenolethers and ar. amines) with $CuBr_2$ in acetonitrile (cf. *17*, 598) s. S. Bhatt, S.K. Nayak, ibid. *37* (8), 1381-8; of phenols and phenolethers with $(BnPh_3P)_2S_2O_8$/KBr s. H. Tajik, I. Mohammadpoor-Baltork, J. Albadi, ibid. *37* (2), 323-8.

Chlorosulfonic acid s. under N-Iodosuccinimide *$ClSO_3H$*
Sodium hypochlorite s. under $CeCl_3$ *NaOCl*

1-Methyl-3-carboxymethylimidazolium halides/ammonium cerium(IV) nitrate ←
α-Monohalogenation of ketones H → Hal
Imidazolium halides as both ionic liquids and radical halogenating agents

292. 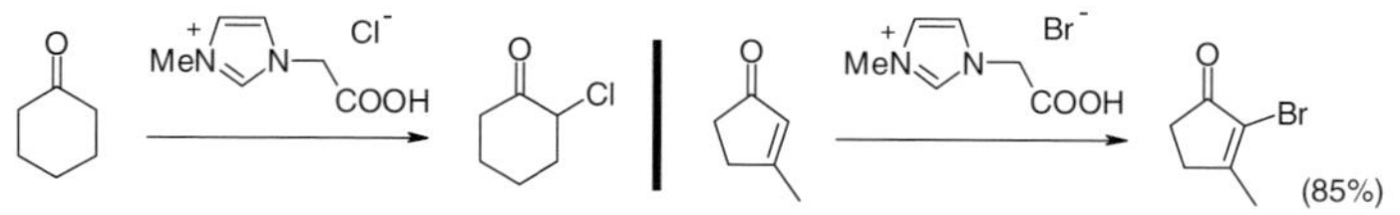

2 Eqs. CAN added to a mixture of 1 eq. cyclohexanone and 1.2 eqs. 1-methyl-3-(carboxymethyl)-imidazolium chloride, stirred at room temp. (25-8°) under N_2 for 4 h (TLC monitoring), quenched with water, extracted with ethyl acetate, and worked up with column chromatographic purification on silica gel → 2-chlorocyclohexanone. Y 90%. The procedure is very clean, simple, fast, high-yielding, highly regioselective and generally applicable to the α-halogenation of a wide variety of acyclic and cyclic ketones, incl. β-keto-esters and cyclic α,β-ethyleneketones. A radical mechanism is invoked, with CAN serving as a one-electron oxidant. F.e. incl. α-bromo-and α-iodo-derivs. s. B.C. Ranu, L. Adak, S. Banerjee, Aust. J. Chem. *2007, 60* (5), 358-62.

Hydrogen iodide/oxygen *HI/O_2*
α,β-Ethylene-α-iodoaldehydes from ethynylcarbinols C(OH)C≡CH → C=C(I)CHO
with pyridinium dichromate cf. *36*, 476; in toluene with aq. HI under O_2 s. S.F. Chen, J.B. Wang, J. Org. Chem. *2007*, *72* (13), 4993-6.

Nickel(II) or palladium(II) phosphine complexes s. under N-Fluorobenzenesulfonimide ←

Palladium(II) chloride/copper(II) chloride *$PdCl_2/CuCl_2$*
3-Halogenoindoles from *o*-acetyleneamines ○
3-bromo-derivs. with NBS or Br_2 cf. *65*, 197s*68*; 1-acetyl-3-bromo- and -3-chloro-derivs. with $PdBr_2/CuBr_2$ and $PdCl_2/CuCl_2$, respectively, s. S. Tang, Y.X. Xie, J.H. Li, N.X. Wang, Synthesis *2007* (12), 1841-7.

*Cyclooctadiene(methoxy)iridium(II) dimer/copper(II) halides/bis(pinacolato)diboron/ 4,4′-di-*tert*-butylbipyridyl* ←

m-Halogenation of 1,3-disubst. arenes via iridium-catalyzed m-boronation ArH → ArHal

293. 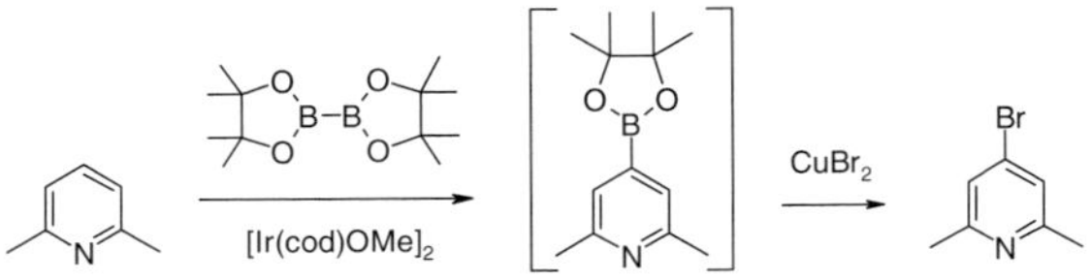

One-pot conversion. 2,6-Dimethylpyridine, 0.7 eqs. bis(pinacolato)diboron, 5 mol% $[Ir(cod)(OMe)]_2$ and 10 mol% 4,4′-di-*tert*-butylbipyridyl added to THF in a glovebox, the mixture heated in a sealed vessel at 80° for 16 h (cf. *65*, 62), cooled to room temp., volatile materials removed *in vacuo*, the residue dissolved in methanol, added to 3 eqs. aq. $CuBr_2$ in a glass bomb, the mixture heated at 80° for 4-6 h (GC monitoring), cooled to room temp., extracted with ether, and the product purified by chromatography on silica → 2,6-dimethyl-4-bromopyridine. Y 72%. The method is successful for electron-rich and -poor 1,3-disubst. benzenes and pyridines (24 examples; 55-81% yield) with *$CuBr_2$* and *$CuCl_2$* (the latter at 90°) as halogenating agent in the presence of ether, chloro, nitrile, ester, amide, pivalate and silyl ether functionality. 3-Methylpyridine gave 3-methyl-5-bromopyridine and the chloro analog in 51 and 46% yield, respectively. Other halogenating agents were less successful. F.e.s. J.M. Murphy, X. Liao, J.F. Hartwig, J. Am. Chem. Soc. *2007, 129* (50), 15434-5.

Oxygen ↾ HalC ⇃↾ O

Without additional reagents *w.a.r.*

4-Iodotetrahydropyrans from 3-ethylenealcohols and ketones via Prins cyclization ○

294. 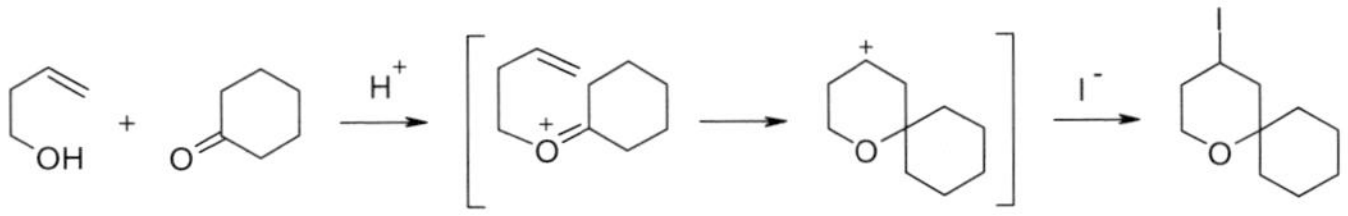

4-Iodotetrahydro-2-spiropyrans. A mixture of 2 eqs. startg. homoallyl alcohol, cyclohexanone, and 1 eq. I_2 in methylene chloride stirred at 23° for 30 min, and quenched with water → 4-iodo-1-oxaspiro[5.5]undecane. Y 96%. The procedure is mild, rapid, clean, generally applicable to cyclic and acyclic ketones, and is highly diastereoselective with subst. homoallyl alcohols. There was no reaction, however, with LiI, *t*-Bu_4NI or NaI. F.e. and **4-iodo-3,6-dihydro-2*H*-pyrans from 3-acetylenealcohols** s. J.S. Yadav, B.V. Subba Reddy, V.H. Krishna, T. Swamy, G.G.K.S. Narayana Kumar, Can. J. Chem. *2007, 85* (6), 412-5.

Lithium chloride or bromide or Sodium iodide *LiCl or LiBr or NaI*

3-Halogeno-1,3-diene-2- from α-allene-β′-hydroxycarboxylic acid esters C=C=C-C(OH) → C=C(Hal)C=C

295. 

A mixture of 2 eqs. NaI and the startg. allenol added to *trifluoroacetic acid*, the mixture heated with stirring at 80° for 1 h (TLC monitoring), cooled to room temp., quenched by slow addition of water, neutralized with $NaHCO_3$, extracted with ether, the extracts washed with satd. aq. $Na_2S_2O_3$ and brine, the organic layer dried over Na_2SO_4, evaporated *in vacuo*, and the residue purified by chromatography → (Z)-2-iodo-3-methoxycarbonyl-1,3-nonadiene. Y 75%. Various iodo-, bromo- and chloro-derivs. were obtained with NaI (Y 50-75%), LiBr·H_2O (Y 44-76%) or LiCl·H_2O (Y 26-59%), respectively, using trifluoroacetic acid as the optimum solvent. Lowest yields were obtained with the phenyl-deriv. but the electron-rich 4-*p*-tolyl deriv. gave good yields for all three halo derivs. in acetic acid. F.e. and prepn. of the startg. m., also subsequent Negishi and Sonogashira coupling, s. Y. Deng, X. Jin, Sh. Ma, J. Org. Chem. *2007*, *72* (15), 5901-4.

Lithium bromide *NaBr*
4-α-Bromo-2-oxazolidones from N-carbo-*tert*-butoxyaziridin-2-ylcarbinols ◌
Regio- and stereo-specific conversion s. *72*, 201

Fluoroboric acid s. under Bis(pyridine)iodine(I) fluoroborate *HBF_4*
Scandium(III) triflate s. under PhCOCl *$Sc(OTf)_3$*
Imidazolium salts s. under Et_2NSF_3 ←
Dimethylformamide s. under PhCOCl *DMF*

Dimethyl sulfoxide/oxalyl chloride/triethylamine *$DMSO/(COCl)_2/Et_3N$*
3,3-ω-Dichlorooxindoles from 3-ω-hydroxyindoles ←

296.

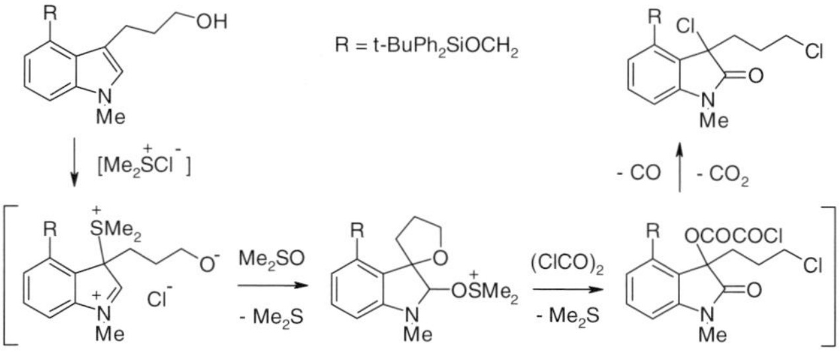

7 Eqs. DMSO added to a soln. of 5 eqs. oxalyl chloride in dry methylene chloride at -78° under argon, stirred for ca. 10 min until effervescence ceased, a soln. of the startg. 3-(hydroxyalkyl)indole (0.77 mmol) in dry methylene chloride added dropwise via cannula, the resulting red soln. stirred for 10 min at -78°, 10 eqs. triethylamine added, left to warm to room temp. for 20 min with stirring, and worked up with rapid column chromatographic purification on silica gel → product. Y 90%. Both oxidation of the indole nucleus and chlorination take place, possibly via spirocyclization of an intermediate iminium ion. F.e. and diastereoselectivity, also 3-(4-chloro-butyl)indole analogs, s. P. Lopez-Alvarado, J. Steinhoff, S. Miranda, C. Avendaño, J.C. Menendez, Synlett *2007* (18), 2792-6.

Benzoyl chloride/dimethylformamide/scandium(III) triflate/bis(trichloromethyl) carbonate ←
α,β-Ethylenehalides from ketones CHCO → C=C(Hal)
with $POCl_3$ or $POBr_3$/DMF cf. *3*, 470s*66*; with catalytic amounts of PhCOCl/DMF/$Sc(OTf)_3$ and 0.4 eqs. triphosgene (to regenerate PhCOCl from PhCOOH) s. W. Su, C. Jin, Org. Lett. *2007*, *9* (6), 993-6; with $(PhO)_3P/Cl_2$ or Br_2/Et_3N, also 1,1-dihalides from non-enolizable ketones or aldehydes (cf. *18*, 608), s. A. Spaggiari, D. Vaccari, P. Davoli, G. Torre, F. Prati, J. Org. Chem. *2007*, *72* (6), 2216-9.

Oxalyl chloride s. under Dimethyl sulfoxide *$(COCl)_2$*

Polymethylhydrosiloxane/bromodimethylsulfonium bromide or N-bromosuccinimide ←
Benzyl bromides from ar. aldehydes ArCHO → $ArCH_2Br$

297. 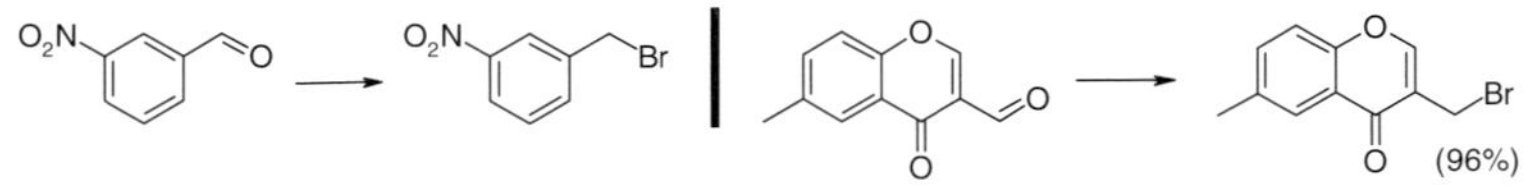

m-Nitrobenzaldehyde (1 mmol) and polymethylhydrosiloxane (1.5 mmol) added to chloroform (5 ml), the mixture stirred at room temp., bromodimethylsulfonium bromide (1 mmol) added, and stirring continued for 4 h → *m*-nitrobenzyl bromide. Y 94%. The reaction is successful with benzaldehyde derivs. bearing electron-donating or -withdrawing substituents, affording excellent yields of the corresponding benzyl bromides in all cases, although the latter examples require longer reaction times. The potential for wider applicability of the procedure is indicated by the successful transformation of the chromone deriv. also shown above. Ketones and aliphatic aldehydes did not react under the present conditions. F.e., also using N-bromosuccinimide as the brominating agent, s. B. Das, Y. Srinivas, H. Holla, K. Laxminarayana, R. Narender, Tetrahedron Lett. *2007*, *48* (38), 6681-3.

Triphenyl phosphite/triethylamine *$(PhO)_3P/Et_3N$*
α,β-Ethylenehalides from ketones s. *3*, 470s*72* CHCO → C=C(Hal)

Diethylaminosulfur trifluoride *Et_2NSF_3*
Halogeno-N-heterocyclics from hydroxy-N-heterocyclics with ring expansion ◌
3-chloro-1,2,3,4-tetrahydroquinolines with Ph_3P/CCl_4 cf. *66*, 233; 3-fluoropiperidines from 2-α-hydroxypyrrolidines with Et_2NSF_3 or, with lower regioselectivity, Deoxo-Fluor s. I. Dechamps, D.G. Pardo, J. Cossy, Synlett *2007* (2), 263-7.

Diethylaminosulfur trifluoride/1-methyl-3-octylimidazolium hexafluorophosphate ←
Replacement of oxygen by fluorine OH → F or CO → CF_2
of hydroxyl cf. *31*, 501s*33*, and 1,1-difluorides from oxo compds. cf. *30*, 365; in $[C_8mim]PF_6$ as recyclable ionic liquid solvent s. S. Das, S. Chandrasekhar, J.S. Yadav, R. Grée, Tetrahedron Lett. *2007*, *48* (30), 5305-7.

Bis(pyridine)iodine(I) fluoroborate/fluoroboric acid *Py_2IBF_4/HBF_4*
Glycosyl fluorides from pent-4-enyl glycosides s. *72*, 302 OR → F

Nitrogen ↑ HalC ↓↑ N

Sodium hydrogen carbonate or Potassium phosphate *$NaHCO_3$ or K_3PO_4*
4-Iodoisoquinolines from *o*-alkynylbenzyl azides ◌
Iodine-mediated electrophilic cyclization

298. 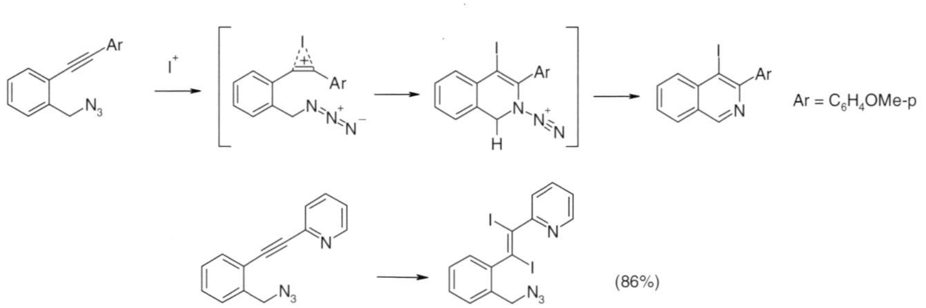

K_3PO_4 (5 eqs.) then I_2 (5 eqs.) added to a soln. of the startg. acetyleneazide in methylene chloride, and stirred *in the dark* for 24 h at room temp. → product. Y 94%. For sec. azides, the use of

$NaHCO_3$ as base was preferred over K_3PO_4. As a general rule, in order for cyclization to occur as described, the alkyne terminal substituent must either be transition-state stabilizing or bulky. However, in the case of the 2-pyridyl example shown, its proximity allows the basic pyridyl N-atom to preferentially coordinate to the intermediate iodonium ion, thereby preventing attack by the azide moiety at C2′ and allowing attack by iodide ion at C1′. In an attempt to widen the scope of the transformation, the authors investigated the use of electrophiles with less-nucleophilic counterions and discovered that both the Barluenga reagent (Py_2IBF_4/HBF_4) and N-iodosuccinimide were able to effect cyclization, albeit only in moderate yields. F.e.s. D. Fischer, H. Tomeba, N.K. Pahadi, N.T. Patil, Y. Yamamoto, Angew. Chem. Int. Ed. Engl. *2007*, *46* (25), 4764-6.

Halogen ↑ HalC ↓↑ Hal

Microwaves s. under KF [\\\\]
Potassium fluoride s.a. under $Pd_2(dba)_3$ *KF*

Potassium fluoride/1-butyl-3-methylimidazolium fluoroborate/microwaves ←
Potassium fluoride/polymeric imidazolium bromide/microwaves ←
Ar. fluorides from chlorides under microwave irradiation Cl → F
with poly(diallyldimethylammonium chloride) cf. *4*, 513s*68*; with a polymeric imidazolium bromide as phase transfer catalyst in DMSO, ar. nitrofluorides, s. Z.Y. Liang, C.X. Lu, J. Luo, L.B. Dong, J. Fluorine Chem. *2007*, *128* (6), 608-11; fluoropyridazines in [bmim]BF_4, also under conventional heating, s. L. Sun, W. Pei, Chin. J. Chem. *2007*, *25* (7), 1005-7.

Hydrogen fluoride/antimony pentafluoride *HF/SbF_5*
1,1,1-Trifluorides from dichloromethylene compds. in superacidic medium $C{=}CCl_2 \rightarrow CHCF_3$

299. 

1,1-Dichloro-2-(3-hydroxy-6-methylphenyl)ethylene (1.6 mmol) added to a soln. of HF (6 ml) and SbF_5 (3 ml) (*superacidic, caution!*) at -40° in a Teflon reactor, the resulting soln. maintained at the same temp. for 15 min, then poured into a vigorously-stirred mixture of Na_2CO_3, water and ice, and extracted with methylene chloride → product. Y 72%. Initial protonation of the alkene affords the α-dichlorocarbenium ion, which is trapped by fluoride ion (SbF_6^-, $Sb_2F_{11}^-$...). Subsequent, sequential protonation of each of the remaining chlorine atoms, followed by loss of HCl and fluorination of the resulting α-halocarbenium ions affords the 1,1,1-trifluoride. In some cases, addition of HF-pyridine (70/30) is required to complete the reaction. No fluorination occurs with substrates that give rise to stable intermediate carbenium ions. Some esters are cleaved under the reaction conditions and 1,1-dichloro-1-decene afforded a complex mixture of products. F.e.s. A.-C. Cantet, J.-P. Gesson, B. Renoux, M.-P. Jouannetaud, Tetrahedron Lett. *2007*, *48* (30), 5255-60.

Tris(dibenzylideneacetone)dipalladium/2-(dicyclohexylphosphino)biphenyl/ hexamethyldisilane/potassium fluoride/iodine monochloride ←
Ar. iodides from ar. chlorides ArCl → ArI

300. 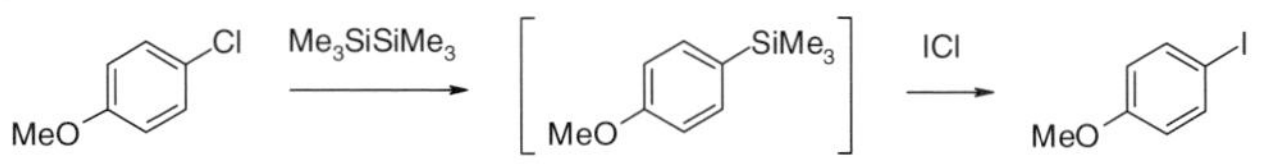

4-Methoxyphenyl chloride and 1.2 eqs. hexamethyldisilane added to 2 mol% tris(dibenzylideneacetone)dipalladium, 6 mol% 2-(dicyclohexylphosphino)biphenyl, 1-2 eqs. water and 5 eqs. dry KF in dioxane in a sealed vial under argon, stirred at 100° for 24 h, cooled to room temp., 1.5 eqs. iodine monochloride added, stirred 4 h, aq. $Na_2S_2O_4$ added, and extracted with hexanes → 4-methoxyphenyl iodide. Y 80%. The intermediate silane can be isolated in 90% yield. F.e.s. E. McNeill, T.E. Barder, S.L. Buchwald, Org. Lett. *2007*, *9* (19), 3785-8.

Sulfur ↑ HalC ↓↑ S

N-Iodosuccinimide/trifluoromethanesulfonic acid *NIS/TfOH*
2-Deoxy-2-iodohexosides from carbohydrate-based 5-hydroxythioenolethers ○
Stereospecific one-pot conversion under mild conditions

301. 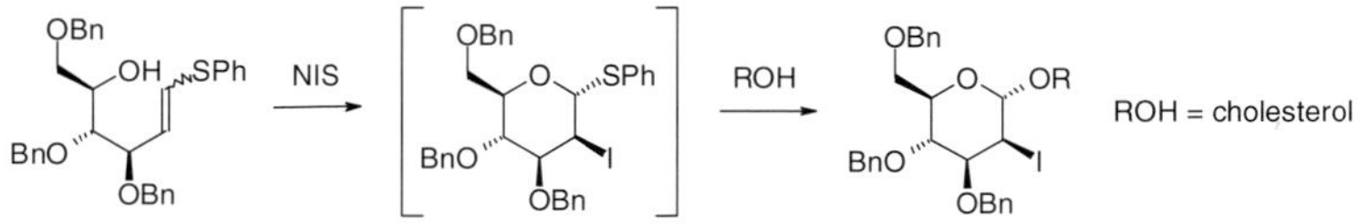

A mixture of the startg. thioenolether, 2 eqs. acceptor alcohol and 4 Å molecular sieves stirred in dry methylene chloride at room temp. for 30 min, cooled to -60°, 3 eqs. N-iodosuccinimide added, stirred at room temp. until cyclization complete (TLC), cooled to -60°, 0.2 eqs. trifluoromethanesulfonic acid added, stirred between -40 and -10° until reaction complete (TLC), quenched with aq. $Na_2S_2O_3$ and $NaHCO_3$, extracted with methylene chloride, the extracts dried over $MgSO_4$, evaporated *in vacuo*, and the residue purified by chromatography → 2-deoxy-2-iodoglycoside deriv. Y 54%. This one-pot method generally gave better yields than the stepwise process with only marginal loss in stereoselectivity (glycosylation occuring predominantly *trans* to the iodo group). Four substrates (*ribo, lyxo, arabino and manno* configuration) and three acceptor alcohols gave 50-71% yields with stereoselectivity generally >80%. Careful TLC monitoring of the reaction is essential for good yields with trifluoromethanesulfonic acid only added after cyclization is complete. F.e.s. M.A. Rodríguez, O. Boutureira, M.I. Matheu, Y. Díaz, S. Castillón, Eur. J. Org. Chem. *2007* (15), 2470-6.

Bis(pyridine)iodine(I) fluoroborate/fluoroboric acid *[(Py)$_2$I]BF$_4$/HBF$_4$*
Bis(pyridine)iodine(I) fluoroborate/hydrogen fluoride-pyridine *[(Py)$_2$I]BF$_4$/HF-Py*
α-Glycosyl fluorides from thioglycosides under mild conditions ←

302. 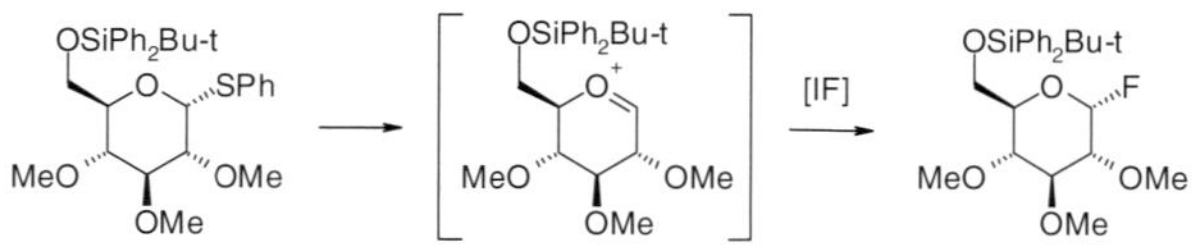

Fully protected α-glycosyl fluorides. 1.2 Eqs. HBF_4 added to 1.2 eqs. bis(pyridine)iodine(I) fluoroborate (*as iodine monofluoride equivalent*) in dry methylene chloride under argon at -40°, the mixture stirred for 5 min, 1 eq. startg. thioglycoside added, the mixture stirred until reaction complete, diluted with methylene chloride, washed with 10% aq. $Na_2S_2O_3$, $NaHCO_3$ and water, the organic layer dried over Na_2SO_4, evaporated *in vacuo*, and the residue purified by chromatography → α-glycosyl fluoride deriv. Y 94%. The α-anomer was formed as sole or major product from fully-protected α- or β-thioglycosides in 64-99% yield (6 examples) in the presence of benzyl, benzoyl, *tert*-butyldiphenylsilyl and benzylidene protecting groups. Other pyridine scavengers [$BF_3 \cdot Et_2O$ and $Yb(OTf)_3$] gave lower yields. The conditions had limited success with partially-protected substrates but replacing HBF_4 with HF-pyridine complex enabled the preparation of α-glycosyl fluorides as major or sole products in the presence of both sec. and prim. hydroxyl groups (10 examples; Y 61-99%). F.e. **and from pent-4-enyl glycosides** s. J.C. López, P. Bernal-Albert, C. Uriel, S. Valverde, A.M. Gómez, J. Org. Chem. *2007*, *72* (26), 10268-71; from pent-4-enyl glycosides (or carbohydrate pent-4-enyl orthoesters) s.a. J.C. López, C. Uriel, A. Guillamon-Martin, S. Valverde, A.M. Gómez, Org. Lett. *2007*, *9* (15), 2759-62.

Remaining Elements ↾ HalC ⇃↾ Rem

Without additional reagents *w.a.r.*

Iodonium salts from borates $B^{+} \leqslant \rightarrow I^{+}R$

from alkali metal tetraarylborates and iodosocarboxylates in acetic acid cf. *52*, 207; ar. iodonium fluoroborates from organo(trifluoro)borates and ar. iododifluorides s. M. Yoshida, K. Osafune, S. Hara, Synthesis *2007* (10), 1542-6.

2-Acylamino-3-halogeno-O-heterocyclics ○
from ω-siloxyenacylamines with asym. induction

303. 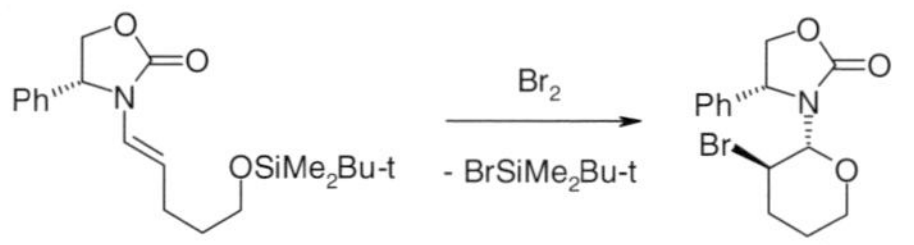

Chiral 2-acylamino-3-halogenotetrahydropyrans. A 0.5 *M* soln. of bromine (0.8 mmol) in methylene chloride added dropwise over 15 min at -78° to a soln. of the startg. chiral enamide (0.53 mmol) in dry toluene, warmed to room temp., stirred for 30 min (TLC monitoring), concentrated *in vacuo*, and worked up with flash column chromatographic purification on silica gel → product. Y 83% (d.r. 7.4:1). The major isomer was separated via trituration with ethyl acetate/hexane (Y 50%). The procedure is applicable to chloro-, bromo- and iodo-derivs., and the corresponding chiral 2-imidazolidone ring can be used as asym. inducer in place of the 2-oxazolidone ring. F.e.s. C. Ko, R.P. Hsung, Z.F. Al-Rashid, J.B. Feltenberger, T. Lu, J.-H. Yang, Y. Wei, C.A. Zificsak, Org. Lett. *2007, 9* (22), 4459-62.

Gold(III) chloride/N-iodosuccinimide *$AuCl_3$/NIS*

4-Iodo-3(2*H*)-furanones from β,γ-acetylene-α-siloxyketones ○
Gold-catalyzed ring closure with 1,2-C-alkyl migration

304. 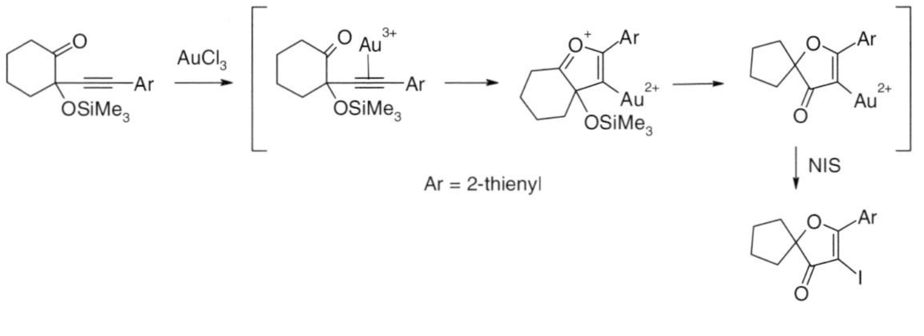

under mild conditions. A soln. of 5 mol% $AuCl_3$ in acetonitrile added to a soln. of the startg. ketone and 1.5 eqs. N-iodosuccinimide in methylene chloride in a reaction vial, the latter sealed, the soln. stirred at room temp. in the absence of light for 30 min (TLC monitoring), quenched with satd. aq. $Na_2S_2O_3$, extracted with methylene chloride, the organic layers washed with brine, dried over Na_2SO_4, concentrated *in vacuo*, and purified by chromatography on silica → 3-iodo-2-thiophen-2-yl-1-oxaspiro[4.4]non-2-en-4-one. Y 99%. The method is rapid for a number of 6- and 7-membered ring analogs bearing alkyl, alkenyl, aryl or benzyloxy substituents giving yields of 60-99% in 20-30 min at room temp. One acyclic analog with diphenyl in place of -$(CH_2)_4$- gave a 91% yield but two diethyl analogs were low-yielding (27 and 43%). The reaction can also be accomplished in the absence of the gold catalyst in comparable yields for cyclic substrates (acyclic analogs being low-yielding) but reaction times are extended (120-300 min). The mechanism is thought to involve an oxonium ion intermediate. F.e. and subsequent **Suzuki and Sonagashira coupling** s. B. Crone, S.F. Kirsch, J. Org. Chem. *2007, 72* (14), 5435-8.

N-Bromosuccinimide *NBS*
α,β-Ethylenehalides from enesilanes C=C(Si⩽) → C=C(Hal)
α,β-ethyleneiodides with NIS cf. *33*, 501s*52*; β,β-dibromo- from β,β-disilyl-styrenes with NBS in acetonitrile at room temp. s. P. Pawluæ, G. Hreczycho, J. Walkowiak, B. Marciniec, Synlett *2007* (13), 2061-4.

Chiral N-chloroimidodicarbonates/samarium(III) triflate ←
α-Halogenooxo compds. from enoxysilanes C=C(OSi⩽) → C(Hal)CO
with Cu(II)-halides cf. *30*, 374s*47*; with chiral N-chloroimidodicarbonates, e.g. the di-(-)-menthyl deriv./Sm(OTf)$_3$, α-chloroketones with low enantioselectivity (e.e. up to 40%), s. S. Hajra, M. Bhowmick, B. Maji, D. Sinha, J. Org. Chem. *2007*, *72* (13), 4872-6.

N-Chloromethyl-N′-fluoro-1,4-diazoniabicyclo[2.2.2]octane bis(fluoroborate) *Selectfluor*
Densely subst. 2-methylenecyclohexyl fluorides ←
from 1-(silylmethyl)cyclohexenes with asym. induction

305. 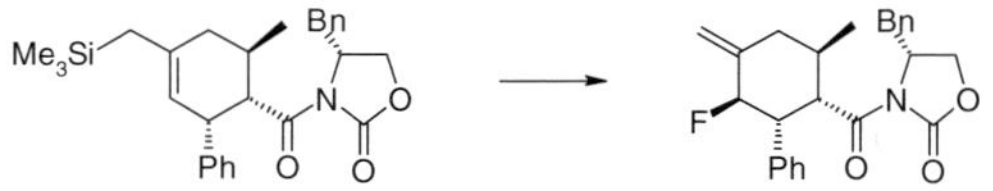

Chiral *syn,anti,syn*-3-fluoro-4-methylenecyclohexanecarboxamides. 1.1-1.3 Eqs. Selectfluor added at room temp. to a stirred soln. of the startg. silylated adduct in acetonitrile under argon, the solvent removed *in vacuo* after 2-4 h (TLC monitoring), the residue fractionated between ether and satd. aq. NaHCO$_3$, and worked up with chromatographic purification on silica gel → product. Y 75% (d.r. >20:1). This is the second stage of a simple 2-step conversion via Diels-Alder reaction, the introduction of fluorine at the latter stage being designed to avoid any possible complications with the formation of fluorinated intermediates. The stereochemical preference for fluorination of both *endo*- and *exo*-adducts can be rationalized by an S_E2' mechanism, with electrophilic fluorine preferentially approaching the silylated cyclohexene axially and *anti* with respect to the silylmethyl group. F.e. incl. *anti,anti,anti*-products and *hexasubst. anti,syn,syn,syn*-products (both from *exo*-adducts) s. Y. Lam, C. Bobbio, I.R. Cooper, V. Gouverneur, Angew. Chem. Int. Ed. Engl. *2007, 46* (27), 5106-10.

Carbon ↾ HalC ⇃↾ C

Copper(II) chloride or bromide/dicyclohexylamine hydro-chloride or -bromide *CuHal$_2$/Cy$_2$NHHal*
Halogenative ring closures of *o*-functionalized acetylene derivs. ○
with demethylation cf. *62*, 203s*67,70*; 4-halogenoisocoumarins from *o*-acetylenecarboxylic acid esters with CuCl$_2$ *or* CuBr$_2$/dicyclohexylamine hydro-chloride or -bromide s. Y. Liang, Y.X. Xie, J.H. Li, Synthesis *2007* (3), 400-6.

*Lead(II) acetate/*tert*-butyl hypochlorite/benzoyl peroxide* ←
Carboxylic acid chlorides from terminal glycols via aldehydes CH(OH)CH$_2$OH → COCl
Degradation with loss of one C-atom

306. 

One-pot conversion. A little benzoyl peroxide stirred with 2 eqs. Pb(OAc)$_2$ and 4 eqs. *tert*-butyl hypochlorite in toluene at room temp. for 40 min, 1,2-octanediol added, and stirring continued for 15 min → product. Y 90%. Initial reaction of Pb(OAc)$_2$ with *tert*-butyl hypochlorite produces *lead tetraacetate* in situ, which cleaves the glycol to the aldehyde prior to chlorination to the acyl chloride. The procedure is simple and mild. F.e. and with isolation of the intermediate aldehydes s. V.N. Telvekar, J.K. Mali, M.G. Dighe, Synth. Commun. *2007, 37* (5), 865-8.

Diphosphorus tetraiodide/tetraethylammonium bromide *P_2I_4/Et_4NBr*
Potassium persulfate/sodium bromide/sodium carbonate ←
α,β-Ethylenehalides from α,β-ethylenecarboxylic acids COOH → Hal
with NaHal/Oxone cf. *62*, 204s*63*; α,β-ethylenebromides with $K_2S_2O_8/NaBr/Na_2CO_3$ s. H. Krawczyk, L. Albrecht, Synthesis *2007* (12), 1877-81; with P_2I_4/Et_4NBr s. V.N. Telvekar, S.N. Chettiar, Tetrahedron Lett. *2007*, *48* (26), 4529-32.

Formation of S-S Bond

Exchange ↓↑

Hydrogen ↑ SS ↓↑ H

Alumina/potassium fluoride/microwaves *$Al_2O_3/KF/$*[\\\\]
Sym. disulfides from mercaptans s. *47*, 468s*72* 2 RSH → $(RS)_2$

Diethyl azodicarboxylate *ROOCN=NCOOR*
Disulfides from two different mercaptan molecules RSH + HSR′→ RSSR′
cf. *47*, 470s*51*; glycosyl disulfides using diethyl azodicarboxylate (cf. *56*, 201) s. G.R. Morais, R.A. Falconer, Tetrahedron Lett. *2007*, *48* (43), 7637-41.

Alkanesulfonyl chloride/pyridine *RSO_2Cl/C_5H_5N*
Bis(trimethylsilyl) chromate/silica ←
Hexamethylenetetramine-bromine complex ←
Sym. disulfides from mercaptans 2 RSH → $(RS)_2$
cf. *47*, 468s*71*; with bis(trimethylsilyl) chromate/silica in acetonitrile s. M.M. Heravi, F. Derikvand, H.A. Oskooie, R.H. Shoar, Synth. Commun. *2007*, *37* (3), 513-7; with alkanesulfonyl halides in a nitrogenous base, e.g. $MeSO_2Cl$/pyridine, s. F.I. Ali, I.A. Hashmi, B.S. Siddiqui, M. Rasheed, ibid. *37* (15), 2461-5; with hexamethylenetetramine-bromine complex s. I. Yavari, A. Rounaqi, L. Moradi, Phosphorus, Sulfur Silicon Relat. Elem. *2006*, *181* (11), 2659-62; with alumina/KF in the absence of solvent under microwave irradiation or by heating s. E.J. Lenardão, R.G. Lara, M.S. Silva, R.G. Jacob, G. Perin, Tetrahedron Lett. *2007*, *48* (43), 7668-70.

Via intermediates *v.i.*
Unsym. disulfides from two different mercaptan molecules RSH + HSR′ → RSSR′

307.

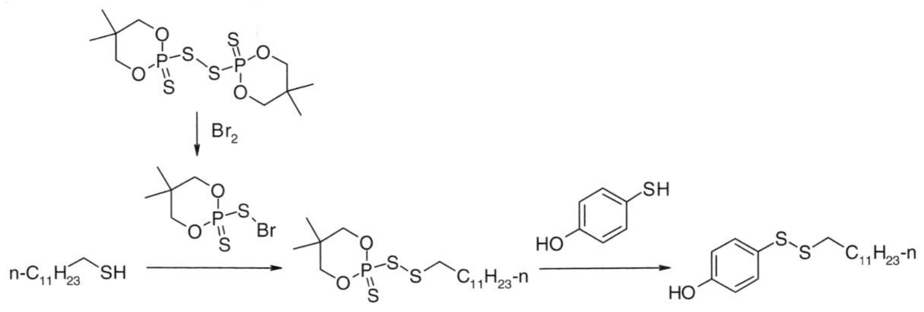

via 2-disulfido-1,3,2-dioxaphosphorinane-2-thiones. Br_2 (6 mmol) added to a soln. of bis(5,5-dimethyl-2-thiono-1,3,2-dioxaphosphorinan-2-yl) disulfide (7 mmol) in dry methylene chloride at -5° under N_2, followed, after 30 min, by a soln. of 1-dodecanethiol (11 mmol) in the same solvent, the mixture stirred at room temp. for 30 min, washed with water, dried, and purified by column chromatography → 5,5-dimethyl-2-thiono-1,3,2-dioxaphosphorinanyl-2-disulfanyldodecane (Y 100%), to a soln. of 0.5 mmol of which in methylene chloride was added a soln. of *p*-mercaptophenol (0.5 mmol) in methylene chloride and triethylamine (2 mmol), and stirred at room temp. for 15 min → disulfide (Y 96%). A variety of prim. aliphatic and aryl mercaptans all gave excellent yields, with alcohols, carboxylic acids, esters, amines and amides all tolerated under the mild reaction conditions. F.e.s. S. Antoniow, D. Witt, Synthesis *2007* (3), 363-6.

Sulfur ↑ SS ↓↑ S

Triethylamine *Et_3N*
Disulfides from 2-disulfido-1,3,2-dioxaphosphorinane-2-thiones $RSSPO(OR')_2 \rightarrow RSSR''$
and mercaptans s. *72*, 307

Bromine *Br_2*
2-Disulfido-1,3,2-dioxaphosphorinane-2-thiones from mercaptans $RSH \rightarrow RSSPO(OR')_2$
s. *72*, 307

Carbon ↑ SS ↓↑ C

Without additional reagents *w.a.r.*
Disulfides from sulfenyl chlorides and 2-(trimethylsilyl)ethyl mercaptans $SCl \rightarrow SSR$

308.

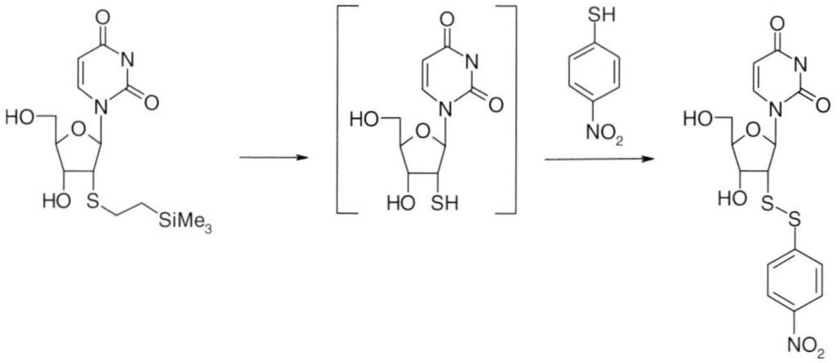

Nucleoside disulfides. *p*-Nitrophenylsulfenyl chloride (3 eqs.) added to a soln. of the startg. 2-(trimethylsilyl)ethyl mercaptan in freshly distilled methanol, stirred at room temp. for 4 h, evaporated, and worked up with chromatographic purification on silica gel → product. Y 92%. There is no need to protect free amino or hydroxyl groups, or isolate the intermediate thiols. F.e. incl. **nucleoside enedisulfides** from the corresponding thioenols s. B. Gerland, J. Désiré, M. Lepoivre, J.-L. Décout, Org. Lett. *2007, 9* (16), 3021-3.

Tetrakis(acetonitrile)copper(I) hexafluorophosphate *$[Cu(MeCN)_4]PF_6$*
Copper(I) chloride *CuCl*
Sym. disulfides from thioethers $2\ RSR' \rightarrow (RS)_2$
cf. *25*, 417s*46,48*; from trityl thioethers with 20 mol% CuCl and 2 eqs. water in methylene chloride, and acceleration under ultrasonication, also with $Cu(MeCN)_4PF_6$, s. M. Ma, X. Zhang, L.L. Peng, J.B. Wang, Tetrahedron Lett. *2007*, *48* (7), 1095-7.

Formation of S-Rem Bond

Exchange ⇃↾

Oxygen ↾ SRem ⇃↾ O

Pivaloyl chloride or Diphenyl phosphorochloridate/pyridine ←
Monothiolphosphorous acid diesters from phosphorous acid monoesters and mercaptans P-OH → P-SR

309. 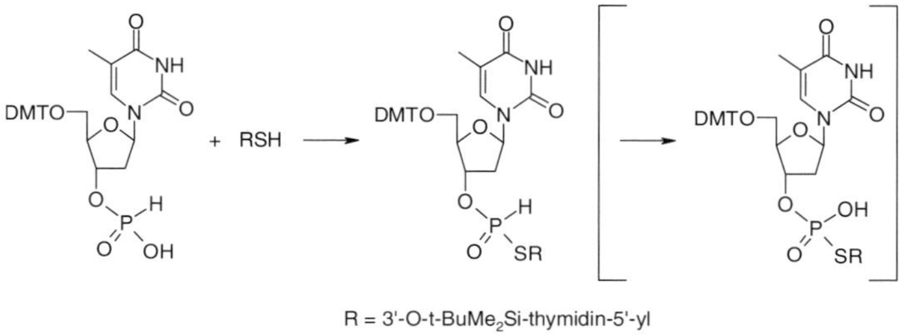

Nucleoside H-thiolphosphonates. The startg. monoester rendered anhydrous by evaporation of added pyridine, the residue dissolved in 4:1 methylene chloride/pyridine, 2 eqs. of the startg. mercaptan and 1 eq. diphenyl phosphorochloridate added, allowed to react for 5 min (^{31}P NMR monitoring), and the product isolated after conversion to the corresponding monothiolphosphoric acid diester by Todd-Atherton oxidation (i.e. by addition of 10 eqs. carbon tetrachloride, 50 eqs. water and 2 eqs. Et_3N, and work-up after a further 5 min) → product. Y 64%. The reactivity of H-thiolphosphonates is significantly different to that of H-phosphonates, notably being highly susceptible to nucleophilic substitution at phosphorus. F.e. and with pivaloyl chloride s. R. Hiresova, J. Stawinski, Synlett *2007* (17), 2748-52.

Formation of S-C Bond

Uptake ⇓

Addition to Oxygen and Carbon SC ⇓ OC

Without additional reagents *w.a.r.*
2-Hydroxydithiourethans from epoxides and amines ▽O → C(OH)C(SC(S)N<)
Regiospecific ring opening

310. 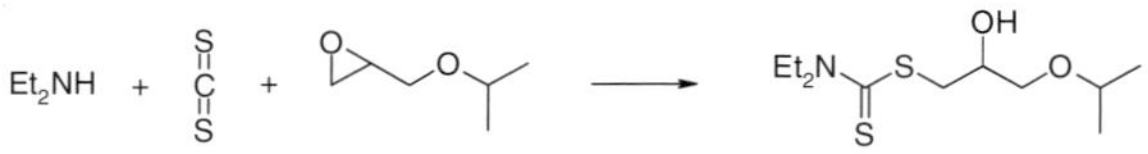

in water. Carbon disulfide (16 mmol) added dropwise over 2 min to a mixture of diethylamine (6 mmol) and water (20 ml), the resulting mixture stirred at room temp. for 20 min, (isopropyloxymethyl)oxirane (8 mmol), added, and the mixture stirred vigorously for 18 h → product. Y 95%. The reaction could also be carried out *in DMF*, with comparable yields, although the addition of anhydrous $LiClO_4$ significantly reduced reaction times to around 90 min. In all cases, apart from styrene oxide (the only aryl-subst. example), regioselectivity of addition of the intermediate dithiocarbamate anion was very high, favouring attack at the least-hindered position to give the prim. dithiocarbamate. With styrene oxide, however, regioselectivity was lower and, notably, reversed depending on the solvent used; furthermore, while the reaction in water was consistent with the aliphatic examples, reaction in DMF was more of an S_N2-type process, favouring attack at the β-position. F.e.s. A. Ziyaei-Halimjani, M.R. Saidi, Can. J. Chem. *2006*, *84* (11), 1515-9; in the absence of solvent s. N. Azizi, B. Pourhasan, F. Aryanasab, M.R. Saidi, Synlett *2007* (8), 1239-42; cf. *48*, 536s*67*.

Magnesium bromide-etherate *$MgBr_2$-OEt_2*
Fluoroboric acid/silica *HBF_4/SiO_2*
Boron fluoride *BF_3*
2-Hydroxythioethers from epoxides and mercaptans ▽O → C(OH)C(SR)
regiospecific ring opening in the absence of solvent with $LiClO_4$ cf. *47*, 479s*71*; with a little $MgBr_2$-OEt_2 cf. M.M. Mojtahedi, M.S. Abaee, M. Bolourtchian, H. Abbasi, Phosphorus, Sulfur Silicon Relat. Elem. *2007*, *182* (4), 905-10; with HBF_4/silica s. B.P. Bandgar, A.V. Patil, O.S. Chavan, V.T. Kamble, Catal. Commun. *2007*, *8* (7), 1065-9; with a little $Gd(OTf)_3$, also β-hydroxysulfoxides by addition of 2 eqs. H_2O_2 (cf. *59*, 204), s. W.K. Su, J.X. Chen, H.Y. Wu, C. Jin, J. Org. Chem. *2007*, *72* (12), 4524-7; with BF_3-OEt_2 in toluene, chiral 3-amino-2-hydroxythioethers, s. J.M. Concellon, V. del Solar, J.R. Suarez, E.G. Blanco, Tetrahedron *2007*, *63* (13), 2805-10.

Indium(III) bromide/(R,R)-6,6′-bis(1-hydroxy-2,2-dimethylpropyl)-2,2′-bipyridyl ←
2-Hydroxythioethers from epoxides and mercaptans
Asym. ring opening
with Zn-L-tartrate cf. *41*, 493; with $InBr_3$/(R,R)-6,6′-bis(1-hydroxy-2,2-dimethylpropyl)-2,2′-bipyridyl s. M.V. Nandakumar, A. Tschöp, H. Krautscheid, C. Schneider, Chem. Commun. *2007* (26), 2756-8.

Scandium(III) triflate/(S,S)-6,6′-bis(1-hydroxy-2,2-dimethylpropyl)-2,2′-bipyridyl ←
2-Hydroxythioethers from *meso*-epoxides with desymmetrization

311.

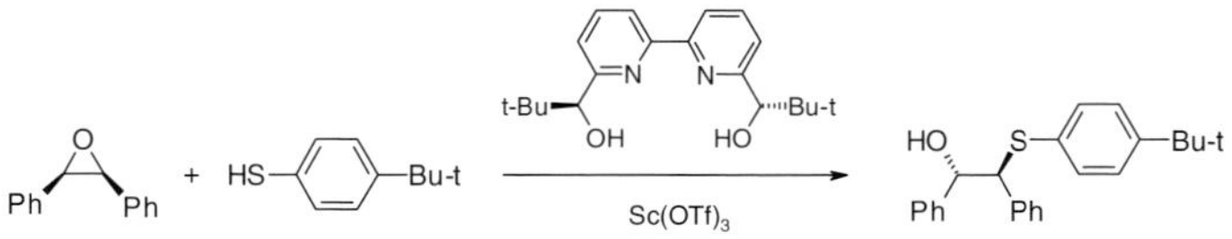

10 Mol% $Sc(OTf)_3$ and 12 mol% (S,S)-6,6′-bis(1-hydroxy-2,2-dimethylpropyl)-2,2′-bipyridyl in methylene chloride stirred at room temp. for 10 min, 1 eq. *cis*-stilbene oxide in methylene chloride and 3 eqs. 4-*tert*-butylbenzenethiol in methylene chloride added successively, stirred at the same temp. for 12 h, quenched with water, and worked up with thin layer chromatographic purification on silica gel → product. Y 87% (e.e. 95%). Reaction is applicable to both aryl and alkyl mercaptans with high enantioselectivity (90-97%), both electron-withdrawing and electron-donating groups on the aromatic ring being tolerated. Enantioselectivity was lower (50%), however, with 2-butene oxide. F.e.s. C. Ogawa, N. Wang, S. Kobayashi, Chem. Lett. *2007, 36* (1), 34-5.

Gadolinium(III) triflate *$Gd(OTf)_3$*
2-Hydroxythioethers from epoxides and mercaptans
s. *47*, 479s*72* – Also β-hydroxysulfoxides by oxidation *in situ* s. *59*, 204s*72*

β-Cyclodextrin/sodium salt *cyclodextrin/Na^+*
Ar. 2-hydroxythioethers from epoxides ▽O → C(OH)C(SR)
s. *64*, 63s*72*

1-(1-Carboxyethyl)-3-methylimidazolium thiocyanate ←
2-Hydroxythiocyanates from epoxides → C(OH)C(SCN)
Regio- and stereo-specific ring opening s. *72*, 279

1-Butyl-3-methylimidazolium bromide *[bmim]Br*
2-Hydroxythioethers from epoxides → C(OH)C(SR)
Regio- and stereo-specific conversion in ionic liquids

312.

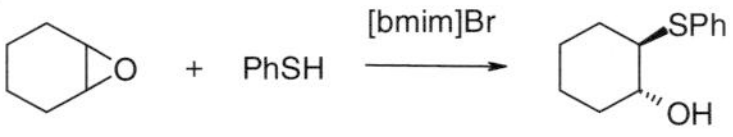

under mild conditions. 1 Eq. phenyl mercaptan added to a mixture of cyclohexene oxide and 0.24 eqs. 1-butyl-3-methylimidazolium bromide, the mixture stirred for 3 h (TLC monitoring), and the product isolated by direct distillation *in vacuo* → *trans*-2-(phenylthio)cyclohexanol. Y 95%. Three other cycloalkene epoxides gave the *trans*-diequatorial 2-hydroxythioethers *stereospecifically* in 80-95% yields, while a series of terminal epoxyalkanes underwent epoxide cleavage by reaction of thiolate at the *least* hindered carbon in similar yields using phenyl-, 4-chlorophenyl and benzyl mercaptan. Styrene oxide, however, gave a 2:1 mixture of adducts. The procedure is environmentally friendly, requiring no additional solvent, and permits *recycling* of the non-toxic ionic liquid (up to six times without appreciable loss in catalytic activity). F.e. and in 1-butyl-3-methylimidazolium hydroxide for the conversion of analogous substrates to ketothioethers (*72*, 349) s. B.C. Ranu, T. Mandal, S. Banerjee, S.S. Dey, Aust. J. Chem. *2007*, *60* (4), 278-83.

Lithium perchlorate *$LiClO_4$*
2-Hydroxydithiourethans from epoxides and amines s. *72*, 310 → C(OH)C(SC(S)N<)

Addition to Nitrogen and Carbon SC ⇓ NC

Without additional reagents *w.a.r.*
Isothiocyanates from nitrile oxides via 1,4,2-oxathiazoles ○
Polymer-based conversion s. *72*, 332

Cesium hydroxide/chiral N-arylmethylcinchoninium halide ←
2-(Sulfonylamino)thioethers → C(NH-)C(SR)
from *meso*-N-sulfonylaziridines and mercaptans
Desymmetrization under mild conditions

313.

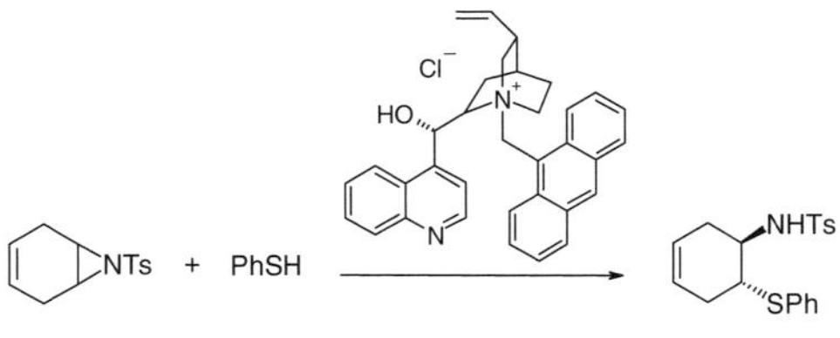

1.2 Eqs. phenyl mercaptan and the startg. aziridine added sequentially to a mixture of 10 mol% cinchoninium catalyst and 2 eqs. $CsOH{\cdot}H_2O$ in carbon tetrachloride at 0° under argon, the mixture stirred for 18 h (TLC monitoring), quenched with water, extracted with methylene chloride, the organic phase washed with brine, dried over Na_2SO_4, evaporated *in vacuo*, and purified by chromatography → 2-aminothioether deriv. Y 88% (e.e. 72%). A series of *meso*-cyclohexane- and -hex-4-ene-fused N-sulfonylaziridines were cleaved with aryl mercaptans in 80-99% yield and variable enantioselectivity (e.e. 40-73%) using one of two effective N-arylmethylcinchoninium halides as catalysts. *o*-Methylphenyl mercaptan gave moderate enantioselectivity (e.e. 28%) as did cyclopentane- and tetrahydrofuran-fused N-sulfonylaziridines (e.e. 40-45%). Absolute stereochemistry was determined by X-ray diffraction analysis in one case. F.e. incl. catalyst development s. Z.-B. Luo, X.-L. Hou, L.-X. Dai, Tetrahedron:Asym. *2007*, *18* (4), 443-6.

Indium(III) chloride *$InCl_3$*
2-Aminothioethers from aziridines and mercaptans → C(NH-)C(SR)
N-sulfonyl- and N-Boc-derivs. with $Bi(OTf)_3$ cf. *68*, 220; N-phenyl-derivs. with 1 mol% $InCl_3$ in methylene chloride s. S. Peruncheralathan, M. Henze, C. Schneider, Tetrahedron Lett. *2007*, *48* (38), 6743-6.

1-(1-Carboxyethyl)-3-methylimidazolium thiocyanate ←
2-(Tosylamino)thiocyanates from N-tosylaziridines → C(NH-)C(SCN)
Regio- and stereo-specific ring opening s. *72*, 279

Thioacetic acid/formic acid/hydrogen peroxide/hydrogen chloride ←
β-Aminosulfonic acids from aziridines → $C(NH-)C(SO_3H)$
with $NaHSO_3$, chiral compds., cf. *65*, 208; with AcSH followed by H_2O_2/formic acid then 10% HCl s. L.B. Hu, H. Zhu, D.M. Du, J.X. Xu, J. Org. Chem. *2007*, *72* (12), 4543-6.

Addition to Sulfur and Carbon SC ⇓ SC

β-Cyclodextrin *cyclodextrin*
Ring opening with mercaptans in protic media ←
of N-tosylaziridines in acetone/water cf. *69*, 228; of thiiranes in 1:15 methanol/water s. M.S. Reddy, K. Surendra, N.S. Krishnaveni, M. Narender, K.R. Rao, Helv. Chim. Acta *2007*, *90* (2), 337-9.

Addition to Carbon-Carbon Bonds SC ⇓ CC

Without additional reagents *w.a.r.*
Uncatalyzed anti-Markovnikov addition of S-nucleophiles to enolethers in the absence of solvent C=C(OR) → C(S-)CH(OR)

314.
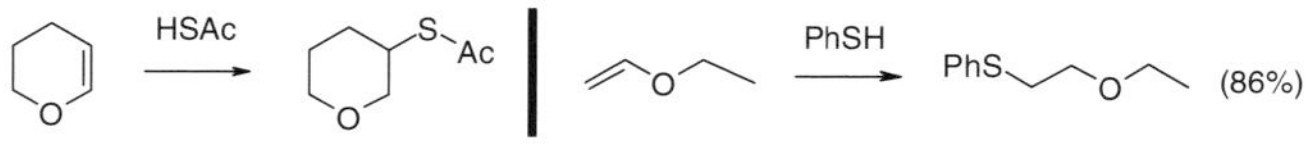

2-(Acylthio)ethers. A mixture of thiolacetic acid and 1.2 eqs. dihydropyran stirred at 30° for 3 h → 3-acetylthiotetrahydropyran. Y 91% (selectivity 99%). The reaction is successful in the absence of solvent or catalyst with alkyl, benzyl and aryl thiols, leading to the corresponding **2-alkoxy-thioethers,** in addition to thiolacetic acid with yields in the range 72-96% and *very high selectivity* (93-100%) for the anti-Markovnikov product (16 examples). Optimum reaction temp. was in the range 30-60° dependant on substrate (higher temps. generally being required for aliphatic thiols). The high selectivity is attributed in part to the presence of the oxygen atom which is expected to interact with the sulfhydryl group leading to C-C activation. F.e.s. F.-W. Lou, J.-M. Xu, B.-K. Liu, Q. Wu, Q. Pan, X.-F. Lin, Tetrahedron Lett. *2007*, *48* (50), 8814-8.

Generation and cycloaddition reactions of singlet sulfur monoxide ○

315.
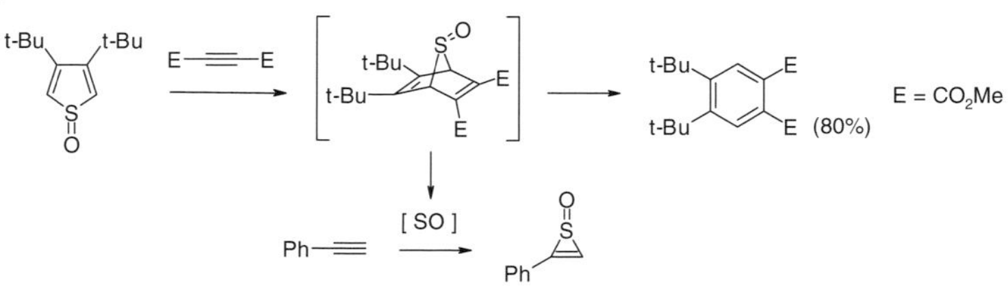

Thiirene 1-oxides from acetylene derivs. Dimethyl acetylenedicarboxylate (0.47 mmol) added to a stirred soln. of 3,4-di-*tert*-butylthiophene 1-oxide (0.47 mmol) and 2-phenylacetylene (4.7

mmol) in methylene chloride, and the mixture stirred at room temp. for 30 min → product. Y 20%. This is the first report of the prepn. of monosubst. thiirene 1-oxides, albeit in low yields. The products are much more stable than expected and can be purified by silica gel chromatography, recrystallization or distillation. The addition reactions of the *in situ* generated SO resemble those of singlet carbenes and it appears that the initially-formed singlet SO is trapped without decay to the more stable triplet form. F.e., **also thiirane 1-oxides from ethylene derivs.**, s. J. Nakayama, Y. Tajima, P. Xue-hua, Y. Sugihara, J. Am. Chem. Soc. *2007*, *129* (23), 7250-1.

Microwaves s. under $Pd(OAc)_2$ [\\\\]

(Triphenylphosphine)gold(I) chloride/silver fluoroborate $(Ph_3P)AuCl/AgBF_4$
Thioethers from activated ethylene derivs. C=C → C(SR)CH
Regiospecific gold(I)-catalyzed addition of mercaptans

316.
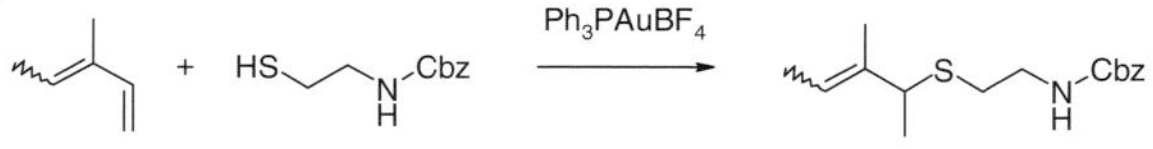

under mild conditions. A mixture of 5 mol% $(Ph_3P)AuCl$ and 5 mol% $AgBF_4$ stirred in methylene chloride for a few min at *room temp.* under N_2, 2-mercapto-N-(benzyloxycarbonyl)ethylamine added, then 1.2 eqs. *cis/trans* 3-methyl-1,3-pentadiene, the mixture stirred overnight, filtered through Celite, the filtrate evaporated *in vacuo*, and the crude mixture purified by chromatography → product. Y 70%. In the absence of the N-protecting group no addition products were obtained. The method is successful with activated alkenes (1,3-dienes, styrene and indene) giving 63-95% yields (by NMR) of the **Markovnikov adducts** in *chlorinated solvents* with *in situ*-generated $(Ph_3P)AuBF_4$ (other counterions being less effective or unreactive). The method is tolerant of sterically-demanding mercaptans (*t*-BuSH giving 85% but Ph_3CSH not reacting), and is *surprisingly tolerant of OH, OMe and* NO_2 *functionality*, but there was no reaction in the presence of prim. amines. Styrene required the use of freshly-distilled thiophenol to prevent polymerization and the product contained some 10% of the anti-Markovnikov product (86% yield). Attempts with unactivated alkenes were largely unsuccessful, with 1-octene suffering isomerization to 2-octene only, while, surprisingly, vinylcyclohexane gave a 1:1 mixture of adducts in unspecified yield. Initial experiments also demonstrated addition of mercaptoacetic, mercaptobenzoic and diphenylphosphinodithioic acids in good yields but this was not explored in detail. F.e.s. C. Brower, R. Rahaman, C. He, Synlett *2007* (11), 1785-9.

Gallium(III) chloride $GaCl_3$
2-Deoxythioglycosides from glycals
with CAN cf. *47*, 106(s*68*); with a little $GaCl_3$ in methylene chloride, α-selectivity, s. J.S. Yadav, B.V.S. Reddy, E.V. Bhasker, S. Raghavendra, A.V. Narsalah, Tetrahedron Lett. *2007*, *48* (4), 677-80.

Ammonium cerium(IV) nitrate *CAN*
Thioethers from ethylene derivs. and mercaptans s. *47*, 488s*72*

Scandium(III) triflate/2,6-bis((4S,5S)-4,5-diphenyl-Δ²-oxazolin-2-yl)pyridine ←
Asym. Michael addition of mercaptans
to 3-(α,β-ethyleneacyl)-2-oxazolidones under Hf(IV)-catalysis cf. *58*, 194s*68*; with 10 mol% $Sc(OTf)_3$/2,6-bis((4S,5S)-4,5-diphenyl-Δ^2-oxazolin-2-yl)pyridine, and improved method for making (4S,5S)-diphenyl-Pybox, s. A.M.M. Abe, S.J.K. Sauerland, A.M.P. Koskinen, J. Org. Chem. *2007*, *72* (14), 5411-3; to (E)-3-crotonoyl-2-oxazolidone with 10 mol% $Fe(BF_4)_2 \cdot 6H_2O$/(S,S)-*ip*-Pybox, (e.e. up to 95%) s. M. Kawatsura, Y. Komatsu, M. Yamamoto, S. Hayase, T. Itoh, Tetrahedron Lett. *2007*, *48* (37), 6480-2.

β-Cyclodextrin *cyclodextrin*
Thioenolethers from acetylene derivs. C≡C → CH=C(SR)
with Cs_2CO_3/TEMPO cf. *43*, 458; with 1 eq. β-cyclodextrin in aq. medium under neutral conditions, (E)-selectivity, s. R. Sridhar, K. Surendra, N.S. Krishnaveni, B. Srinivas, K.R. Rao, Synlett *2006* (20), 3495-7; with 1 mol% PhSeBr in the absence of solvent s. F. Manarin, J.A. Roehrs, M. Prigol, D. Alves, C.W. Nogueira, G. Zeni, Tetrahedron Lett. *2007*, *48* (28), 4805-8; (Z)-2-phosphino-

thioenolethers with 5 mol% $Pd(OAc)_2$ in ethanol s. A. Kondoh, H. Yorimitsu, K. Oshima, Org. Lett. *2007*, *9* (7), 1383-5.

*2(S)-[Bis[3,5-bis(trifluoromethyl)phenyl]trimethylsiloxymethyl]-4(R)-(*tert*-butyldimethylsiloxy)pyrrolidine/benzoic acid* ←
Organocatalyzed asym. Michael addition of mercaptans to α,β-ethylenealdehydes C=C → C(SR)CH
chiral 3-hydroxythioethers with 2(S)-[bis[3,5-bis(trifluoromethyl)phenyl]trimethylsiloxymethyl]pyrrolidine cf. *70*, 228; with the 4(R)-(*tert*-butyldimethylsiloxy)-deriv. s. T. Ishino, T. Oriyama, Chem. Lett. *2007*, *36* (4), 550-1.

1-Butyl-3-methylimidazolium hexafluorophosphate s. under $Pd(PPh_3)_4$ *$[bmim]PF_6$*
Chiral bis(Δ^2-oxazolin-2-yl)pyridines s. under $Sc(OTf)_3$ and $Fe(BF_4)_2$ *Pybox*

Phenylselenyl bromide *PhSeBr*
Thioenolethers from acetylene derivs. s. *43*, 458s*72* C≡C → CH=C(SR)

1,3-Bis(diphenylphosphino)propane *dppp*
3-Sulfonylthiazolidines from electron-deficient acetylene derivs.
via double Michael addition s. *72*, 177

Bromine/sodium thiosulfate *$Br_2/Na_2S_2O_3$*
2-(α-Sulfonylalkylidene)thiiranes from α,β-allenesulfones

317. 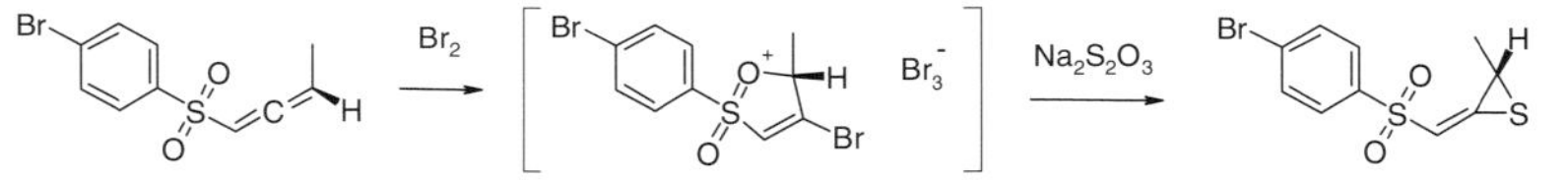

with retention of chirality. A soln. of Br_2 (0.75 mmol; 0.5 *M* in acetonitrile) added dropwise, with stirring, to a soln. of startg. (R)-α-allenylsulfone (0.29 mmol) in acetonitrile at 23°, the resulting soln. stirred for 15 min, then quenched sequentially with water (6 ml) and satd. aq. $Na_2S_2O_3$ (3 ml) → product. Y 74%. The formation of the intermediate Br_3^- salt was confirmed by isolation and X-ray diffraction analysis of its *trans* isomer. Subsequent nucleophilic attack by thiosulfate anion at the asymmetric carbon centre causes ring opening, followed by loss of SO_2, intramolecular Michael addition of the resulting sulfide anion to the vinyl sulfone and elimination of bromide. Alkylidenethiirane formation was initially observed as a side-reaction during attempted bromohydrin preparation using aq. Br_2. The neighbouring group participation effect of the sulfone moiety had not been expected but was exploited and developed by the authors as a general method for the synthesis of 2-(α-sulfonylalkylidene)thiiranes in excellent yields. F.e.s. C. Zhou, C. Fu, S. Ma, Angew. Chem. Int. Ed. *2007*, *46* (23), 4379-81.

Iron(II) fluoroborate/chiral 2,6-bis(Δ^2-oxazolin-2-yl)pyridine ←
Asym. Michael addition of mercaptans C=C → CHC(SR)
to 3-(α,β-ethyleneacyl)-2-oxazolidones s. *58*, 194s*72*

Palladium(II) acetate *$Pd(OAc)_2$*
Thioenolethers from acetylene derivs. C≡C → CH=C(SR)
(Z)-2-Phosphinothioenolethers s. *43*, 458s*72*

Palladium(II) acetate/microwaves *$Pd(OAc)_2$/\[////]*
Thioenolethers from terminal acetylene derivs. CH≡C → CH_2=C(SR)
Catalysis with ligand-modified palladium nanoparticles

318. 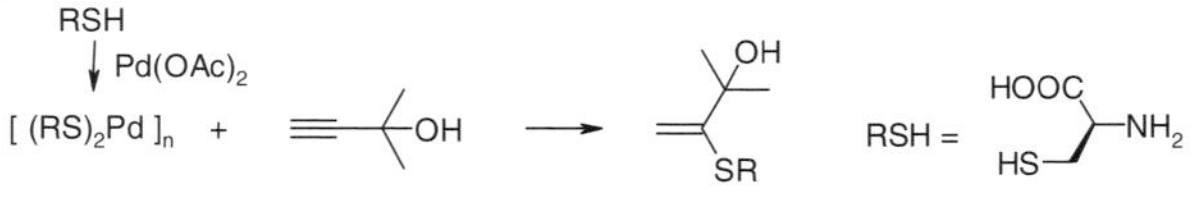

Anti-Markovnikov addition of alkyl mercaptans. A mixture of the startg. alkyne (0.001 mol) and 1 eq. γ-terpinene (to suppress side-reactions) added to 5 mol% $Pd(OAc)_2$, shaken at room temp. until a homogeneous dark-brown soln. formed, 1.1 eqs. of the startg. mercaptan added at room temp. (resulting in the slow formation of a dark-brown precipitate), and the mixture subjected to microwave heating in a screw-capped tube for 20 min → product. Y 92% (regioselectivity >99:1). The nanocatalyst is formed as a network of *1-dimensional nanobelts* with a high degree of 3-dimensional networking possessing sufficient void spaces to include other molecules. Catalytic selectivity is unusually high, leaving a number of functional groups unaffected (e.g. OH, OAc, NMe_2). However, the nature of the palladium salt appears critical, as $PdCl_2$ led to less active μm-scale species. Mechanistically, reaction involves initial coordination of the alkyne to palladium, followed by alkyne insertion into the Pd-S bond prior to protonolysis with a molecule of the mercaptan. F.e. and thermal procedure s. V.P. Ananikov, N.V. Orlov, I.P. Beletskaya, V.N. Khrustalev, M.Y. Antipin, T.V. Timofeeva, J. Am. Chem. Soc. *2007*, *129* (23), 7252-3.

Tetrakis(triphenylphosphine)palladium(0)/1-butyl-3-methylimidazolium hexafluorophosphate *$Pd(PPh_3)_4$/[bmim]PF_6*

Ionic liquid-mediated, stereospecific addition of diaryl disulfides to terminal acetylene derivs. C≡C → C(SAr)=C(SAr)

319. 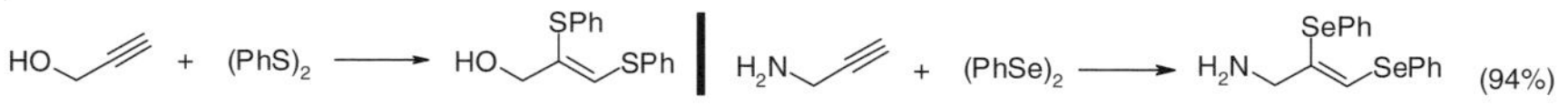

(Z)-1,2-Bis(arylthio)ethylenes. Propargyl alcohol (1.2 mmol), diphenyl disulfide (1 mmol), $Pd(PPh_3)_4$ (2 mol%) and [bmim]PF_6 (1.5 ml) stirred at 60° under argon for 2 h, then cooled to 30°, and extracted with light petroleum ether → product. Y 98%. In addition to its high stereoselectivity, the present method offers the advantages that it avoids the use of organic solvents and permits recycling of the catalyst – indeed, the recovered ionic liquid, containing the Pd-catalyst, could simply be concentrated *in vacuo* and reused. F.e., also **(Z)-1,2-bis(arylseleno)ethylenes from diaryl diselenides**, s. M. Cai, Y. Wang, W. Hao, Green Chem. *2007*, *9* (11), 1180-4.

Rearrangement ∩

Nitrogen/Sulfur Type SC ∩ NS

Gold(III) bromide or Indium(III) bromide *$AuBr_3$ or $InBr_3$*

Sulfonylindoles from *o*-acetylene-N-sulfonylamines via N→C-sulfonyl group migration

320. 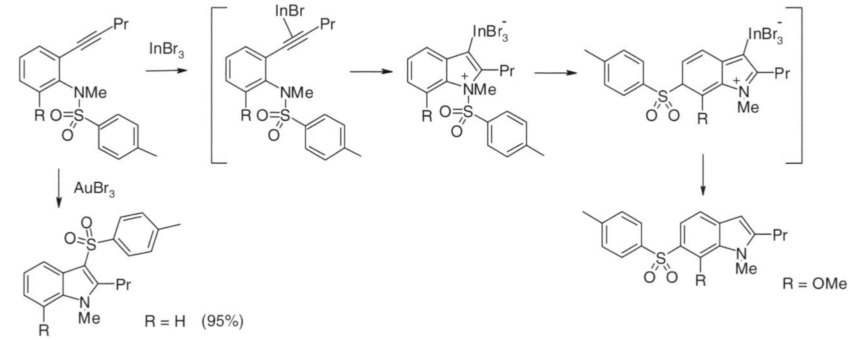

6-Sulfonylindoles under mild conditions. Toluene added to a mixture of 5 mol% $InBr_3$ and 2-methoxy-N-methyl-6-(1-pentynyl)-N-tosylaniline at room temp., warmed at 80° for 2 h (TLC monitoring), and the products isolated by chromatography → 1-methyl-2-*n*-propyl-6-tosyl-7-

methoxyindole. Y 95% (87:13 mixture with the 3-tosyl isomer). The presence of the methoxy group is essential for the unprecedented 1,7-sulfonyl migration to occur giving *p*-subst. 6-(benzenesulfonyl)indoles as major products (substituted by H, Me, MeO or NO_2) in 88-99% yields (90:10 to 66:34 mixtures with the 3-sulfonyl isomers) from internal alkynes (diaryl and aryl(alkyl)). A further example, utilizing a terminal alkyne, was less selective (54:46; 73% yield). Use of $AuBr_3$ as catalyst under similar conditions gave the **3-sulfonylindoles** as major products in variable yields (44-95%) accompanied by 0-25% of the 4- and 6-sulfonyl isomers. A series of experiments established that reaction occurred via intramolecular migration of the sulfonyl group and a mechanism was proposed. F.e. and prepn. of the startg. alkynes s. I. Nakamura, U. Yamagishi, D. Song, S. Konta, Y. Yamamoto, Angew. Chem. Int. Ed. Engl. *2007*, *46* (13), 3284-7.

Sulfur/Remaining Elements Type SC ∩ SRem

Gold(I) chloride *AuCl*

Gold-catalyzed ring closure of (*o*-alkynylphenylthio)silanes ○

321. 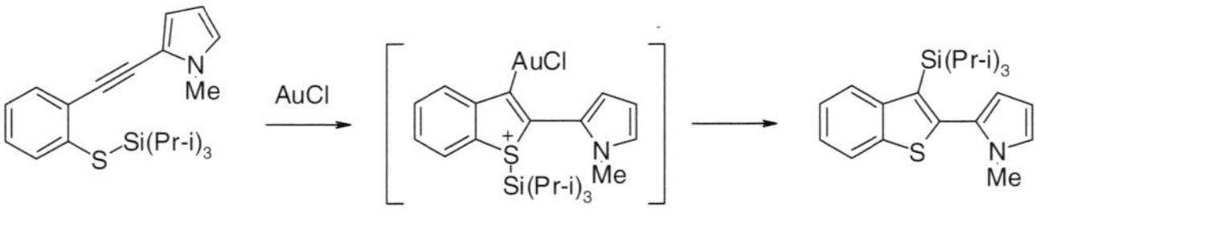

3-Silylbenzo[*b*]thiophenes. The startg. silane heated with 10 mol% AuCl in toluene under argon at 45° for 18 h → product. Y 90%. The reaction occurs via cyclization of an initial Au-alkyne *pi*-complex and subsequent **silyldemetalation**. Using mixtures of different silanes produced crossover products via silyl group migration; no silyl migration to pyrrole occurred in the illustrated example. The reaction is successful with catalyst loadings of 2-25 mol% with higher quantities required for bulky alkyne substituents. Other Au-based catalysts were less effective while catalysts based on Cu, Pt, Pd, In and Ag did not promote the reaction. The method gives lower yields with electron-withdrawing groups on the alkyne moiety and with $SiPh_3$ replacing Si(Pr-*i*)$_3$. Benzene, hexane, methylene chloride and ethyl acetate can be used in place of toluene as solvent but not acetonitrile or THF. Full experimental details are not given in the paper or supporting information. F.e.s. I. Nakamura, T. Sato, M. Terada, Y. Yamamoto, Org. Lett. *2007*, *9* (20), 4081-3.

Carbon/Carbon Type SC ∩ CC

Copper(II) hexafluoroacetoacetonate *Cu(hfacac)$_2$*

2,5-Dihydrothiophenes from 2-vinylthiiranes ◌

Copper-catalyzed ring expansion

322. 

A soln. of 1,2-epithio-3-hexyl-2-methyl-3-butene and 10 mol% Cu(hfacac)$_2$ refluxed (80°) in benzene under N_2 for 20 h, the mixture cooled, concentrated *in vacuo*, the residue filtered off, and purified by chromatography → 3-*n*-hexyl-4-methyl-2,5-dihydrothiophene. Y 75%. Only the selected copper catalyst promoted the desired reaction while suppressing the anticipated side-reactions (diene or dihydrodithiin formation), the conversion being successful at loadings as low as *0.1 mol%*. The reactions are complete in 2-20 h with 10 mol% catalyst in refluxing benzene or toluene giving 2-, 3-, 2,3- and 2,4-subst. products in 65-95% yield (9 examples). The formation of 2-(1-cyclohexenyl)thiirane, however, required the less electrophilic copper(II) tetrafluoroaceto-aceton-onate as catalyst (30 mol%) giving only 35% of the hexahydrobenzothiophene in 24 h. F.e., also prepn. of the startg. m., and application of the method to synthesis of biotin and plavix, s. E. Rogers, H. Araki, L.A. Batory, C.E. McInnis, J.T. Njardarson, J. Am. Chem. Soc. *2007*, *129* (10), 2768-9.

Exchange ⇃↾

Hydrogen ↾ SC ⇃↾ H

Potassium thiocyanate or thioacetate/microwaves *KSCN or KSAc/*[\\\\]
Δ^4-Imidazoline-2-thiones from imidazolium salts ←
with K_2CO_3 cf. *42*, 507; with K-thiocyanate or -thioacetate under microwave irradiation in the absence of solvent s. X.-L. Tao, M. Lei, Y.-G. Wang, Synth. Commun. *2007*, *37* (3), 399-408.

Ammonium thiocyanate s. under [Me_2SBr]Br, $K_2S_2O_8$ or I_2 *NH_4SCN*
Sodium azide s. under 1,1,1-Triacetoxy-1,2-benziodoxol-3(1H)-one *NaN_3*
Copper(II) sulfate s. under $K_2S_2O_8$ *$CuSO_4$*

1,1,1-Triacetoxy-1,2-benziodoxol-3(1H)-one/sodium azide ←
Thiolic acid aryl esters from aldehydes and arylmercaptans RCHO → RC(O)SAr

323. 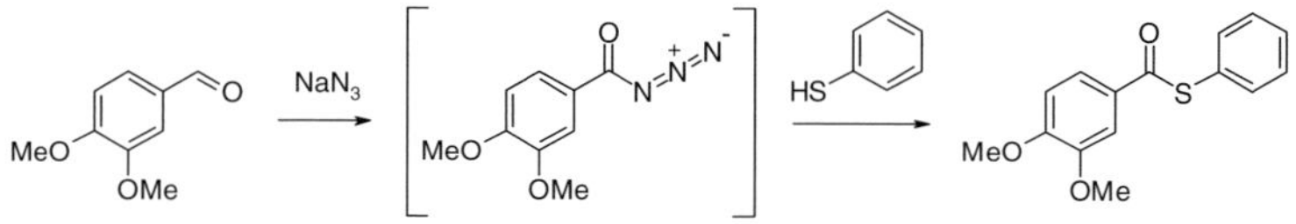

3,4-Dimethoxybenzaldehyde (1 mmol) treated with Dess-Martin periodinane (6 mmol), NaN_3 (6.5 mmol) and thiophenol (1 mmol) in methylene chloride at 0°, then stirred at 25° for 30 min → thioester. Y 94%. High yields were obtained with aryl, heteroaryl, alkyl and α,β-unsatd. aldehydes, although the presence of electron-withdrawing ring-substituents leads to slightly lower yields and longer reaction times. Replacing thiophenol with diphenyl disulfide also gave consistently high yields of the corresponding thioesters. The reaction proceeds via an initially-formed azidohydrin which is further oxidized to an acyl azide prior to nucleophilic displacement by the thiol. Use of NaCN in one example, in the expectation that NaCN would prove to be a better activator than NaN_3, was unexpectedly detrimental to the yield obtained and required a longer reaction time. F.e.s. S.B. Bandgar, B.P. Bandgar, B.L. Korbad, S.S. Sawant, Tetrahedron Lett. *2007*, *48* (43), 1287-90.

N-Bromosuccinimide/1,8-diazabicyclo[5.4.0]undec-7-ene *NBS/DBU*
Thioenolethers from ethylene derivs. C=CH → C=C(SR)
Regiospecific one-pot conversion via addition-elimination

324. 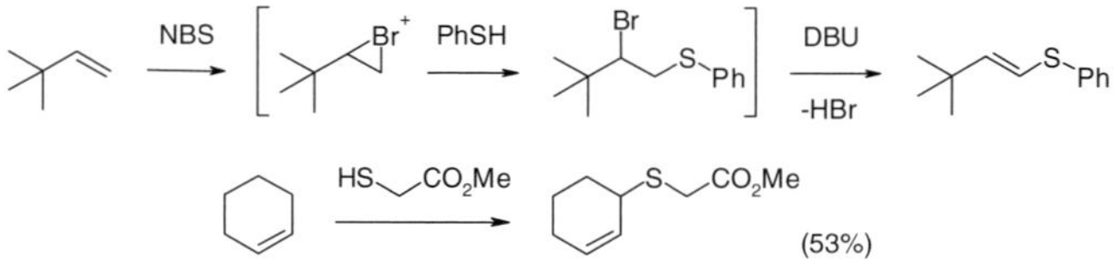

under mild conditions. 1 Eq. phenyl mercaptan and 1.6 eqs. DBU added dropwise to an equimolar mixture of 3,3-dimethyl-1-butene and NBS in THF at 0°, the mixture stirred at room temp. for 4 h, diluted with ether, adjusted to pH 5 with aq. 2 *N* H_2SO_4, the organic layer washed with aq. K_2CO_3, water, dried over Na_2SO_4, the solvent evaporated *in vacuo*, and the residue purified by chromatography → (E)-1-phenylthio-3,3-dimethyl-1-butene. Y 82% (exclusively E). Styrene and 3,3-dimethyl-1-butene gave the (E)-isomers as sole products while products from 1-pentene also contained some of the (Z)-isomer (85:15 to 65:35) with yields in the range 74-87% (7 examples). Similar treatment of cyclohexene afforded **3-alkylthiocyclohexenes** as major products

(3 examples; 53-62%). F.e.s. H. Zoghlami, I. Chehidi, M. Romdhani, M.M. Chaabouni, A. Boulouti, Tetrahedron Lett. *2007*, *48* (32), 5645-7.

N-Chloromethyl-N′-fluoro-1,4-diazoniabicyclo[2.2.2]octane bis(fluoroborate) ←
Oxidative 3-sulfenylation of indoles H → SR

325. 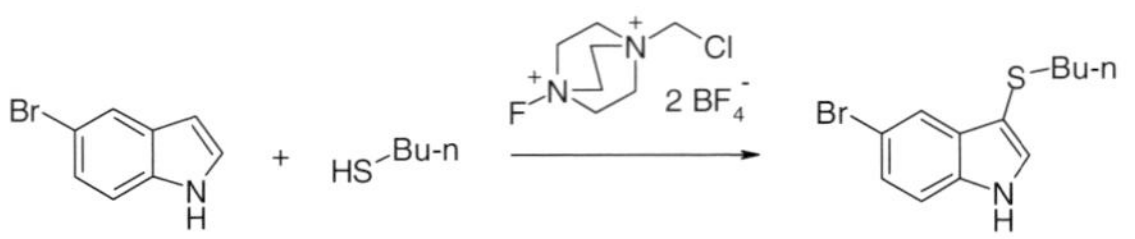

Rapid conversion under mild conditions. An equimolar mixture of 5-bromoindole, butanethiol and Selectfluor stirred in *acetonitrile* at room temp. for 30 min (TLC monitoring), the reaction quenched with water, extracted with methylene chloride, the extracts dried over Na_2SO_4, evaporated *in vacuo*, and purified by chromatography → 3-butylthio-5-bromoindole. Y 78%. The reaction is rapid and atom-efficient with aliphatic, benzylic and aryl thiols (incl. electron-deficient 4-nitrothiophenol and sterically hindered 2-thionaphthol), being complete within 20-30 min and giving 78-96% yield (20 examples). The reaction is tolerant of bromo and methoxy ring substituents, and is also applicable to N-benzylindoles. No bis-sulfenation products are observed (indeed with 3-methylindole no sulfenation occurs) and no fluorination of the indole ring. F.e.s. J.S.Yadav, B.V. Subba Reddy, Y. Jayasudhan Reddy, Tetrahedron Lett. *2007*, *48* (39), 7034-7.

Bromodimethylsulfonium bromide/ammonium thiocyanate *$[Me_2SBr]Br/NH_4SCN$*
Ar. thiocyanation H → SCN
cf. *50*, 307s*69*; with bromodimethylsulfonium bromide in place of $PhI(OAc)_2$ in acetonitrile at room temp. s. D.S. Bhalerao, K.G. Akamanchi, Synlett *2007* (19), 2952-6.

Potassium persulfate/copper(II) sulfate/ammonium thiocyanate *$K_2S_2O_8/CuSO_4/NH_4SCN$*
Iodine/ammonium thiocyanate *I_2/NH_4SCN*
α-Thiocyanation of carbonyl compds.
of β-dicarbonyl compds. with $PhICl_2/Pb(SCN)_2$ cf. *42*, 456s*61*; of carbonyl and dicarbonyl compds. with $K_2S_2O_8/CuSO_4/NH_4SCN$ s. A. Kumar, P. Ahamd, R.A. Maurya, Tetrahedron Lett. *2007*, *48* (8), 1399-401; of ketones with I_2/NH_4SCN in refluxing methanol s. J.S. Yadav, B.V. Subba Reddy, U.V. Subba Reddy, A.D. Krishna, ibid. *48* (30), 5243-6.

Oxygen ↑ SC ↓↑ O

Without additional reagents *w.a.r.*
S-Acylation in the absence of solvent s. *10*, 295s*72* SH → SAc

Sodium hydroxide *NaOH*
α-Functionalized β,γ-ethylenecarboxylic acids from 2′-ethylene-2,2,2-trichloroalcohols ←

326. 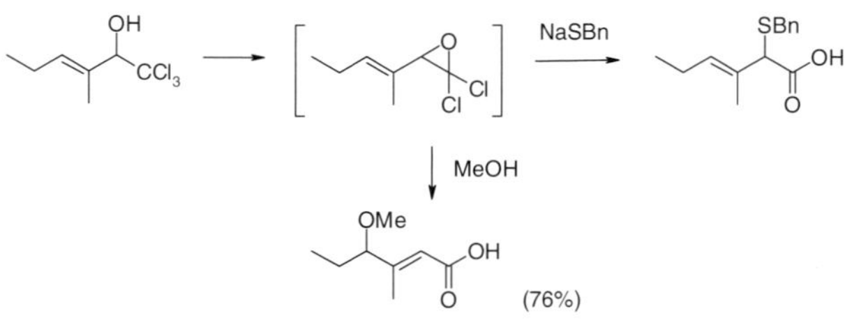

3 Eqs. benzyl mercaptan and 6 eqs. powdered NaOH added to the startg. alcohol in methanol, the mixture stirred at room temp. for 5 min then at 55° for 12 h, cooled, the methanol evaporated *in*

vacuo, the residue acidified to pH 1 (2 *N* HCl), extracted with ethyl acetate, dried over anhydrous Na_2SO_4, and concentrated *in vacuo* → product. Y 85%. The reaction involves an intermediate ethenyl *gem*-dichloroepoxide which reacts with the soft mercaptan nucleophile by an S_N2-type mechanism to afford the α-thio-β,γ-enoic acid. For harder alkoxide nucleophiles the selectivity is dependent on the level of ethene substitution, with trisubstituted ones affording mainly γ-alkoxy-α,β-ethylenecarboxylic acids by S_N2'-type substitution, whereas 2,2-disubstituted ethenes afford α-alkoxy derivs. exclusively, and 1,2-disubstituted ones give mixtures of α- and γ-alkoxy isomers. Yields were generally >75%. The reaction requires a protic solvent to form the intermediate epoxide. F.e.s. J.L. Shamshina, T.S. Snowden, Org. Lett. *2006*, *8* (25), 5881-4.

*Sodium hydroxide/tetra-*n*-butylammonium iodide* *NaOH/Bu₄NI*
2-Ethylenethioethers from acoxy-2-ethylenes C=C-C(OAc) → C(SR)C=C
Stereospecific conversion with double bond shift

327. 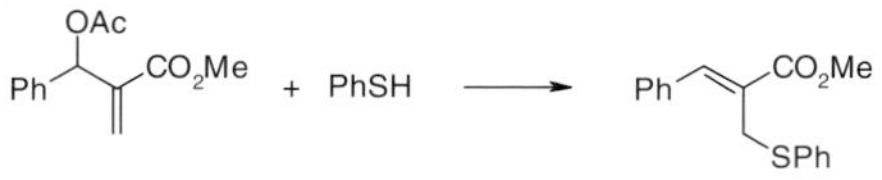

(Z)-α,β-Ethylene-α-(organothiomethyl)carboxylic acid esters. A little tetrabutylammonium iodide and 15% aq. NaOH added to a soln. of thiophenol in DMSO, stirred at room temp. for 30 min, 1 eq. Baylis-Hillman acetate added, stirring continued for 30-60 min, and quenched with water → (Z)-product. Y 92%. cf. *67*, 233. F.e. and (E)-α,β-ethylene-α-(organothiomethyl)nitriles s. B. Das, N. Chowdhury, K. Damodar, J. Banerjee, Chem. Pharm. Bull. *2007*, *55* (8), 1274-6.

Sodium thiocyanate/oxalic or trichloroacetic acid/sodium iodide/iodine ←
Thiocyanates from alcohols OH → SCN
with $NH_4SCN/Ph_3P/Br_2$ cf. *33*, 542s*58*; with NaSCN/oxalic (or trichloroacetic) acid/NaI/I_2 in nitromethane, also isothiocyanates in the absence of NaI and/or I_2, s. H. Miyake, Y. Nakao, M. Sasaki, Chem. Lett. *2006*, *35* (11), 1262-3.

Lithium mercaptides ←
Thioenolethers from enol phosphates or tosylates C=C(SR)
(E)-α,β-Bis(arylthio)- or (E)-α-arylseleno-β-arylthio-α,β-ethylenecarbonyl compds. s. *37*, 519s*72*

Sodium sulfide *Na_2S*
Improved Willgerodt-Kindler reaction under mild conditions CHO → C(S)N<

328. 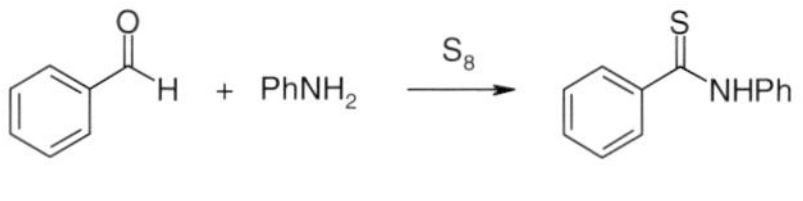

Arylcarboxylic acid N-(aryl)thioamides. *15 Mol%* $Na_2S{\cdot}9H_2O$ added to a mixture of 5 mmol S_8 and 6 mmol aniline in DMF, the suspension stirred at 115° for 30 min under N_2, cooled to room temp., 4 mmol benzaldehyde added, stirring continued for 12 h at 115°, cooled to room temp., quenched with satd. aq. NH_4Cl, and worked up with column chromatographic purification on silica gel → N-phenylthiobenzamide. Y 91%. The catalyst facilitates generation of reactive polysulfide anion from S_8. F.e. and with pyridylamines, also N-unsubst. thioamides (with $S_8/NH_3/Na_2S$), s. K. Okamoto, T. Yamamoto, T. Kanbara, Synlett *2007* (17), 2687-90; Willgerodt-Kindler reaction in water, also from aldimines, s. K. Aghapoor, F. Mohsenzadeh, G. Khanalizadeh, H.R. Darabi, Monatsh. Chem. *2007*, *138* (1), 61-5.

Aluminum triflate *$Al(OTf)_3$*
S-Acetylation in the absence of solvent s. *54*, 79s*72* SH → SAc

1-Butyl-3-methylimidazolium hydroxide *[bmim]OH*
Cyclic α,β-ethylene-α-(organothio)ketones ▽O → C=C(SR)
from α,β-epoxyketones s. *72*, 341

Amberlyst 15 ←
2,3-Unsatd. thioglycosides from O³-acylglycals s. *24*, 240s*72* OAc → SR

Oxalic acid (s.a. under NaSCN) *$(COOH)_2$*
Mercaptals from oxo compds. $CO \rightarrow C(SR)_2$
cf. *33*, 540s*71*; with oxalic acid (2.4 eqs.) in nitromethane s. H. Miyake, Y. Nakao, M. Sasaki, Chem. Lett. *2007*, *36* (1), 104-5; with a little benzyltriphenylphosphonium tribromide by grinding in a pestle and mortar in the absence of solvent s. A.R. Hajipour, S.A. Pourinousavi, A.E. Ruoho, Phosphorus, Sulfur Silicon Relat. Elem. *2007*, *182* (5), 921-37; chemoselectivity under continuous flow by optimizing a reagent's residence time within a packed-bed reactor s. C. Wiles, P. Watts, S.J. Haswell, Tetrahedron Lett. *2007*, *48* (41), 7362-5.

Carbon tetrabromide *CBr_4*
S-Acetylation in the absence of solvent s. *72*, 102 $SH \rightarrow SAc$

Silica s. under $HClO_4$ *SiO_2*

Silica chloride ←
2-Aminothiazoles from ketones ○
and thiourea with I_2 cf. *27*, 607s*71*; with silica chloride as catalyst in refluxing acetonitrile s. H. Karade, M. Sathe, M.P. Kaushik, Catal. Commun. *2007*, *8* (4), 741-6.

*Tri-*n*-butylphosphine* *Bu_3P*
(Z)-2-Arylidene-3-oxo-S-heterocyclics
from arylpropiolic acid esters and bifunctional S-nucleophiles

329. 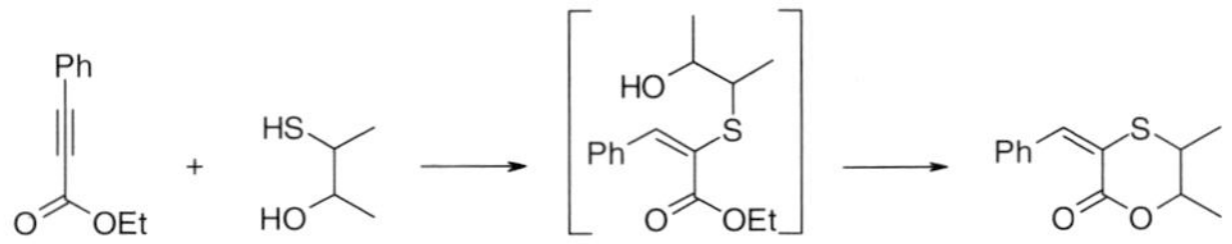

(Z)-3-Arylidene-1,4-oxathian-2-ones. Ethyl phenylpropiolate added dropwise to a mixture of 1.1 eqs. racemic 2-mercapto-3-butanol and 0.2 eqs. tri-*n*-butylphosphine in dry toluene under argon, heated at 25° for 12 h, evaporated under reduced pressure, and purified by flash chromatography → (Z)-3-benzylidene-5,6-dimethyl-1,4-oxathian-2-one. Y 77% (*syn/anti* 40:60). 2-Aminomercaptans, α-mercaptocarboxamides and thioureas also served as the dinucleophile to give the corresponding **(Z)-2-arylidene-5,6-dihydro-1,4-thiazin-3(4*H*)-ones**, 3,5-dione analogs and **(Z)-5-arylidene-2-imino-4-thiazolidones**, respectively, reaction being tolerant of chloro, *tert*-butyl and methoxy groups in the phenyl ring. In some cases catalytic *p*-toluenesulfonic acid was added to the mixture to force closure of the intermediate S-adducts. This is another application of phosphine-catalysis initiated by 1,4-addition, in this instance reversing the polarity of the acetylene group to achieve anti-Michael addition of the S-nucleophile prior to ring closure. F.e., also formation of **(Z)-5-arylidene-4-thiazolidone-2-thiones** (rhodanine derivs.) by reaction of the arylpropiolates with dithiocarbamates (generated *in situ* from primary amines and CS_2), s. S. Gabillet, D. Lecerclé, O. Loreau, M. Carboni, S. Dézard, J.-M. Gomis, F. Taran, Org. Lett. *2007*, *9* (20), 3925-7.

Triphenylphosphine/diethyl azodicarboxylate *$Ph_3P/ROOCN{=}NCOOR$*
Xanthates from two different alcohol molecules ROC(S)SR′
One-pot conversion under mild conditions

330. 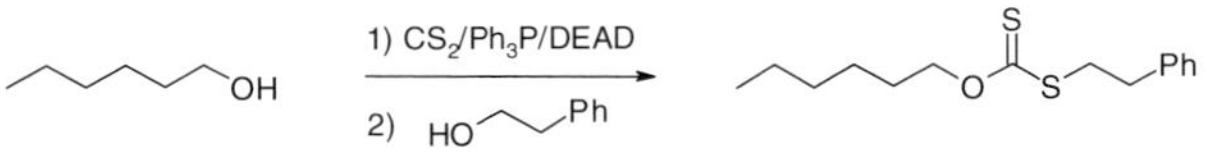

Triphenylphosphine (9 mmol) then diethyl azodicarboxylate (9 mmol) (*slowly, in 2-3 portions*) added to a soln. of *n*-hexanol (9 mmol) and carbon disulfide (4 ml, excess) in dry DMSO, 2-

phenylethanol (9 mmol) added, the mixture stirred at room temp. for 4 h, then poured into distilled water, and extracted with ethyl acetate → O-hexyl S-phenethyl xanthate. Y 97%. A variety of prim., sec. and tert. alcohols afford the corresponding xanthates in good to excellent yields. The method offers the advantages that it avoids long reaction times, use of expensive, strongly-basic reagents and tedious work-up procedures. F.e.s. D. Chaturvedi, S. Ray, Tetrahedron Lett. *2007*, *48* (1), 149-151.

*2,4-Bis(*p*-methoxyphenyl)-1,3,2,4-dithiadiphosphetane 2,4-disulfide/triethylamine or pyridine* ←
***o*-Halogenothioacylamines from *o*-halogenacylamines** s. *72*, 337 CO → CS

Bismuth trichloride *$BiCl_3$*
2-Acetylenethioethers from 2-acetylenealcohols s. *72*, 106 OH → SR

Iodine s. under NaSCN *I_2*

Magnesium perchlorate *$Mg(ClO_4)_2$*
Perchloric acid/silica *$HClO_4/SiO_2$*
2,3-Dihydro-1,5-benzothiazepines from α,β-ethyleneketones ○
and *o*-aminomercaptans by heating on silica cf. *13*, 896s*68*; with 1 mol% $HClO_4$/silica in refluxing methanol s. G.L. Khatik, G. Sharma, R. Kumar, A.K. Chakraborti, Tetrahedron *2007*, *63* (5), 1200-10; with 5 mol% anhydrous Mg-perchlorate in refluxing 1,2-dichloroethane s. G.L. Khatik, R. Kumar, A.K. Chakraborti, Synthesis *2007* (4), 541-6.

Hydrochlorides ←
2-Subst. benzothiazoles or benzothiazolines from *o*-aminomercaptans and aldehydes
Chemoselective conversion in water

331. 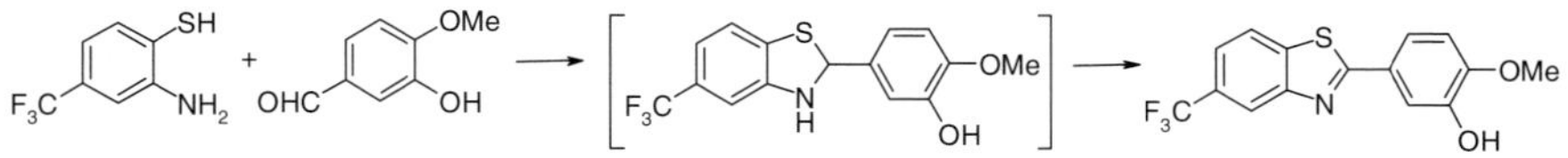

A mixture of 3-hydroxy-4-methoxybenzaldehyde (3 mmol) and 2-amino-4-(trifluoromethyl)-thiophenol (3 mmol) (as its hydrochloride) in *tap water* (10 ml) stirred and heated at 110° for 5 h, the water decanted off, and the residue extracted with ether → 2-(3-hydroxy-4-methoxyphenyl)-5-(trifluoromethyl)benzothiazole. Y 92%. Aryl, heteroaryl and styryl aldehydes afforded benzothiazoles in excellent yields, while alkyl and aralkyl aldehydes gave the corresponding benzothiazolines or product mixtures. Shorter reaction times for benzothiazole formation were achieved by bubbling O_2 through the reaction mixture, indicating that the reaction proceeds via an intermediate benzothiazoline, further dehydrogenation of which is dependent on the nature of the substituent at the 2-position. Studies using alternative solvents showed the reaction rate to be fastest in water, supporting a postulated mechanism in which water coordinates to both thiol sulfur and aldehyde oxygen moieties in an initial transition state complex. Further mechanistic studies showed the alternative pathway of imine formation followed by intramolecular nucleophilic attack by sulfur to be unlikely. Tap water was used for reasons of expedience, the reaction being equally successful using either distilled water or satd. brine. F.e.s. A.K. Chakraborti, S. Rudrawar, K.B. Jadhav, G. Kaur, S.V. Chankeshwara, Green Chem. *2007*, *9* (12), 1335-40.

*Tetra-*n*-butylammonium iodide s. under NaOH* *Bu_4NI*

Benzyltriphenylphosphonium tribromide *$BnPh_3PBr_3$*
Mercaptals from oxo compds. CO → $C(SR)_2$
by grinding in the absence of solvent s. *33*, 540s*72*

Via intermediates *v.i.*
Isothiocyanates from nitrile oxides RC≡N→O → RN=C=S
Polymer-based conversion via 1,4,2-oxathiazoles

332.
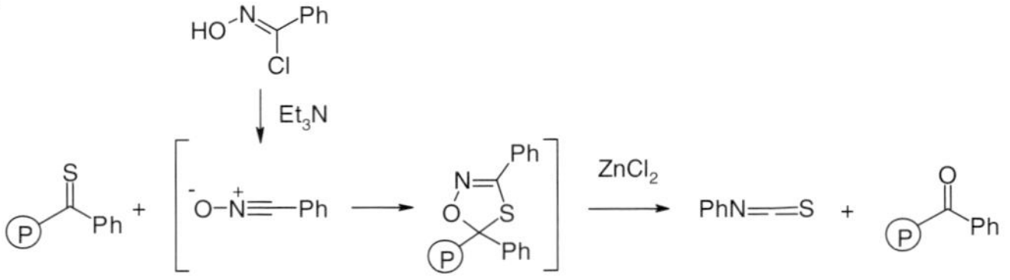

'Catch-and-release' methodology. 1.1 Eqs. phenylhydroximidoyl chloride added to polymer-supported thiobenzophenone (assumed loading 2 mmol/g) in methylene chloride, the mixture stirred under inert atm. for 5 min, 1.1 eqs. Et_3N added over several min, stirring continued until the blue colour of the thione became colourless (2 h), the resulting polymer-supported 1,4,2-oxathiazole filtered off, washed with boiling toluene, acetone, ether and methylene chloride, dried *in vacuo*, the beads swollen in dry toluene under inert atm. for 10 min, 0.2 eqs. of anhydrous $ZnCl_2$ added, the mixture refluxed 24 h, the polymer filtered off, washed with methylene chloride, and the filtrate evaporated *in vacuo* → phenyl isothiocyanate. Y 84%. The initial **1,3-dipolar cycloaddition** is *self-indicating* by the described colour change, but no yield is given for the resulting supported 1,4,2-oxathiazole. Zinc chloride was the Lewis acid of choice for the fragmentation step since competing cycloreversion, promoted by other catalysts, was not observed, suggesting that zinc ion coordinates to the sulfur of the oxathiazole reducing the C_5-S bond order. Toluene is an effective solvent for this step due to increased swelling of the resin compared with other solvents. F.e.s. B.A. Burkett, J.M. Kane-Barber, R.J. O'Reilly, L. Shi, Tetrahedron Lett. *2007, 48* (31), 5355-8.

Halogen ↑ SC ↓↑ Hal

Microwaves s. under K-salt, 1,8-Diazabicyclo[5.4.0]undec-7-ene and Pd$_2$(dba)$_3$ [\\\\]
Potassium carbonate/sodium hypochlorite/acetic acid *K_2CO_3/NaOCl/AcOH*
***o*-Arylsulfonyl- from *o*-chloro-N-heterocyclics and arylmercaptans** Cl → SO_2Ar

333.
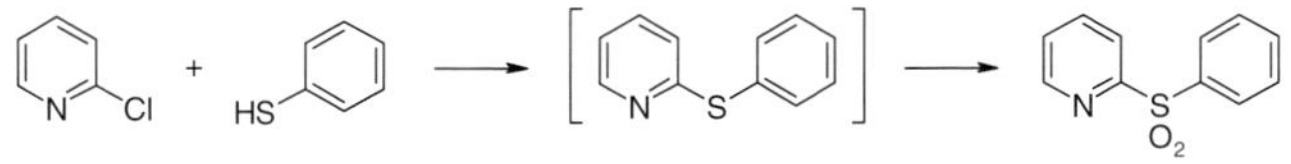

Large-scale one-pot procedure. 962 mmol K_2CO_3 added to a soln. of 790 mmol 2-chloropyridine and 852 mmol thiophenol in 450 ml DMF, heated to 110°, stirred for 20 h, filtered through polypropylene, the resulting soln. of 2-phenylthiopyridine (crude) treated with 995 mmol glacial acetic acid, heated to 40°, 1.7 mol 13 wt% NaOCl soln. added slowly over 2 h, water added when the sulfide had been consumed (HPLC monitoring), heating discontinued, the pH raised to 9 with 20% NaOH, the slurry cooled to <5°, stirred cold for 1.5 h, and worked up → 2-benzene-sulfonylpyridine. Y 86% (purity >99%). This tandem S_NAr/oxidation is clean, high-yielding on the large scale, environmentally friendly, and proceeds without any detectable isolation of the intermediate sulfoxide. The method was notably successful for the preparation of **2-aryl-sulfonylpyridines,** and also applied to 2-chloro-quinoxaline and -quinoline (after re-oxidation to convert all of the intermediate sulfoxide to sulfone). Electron-withdrawing groups in the aromatic ring reduced the reaction rate. F.e. and with retention of cyano groups, also comparison of oxidants, s. W.G. Trankle, M.E. Kopach, Org. Proc. Res. Dev. *2007, 11* (5), 913-7.

Potassium salt/microwaves *K⁺/[\\\\]*
2(3*H*)-Benzothiazolethiones from *o*-halogenamines ○
and K-O-ethyl xanthate cf. *54*, 221; *69*, 238; under microwave irradiation in DMF s. W. Huang, Y. Tan, M.W. Ding, G.F. Yang, Synth. Commun. *2007*, *37* (3), 369-76.

1,8-Diazabicyclo[5.4.0]undec-7-ene/microwaves *DBU/[\\\\]*
Mercaptals from 1,1-dihalides and mercaptans $CHal_2 \rightarrow C(SR)_2$
in the absence of solvent with KOH/sand cf. *42*, 521; with DBU under microwave irradiation, incl. 1,3-dithiolanes from 1,2-dithiols, s. Y.J. Cao, Y.Y. Lai, H. Cao, X.N. Xing, X. Wang, W.J. Xiao, Can. J. Chem. *2006*, *84* (11), 1529-33.

Copper(II) oxide nanoparticles/potassium hydroxide *CuO/KOH*
Ar. thioethers from ar. iodides and mercaptans ArI → ArSR
under mild conditions with readily recyclable copper(II) oxide nanoparticles

334.

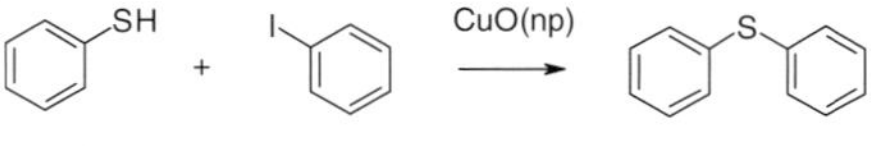

A mixture of 1.1 eqs. iodobenzene, 1 eq. benzenethiol, 1.26 mol% CuO nanoparticles and 1.5 eqs. KOH stirred *at 80°* under N_2 in DMSO for 10 h (TLC monitoring), treated with water and ethyl acetate, the aq. layer extracted with ethyl acetate, the organic solutions combined, and worked up with purification through a short pad of Celite → diphenyl sulfide.Y 95%. The catalyst is relatively inexpensive, air-stable, easily recovered by centrifugation, and retained its activity over three cycles. The procedure is notably milder than existing methods, and generally applicable in high yield to alkyl and aryl mercaptans, substrates with electron-donating groups in the aromatic ring being more reactive than those with electron-withdrawing groups. The fact that 1-iodo-4-nitrobenzene was more reactive than 4-iodoanisole supports a mechanism involving oxidative addition of CuO to the C-I bond, followed by reaction with the mercaptan then reductive elimination to regenerate the catalyst. Other copper salts were less effective. F.e.s. L. Rout, T.K. Sen, T. Punniyamurthy, Angew. Chem. Int. Ed. Engl. *2007, 46* (29), 5583-6.

*Copper(I) chloride/*trans-*1,2-diaminocyclohexane* ←
Diaryl sulfides from ar. halides and arylmercaptans in water ArI → ArSAr′

335.

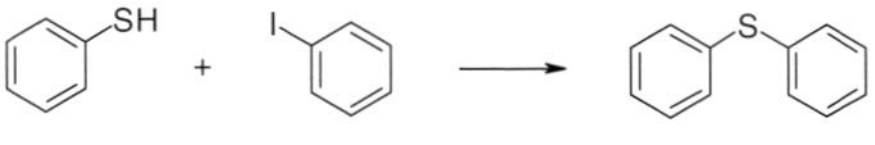

A Schlenk flash containing 0.048 mmol CuCl, 1.05 mmol iodobenzene, 2.21 mmol *trans*-1,2-diaminocyclohexane, *6.3 ml* water and 0.57 mmol thiophenol sealed under a positive pressure of argon, the resulting violet soln. heated overnight at 120°, and worked up with flash chromatographic purification → diphenyl sulfide. Y 97%. The procedure is safe, cost-effective, environmentally friendly, applicable to a wide range of thiophenols, aryl iodides and aryl bromides (the latter, where necessary, being converted *in situ* to the more reactive iodides with NaI), and leaves unprotected amino and hydroxyl groups unaffected. Furthermore, the aqueous medium after extraction of the product can be used up to three times without significant loss of activity (although the actual volume of water used appears to be critical). The diamine serves as both ligand and base. F.e. and comparison of copper salts and diamines, also from *o*-subst. ar. halides and arylmercaptans, s. M. Carril, R. SanMartin, E. Dominguez, I. Tellitu, Chem. Eur. J. *2007, 13* (18), 5100-5.

Copper(I) iodide/1-butyl-3-methylimidazolium N,N-dimethylglycinate ←
Thioenolethers from α,β-ethylenehalides and mercaptans C=C(Hal) → C=C(SR)
(E)-isomers with CuBr/L-proline/K_2CO_3/[bmim]BF_4 cf. *68*, 230s*70*; (Z)-isomers with CuI/1-butyl-3-methylimidazolium N,N-dimethylglycinate at 90°, also eneselenides with selenophenol (cf. *41*, 584), s. Z.M. Wang, H.J. Mo, W.L. Bao, Synlett *2007* (1), 91-4.

Copper(I) iodide/1-ethyl-3-methylimidazolium (S)-2-amino-3-methylbutyrate/sodium salt ←
Ar. sulfones from halides and sulfinic acids Hal → SO_2R
with $CuOTf{\cdot}C_6H_6$/DMEDA cf. *42*, 522s*65*; from Na-sulfinates with CuI in 1-ethyl-3-methylimidazolium (*S*)-2-amino-3-methylbutyrate/DMSO, also α,β-ethylene-sulfones from -bromides, s. M. Bian, F. Xu, C. Ma, Synthesis *2007* (19), 2951-6.

Copper(I) iodide/benzotriazole/potassium tert-*butoxide* *CuI/BtH/KOBu*-t
Copper(I) iodide/L-proline/potassium carbonate or phosphate *CuI/Pro-OH/K_2CO_3 or K_3PO_4*
Ar. thioethers from ar. halides and mercaptans Hal → SR
with CuI/KOH cf. *31*, 522s*69*; diaryl sulfides from ar. bromides and arylmercaptans with CuI (0.5 mol%)/benzotriazole (1 mol%)/KOBu-*t* in DMSO at 100° s. A.K. Verma, J. Singh, R. Chaudhary, Tetrahedron Lett. *2007*, *48* (40), 7199-202; from ar. iodides with CuI/L-proline and K_2CO_3 (for ar. mercaptans) or K_3PO_4 (for aliphatic mercaptans) in glyme at 80° s. H. Zhang, W. Cao, D. Ma, Synth. Commun. *2007*, *37* (1), 25-35.

Copper(I) iodide/N,N-dimethylglycine/sodium salt *$CuI/Me_2NCH_2COOH/Na^+$*
Dithiocarbamic acid esters from halides Hal → SC(S)N<
cf. *18*, 689; aryl dithiocarbamates from ar. iodides and vinyl dithiocarbamates from vinyl bromides (with retention of configuration) using Na-dimethyldithiocarbamate in the presence of CuI/N,N-dimethylglycine in DMF at 110° s. Y.Y. Liu, W.L. Bao, Tetrahedron Lett. *2007*, *48* (27), 4785-8.

Tributyl(2-hydroxyethyl)ammonium thiocyanate *$[Bu_3(HOCH_2CH_2)N]SCN$*
Thiocyanates from halides in an ionic liquid as reagent and solvent Hal → SCN
using [bmim]SCN cf. *4*, 566s*69*; using tributyl(2-hydroxyethyl)ammonium thiocyanate s. F. Mohanazadeh, M. Aghvami, Tetrahedron Lett. *2007*, *48* (40), 7240-2.

Imidazolium salts s. under CuI ←
Benzotriazole s. under CuI *BtH*
Acetic acid s. under K_2CO_3 *AcOH*
Amino acids s. under CuI ←
9,9-Dimethyl-4,5-bis(diphenylphosphino)xanthene s. under $Pd_2(dba)_3$ *Xantphos*

Sodium thiophosphate *Na_3PO_3S*
1,2-*trans*-Thioaldoses from glycosyl halides Hal → SH
One-pot conversion with high stereoselectivity

336. 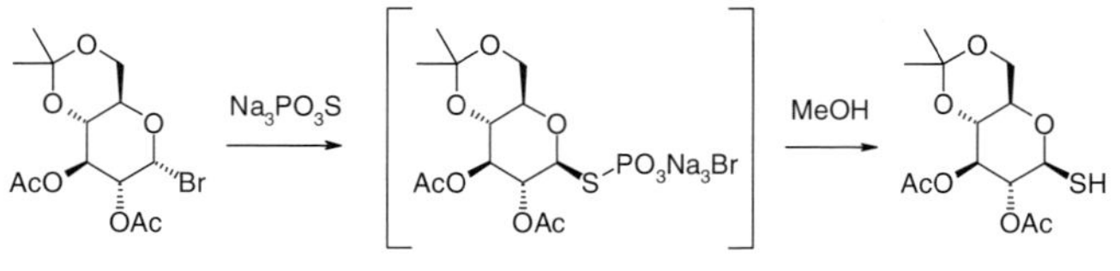

under mild conditions. 1.5 eqs. $Na_3PO_3S{\cdot}10H_2O$ added to a soln. of the startg. glycosyl halide in methanol under argon, the mixture stirred at room temp. for 4-7 h (TLC), added to water, extracted with methylene chloride, the extracts washed with brine, dried over Na_2SO_4, evaporated *in vacuo*, and the residue purified by chromatography → 1,2-*trans*-thioaldose. Y 92%. The substitution occurs with *almost complete inversion of configuration* for both α- and β-glycosyl chlorides and bromides in 88-97% yields (9 examples) in the presence of common carbohydrate protecting groups (acetate/pivalate esters, benzyl ethers, acetals and N-phthalimides) with an inexpensive reagent. F.e.s. W. Xue, X. Cheng, J. Fan, H. Diao, C. Wang, L. Dong, Y. Luo, J. Chen, J. Zhang, Tetrahedron Lett. *2007*, *48* (35), 6092-5.

*2,4-Bis(*p*-methoxyphenyl)-1,3,2,4-dithiadiphosphetane 2,4-disulfide/cesium carbonate* ←
2-Subst. benzothiazoles from *o*-halogenacylamines ○
One-pot conversion via *o*-halogenothioacylamines

337. 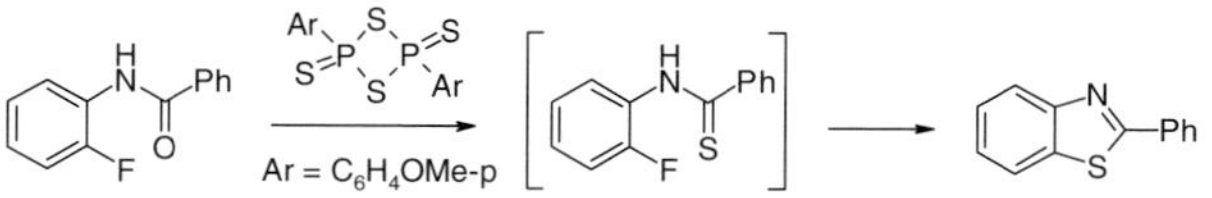

A mixture of N-benzoyl-2-fluoroaniline and 0.6 eqs. Lawesson's reagent stirred/heated under argon at 105-115° for 2.5 h, 2 eqs. Cs_2CO_3 added, the mixture heated under reflux (140°) for 3 h, cooled, evaporated to dryness *in vacuo*, and purified by chromatography → 2-phenylbenzothiazole. Y 84%. 2-Methyl- and 2-phenyl-benzothiazole were prepared by this method in 68-84% yields from *o*-fluoro- and *o*-iodo-derivs., and yields and reaction times were acceptable with the less expensive and more readily available bromo- and chloro-analogs. In all cases, thionation was complete within 2-3 h, and cyclization of the intermediate thioamides in 3-5 h. The intermediates could be obtained in 83-85% yields by heating the reactants in pyridine or triethylamine. The scope of the reaction is under investigation. F.e.s. D. Bernardi, L.A. Ba, G. Kirsch, Synlett *2007* (13), 2121-3.

Sodium hypochlorite s. under K_2CO_3 *NaOCl*

Bis(1,3-dibenzylimidazol-2-ylidene)nickel(0)/potassium tert-*butoxide* ←
Ar. thioethers from ar. halides ArHal → ArSR

338. 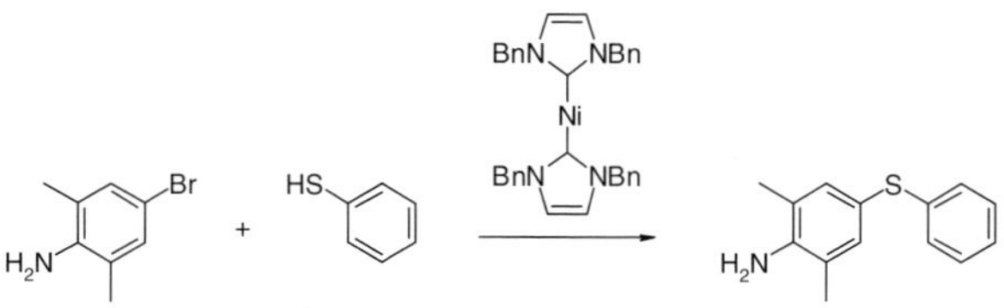

A mixture of the startg. ar. bromide, 1.05 eqs. thiophenol, 1.1 eqs. KOBu-*t* and 4 mol% bis(1,3-di-dibenzylimidazol-2-ylidene)nickel(0) heated in DMF at 110° for 16 h in a capped vial → 2,6-dimethyl-4-phenylthioaniline. Y 96% (isolated yield). Such nickel(0) N-heterocyclic carbene catalysts are inexpensive, environmentally friendly, readily prepared by reaction of bis(1,5-cyclooctadiene)nickel(0) with 1,3-disubst. imidazolium bromides, and useful alternatives to traditional nickel and palladium phosphine complexes. A series of diaryl sulfides and one aryl alkyl sulfide were prepared by this method in 80-99% yield. Different ligand/base combinations can accommodate both electron-rich and electron-deficient aryl bromide and iodides. The mechanism is believed to follow the oxidative addition/reductive elimination cycle identified for palladium phosphine-catalyzed coupling (cf. *36,* 561s*69*). F.e. and with bridged bis(imidazol-2-ylidenes) as ligand s. Y. Zhang, K.C. Ngeow, J.Y. Ying, Org. Lett. *2007*, *9* (18), 3495-8.

Tris(dibenzylideneacetone)dipalladium/9,9-dimethyl-4,5-bis(diphenylphosphino)xanthene/ ethyldiisopropylamine/potassium salt/microwaves ←
Thiolic acid aryl esters from thiolic acids and ar. halides RC(O)SAr
with a Cu(I)-salt and ar. iodides cf. *38*, 529; with K-thioacetate and ar. bromides (or one example of an aryl triflate) in dioxane under microwave irradiation in the presence of 2.5 mol% $Pd_2(dba)_3$, 5 mol% Xantphos and 2 eqs. *i*-Pr_2NEt s. C.Q. Lai, B.J. Backes, Tetrahedron Lett. *2007*, *48* (17), 3033-7.

Sulfur ↑ SC ↓↑ S

*Sodium carbonate/tetra-*n*-butylammonium hydrogen sulfate/triphenylphosphine/ hydrogen chloride* ←

Thioglycosides from glycosyl halides and disulfides Hal → SR

by cleavage of the disulfides with Ph_3P then S-glycosylation in the presence of $NaOH/BnEt_3NCl$ cf. *7*, 634s*71*; cleavage with Ph_3P/HCl in toluene then S-glycosylation in the presence of satd. aq. Na_2CO_3/Bu_4NHSO_4, also Michael additions, s. C. Mukherjee, A.K. Misra, J. Carbohydr. Chem. *2007, 26* (4), 213-21.

Cesium carbonate Cs_2CO_3 ←

3-Component reactions of monothioldicarboxylic acid anhydrides with amines and sulfonic acid amides

339. 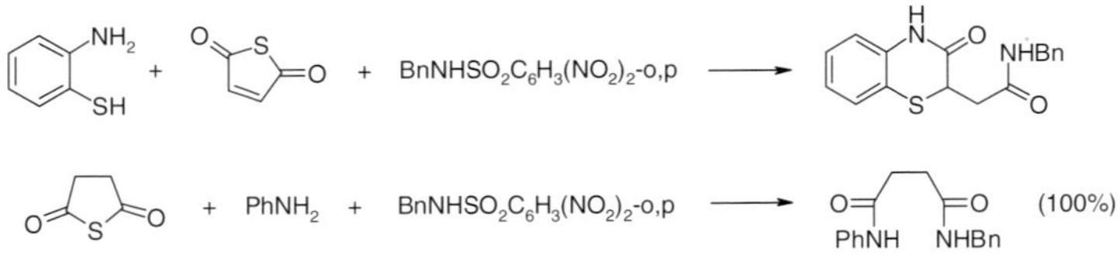

4-6-Membered dicarboxylic acid thioanhydrides react with amides and sulfonamides to give **mixed N,N′-disubst. dicarboxylic acid amides,** while functionalized amines present another point of molecular diversity to effect simultaneous ring closure with thiomaleic anhydride. **E: 2*H*-1,4-Benzothiazin-3(4*H*)-one-2-acetamides from *o*-aminomercaptans.** 2-Aminothiophenol added to a stirred soln. of 1.2 eqs. maleic thioanhydride in DMF at 0°, the dark purple mixture warmed to room temp., stirred for 30 min, cooled to 0°, Cs_2CO_3 added, followed immediately by 1 eq. N-benzyl-(2,4-dinitrophenyl)sulfonamide, warmed to room temp., stirring continued for 30 min, solvent removed under vacuum, and the residue worked up → 3,4-dihydro-3-oxo-N-benzyl-2*H*-1,4-benzothiazine-2-acetamide. Y 96%. F.e., also mixed α-(carbobenzoxy-amino)succinic acid amides from N-(benzoxycarbonyl)-L-aspartic acid thioanhydride, and 3-component reactions with azides, s. D. Crich, A.A. Bowers, Org. Lett. *2007, 9* (25), 5323-5.

*1,1,1-Triacetoxy-1,2-benziodoxol-3(1*H*)-one/sodium azide* ←

Thiolic acid aryl esters from aldehydes RCHO → RC(O)SAr

and diphenyl disulfide s. *72*, 323

Remaining Elements ↑ SC ↓↑ Rem

Copper(II) acetate/N-benzylimidazole/potassium carbonate/sodium salt ←

Ar. sulfones from arylboronic acids and sulfinic acid salts $ArB(OH)_2 \rightarrow ArSO_2R$

340. 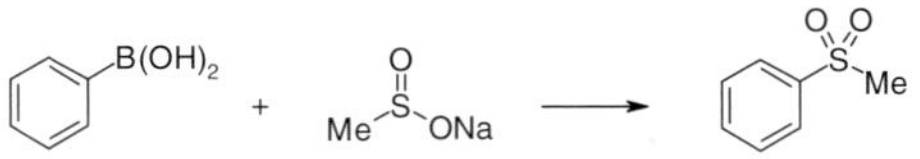

20 Mol% $Cu(OAc)_2$, 40 mol% N-benzylimidazole, 1.5 eqs. Na-methylsulfinate, 2 eqs. K_2CO_3 and 4 Å molecular sieves added to a soln. of the startg. arylboronic acid in DMSO, stirred for 22 h at 60°, filtered, the filtrate diluted with ethyl acetate and aq. NH_4Cl, the aq. layer extracted with ethyl acetate, and worked up with column chromatographic purification → methyl phenyl sulfone. Y 64%. The aryl residue may possess electron-withdrawing or -donating groups, and reaction leaves chlorine and bromine unaffected. However, it appears that a *monodentate* ligand is essential for the conversion. F.e. incl. diaryl sulfones, hetaryl methyl sulfones, **also α,β-ethylenesulfones** from α,β-ethyleneboronic acids, s. A. Kar, I.A. Sayyed, W.F. Lo, H.M. Kaiser, M. Beller, M.K. Tse, Org. Lett. *2007, 9* (17), 3405-8.

Carbon ↑ **SC ↓↑ C**

1-Butyl-3-methylimidazolium hydroxide *[bmim]OH*
2-Ketothioethers from α,β-epoxyketones ▽O → CHSR
Regiospecific conversion in ionic liquids under mild conditions

341.

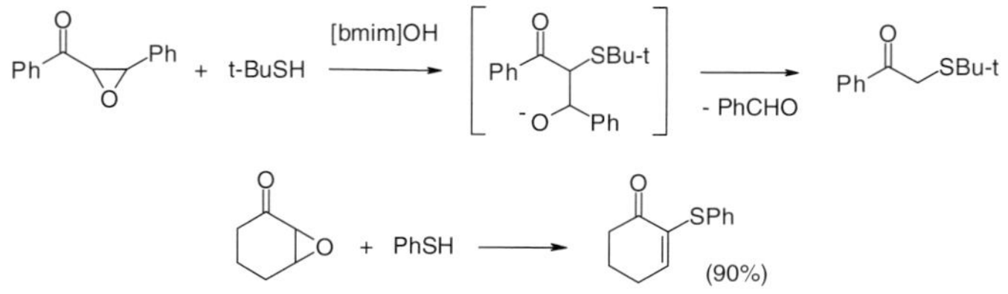

via retro-aldol cleavage. 1 Eq. *tert*-butyl mercaptan added to a mixture of 2-benzoyl-3-phenyloxirane and 0.24 eqs. [bmim]OH, stirred for 20 min (TLC), and the product isolated by direct distillation *in vacuo* → α-(*tert*-butylthio)acetophenone. Y 80%. Sixteen acyclic α,β-epoxyketones gave analogous products in 75-90% yields with aryl, alkyl and benzyl mercaptans. By contrast, cyclic analogs suffered dehydration (rather than retro-aldol reaction) after thiolysis to produce **cyclic α,β-ethylene-α-(organothio)ketones** in 70-90% yields (4 examples). The procedure is environmentally friendly, requiring no additional solvent, and permits *recycling* of the non-toxic ionic liquid (up to six times without appreciable loss in catalytic activity). F.e. and in 1-butyl-3-methylimidazolium bromide for the conversion of epoxides to 2-hydroxythioethers (*72*, 312) s. B.C. Ranu, T. Mandal, S. Banerjee, S.S. Dey, Aust. J. Chem. *2007*, *60* (4), 278-83.

Tris(dibenzylideneacetone)dipalladium/9,9-dimethyl-4,5-bis(diphenylphosphino)xanthene/ potassium hydroxide ←
Sulfoxide exchange via sulfenate anions Hal → S(O)R
from methyl 2-sulfinylacrylates and alkyl halides with Li-cyclohexoxide cf. *48*, 529s*66*; hetaryl *p*-tolyl sulfoxides from *tert*-butyl 3-(*p*-tolylsulfinyl)propionate and hetar. bromides with $Pd_2(dba)_3$/Xantphos/aq. KOH in toluene s. F. Colobert, R. Ballesteros-Garrido, F.R. Leroux, R. Ballesteros, B. Abarca, Tetrahedron Lett. *2007*, *48* (39), 6896-9.

Tris(dibenzylideneacetone)dipalladium/chiral ferrocenyldi(phosphines)/cesium carbonate ←
Ar. sulfoxides from ar. iodides and β-sulfinylcarboxylic acid esters ArI → ArS(O)R
via palladium-catalyzed asym. S-arylation of sulfenate anions

342.

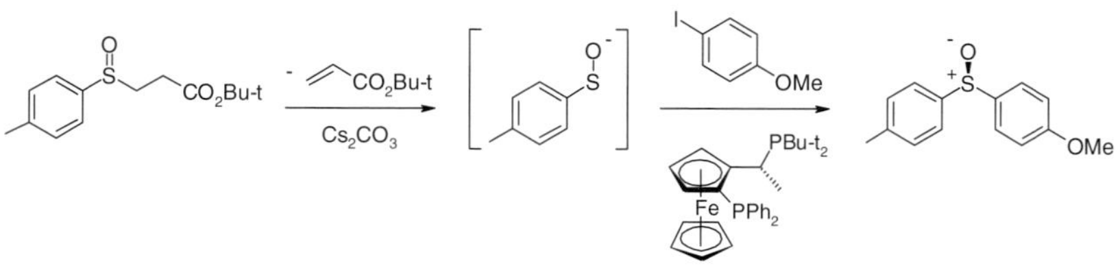

1 Mol% $Pd_2(dba)_3$ in toluene added to 2 mol% chiral ferrocenyldi(phosphine), stirred at room temp. for 5 min, a soln. of 1.2 eqs. startg. aryl iodide in the same solvent added, followed by the startg. sulfinylester in toluene and 4 eqs. Cs_2CO_3, stirred vigorously at reflux for 16 h, cooled to room temp., quenched with satd. aq. NH_4Cl, and worked up with flash chromatographic purification → product. Y 83% (e.e. of (R)-isomer 95%). Enantioselectivity was moderate to good for the reaction of 2-(carbo-*tert*-butoxy)ethyl tolyl, 2-naphthyl and benzyl sulfoxides with aryl or hetaryl iodides, incl. *p*-iodonitrobenzene, but *o*-iodoanisole gave the racemic product. F.e. and chiral [Josiphos-type] ligands s. G. Maitro, S. Vogel, M. Sadaoui, G. Prestat, D. Madec, G. Poli, Org. Lett. *2007*, *9* (26), 5493-6.

Elimination ⇑

Hydrogen ↑ SC ⇑ H

2,3-Dichloro-5,6-dicyanoquinone *DDQ*
Benzothiazoles from ar. thioacylamines ○
cf. *23*, 630s*69*; with 1.1 eqs. DDQ in methylene chloride s. D.S. Bose, M. Idrees, Tetrahedron Lett. *2007*, *48* (4), 669-72.

Oxygen ↑ SC ⇑ O

Dicyclohexylcarbodiimide *DCC*
2-Amino-Δ^2-thiazolines from 2-hydroxythioureas ○
with HCl cf. *25*, 458s*69*; polymer-based synthesis with DCC (and cleavage of the ArgoGel-MB-CHO resin with 95% TFA/water) s. H.S. Jeon, J.N. Kim, T.H. Kim, J. Comb. Chem. *2006*, *8* (6), 799-801.

Acetic acid/microwaves *AcOH/*[\\\\]
2-Acylamino-1,3,4-thiadiazoles from 1,4-diacylthiosemicarbazides
s. *12*, 690s*60*; s.a. T.-B. Wei, H. Liu, J.-H. Hu, M.-L. Li, W.-X. Xu, L.-Z. Yang, Y.-M. Zhang, Indian J. Chem. *2006*, *45B* (12), 2754-6.

Trifluoromethanesulfonic anhydride/potassium carbonate/ethanolamine ←
Benzo[*b*]thiophenes from β-styryl sulfoxides by Pummerer-type ring closure

343.

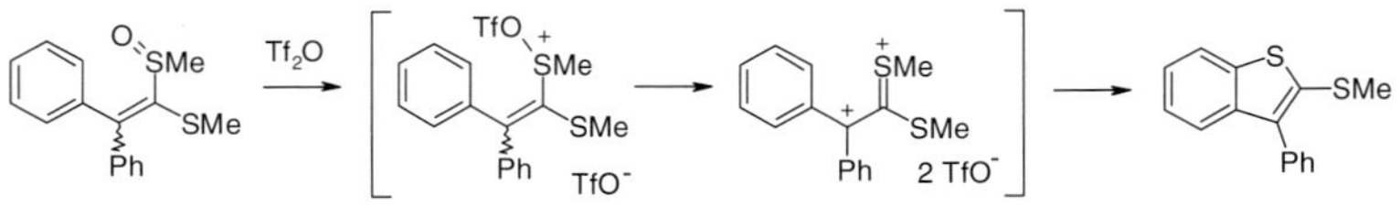

2-Alkylthiobenzo[*b*]thiophenes. Methyl 1-methylsulfinyl-2,2-diphenylethenyl sulfide (0.19 mmol), K_2CO_3 (0.66 mmol) and toluene treated under argon with triflic anhydride (0.27 mmol), stirred at 25° for 3 h, ethanolamine (1 mmol) added, stirring continued at 25° for 3 h, quenched with satd. aq. $NaHCO_3$, and worked up with chromatographic purification on silica gel → 2-methylthio-3-phenylbenzo[*b*]thiophene. Y 86%. Reaction is thought to involve ring closure of an intermediate dication which is stabilized by the methylthio group. F.e. incl. 2-phenyl derivs. s. S. Yoshida, H. Yorimitsu, K. Oshima, Org. Lett. *2007, 9* (26), 5573-6.

*Dichlorobis(*p*-cymene)ruthenium(II)* *$RuCl_2$(*p*-cymene)$_2$*
2-(Arylthiomethyl)furans from α-allenyl-α-(arylthio)carboxylic acid esters via ruthenium(II)-catalyzed 1,4-arylthio group migration

344.

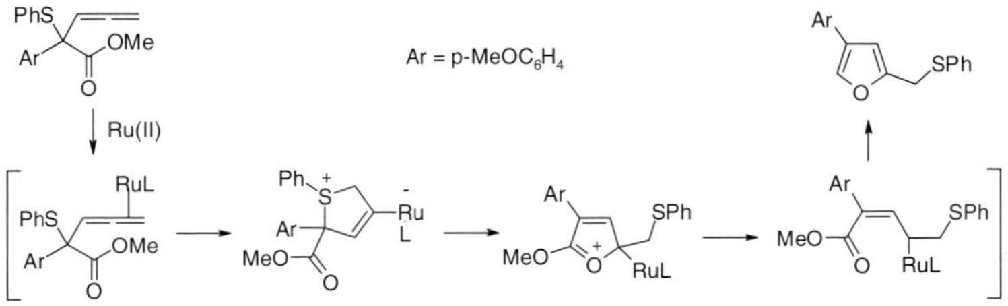

Startg. allenyl thioether and 5 mol% [$RuCl_2$(*p*-cymene)]$_2$ mixed in dry toluene, the soln. heated at 80° for 1 h (TLC monitoring), cooled, the solvent evaporated *in vacuo*, and the residue purified by chromatography on alumina → 4-(*p*-methoxyphenyl)-2-(phenylthiomethyl)furan. Y 98%. The reaction is successful in the presence of keto, ether, ester and halo functionality affording 60-95% yields of the corresponding 4,5-disubst. 2-(phenylthiomethyl)furans. The method was further developed for rapid generation and cyclization of the allenyl thioethers *in situ* via ruthenium(II)-catalyzed reaction of the corresponding α-diazoesters with (phenylthiomethyl)acetylene using $RuCl_2$(*p*-cymene)$_2$ for both steps (6 examples; 45-67% yields) or $Rh_2(OAc)_4$ for thioether generation (9 examples; 50-90%). Crossover experiments for the rearrangement step showed the thioether migration to be intramolecular via a five-membered thiolanium ion. F.e. incl. preparation of the startg. allenyl thioethers s. L. Peng, X. Zhang, M. Ma, J. Wang, J. Am. Chem. Soc. *2007*, *46* (11), 1905-8.

Halogen ↑ SC ⇑ Hal

*Tris(dibenzylideneacetone)dipalladium/2-(di-*tert-*butylphosphino)biphenyl/ cesium carbonate/sodium* tert-*butoxide* ←

Benzothiazoles from N-(*o*-bromoaryl)thioacylamines ○

cf. *13*, 631s*66*; aminobenzothiazoles with concomitant replacement of ar. bromine by amino with added NaOBu-*t* (2 eqs.) in toluene/dioxane s. M.D. Vera, J.C. Pelletier, J. Comb. Chem. *2007*, *9* (4), 569-70.

Carbon ↑ SC ⇑ C

Lithium hydroxide *LiOH*

Thiiranes from 2-acoxythiocyanates with inversion of configuration

345.

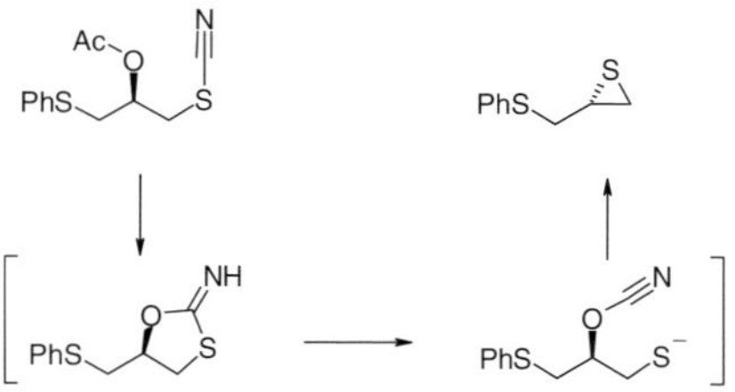

in a 2-phase medium. 4 Eqs. $LiOH{\cdot}H_2O$ in water added to the startg. 2-acoxythiocyanate (e.e. 90%) in THF, the mixture stirred at room temp. for 2 h (TLC monitoring), the organic layer separated, evaporated *in vacuo*, and the residue purified by chromatography → product. Y 79% (e.e. 86%). Six examples bearing pendant aryloxy or arylthio groups gave 67-87% yields of chiral thiiranes *without loss of optical activity*. The key features of the method are use of the *small* lithium cation in place of, say, potassium, adoption of a 2-phase aq. LiOH/THF system to limit the contact time of the thiirane product, and use of chiral acetates as substrates. Significantly, attempted cyclization of the corresponding chiral alcohols afforded chiral polymers. The configuration of the product is assumed based on the proposed mechanism. F.e., also prepn. of the startg. m. via treatment of the corresponding oxiranes with NH_4SCN (cf. *59*, 203), acetylation of the resulting alcohols and subsequent lipase-catalyzed resolution, s. E. Lukowska, J. Plenkiewicz, Tetrahedron: Asym. *2007*, *18* (4), 493-9.

Zinc chloride *$ZnCl_2$*

Isothiocyanates from nitrile oxides via 1,4,2-oxathiazoles

Polymer-based conversion s. *72*, 332

Formation of Rem-Rem Bond

Exchange ↓↑

Nitrogen ↑ RemRem ↓↑ N

Pyridine C_5H_5N

Monoselenophosphoric acid O,O-diesters ←

from phosphorous acid diesters [H-phosphonates] – Dinucleoside monoselenophosphates s. *72*, 346

Halogen ↑ RemRem ↓↑ Hal

Samarium/samarium diiodide Sm/SmI_2

Disilanes from chlorosilanes 2 ⩾Si-Cl → ⩾Si-Si⩽

with Li cf. *8*, 695s*52*; with Sm/SmI_2 in refluxing DME s. Z. Li, K. Iida, Y. Tomisaka, A. Yoshimura, T. Hirao, A. Nomoto, A. Ogawa, Organometallics *2007*, *26* (5), 1212-6.

Remaining Elements ↑ RemRem ↓↑ Rem

Pyridine C_5H_5N

Monoselenophosphoric acid O,O-diesters ←
from phosphorous acid diesters [H-phosphonates]
via Se-de(2-cyanoethyl)ation

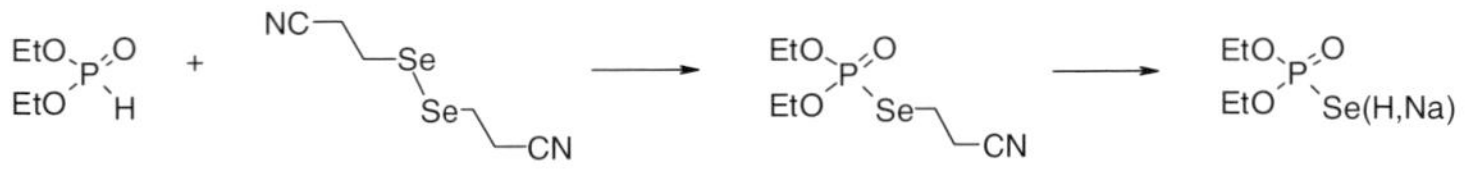

346. 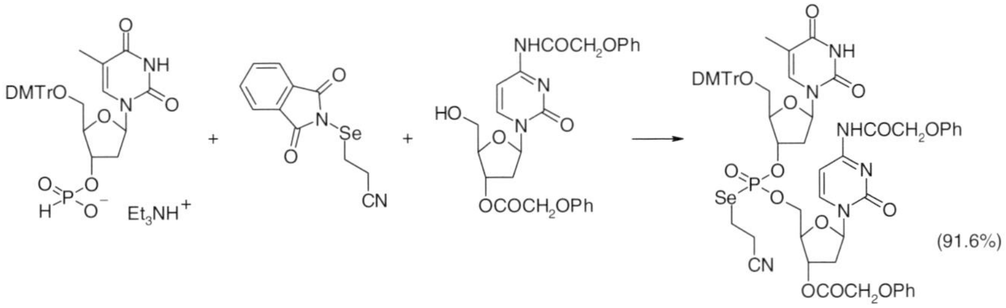

Se-(2-Cyanoethyl) diselenide added to 1.2 eqs. diethyl phosphite in pyridine, stirred overnight at room temp. with exclusion of light, concentrated under reduced pressure, and worked up with flash column chromatographic purification → Se-(2-cyanoethyl) diethyl phosphoroselenoate (Y 76.7%), dried by evaporation with dry toluene, the residue dissolved in methylene chloride, excess of trimethylsilyl chloride added, followed by DBU, after 30 min the mixture concentrated under reduced pressure, and worked up with purification on an Amberlite column (Na^+ form) → diethyl phosphoroselenoate (Y 68%). F.e. and with N-(2-cyanoethylseleno)phthalimide in place of the diselenide, especially for **dinucleoside monoselenophosphates**, s. K. Tram, X. Wang, H. Yan, Org. Lett. *2007*, *9* (24), 5103-6.

Formation of Rem-C Bond

Uptake ⇓

Addition to Oxygen and Carbon RemC ⇓ OC

Copper(I) triflate *CuOTf*
Si-Alkoxy-2-ethylenesilanes from alkoxy-2-ethylenes by metal-catalyzed silylene insertion OR → Si(OR)<

347. 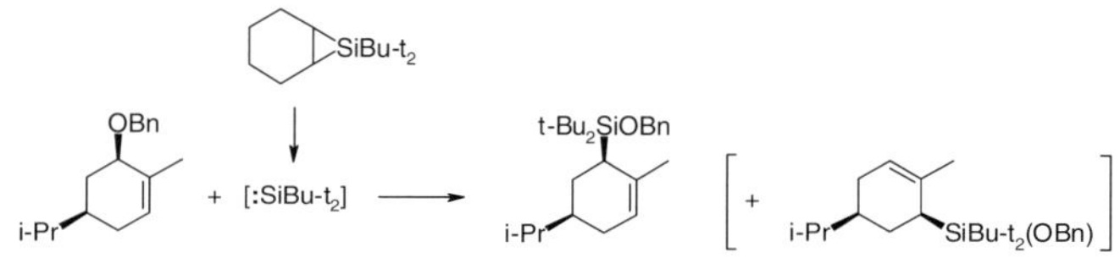

with retention of configuration. 5 Mol% CuOTf·toluene (2:1 complex) added to a Schlenk flask containing the startg. chiral benzyloxy-2-ethylene and 1.5 eqs. startg. silacyclopropane in toluene, stirred under argon for 12 h, concentrated *in vacuo*, and worked up with flash chromatographic purification → product. Y 79%. An allylic rearranged by-product was also formed, presumably as a result of a [2.3]-sila-Wittig reaction, while the retention of configuration is in line with a [1.2]-Stevens rearrangement of an intermediate oxonium ylid. Silylene insertion also takes place in good yield (using CuOTf or AgOTf) with acoxy-2-ethylenes and alkyl allyl ethers, reaction tolerating 1,2-di-, tri- and 1,1-di-substitution of the alkene group (the latter with allyl shift). Substrates with mono-subst. alkene groups, however, underwent silacyclopropanation. F.e.s. L.E. Bourque, P.A. Cleary, K.A. Woerpel, J. Am. Chem. Soc. *2007*, *129* (42), 12602-3.

Indium *In*
2-Hydroxyselenides from epoxides ▽O → C(OH)C(SeR)
and diselenides with Zn/aq. NaOH cf. *59*, 219s*69*; with In in aq. medium, regiospecificity, s. X. Chen, H. Wu, W. Su, R. Xu, M. Liu, J. Ding, J. Chem. Res. *2007* (6), 325-7.

Triethylaluminum *Et_3Al*
1,1-Siloxyphosphines from aldehydes CHO → CH(OSi≤)P<
cf. *68*, 239s*69*; with 10 mol% Et_3Al, also 2-siloxyphosphines from epoxides (with 1 eq. of Et_3Al), s. M. Hayashi, Y. Matsuura, Y. Nishimura, T. Yamasaki, Y. Imai, Y. Watanabe, J. Org. Chem. *2007*, *72* (20), 7798-800; **asym. conversion** with Ti(OPr-*i*)$_4$/diisopropyl L-tartrate/*n*-BuLi (e.e. up to 82%), also oxidation to chiral α-siloxyphosphine oxides, s. Y. Matsuura, T. Yamasaki, Y. Watanabe, M. Hayashi, Tetrahedron:Asym. *2007*, *18* (18), 2129-32.

Chiral aluminum salalen complex *Al(III)-salalen*
α-Hydroxyphosphonic acid esters from aldehydes CHO → CH(OH)PO(OR)$_2$
Aluminum-catalyzed asym. hydrophosphonylation

348. 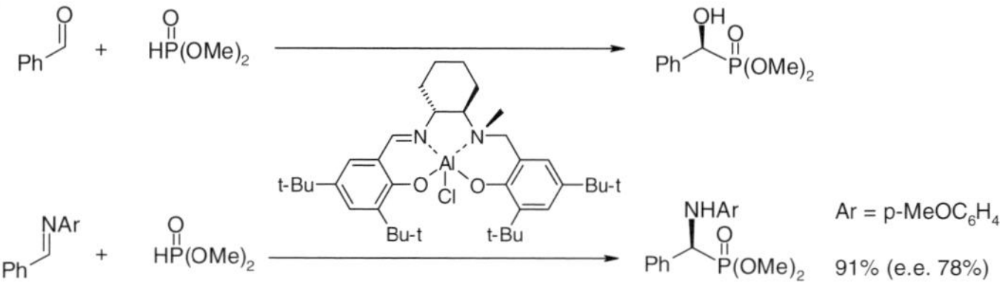

A soln. of 2.05 eqs. dimethyl phosphite and 10 mol% chiral Al-salalen complex in THF stirred under N_2 for 10 min at 0°, cooled to -15°, startg. aldehyde added, stirring continued for 48 h, quenched with 1 *M* HCl, and worked up with chromatographic purification on silica gel → (S)-dimethyl phenyl(hydroxy)methylphosphonate. Y 87% (e.e. 90%). High enantioselectivity was effected with *both aliphatic and ar.* aldehydes, and the catalyst is equally effective for the **asym. hydrophosphonylation of** aromatic, heteroaromatic and *branched* aliphatic **N-arylaldimines**. The complex possesses a unique chiral trigonal bipyramidal configuration, providing an effective asym. reaction site in which nitrogen is bound to the metal ion and the N-methyl group is *cis* to the chloro ligand. F.e. and comparison of N-protective groups on the aldimines s. B. Saito, H. Egami, T. Katsuki, J. Am. Chem. Soc. *2007*, *129* (7), 1978-86.

Titanium tetraisopropoxide/diisopropyl L-*tartrate/*n-*butyllithium* ←
1,1-Siloxyphosphines from aldehydes CHO → CH(OSi≤)P<
Asym. conversion s. *68*, 239s*72*

Addition to Nitrogen and Carbon RemC ⇓ NC

Silver triflate/phenyltrimethylsilane *AgOTf/PhSiMe$_3$*
Syntheses via metal-catalyzed silylene transfer to azomethines ←

349. 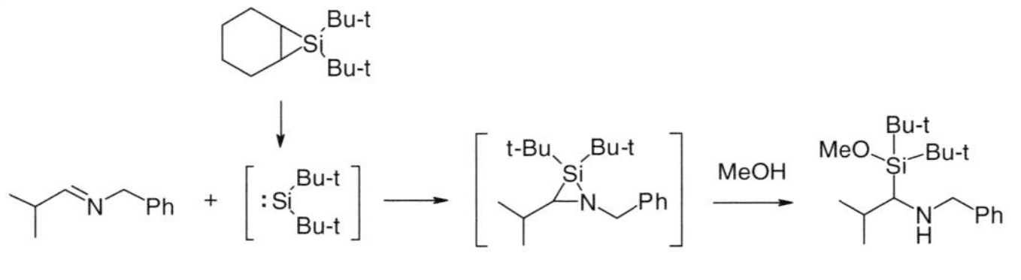

Si-Alkoxy-1,1-aminosilanes via silaaziridines. 1 Mol% Ag-triflate added to a mixture of the startg. imine, 1 eq. of the startg. silacyclopropane (as silylene precursor), and 20 mol% phenyltrimethylsilane in deuteriotoluene, stirred at 23° for 12 h, 42 eqs. methanol added, and stirring continued for 1 h → N,Si-acetal. Y 53%. The intermediate silaaziridine is sensitive to air and water but can be isolated in 84% yield (s. *72*, 458). Reaction under thermal conditions at 100° results in formation of significant amounts of ring-expanded **2,1,3-siladiazolidines** via dimerization of the intermediate silaaziridine. The reaction is successful with N-benzyl- and N-aryl-imines, incl. sterically-hindered ketimines. F.e. and treatment of the intermediate silaaziridines with electrophiles to afford ring-expanded silicon-containing heterocycles s. Z. Nevárez, K.A. Woerpel, Org. Lett. *2007*, *9* (19), 3773-6.

Chiral aluminum salalen complex *Al(III)-salalen*
Asym. hydrophosphonylation of N-arylaldimines CH=NAr → CH(NHAr)PO(OR)$_2$
s. *72*, 348

Tetraethylammonium tetraselenotungstate *[Et$_4$N]$_2$WSe$_4$*
Tetraethylammonium tetraselenotungstate as selenium transfer agent ←
sym. diselenides from halides or alcohols cf. *47*, 563s*65*; sym. N-protected 2,2′-diaminodiselenides from aziridines in CH_2Cl_2 at room temp., regio- and stereo-specificity, s. D. Sureshkumar, T. Gunasundari, V. Saravanan, S. Chandrasekaran, Tetrahedron Lett. *2007*, *48* (4), 623-6.

Addition to Carbon-Carbon Bonds RemC ⇓ CC

Without additional reagents *w.a.r.*
(3E)-2-Arylseleno-6-halogeno-1,3-dienes from cyclopropylallenes

350. 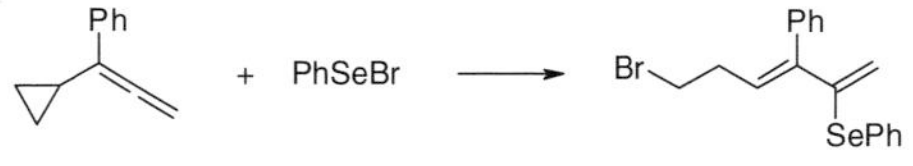

A soln. of phenylselenyl bromide in methylene chloride added slowly to a soln. of 1-phenyl-1-cyclopropylallene in the same solvent at 0° under N$_2$, stirred for a further 1 min, solvent removed *in vacuo*, and the residue worked up with purification by TLC → 2-phenylselenyl-3-phenyl-6-bromo-1,3(E)-hexadiene. Y 78%. Reaction is presumed to involve initial attack of the allene by a phenylselenium cation, followed by quenching of the generated ring-opened carbocation by bromide ion. F.e. and with PhSeSePh/PhI$(OAc)_2$ as reagent s. L. Yu, B. Meng, X. Huang, Synlett *2007* (18), 2919-21.

Irradiation ///

Phosphine selenides C=C → CHC-P(Se)<

from ethylene derivs. and selenophosphinous acids

Regiospecific radical addition

351. 

Equivalent amounts of the startg. sec. phosphine selenide and styrene in dioxane irradiated in a quartz ampoule under argon with a 200 W Hg-arc lamp for 4 h → triphenethylphosphine selenide. Y 95%. Reaction is facile, easy to manage, environmentally friendly, and generally applicable to alkenes, styrenes, enolethers and thioenolethers. F.e.s. N.K. Gusarova, S.F. Malysheva, N.A. Belogorlova, B.G. Sukhov, B.A. Trofimov, Synthesis *2007* (18), 2849-52.

Chiral pyrrolidines s. under 2(S)-α-Siloxypyrrolidine-carboxylic acid ←

(-)-Quinine ←

β-Nitrophosphonic acid esters $C{=}C(NO_2) \rightarrow C(PO(OR)_2)CH(NO_2)$

from 1,1-nitroethylene derivs. – Organocatalyzed asym. Michael addition s. *72*, 356

Copper(I) iodide/cesium carbonate *CuI/Cs_2CO_3*

(Z)-Ene-1,2-di(phosphines) from ynephosphines C≡C → CH=C(P<)

Regio- and stereo-specific hydrophosphination

352. 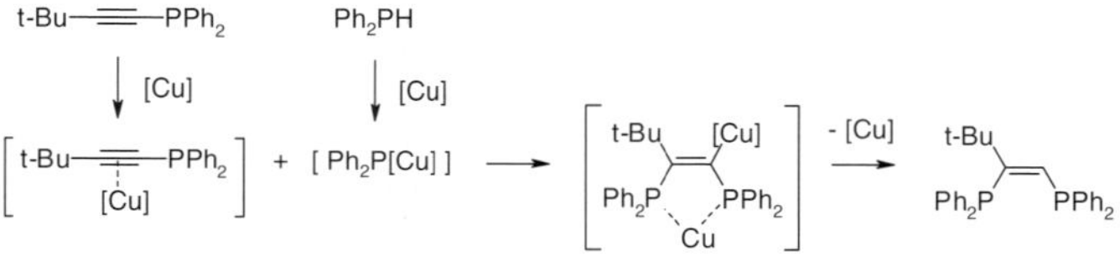

DMF, the startg. ynephosphine and ca. 1 eq. diphenylphosphine added sequentially to a mixture of 2 mol% CuI and 3.2 mol% *Cs_2CO_3* under argon, heated at 90° for 4 h, cooled to room temp., passed through a pad of Florisil, and worked up with purification on silica gel → product. Y 69%. The products, which are obtainable on a gram scale, are sensitive to air, and were more conveniently isolated as the corresponding bis(phosphine sulfides) from which they may be regenerated simply (with tris(trimethylsilyl)silane). Perfect regio- and stereo-selectivity were recorded as a result of *anti*-addition of an intermediate copper(I) phosphide. AgI and CuCl were also effective catalysts, but other copper salts as well as gold, nickel, palladium and platinum salts were not. The choice of a polar solvent and the base is also critical. F.e. and in aq. medium s. A. Kondoh, H. Yorimitsu, K. Oshima, J. Am. Chem. Soc. *2007*, *129* (13), 4099-104.

Copper(I) iodide/ethylenediamine *CuI/$(CH_2NH_2)_2$*

(E)-α,β-Ethylenephosphine oxides from acetylene derivs. C≡C → CH=C(P(O)<)

Regio- and stereo-selective copper(I)-catalyzed hydrophosphinylation

353. 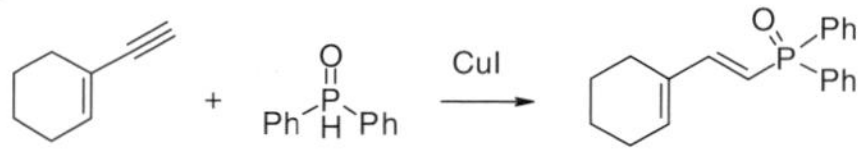

Diphenylphosphine oxide, 1.5 eqs. startg. alkyne, 10 mol% CuI and 15 mol% ethylenediamine stirred at 60° in *dimethyl sulfoxide* under N_2 for 10 h, the soln. cooled to room temp., chloroform and brine added, the organic layer washed with brine, dried over Na_2SO_4, concentrated *in vacuo*, and the residue purified by chromatography → (E)-vinylphosphine oxide deriv. Y 70% (>99%

regioselectivity). Both terminal and internal alkynes reacted with dibenzylphosphine oxide, diphenylphosphine oxide or ethyl hydrogen (phenyl)phosphonate to give predominantly anti-Markovnikov adducts with (E)-configuration, leaving hydroxyl and conjugated ethylene groups unaffected (11 examples; Y 68-99%). Reactions with terminal alkynes were generally complete within 18 h at 60° using commercially available and inexpensive CuI, but internal alkynes were less reactive, requiring heating at 100° to effect completion. F.e. and optimization s. M. Niu, H. Fu, Y. Jiang, Y. Zhao, Chem. Commun. *2007* (3), 272-4.

Zinc/ruthenium trichloride *Zn/RuCl₃*

β-(Organoseleno)- from α,β-ethylene-carbonyl compds. C=C → C(SeR)CH

from diaryl diselenides with In/Me_3SiCl cf. *55*, 229s*69*; with Zn/$RuCl_3$ in acetonitrile/water, also from dibenzyl diselenides, s. B. Movassagh, A. Tatar, Synlett *2007* (12), 1954-6.

Sodium tetrahydridoborate/1-butyl-3-methylimidazolium fluoroborate *$NaBH_4$/[bmim]BF_4*

Eneselenides from acetylene derivs. C≡C → CH=C(SeR)

and diphenyl diselenide with *i*-Bu_2AlH cf. *48*, 549s*52*; with $NaBH_4$ (cf. *50*, 343s*58,70*) in [bmim]BF_4 s. E.J. Lenardão, L.G. Dutra, M.T. Saraiva, R.G. Jacob, G. Perin, Tetrahedron Lett. *2007*, *48* (45), 8011-3.

Bis(pentafluorophenyl)borane-dimethyl sulfide *$(C_6F_5)_2BH$-SMe_2*

(E)-α,β-Ethyleneboronic acid esters from acetylene derivs. C≡C → CH=C-$B(OR)_2$

with diisopinocampheylborane cf. *43*, 516s*49*; from 1-alkynes with pinacolborane in the presence of $(C_6F_5)_2BH$-SMe_2 [obtained from $(C_6F_5)_3B$ and BH_3-SMe_2 in hexane] s. M. Hoshi, K. Shirakawa, M. Okimoto, Tetrahedron Lett. *2007*, *48* (48), 8475-8.

Triethylaluminum *Et_3Al*

(Z)-α,β-Ethylene-β-phosphino-α-silylcarboxylic from α,β-acetylenecarboxylic acid esters C≡C → C(Si≤)=C(P<)

Regio- and stereo-specific silylphosphination

354.

1.2 Eqs. startg. silylphosphine added at room temp. to a stirred mixture of the startg. alkyne and 1 eq. triethylaluminum (1 *M* in *n*-hexane) in methylene chloride, stirring continued for 2 h, the solvent removed *in vacuo*, and the residue worked up with column chromatographic purification on silica gel → product. Y 78%. The procedure is applicable to alkyl-, alkenyl-, aryl- and hetaryl-subst. propiolic acid esters, but yields were low with less hindered silylphosphines. Other Lewis acids, except Et_2AlCl, were ineffective. F.e.s. M. Hayashi, Y. Matsuura, K. Kurihara, D. Maeda, Y. Nishimura, E. Morita, M. Okasaki, Chem. Lett. *2007*, *36* (5), 634-5.

Ammonium cerium(IV) nitrate *CAN*

Selenides from ethylene derivs. and selenols C=C → C(SeR)CH

with NaOMe cf. *18*, 704; addition of benzeneselenol to enones with a little CAN in the absence of solvent, also addition of mercaptans (cf. *47*, 488s*68*), s. C.-M. Chu, S. Gao, M.N.V. Sastry, C.-W. Kuo, C. Lu, J.-T. Liu, C.-F. Yao, Tetrahedron *2007*, *63* (8), 1863-71.

2(S)-[Bis[3,5-bis(trifluoromethyl)phenyl](trimethylsiloxy)methyl]pyrrolidine/ p-nitrobenzoic acid or o-fluorobenzoic acid ←

Organocatalyzed asym. hydrophosphination of α,β-ethylenealdehydes C=C → CHC-P<

355.

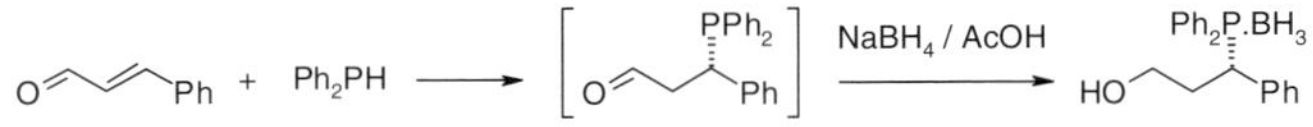

Chiral 3-hydroxyphosphine-borane complexes. The startg. enal added to a soln. of 10 mol% each of 2(S)-[bis[3,5-bis(trifluoromethyl)phenyl](trimethylsiloxy)methyl]pyrrolidine and *p*-nitrobenzoic acid in dry ether, the reaction vial capped, stirred at room temp. for 5 min, cooled to -10°, after 10 min 1 eq. of diphenylphosphine added, stirring continued for 24 h, cooled to -40°, diluted with dry THF, 3.5 eqs. solid $NaBH_4$ added in one portion, followed after 20 min by a soln.

of 3 eqs. glacial acetic acid in THF, stirred for a further 30 min, quenched with a few drops of water, and worked up with flash chromatographic purification → (R)-product. Y 72% (e.e. 94%). Isolation of the product as the borane complex facilitates the purification process as the adduct is stable for a long time. This is the first example of such an organocatalyzed asym. transformation, which is applicable to a range of aryl-, hetaryl-, alkyl- and alkenyl-subst. enals. The formation of (R)-products is consistent with approach of the phosphine from the *re* face of the intermediate iminium ion. F.e. and comparison of chiral reagents and acids s. A. Carlone, G. Bartoli, M. Bosco, L. Sambri, P. Melchiorre, Angew. Chem. Int. Ed. Engl. *2007*, *46* (24), 4504-6; with *o*-fluorobenzoic acid as additive cf. I. Ibrahem, R. Rios, J. Vesely, P. Hammar, L. Eriksson, F. Himo, A. Córdova, ibid. 4507-10.

1-Butyl-3-methylimidazolium fluoroborate s. under $NaBH_4$ *[bmim]*BF_4

Phenyl iodosoacetate *PhI(OAc)*$_2$

(3E)-2-Arylseleno-6-halogeno-1,3-dienes from cyclopropylallenes C̨
s. *72*, 350

Axially-chiral cyclic guanidines ←

β-Nitrophosphonic acid esters $C{=}C(NO_2) \rightarrow C(PO(OR)_2)CH(NO_2)$
from 1,1-nitroethylene derivs.
Organocatalyzed asym. Michael addition

356.

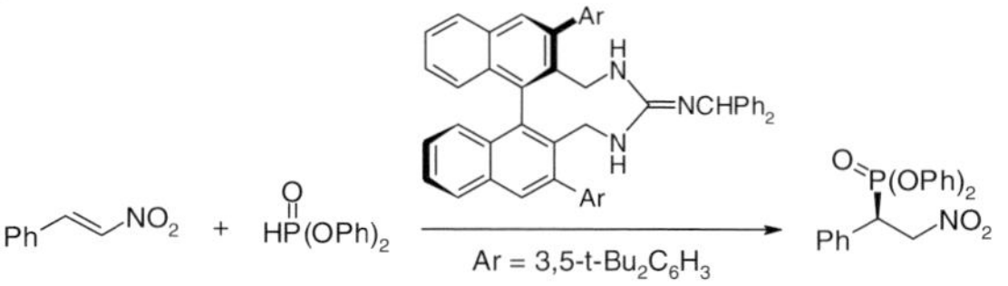

1.1 Eqs. diphenyl phosphite added at room temp. to a mixture of *trans*-β-nitrostyrene and 4 Å molecular sieves *in* tert-*butyl methyl ether*, stirred for 10 min at -40°, *1 mol%* chiral cyclic guanidine added, stirring continued for 2 h, quenched with satd. aq. NH_4Cl, filtered, and worked up with flash column chromatographic purification → (R)-diphenyl 2-nitro-1-phenylethyl-phosphonate. Y 94% (e.e. 92%). The procedure, based on activation of the P-nucleophile by the chiral reagent, is applicable *at low catalyst loading* to a range of aryl-, hetaryl- [with 5 mol% catalyst] and alkyl-subst. nitroolefins in high yield and enantioselectivity (80-94%). F.e. and optimization of the catalyst by varying the aryl- and N-substitution s. M. Terada, T. Ikehara, H. Ube, J. Am. Chem. Soc. *2007*, *129* (46), 14112-3; with (-)-quinine in xylene, also conversion of the adducts to chiral β-aminophosphonic acid esters, s. J. Wang, L.D. Heikkinen, H. Li, L. Zu, W. Jiang, H. Xie, W. Wang, Adv. Synth. Catal. *2007*, *349* (7), 1052-6.

Phosphines s. under Ni(cod)$_2$ *and Pd(OAc)*$_2$ ←

Bis(cyclooctadiene)nickel(0)/tricyclohexylphosphine/cesium carbonate ←

β-Boronyl- from α,β-ethylene-carboxylic acid derivs. $C{=}C \rightarrow CHC\text{-}B(OR)_2$

357.

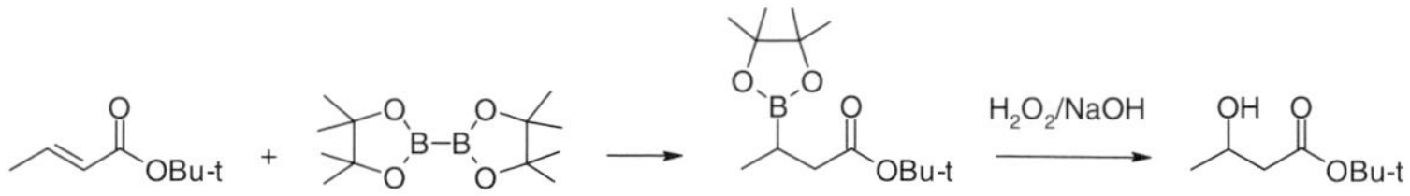

β-Boronylcarboxylic acid esters. Toluene and a soln. of 10 mol% tricyclohexylphosphine in the same degassed solvent (0.5 *M*) added dropwise to a mixture of 1 mol% Ni(cod)$_2$ and 1.5 eqs. Cs_2CO_3 under argon, the suspension stirred for 10 min at 0°, a soln. of *tert*-butyl (E)-crotonate and 1.5 eqs. bis(pinacolato)diborane in toluene added, treated with methanol and 1.5 eqs. water, allowed to warm to room temp., stirring continued for 11 h, poured into water, and worked up with column chromatographic purification on silica gel → *tert*-butyl 3-(4,4,5,5-tetramethyl-1,3,2-dioxaborolan-2-yl)butanoate. Y 77%. The products were readily converted to the corresponding **β-hydroxycarboxylic acid derivs.** (with H_2O_2/aq. NaOH in ethanolic THF). F.e. **and β-boronylcarboxylic acid amides** s. K. Hirano, H. Yorimitsu, K. Oshima, Org. Lett. *2007*, *9* (24), 5031-3.

Nickel(II) acetoacetonate *Ni(acac)$_2$*
Eneselenides from acetylene derivs. C≡C → CH=CH(SeR)
and selenols in the presence of $(PhSe)_2$ under irradiation cf. *46*, 530s*47*; with a self-organized nanosized catalyst generated from $Ni(acac)_2$ s. V.P. Ananikov, N.V. Orlov, I.P. Beletskaya, Organometallics *2007*, *26* (3), 740-50.

Tris(acetonitrile)(η⁵-pentamethylcyclopentadienyl)ruthenium hexafluorophosphate *[Cp*Ru(MeCN)$_3$]PF$_6$*
Ruthenium-catalyzed hydrosilylation of acetylene derivs. CH=C(Si≤)
1,1-disubst. enesilanes cf. *62*, 230; **silacyclopentadienes** [siloles] from 1,3-diynes, 2,5-diaryl-derivs., especially spiro-compds. with 9-silafluorene, s. T. Matsuda, S. Kadowaki, M. Murakami, Chem. Commun. *2007* (25), 2627-9.

Ruthenium trichloride s. under Zn *RuCl$_3$*

Rhodium(I) complex *[Rh(I)]*
Regiospecific hydrophosphination with silylphosphines C=C → CHC(P<)
of ethylene derivs. with Bu_4NF cf. *68*, 239; of cyclic enones under Rh(I)-catalysis, also oxidation to cyclic β-phosphinylketones, s. V.T. Trepohl, M. Oestreich, Chem. Commun. *2007* (31), 3300-2.

Palladium(II) acetate/1,2-bis(diphenylphosphino)ethane *Pd(OAc)$_2$/dppe*
α-Methylenephosphinic acid esters from terminal acetylene derivs. C≡CH → C(PO(R)(OR′))=CH$_2$

358. 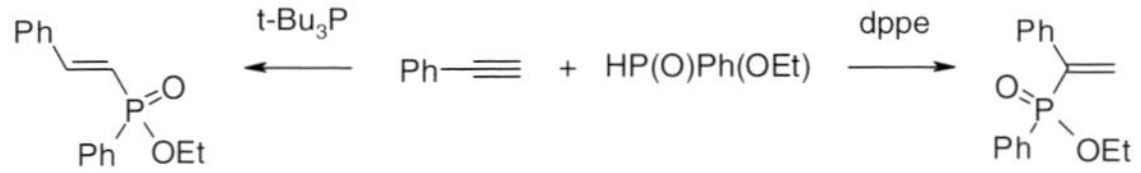

Markovnikov addition. 1.33 Mmol ethyl phenylphosphinate and 1.39 mmol phenylacetylene added under N_2 to a mixture of 0.066 mmol $Pd(OAc)_2$ and 1.5 eqs. [relative to Pd] dppe in dry toluene, heated at 100° for 3 h, evaporated, and worked up with column chromatographic purification on silica gel → product. Y 78.5% (99% Markovnikov adduct). No acidic additives were required. There is a significant **ligand effect on regioselectivity,** the bulky, electron-rich tri-*tert*-butylphosphine affording predominantly the anti-Markovnikov adduct (**(E)-α,β-ethylenephosphinic acid esters**). Bulky substrates and internal acetylene derivs., however, were unreactive, as also was propargyl methyl ether. Cyano and hydroxyl groups were unaffected. F.e. and with terminal aliphatic alkynes s. S.K. Nune, M. Tanaka, Chem. Commun. *2007* (27), 2858-60; **α-methylenephosphine oxides** from diphenylphosphine oxide in propionitrile (cf. *61*, 225s*62*), also ene-1,2-di(phosphine oxides) with a triarylphosphine, such as phenyl(di-*o*-tolyl)phosphine, in place of dppe, s. N. Dobashi, K. Fuse, T. Hoshino, J. Kanada, T. Kashiwabara, C. Kobata, S.K. Nune, M. Tanaka, Tetrahedron Lett. *2007*, *48* (27), 4669-73.

Tris(dibenzylideneacetone)dipalladium/9,9-dimethyl-4,5-bis(diphenylphosphino)xanthene ←
One-pot palladium-catalyzed synthesis of phosphonic acids from hypophosphorous acid ←

359. 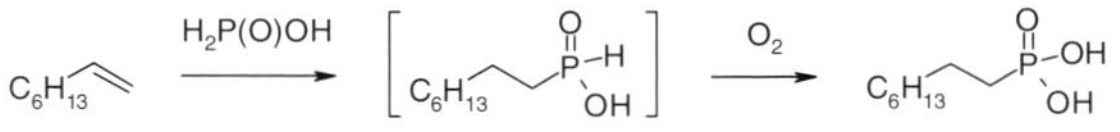

and ethylene derivs. Freshly concentrated H_3PO_2 (prepared by rotary evaporation of the commercial 50% aq. soln.) in *undried, reagent-grade* DMF heated with 1-octene in the presence of 1 mol% $Pd_2(dba)_3$ and 2 mol% Xantphos *under air* at 110° for 20-24 h → product. Y 100%. The intermediate H-phosphinic acid was isolated as the main product in dry DMF under N_2 or by lowering the catalyst loading (to 0.05 mol%). The procedure is efficient, atom-economical and environmentally friendly. Acetylene derivs., protected enamines and allylamines, allyl alcohols, and ar. iodides and bromides may also be used in place of ethylene derivs., preferably by a step-wise procedure whereby the H-phosphinic acid is preformed, followed by *in situ* Pd-catalyzed oxidation under air. F.e. and with the 50% aq. soln. of H_3PO_2 in DMF s. K. Bravo-Altamirano, J.-L. Montchamp, Tetrahedron Lett. *2007*, *48* (33), 5755-9.

Tetrakis(triphenylphosphine)palladium(0) *$Pd(PPh_3)_4$*
Enestannanes from acetylene derivs. C≡C → CH=C(Sn≤)
(E)-α,β-ethylene-α-stannylphosphonic acid esters cf. *43*, 520s*66*; (E)-α,β-ethylene-α-stannyl- from α,β-acetylene-sulfones (and Bu_3SnH) in benzene at room temp. s. M. Cai, G. Chen, W. Hao, D. Wang, Synlett *2007* (20), 3492-4.

Tetrakis(triphenylphosphine)palladium(0)/1-butyl-3-methyl-imidazolium hexafluorophosphate *$Pd(PPh_3)_4$/[bmim]PF_6*
(Z)-1,2-Bis(arylseleno)ethylenes from terminal acetylene derivs. C(SeAr)=C(SeAr)
and diaryl diselenides in an ionic liquid solvent s. *72*, 319

Exchange ⇃↾

Hydrogen ↾ RemC ⇃↾ H

1,5,7-Triazabicyclo[4.4.0]dec-5-ene/deuteriochloroform ←
Deuteriation H → D
of ketones with Et_3N/D_2O cf. *17*, 705s*56*; α,α-dideuteriation with 10 mol% 1,5,7-triazabicyclo-[4.4.0]dec-5-ene in deuteriochloroform (deuterium incorporation 77-97%), and selectivity for ketones over esters, s. C. Sabot, K.A. Kumar, C. Antheaume, C. Mioskowski, J. Org. Chem. *2007*, *72* (13), 5001-4.

Palladium(II) acetate/triarylphosphines *$Pd(OAc)_2/Ar_3P$*
Ene-1,2-di(phosphine oxides) from acetylene derivs. C≡C → C(P(O)<)=C(P(O)<)
s. *72*, 358

*(1,5-Cyclooctadiene)(methoxy)iridium(I) dimer/4,4′-di-*tert*-butyl-2,2′-bipyridyl/bis(pinacolato)diboron/sodium periodate or potassium hydrogen difluoride* ←
Arylboronic acids or potassium aryl(trifluoro)borates from arenes H → $B(OH)_2$ or BF_3K
via arylboronic acid esters s. *64*, 219s*72*

Oxygen ↾ RemC ⇃↾ O

Microwaves s. under SiO_2 [\\\\]

Organolithium compds. or Magnesium *RLi or Mg*
Phosphine oxides from phosphinic acid esters with retention of configuration P-OR → P-R′
cf. *49*, 508; sec. alkyl- and aryl-(phenyl)phosphine oxides from (R_P)-(-)-menthyl hydrogeno-(phenyl)phosphinate and organolithium or Grignard compds. (e.e. up to 95%) s. A. Leyris, J. Bigeault, D. Nuel, L. Giordano, G. Buono, Tetrahedron Lett. *2007*, *48* (30), 5247-50.

Lithium organochalcogenides *LiSeR or LiTeR*
Vinyl chalcogenides from enol phosphates C=C-$OPO(OR)_2$ → C=C(Se,TeR′)
thioenolethers with LiSR cf. *37*, 519 and (Z)-enetellurides with RLi/Te via Li-organotellurides cf. *37*, 519s*58*; also (Z)-eneselenides, and from enol tosylates, (E)-α-aryl-thio- or -seleno-β-aryl-thio- or -seleno- or -organotelluro-α,β-ethylenecarbonyl compds., s. C.C. Silveira, R.B. Guerra, J.V. Comasseto, Tetrahedron Lett. *2007*, *48* (27), 5121-4.

β-*Cyclodextrin* *cyclodextrin*
3-Component synthesis of α-aminophosphonic acid esters CO → C(N<)$PO(OR)_2$
Kabachnik-Fields reaction s. *33*, 593s*72*

Oxalyl chloride/lithium tetrahydridoaluminate *$(COCl)_2/LiAlH_4$*
Carboxylic acid telluroamides from amides C(O)N< → C(Te)N<
with $(Me_2Al)_2Te$ cf. *45*, 347s*53*; via chloroimidium chlorides generated *in situ* with oxalyl chloride, followed by treatment with $LiAlH_4$/Te, also from 1-(alkylseleno)azomethinium triflates, s. Y. Mutoh, T. Murai, S. Yamago, J. Organomet. Chem. *2007*, *692* (1-3), 129-35.

Silica/microwaves *SiO_2/[\\\\]*
3-Component synthesis of α-aminophosphonic acid esters CO → C(N<)PO$(OR)_2$
s. *33*, 593s*72*

Stannyllithium compds./methanesulfonyl chloride/triethylamine *⩾SnLi/$MeSO_2Cl/Et_3N$*
Enestannanes C=C(Sn⩽)
from enol triflates with higher-order lithium stannylcuprates cf. *44*, 521; **from ketones** with Bu_3SnLi followed by addition of MsCl/Et_3N to the generated alkoxide, especially for cyclic enestannanes, s. A. Darwish, J.M. Chong, J. Org. Chem. *2007*, *72* (4), 1507-9.

Bismuth(III) nitrate *$Bi(NO_3)_3$*
3-Component synthesis of α-aminophosphonic acid esters CO → C(N<)PO$(OR)_2$
in the absence of solvent s. *33*, 593s*72*

Methanesulfonyl chloride s. under Stannyllithium compds. *$MeSO_2Cl$*

Bismuth(III) nitrate *$Bi(NO_3)_3$*
Magnesium perchlorate *$Mg(ClO_4)_2$*
3-Component synthesis of α-aminophosphonic acid esters CO → C(N<)PO$(OR)_2$
Kabachnik-Fields reaction with aldehydes using a little $Mg(ClO_4)_2$ in the absence of solvent cf. *33*, 593s*71*; also with ketones, and with trialkyl phosphites (cf. *59*, 234s*67*), s. S. Bhagat, A.K. Chakraborti, J. Org. Chem. *2007*, *72* (4), 1263-70; from aldehydes with β-cyclodextrin in water at reflux s. B. Kaboudin, M. Sorbiun, Tetrahedron Lett. *2007*, *48* (51), 9015-7; with 10 mol% $Bi(NO_3)_3 \cdot 5H_2O$ in the absence of solvent s. A.K. Bhattacharya, T. Kaur, Synlett *2007* (5), 475-8; from fluorinated ar. aldehydes under microwave irradiation using silica as support in the absence of solvent s. X. Lv, J. Zhang, C. Xing, W. Du, S. Zhu, Synth. Commun. *2007*, *37* (5), 743-57.

Dichloro[1,3-bis(diphenylphosphino)propane]palladium(II)/potassium carbonate ←
Enephosphine-borane complexes from enol triflates OTf → $PR_2 \cdot BH_3$
and sec. phosphine-borane complexes with $Pd(PPh_3)_4$ cf. *45*, 350s*56*; with $PdCl_2$(dppe), incl. chiral compds., s. D. Julienne, J.F. Lohier, O. Delacroix, A.C. Gaumont, J. Org. Chem. *2007*, *72* (6), 2247-50.

Nitrogen ↑ RemC ↓↑ N

Chiral 2-[(diaryl)(trimethylsiloxy)methyl]pyrrolidine ←
*Chiral 2-[(diaryl)(trimethylsiloxy)methyl]pyrrolidine/*p*-nitrobenzoic acid* ←
Organocatalyzed α-selenylation of aldehydes H → SeR
with L-prolinamide cf. *68*, 245; **asym. α-selenylation** with a chiral 2-[(diaryl)(trimethylsiloxy)-methyl]pyrrolidine (e.e. up to >99%) s. H. Sundén, R. Rios, A. Córdova, Tetrahedron Lett. *2007*, *48* (44), 7865-9; with added *p*-nitrobenzoic acid s. M. Tiecco, A. Carlone, S. Sternativo, F. Marini, G. Bartoli, P. Melchiorre, Angew. Chem. Int. Ed. Engl. *2007*, *46* (36), 6882-5.

Hexamethylselenophosphoramide *$(Me_2N)_3PSe$*
2-(Arylseleno)lactones from ethylenecarboxylic acids ○
Stereoselective Lewis base-catalyzed conversion

360.
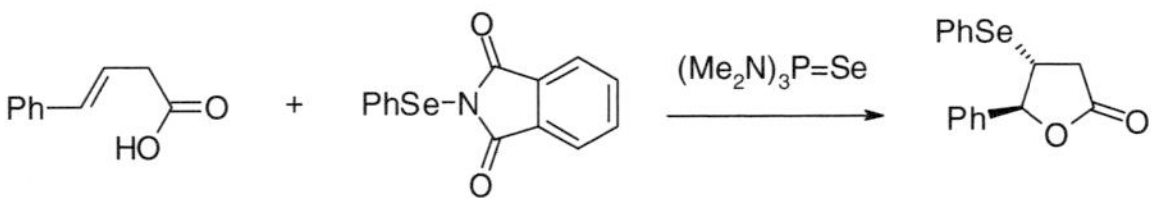

***trans*-β-(Arylseleno)-γ-lactones.** 0.1 Eq. hexamethylselenophosphoramide added to (E)-4-phenyl-3-butenoic acid in methylene chloride, 1 eq. N-(phenylseleno)phthalimide added over 5 min, stirred for 3 h, quenched with satd. aq. $Na_2S_2O_3$, stirred for a further 1 h, diluted with water, and extracted with methylene chloride → *rel*-(4R,5S)-dihydro-4-(phenylseleno)-5-phenylfuran-2(3*H*)-one. Y 95%. The paper explores the use of several selenylating agents with a view to reducing the usual reactivity of such compds. (e.g. benzeneselenyl chloride, cf. *33*, 596). The reduced reactivity coupled with activation of these agents by a suitable Lewis base containing soft donor atoms (S, P, Se) was expected to allow 'fine tuning' of the reaction and increased stereoselectivity. N-(Phenylseleno)phthalimide and the analogous N-(phenylseleno)succinimide were the most useful reagents. Successful Lewis bases included hexamethylphosphoramide and its deoxy, thio and illustrated seleno analogs, triphenyl- and tris(cyclohexyl)-phosphine sulfides, and N,N′-dimethyltrimethylenethiourea. F.e.s. S.E. Denmark, W.R. Collins, Org. Lett. *2007*, *9* (19), 3801-4.

Halogen ↑ RemC ↓↑ Hal

Without additional reagents *w.a.r.*

α-(Phosphorylamino)phosphonic acid esters CO → $C(NHPO(OR)_2)PO(OR)_2$
from oxo compds., phosphoromonoamidates and phosphoromonochloridites
3-Component Mannich-type reaction

361.

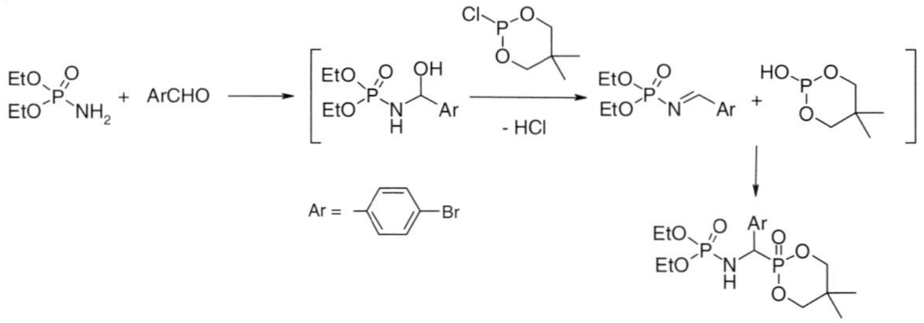

under solvent- and catalyst-free conditions. An equimolar mixture of 4-bromobenzaldehyde, diethyl phosphoramidate and 2-chloro-5,5-dimethyl-1,3,2-dioxaphosphorinane stirred at 50-60° for 30 s → 2-[(N-diethoxyphosphorylamino)(4-bromophenyl)methyl]-5,5-dimethyl-1,3,2λ^5-dioxaphosphorinane-2-oxide. Y 83%. The method is convenient, rapid, clean and environmentally benign, affording the products in 70-94% yield in less than 5 min for aromatic aldehydes only (12 examples) without any side-reactions. A single ketone example (acetophenone) gave a 77% yield in 90 min, while aliphatic and conjugated aldehydes gave little or no reaction under these conditions, as did the use of solvents (THF, benzene or acetyl chloride). The mechanism is assumed to involve dehydration of an initial addition product generating a phosphite and HCl, with the latter possibly activating the phosphite to further reaction. F.e.s. J. Zhang, Z. Cui, F. Wang, Y. Wang, Z. Miao, R. Chen, Green Chem. *2007*, *9* (12), 1341-5.

Vinyllithium compds. ←

Phosphines from chlorophosphines P-Hal → P-R

1-acetylenephosphines with Mg cf. *7*, 670s*10*; 1,2-difluoro-2-[dimethyl(phenyl)silyl]vinylphosphines from 1,2-difluoro-2-[dimethyl(phenyl)silyl]vinyllithium, also oxidation to the corresponding phosphine oxides (as *trans*-isomers) with H_2O_2 or *t*-BuOOH, s. H.-J. Tsai, C.-W. Hsieh, S.-C. Wu, Phosphorus, Sulfur Silicon Relat. Elem. *2007*, *182* (2), 491-501.

Copper(I) iodide/1,8-diazabicyclo[5.4.0]undec-7-ene *CuI/DBU*

Reactions of isonitriles via isoselenocyanates ←

monoselenoiminocarbonic acid esters by reaction with Se, alcohols and halides cf. *68*, 235; 2-alkoxy-, 2-amino- or 2-organothio-1,3-benzoselenazoles from *o*-halogenisonitriles and alcohols, amines or mercaptans, respectively, in the presence of CuI/DBU, also 2-amino-1,3-benzotellurazoles, s. S. Fujiwara, Y. Asanuma, T. Shin-ike, N. Kambe, J. Org. Chem. *2007*, *72* (21), 8087-90.

Copper(I) iodide/1-butyl-3-methylimidazolium N,N-dimethylglycinate ←
Eneselenides from α,β-ethylenehalides C=C(Hal) → C=C(SeR)
s. *41*, 584s*72; 68*, 230s*72*

Zinc s. under $CoBr_2$ *Zn*

Lithium tetrahydridoaluminate *$LiAlH_4$*
Carboxylic acid telluroamides from chloroimidium chlorides C(Cl)=N⁺< → C(Te)N<
s. *45*, 347s*72*

B-Bromocatecholborane s. under $CoBr_2$ ←

Carbon monoxide/water/sodium hydroxide *$CO/H_2O/NaOH$*
Sym. diselenides from halides 2 RHal → $(RSe)_2$
with NaOH/PEG-400 under microwave irradiation cf. *47*, 563s*58*; with $CO/Se/H_2O/NaOH$ s. Y. Chen, F. Tian, M. Song, S. Lu, Synth. Commun. *2007*, *37* (16), 2687-92.

β-*Cyclodextrin* *cyclodextrin*
2-Aminoselenazoles from α-bromoketones ○
and selenourea with K_2CO_3 cf. *2*, 580; in water containing β-cyclodextrin at 50°, 4-aryl-derivs., s. M. Narender, M. Somi Reddy, V. Pavan Kumar, V. Prakash Reddy, Y.V.D. Nageswar, K. Rama Rao, J. Org. Chem. *2007*, *72* (5), 1849-51.

Trifluoroacetic acid s. under $CoBr_2$ *CF_3COOH*
Allyl chloride s. under $CoBr_2$ *$H_2C{=}CHCH_2Cl$*

Trialkylgermyllithium compds./hexamethylphosphoramide *$R_3GeLi/HMPA$*
Enegermanes from 1-functionalized ethylene derivs. C=C(X) → C=C(Ge≤)
from thioenolethers or enestannanes by Et_3B-catalyzed radical substitution cf. *43*, 535; from α,β-ethylenebromides and Me_3GeLi in HMPA, (E)-α,β-ethylene-α-germylboronic acid esters, also their conversion to acylgermanes (with trimethylamine oxide/water) or to carboxylic acids (with $H_2O_2/NaOH$), s. N.G. Bhat, N. Varghese, Synlett *2007* (13), 2023-4.

Cobalt(II) bromide/zinc/trifluoroacetic acid/B-bromocatecholborane/allyl chloride ←
Arylboronic acid esters from ar. halides Hal → $B(OR)_2$

362. 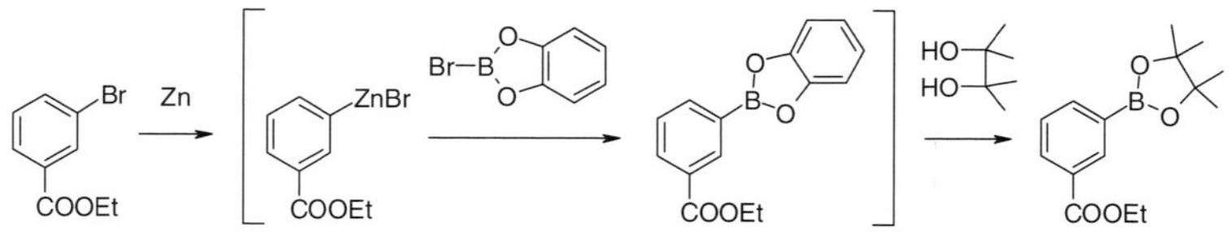

A mixture of 0.1 eq. $CoBr_2$, 1.5 eqs. Zn dust and 0.3 eq. allyl chloride in acetonitrile stirred at room temp., the catalyst activated with trifluoroacetic acid, stirred for 5 min and the temp. raised, the resulting grey soln. treated successively with equivalent amounts of the startg. aryl halide and B-bromocatecholborane, stirred for ca. 1 h (GC monitoring), THF added, followed by 1 eq. pinacol, stirred until the more stable pinacol ester was formed, hydrolyzed with NH_4Cl, and worked up with flash chromatographic purification → product. Y 95%. This is the first example of **transmetalation between an organozinc and a boron compd.** The procedure is generally applicable to aryl bromides or iodides possessing electron-withdrawing or -donating groups (substrates with the latter reacting more efficiently with B-chlorocatecholborane). Certain hetaryl bromides (but not pyridyl bromides) also participated efficiently, but yields were low with *o*-subst. aryl halides. F.e.s. S. Claudel, C. Gosmini, J.M. Paris, J. Périchon, Chem. Commun. *2007* (35), 3667-9.

Tris(dibenzylideneacetone)dipalladium/2-(dicyclohexylphosphino)biphenyl ←
Arylsilanes from ar. chlorides ArCl → ArSi⩽

363.

4-Methoxyphenyl chloride and 1.2 eqs. hexamethyldisilane added to 2 mol% tris(dibenzylideneacetone)dipalladium, 6 mol% 2-(dicyclohexylphosphino)biphenyl, 1-2 eqs. water and 5 eqs. dry KF in dioxane stirred in a sealed vial under argon at 100° for 24 h, cooled to room temp., and extracted with hexanes → 4-methoxyphenyl(trimethyl)silane. Y 90%. By variation of the reaction conditions and biphenyl catalyst the reaction was successful with both electron-rich and -poor aryl chlorides bearing amine, ether, thioether, phenol, acyl, ester, nitrile or nitro groups. The product could be converted to 4-methoxyphenyl iodide by treatment *in situ* with iodine monochloride. F.e.s. E. McNeill, T.E. Barder, S.L. Buchwald, Org. Lett. *2007*, *9* (19), 3785-8.

Tris(dibenzylideneacetone)dipalladium/9,9-dimethyl-4,5-bis(diphenylphosphino)xanthene ←
One-pot palladium-catalyzed synthesis of arylphosphonic acid esters Hal → $PO(OR)_2$
from hypophosphorous acid and ar. halides s. *72*, 359

*Bis(tri-*tert*-butylphosphine)palladium(0)/triethylamine* $Pd(PBu\text{-}t_3)_2/Et_3N$
Palladium-catalyzed Si-arylation Si-H → Si-Ar
trialkyl(aryl)silanes with $Pd(PBu\text{-}t_3)_2/K_3PO_4$ cf. *55*, 227s*62*; mono-Si-arylation of dihydrosilanes with (het)aryl iodides and Et_3N as base s. Y. Yamanoi, T. Taira, J. Sato, I. Nakamula, H. Nishihara, Org. Lett. *2007*, *9* (22), 4543-6; of 1,1,1,3,5,5,5-heptamethyltrisiloxane, comparison of substrate scope under Pd(0)-, Pt(0)-, or Rh(I)-catalysis, s. M. Murata, K. Ota, H. Yamasaki, S. Watanabe, Y. Masuda, Synlett *2007* (9), 1387-90.

Chiral platinum complexes ←
Tert. from sec. phosphines via asym. P-alkylation P-H → P-R
s. *71*, 245; also asym. P-arylation (cf. *65*, 230s*67*), and Michael addition s. D.S. Glueck, Synlett *2007* (17), 2627-34.

Sulfur ↑ RemC ↓↑ S

Sodium or potassium bis(trimethylsilyl)amide $NaN(SiMe_3)_2$ *or* $KN(SiMe_3)_2$
Potassium tert*-butoxide* *KOBu*-t
Cyclic α,β-ethylenephosphonic acid esters $C(SO_2R){=}C \rightarrow C{=}C\text{-}PO(OR')_2$
from α,β-ethylenesulfones with position interchange

364.

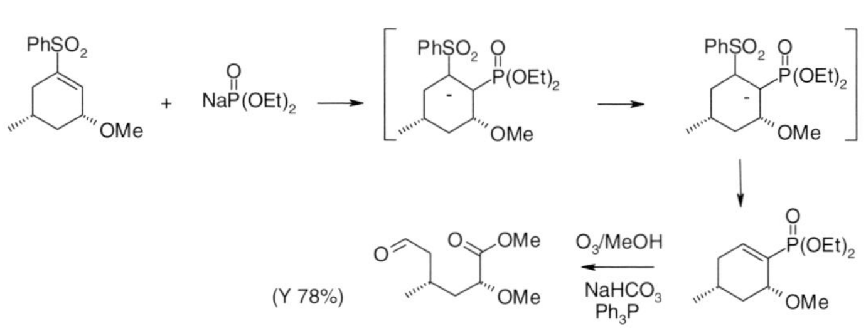

Reversal of polarity of electron-deficient ethylene derivs. may be achieved under milder, safer conditions using a P-nucleophile in place of KCN/18-crown-6 (cf. *37*, 837). **E:** Freshly distilled, dry THF added at room temp. to 2.3 eqs. diethyl phosphite in a flame-dried, 3-necked flask, stirred for 1 min, 2.2 eqs. 2 *M* $NaN(SiMe_3)_2$ in THF added dropwise via syringe, stirring continued for 30 min, stirring stopped, dry solid vinyl sulfone (1 eq.) added in one portion under positive dry N_2 pressure, stirring resumed for 1-30 min (TLC monitoring) or until a heavy yellowish-white precipitate formed, then ether and brine soln. added, followed by a few drops of

DI-water, stirred for 10 min, the organic layer dried over Na_2SO_4 for 30 min, filtered, concentrated *in vacuo*, and purified by column chromatography → product. Y 93%. The products may be cleaved to **aldehydocarboxylic acid esters** by ozonolysis in 4:1 methylene chloride/methanol. The methodology is fast, high-yielding, and tolerates a variety of functional groups, such as ethers, silyl ethers and esters. F.e., incl. cyclo-pentene and -heptene analogs, and application to the synthesis of segments of Aplyronine A, also generation of the phosphite anions *in situ* using $KN(SiMe_3)_2$ or KOBu-*t*, s. M.N. Noshi, A. El-awa, E. Torres, P.L. Fuchs, J. Am. Chem. Soc. *2007*, *129* (36), 11242-7.

Remaining Elements ↑ RemC ↓↑ Rem

Copper(I) tert-*butoxide/(R,R)-2,3-bis[*tert-*butyl(methyl)phosphino]quinoxaline* ←
β,γ-Ethyleneboronic from (Z)-2-ethylenecarbonic acid esters ←
Regiospecific asym. conversion with double bond shift

365.

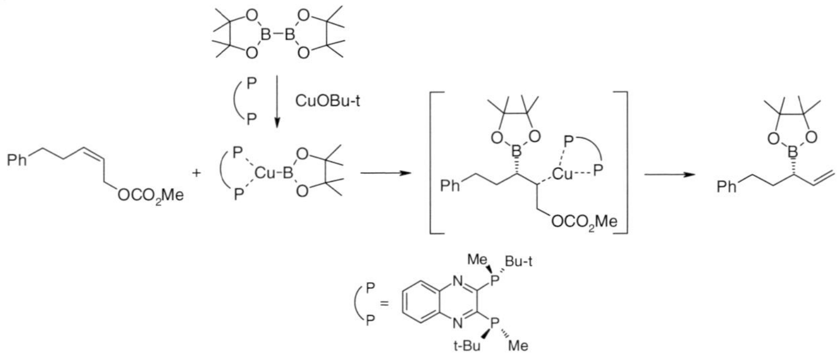

5 Mol% CuOBu-*t* and 5 mol% (R,R)-2,3-bis[*tert*-butyl(methyl)phosphino]quinoxaline in dry THF stirred [in a glove box] to give a yellow suspension, 2 eqs. bis(pinacolato)diboron added [with a colour change to dark red], the reaction vial sealed with a rubber septum, the startg. allyl carbonate added under argon at 0° (*caution: possible evolution of CO with the liberated gas*), stirred for 20 h, evaporated under reduced pressure, and worked up with flash chromatographic purification within 10 min → (S)-4,4,5,5-tetramethyl-2-[(3S)-5-phenyl-1-penten-3-yl]-1,3,2-dioxaborolane. Y 77% (e.e. 94%). This is the first such copper-catalyzed asym. substitution with a non-carbon nucleophile, reaction being applicable to various allylic carbonates possessing an alkyl group at the 3-position (although substrates with branched alkyl groups required a higher catalyst loading, while the 3-isopropyl deriv. was unreactive). Enantioselectivity was low, however, from the corresponding (E)-allyl carbonate. The key step in the transformation is presumed to be the regiospecific addition of a borylcopper complex to the ethylene group through a four-centred rigid diastereoisomeric transition state which is responsible for the high face selectivity. F.e. and comparison of chiral ligands s. H. Ito, S. Ito, Y. Sasaki, K. Matsuura, M. Sawamura, J. Am. Chem. Soc. *2007*, *129* (48), 14856-7.

Sodium tetrahydridoborate *$NaBH_4$*
Selenides from sym. diselenides and halides $(RSe)_2$ → RSeR′
cf. *33*, 603s*71*; (phenyl- or benzyl-seleno)carboxylic from chlorocarboxylic acid esters, also hydrolysis to the corresponding acids, s. A. Bhalla, S. Sharma, K.K. Bhasin, S.S. Bari, Synth. Commun. *2007*, *37* (5), 783-93.

Lithium tetrahydridoaluminate *$LiAlH_4$*
Carboxylic acid telluroamides C(SeR)=N^+< → C(Te)N<
from 1-(alkylseleno)azomethinium triflates s. *45*, 347s*72*

Palladium(II) acetate/methanol *$Pd(OAc)_2/MeOH$*

Exocyclic (E)-γ,δ-ethyleneboronic acid esters from 1,6-enynes ○

366.

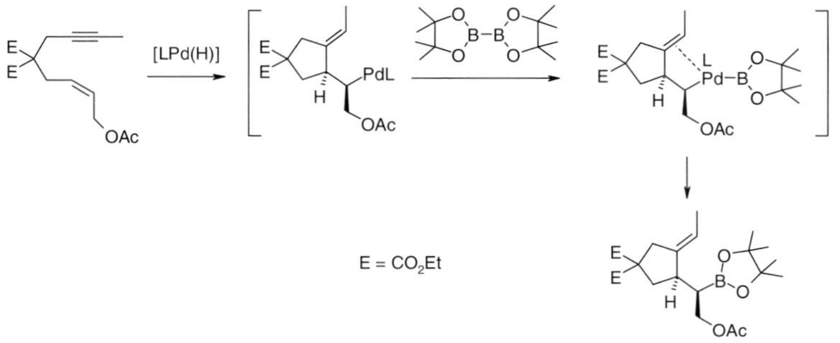

Dry toluene and 1 eq. methanol added to a soln. of the startg. enyne (ca. 100 mg), 1.1 eqs. bis(pinacolato)diboron and 5 mol% $Pd(OAc)_2$ under argon, heated at 50° for 4 h, cooled to room temp., Celite added, the solvent evaporated, and the residue worked up with column chromatographic purification → product. Y 95%. The allylic functionality can be acetoxy, benzoyloxy, methoxy or benzyloxy, and the tethered functionality may be geminal carbon, oxygen or NTs, yields being highest with internal alkyne derivs. Reaction may involve a cyclopalladation, or perhaps an initial hydridopalladation of the alkyne followed by intramolecular carbopalladation, transmetalation and reductive elimination. The ligandless conditions are presumed to obviate undesirable β-hydride and β-oxygen elimination as predominates in the presence of triphenylphosphine. F.e. incl. reaction with hindered enynes, also comparison of catalysts and additives, s. J. Marco-Martínez, V. López, E. Buñuel, R. Simancas, D.J. Cárdenas, J. Am. Chem. Soc. *2007*, *129* (7), 1874-5.

Tris(dibenzylideneacetone)dipalladium/2-(dicyclohexylphosphino)biphenyl ←

Arylsilanes from ar. chlorides s. *72*, 363 ArCl → ArSi≤

Dichloro[1,1′-bis((R,R)-2,4-diethylphosphetidin-1-yl)ferrocene]palladium(II) ←

Tert. ar. phosphines from ar. iodides and silylphosphines I → P<

Dynamic kinetic asym. phospha-Stille coupling

367.

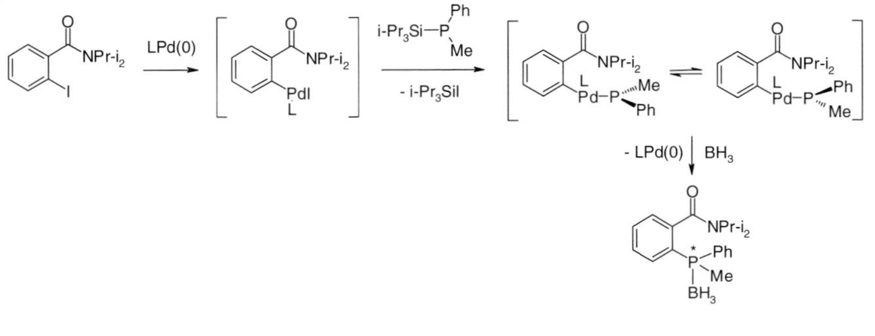

Enhancement of enantioselectivity with *o*-carbamyl groups. 1.1 Eqs. startg. silylphosphine added via syringe to a J-Young NMR tube containing a soln. of 1.25 eqs. startg. ar. iodide and 5 mol% dichloro[1,1′-bis((R,R)-2,4-diethylphosphetidin-1-yl)ferrocene]palladium(II) in DMPU under N_2 (in a glove box), the tube sealed, heated at 60° for 1 h, and the resulting phosphine protected by formation of the borane complex (by addition of 10 eqs. 1 *M* BH_3-THF, and stirring for 30 min at room temp.) → product. Y 53% (e.e. 98%). Enantioselectivity was noticeably lower with other *o*-subst. ar. iodides, and low (32%) with N,N-diisopropyl-*o*-iodoarylacetamide. The dynamic kinetic conversion is facilitated by the low barrier to pyramidal inversion of metal-phosphide complexes which ensures rapid epimerization of the intermediate palladium(II) phosphite (relative to reductive elimination). With the *o*-carbamyl substituent, further enhancement of enantioselectivity is presumed to be the result of coordination of the amide to palladium in a 5-membered palladacycle. F.e. and with hetaryl iodides, incl. 2-iodo-1-(N,N-diisopropylcarbamyl)-indole, s. V.S. Chan, R.G. Bergman, F.D. Toste, J. Am. Chem. Soc. *2007*, *129* (49), 15122-3.

*(1,5-Cyclooctadiene)(methoxy)iridium(I) dimer/4,4′-di-*tert*-butyl-2,2′-bipyridyl* ←
Arylboronic acid esters from arenes H → $B(OR)_2$
and bis(pinacolato)diboron cf. *64*, 219; with conversion of the crude boronates to **arylboronic acids** (using $NaIO_4$) **or potassium aryl(trifluoro)borates** (using KHF_2) s. J.M. Murphy, C.C. Tzschucke, J.F. Hartwig, Org. Lett. *2007*, *9* (5), 757-60; indol-7-ylboronic acid esters (useful intermediates for the synthesis of a variety of 2,7-di- and 2,4,7-tri-subst. indoles) s. W.F. Lo, H.M. Kaiser, A. Spannenberg, M. Beller, M.K. Tse, Tetrahedron Lett. *2007*, *48* (3), 371-5.

Dichloro(cyclooctadiene)platinum(II) *$PtCl_2$(cod)*
1-Oxa-2-silacyclohex-5-enes from α,β-ethyleneketones ○
Platinum silylene complexes as intermediates

368.

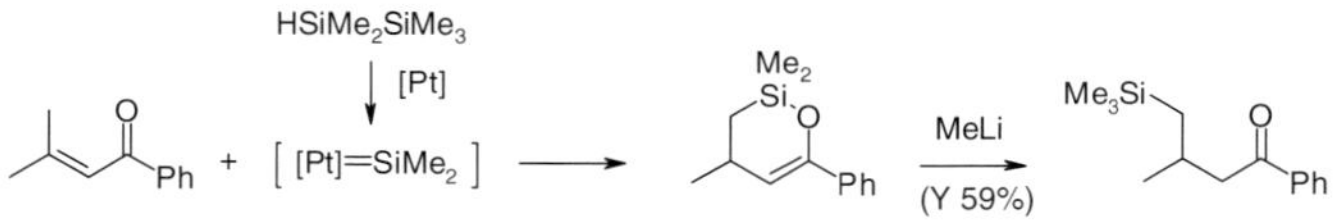

The startg. enone and 1.2 eqs. pentamethyldisilane added to a soln. of 3 mol% $PtCl_2$(cod) in toluene, stirred at 40° for 12 h, and worked up with purification by Kugelrohr distillation → 6-phenyl-2,2,4-trimethyl-1-oxa-2-silacyclohex-5-ene. Y 79% (90% regio-purity). Subsequent work-up by treatment with 1.13 *M* MeLi (1.5 eqs.) in ether at 0° afforded the corresponding **γ-silyl-ketones,** while subsequent treatment of the oxasilacyclohexene with KHF_2 in THF, followed by Tamao oxidation (30% aq. H_2O_2/KF/$KHCO_3$; cf. *39*, 200s*47*), gave the corresponding **γ-hydroxy-ketones.** Reaction is presumed to involve intermediate formation of a platinum silylene complex as the reactive species. F.e.s. K. Okamoto, T. Hayashi, Org. Lett. *2007*, *9* (24), 5067-9.

Carbon ↾ RemC ⇃↾ C

Magnesium perchlorate *$Mg(ClO_4)_2$*
Trimethyl(phenyl)ammonium chloride *Me_3PhNCl*
3-Component synthesis of α-aminophosphonic acid esters CO → C(N<)$PO(OR)_2$
from trialkyl phosphites with a little $Mg(ClO_4)_2$ in the absence of solvent s. *59*, 234s*72*; with 1 mol% Me_3PhNCl in refluxing methylene chloride, also α-hydrazino- and α-hydroxylamino-derivs. (with added Me_3SiCl), s. A. Heydari, A. Arefi, Catal. Commun. *2007*, *8* (7), 1023-6.

Carbonyl(chloro)hydridobis(tricyclohexylphosphine)ruthenium(II) *$[RuH(CO)Cl(PCy_3)_2]$*
Germylacetylenes from terminal acetylene derivs. and vinylgermanes H → Ge≤

369.

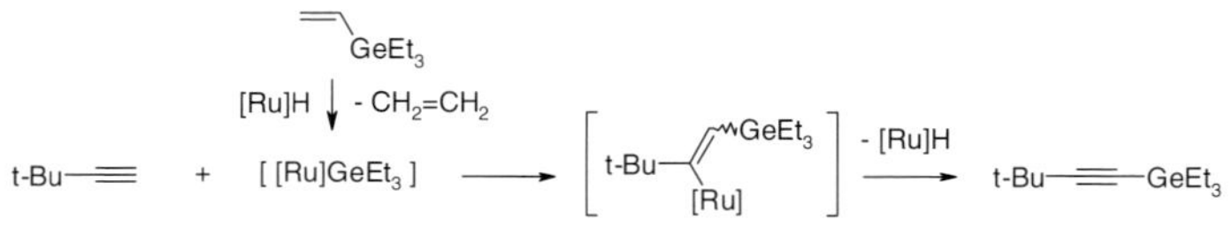

tert-Butylethyne and 2 eqs. of triethyl(vinyl)germane added to a glass ampoule containing 2 mol% $[RuH(CO)Cl(PCy_3)_2]$ in toluene under argon (with decane as internal standard), and the sealed vessel heated at 100° for 24 h → 1-(*tert*-butyl)-2-(triethylgermyl)ethyne. Y 99%. This is the first direct transition metal-catalyzed route to this class of compound, being applicable to the germylation of terminal alkyl-, cycloalkyl-, silyl- and germyl-acetylenes. There was no reaction, however, with phenylacetylene. The conversion is thought to involve initial hydridoruthenation of the vinylgermane, followed by elimination of ethylene and insertion of the alkyne into the resulting germylruthenium species; β-hydrogen elimination completes the cycle. F.e.s. B. Marciniec, H. Lawicka, B. Dudziec, Organometallics *2007*, *26* (21), 5188-92; **silylacetylenes** from terminal acetylene derivs. and vinylsilanes (cf. *60*, 219) s. B. Marciniec, B. Dudziec, I. Kownacki, Angew. Chem. Int. Ed. Engl. *2006*, *45* (48), 8180-4.

Formation of C-C Bond

Uptake ⇓

Addition to Oxygen and Carbon CC ⇓ OC

Potassium hydroxide/1-butyl-3-methylimidazolium hexafluorophosphate *KOH/[bmim]PF$_6$*
2-Acetylenealcohols from oxo compds. CO → C(OH)C≡C
with KOH/liq. NH_3 cf. *19*, 734; with KOH in [bmim]PF_6 as ionic liquid solvent s. S.R. Jagtap, B.M. Bhanage, J. Chem. Res. *2007* (6), 370-2; in the presence of 10 mol% CuOTf-KOBu-*t*/Xantphos or $Cu(OTf)_2$-2KOBu-*t*/phenanthroline, 2-acetylene-2′,2′,2′-trifluoroalcohols from trifluoromethyl ketones, s. R. Motoki, M. Kanai, M. Shibasaki, Org. Lett. *2007*, *9* (16), 2997-3000; ligand effects in addition of alk-1-ynes to aldehydes or diketones using $Rh(CO)_2$(β-diketonates)/*o*-(*t*-Bu_2P)biphenyl s. P.K. Dhondi, P. Carberry, J.D. Chisholm, Tetrahedron Lett. *2007*, *48* (49), 8743-6.

Lithium diisopropylamide/zinc chloride i-*Pr_2NLi/ZnCl$_2$*
Asym. aldol condensation with sulfonic acid esters CHO → CH(OH)C(SO_3R)

370. 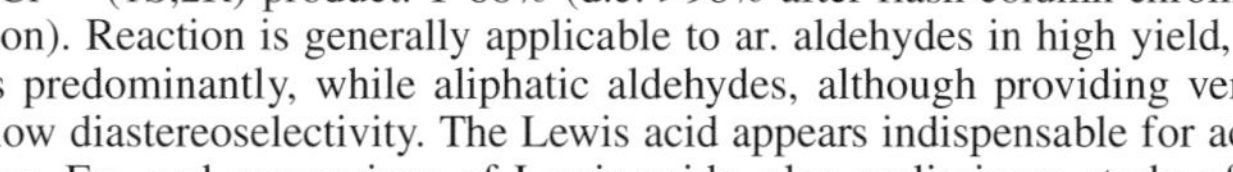

Chiral *syn*-β-hydroxysulfonic acid esters. A soln. of the startg. chiral sulfonic acid ester [based on 1,2:5,6-di-O-isopropylidene-α-D-allofuranose (cf. *63*, 377)] in THF added dropwise under N_2 to a freshly prepared soln. of 1 eq. LDA in dry THF at -78°, stirred for 1 h, 2 eqs. benzaldehyde and 1.2 eqs. 1 *M* $ZnCl_2$ in THF added dropwise, stirred again at -78° for 1 h, and quenched with satd. aq. NH_4Cl → (1S,2R)-product. Y 88% (d.e. >98% after flash column chromatography and recrystallization). Reaction is generally applicable to ar. aldehydes in high yield, affording *syn*-diastereomers predominantly, while aliphatic aldehydes, although providing very high yields, reacted with low diastereoselectivity. The Lewis acid appears indispensable for activation of the carbonyl group. F.e. and comparison of Lewis acids, also preliminary study of removing the chiral ligand, s. W. Harnying, W. Kitisriworaphan, M. Pohmakotr, D. Enders, Synlett *2007* (16), 2529-32.

Potassium cyanide/18-crown-6 polyether *KCN/crown*
α-Cyano-α-phosphoryloxycarboxylic from acylphosphonic acid esters ←

371. 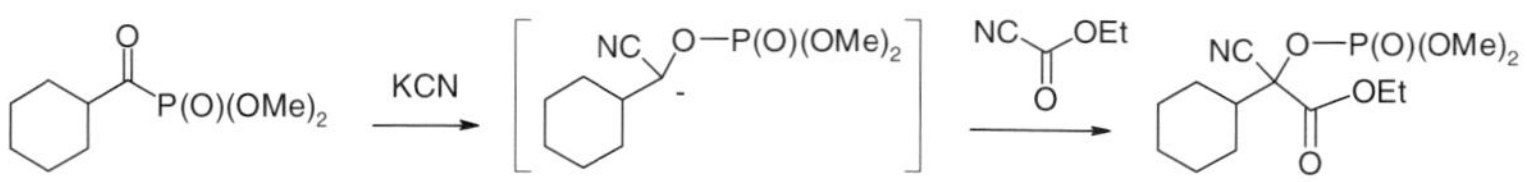

Dry THF (1 ml), ethyl cyanoformate (0.6 mmol), 18-crown-6 (0.025 mmol) and KCN (2 mg) added to the startg. acylphosphonate (0.5 mmol) in an oven-dried Schlenk flask, and stirred under argon for 15-20 min at room temp. → product. Y 90%. The acylphosphonates act as acyl anion precursors and, unlike the analogous acylsilanes, are readily available in multigram quantities (from acyl chlorides and trialkyl phosphites via an Arbuzov reaction). The catalytic cycle is presumed to involve initial addition of cyanide ion to the carbonyl group of the acylphosphonate, followed by C→O-phosphonyl group migration prior to addition of the carbethoxy group. F.e.s. A.S. Demir, B. Reis, Ö. Reis, S. Eymür, M. Göllü, S. Tural, G. Saglam, J. Org. Chem. *2007*, *72* (19), 7439-42; **cyanohydrin phosphates** from acylphosphonic acid esters with 1.1 eqs. KCN in an ethereal solvent/water s. A.S. Demir, Ö. Reis, I. Esiringü, B. Reis, S. Baris, Tetrahedron *2007*, *63* (1), 160-5.

Sodium iodide *NaI*
2-Nitroalcohols from aldehydes CO → C(OH)C(NO_2)
with KF cf. *34*, 610; with a little NaI, 1-nitro-1,2-bromohydrins, and conversion to (E)-1,1-nitroethylene derivs. with SmI_2 (cf. *59*, 484) s. J.M. Concellón, P.L. Bernad, H. Rodríguez-Solla, C. Concellón, J. Org. Chem. *2007*, *72* (14), 5421-3.

Ethyldiisopropylamine or 1,8-Diazabicyclo[5.4.0]undec-7-ene i-Pr_2NEt *or DBU*
Stereospecific tert. amine-catalyzed aldol condensation CHO → CH(OH)C-CO

372. 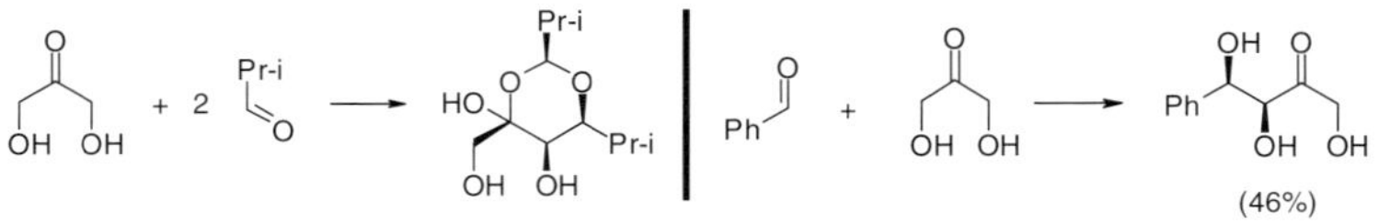

under mild, solvent-free conditions. The first generalized tert. amine-catalyzed aldol condensation of aldehydes with ketones is reported, as illustrated with the exclusively ***syn*-selective reaction** with dihydroxyacetone. **E:** 5 Mol% *ethyldiisopropylamine* added to a mixture of dihydroxyacetone and 2 eqs. 2-methylpropionaldehyde at 0°, stirred at room temp. for 6-48 h (TLC monitoring), the mixture absorbed onto Celite, and purified by chromatography → product. Y 92%. Two molecules of aliphatic aldehydes condense to provide the O,O-alkylidene derivs. (Y 68-94%) with only traces (<5%) of unprotected adducts, whereas benzaldehyde gave the anticipated unprotected aldol adduct with the same high (99%) *syn*-selectivity (Y 46%). Significantly, only *5 mol%* of the catalyst is required *without* any other additives. Similarly, monohydroxyacetone underwent condensation regiospecifically with DBU as catalyst to give ***syn*-α,β-dihydroxyketones** (Y 83-92%; 68-91% *syn*-selectivity). F.e. and application of the method to the synthesis of carbohydrate derivs., e.g. 6-deoxy-L-sorbose, s. M. Markert, M. Mulzer, B. Schetter, R. Mahrwald, J. Am. Chem. Soc. *2007*, *129* (23), 7258-9.

Chiral 1,2-diamines s. (1R,2R)-N,N-Dipropylcyclohexane-1,2-diaminetrifluoromethanesulfonic acid ←
Chiral 2-aminoalcohols s. under Cu(OAc)$_2$ and R_2Zn ←
Phenanthroline s. under Cu(OTf)$_2$ ←

Copper(I) tert-*butoxide/chiral ferrocenyldi(phosphine)* ←
Asym. synthesis of 2-acetylenealcohols from aldehydes CHO → CH(OH)C≡C
s. *58*, 236s*72*

Chiral copper(I) salan complex *Cu(II)-salan*
Copper(II) acetate/chiral 2-[N-(thien-2-ylmethyl)amino]alcohols or bicyclic N-(pyrid-2-ylmethyl)imines/ethyldiisopropylamine ←
Copper(I) triflate-benzene/chiral 2-(2-hydroxy-5-methylbenzylideneamino)-2′-diphenylphosphino-1,1′-binaphthyl ←
Copper(II) triflate/2,6-bis[(S)-4-isopropyl-1-phenyl-Δ^2-imidazolin-2-yl]pyridine/triethylamine ←
Asym. Henry [nitroaldol] reaction CO → C(OH)C(NO_2)
with chiral phenylalanine-derived Cu(II)-salen complexes cf. *62*, 250s*71*; with a chiral Cu(I)-salan catalyst (e.e. up to 96%) and application to (S)-norphenylephrine synthesis s. Y. Xiong, F.

Wang, X. Huang, Y. Wen, X. Feng, Chem. Eur. J. *2007, 13* (3), 829-33; under dual Lewis acid/Brønsted base catalysis with $Cu(OTf)_2$/2,6-bis[(S)-4-isopropyl-1-phenyl-Δ^2-imidazolin-2-yl]-pyridine/Et_3N s. K. Ma, J. You, Chem. Eur. J. *2007, 13* (6), 1863-71; with $Cu(OAc)_2$/chiral 2-[N-(thien-2-ylmethyl)amino]alcohols (e.e. up to 88%) s. W. Mansawat, I. Saengswang, P. U-prasitwong, W. Bhanthumnavin, T. Vilaivan, Tetrahedron Lett. *2007, 48* (24), 4235-8; with chiral bicyclic N-(pyrid-2-ylmethyl)imines as the ligand and *i*-Pr_2NEt s. G. Blay, E. Climent, I. Fernández, V. Hernández-Olmos, J.R. Pedro, Tetrahedron:Asym. *2007, 18* (13), 1603-12; with $(CuOTf)_2$-C_6H_6/chiral 2-(2-hydroxy-5-methylbenzylideneamino)-2′-diphenylphosphino-1,1′-bi-naphthyl (e.e. 65-80%) s. J.-J. Jiang, M. Shi, ibid. *18* (11), 1376-82; with a chiral ClCr(III)-salen complex/*i*-Pr_2NEt s. R. Kowalczyk, L. Sidorowicz, J. Skarzewski, ibid. *18* (21), 2581-6; under organocatalysis with a chiral 2,2′-bis(thioureido)octahydro-1,1′-binaphthyl/*i*-Pr_2NEt s. X.-G. Liu, J.-J. Jiang, M. Shi, ibid. *18* (23), 2773-81; reaction of trifluoromethyl ketones with a chiral lanthanum(III) triflate complex/Proton Sponge s. F. Tur, J.M. Saá, Org. Lett. *2007, 9* (24), 5079-82.

Copper(I) triflate/potassium tert-*butoxide/9,9-dimethyl-4,5-bis(diphenylphosphino)-xanthene* ←
Copper(II) triflate/potassium tert-*butoxide/phenanthroline* ←
2-Acetylenealcohols from oxo compds. s. *19*, 734s*72* CO → C(OH)C≡C

Copper(II) triflate/chiral fluorinated bis(Δ^2-oxazolines) ←
Asym. carbonyl-ene reaction ←
with Cu(II)-perchlorate/chiral aminosulfoximines cf. *56*, 242s*70*; asym. glyoxylate-ene reaction with $Cu(OTf)_2$/chiral fluorinated bis(Δ^2-oxazolines) s. R. Kolodziuk, C. Goux-Henry, D. Sinou, Tetrahedron:Asym. *2007, 18* (23), 2782-6; with a chiral silylated Co(III)-salen complex s. G.E. Hutson, A.H. Dave, V.H. Rawal, Org. Lett. *2007, 9* (20), 3869-72.

Silver hexafluoroantimonate s. under $PdCl_2$[(S)-SegPhos] *$AgSbF_6$*

Magnesium/magnesium bromide *$Mg/MgBr_2$*
Asym. induction with chiral (1-methoxy-2,2,2-triphenylethyl)dimethylsilyl ethers ←
Synthesis of glycol or 1,3-diol monosilyl ethers from α- or β-hydroxyoxo compds. s. *72*, 1

Magnesium/iron(III) chloride *$Mg/FeCl_3$*
Low-valent iron-catalyzed Reformatsky reaction with aldehydes under mild conditions CHO → CH(OH)C-CO

373.

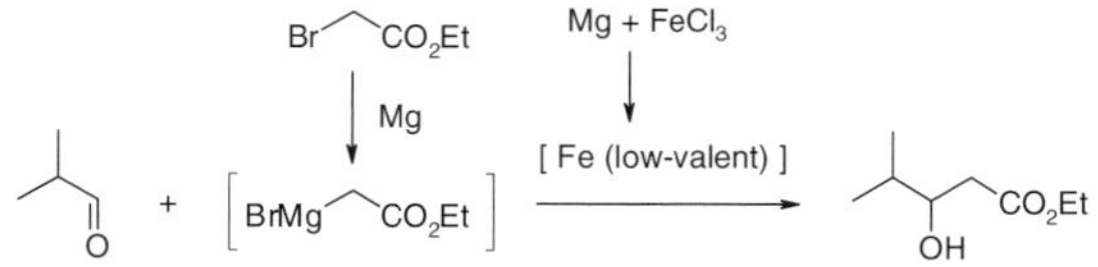

β-Hydroxycarboxylic acid esters. 0.32 Eqs. Mg turnings added in one portion to a stirred mixture of isobutanal, 2 eqs. ethyl 2-bromoacetate and 3 eqs. $FeCl_3$ in *moist THF*, the mixture stirred at room temp. for 4 h, treated sequentially with water and ethyl acetate, stirred for 10 min, filtered, the filtrate treated with 2% aq. HCl, the organic layer separated, the aq. layer extracted with ethyl acetate, the combined organic layers washed with water and brine, dried, evaporated to dryness *in vacuo*, and the residue purified by chromatography → ethyl 3-hydroxy-4-methyl-pentanoate. Y 82.6%. This simple and efficient method, based on inexpensive reagents, employs a **bimetal redox strategy** to generate low-valent iron *in situ* and is assisted by the presence of moisture which is believed to aid dissolution of the metal salts. The conversion is successful with aliphatic and aromatic aldehydes (Y 71-85%) but shows little stereoselectivity with chiral aldehydes. The reaction can also be accomplished with $CuCl_2 \cdot 2H_2O$ but is slower (6-8 h cf. 3-4 h) and yields are ca. 10% lower, while Zn dust is a poor substitute for Mg turnings despite its greater surface area. F.e.s. A. Chattopadhyay, A.Kr. Dubey, J. Org. Chem. *2007, 72* (24), 9357-9.

Nanocrystalline magnesium oxide *MgO*
Magnesium-lanthanum mixed oxide ←
α-Diazo-β-hydroxycarboxylic acid esters from oxo compds. CO → C(OH)C(N_2)COOR
with Bu_4NOH in DMSO cf. *26*, 668s*70*; with nanocrystalline MgO in DMSO s. M. Lakshmi Kantam, L. Chakrapani, T. Ramani, Tetrahedron Lett. *2007*, *48* (35), 6121-3; with highly-basic magnesium-lanthanum mixed oxide in water, also β-sulfonylamino-derivs. from N-tosylimines, s. M. Lakshmi Kantam, V. Balasubrahmanyam, K.B. Shiva Kumar, G.T. Venkanna, F. Figueras, Adv. Synth. Catal. *2007*, *349* (11-12), 1887-90.

*Diorganozinc compds. (or zinc)/chloro(norbornadiene)rhodium(I) dimer/ (4S)-*tert*-butyl-2-[*o*-(diphenylphosphino)phenyl]-Δ^2-oxazoline* ←
Synthesis of δ-ketocarboxylic acids from glutaric anhydrides with desymmetrization C

374. 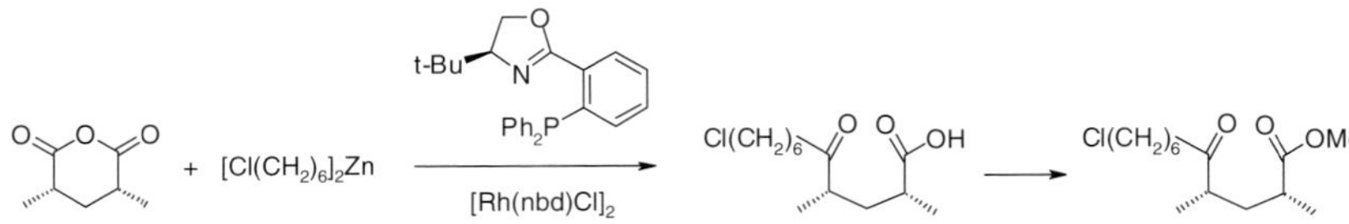

A mixture of 0.0088 mmol [Rh(nbd)Cl]$_2$ and 0.0176 mmol (4S)-*tert*-butyl-2-[*o*-(diphenylphosphino)phenyl]-Δ^2-oxazoline purged with argon for 15 min in a flask sealed with a septum, THF introduced, followed via syringe by 0.3 mmol freshly prepared bis(6-chlorohexyl)zinc (0.3 *M* in THF), heated to 50°, 0.176 mmol dimethylglutaric anhydride added, stirred overnight at 50°, the free acid worked up, and isolated as the methyl ester (on treatment with trimethysilyldiazomethane in 1:1 toluene/methanol for 10 min) → product. Y 76% (e.e. 94%). The procedure is applicable to a range of *freshly prepared* dialkylzincs, diarylzincs, and organozinc halides (notably benzylzinc bromides), as well as zinc homoenolates, leaving alkyl chlorides and ester groups unaffected. F.e. and comparison of chiral auxiliaries s. M.J. Cook, T. Rovis, J. Am. Chem. Soc. *2007*, *129* (30), 9302-3.

Dialkylzinc/chiral polymer-based 2-aminoalcohols or o*-α-aminophenols or dihydrofuran-2-yl-, tetrahydrofuran-2-yl-, or furan-2-yl-2-(dialkylamino)ethanols or 1-(perfluoroalkyl)-2-(naphth-1-ylmethyleneamino)alcohols or dendritic 1,1′-bi-2-naphthols or 3-aryl-octahydro-1,1′-bi-2-naphthols/titanium tetraisopropoxide or 2,2-dimethyl-α,α,α′-tris(perfluorooctyl)-1,3-dioxolane-4,5-dimethanol* ←
Asym. synthesis of sec. alcohols from aldehydes CHO → CH(OH)R
with chiral 1,1′-bi-2-naphthols with or without Ti(OPr-*i*)$_4$ cf. *42*, 616s*70*; with chiral dendritic BINOLS/Ti(OPr-*i*)$_4$ cf. L. Yin, R. Li, F. Wang, H. Wang, Y. Zheng, C. Wang, J. Ma, Tetrahedron: Asym. *2007*, *18* (11), 1383-9; with 2 mol% (R)-3-(3,5-diphenylphenyl)-H_8-BINOL and 1.4 eqs. Ti(OPr-*i*)$_4$ (e.e. 94-8%) s. T. Harada, T. Ukon, ibid. *18* (21), 2499-502; with chiral *o*-α-aminophenols s. X.-F. Yang, Z.-H. Wang, T. Koshizawa, M. Yasutake, G.-Y. Zhang, T. Hirose, ibid. *18* (10), 1257-63; with chiral 1-(perfluoroalkyl)-2-(naphth-1-ylmethyleneamino)alcohols s. M. Omote, Y. Eto, A. Tarui, K. Sato, A. Ando, ibid. (23), 2768-72; with chiral dihydrofuran-2-yl, tetrahydrofuran-2-yl-, and furan-2-yl-2-(dialkylamino)ethanols s. C. Paolucci, G. Rosini, ibid. *18* (24), 2923-46; with a chiral 2,2-dimethyl-α,α,α′-tris(perfluorooctyl)-1,3-dioxolane-4,5-dimethanol s. Y.S. Sokeirik, H. Mori, M. Omote, K. Sato, A. Tarui, I. Kumadaki, A. Ando, Org. Lett. *2007*, *9* (10), 1927-9; with chiral polymer-based 2-aminoalcohols s. S.-X. Wang, F.-E. Chen, Chem. Pharm. Bull. *2007*, *55* (7), 1011-3.

Dialkylzinc/chiral sec. alcohols *$R_2Zn/R'CH(OH)R''$*
Asym. synthesis of alcohols from aldehydes CHO → CH(OH)R
chiral 5-α-hydroxypyrimidines cf. *51*, 271, 271s*69*; correlation between the absolute configuration of the chiral initiators and the product in sec. alcohol-induced asym. autocatalysis s. T. Shibata, K. Iwahashi, T. Kawasaki, K. Soai, Tetrahedron:Asym. *2007*, *18* (15), 1759-62.

Dialkylzinc/chiral nickel(II) α-aminocarboxamide complexes ←
Nickel-catalyzed asym. synthesis of sec. alcohols from aldehydes CHO → CH(OH)R
Effect of catalyst/ligand ratio on face selectivity

375. 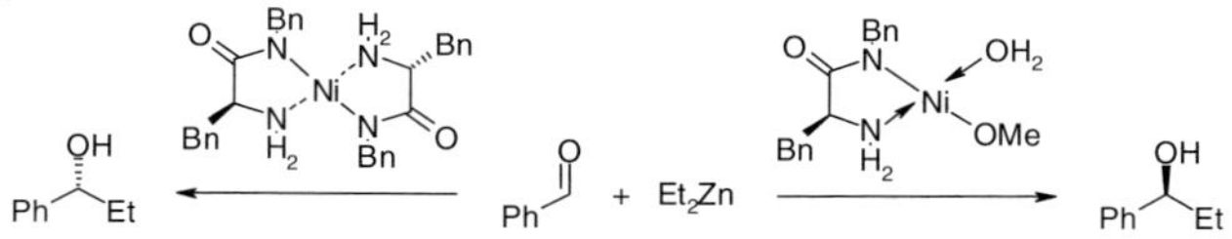

For the first time, a single ligand has been found to control face selectivity in the asym. addition of dialkylzincs to aldehydes. **E:** A soln. of *1 mol%* chiral nickel(II) α-aminocarboxamide *1:1* complex in dry toluene stirred in a Schlenk tube at 0° for 30 min, 1.1 *M* diethylzinc (21 mmol) in the same solvent added, stirred for a further 30 min at room temp., a soln. of the startg. aldehyde (10 mmol) in toluene added slowly, stirring continued for 18 h, quenched with 1 *M* HCl, and worked up with column chromatographic purification on silica gel → (S)-product. Conversion 99% (selectivity 97%; e.e. 97%). Remarkably, the same reaction with the more stable chiral nickel(II) α-aminocarboxamide *1:2* complex gave the (R)-enantiomer (conversion 99%; selectivity 96%; e.e. 96%). The procedure is quite general and can be performed at catalyst loadings as low as 0.1 mol% without affecting the selectivity and enantioselectivity. It is possible that the (S)-product arises from the 1:1 complex via an *anti-trans*-disposed tricyclic transition state, while a *syn-trans*-tricyclic transition state favours the (R)-enantiomer with the 1:2 complex. F.e.s. M.I. Burguete, M. Collado, J. Escorihuela, S.V. Luis, Angew. Chem. Int. Ed. Engl. *2007*, *46* (47), 9002-5.

Dialkylzinc/dichloro(cyclooctadiene)palladium(II)/P-chiral P,P′-di(alk-1-ynyl)-1,2-di(phosphines) ←
Asym. transition metal-catalyzed reactions ←
with P-chiral P,P′-di(alk-1-ynyl)-1,2-di(phosphines) as ligand

376. 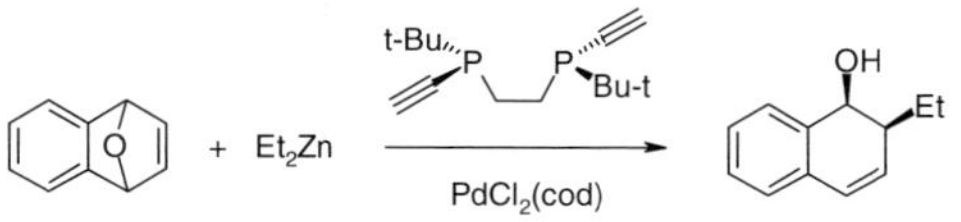

Easily prepared P-chiral P,P′-di(alk-1-ynyl)-1,2-di(phosphines) display *almost perfect* enantioselectivity in a number of familiar asym. transition metal-catalyzed conversions. **E: Asym. alkylative ring opening of 1,4-oxido-1,4-dihydronaphthalenes.** A soln. of 2 mol% $PdCl_2$(cod) and 2.2 mol% (S,S)-1,2-bis[1-ethynyl(*tert*-butyl)phosphino]ethane in methylene chloride stirred at 40° under N_2 for 15 min, the startg. oxidodihydronaphthalene in the same solvent added, followed by 1 *M* dimethylzinc in hexane, stirred at room temp. for 6 h, quenched with a few drops of water, filtered, and worked up with preparative thin layer chromatographic purification → product. Y 90% (*e.e. 99.9%*). F.e. and comparison of chiral reagents, also application in rhodium-catalyzed asym. homogeneous hydrogenation and 1,4-addition of arylboronic acids to cyclic enones, s. T. Imamoto, Y. Saitoh, A. Koide, T. Ogura, K. Yoshida, Angew. Chem. Int. Ed. Engl. *2007*, *46* (45), 8636-9.

Di(m)ethylzinc/chiral 2-aminoalcohols or 2-(sulfonylamino)alcohols or (R)-1,1′-bi-2-naphthol-N-methylimidazole/titanium tetraisopropoxide or chiral macrocyclic bis[3,3′-bis(alkylideneamino)-1,1′-bi-2-naphthol] or o*-α-aminophenol* ←
Asym. synthesis of 2-acetylenealcohols from aldehydes CHO → CH(OH)C≡C
with Et_2Zn and chiral 2-aminoalcohols cf. *58*, 236s*68,70* and with Me_2Zn and (R)-BINOL/phenol/$Ti(OPr\text{-}i)_4$ cf. *65*, 249; with Et_2Zn and a chiral aziridin-2-yl(ferrocenyl)carbinol/$Ti(OPr\text{-}i)_4$ (e.e. up to 96%) s. H. Koyuncu, Ö. Dogan, Org. Lett. *2007*, *9* (17), 3477-9; with a chiral 2-(sulfonylamino)alcohol/DME/$Ti(OPr\text{-}i)_4$, chiral α,β-acetylene-γ-hydroxy esters (e.e. up to 94%), s. L. Lin, X. Jiang, W. Liu, L. Qiu, Z. Xu, J. Xu, A.S.C. Chan, R. Wang, ibid. *9* (12), 2329-32; with

CuOBu-*t*/chiral bis(ferrocenyl)di(phosphine) [(S,S)-(R,R)-Ph-TRAP] without R_2Zn (e.e. up to 56%) s. Y. Asano, K. Hara, H. Ito, M. Sawamura, ibid. *9* (20), 3901-4; with (R)-BINOL/N-methylimidazole/Ti(OPr-*i*)$_4$ (e.e. up to 94%) s. F. Yang, P. Xi, L. Yang, J. Lan, R. Xie, J. You, J. Org. Chem. *2007*, *72* (14), 5457-60; addition to aliphatic and α,β-ethylenic aldehydes with Me_2Zn and a chiral macrocyclic bis[3,3′-bis(alkylideneamino)-1,1′-bi-2-naphthol] (e.e. 89-96%) s. Z.-B. Li, T.-D. Liu, L. Pu, ibid. *72* (12), 4340-3; with a chiral *o*-α-aminophenol (e.e. up to 99%) s. X.-F. Yang, T. Hirose, G.-Y. Zhang, Tetrahedron:Asym. *2007*, *18* (22), 2668-73; with a chiral 2-α-hydroxypyridine s. S. Liebehentschel, J. Cvengroš, A.J. von Wangelin, Synlett *2007* (16), 2574-8.

Magnesium bromide s. under Mg *$MgBr_2$*
Zinc chloride s. under i-*Pr_2NLi* *$ZnCl_2$*

Indium *In*
3-Ethylenealcohols CO → C(OH)C-C=C
from oxo compds. and β,γ-ethylenehalides
with In cf. *43*, 564s*71*; diastereoselective synthesis of quaternary C-atoms from ketones and γ-subst. allyl halides and regioselectivity in dry THF or THF/water s. S.A. Babu, M. Yasuda, A. Baba, J. Org. Chem. *2007*, *72* (26), 10264-7; In-mediated allylation of trichloromethyl ketones in aq. media s. Z.J. Bao, J. Lu, S.J. Ji, Chin. J. Chem. *2007*, *18* (9), 1061-3.

Triorganoaluminum compds./cyclooctadiene(methoxy)rhodium(I) dimer/ (S)-2,2′-bis(diphenylphosphino)-1,1′-binaphthyl ←
Rhodium-catalyzed asym. synthesis of cyclic 2-ethylene-*tert*-alcohols CO → C(OH)R
from cyclic α,β-ethyleneketones

377.

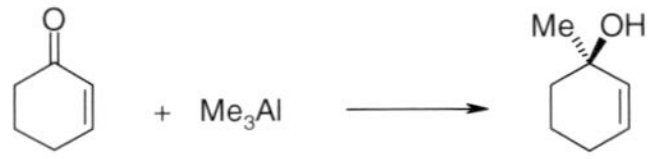

A slurry of 2.5 mol% $[Rh(cod)OMe]_2$ and 6 mol% (S)-BINAP in THF stirred for 30 min at room temp., cooled to 0°, equivalent amounts of trimethylaluminum (2 *M* in hexane) and cyclohex-2-enone added, stirred for 2 h, and worked up with column chromatographic purification on silica gel → product. Y 84% (e.e. 98%). This is the first example of a rhodium-catalyzed asym. 1,2-addition of a triorganoaluminum compd. to a cyclic enone [with no evidence of 1,4-addition!]. Yields and enantioselectivity for this inexpensive procedure were high with cyclohexenone and cycloheptenone derivs., but cyclopentenone derivs. and acyclic enones were low yielding. The corresponding chiral 1-arylcyclohex-2-enols were also obtained with freshly prepared aryl-(dimethyl)aluminum (*without* addition of the methyl residue!). Dimethylaluminum chloride and methoxide, however, gave inferior results. F.e.s. J. Siewert, R. Sandmann, P. von Zezschwitz, Angew. Chem. Int. Ed. Engl. *2007*, *46* (37), 7122-4.

Trimethylaluminum *Me_3Al*
α,β-Dialkoxy-β-siloxycarboxylic acid esters ←
from carboxylic acid esters and O-silyl O-alkyl alkoxyketeneacetals
Aluminum-mediated aldol-type condensation

378.

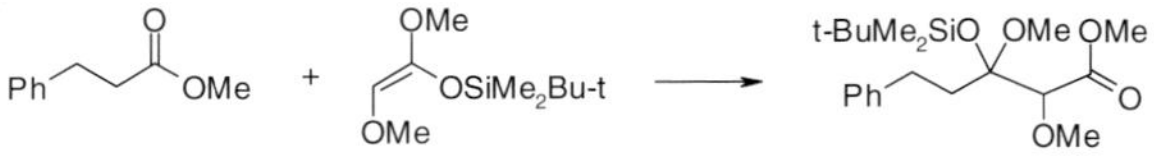

A mixture of the startg. methyl ester and 2 eqs. of the startg. O-silyl O-alkyl keteneacetal treated with 1.1 eqs. trimethylaluminum in toluene under argon at -78°, stirred for 6 h at 0°, and quenched with aq. $NaHCO_3$ → product. Y 95% (as a mixture of diastereoisomers). The procedure is applicable to aliphatic, aromatic and α,β-acetylenic carboxylic acid *methyl* esters, but methylaluminoxane was preferred for the reactions of methyl *p*-chlorobenzoate and 5-bromopentanoate. $TiCl_4$ was less effective, but there was no reaction with BF_3, $Sc(OTf)_3$ or *t*-BuMe$_2$SiOTf. F.e.s. S. Iwata, T. Hamura, T. Matsumoto, K. Suzuki, Chem. Lett. *2007*, *36* (4), 538-9.

Dibutylborinyl triflate/ethyldiisopropylamine *Bu_2BOTf/i-Pr_2NEt*
Aldol condensation with asym. induction CHO → CH(OH)C–CO
Effect of the order of addition on diastereoselectivity

379.

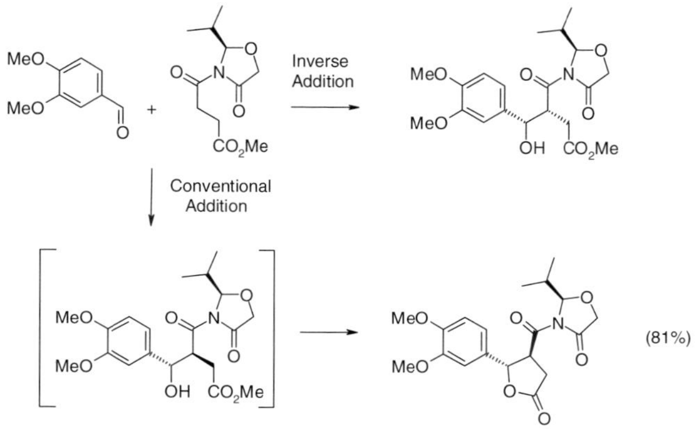

2.5 Eqs. dibutylboron triflate, 1.4 eqs. 3,4-dimethoxybenzaldehyde and 2.6 eqs. ethyldiisopropylamine added *in this order* to the startg. chiral N-acyl-2-oxazolidone in methylene chloride at -78°, and worked up when reaction complete → *anti*-aldol adduct. Y 92% (*anti/syn* >95:5). The presence of an additional chelating group in the N-acyl-2-oxazolidone has an impact on stereoselectivity, with the conventional order of addition (Lewis acid, then amine, then aldehyde) giving the *syn*-aldol (isolated as the corresponding γ-butyrolactone in 70-81% yield; *trans/cis* >95:5). The Lewis acid can be $TiCl_4$ (Y 41-88%) or *n*-Bu_2BOTf (Y 49-92%), although reaction was incomplete with cinnamaldehyde (37-48% recovery), and hexanal gave mixtures of unspecified products with each catalyst. Crotonaldehyde, however, underwent direct cyclization of the *anti*-aldol adduct to afford the *cis*-γ-butyrolactone in 56 or 66% yield with $TiCl_4$ or *n*-Bu_2BOTf, respectively. Stereochemistry was confirmed by X-ray diffraction. F.e.s. S. Hajra, A.K. Giri, A. Karmakar, S. Khatua, Chem. Commun. *2007* (23), 2408-10.

5,10,15,20-Tetrakis(4-chlorophenyl)porphyrinatobis(tetrahydrofuran)aluminum tetracarbonylcobaltate ←
Succinic anhydrides from epoxides
Biscarbonylation via β-lactones

380.

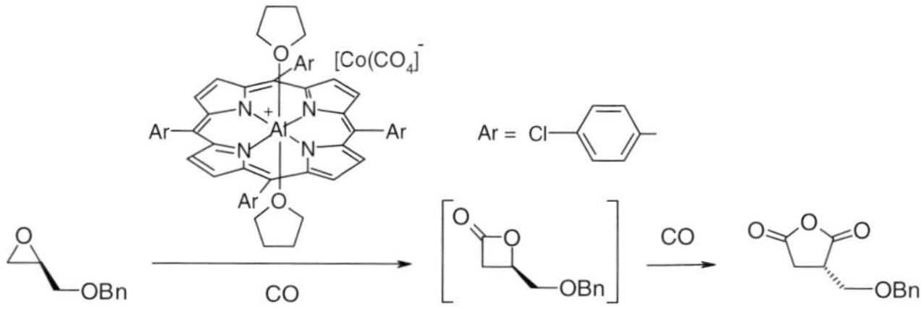

1 Mol% 5,10,15,20-tetrakis(4-chlorophenyl)porphyrinatobis(tetrahydrofuran)aluminum tetracarbonylcobaltate in *dioxane* stirred in a dry Parr high-pressure reactor under 200 psi CO for 10 min, the mixture vented down to 20 psi without stirring, startg. chiral epoxide injected at room temp., the reactor repressurized to 850 psi CO, the mixture stirred rapidly and heated at 90° for

24 h, the reactor cooled to <0°, slowly vented, solvents evaporated *in vacuo*, and the resulting oil distilled → 2-(benzyloxymethyl)succinic anhydride. Y 96% (e.e. 97%) (Y 87%; e.e. >99% after recrystallization from ether). The conversions were complete within 3 h at 90° for most examples using as little as *0.07 mol%* catalyst, but in the illustrated case the reaction is run at lower temp. to ensure clean **inversion of configuration**. Over 25 examples are given (Y 93-9%) using mainly racemic mono-subst. epoxides, while *cis/trans*-dialkyl-subst. epoxides undergo clean and near-complete inversion at both sites under these conditions. The reaction is compatible with ester, nitrile, amide, silyl ether, alkene, phenyl, ether and *alcohol* functionality, and work-up is straightforward requiring little more than removal of catalyst and solvent. The mechanism was studied in some detail and it was determined that formation of the intermediate β-lactone, via nucleophilic ring opening of the epoxide with tetracarbonylcobaltate anion, is essentially complete before the second carbonylation occurs. F.e. and optimization s. J.M. Rowley, E.B. Lobkovsky, G.W. Coates, J. Am. Chem. Soc. *2007*, *129* (16), 4948-60.

Aluminum fluoride s. under $Pd_2(dba)_3$ — *AlF_3*
Lanthanum oxide s. under Magnesium-lanthanum mixed oxide ←

Chiral lanthanum(III) triflate complex/1,8-bis(dimethylamino)naphthalene ←
Asym. nitro-Henry reaction s. *62*, 250s*72* — CO → C(OH)C(NO_2)

Alcohols and diols s. under R_2Zn ←
18-Crown-6 polyether s. under KCN — *crown*

*Chiral 3,3′-bimorpholine trifluoroacetates or triflates or N-(2′-Hydroxy-1,1′-binaphthyl-2-yl)-L-prolinamide or L-Proline diamides-acetic acid or Methyl 12-[prolinoyl-amino]cholate or Polyleucines or L-Proline or Polymer-based L-proline or O-*tert*-Butyl-L-threonine/5-methyltetrazole* ←
Organocatalyzed asym. aldol condensation — CO → C(OH)C-CO
update s. *58*, 245s*70*; with chiral 4-monosubst. 3,3′-bimorpholine triflates or trifluoroacetates (e.e. up to 94%), also intramolecular reaction (e.e. up to 95%), s. T. Kanger, K. Kriis, M. Laars, T. Kailas, A.-M. Müürisepp, T. Pehk, M. Lopp, J. Org. Chem. *2007*, *72* (14), 5168-73; with chiral N-(2′-hydroxy-1,1′-binaphthyl-2-yl)-L-prolinamide (e.e. up to 77%) s. A. Russo, G. Botta, A. Lattanzi, Synlett *2007* (5), 795-9; with L-proline diamides derived from chiral 1,1′-binaphthyl-2,2′-diamines/AcOH (e.e. up to 98%) s. G.-N. Ma, Y.-P. Zhang, M. Shi, Synthesis *2007* (2), 197-208; of isocyclic α-(arylthio)aldehydes with L-proline in DMSO s. A.M. Bernard, A. Frongia, R. Guillot, P.P. Piras, F. Secci, M. Spiga, Org. Lett. *2007*, *9* (3), 541-4; dynamic kinetic resolution of 2-oxo-3-arylsuccinates by L-proline-catalyzed addition of acetone in acetonitrile s. Y. Wang, Z. Shen, B. Li, Y. Zhang, Y. Zhang, Chem. Commun. *2007* (12), 1284-6; with polystyrene-supported or dendronized resin-based L-proline s. T. Kehat, M. Portnoy, ibid. *2007* (27), 2823-5; of unprotected dihydroxyacetone with O-*tert*-butyl-L-threonine/5-methyltetrazole (as aldolase mimics) in DMF (e.e. up to 98%, d.r. up to 15:1) s. S.S.V. Ramasastry, K. Albertshofer, N. Utsumi, F. Tanaka, C.F. Barbas III, Angew. Chem. Int. Ed. Engl. *2007*, *46* (29), 5572-5; reactions with protected dihydroxyacetone cf. N. Utsumi, M. Imai, F. Tanaka, S.S.V. Ramasastry, C.F. Barbas III, Org. Lett. *2007*, *9* (17), 3445-8; with polyleucines cf. G. Carrea, G. Ottolina, A. Lazcano, V. Pironti, S. Colonna, Tetrahedron:Asym. *2007*, *18* (11), 1265-8; with proline derivs. of bile acids, especially a methyl 12-[prolinoylamino]cholate, s. G.L. Puleo, M. Masi, A. Iuliano, ibid. *18* (11), 1364-75; of cyclic ketones in water or methylene chloride s. G.L. Puleo, A. Iuliano, ibid. *18* (24), 2894-900; in the presence of water (cf. *70*, 276) with a linear polystyrene-anchored L-proline s. Y.-X. Liu, Y.-N. Sun, H.-H. Tan, W. Liu, J.-C. Tao, ibid. *18* (22), 2649-56; with O-*tert*-butyldiphenylsilyl-L-proline s. Y.-C. Teo, ibid. *18* (10), 1155-8; with α-aminoacyl derivs. of glucose, e.g. methyl 2-(L-prolinoylamino)-2-deoxyglucosides s. A. Tsutsui, H. Takeda, M. Kimura, T. Fujimoto, T. Machinami, Tetrahedron Lett. *2007*, *48* (30), 5213-7; with L-prolinamide s. Y.-Y. Peng, H. Liu, M. Cui, J.-P. Cheng, Chin. J. Chem. *2007*, *25* (7), 962-7; hydrophobically directed

aldol reactions with [recyclable] polystyrene-supported L-proline in the presence of water (or methanol) s. M. Gruttadauria, F. Giacalone, A.M. Marculescu, P. L. Meo, S. Riela, R. Noto, Eur. J. Org. Chem. *2007* (28), 4688-98.

(2-Hydroxyethyl)trimethylammonium L*-prolinate* *[Choline][Pro]*
Imidazolium-based L*-proline/1-butyl-3-methylimidazolium fluoroborate* ←
Organocatalyzed asym. aldol condensation CO → C(OH)C-CO
in ionic liquids cf. *68*, 259; with a [recyclable] imidazolium-based L-proline in [bmim]BF_4 s. L. Zhou, L. Wang, Chem. Lett. *2007*, *36* (5), 628-9; with [reusable] (2-hydroxyethyl)trimethylammonium L-prolinate in water s. S. Hu, T. Jiang, Z. Zhang, A. Zhu, B. Han, J. Song, Y. Xie, W. Li, Tetrahedron Lett. *2007*, *48* (32), 5613-7.

1-Butyl-3-methylimidazolium hexafluorophosphate s. under KOH *[bmim]PF_6*
2,6-Bis[(S)-4-isopropyl-1-phenyl-Δ^2-imidazolin-2-yl]pyridine s. under Cu(OTf)$_2$ ←
Chiral fluorinated bis(Δ^2-oxazolines) s. under Cu(OTf)$_2$ ←

(S)-2-Amino-N-[(S)-1,1-bis[3,5-bis(trifluoromethyl)phenyl]-1-hydroxy-3-methylbutan-2-yl]-4-methylpentanamide ←
***syn*-α,β-Dihydroxyketones from aldehydes and α-hydroxyketones** CHO → CH(OH)C-CO
Organocatalyzed asym. aldol condensation

381.

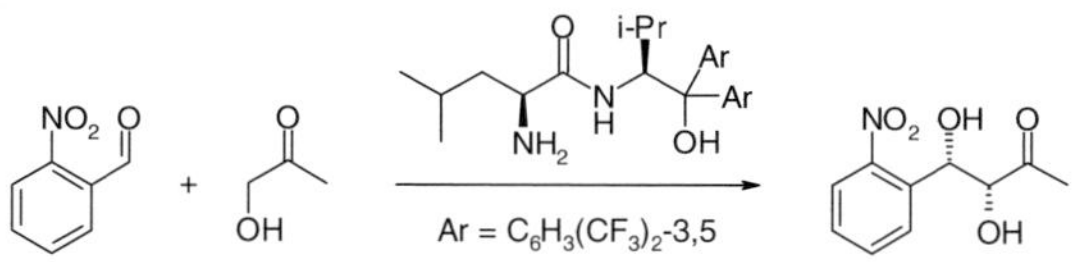

5 Mol% L-leucine-based catalyst added to a soln. of 0.3 mmol 2-nitrobenzaldehyde and 206 μL hydroxyacetone in *m*-xylene, stirred at 25° for 24 h, the reaction quenched with satd. aq. NH_4Cl, extracted with ethyl acetate, washed with brine, and dried → (3S,4R)-3,4-dihydroxy-4-(2-nitrophenyl)butan-2-one. Y 92% (*syn/anti* >20:1; e.e. 98%). The reaction is general for both ar. and aliphatic aldehydes and is independent of the size and electronic nature of ar. substituents affording *syn*-diols in high yields (generally >80%) and selectivities (e.e. >90%; *syn/anti* >10:1). Lowest yields were obtained with aliphatic aldehydes (*i*-PrCHO 45%; c-C_6H_{11}CHO 68%) but enantio- and diastereo-selectivity were still excellent. A second catalyst (with *gem*-diphenyl in place of *gem*-diaryl) promoted the reaction between benzaldehydes and fluoroacetone, chloroacetone or the less reactive 3-pentanone but with reduced selectivity. F.e.s. X.-Y. Xu, Y.-Z. Wang, L.-Z. Gong, Org. Lett. *2007*, *9* (21), 4247-9.

(S)-N-((S)-1-Hydroxy-4-methyl-1,1-diphenylpent-2-yl)pyrrolidine-2-carboxamide/brine ←
(S)-Pyrrolidine-2-carboxylic acid ((S)-2-hydroxy-1,1'-binaphthyl-2-yl)amide/trifluoroacetic acid ←
Stoichiometric organocatalyzed asym. aldol condensation in water
under inverse phase transfer catalysis s. *72*, 383

Amberlyst A-21 s. under Amberlite IRA-402 (nitrite) ←
Phenol s. under Bu$_3$P *PhOH*
1,1'-Bi-2-naphthols s. under R$_2$Zn *BINOLs*
m*-Nitrobenzoic acid s. under (1R,2R)-N,N-Dipropylcyclohexane-1,2-diamine-trifluoromethanesulfonic acid* *ArCOOH*

Mesoporous silica-supported (S)-proline ←

Heterogeneous organocatalyzed asym. aldol condensation CO → C(OH)C-CO

Improved procedure in hydrophilic polar solvents

382. 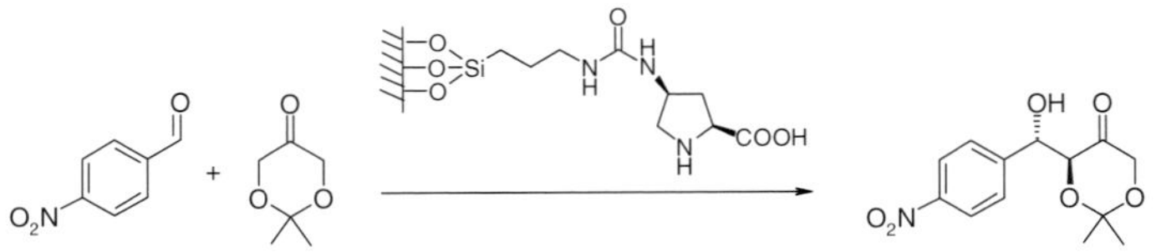

p-Nitrobenzaldehyde (10 mg) added to a suspension of startg. ketone (16.9 mg) and MCM41-Pro (30 mol%; 27.8 mg, loading: 0.7 mmol/g) *in DMF*, stirred *at room temp.* for 61 h, the catalyst removed by centrifugation, quenched with satd. aq. NH_4Cl, and worked up → product. Conversion 80% (*anti/syn* 5:1; e.e. 82%). With (S)-proline itself under homogeneous conditions in the same solvent the *anti/syn* selectivity was 2:1 (e.e. 63%). Reaction is notably efficient in *hydrophilic polar* solvents, as well as in hydrophobic toluene containing a little water. F.e. and comparison of solvents and support linkages, also reaction with chiral aldehydes to give adducts *en route* to azasugars, s. E.G. Doyaguez, F. Calderon, F. Sanchez, A. Fernandez-Mayoralas, J. Org. Chem. *2007*, *72* (24), 9353-6.

trans-*4-(4-*tert-*Butylphenoxy)-*L-*proline/sulfated β-cyclodextrin* ←

Stoichiometric organocatalyzed asym. aldol condensation in water under inverse phase transfer catalysis

383. 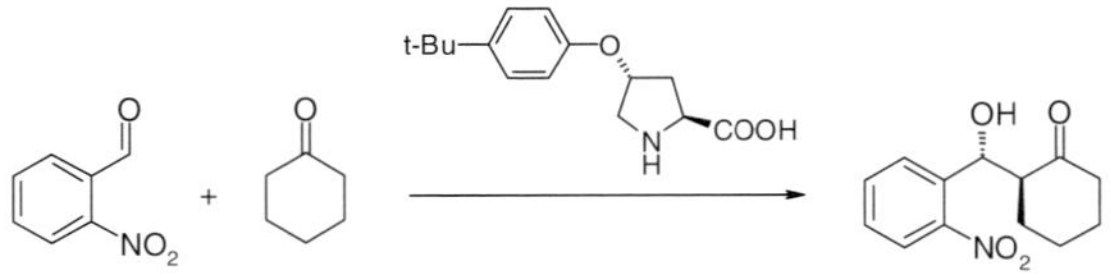

An asym. aldol condensation of ar. aldehydes with cyclohexanone *in stoichiometric* amounts can be achieved in water with the aid of a water-insoluble chiral proline deriv. as organocatalyst and water-soluble sulfated β-cyclodextrin *as inverse phase transfer catalyst*, reaction taking place *within a hydrophobic chiral pocket* embracing all reactants and catalysts in close proximity and diminishing contact between bulk water and the transition states. **E:** Cyclohexanone (5 mmol) and 1 eq. *o*-nitrobenzaldehyde added to a mixture of 2 mol% *trans*-4-(4-*tert*-butylphenoxy)-L-proline and 10 mol% sulfated β-cyclodextrin in water, stirred at room temp. for 48 h, the product phase separated, extracted with ethyl acetate, and worked up with flash chromatographic purification on silica → product. Y 97% (*anti/syn* >99:1; e.e. >99%). The procedure is simple in operation, environmentally friendly and affords adducts with very high *anti/syn* diastereoselectivity (88:12 to 96:4) and 96->99% enantioselectivity. The diastereo- and enantio-selectivity were generally lower for the same conversion **on water** (without β-cyclodextrin). F.e.s. J. Huang, X. Zhang, D.W. Armstrong, Angew. Chem. Int. Ed. Engl. *2007*, *46* (47), 9073-7; with a L-proline-functionalized β-cyclodextrin cf. K. Liu, D. Häussinger, W.-D. Woggon, Synlett *2007* (14), 2298-300; with (S)-pyrrolidine-2-carboxylic acid ((S)-2-hydroxy-1,1′-binaphthyl-2-yl)amide *and 10 mol% trifluoroacetic acid* with a ketone/aldehyde ratio of 2:1 cf. C. Wang, Y. Jiang, X.X. Zhang, Y. Huang, B.G. Li, G.L. Zhang, Tetrahedron Lett. *2007*, *48* (24), 4281-5; **in brine** with (S)-N-((S)-1-hydroxy-4-methyl-1,1-diphenylpent-2-yl)pyrrolidine-2-carboxamide and a ketone/aldehyde ratio of 4:1 cf. V. Maya, M. Raj, V.K. Singh, Org. Lett. *2007*, *9* (13), 2593-5.

Chiral 2,2′-bis(thioureido)-5,5′,6,6′,7,7′,8,8′-octahydro-1,1′-binaphthyl/ethyldiisopropylamine ←

Asym. nitro-Henry reaction s. *62*, 250s*72* CO → C(OH)C(NO_2)

Silica s. under (S)-Proline *SiO_2*

1-Methoxy-1-trimethylsiloxy-2,2-dimethylethylene *$Me_2C{=}C(OSiMe_3)OMe$*

α-Siloxynitriles from oxo compds. CO → C(OSi≤)CN

with N-oxides cf. *65*, 253s*67,68*; with 1-methoxy-1-trimethylsiloxy-2,2-dimethylethylene, retention of acid-sensitive groups, also cyanohydrin carbonates (cf. *39*, 895), s. X. Wang, S.-K. Tian, Synlett *2007* (9), 1416-20; from aldehydes with a little NbF_5 in the absence of solvent s. S.S. Kim, G. Rajagopal, Synthesis *2007* (2), 215-8.

Titanium tetraisopropoxide s. under R_2Zn *Ti(OPr-*i*)$_4$*

Chiral titanium(IV) N-β-hydroxysalen complex/cinchonine ←

Cyanohydrin carbonates from aldehydes – Asym. conversion CHO → CH(OCOOR)CN

and ethyl cyanoformate under Ti(IV) catalyst cf. *63*, 253s*66,70,71*; with a self-assembled Ti-catalyst from Ti(OPr-*i*)$_4$/chiral N-β-hydroxysalen/cinchonine s. W. Wang, S. Gou, X. Liu, X. Feng, Synlett *2007* (18), 2875-8.

Germanium dichloride-dioxane *$GeCl_2$-dioxane*

Reformatsky-type synthesis CHO → CH(OH)C-CO

with SmI_2 cf. *49*, 591s*65*; β-hydroxyaldehydes with $GeCl_2$-dioxane in THF, diastereoselectivity, also one-pot conversion to 1,3-diols or 3-aminoalcohols, s. S. Tanaka, M. Yasuda, A. Baba, Synlett *2007* (11), 1720-4.

Amberlite IRA-402 (nitrite)/Amberlyst A-21 ←

Heterogeneous Henry reaction CHO → CH(OH)CH(NO_2)

with prim. nitro compds. generated *in situ* from prim. halides

384.

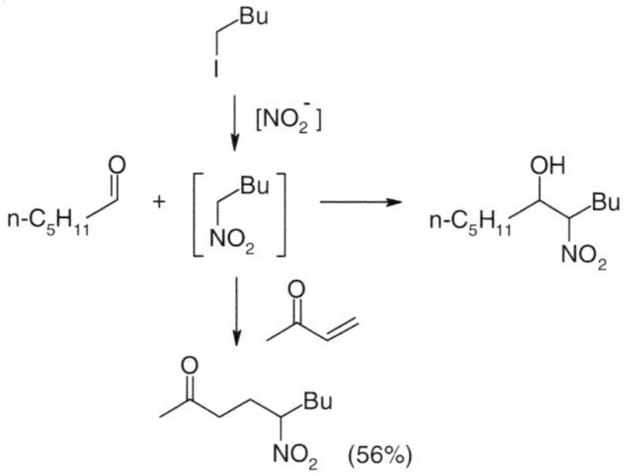

under mild conditions. Amberlite IRA-402 (nitrite) (800 mg) and Amberlyst A-21 (600 mg) added to a soln. of pentyl iodide (1 mmol) and the startg. aldehyde (1 mmol) in DMF, stirred for 8 h, filtered, and the filtrate worked up with flash chromatographic purification → product. Y 63%. The procedure avoids a complicated work-up, and both waste and energy consumption are reduced in comparison to existing methods. The weakly basic Amberlyst A-21 is required to deprotonate the generated aliphatic nitro compds. to form the corresponding reactive nitronates as nucleophiles. **Michael addition of *in situ*-generated prim. nitro compds.** was also effected under the same conditions. F.e. and with alkyl bromides (more slowly) s. R. Ballini, L. Barboni, A. Palmieri, Synlett *2007* (19), 3019-21.

Benzyltriphenylphosphonium chloride *$[BnPPh_3]Cl$*

α-Siloxynitriles from ketones CO → C(OSi≤)CN

Catalysis with phosphonium salts under mild conditions

385.

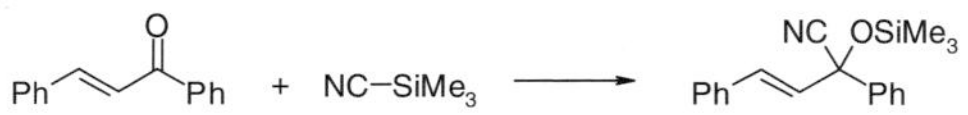

1.5 Eqs. trimethylsilyl cyanide and 5 mol% benzyltriphenylphosphonium chloride added to chalcone in dry chloroform, the mixture stirred at 50° for 24 h (monitored by GC/TLC), and purified directly by chromatography → cyanohydrin silyl ether. Y 98%. Eighteen ketones and enones were converted to the desired products in 90-99% yield at 25° or 50° with *1-5 mol%* of the catalyst (prepared from inexpensive triphenylphosphine and benzyl chloride), alkene, ester and acetal functionality being tolerated. The catalyst is thought to facilitate the reaction by **dual activation,** the phosphonium ion serving as Lewis acid to activate the ketone with the chloride counterion possibly serving to activate the silyl cyanide. F.e.s. X. Wang, S.-K. Tian, Tetrahedron Lett. *2007*, *48* (34), 6010-3.

Tributylphosphine/phenol *Bu_3P/PhOH*
Baylis-Hillman reaction CHO → CH(OH)C(=C)CO
with tert. phosphines cf. *39*, 593s*70*; of masked 5-alkylidene-2-cyclopentenones with Bu_3P/PhOH (followed by unmasking by flash vacuum pyrolysis) s. M. Pohmakotr, S. Thamapipol, P. Tuchinda, S. Prabpai, P. Kongsaeree, V. Reutrakul, J. Org. Chem. *2007*, *72* (14), 5418-20.

Tris(2,4,6-trimethoxyphenyl)phosphine *Ar_3P*
1-α-Siloxycyclopropenes from 1-silylcyclopropenes and oxo compds. ←
Sila-Baylis-Hillman reaction under mild conditions

386. 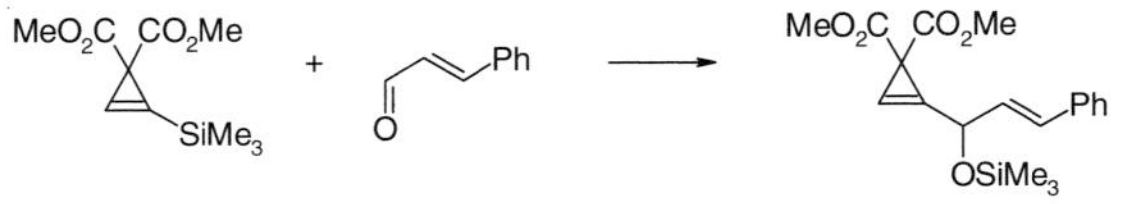

A mixture of startg. silylcyclopropene, 1.2 eqs. cinnamaldehyde and 1 mol% tris(2,4,6-trimethoxyphenyl)phosphine stirred in dry dioxane under N_2 until reaction complete (GC/MS), and purified directly by chromatography → dimethyl 1-[(2E)-1-trimethylsilyloxy-3-phenylprop-2-en-1-yl]cyclopropene-3,3-dicarboxylate. Y 76%. This first direct method for preparing such compds. is successful with aromatic and heteroaromatic aldehydes, *cinnamaldehyde and aliphatic aldehydes* (incl. sterically hindered α,α-dimethylphenylacetaldehyde) as well as α,α,α-trifluoroacetophenone and benzil with yields in the range 50-86%. The reaction is tolerant of halo, ester and ether functionality but cyclopropene substitution is limited to electron-withdrawing groups, unsymmetrically substituted cyclopropenes giving 1:1 mixtures of diastereomers. F.e. and prepn. of the startg. m. s. S. Chuprakov, D.A. Malyshev, A. Trofimov, V. Gevorgyan, J. Am. Chem. Soc. *2007*, *129* (48), 14868-9.

*2-(Di-*tert*-butylphosphino)biphenyl s. under Dicarbonylrhodium β-diketonates* *2-(*t*-Bu_2P)biphenyl*
*Tri-*o*-tolylphosphine s. under $Pd_2(dba)_3$* *Tol_3P*
*(4S)-*tert*-Butyl-2-[*o*-(diphenylphosphino)phenyl]-Δ^2-oxazoline s. under R_2Zn* ←
Chiral 2-(2-hydroxy-5-methylbenzylideneamino)-2′-diphenylphosphino-1,1′-binaphthyl s. under CuOTf ←
P-Chiral P,P′-di(alk-1-ynyl)-1,2-di(phosphines) s. under R_2Zn ←
(S)-2,2′-Bis(diphenylphosphino)-1,1′-binaphthyl s. under R_3Al *(S)-BINAP*
9,9-Dimethyl-4,5-bis(diphenylphosphino)xanthene s. under CuOTf *Xantphos*

Chiral tetraaminophosphonium chloride/potassium tert*-butoxide* *[(>N)$_4$P]Cl/KOBu-*t
Asym. Henry reaction under mild conditions CHO → CH(OH)C(NO_2)
with a chiral tetraaminophosphonium salt as catalyst

387. 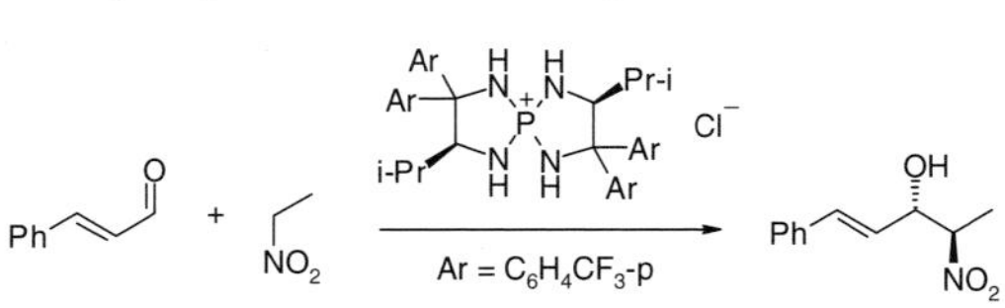

Chiral *anti*-2-nitroalcohols. 5 Mol% KOBu-*t* added to 5.5 mol% chiral tetraaminophosphonium chloride and nitroethane in THF under argon at -78°, the mixture stirred for 30 min, 1 eq. (E)-

cinnamaldehyde added dropwise, stirring continued for 21 h, the reaction quenched with 0.5 *M* trifluoroacetic acid in toluene, poured into ice-cold water, the aq. phase extracted with ethyl acetate, the extracts dried over Na_2SO_4, evaporated *in vacuo*, and the residue purified by chromatography → 1-phenyl-4-nitro-1-hexen-3-ol. Y 74% (*anti/syn* >19:1; e.e. 99%). The reaction is *anti*-selective for aromatic, heteroaromatic, unsaturated and aliphatic aldehydes with enantiomeric excesses of 93-99%, *anti/syn*-selectivity generally >19:1 (4:1 for aliphatic aldehydes) and yields of 74-96%. The high degree of selectivity is attributed to the rigid spirocyclic core of the phosphorus-centred catalyst. F.e. and catalyst development s. D. Uraguchi, S. Sakaki, T. Ooi, J. Am. Chem. Soc. *2007, 129* (41), 12392-3.

Niobium pentafluoride *NbF_5*
α-Siloxynitriles from oxo compds. s. *65*, 253s*72* CO → C(OSi<)CN

*Chiral 2-trifluoromethanesulfonylamino-4,5-dihydro-3*H*-dinaphtho[2,1-*c*:1′,2′-*e*]azepine* ←
Organocatalyzed asym. cross-aldol condensation of aldehydes CHO → CH(OH)C-CHO

388. 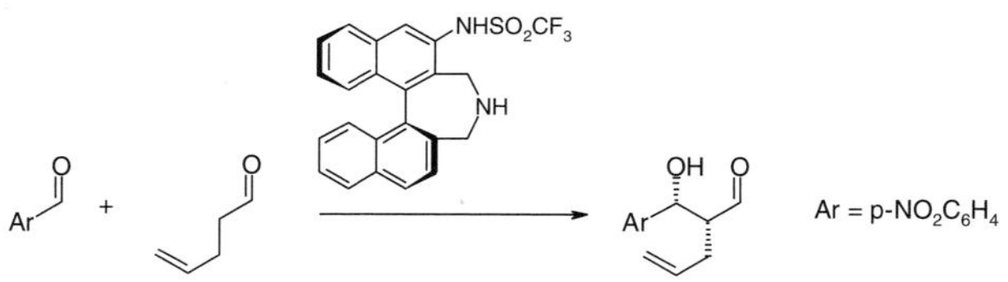

A rare example of an asym. syn-*selective* cross-aldol condensation of aldehydes is reported with the aid of an *axially-chiral N-sulfonyldiamine* as organocatalyst. **E:** *p*-Nitrobenzaldehyde and 2 eqs. 4-pentenal added sequentially to a stirred soln. of 5 mol% chiral 2-trifluoromethane-sulfonylamino-4,5-dihydro-3*H*-dinaphtho[2,1-*c*:1′,2′-*e*]azepine in NMP *at room temp.*, and stirred for 36 h → product. Y 79% (*syn*-selectivity 95%; e.e. 98%). Reaction is also applicable to fluoroar. and hetar. aldehydes as well as phenylglyoxal as the acceptor aldehyde. Intermediate formation of a *syn*-enamine is presumed responsible for the unusual stereoselectivity. F.e.s. T. Kano, Y. Yamaguchi, Y. Tanaka, K. Maruoka, Angew. Chem. Int. Ed. Engl. *2007, 46* (10), 1738-40.

(1R,2R)-N,N-Dipropylcyclohexane-1,2-diamine-trifluoromethanesulfonic acid/ ←
m-*nitrobenzoic acid*
Organocatalyzed asym. aldol condensation in the absence of solvent CHO → CH(OH)C-CO

389. 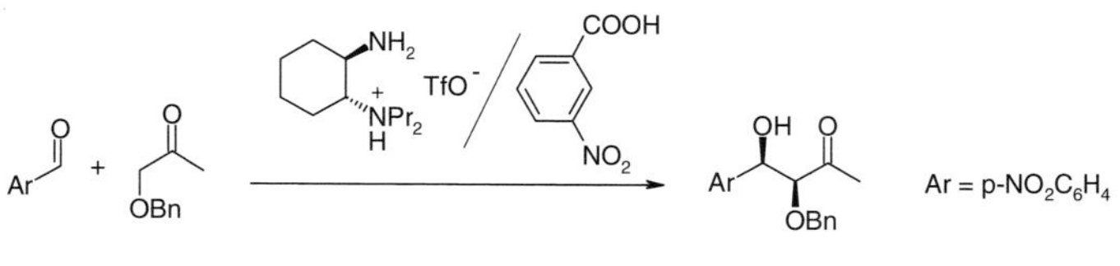

Chiral *syn*-β-hydroxyketones. The startg. aldehyde, 10 mol% *m*-nitrobenzoic acid and 10 mol% (1R,2R)-N,N-dipropylcyclohexane-1,2-diamine-trifluoromethanesulfonic acid complex added to 20 eqs. startg. anhydrous ketone, stirred *at room temp.* for 20-72 h, and worked up with flash column chromatographic purification on silica gel → product. Y 99% (*syn/anti* 10:1; e.e. >99%). Notable features include the high enantioselectivity, regioselectivity and unprecedented *syn*-selectivity, as well as the use of a chiral *prim-tert*-1,2-diamine as auxiliary in the presence of a strong acid and a secondary acidic catalyst, which is thought to facilitate the catalytic cycle. Significantly, the procedure is also applicable to *small* aliphatic ketones, but limited to condensation with ar. aldehydes (with either electron-donating or -withdrawing substituents). The formation of an intermediate (Z)-enamine is presumed responsible the high *syn*-selectivity. F.e.s. S. Luo, H. Xu, J. Li, L. Zhang, J.-P. Cheng, J. Am. Chem. Soc. *2007, 129* (11), 3074-5; in the absence of solvent with recoverable (S_a)-binam-L-prolinamide/benzoic acid as catalyst cf. G. Guillena, M. del Carmen Hita, C. Najera, S.F. Viozquez, Tetrahedron:Asym. *2007, 18* (19), 2300-4.

Sulfated β-cyclodextrin s. under trans-*4-(4-*tert-*Butylphenoxy)-*L*-proline* ←

Chiral chromium(III) salen complex/ethyldiisopropylamine *[Cr]*
Asym. nitro-Henry reaction s. *62*, 250s*72* CO → C(OH)C(NO_2)

*Tetra-*n*-butylammonium bromide s. under $Pd_2(dba)_3$* *Bu_4NBr*
*Chiral ferrocenyldi(phosphine) s. under CuOBu-*t ←
Iron(III) chloride s. under Mg *$FeCl_3$*
Tetracarbonylcobaltates s.a. 5,10,15,20-Tetrakis(4-chlorophenyl)porphyrinato-bis(tetrahydrofuran)aluminum tetracarbonylcobaltate ←

Tetracarbonylhydridocobalt(I) *$HCo(CO)_4$*
β-Hydroxy-δ-lactones from 3,4-epoxyalcohols via carbonylation OC

390. 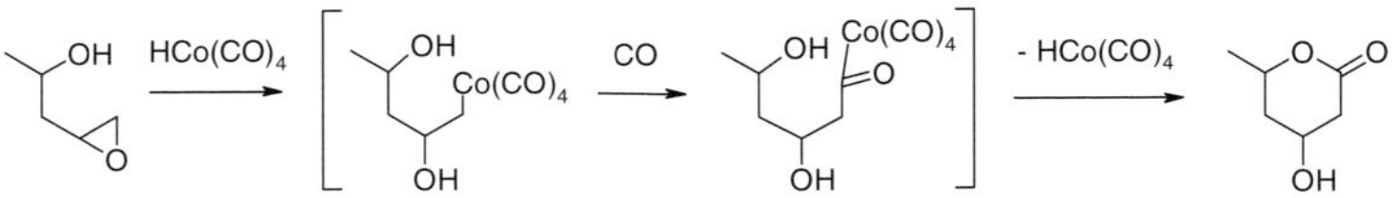

2 Mol% $HCo(CO)_4$ (0.05 *M* in DME) added via syringe to a vial containing the startg. epoxide (in a high-pressure stainless steel reactor), the reactor sealed, pressurized to 800 psi CO, heated at 60° for 24 h with stirring, and worked up with chromatographic purification → product. Y 73%. Competing β-lactone ring closure was only evident to a small degree with the more hindered 3,4-epoxy-*tert*-alcohols. Reaction is presumed to involve initial protonation and ring opening of the epoxide with the catalyst, followed by carbonylation of the generated alkylcobalt complex prior to ring closure with elimination of cobalt. F.e. incl. δ-aryl-derivs. s. J.W. Kramer, D.Y. Joh, G.W. Coates, Org. Lett. *2007*, *9* (26), 5581-3.

Chiral silylated cobalt(III) salen complex *[Co(III)]*
Asym. carbonyl-ene reaction ←
s. *56*, 242s*72*

Chiral nickel(II) α-aminocarboxamide complexes s. under R_2Zn *[Ni(II)]*

*Dicarbonylrhodium β-diketonates/2-(di-*tert-*butylphosphino)biphenyl* ←
2-Acetylenealcohols from oxo compds. s. *19*, 734s*72* CO → C(OH)C≡C

Cyclooctadiene(methoxy)rhodium(I) dimer s. under R_3Al *$[Rh(cod)OMe]_2$*
Chloro(norbornadiene)rhodium(I) dimer s. under R_2Zn *$[Rh(nbd)Cl]_2$*

*Tris(dibenzylideneacetone)dipalladium/tri-*o*-tolylphosphine/aluminum fluoride/tetra-*n*-butylammonium bromide* ←
α-Acylaminocarboxylic acid aryl esters from 1,1-aryloxyacylamines by carbonylation OAr → COOAr

391. 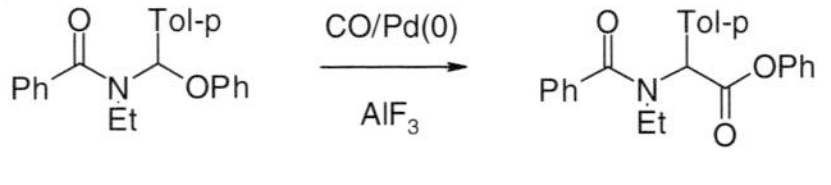

under mild conditions. Carbon monoxide (60 psi) added to a reaction bomb containing a degassed soln. of N-ethyl-N-[phenoxy(*p*-tolyl)methyl]benzamide, 10 mol% $Pd_2(dba)_3$, 0.5 eq. Bu_4NBr and 15 mol% tri-*o*-tolylphosphine in acetonitrile, stirred at 65° for 14 h, solvent removed *in vacuo*, and the resulting oil chromatographed → phenyl N-ethyl-α-benzamido-*p*-tolylacetate. Y 83%. Initial experiments using BF_3 as the Lewis acid allowed the isolation of a stable palladacycle as the fluoroborate salt which *did not* undergo the illustrated insertion reaction. With the less acidic AlF_3 as Lewis acid, however, no such complex could be formed and the insertion reaction was successful with substrates possessing a range of bulky and electron-withdrawing substituents in generally good yields (10 examples; 45-94%). F.e.s. Y. Lu, B.A. Arndtsen, Org. Lett. *2007*, *9* (21), 4395-7.

Dichloro(cyclooctadiene)palladium(II) s. under R_2Zn *$PdCl_2(cod)$*

Dichloro[(S)-2,3:2′,3′-bis(methylenedioxy)-6,6′-bis(diphenylphosphino)biphenyl]-palladium(II)/silver hexafluoroantimonate ←

Asym. carbonyl-ene reaction with enoxysilanes ←

392.

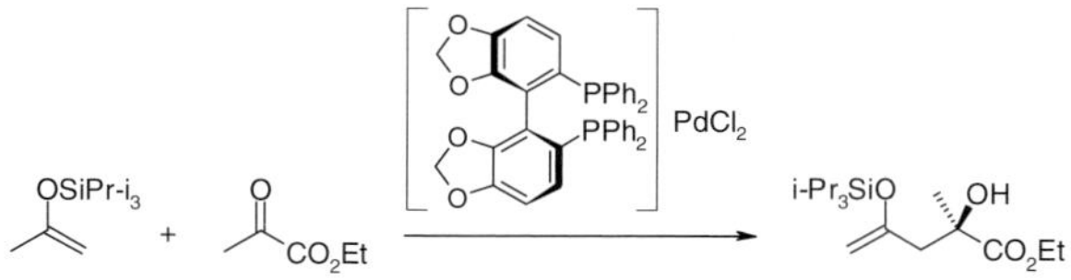

Chiral α-subst. γ,δ-ethylene-α-hydroxy-γ-siloxycarboxylic acid esters. *0.022 Mol%* $AgSbF_6$ added at room temp. to *0.01 mol%* $PdCl_2$[(S)-SegPhos] in methylene chloride, stirred for 30 min, 2.2 eqs. ethyl pyruvate and the startg. enoxysilane added at 0°, stirring continued for 48 h, treated with triethylamine, and worked up with chromatographic purification on silica gel → product. Y 85% (e.e. 90%). This is the first such carbonyl-ene reaction which generates **chiral quaternary hydrocarbon groups** at the α-position. The method is applicable *at very low catalyst loading* to both α-ketocarboxylic acid esters, ethyl glyoxylate and α-diketones, symmetrical derivs. of the latter participating **with desymmetrization** while unsymmetrical derivs. reacted with complete regioselectivity. An **asym. two-directional carbonyl-ene reaction** was also effected by subsequent addition of the product to a second α-ketocarbonyl compd. F.e. and comparison of catalysts and siloxy groups on the enoxysilane, also hydrolysis of the products to the corresponding **chiral α-hydroxy-γ-ketocarbonyl compds.**, s. K. Mikami, Y. Kawakami, K. Akiyama, K. Aikawa, J. Am. Chem. Soc. *2007*, *129* (43), 12950-1.

Addition to Nitrogen and Carbon CC ⇓ NC

Irradiation s. under $Mn_2(CO)_{10}$ ///

Electrolysis ϟ

N-Protected 2-nitramines C=N- → C(NH-)C(NO_2)

from imines and aliphatic nitro compds.

Electrochemical aza-Henry reaction in the absence of solvent

393.

Nitromethane (1 ml) electrolyzed *without supporting electrolyte* in the cathode compartment of a diaphragm-divided cell fitted with a spiral Pt cathode and a cylindrical Pt spiral anode under galvanostatic control (*J* 40 mA/cm^2) under inert atmosphere until *0.1 F/mol* (relative to imine) passed, the imine (0.2 mmol) added, stirred for 5 min at room temp., and worked up after 5 min → product. Y 92%. Cf. *39*, 578s*47,49*. The method is mild, clean, high-yielding, and avoids classical solvent and catalyst. It is applicable to the reaction of aromatic or heteroaromatic N-(diphenylphosphinoyl)- or N-tosyl-aldimines with a number of nitroalkanes, and is unaffected by the electronic nature of substituents on the aromatic ring. Only a catalytic quantity of current is required to generate a stabilized nitroalkane carbanion which adds to the imine, the charged product then deprotonating another molecule of the nitroalkane to continue the cycle. F.e.s. L. Rossi, G. Bianchi, M. Feroci, A. Inesi, Synlett *2007* (16), 2505-8.

Lithium p-*methoxyphenoxide s. under Lanthanum(III)* p-*methoxyphenoxide* *LiOAr*

n-*Butyllithium* *BuLi*
Lithium bis(trimethylsilyl)amide *$LiN(SiMe_3)_2$*
Asym. synthesis of sulfinylamines from sulfinylimines C=NS(O)R → C(R′)NHS(O)R
cf. *50*, 380s*71*; β-(sulfinylamino)hydroxamic acid esters (as building blocks for β-amino-carboxylic acids or -oxo compds.) with $LiN(SiMe_3)_2$ s. F.A. Davis, M. Song, Org. Lett. *2007*, *9* (12), 2413-6; 2-α-(sulfinylamino)indoles from indoles using *n*-BuLi s. L. Cheng, L. Liu, Y. Sui, D. Wang, Y.-J. Chen, Tetrahedron:Asym. *2007*, *18* (15), 1833-43.

Chiral copper(I) N,N′-propylenebis(piperidine N-oxide) complexes/ethyldiisopropylamine ←
[(1R,2R)-N,N′-Bis(3-hydroxysalicylidenato)-1,2-cyclohexanediamine]copper(II)/ ←
*samarium triisopropoxide/4-*tert*-butylphenol*
N-Protected *syn*-2-nitramines C=N- → C(NH-)C(NO_2)
from imines and aliphatic nitro compds.
Asym. aza-Henry reaction

394. 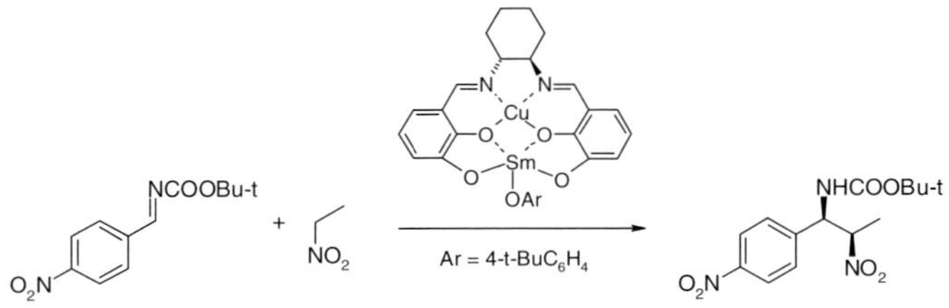

The first example of a highly enantiospecific syn-*selective* aza-Henry reaction is reported **under heterobimetallic catalysis. E:** 10 Mol% Sm(OPr-*i*)$_3$ (0.2 *M* in THF) added to a suspension of 10 mol% [(1R,2R)-N,N′-bis(3-hydroxysalicylidenato)-1,2-cyclohexanediamine]copper(II) in THF, stirred at 80° for 1 h, the resulting bimetallic complex cooled to room temp., 10 mol% 4-*tert*-butylphenol (0.2 *M* in THF) added, stirring continued for 15 min, cooled to -40°, 10 eqs. nitroethane added, stirred again for 1 h, a soln. of the startg. protected imine in THF added, stirring continued for 48 h, diluted with ether, filtered, and the filtrate worked up with flash column chromatographic purification on silica gel → product. Y 90% (d.e. >20:1; e.e. 98%). The ratio of the two metals in the complex is critical and there was no reaction with copper(II) or samarium(III) alone. The procedure is applicable to ar. or hetar. N-carbo-*tert*-butoxyimines, and both electron-withdrawing and -donating substituents were tolerated. F.e.s. S. Handa, V. Gnanadesikan, S. Matsunaga, M. Shibasaki, J. Am. Chem. Soc. *2007*, *129* (16), 4900-1; asym. aza-Henry reaction **with N-tosylimines** in the presence of chiral copper(I) N,N′-propylene-bis(piperidine N-oxide) complexes and ethyldiisopropylamine as base cf. H. Zhou, D. Peng, B. Qin, Z. Hou, X. Liu, X. Feng, J. Org. Chem. *2007*, *72* (26), 10302-4.

*Copper(II) triflate/chiral N-(*o*-hydroxybenzylidene)-N′-tosyl-1,2-diamines/dimethylzinc* ←
Asym. synthesis of 2-acetyleneamines from azomethines C=NR → C(NHR)C≡C
with CuOTf/2,2′-diamino-1,1′-binaphthyl cf. *66*, 290s*69*; with Cu(OTf)$_2$/chiral [tridentate] N-(*o*-hydroxybenzylidene)-N′-tosyl-1,2-diamine/Me$_2$Zn (e.e. up to 92%) s. B. Liu, L. Huang, J. Liu, Y. Zhong, X. Li, A.S.C. Chan, Tetrahedron:Asym. *2007*, *18* (24), 2901-4; effects of the ligand-to-metal ratio and additives on stereofacial selection in the Cu(I)-catalyzed asym. addition of arylacetylenes to α-imino esters (e.e. up to 74%) s. Z. Shao, J. Wang, K. Ding, A.S.C. Chan, Adv. Synth. Catal. *2007*, *349* (14-15), 2375-9.

Silver acetate/chiral 3,3′-diaryl-1,1′-bi-2-naphthol cyclic phosphates ←
Asym. synthesis with organometallics ←
by simultaneous activation with a chiral Brønsted acid and metal salt in catalytic amount

395.

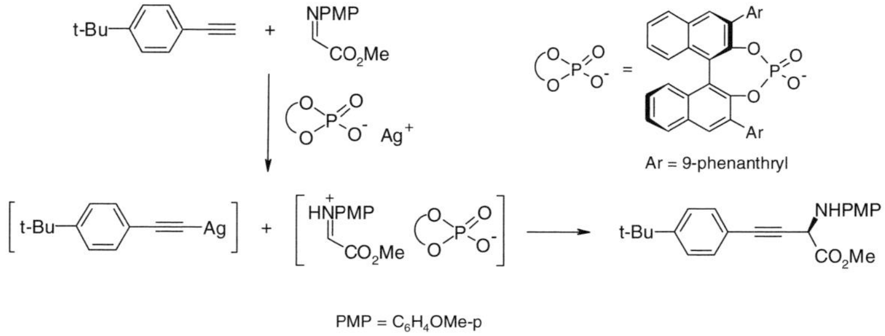

Dual catalysis is exemplified for the first time by simultaneous electrophilic activation with a chiral Brønsted acid and nucleophilic activation with a metal salt, each in catalytic amount. **E: Asym. synthesis of β,γ-acetylene-α-aminocarboxylic acid esters from terminal acetylene derivs.** A suspension of the startg. methyl 2-(arylimino)acetate, 2 eqs. startg. alkyne, 10 mol% chiral 3,3′-diaryl-1,1′-bi-2-naphthol phosphate and 5 mol% AgOAc in toluene stirred for 10-12 h in a screw-capped vial → product. Y 80% (e.r. 96:4). The procedure is mild and simple, and does not require pre-formation of catalyst. It is thought that reaction involves intermediate formation of a chiral silver(I) BINOL phosphate complex, wherein the *chiral counterion* induces the asymmetry. F.e. and comparison of silver and copper salts, also solvent effect and with alternative N-protecting groups, s. M. Rueping, A.P. Antonchick, C. Brinkmann, Angew. Chem. Int. Ed. Engl. *2007*, *46* (36), 6903-6.

Silver hexafluorophosphate *$AgPF_6$*
2-Aryltosylamines from N-tosylaziridines and arenes N-aziridine → C(NH)C(Ar)
Regiospecific silver-catalyzed ring opening under mild conditions

396.

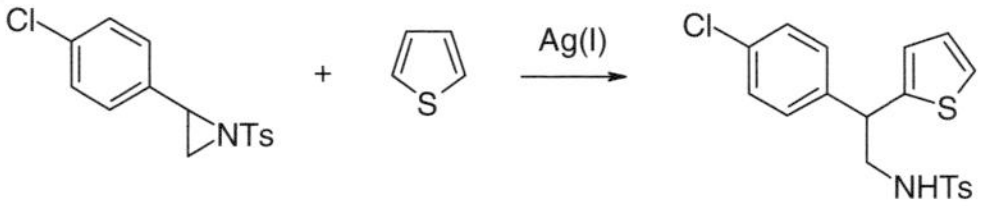

2,2-Diaryltosylamines. A mixture of 2-(4-chlorophenyl)-N-tosylaziridine, 2 eqs. thiophene and 2 mol% *$AgPF_6$* stirred in dry *dichloroethane* at room temp. for 1 h (TLC monitoring), the mixture diluted with water, extracted with methylene chloride, the extracts dried over Na_2SO_4, concentrated *in vacuo*, and the residue purified by chromatography → N-tosyl-2-(4-chlorophenyl)-2-(2-thienyl)ethylamine. Y 85%. The method is clean, rapid and simple in execution for ring-activated arenes and heteroarenes reacting exclusively at the benzylic position giving 75-85% yield and reaching completion in 0.5-3 h at room temp. However, the reaction fails with alkyl groups in place of aryl on the aziridine ring as well as in the absence of added dichloroethane. The catalytic activity is attributed to the known binding ability of Ag(I) towards both arenes and aziridines. F.e.s. M. Bera, S. Roy, Tetrahedron Lett. *2007*, *48* (40), 7144-6.

Barium aroxides *Ba(OAr)₂*

Stereospecific barium-catalyzed Mannich reaction ←
of N-protected carboxylic acid amides with N-phosphinylimines under mild conditions

397. 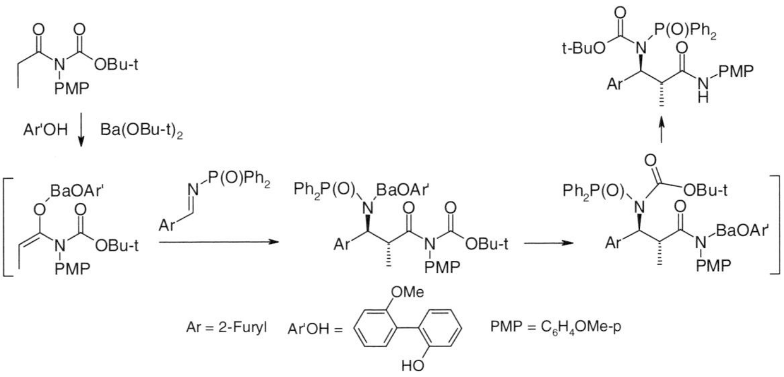

N$^{\beta}$-Diprotected *anti*-β-aminocarboxylic acid amides. A mixture of startg. N-phosphinylimine and Boc-protected N-*p*-methoxyphenylamide stirred with 10 mol% Ba(II)-*tert*-butoxide, 22 mol% 2-hydroxy-2′-methoxybiphenyl and 5 Å molecular sieves in DMF at 0° for 48 h → product. Y 81% (*anti/syn* 91:9). The bulky biphenyl ligand gave the highest stereoselectivity of several examined, giving 84:16 to 91:9 *anti*-selectivity (Y 76-89%) for N-diphenylphosphinylimines derived from aryl or hetaryl aldehydes. In initial experiments benzenesulfonyl was also found to be a useful protecting group, but no conversion was observed with phenyl- or benzyl-protected analogs. Reactions are successful in polar solvents (DMF or DMSO) and are generally complete within 30 min at 0° with acetamide derivs. using *p*-methoxyphenol as ligand (Y 63-95%). F.e.s. S. Saito, T. Tsubogo, S. Kobayashi, Chem. Commun. *2007* (12), 1236-7.

α,β-Ethylene-β′-(phosphinylamino)carboxylic ←
from β,γ-ethylenecarboxylic acid esters and N-phosphinylimines
Regiospecific barium-catalyzed Mannich reaction with double bond shift

398. 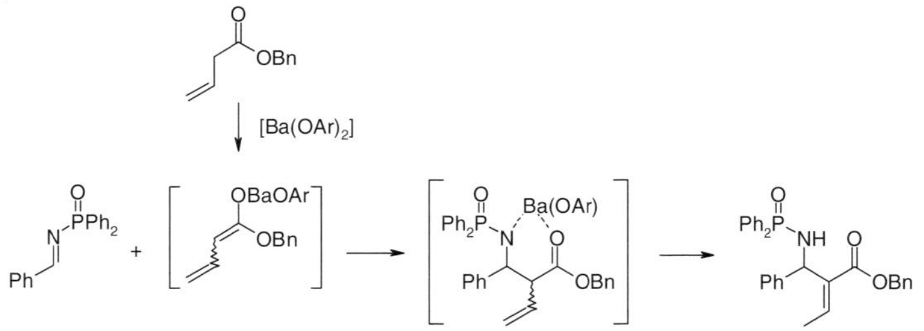

10 Mol% Ba(OPr-*i*)$_2$ (0.2 *M* in THF) added to a soln. of 20 mol% 4-methoxyphenol in THF at 0°, stirred for 5 min, the startg. imine and 1.3 eqs. benzyl 3-butenoate in THF added, stirring continued for 17 h at 0°, quenched with satd. aq. NH_4Cl, and worked up with flash column chromatographic purification on silica gel → product. Y 74% (α/γ-substitution >15:1). This is a useful alternative route to *β-subst.* aza-Baylis-Hillman adducts which are difficult to prepare by the conventional Baylis-Hillman route, reaction taking place via regiospecific α-addition of a barium dienolate to the imine. Lithium aroxides also promoted the α-addition-isomerization, but

other alkaline earth salts and rare earth metal salts were ineffective. The procedure is applicable to N-diphenylphosphinylimines derived from aryl, hetaryl and aliphatic aldehydes, but yields were low with the corresponding N-tosyl- and N-carbo-*tert*-butoxy-imines. An **asym. variant** was also devised based on a chiral (S)-2,2′-dihydroxybiphenyl as auxiliary (with e.e. up to 80%). F.e.s. A. Yamaguchi, N. Aoyama, S. Matsunaga, M. Shibasaki, Org. Lett. *2007*, *9* (17), 3387-90.

Dimethylzinc s. under $Cu(OTf)_2$ *Me_2Zn*

Magnesium-lanthanum mixed oxide ←
α-Diazo-β-(sulfonylamino)carboxylic acid esters C=NTs → $C(NHTs)C(N_2)COOR$
from tosylimines s. *26*, 668s*72*

Magnesium perchlorate/chiral bis(Δ^2-oxazolines)/ethyldiisopropylamine ←
Imidazolidine-2-thione-4-carboxylic ○
from α-isothiocyanatocarboxylic acid derivs. and tosylimines
Catalytic *anti*-selective asym. Mannich reaction

399. 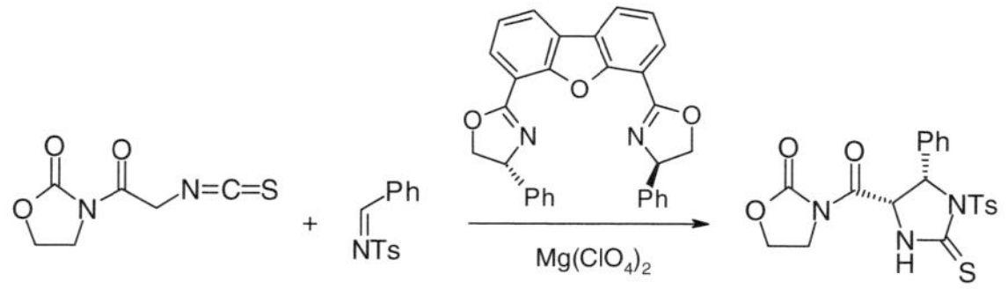

A novel asym. synthesis of **chiral N-protected *anti*-α,β-diaminocarboxylic acid derivs.** is reported. **E:** A mixture of 10 mol% $Mg(ClO_4)_2$, 11 mol% (R,R)-4,6-dibenzofurandiyl-2,2′-bis(4-phenyl-Δ^2-oxazoline) and 3-(2-isothiocyanatoacetyl)-2-oxazolidone in dry methylene chloride stirred for 1 h with activated powdered 4 Å molecular sieves under N_2 at room temp., cooled to -78°, after 15 min 2 eqs. N-tosylbenzaldimine and 20 mol% ethyldiisopropylamine added, stirring continued for 24 h at the same temp., quenched with satd. aq. NH_4Cl, and worked up with flash chromatographic purification on silica gel → product. Y 94% (*anti/syn* 88:12; e.e. 96%). The procedure is high-yielding, predominantly *anti*-selective [unlike the corresponding conversion with aldehydes] and highly enantioselective for the reaction of N-tosylaldimines derived from ar., hetar. and aliphatic aldehydes, both electron-donating and -withdrawing groups being tolerated in the aromatic ring. The diastereoselectivity arises from the (E)-configuration of the startg. N-tosylaldimines which forces coordination of the Lewis acid *syn* to the imino group. F.e.s. G.A. Cutting, N.E. Stainforth, M.P. John, G. Kociok-Köhn, M.C. Willis, J. Am. Chem. Soc. *2007*, *129* (35), 10632-3.

Triethylborane/air *Et_3B/air*
α-*tert*-Aminocarboxylic acid amides from isocyanates and tert. amines H → CONHR
Radical α-carbamylation

400. 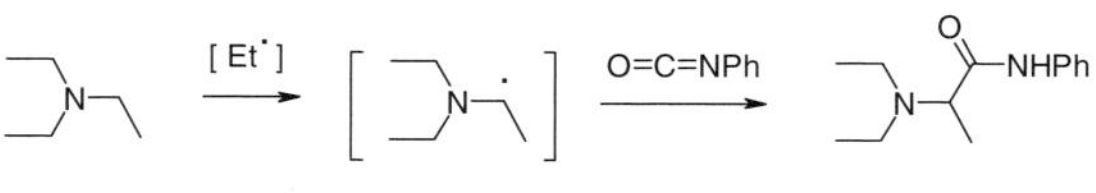

Triethylborane (9 mmol) added to a soln. of phenyl isocyanate (1.5 mmol) in triethylamine (52.5 mmol) at room temp. under argon, the argon balloon removed, the mixture stirred with continuous bubbling of dry air for 7 h, methanol added, stirring continued for a further 1 h with continuation of the air supply until the trialkylborane no longer remained, concentrated under reduced pressure, and worked up with flash column chromatographic purification on silica gel → 2-(diethylamino)-N-phenylpropanamide. Y 82%. The procedure is applicable to aliphatic tert. amines and dialkyl(aryl)amines, but yields decreased with increase in the alkyl chain length. Low yields were recorded, however, with triisopropylamine, while 5- and 6-membered cyclic tert. amines gave a mixture of carbamylation products. F.e.s. T. Yoshimitsu, K. Matsuda, H. Nagaoka, K. Tsukamoto, T. Tanaka, Org. Lett. *2007*, *9* (24), 5115-8.

Trimethylaluminum or Ynalanes *Me₃Al or RC≡AlR'₂*

2-Acetylenehydroxylamines from nitrones and terminal acetylene derivs. C(N(R)OH)C≡C with $Zn(OTf)_2$/*i*-Pr_2NEt cf. *58*, 259; with 1 eq. Me_3Al in methylene chloride s. T. Bunlaksananusorn, T. Lecourt, L. Micouin, Tetrahedron Lett. *2007*, *48* (8), 1457-9; from ynalanes and carbohydrate-derived nitrones in toluene, diastereoselectivity, s. C. Pillard, V. Desvergnes, S. Py, ibid. *48* (35), 6209-13.

*2-Diisopropylamino-4*H*-benzo[1,3,2]dioxaborinane* ←

α-*tert*-Aminocarboxylic acid amides CHO → CH(N<)CONHR
from aldehydes, isonitriles and sec. amines
Ugi-type reaction without carboxylic acids

401. 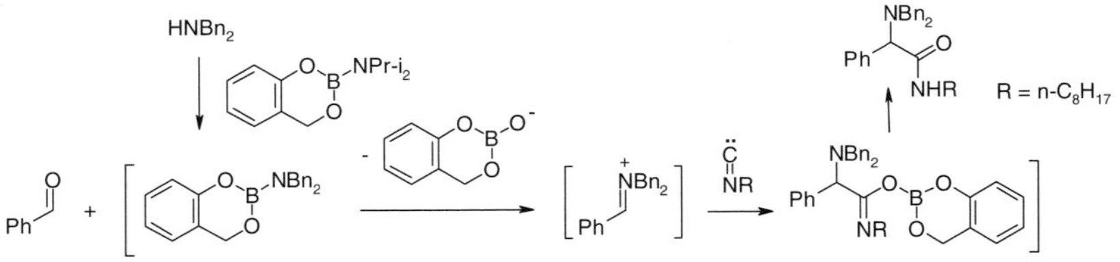

Dibenzylamine (0.6-0.8 mmol), benzaldehyde (0.4 mmol) and *n*-octyl isocyanide added at room temp. to a soln. of 2-diisopropylamino-4*H*-benzo[1,3,2]dioxaborinane (0.6 mmol) in THF, stirred for 12-16 h, quenched with water, and worked up → product. Y 85%. This is the first example of a Ugi-type reaction with *sec.* amines which, significantly, does not require a carboxylic acid, the cyclic aminoboronate effectively serving as an **aldiminium ion generator**. The procedure is generally applicable to the condensation of aromatic, heteroaromatic and aliphatic aldehydes (irrespective of the electronic nature of substituents in the aromatic ring) with aliphatic or cyclic sec. amines. Prim. amines simply gave the corresponding imines under these conditions. F.e.s. Y. Tanaka, T. Hasui, M. Suginome, Org. Lett. *2007*, *9* (22), 4407-10.

Aluminum chloride *$AlCl_3$*

Ar. aldimines from electron-rich arenes and isonitriles H → CH=NR
Lewis acid-catalyzed ar. imidoylation

402. 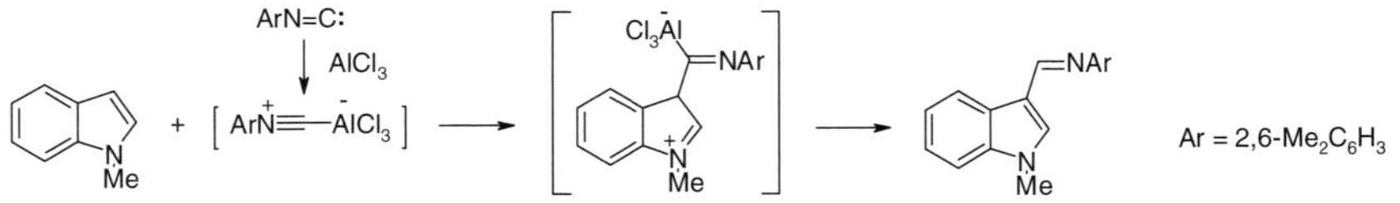

$AlCl_3$ (1.2 mmol) added under a gentle stream of N_2 to a soln. of N-methylindole (1 mmol) and 2,6-xylyl isocyanide in toluene, the reaction vial capped, stirred at room temp. under N_2 for 15 min, diluted with ethyl acetate, and quenched with satd. aq. $NaHCO_3$ → N-(1-methylindol-3-yl-methylene)-2,6-dimethylaniline. Y 88%. This inexpensive procedure is applicable to the reaction of electron-rich arenes (N-methylindoles, pyrroles, thiophenes, 1,3,5-trimethoxybenzene) with aliphatic or aromatic isonitriles (the latter of which may possess electron-donating or -withdrawing groups). Reaction appears to involve nucleophilic attack of the arene on an intermediate Lewis acid-isonitrile adduct, followed by aromatization and protonolysis. Ethers, bromides and esters remained unaffected. F.e.s. M. Tobisu, S. Yamaguchi, N. Chatani, Org. Lett. *2007*, *9* (17), 3351-3.

Indium(III) chloride s. under $Mn_2(CO)_{10}$ *$InCl_3$*
Lanthanum oxide s. under Magnesium-lanthanum mixed oxide ←
Samarium triisopropoxide s. under [(1R,2R)-N,N'-Bis(3-hydroxysalicyl-idenato)-1,2-cyclohexanediamine]copper(II) *Sm(OPr-i)₃*

Lanthanum(III) p-*methoxyphenoxide/chiral bis(Δ^2-oxazolin-2-yl)pyridines/lithium* p-*methoxyphenoxide* ←
Catalytic asym. Mannich reaction with N-sulfonylimines ←
with N-tosyl-derivs. using $CuOCl_4$ and a chiral phosphino-Δ^2-oxazoline cf. *62*, 272s*65*; with N-(thien-2-ylsulfonyl)-derivs. and trichloromethyl ethyl ketone (as propionate equivalent) using $La(OC_6H_4OMe$-*p*$)_3$/*i*-Pr-Pybox/$LiOC_6H_4OMe$-*p* s. H. Morimoto, G. Lu, N. Aoyama, S. Matsunaga, M. Shibasaki, J. Am. Chem. Soc. *2007*, *129* (31), 9588-9.

[Ethylenebis(η^5-indenyl)][bis(trimethylsilyl)amido]lanthanide(III) complexes *[Ln(III)]*
α,β-Acetyleneamidines from carbodiimides RN=C=NR′ → RN=C(NHR′)C≡C
and terminal acetylene derivs. s. *28*, 617s*72*

Chiral bis(Δ^2-oxazolines) s. under $Mg(ClO_4)_2$ *Box*
Chiral bis(Δ^2-oxazolin-2-yl)pyridines s. under Lanthanum(III) p-*methoxyphenoxide* *Pybox*
*4-*tert-*Butylphenol s. under [(1R,2R)-N,N′-Bis(3-hydroxysalicylidenato)-1,2-cyclohexanediamine]copper(II)* *ArOH*
Benzoic acid s. under Chiral 2-(diphenylphosphino)-2′-hydroxy-3′-[2-(o-hydroxy-phenyl)ethoxymethyl]-1,1′-binaphthyl *PhCOOH*

L-*Proline/triethylenediamine or imidazole* ←
Organocatalyzed asym. aza-Baylis-Hillman reaction with N-protected imines ←
s. *68*, 272s*72*

L-*Threonine* tert-*butyldimethylsilyl ether* ←
***anti*-α-Siloxy-β-(sulfonylamino)ketones from N-sulfonylimines** $C{=}NSO_2R \rightarrow C(NHSO_2R)CH(OSi{\leqslant})CO$
Organocatalyzed asym. Mannich reaction with addition of three C-atoms

403.

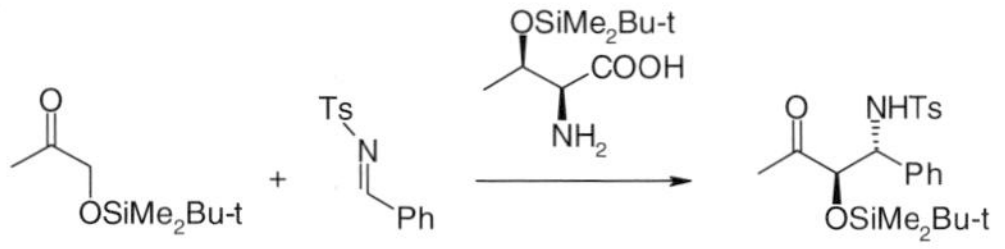

A mixture of phenyl-N-tosylmethanimine, hydroxyacetone *tert*-butyldimethylsilyl ether and 10 mol% L-threonine *tert*-butyldimethylsilyl ether *in toluene* stirred at room temp. for 40 h → product. Y 76% (*anti/syn* 27:1; e.e. 99% after column chromatography). Remarkably high *anti*-selectivity (up to 120:1) and enantioselectivity (98-99%) were achieved with this organocatalyst irrespective of the electronic nature of substituents in the aromatic ring. The N-sulfonyl group, which probably stabilizes an intermediate (Z)-enamine group through H-bonding with the sulfonyl oxygens, appears critical as both diastereo- and enantio-selectivity were low with N-PMP-, N-Bn- or N-Boc-imines, while aliphatic aldehydes gave a complex mixture of products. Enantioselectivity was also ca. 100% *in water*, but diastereoselectivity was lower. F.e.s. L. Cheng, X. Han, H. Huang, M.W. Wong, Y. Lu, Chem. Commun. *2007* (40), 4143-5.

(R)-β-Homovaline/brine ←
Organocatalyzed asym. Mannich-type reaction C=NR → C(R′)C(NHR)
of ketones in DMSO with proline under ultrasonication cf. *63*, 266s*71*; with 30 mol% (R)-β-homovaline and 5 eqs. brine (0.5-0.9 *M* NaCl), chiral α-amino-γ-keto esters, s. P. Dziedzic, A. Córdova, Tetrahedron:Asym. *2007*, *18* (9), 1033-7.

(R)-2-(Diphenylphosphino)-2′-hydroxy-1,1′-binaphthyl or (S)-2-(Diphenylphosphino)-2′-hydroxy-5,5′,6,6′,7,7′,8,8′-octahydro-1,1′-binaphthyl ←
Chiral 2-(diphenylphosphino)-2′-hydroxy-3′-[2-(o-hydroxyphenyl)ethoxymethyl]-1,1′-binaphthyl/benzoic acid ←
Chiral 2-(N′-phenylthioureido)-2′-diphenylphosphino-1,1′-binaphthyl ←
Organocatalyzed asym. aza-Baylis-Hillman reaction with N-protected imines ←
using (R)-2-(dimethylphosphino)-2′-hydroxy-1,1′-binaphthyl cf. *68*, 272s*70*; of N-tosylaldimines using chiral 2-(diphenylphosphino)-2′-hydroxy-3′-[2-(*o*-hydroxyphenyl)ethoxymethyl]-1,1′-

binaphthyl s. K. Ito, K. Nishida, T. Gotanda, Tetrahedron Lett. *2007*, *48* (35), 6147-9; using chiral 2-(N′-phenylthioureido)-2′-diphenylphosphino-1,1′-binaphthyl/PhCOOH s. Y.-L. Shi, M. Shi, Adv. Synth. Catal. *2007*, *349* (13), 2129-35; of N-aryliminoacetates and enones using (R)-2-(diphenylphosphino)-2′-hydroxy-1,1′-binaphthyl or the (S)-octahydro-deriv. s. M. Shi, G.-N. Ma, J. Gao, J. Org. Chem. *2007*, *72* (25), 9779-81; of N-(*p*-methoxyphenyl)iminoacetates and enals using L-proline/imidazole s. N. Utsumi, H. Zhang, F. Tanaka, C.F. Barbas III, Angew. Chem. Int. Ed. Engl. *2007*, *46* (11), 1878-80; of ar. N-Boc-aldimines and enals using L-proline/DABCO s. J. Vesely, P. Dziedzic, A. Córdova, Tetrahedron Lett. *2007*, *48* (39), 6900-4.

Chiral phosphorous acid diesters/lithium bis(trimethylsilyl)amide ←
Asym. synthesis of α-(siloxylamino)ketones C(O)C(N-OSi≤)
from nitrones and acylsilanes

404. 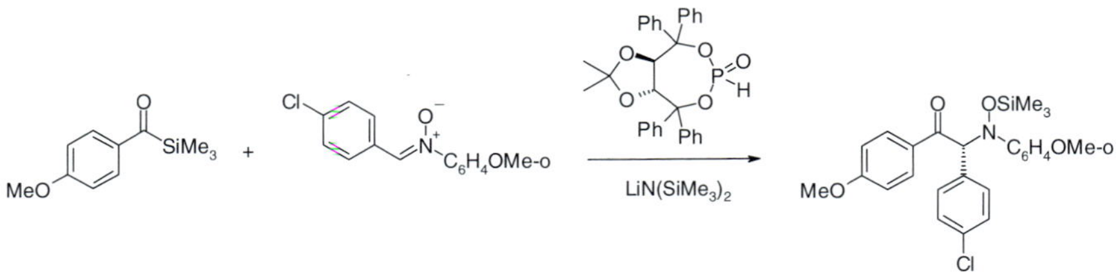

under mild conditions. The startg. nitrone, 1.5 eqs. startg. acylsilane, 0.25 eq. (R,R)-TADDOL-phosphite, and 0.25 eq. $LiN(SiMe_3)_2$ mixed (in a glove box), *2-methyltetrahydrofuran* added, the soln. stirred at room temp. under N_2 for 5 min (TLC), the solvent removed *in vacuo*, and the residue purified by chromatography on silica (deactivated with 5% Et_3N/hexanes to minimize $HOSiMe_3$ elimination) → 2-(4-chlorophenyl)-1-(4-methoxyphenyl)-2-[N-(2-methoxyphenyl)-N-trimethylsilyloxy]aminoethanone. Y 89% (e.r. 97:3). The illustrated example was carried out on a 16 *g* scale and the absolute stereochemistry was determined by X-ray diffraction. Twelve examples gave yields in the range 77-94% but in three cases these were somewhat reduced by elimination of $HOSiMe_3$ to form achiral α-ketoimines. This novel C-acylation requires a slight excess of acylsilane to minimize irreversible addition of the chiral metallophosphite to the nitrone and, although limited to startg. ms. derived from aromatic aldehydes, accommodates both electron-rich and -poor systems. It is considered that the nitrones are good electrophiles due to re-establishment of thermoneutrality by the **O-silyl transfer**. F.e. incl. reaction optimization and reductive cleavage of the O-N bond using Zn/NH_4Cl (cf. *61*, 16) s. M.R. Garrett, J.C. Tarr, J.S. Johnson, J. Am. Chem. Soc. *2007*, *129* (43), 12944-5.

Chiral 1,1′-binaphthyl-2,2′-diyl hydrogen phosphates (s.a. under AgOAc) ←
2-α-Acylaminopyrroles from pyrroles and N-acylimines C=NAc → C(NHAc)Ar
Organocatalyzed asym. Friedel-Crafts α-acylaminoalkylation

405. 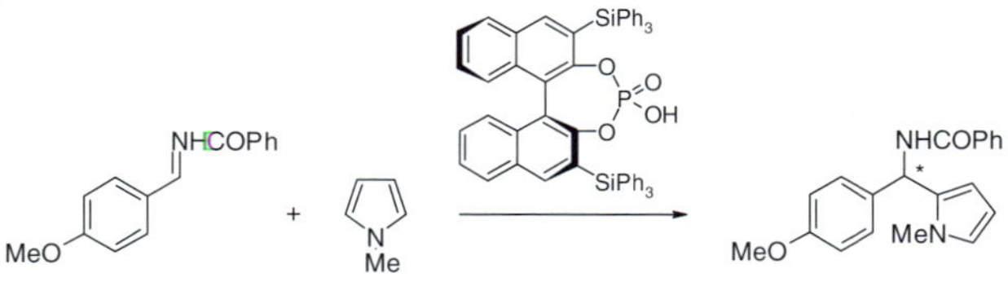

A mixture of startg. N-acylimine and 5 mol% chiral 3,3′-bis(triphenylsilyl)-1,1′-binaphthyl-2,2′-diyl hydrogen phosphate stirred in dry chloroform under argon at -60° for 10 min, 2 eqs. N-methylpyrrole added, and stirring continued for 20 h → N-[(4-methoxyphenyl)(1-methyl-1*H*-pyrrol-2-yl)methyl]benzamide. Y 95% (e.e. 96%). High enantioselectivity is favoured by electron-rich imines but the reaction is tolerant of mildly electron-withdrawing groups. The use of polar solvents (e.g. acetonitrile) reduces enantioselectivity and yields as does the presence of bulky groups on the pyrrole nitrogen. However, enantioselectivity is high (e.e. >90%) with N-methyl-, N-allyl-, N-bromoethyl- and N-cyanoethyl-pyrroles but lower for N-propyl-, N-hexyl- and N-benzyl-pyrroles, while N-isopropyl and N-unsubst. pyrroles gave only a trace of product. The absolute configuration of the products was not discussed. F.e. and catalyst optimization, also double addition of the imine at room temp., s. G. Li, G.B. Rowland, E.B. Rowland, J.C. Antilla, Org. Lett. *2007*, *9* (20), 4065-8.

Chiral 3,3′-bis(triphenylsilyl)-1,1′-binaphthyl-2,2′-diyl hydrogen phosphate ←
Asym. Friedel-Crafts aminoalkylation of indoles H → $CH(NHSO_2R)R'$
under organocatalysis with chiral quini[di]ne-derived thioureas cf. *71*, 271; chiral 3-α-acylamino-indoles with 3,3′-bis(triphenylsilyl)-1,1′-binaphthyl-2,2′-diyl hydrogen phosphate s. G.B. Rowland, E.B. Rowland, Y. Liang, J.A. Perman, J.C. Antilla, Org. Lett. *2007*, 9 (14), 2609-11.

Air s. under Et_3B *air*
Chiral N-(o-hydroxybenzylidene)-N′-tosyl-1,2-diamines s. under $Cu(OTf)_2$ ←

Dimanganese decacarbonyl/indium(III) chloride/irradiation *$Mn_2(CO)_{10}/InCl_3$/hν*
Manganese-mediated radical addition to N-acylhydrazones C=NN< → C(R)NHN<

406.
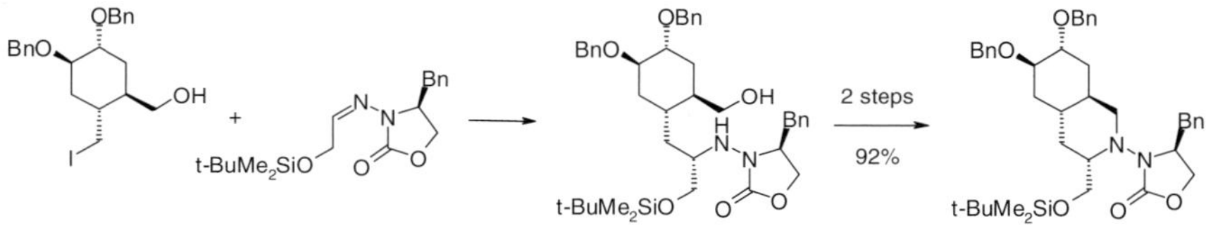

with asym. induction. A soln. of the startg. hydrazone in methylene chloride deoxygenated (with nitrogen or argon) for 30 min, 2.2 eqs. $InCl_3$ added, stirred 40 min, 1.25 eqs. of the startg. alkyl iodide and 1.2 eqs. $Mn_2(CO)_{10}$ added, the mixture irradiated (300 nm, Rayonet photochemical reactor) for 10-20 h, 5 eqs. Et_3N added, the mixture stirred for 40 min, and concentrated → product. Y 93%. The reaction forms part of an investigation of a radical-ionic annulation approach to functionalized **perhydroisoquinolines** with the ultimate aim of achieving cyclization of the illustrated intermediate *in situ*. This radical addition to N-acylhydrazones is versatile, allowing incorporation of the required asymmetry in either alkyl iodide or hydrazone components. The illustrated reaction was accomplished on a 1 mmol scale and utilizes near stoichiometric quantities of the two components. A second example with hydrogens replacing the benzyl ethers gave a 67% yield while a series of compounds with a 3,4-double bond gave modest yields (26-49%) due to catalyst interference by the alkene moiety. F.e. incl. preparation of starting materials and cyclization of the illustrated product via O-tosylation and treatment with NaI (2 steps; 92%) s. C.S. Korapala, J. Qin, G.K. Friestad, Org. Lett. *2007*, *9* (21), 4243-6.

Addition to Carbon-Carbon Bonds CC ⇓ CC

Without additional reagents *w.a.r.*
Diels-Alder reaction with quinones ○
cf. *17*, 198; with *o*-alkoxy-*p*-quinones in methylene chloride, quinoflavanoid synthesis, s. M.-F. Laroche, A. Marchand, A. Duflos, G. Massiot, Tetrahedron Lett. *2007*, *48* (50), 9056-8.

Diels-Alder reaction with vinylheterocyclics
under microwave irradiation cf. *37*, 646s*71*; of 4-vinylthiazoles without enhancement s. M. Alajarín, J. Cabrera, A. Pastor, P. Sánchez-Andrada, D. Bautista, J. Org. Chem. *2007*, *72* (6), 2097-105.

1,3-Dipolar cycloaddition with nitrile oxides and *exo*-glycals
cf. *16*, 735s*70*; 5-spiro-Δ^2-isoxazoline C-disaccharides s. P.-Z. Zhang, X.-L. Li, H. Chen, Y.-N. Li, R. Wang, Tetrahedron Lett. *2007*, *48* (44), 7813-6.

3-Arylidene-5-sulfonyloxytetrahydropyrans
from 2-arylidenecyclopropylcarbinols, aldehydes and sulfonic acids
Regio- and stereo-specific Prins-type reaction

407. 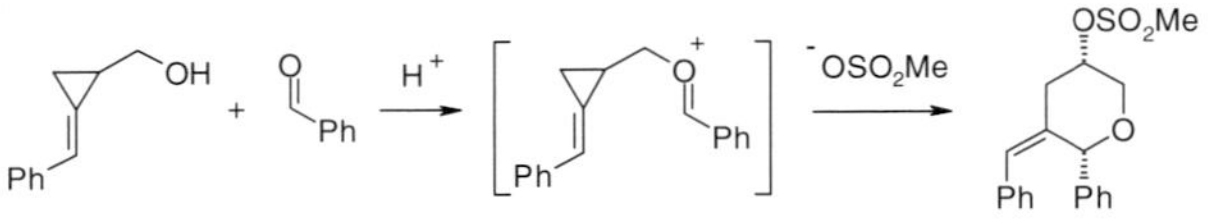

The startg. 2-arylidenecyclopropylcarbinol added to a soln. of 1.2 eqs. benzaldehyde and 1.4 eqs. methanesulfonic acid in methylene chloride, stirred at room temp. for 20 min, solvent removed under reduced pressure, and the residue worked up with column chromatographic purification on silica gel → product. Y 85% (12:1 mixture of geoisomers). The procedure is applicable to both aliphatic and aromatic aldehydes, irrespective of ring substitution, and is equally facile with *p*-toluenesulfonic acid. A Prins-type mechanism is invoked. F.e.s. G.Q. Tian, M. Shi, Org. Lett. *2007*, *9* (12), 2405-8.

Irradiation (s.a. under CuOTf)
5-Methylene-5,6-dihydro-2-pyridones
from 4-methylenecyclobutenones and azomethines
via hetero-Diels-Alder reaction with allenylketenes

408. 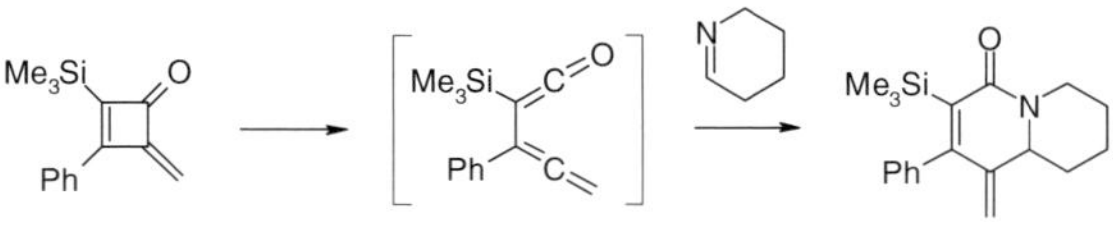

5-Methylene-3-silyl-5,6-dihydro-2-pyridones. The startg. 4-methylenecyclobutenone added to a soln. of 3 eqs. startg. azomethine in ether, and the mixture irradiated with a high-pressure Hg-lamp at room temp. for 8 h under N_2 → product. Y 71%. The procedure is applicable to the preparation of the isolated ring system as well as quinolizidinone and indolizidinone analogs, the silyl group being essential for the conversion. Reaction with N-benzylbenzaldimine, however, was sluggish, while ketimine derivs. were unreactive. F.e.s. S. Aoyagi, K. Kikuchi, K. Shimada, Y. Takikawa, Synlett *2007* (16), 2553-6.

Microwaves s.a. under Ruthenium complexes [\\\\]

*Microwaves/trimethylsilyl cyanide/tetra-*n*-butylammonium fluoride* ←
β-Cyano- from α,β-ethylene-ketones C=C → CHC(CN)
1,4-Addition under microwave irradiation in the absence of solvent

409. 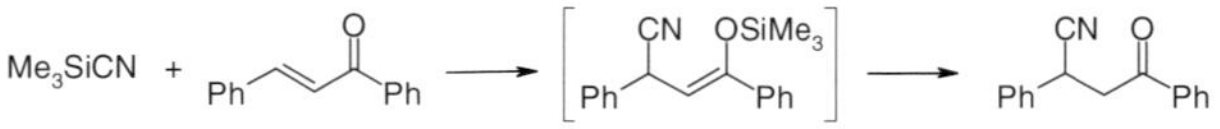

A mixture of (E)-chalcone and 2 eqs. trimethylsilyl cyanide irradiated in a microwave oven (CEM Discover) at 200° for 5 min, cooled to room temp., evaporated *in vacuo*, the residue dissolved in THF, treated at room temp. with 10% tetra-*n*-butylammonium fluoride, stirred for 5 min, water added, and the mixture worked up with short column chromatographic purification on silica gel → product. Y 87%. The procedure is rapid, does not require a Lewis or Brønsted acid, and is not complicated by 1,2-addition. 3-Nonen-2-one reacted similarly to give 65% product (with 24% recovery of startg. m.). However, there was no reaction with enoates. F.e.s. H. Iida, T. Moromizato, H. Hamana, K. Matsumoto, Tetrahedron Lett. *2007*, *48* (11), 2037-9.

Alkyllithiums s. under Cp_2ZrCl_2 *RLi*
Methyllithium s. under $Cp_2Zr(H)Cl$ *MeLi*
n-*Butyllithium s. under Ti(OPr*-i*)$_4$* *BuLi*
Sodium azide s. under L-*Proline* *NaN_3*
Lithium chloride s. under $Cp_2Zr(H)Cl$ *LiCl*
Lithium iodide s. under $CeCl_3$ *LiI*
Chiral 1,2-diamines s.a. under $Cu(OTf)_2$ ←

Chiral diamine/trifluoroacetic acid ←
Organocatalyzed asym. Michael addition s. *62*, 282s*72* C=C → CHC(R)

Chiral cyclic 1,2-diamines/trifluoroacetic acid ←
*Axially-chiral 1,2-diamines/*p*-toluenesulfonic acid* ←
Organocatalyzed asym. Diels-Alder reaction ○
with α,β-ethylenealdehydes using chiral 1,2-diamines/$C_6F_5SO_3H$ cf. *59*, 301s*69*; using chiral cyclic 1,2-diamines/CF_3COOH s. C. Biaggi, M. Benaglia, S. Rossi, S. Proto, R. Annunziata, Tetrahedron Lett. *2007*, *48* (48), 8521-5 ; with an axially-chiral diamine/TsOH s. T. Kano, Y. Tanaka, K. Maruoka, Chem. Asian J. *2007*, *2* (9), 1161-5; with 2(S)-[bis[3,5-bis(trifluoromethyl)phenyl]-triethylsiloxymethyl]pyrrolidine/CF_3COOH s. H. Gotoh, Y. Hayashi, Org. Lett. *2007*, *9* (15), 2859-62; of 3-hydroxy-2-pyrones using bifunctional cinchona-derived catalysts, e.g. a thiourea, s. Y. Wang, H. Li, Y.-Q. Wang, Y. Liu, B.M. Foxman, L. Deng, J. Am. Chem. Soc. *2007*, *129* (20), 6364-5.

Chiral triamine/trifluoromethanesulfonimide ←
1-Acoxycyclobutanecarboxaldehydes □
from unactivated ethylene derivs. and α-acoxyacroleins
Organocatalyzed asym. [2+2]-cycloadditions

410.

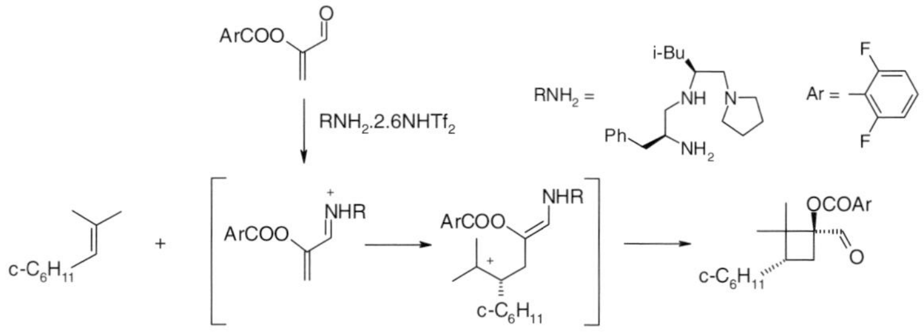

under mild conditions. 1,1-Dimethyl-2-cyclohexylethene added to a soln. of 10 mol% chiral triamine (derived by reduction of the corresponding L-Phe-L-Leu amide) and 26 mol% trifluoromethanesulfonimide in *nitropropane*, the mixture cooled to 0°, 2 eqs. startg. acyloxyacrolein added, the mixture stirred at 0° for 48 h, the reaction quenched with $NaHCO_3$, diluted with ethyl acetate, washed with brine and water, the organic layer dried over $MgSO_4$, concentrated *in vacuo*, and the residue purified by chromatography → cyclobutane deriv. Y 89% (*syn/anti* 8:92; e.e. 82%). The reaction is successful at 0° or -20° with cyclic and acyclic trisubst. ethylene derivs. and electronically diverse acyloxyacroleins giving variable yields (24-89%; e.e. 80-95%) and affording predominantly *anti* products (selectivity 83-96%). 1,1- and 1,2-Dialkylethylenes were unreactive but α-methylstyrene gave a 20% yield of product (selectivity 84%; e.e. 80%). F.e. and subsequent ring expansion to **chiral 2-acoxycyclopentanones** (with $AlCl_3$) and 2-hydroxy-derivs. (with Bu_4NF) s. K. Ishihara, K. Nakano, J. Am. Chem. Soc. *2007*, *129* (29), 8930-1.

Chiral proline-derived triamine ←
Organocatalyzed asym. Michael addition s. *62*, 282s*72* C=C → CHC(R)

α,α-Diphenylprolinol ←
Pyrrolidines from ethylene derivs. and azomethines ○
via asym. 1,3-dipolar cycloaddition
under Cu-catalysis cf. *67*, 301s*71*; under organocatalysis with α,α-diphenylprolinol, chiral 5,5-dicarbalkoxypyrrolidine-3-carboxaldehydes, s. J.L. Vicario, S. Reboredo, D. Badía, L. Carrillo, Angew. Chem. Int. Ed. Engl. *2007*, *46* (27), 5168-70; with 5 mol% [(R)- or (S)-binap]$AgClO_4$/Et_3N, chiral 2-carbalkoxypyrrolidine-3,4-dicarboxylic acid imides, s. C. Nájera, M. de Gracia Retamosa, J.M. Sansano, Org. Lett. *2007*, *9* (20), 4025-8.

Chiral 2-α-siloxy- or 2-α-(sulfonylamino)-pyrrolidines ←
Organocatalyzed asym. Michael addition s. *62*, 282s*72* C=C → CHC(R)

2(S)-[Diphenyl(trimethylsiloxy)methyl]pyrrolidine ←
Organocatalyzed asym. Michael addition
of aldehydes to 1,2-ethylene-1,1-di(phosphonic acid esters)

411. 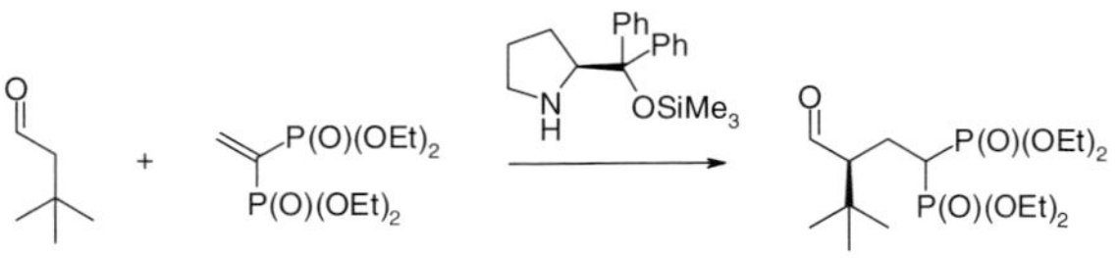

A soln. of 20 mol% 2(S)-[diphenyl(trimethylsiloxy)methyl]pyrrolidine in chloroform stirred for 12 h at room temp. with 1 eq. tetraethyl ethylidenebis(phosphonate) and 10 eqs. 3,3-dimethylbutyraldehyde, the mixture hydrolyzed with satd. NH_4Cl, and extracted with methylene chloride → (S)-2-*tert*-butyl-4,4-bis(diethoxyphosphinyl)butanal. Y 85% (e.e. 97%). The catalyst is the most effective of seven proline-based catalysts tested, giving title products in >75% yields with 75-97% enantiomeric excess from aliphatic aldehydes. Lower selectivity and conversion were achieved with 2-pentenal or by heating the reactions or replacing chloroform with aq. ethanol. The reaction is successful with catalyst loadings down to 10 mol%. The enantioselectivity is believed to result from *si*-face attack of an intermediate (E)-enamine on the Michael acceptor. The products are converted to synthetically useful **chiral 3-subst. 1,1-di(phosphonic acid esters)** without loss of optical activity. F.e.s. S. Sulzer-Mossé, M. Tissot, A. Alexakis, Org. Lett. *2007*, *9* (19), 3749-52.

2(S)-[Diphenyl(trimethylsiloxy)methyl]pyrrolidine/benzoic acid ←
4-Formyltetrahydrothiophene-3-acetic acid esters ○
from α,β-ethylenealdehydes and α,β-ethylene-γ-mercaptocarboxylic acid esters
Ring closure via organocatalyzed asym. double Michael addition

412. 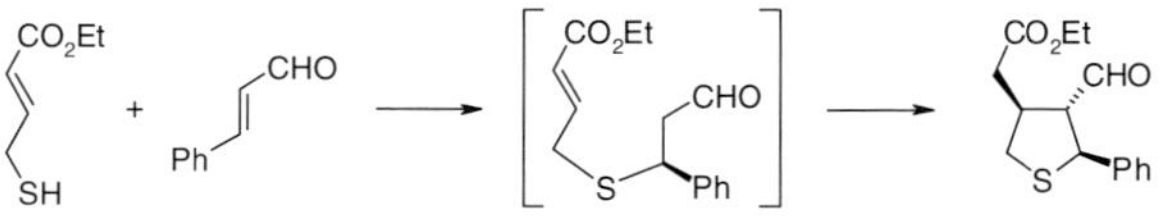

under mild conditions. A soln. of *trans*-cinnamaldehyde, 10 mol% 2(S)-[diphenyl(trimethylsiloxy)methyl]pyrrolidine and 10 mol% benzoic acid in *toluene* treated with 1 eq. 4-mercaptobut-2-enoic acid ethyl ester, stirred for 3 days at room temp., and worked up with column chromatographic purification → product. Y 76% (e.e. >99%; d.e. 15:1). High enantioselectivity (94-9%) and diastereoselectivity (6:1 to 18:1) were recorded for the generation of *three* contiguous chiral centres by reaction with cinnamaldehydes (incl. *o*-subst. derivs.) and hetaryl analogs (irrespective of the nature of ring substitution). Reaction is presumed to involve initial Michael addition of the mercaptan to the enal, followed by intramolecular Michael addition onto the enoate. F.e and comparison of organocatalysts s. H. Li, L. Zu, H. Xie, J. Wang, W. Jiang, W. Wang, Org. Lett. *2007*, *9* (9), 1833-5.

2(S)-[Diphenyl(trimethylsiloxy)methyl]pyrrolidine/3-benzyl-4,5-dimethyl-thiazolium chloride/triethylamine/ethyldiisopropylamine ←

Asym. synthesis of β-functionalized carboxylic acid esters from α,β-ethylenealdehydes ←
Sequential catalysis in one pot with two different organocatalysts

413.

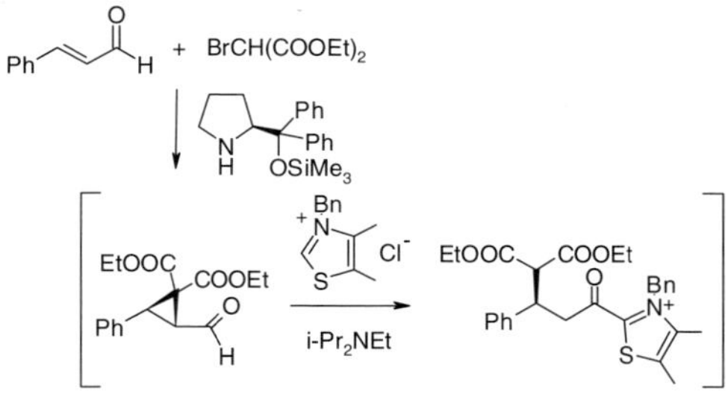

trans-Cinnamaldehyde (0.3 mmol), diethyl bromomalonate (0.25 mmol), triethylamine (0.25 mmol) and 3 eqs. ethanol added to a stirred soln. of 20 mol% (S)-2-[diphenyl(trimethylsiloxy)-methyl]pyrrolidine in chloroform, stirred vigorously at room temp. for 3 h, 40 mol% ethyldiiso-propylamine and 20 mol% 3-benzyl-4,5-dimethylthiazolium chloride added, warmed to 30°, stirring continued for 15 h, treated with satd. aq. NH_4Cl, and worked up with flash chromatographic purification → (R)-2-ethoxycarbonyl-3-phenylpentanedioic diethyl ester. Y 69% (e.e. 97%). Chirality is introduced by the initial amine-catalyzed 1,4-addition of the malonate, which is followed by *3-exo-tet* cyclization to give an intermediate cyclopropanecarboxaldehyde; this then undergoes N-heterocyclic carbene-catalyzed ring opening with concomitant oxidation of the formyl group to the corresponding ester. **Chiral β-hydroxy- and β-carbalkoxyamino-carboxylic acid esters** were obtained similarly (with H_2O_2 and CbzNHOAc, respectively). F.e. (e.e. 91-7%) s. G.-L. Zhao, A. Cordova, Tetrahedron Lett. *2007*, *48* (34), 5976-80.

2(S)-[Bis[3,5-bis(trifluoromethyl)phenyl]trimethylsiloxymethyl]pyrrolidine/ triethylenediamine ←

2,4-Dinitrocyclohexanols from α,β-ethylenealdehydes and 1,3-dinitro compds. via organocatalyzed asym. Michael addition-intramolecular Henry reaction ○

414.

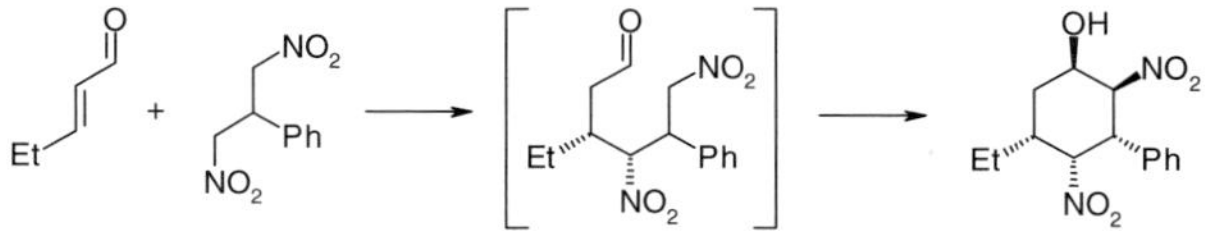

Chiral pentasubst. cyclohexanols. Startg. 1,3-dinitro compd. and 1 eq. startg. enal added to a vial containing 20 mol% 2(S)-[bis[3,5-bis(trifluoromethyl)phenyl]trimethylsiloxymethyl]pyr-rolidine and 10 mol% triethylenediamine in methylene chloride, stirred at room temp. until reaction complete, and worked up with flash chromatographic purification on silica gel → (1R,2S,3R,4R,5R)-5-ethyl-2,4-dinitro-3-phenylcyclohexanol. Y 45% (e.e. 90%). Significantly, and for the first time, *two* new carbon-carbon bonds and *five* new chiral centres are created in one operation. Reaction is generally applicable to the reaction of β-alkyl- and β-alkenyl-subst. enals with 2-aryl- and 2-hetaryl-1,3-dinitro compds. (irrespective of the nature of ring substituents). For simplification, the iminium activation of the enal is not indicated in the scheme. F.e. and formation of the opposite enantiomers with the same reagent lacking the trimethylsilyl group s. E. Reyes, H. Jiang, A. Milelli, P. Elsner, R.G. Hazell, K.A. Jørgensen, Angew. Chem. Int. Ed. Engl. *2007*, *46* (48), 9202-5.

2(S)-[Bis[3,5-bis(trifluoromethyl)phenyl]triethylsiloxymethyl]pyrrolidine/ trifluoroacetic acid ←
Organocatalyzed asym. Diels-Alder reaction s. *59*, 301s*72* ○

*(R)-2-(Trifluoromethanesulfonylaminomethyl)pyrrolidine/*p*-methoxybenzoic acid* ←
Bridged 3-hydroxycyclohexanone ring from cyclic ketones and α,β-ethyleneketones via organocatalyzed asym. Michael addition-intramolecular aldol condensation

415. 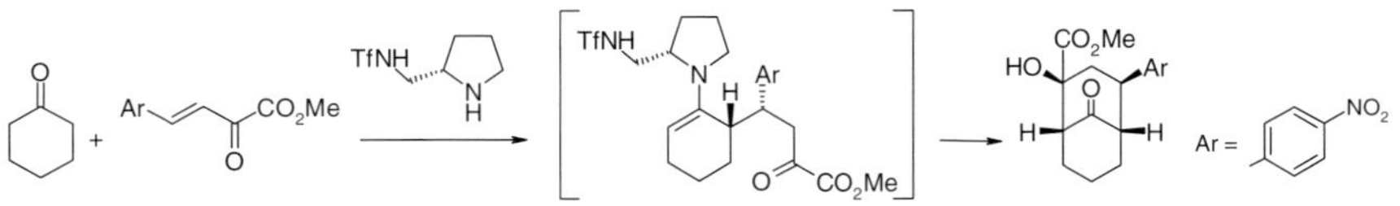

A mixture of cyclohexanone, 20 mol% (R)-2-(trifluoromethanesulfonylaminomethyl)pyrrolidine and 20 mol% 4-methoxybenzoic acid stirred for 10 min, 1 eq. startg. enone added, stirred for 120 h, concentrated under reduced pressure, and the residue purified by flash chromatography → [3+3]-adduct. Y 90% (e.e. 90%). The method is successful with a range of β-aryl-subst. enones, tolerating both electron-rich and -poor systems, and giving similar yields and selectivities for Me, Et or Bn esters (Y 74-90%; e.e. 87-99%). Cyclohexanone may be replaced with 4-pyrone, 1-methylpiperidin-4-one and cyclopentanone, but acetone gave the aldol product in 99% yield (e.e. 14%). A single example of a β-methyl-subst. enone gave 56% yield (e.e. 94%). The proposed mechanism involves Michael-type addition of an intermediate chiral enamine to the enone, followed by proton transfer and intramolecular aldol condensation. F.e.s. C.-L. Cao, X.-L. Sun, Y.-B. Kang, Y. Tang, Org. Lett. *2007*, *9* (21), 4151-4.

Chiral prolinal mercaptals/benzoic acid ←
Chiral prolinal mercaptals as organocatalysts ←

416. 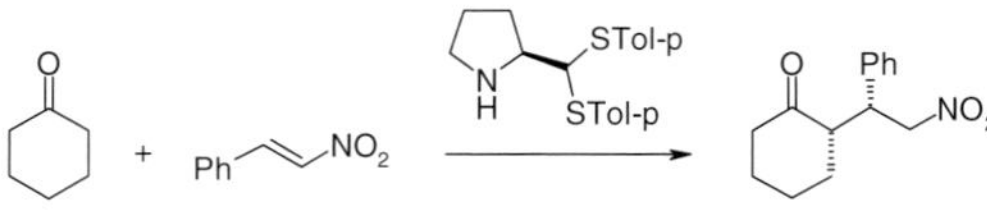

Chiral prolinal mercaptals, readily prepared from N-Boc-(S)-proline and the appropriate mercaptan, are stable, readily tunable organocatalysts, offering ca. 100% diastereo- and enantio-selectivity in the **asym. Michael addition of oxo compds. to 1,1-nitroethylene derivs. E:** A soln. of *trans*-β-nitrostyrene and 3 eqs. cyclohexanone *in methylene chloride* allowed to react with 10 mol% chiral prolinal mercaptal and 10 mol% benzoic acid at room temp. for 28 h → *syn*-product. Y 88% (*syn/anti* 99:1; e.e >99%). Reaction is *syn*-selective and generally applicable to 5- to 6-membered cyclic ketones and enolizable aldehydes, although hindered substrates required more catalyst. F.e.s. T. Mandal, C.-G. Zhao, Tetrahedron Lett. *2007*, *48* (33), 5803-6; **asym. hetero-Diels-Alder reaction** of α,β-ethyleneacylphosphonic acid esters with enolizable aldehydes (e.e. up to 94%) s. S. Samanta, J. Krause, T. Mandal, C.-G. Zhao, Org. Lett. *2007*, *9* (14), 2745-8.

Cinchona alkaloids ←
Organocatalyzed asym. Michael addition C=C → CHC(R)
to 1,1-nitroethylene derivs. cf. *62*, 282s*71*; of anthrone using cinchona alkaloids, e.g. O-benzoylcupreine (5 mol%), s. M. Shi, Z.-Y. Lei, M.-X. Zhao, J.-W. Shi, Tetrahedron Lett. *2007*, *48* (33), 5743-6; of malonates using demethylquinine alkali metal salts (1 mol%) in water s. F.-X. Chen, C. Shao, Q. Wang, P. Gong, D.-Y. Zhang, B.-Z. Zhang, R. Wang, ibid. *48* (48), 8456-9; of cyclohexanones using chiral N-(pyrrolidin-2-ylmethyl)pyridinium salts s. D.-Q. Xu, B.-T. Wang, S.-P. Luo, H.-D. Yue, L.-P. Wang, Z.-Y. Xu, Tetrahedron:Asym. *2007*, *18* (15), 1788-94; using (3′S,5′S)-2-(pyrrolidinomethyl)-1,3′-bipyrrolidine s. H. Chen, Y. Wang, S. Wei, J. Sun, ibid. *18*

(11), 1308-12; of ketones using 2(S)-[3,5-bis(trifluoromethyl)benzenesulfonylaminomethyl]-pyrrolidine s. B. Ni, Q. Zhang, A.D. Headley, ibid. *18* (12), 1443-7; using 1-methyl-3-[*p*-[1-((2S)-pyrrolidin-2-ylmethyl)-1,2,3-triazol-4-yl]benzyl]imidazolium fluoroborate/CF_3COOH s. L.-Y. Wu, Z.-Y. Yan, Y.-X. Xie, Y.-N. Niu, Y.-M. Liang, ibid. *18* (17), 2086-90; of α-lithiated phosphonates (with *i*-Pr_2NLi) and cinchonine s. V. Rai, S.M. Mobin, I.N.N. Namboothiri, ibid. *18* (22), 2719-26; of ketones using 2-((2S)-pyrrolidin-2-ylmethylthio)imidazolium benzoates in polyethylene glycol s. D.Q. Xu, S.P. Luo, Y.F. Wang, A.B. Xia, H.D. Yue, L.P. Wang, Z.Y. Xu, Chem. Commun. *2007* (42), 4393-5; using bifunctional prolinamides derived from L-proline and chiral 2-aminoalcohols s. D. Almai, D.A. Alonso, E. Gómez-Bengoa, Y. Nagel, C. Nájera, Eur. J. Org. Chem. *2007* (14), 2328-43; using a chiral diamine/CF_3COOH in brine s. Vishnumaya, V.K. Singh, Org. Lett. *2007*, *9* (6), 1117-9; of aldehydes using chiral 3,3'-bimorpholines, also intramolecular aldol reaction, s. S. Sulzer-Mossé, M. Laars, K. Kriis, T. Kanger, A. Alexakis, Synlett *2007* (11), 1729-32; of aldehydes or ketones with L-proline and Et_3N s. B.M. Choudary, Ch.V. Rajasekhar, G. Gopi Krishna, K. Rajender Reddy, Synth. Commun. *2007*, *37* (1), 91-8; of nitromethane, dibenzyl malonate or aldehydes to enals using chiral 2-α-siloxypyrrolidines in water s. C. Palomo, A. Landa, A. Mielgo, M. Oiarbide, A. Puente, S. Vera, Angew. Chem. Int. Ed. Engl. *2007*, *46* (44), 8431-5; of ethyl α-cyanoacetate to chalcones using a cinchona alkaloid-derived thiourea (10 mol%) s. C. Gu, L. Liu, Y. Sui, J.-L. Zhao, D. Wang, Y.-J. Chen, Tetrahedron:Asym. *2007*, *18* (4), 455-63; of α-subst. cyanoacetates to vinyl ketones using chiral 2-aminothioureas/tert. amines s. T.-Y. Liu, R. Li, Q. Chai, J. Long, B.-J. Li, Y. Wu, L.-S. Ding, Y.-C. Chen, Chem. Eur. J. *2007*, *13* (1), 319-27.

9-Amino-O-demethyl-9-deoxyepiquinine/2,4,6-triisopropylbenzenesulfonic acid ←
Asym. 1,3-dipolar cycloaddition ○
of cyclic α,β-ethyleneketones to 1-alkylidene-3-pyrazolidonium ylids
with a multifunctional chiral prim. amine as organocatalyst

417. 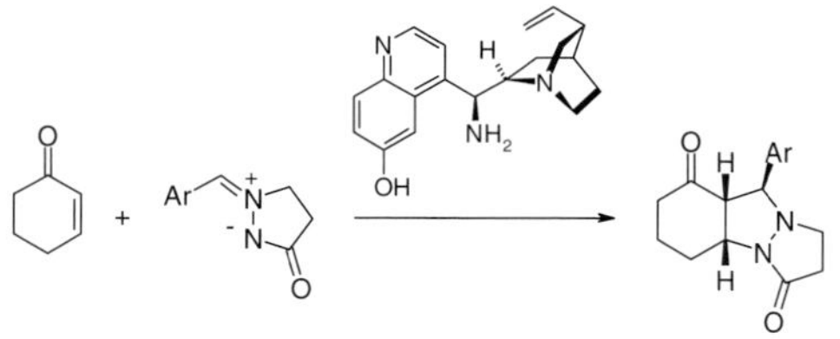

2 Eqs. 2-cyclohexenone added to a mixture of the startg. azomethinium imine, 4 Å molecular sieves, 10 mol% 9-amino-O-demethyl-9-deoxyepiquinine and 20 mol% 2,4,6-triisopropylbenzenesulfonic acid in dry THF at 40°, kept at this temp. for 48 h (TLC monitoring), and worked up with column chromatographic purification on silica gel → product. Y 99% (e.e. 93%). Yields were low and enantioselectivity only moderate (50-60%) with 9-amino-9-deoxyepiquinine as catalyst suggesting that *hydrogen-bonding* between the free phenolic group of the O-demethylated catalyst and the carbonyl group of the dipole has a synergistic action supporting the initial formation of ketiminium ion, thereby facilitating *re*-face attack of the dipolarophile and *endo*-selectivity. Enantioselectivity was also high with 5- and 7-membered cyclic enones. However, addition of a bulky additive, such as 2,4,6-triisopropylbenzenesulfonic acid, is important for achieving high enantioselectivity, as also is the addition of molecular sieves to remove traces of water eliminated in the formation of the ketiminium ion. F.e. and comparison of chiral auxiliaries and additives s. W. Chen, W. Du, Y.-Z. Duan, Y. Wu, S.-Y. Yang, Y.-C. Chen, Angew. Chem. Int. Ed. Engl. *2007*, *46* (40), 7667-70.

Copper(I) thiophene-2-carboxylate s. under R_2Zn *CuOCOR*
Copper(II) acetate s. under $[(Cp^*)RhCl_2]_2$ *$Cu(OAc)_2$*
Copper(I) cyanide s. under $Cp_2Zr(H)Cl$ *CuCN*

(4,4′-Dimethyl-2,2′-bipyridyl)copper(II) nitrate/DNA ←
Asym. copper(II)-catalyzed Michael addition in water with DNA as ligand C=C → CHC(R)

418.

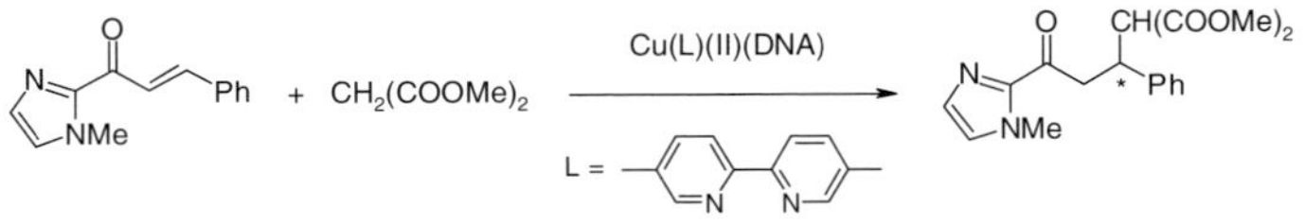

Startg. Michael acceptor (1 mM) added (as a 30 μL aliquot in acetonitrile) to a soln. of (4,4′-dimethyl-2,2′-bipyridyl)copper(II) nitrate (0.3 mM) and salmon testes DNA (1.3 mg/mL) in 20 mM MOPS buffer (pH 6.5), followed by 100 eqs. dimethyl malonate at 0°, mixed for 3 days by continuous inversion at 5°, and worked up → dimethyl 2-[3-(1-methyl-1*H*-imidazol-2-yl)-3-oxo-1-phenylpropyl]malonate. Conversion 100% (e.e. 85%). Enantioselectivities of 82-99% were recorded for the addition of dimethyl malonate or nitromethane (the latter requiring a larger catalyst loading and in 1000-fold excess), these being the highest levels to date for a Michael addition in water. The bipyridyl ligand enhances both enantioselectivity and accelerates the conversion, while imidazole nitrogen of the substrate appears critical in binding to the copper thereby providing the platform for DNA to direct nucleophilic attack on the *si*-face. The procedure may be performed on the preparative scale and the catalyst may be recycled. F.e.s. D. Coquière, B.L. Feringa, G. Roelfes, Angew. Chem. Int. Ed. Engl. *2007*, *46* (48), 9308-11.

Chiral copper(II) bis(Δ²-oxazoline) complexes/1-butyl-3-methyl-imidazolium hexafluoroantimonate *[Cu(II)-box]/[bmim]SbF$_6$*
Asym. Diels-Alder reaction ○
with chiral Cu(II)-bis(Δ²-oxazoline) complexes in ionic liquids cf. *46*, 662s*67*; in [bmim]SbF$_6$ as the solvent (recyclable up to 18 times) s. C.-E. Yeom, H.W. Kim, Y.J. Shin, B.M. Kim, Tetrahedron Lett. *2007*, *48* (52), 9035-9.

Copper(II) triflate/chiral 1,2-diamines ←
Asym. Michael addition of enacylamines to alkylidenemalonic acid esters C=C → CHC(R)

419.

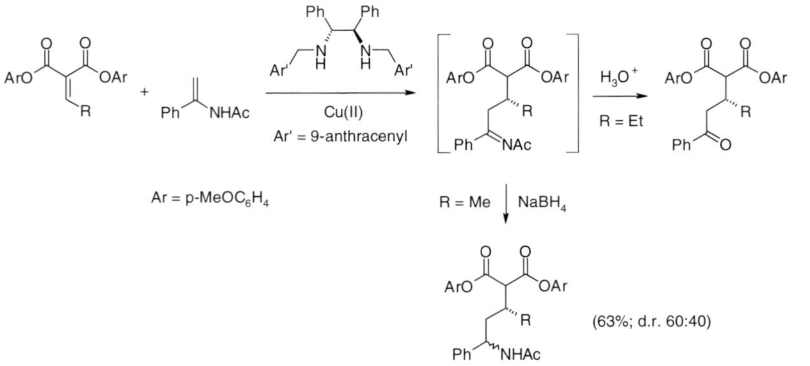

Chiral δ-ketomalonic acid esters. 5.5 Mol% chiral 1,2-diamine in methylene chloride added to 5 mol% Cu(OTf)$_2$ under argon, the resulting yellow soln. stirred for over 12 h as the colour gradually turned to brown, cooled to -78°, solns. of the startg. alkylidenemalonate (1.1 eqs.) and the enacylamine (1 eq.) in methylene chloride added sequentially, stirred for 24 h at -78°, quenched with satd. aq. NaHCO$_3$, allowed to warm to room temp., extracted with methylene chloride, dried, the solvent removed, the residue taken up in THF, treated with 48% aq. HBr, stirred at room temp. for 1.5 min (to hydrolyze the adduct to the ketone), quenched again with satd. aq. $NaHCO_3$, and worked up with column chromatographic purification on silica gel → (R)-bis(4-methoxyphenyl)-2-(1-oxo-1-phenylpentan-3-yl)malonate. Y 85% (e.e. 94%). High yields and enantioselectivities were recorded for the asym. Michael addition of N-acetyleneamines (especially those derived from aromatic ketones), from which a proton is smoothly transferred intramolecularly so that addition of an exogenous proton source is not required. Reduction of the initially formed adduct with $NaBH_4$ affords the corresponding **chiral δ-(acylamino)malonic acid esters.** Reaction appears to involve initial complexation of the chiral copper(II) diamine catalyst to the two carbonyl groups of the malonate, followed by addition of the enacylamine via a 6-membered transition state, either in a concerted or stepwise fashion, the bulky anthracenyl ligands of the chiral diamine being largely responsible for the high level of enantioselectivity. Addition of enecarbamates was less efficient. F.e. and comparison of chiral 1,2-diamines s. F. Berthiol, R. Matsubara, N. Kawai, S. Kobayashi, Angew. Chem. Int. Ed. Engl. *2007*, *46* (41), 7803-5.

Copper(I) triflate/butyl(trimethyl)ammonium triflimide/irradiation ←
[2+2]-Cycloaddition in ionic liquid solvents s. *22*, 705s*72* □

Copper(I) bromide-dimethyl sulfide or iodide s. under Mg *CuHal*
Copper(II) chloride s. under R_2Zn *$CuCl_2$*
Silver triflate s. under K_2PtCl_4 *AgOTf*

[(R)- or (S)-2,2′-Bis(diphenylphosphino)-1,1′-binaphthyl]-silver perchlorate/triethylamine *[(R)- or (S)-binap]$AgClO_4$/Et_3N*
Pyrrolidines from ethylene derivs. and azomethines ○
via asym. 1,3-dipolar cycloaddition s. *67*, 301s*72*

*Acetonitrile[di-*tert*-butyl(biphenyl-1-yl)phosphine]gold(I) hexafluoroantimonate* ←
Regio- and stereo-specific synthesis of enolesters C(R)C(OCOR′)=C
from terminal acoxy-2-acetylenes with 1,2(3)-acoxy group migration

420.
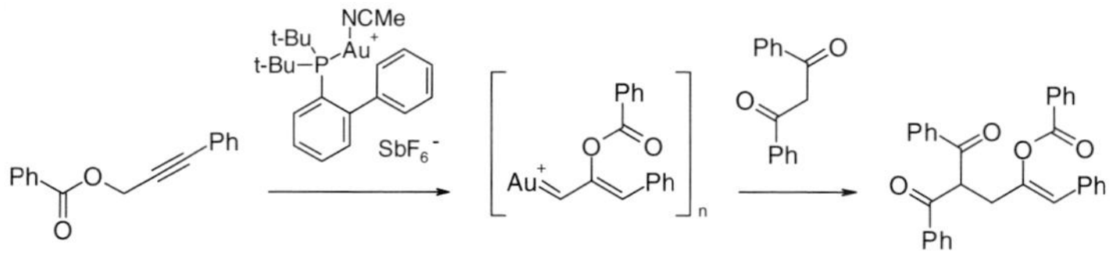

Gold(I) α-acoxy-α,β-ethylenecarbene complexes as intermediates. 3-Phenylpropargyl benzoate and 1 eq. dibenzoylmethane in methylene chloride added to 2-5 mol% freshly prepared acetonitrile[di-*tert*-butyl(biphenyl-1-yl)phosphine]gold(I) hexafluoroantimonate, stirred for 1 h, filtered through silica gel, and evaporated → enol benzoate deriv. Y 92%. The reaction is generally clean and rapid with gold complexes, affording pure (Z)-isomers in variable yields from 3-phenyl- and 3,3-dimethyl-propargyl carboxylates. 3-Methylpropargyl and propargyl carboxylates, however, gave ***trans*-enolesters with 1,3-acoxy group migration via gold(I)-coordinated allenolesters.** Simple gold catalysts (AuCl or $AuCl_3$) gave mixtures of E/Z isomers while $Sc(OTf)_3$ and $Cu(OTf)_2$ promoted simple nucleophilic substitution of the acoxy group. The C-nucleophiles may be β-diketones, β-keto-esters, electron-rich arenes (e.g. 1,3,5-trimethoxybenzene) and indole, and the acoxy group may be acetate, benzoate or pivalate. In some cases 2-5 mol% $Cu(OTf)_2$, $Sc(OTf)_3$ or $AgSbF_6$ were also added to complete the reaction. By-products formed by 1,4-addition to the intermediate gold complex were also isolated in low yields in some cases. F.e.s. C.H.M. Amijs, V. López-Carrillo, A.M. Echavarren, Org. Lett. *2007*, *9* (20), 4021-4.

(Triphenylphosphine)gold(I) chloride/silver(I) hexafluoroantimonate *$(Ph_3P)AuCl/AgSbF_6$*
Cyclic 3-ethyleneamines from 1,6-enynes
vis stereospecific cycloisomerization-hydroamination ○

421. 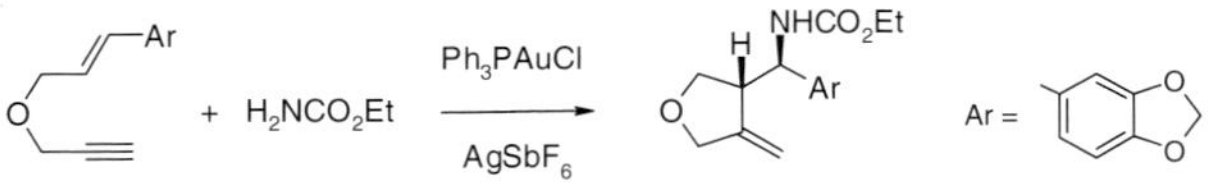

N-Protected *anti*-3-α-amino-4-methylenetetrahydrofurans. A mixture of 3 mol% $(Ph_3P)AuCl$ and 3 mol% $AgSbF_6$ stirred in dry THF under argon for 3 min, 3 eqs. ethyl carbamate and 1 eq. enyne added, stirred 15 h, filtered, and evaporated → ethyl [benzo[1,3]dioxol-5-yl-(4-methylene-tetrahydrofuran-3-yl)methyl]carbamate. Y 76%. The reaction is clean and diastereospecific giving the *anti* isomer exclusively with a range of carbamates and substituted anilines in the presence of nitro, cyano and trifluoromethyl groups. Use of tosylamine as the amine component, however, promotes classical cycloisomerization of the enyne to form significant amounts of cyclodiene by-products. 1,6-Enynes based on dimethyl malonate and bis(benzenesulfonyl)methane reacted similarly to give the cyclopentane analogs. Activation of the alkyne group by Au(I) initiates the ring closure via a transient bicyclic gold cyclopropylcarbene prior to ring opening with the nucleophile. F.e. and with N-benzyloxycarbonyl derivs., also cleavage of the N-allyloxycarbonyl group from the furan product under palladium catalysis, s. L. Leseurre, P.Y. Toullec, J.-P. Genêt, V. Michelet, Org. Lett. *2007*, *9* (20), 4049-52.

4,5-Condensed 2-oxabicyclo[3.1.0]hexane ring from enynes and oxo compds.

422. 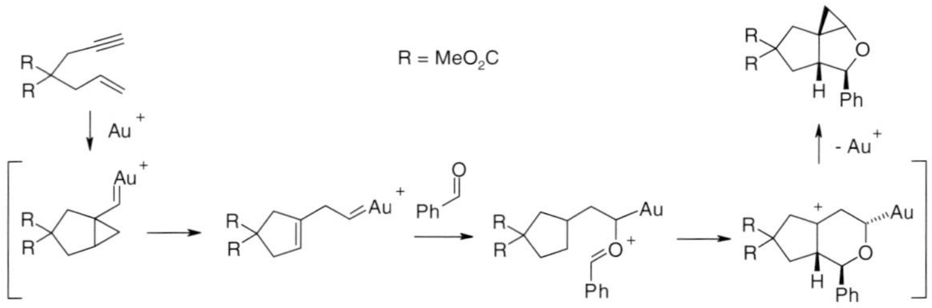

A soln. of the startg. enyne, 5 eqs. benzaldehyde and 5 mol% $AgSbF_6$ in dry methylene chloride under argon treated at -45° with 5 mol% $(Ph_3P)AuCl$, the resulting orange soln. kept at the same temp. for 6 h, allowed to warm slowly to room temp. (22 h), and worked up with flash chromatographic purification on silica → product. Y 68%. Reaction, which is presumed to involve addition of the oxo compd. to an intermediate cyclic gold β,γ-ethylenecarbene complex, is highly diastereoselective and applicable to aliphatic or aromatic aldehydes and ketones as well as cyclic ketones. F.e. and comparison of transition metal catalysts s. M. Schelwies, A.L. Dempwolff, F. Rominger, G. Helmchen, Angew. Chem. Int. Ed. Engl. *2007*, *46* (29), 5598-601.

Magnesium/copper(I) bromide-dimethyl sulfide or iodide/chiral di(phosphines) ←
Catalytic asym. 1,4-addition with Grignard compds. C=C → CHC(R)
to enones or enoates cf. *49*, 644s*69*; to α,β-ethylenethiolic acid esters with $CuBr\text{-}SMe_2$/Josiphos or CuI/Tol-BINAP s. B.M. Ruiz, K. Geurts, M.Á. Fernández-Ibáñez, B. ter Horst, A.J. Minnaard, B.L. Feringa, Org. Lett. *2007*, *9* (24) 5123-6.

Magnesium/iron(III) acetoacetonate *Mg/Fe(acac)$_3$*
Regio- and stereo-specific synthesis of β,γ-ethylenecarboxylic from α-allenecarboxylic acid esters C=C=C → C=C(R)CH
Iron(III)-catalyzed 1,4-addition via magnesium dienolates

423. 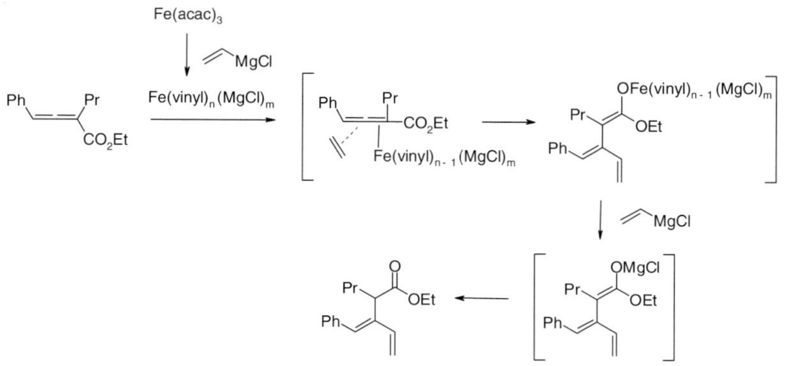

under mild conditions. 5 Mol% Fe(acac)$_3$, the starg. allenoate and *toluene* added sequentially to a dry Schlenk tube under N_2 at room temp., the mixture cooled to -78°, a soln. of 3 eqs. vinylmagnesium chloride in THF added by syringe over 3-5 min, the mixture stirred at -78° for 1 h (TLC), quenched by slow addition of satd. aq. NH_4Cl at -78°, warmed to room temp., diluted with water, extracted with ether, the organic layer washed with 1% aq. HCl, satd. aq. $NaHCO_3$, and brine, dried over Na_2SO_4, evaporated *in vacuo*, and the residue purified by chromatography → ethyl (Z)-4-phenyl-2-propyl-3-vinylbut-3-enoate. Y 82%. The reaction is successful with Grignard compounds possessing alkyl (prim. and sec.; Y 70-94%), vinyl (Y 82 and 86%) and phenyl (Y 63 and 63%) groups, affording *trans*-β,γ-unsatd. esters almost exclusively. Allenyl substituents may be H, alkyl, phenyl or benzyl, and the intermediate magnesium dienolates may be reacted further with electrophiles to generate **allylic quaternary hydrocarbon groups** alpha to the ester group. F.e.s. Z. Lu, G. Chai, S. Ma, J. Am. Chem. Soc. *2007*, *129* (47), 14546-7.

Zinc s. under [Co(dppe)Br$_2$] and (dme)NiBr$_2$ *Zn*
*Cyclopentylmagnesium chloride s. under Ti(OPr-*i*)$_4$* *c-C$_5$H$_9$MgCl*

Dialkylzinc/copper(I) thiophene-2-carboxylate/chiral diaryl(dibenzylamino)phosphines ←
Asym. copper-catalyzed reactions ←
with chiral monodentate diaryl(dibenzylamino)phosphines as ligand

424. 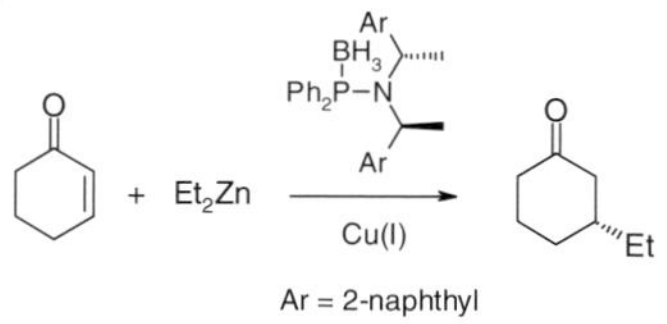

Readily prepared and *highly modular* monodentate chiral diaryl(dibenzylamino)phosphines (*SimplePhos* derivs.) are excellent ligands for a number of highly enantioselective copper-catalyzed reactions, on a par and, in certain instances, superior to established chiral ligands, such as BINOL derivs. **E: Asym. 1,4-addition of dialkylzincs.** Diethylzinc (1 *M* in hexane; 2 eqs.) added slowly to an ethereal soln. of 5 mol% Cu(I)-thiophene-2-carboxylate and 10 mol% chiral aminophosphine (preliminarily stirred for 30 min at room temp., then cooled to -10°), stirred again for 15 min at the same temp., a soln. of the Michael acceptor in dry ether added dropwise in 1 min, stirred at -30° overnight, and quenched with methanol followed by satd. aq. NH_4Cl → product. Conversion >99% (e.e. 95%). Contrary to BINOL-based ligands, the chirality of diaryl(dibenzylamino)phosphines is induced by the helicity of the two aryl groups on phosphorus, which are also chirally flexible as *tropos* ligands. F.e. and other asym. copper-catalyzed reactions [asym. allylic substitution of, e.g., 1,4-dibromobut-2-ene with Grignards, and asym. ring opening of bicyclic hydrazines with trialkylaluminum compds. (cf. *69*, 289)] s. L. Palais, I.S. Mikhel, C. Bournaud, L. Micouin, C.A. Falciola, M. Vuagnoux-d'Augustin, S. Rosset, G. Bernardinelli, A. Alexakis, Angew. Chem. Int. Ed. Engl. *2007*, *46* (39), 7462-5.

Dialkylzinc/copper(II) chloride/chiral 2,2′-diamino-1,1′-binaphthyl/2,6-diphenylaniline ←
Dialkylzinc/copper salts/chiral mannitol-derived bis(phosphites) ←
Catalytic asym. 1,4-addition C=C → CHC(R)
of dialkylzinc compds. cf. *52*, 297s*71*; to cyclic enones (e.e. up to 76%) using $CuCl_2$ (5 mol%), chiral 2,2′-diamino-1,1′-binaphthyl (5.5 mol%) and 2,6-diphenylaniline (10 mol%) s. M. Hatano, T. Asai, K. Ishihara, Tetrahedron Lett. *2007*, *48* (49), 8590-4; with Cu-salts and chiral bis-(phosphites) derived from 1,2:5,6-di-O-cyclohexylidene-D-mannitol s. Q.-L. Zhao, L.-L. Wang, F.Y. Kwong, A.S.C. Chan, Tetrahedron:Asym. *2007*, *18* (16), 1899-905.

Magnesium bromide *$MgBr_2$*
Diels-Alder reaction ○
with chiral α,β-ethylenecarboxylic acid (1-methoxy-2,2,2-triphenylethyl)dimethylsilyl esters with asym. induction s. *72*, 1

Zinc chloride s. under $Cp_2Zr(H)Cl$ *$ZnCl_2$*
Zinc iodide s. under $[Co(dppe)Br_2]$ *ZnI_2*
Sodium tetrahydridoborate s. under $Rh(CO)_2(acac)$ *$NaBH_4$*

Triethylborane/air *Et_3B/air*
Addition of polyhalides to ethylene derivs. s. *18*, 776s*72* C=C → C(Hal)C(R_{Hal})

Chiral 1,3,2-oxazaborolidinium salts ←
Asym. Diels-Alder reaction ○
with 10 mol% chiral 4,5-diaryl-2-bromo-1,3,2-dioxaborolidines cf. *48*, 658*s50*; with chiral 1,3,2-oxazaborolidinium triflates s. D.H. Ryu, K.H. Kim, J.Y. Sim, E.J. Corey, Tetrahedron Lett. *2007*, *48* (33), 5735-7; of 1- or 2-subst. cyclopentadienes with chiral 1,3,2-oxazaborolidinium pentafluorophenylbis(trifluoromethanesulfonyl)methides cf. J.N. Payette, H. Yamamoto, J. Am. Chem. Soc. *2007*, *129* (31), 9536-7.

Chiral aluminum bis(salen) triflate ←
(E)-α,β-Epoxy-γ,δ-ethylene-γ-silylcarboxylic acid amides ←
from terminal 2-acetylenesilanes
Asym. synthesis with addition of two C-atoms

425.

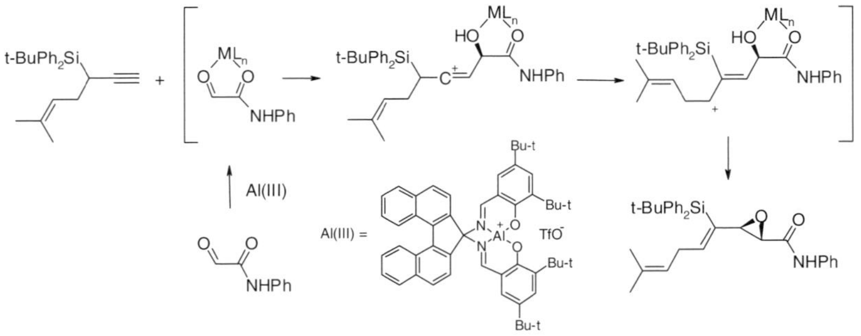

under mild conditions. N-Phenylglyoxamide added to an oven-dried vial containing 10 mol% Al[(S)-sal-BINAM]Cl, 11 mol% AgOTf and 4 Å molecular sieves in a glove box, the vial capped then wrapped in tin foil (to minimize exposure to strong light), removed from the inert atmosphere box, stirred with toluene for 2 min at -78°, 1.5 eqs. startg. propargylsilane added, the tin foil removed, the soln. stirred at -55° for 30 h (TLC), the mixture diluted with pre-chilled 5% ethyl acetate in hexane, filtered through alumina (neutral Brockmann activity I), washed with ether, the filtrate concentrated *in vacuo*, and purified by chromatography → product. Y 83% (e.e. 85%). Eight examples gave 49-97% yield (e.e. 85-94%) from substrates with or without polar substituents or alkene groups. The complex was inactive without intermediate formation of the triflate, while the SbF_6 salt gave a lower conversion in most cases. F.e. incl. reaction optimization and some subsequent conversions without loss of enantiomeric purity s. D.A. Evans, Y. Aye, J. Am. Chem. Soc. *2007*, *129* (31), 9606-7.

Boron fluoride *BF_3*

Ring closures of vinylidenecyclopropanes with N-arylaldimines

426.

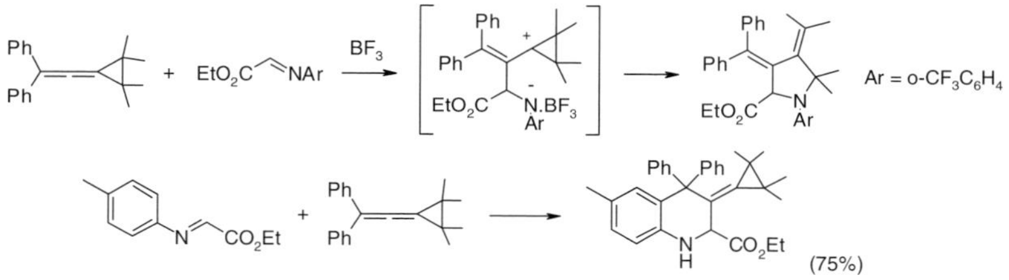

3,4-Di(alkylidene)pyrrolidine-2-carboxylic acid esters. A mixture of the startg. vinylidenecyclopropane and 1.5 eqs. startg. ethyl (arylimino)acetate in methylene chloride stirred in a Schlenk tube under argon for 1 h at 60° with 10 mol% BF_3-etherate, solvent removed under reduced pressure, and the residue worked up with flash chromatographic purification → product. Y 99%. With N-arylimines possessing electron-donating groups, however, the corresponding **3-cyclopropylidenetetrahydroquinoline-2-carboxylic acid esters** were formed. F.e.s. J.-M. Lu, M. Shi, Org. Lett. *2007*, *9* (9), 1805-8.

Gallium(III) chloride *$GaCl_3$*

1,2,3,4-Tetrahydroquinolines from N-arylaldimines

with $ZrCl_4$ cf. *53*, 301s*68*; with $GaCl_3$ s. T. Hirashita, D. Kawai, S. Araki, Tetrahedron Lett. *2007*, *48* (31), 5421-4; with $InCl_3$, 1,2,3,4-tetrahydroquinoline-4-carboxaldehyde hydrazones from α,β-ethylenealdehyde hydrazones, s. V. Sridharan, P.T. Perumal, C. Avendaño, J.C. Menéndez, Org. Biomol. Chem. *2007*, *5* (9), 1351-3.

Scandium(III) triflate *$Sc(OTf)_3$*

Oxidative 1,4-addition s. *72*, 592 C=C → C(R)CH

Samarium(III) triflate/N-bromosuccinimide *$Sm(OTf)_3$/NBS*

2-Arylhalides from ethylene derivs. and electron-rich arenes C=C → C(Ar)C(Hal)

Regio- and stereo-specific halogenative Friedel-Crafts reaction

427.

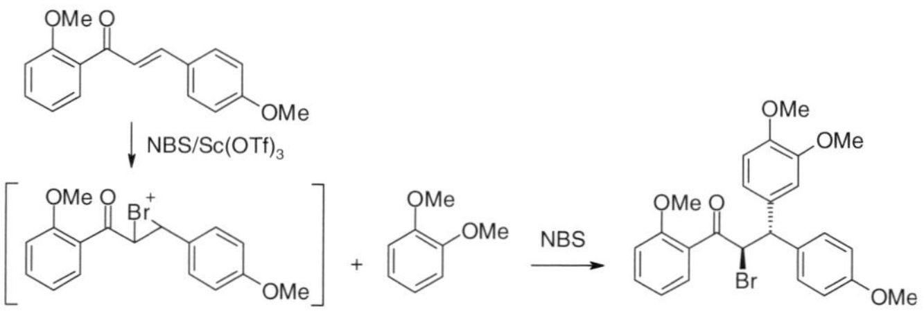

***anti*-β-Aryl-α-bromoketones.** 5 Mol% $Sm(OTf)_3$ and 1.2 eqs. NBS added to a well-stirred suspension of the startg. alkene and 1.2 eqs. startg. electron-rich arene in acetonitrile containing

4 Å molecular sieves, stirred at -15° under argon for 4 h (TLC monitoring), quenched with water, and worked up with flash column chromatographic purification → product. Y 90%. The procedure is inexpensive and generally applicable to styrenes, enones and enoates. The triflate activates NBS in the presence of the alkene to generate a 3-membered bromonium ion which undergoes stereospecific addition of the arene to give *anti*-products. F.e. and comparison of metal triflates, also iodinative Friedel-Crafts reaction with I_2/$NaHCO_3$, s. S. Hajra, B. Maji, S. Bar, Org. Lett. *2007*, *9* (15), 2783-6.

Cerium(III) chloride/lithium iodide *$CeCl_3$/LiI*
γ-(3-Indolyl)ketones from indoles and cyclopropyl ketones C

428. 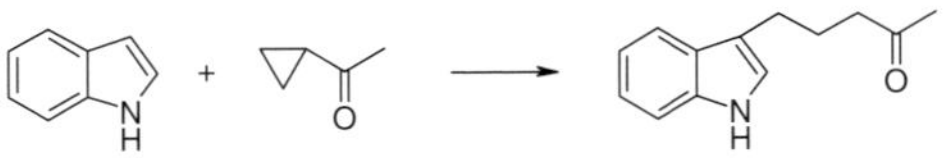

under neutral conditions. Equivalent amounts of indole, startg. cyclopropyl ketone and $CeCl_3 \cdot 3H_2O$ in acetonitrile containing 1.5 eqs. LiI stirred at reflux for 6 h → 5-(1*H*-3-indolyl)-2-pentanone. Y 85%. The procedure is mild, simple, clean, highly selective, economical, non-corrosive and does not require exclusion of atmospheric moisture. Trimethylsilyl iodide and InI_3 were also effective as reagent, but yields were lower with GaI_3, AlI_3 and MgI_2. Significantly, no chloro- or iodo-alkanes were formed as by-products, but yields were slightly lower with indoles possessing electron-withdrawing groups. F.e.s. J.S. Yadav, B.V. Subba Reddy, D. Chandrakanth, G. Satheesh, Tetrahedron Lett. *2007*, *48* (45), 8040-2.

1-Octanol s. under [Ru(η^6-cot)(η^2-dmfm)$_2$] *ROH*
Diglyme s. under Polymer-based 1-[(S)-pyrrolidin-2-ylmethyl]-1,2,3-triazole ←
Polyethylene glycol s. under 2-((2S)-Pyrrolidin-2-ylmethylthio)imidazolium benzoates *PEG*
Cyclic sec. amine-carboxylic acid salts s. under Pyrrolidine derivs. ←

N-Hydrogeno-2-aminopyridinium salts ←
Diels-Alder reaction with 1,1-nitroethylene derivs. ○
cf. *35*, 454, 454s*68*; activation via hydrogen bonding with 20 mol% N-hydrogeno-2-amino-pyridinium salts, e.g. 7-azaindolium tetrakis[3,5-bis(trifluoromethyl)phenyl]borate, also 1,4-addition of heteroarenes (with 0.05-0.5 mol% of the reagent), s. N. Takenaka, R.S. Sarangthem, S.K. Seerla, Org. Lett. *2007*, *9* (15), 2819-22.

*1-*n*-Butyl-3-methylimidazolium salts s. under Chiral Cu(II)-bis(Δ^2-oxazoline) complexes, Fluoroantimonates and $PdCl_2(PPh_3)_2$* *[bmim]$^+$*

Chiral bifunctional prolinamides ←
Organocatalyzed asym. Michael addition C=C → CHC(R)
to 1,1-nitroethylene derivs. s. *62*, 282s*72*

1,3-Disubst. imidazol-2-ylidenes or imidazolin-2-ylidenes s. under $Ni(cod)_2$ ←
Chiral bis(Δ^2-oxazolines) s. under $Ni(ClO_4)_2$ *box*

Polymer-based 1-[(S)-pyrrolidin-2-ylmethyl]-1,2,3-triazole/diglyme ←
Heterogeneous organocatalyzed asym. Michael addition of ketones to 1,1-nitroethylene derivs. in water

429. 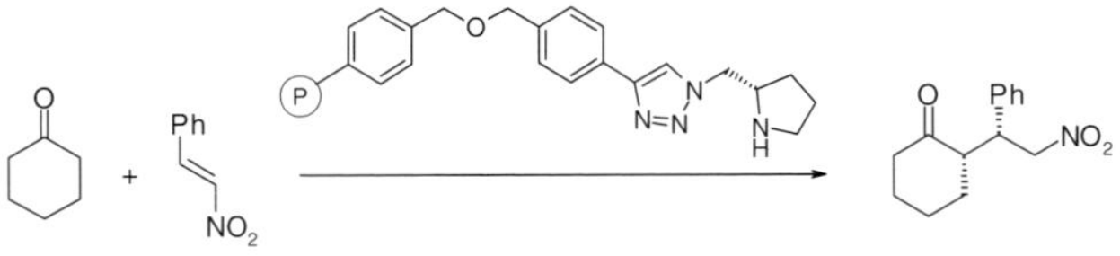

trans-Nitrostyrene, 10 mol% polystyrene-supported 1-[(S)-pyrrolidin-2-ylmethyl]-1,2,3-triazole and 10 mol% diglyme stirred at room temp. with 20 eqs. cyclohexanone in water for 24 h, filtered, and washed with ethyl acetate → (S)-2-((R)-2-nitro-1-phenylethyl)cyclohexanone. Y 85% (conversion 99%; *syn/anti* 95:5; e.e. of 90% raised to 99.9% after a single recrystallization). The illustrated catalyst gave the highest conversion and selectivity of two catalysts tried at room temperature. A range of β-nitrostyrenes containing electron-withdrawing and -releasing groups on the phenyl ring were successfully reacted with cyclohexanone, methyl ethyl ketone and 4-pyrone. Acetone gave a low yield and selectivity due to formation of some double addition product. Selectivity with aliphatic aldehydes (2 examples) was also lower. F.e. incl. preparation of the catalysts on a Merrifield resin s. E. Alza, X.C. Cambeiro, C. Jimeno, M.A. Pericàs, Org. Lett. *2007, 9* (19), 3717-20.

*1-Methyl-3-[*p*-[1-((2S)-pyrrolidin-2-ylmethyl)-1,2,3-triazol-4-yl]benzyl]imidazolium fluoroborate/trifluoroacetic acid* ←
Organocatalyzed asym. Michael addition C=C → CHC(R)
to 1,1-nitroethylene derivs. s. *62*, 282s*72*

2,6-Diisopropylphenyl isocyanide s. under [ReBr(CO)$_3$(thf)]$_2$ *ArNC:*
Amberlyst A-21 s. under Amberlite IRA-402 (nitrite) ←
Formic acid s. under Pd(PPh$_3$)$_4$ *HCOOH*
Acetic acid s. under Platinum(II) bis(phosphinous acid) complexes *AcOH*
Triphenylacetic acid s. under [Ir(cod)$_2$]BAr$_4$ *Ph$_3$CCOOH*
Arylcarboxylic acids s. under Chiral pyrrolidine derivs. *ArCOOH*
L-Proline s.a. under Rh(CO)$_2$(acac) *Pro-OH*

L-Proline/sodium azide *Pro-OH/NaN$_3$*
(Z)-γ-Nitro-δ,ε-ethylenecarbonyl C=C → CHC-C(NO$_2$)C=C
from α,β-ethylenecarbonyl compds. and 1,1-nitroethylene derivs.
via Lewis base-catalyzed double Michael addition

430.

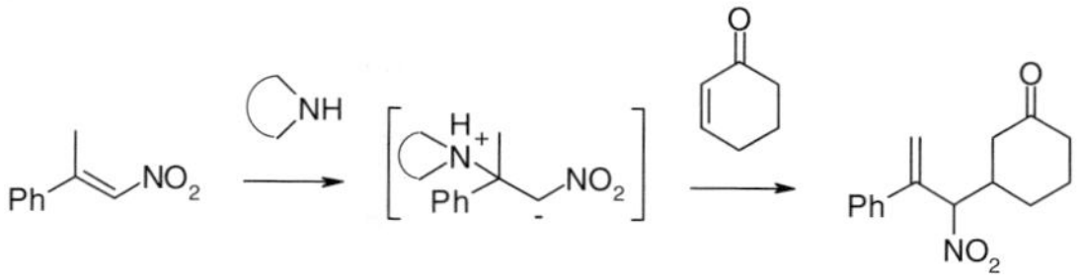

The startg. nitroalkene added to a soln. of 2 eqs. cyclohex-2-enone, 20 mol% L-proline and 1 eq. NaN$_3$ in DMSO, stirred at room temp. for 5 h, and worked up with flash column chromatographic purification on silica gel → 3-(1-nitro-2-phenylallyl)cyclohexanone. Y 87% (e.e. <5%). This *'deconjugative'* Michael addition plays on the enhanced electrophilic activity of nitroolefins over enones (or enoates), reaction involving initial Michael addition of the Lewis base to the former, followed by a second Michael addition to the less reactive enone (or enoate) prior to elimination of the base. Sodium azide serves as a second Lewis base and is thought to fine-tune the acidity. Only sec. amines appear to catalyze the conversion and prevent polymerization. F.e. incl. reaction with acyclic enones as well as cyclic and acyclic enoates s. X. Sun, S. Sengupta, J.L. Petersen, H. Wang, J.P. Lewis, X. Shi, Org. Lett. *2007, 9* (22), 4495-8.

*Trifluoroacetic acid s. under Chiral diamines, Chiral pyrrolidine derivs., 1-Methyl-3-[*p*-[1-((2S)-pyrrolidin-2-ylmethyl)-1,2,3-triazol-4-yl]benzyl]imidazolium fluoroborate and K$_2$PtCl$_4$* *CF$_3$COOH*

Cinchona alkaloid-derived thiourea ←
Organocatalyzed asym. Michael addition s. *62*, 282s*72* C=C → CHC(R)

Organocatalyzed asym. Diels-Alder reaction s. *59*, 301s*72* ○

2-((2S)-Pyrrolidin-2-ylmethylthio)imidazolium benzoates/polyethylene glycol ←
Chiral 2-aminothiourea/tert. amine ←
Organocatalyzed asym. Michael addition s. *62*, 282s*72* C=C → CHC(R)

3-Benzyl-4,5-dimethylthiazolium chloride s. under 2(S)-[Diphenyl(trimethylsiloxy)-methyl]pyrrolidine ←

N-Bromosuccinimide s. under Sm(OTf)$_3$ *NBS*

Trimethylsilyl cyanide s. under Microwaves *Me_3SiCN*

Silica gel *SiO_2*

4-Component synthesis of (Z)-α-(α-aminoarylidene)-α′-carbamylsuccinic acid esters ←

431.

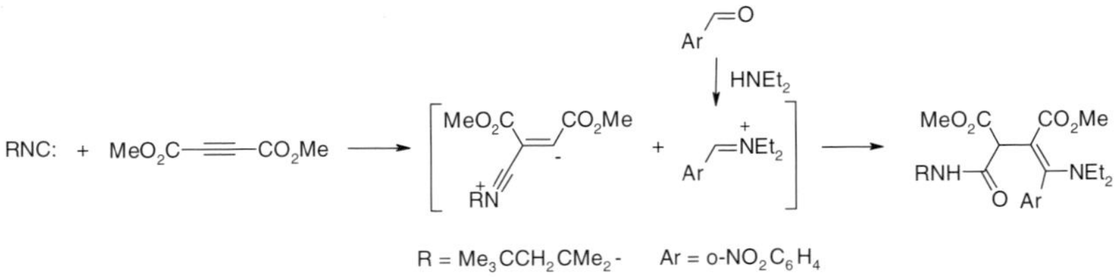

A mixture of diethylamine and 1 eq. *o*-nitrobenzaldehyde in dry methylene chloride stirred at room temp. for 30 min, a soln. of 1.25 eqs. dimethyl acetylenedicarboxylate and 1.25 eqs. 1,1,3,3-tetramethylbutyl isocyanide in the same solvent at -10° added rapidly, stirred for 5 min, silica gel powder added quickly, allowed to warm to room temp., solvent removed under reduced pressure, the residue kept for 24 h, and worked up → dimethyl 2-[(Z)-1-(2-nitrophenyl)-1-(diethylamino)methylene]-3-[[(1,1,3,3-tetramethylbutyl)amino]carbonyl]succinate. Y 77%. Reaction is presumed to involve formation of an intermediate zwitterion by addition of the isonitrile to the acetylene deriv., followed by reaction with *in situ*-generated aldiminium ion. F.e.s. E. Ahmadi, A. Ramazani, M.N. Haghighi, Tetrahedron Lett. *2007*, *48* (39), 6954-7.

*Titanium tetraisopropoxide/cyclopentylmagnesium chloride/*n*-butyllithium* ←

Regio- and stereo-specific synthesis of sec. amines CH=NR → CH(NHR)C-CH
from aldimines and ethylene derivs.
Alkoxide-directed titanium(II)-mediated reductive cross-coupling

432.

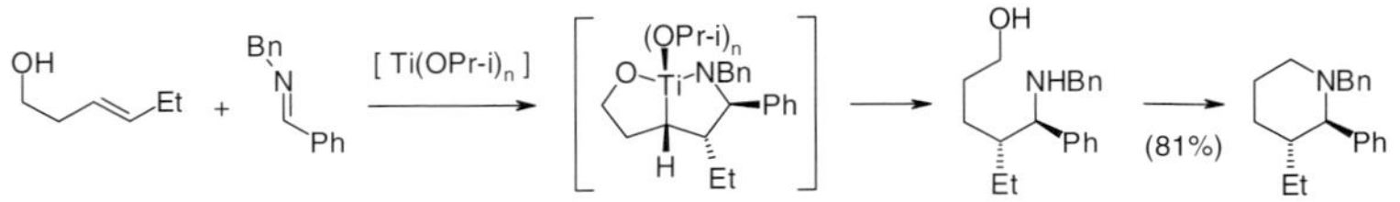

***anti*-5-Aminoalcohols.** 1.2 Mmol 2.15 *M* cyclopentylmagnesium chloride in ether added dropwise to a soln. of 0.403 mmol startg. imine and 0.609 mmol Ti(OPr-*i*)$_4$ in the same solvent at -70°, warmed to -40° over 1 h, stirred for a further 1 h, a soln. of the startg. Li-alkoxide [prepared by treatment of an ethereal soln. of 0.604 mmol of the corresponding alcohol with 0.650 mmol *n*-BuLi (2.6 *M* in hexanes) at -78°, followed by warming to 0° over 20 min] added dropwise to the resulting titanium-imine complex at -40° via cannula, warmed to 0° over 1 h, stirred for 6 h, quenched with water, stirred rapidly until the precipitate became white, and worked up with column chromatographic purification on silica gel → (R*)-4-[(R*)-(N-benzylamino)(phenyl)-methyl]hexan-1-ol. Y 73% (regioselectivity >95:5; d.r. >95:5). The procedure is generally applicable to the coupling of aromatic aldimines with terminal, 1,1-disubst. and (E)- or (Z)-configured disubst. olefins, the latter affording *anti*-products irrespective of the stereochemistry of the double bond. Such stereo-convergence is dependent on the presence of the hydroxyl group and in keeping with the preferential generation of an intermediate *cis*-fused O,Ti,N-bicyclic complex so as to avoid 1,2-steric interactions. F.e. and from chiral aldimines **with asym. induction**, or from chiral 3-ethylenealcohols with **double asym. induction**, also conversion of the products to the corresponding **piperidines** via intramolecular Mitsunobu N-alkylation, s. M. Takahashi, G.C. Micalizio, J. Am. Chem. Soc. *2007*, *129* (24), 7514-6.

*Titanium tetra-*tert*-butoxide s. under Ni(cod)$_2$* *Ti(OBu-*t*)$_4$*

Chlorobis(cyclopentadienyl)hydridozirconium/methyllithium/zinc chloride/ copper(I) cyanide/lithium chloride/trimethylsilyl chloride ←
Synthesis of 2-methylene-*sec*-alcohols from sec. propargyl alcohols via regiospecific hydrozirconation of sec. lithium propargyloxides ←

433. 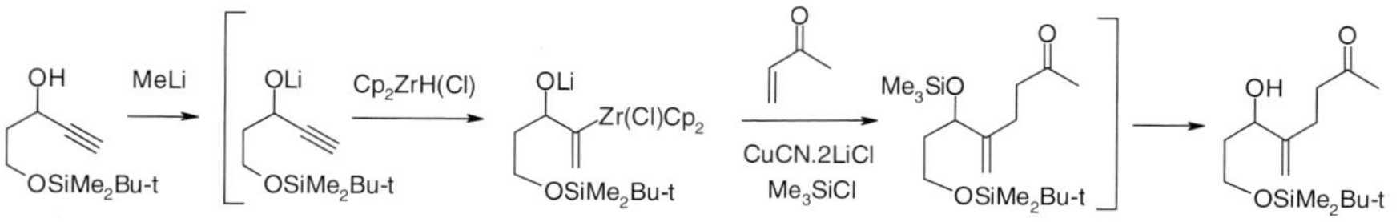

1 Eq. *methyllithium* soln. (1.6 *M*) added to a soln. of the startg. propargyl alcohol in THF at -78°, warmed to 23° after 20 min, the resulting propargylic alkoxide soln. added (with THF washings) via cannula to a mixture of 6 eqs. vacuum-fused *$ZnCl_2$* and 2 eqs. $Cp_2Zr(H)Cl$ in THF at 23° (preliminarily stirred for 3 min to solubilize the zirconium complex), stirred for a further 2 h, cooled to -78°, a soln. of 1 eq. CuCN·2LiCl in THF added, treated sequentially after 20 min with 5 eqs. HMPA, a soln. of 2.5 eqs. methyl vinyl ketone in THF, and 5 eqs. trimethylsilyl chloride, stirred at -30° for 2.5 h, quenched with aq. NH_4Cl, worked up, the crude bis(silyl ether) dissolved in methanol, treated with K_2CO_3 (to remove the trimethylsilyl group selectively), and worked up with flash chromatographic purification on silica gel → product. Y 68%. Significantly, hydrozirconation of the lithium propargyloxide in the presence of $ZnCl_2$ gave the *branched* hydrozirconation adduct *exclusively*, such regioselectivity being the total reverse of that for conventional hydrozirconation of acetylene derivs. F.e. and electrophiles s. D. Zhang, J.M. Ready, J. Am. Chem. Soc. *2007*, *129* (40), 12088-9.

Zirconocene dichloride/alkyllithiums *Cp_2ZrCl_2/RLi*
Regio- and stereo-specific synthesis of 2-arylcyclohexanols from 2-aryl-2,3,4,5-tetrahydrooxepins

434. 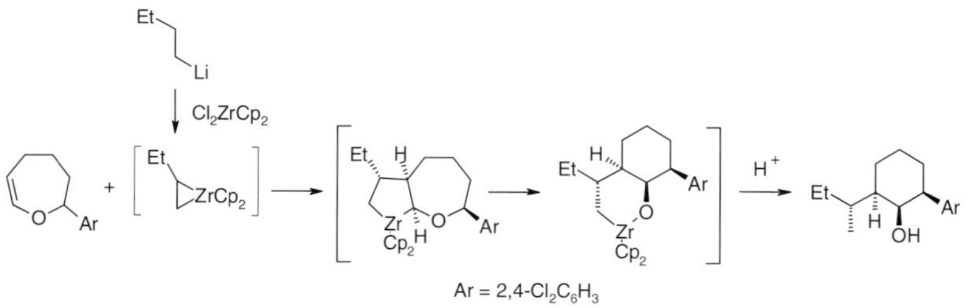

2 Eqs. *n*-butyllithium in ether added to a soln. of Cp_2ZrCl_2 in dry THF at -78° under argon, stirred at the same temp. for 45 min, 0.67 eqs. of the startg. enolether added, allowed to warm to room temp., stirred for 5-8 h, cooled to 0°, excess of water added, and the mixture worked up with column chromatographic purification on silica gel → (1S*,2R*,6S*)-2-[(R*)-*sec*-butyl]-6-(2,4-dichlorophenyl)cyclohexanol. Y 86%. This unprecedented one-pot ring contraction generates *three new carbon-carbon bonds and four contiguous chiral centres* in one operation. The aryl-substituent appears essential, and the conversion is favoured by electron-withdrawing groups in the aromatic ring. Reaction is presumed to involve initial insertion of the intermediate zirconocene η^2-olefin complex, which can also be generated from Grignard reagents. F.e. and electrophiles s. J. Barluenga, L. Álvarez-Rodrigo, F. Rodríguez, F.J. Fañanás, T.L. Sordo, P. Campomanes, Angew. Chem. Int. Ed. Engl. *2007*, *46* (15), 2607-9.

Amberlite IRA-402 (nitrite)/Amberlyst A-21 ←
Heterogeneous Michael addition of prim. nitro compds. C=C → CHC-C(NO_2)
generated *in situ* from prim. halides s. 72, 384

Tricyclopentylphosphine *s. under Ni(cod)$_2$* *$(c\text{-}C_5H_{11})_3P$*

Diphenylphosphinoalanine deriv. ←
Organocatalytic regiospecific asym. [3+2]-cycloaddition ○
4-Acyl-1-cyclopentene- from α-allene-carboxylic acid esters and α,β-ethyleneketones

435. 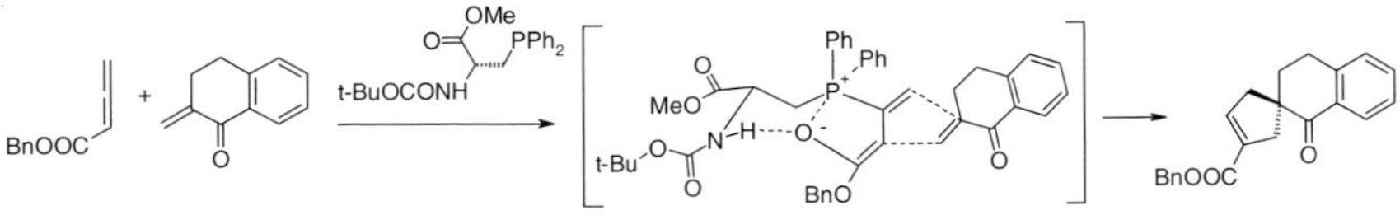

1.5 Eqs. startg. allenoate added to 2-methylene-1-tetralone in toluene, the tube sealed with a rubber septum, cooled to -25°, purged with dry N_2 for 30 min, a soln. of 10 mol% diphenylphosphinoserine deriv. [Boc-PPs-OMe] in toluene added dropwise, the soln. stirred at -25° for 30 h, then loaded directly onto silica gel, and subjected to flash chromatography → product. Y 96% (α/γ >99:1; e.e. 80%). The regioselectivity was lower with acyclic enones; however, a 'deracemization' reaction of γ-subst. allenoates with chalcones proceeded with complete regio- and diastereo-selectivity (e.e. up to 93%). It is believed the transition state may involve hydrogen bonding between the amino acid NH and the allenoate carbonyl, with the enone approaching the zwitterion from the π-face opposite one of the phenyl rings of the catalyst. F.e.s. B.J. Cowen, S.J. Miller, J. Am. Chem. Soc. *2007, 129* (36), 10988-9.

Triphenylphosphine s. under Rh(CO)$_2$(acac) *Ph_3P*
Tri-2-furylphosphine s. under [Rh(cod)$_2$]OTf *Fu_3P*
Chiral phosphines s. under Mg ←

1,3-Bis(diphenylphosphino)propane [s.a. under [Rh(cod)Cl]$_2$ and PdCl$_2$(PPh$_3$)$_2$] *dppp*
Satd. 5-membered N-sulfonyl-N-heterocyclics from electron-deficient acetylene derivs.
via di(phosphine)-catalyzed stereospecific double Michael addition of 2-functionalized sulfonylamines s. 72, 177

1,4-Bis(diphenylphosphino)butane s. under Ni(cod)$_2$ dppb
Chiral diaryl(dibenzylamino)phosphines s. under R$_2$Zn *$Ar_2P(NBn_2)$*
(R)-3,3′-Dichloro-2,2′-dimethoxy-6,6′-bis(diphenylphosphino)biphenyl ←
s. under [Ir(cod)$_2$]BAr$_4$

Tris(dimethylamino)phosphine *$(Me_2N)_3P$*
1-Cyclohexenecarboxylic
from α-allenecarboxylic acid esters and electron-deficient ethylene derivs.
Regio- and stereo-specific phosphorus(III)-catalyzed [4+2]-cycloaddition

436. 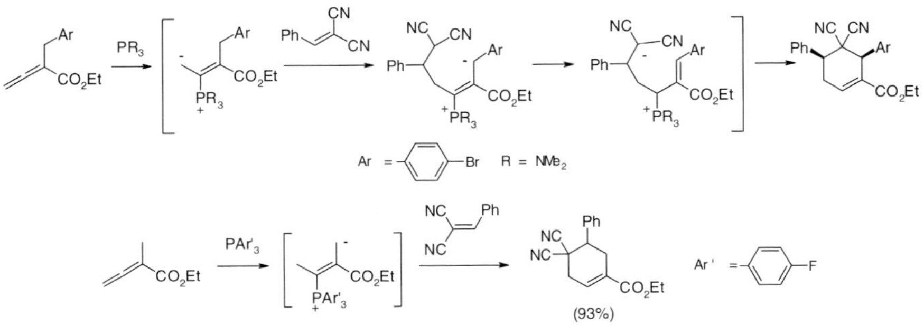

1.2 Eqs. startg. allenoate added via syringe over 3 h to a soln. of benzylidenemalononitrile and 20 mol% $(Me_2N)_3P$ in refluxing benzene (80°) under argon, additional allenoate (0.2 eq.) added over 5 min as required after 8 h (TLC), after 14 h the mixture concentrated, and the residue purified by chromatography → ethyl 3-(4-bromophenyl)-4,4-dicyano-5-phenylcyclohexene-2-carboxylate. Y 96% (*cis/trans* 84:16). With *tris(dimethylamino)phosphine* as catalyst, the method is successful for the reaction of β-hydro, -aryl, -ester, -vinyl and -alkyl subst. allenoates with arylidenemalonates possessing electron-withdrawing or -donating groups (13 examples), affording predominantly *cis*-products in 86-98% yield (the β-isopropylallenoate giving a 66:34 mixture of *trans/cis*-isomers in 77% yield). In the presence of *electron-deficient triarylphosphines*, however, the regioisomer is obtained exclusively in 80-93% yield (6 examples), the reversal in regio-selectivity being rationalized in terms of the inverted polarization of the intermediate 1,4-dipolar synthon. F.e.s. Y.S. Tran, O. Kwon, J. Am. Chem. Soc. *2007*, *129* (42), 12632-3.

Chiral bis(phosphites) s. under R_2Zn ←

1-Butyl-3-methylimidazolium undecafluorodiantimonate *[bmim][Sb_2F_{11}]*
Regio- and stereo-selective hydroarylation of acetylene derivs. C≡C → CH=C(Ar)
under Pd catalysis cf. *59*, 311s*70*; using the *superacidic* ionic liquid, [bmim][Sb_2F_{11}] [freshly prepared by mixing [bmim][SbF_6] and SbF_5 under argon], as catalyst, (Z)-selectivity in the presence of 5 mol% at 90° or (E)-selectivity in the presence of 20 mol% at -25° (or -78°), s. D.S. Choi, J.H. Kim, U.S. Shin, R.R. Deshmukh, C.E. Song, Chem. Commun. *2007* (33), 3482-4; double thienylation of propiolic acid or its ethyl ester in the presence of K_2PtCl_4/AgOTf in TFA s. M.L. Keita, T. Mizuhara, J. Oyamada, T. Kitamura, Chem. Lett. *2007*, *36* (9), 1150-1.

Air s. under [(Cp)$RhCl_2$]$_2$* *air*
Trifluoromethanesulfonimide s. under Chiral triamine *Tf_2NH*
p-*Toluenesulfonic acid s. under Axially-chiral 1,2-diamines* *TsOH*
2,4,6-Triisopropylbenzenesulfonic acid s. under 9-Amino-O-demethyl-9-deoxyepiquinine *$ArSO_3H$*

Tungsten complexes ←
Regio- and stereo-specific [2+2]-cycloaddition with tungsten-complexed 2,4-cyclohexadienones [2*H*-phenols] □

437.

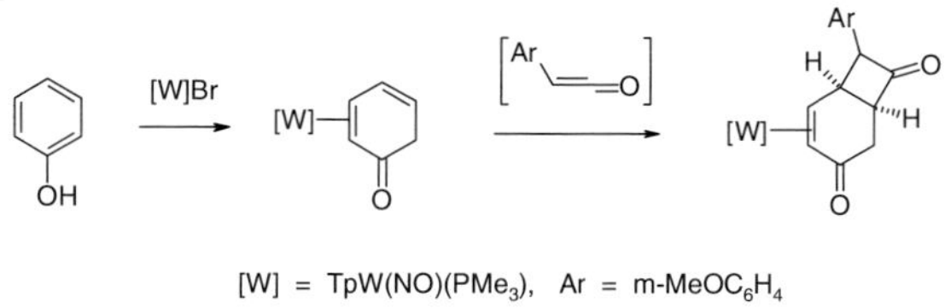

Tungsten-complexed bicyclo[4.2.0]oct-4-ene-3,8-diones. A soln. of 38 eqs. oxalyl chloride in benzene added to 2.3 eqs. 3-methoxyphenylacetic acid in a glovebox under N_2, after 4 h the solvent removed *in vacuo*, benzene added then evacuated (to remove excess oxalyl chloride), the resulting oil dissolved in methylene chloride, added slowly dropwise over ca. 30 min to a well-stirred soln. of TpW(NO)(Me_3P)(η^2-phenol) (single isomer, obtained by precipitation from methanol in the presence of 0.25 eq. DBU) and 2.35 eqs. *i*-Pr_2NEt in methylene chloride, the mixture removed from the glovebox, quenched with 1 *M* HCl, the organic layer extracted, dried, filtered, evaporated, the residue taken up in acetone, stirred vigorously, and filtered after 15 min, washing with acetone → the *endo*-isomer (Y 29%), while the filtrate was added to hexane and the precipitate collected → the *exo*-isomer (Y 52%). F.e. incl. cycloaddition with dimethyl acetylene-dicarboxylate, also reductions to the corresponding cyclobutanols, ring contraction of the 7-chloro-analog, and demetalation with NBS, s. M.A. Todd, M. Sabat, W.H. Myers, W.D. Harman, J. Am. Chem. Soc. *2007*, *129* (36), 11010-11; large scale syntheses of TpW(NO)(PMe_3)(η^2-arene) complexes (arene = benzene, anisole, dimethoxybenzene, trifluorotoluene and naphthalene) and their precursors such as TpW(NO)(PMe_3)Br s. K.D. Welch, D.P. Harrison, E.C. Lis, Jr., W. Liu, R.J. Salomon, W.D. Harman, Organometallics *2007*, 26 (10), 2791-4.

*Tetra-*n*-butylammonium fluoride s. under Microwaves* *Bu_4NF*

Bromotricarbonyl(tetrahydrofuran)rhenium dimer/ 2,6-diisopropylphenyl isocyanide *$[ReBr(CO)_3(thf)]_2$/ArNC:*

[2+2]-Cycloaddition with acetylene derivs. □

to norbornadienes under Ru-catalysis cf. *60*, 288s*66,67*; with a little $[ReBr(CO)_3(thf)]_2$/ 2,6-*i*-$Pr_2C_6H_3NC$: s. Y. Kuninobu, P. Yu, K. Takai, Chem. Lett. *2007, 36* (9), 1162-3; to oxanorbornadienes with Cp*RuCl(cod) s. R.R. Burton, W. Tam, J. Org. Chem. *2007, 72* (19), 7333-6.

Iron(III) acetoacetonate s. under Mg *$Fe(acac)_3$*

Iron(III) chloride *$FeCl_3$*

Diels-Alder reaction ○

using $AlCl_3$ cf. *16*, 746; of N-acryloyl-2-pyrrolidones using $FeCl_3 \cdot 6H_2O$ at room temp. in the absence of solvent s. W. Pei, Y.-J. Wang, C.-Q. Yu, Chin. J. Chem. *2007, 25* (6), 814-7.

Dibromo[1,3-bis(diphenylphosphino)propane]cobalt(II)/zinc/zinc iodide ←

1,4-Dienes C≡C → CH=C-CH_2CH=CHR

from acetylene derivs. and terminal ethylene derivs.

Regio- and stereo-specific cobalt(II)-catalyzed Alder-ene reaction

438. 

Allyltrimethylsilane and 1.5 eqs. ethyl 2-butynoate added to a suspension of 10 mol% $[Co(dppe)Br_2]$, 20 mol% ZnI_2 and 20 mol% Zn dust in dry methylene chloride under N_2, stirred for 16 h at room temp., filtered, and worked up with column chromatographic purification on silica gel → product. Y 100% (E/Z >99:1). The procedure is inexpensive, atom-economical and provides good yields with high chemo-, regio- and stereo-selectivity from *internal* acetylene derivs. and a range of terminal ethylene derivs. Regioselectivity is very high with unsym. alkynes, the alkene group generated therefrom being exclusively (Z)-configured, while high levels of *trans*-selectivity are recorded for the migrated alkene group. Terminal acetylene derivs. favour [2+2+2]-cycloaddition. F.e.s. G. Hilt, J. Treutwein, Angew. Chem. Int. Ed. Engl. *2007, 46* (44), 8500-2.

Stilbenes from 1,3-dienes, 2-acetylenephosphonium salts and ar. aldehydes ○

One-pot conversion via cobalt(I)-catalyzed Diels-Alder reaction and Wittig synthesis

439. 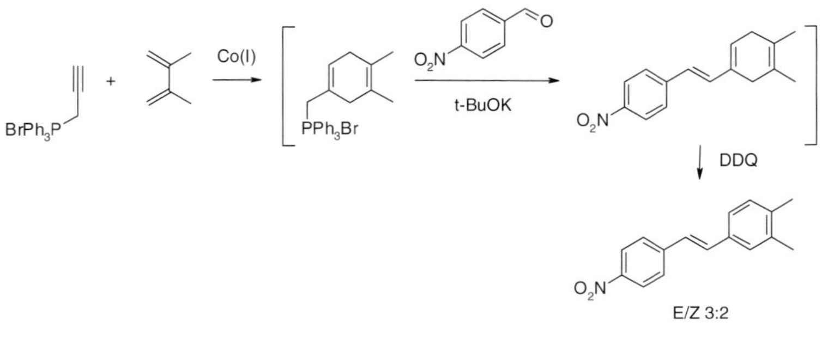

ZnI_2 (1 mmol), Zn dust (0.6 mmol) and dibromo[1,2-bis(diphenylphosphino)ethane]cobalt(II) suspended in dry methylene choride (5 ml) to generate the active catalytic species (with a colour change from green to brown), triphenyl(prop-2-ynyl)phosphonium bromide (0.5 mmol) and

2,3-dimethyl-1,3-butadiene (0.75 mmol) added, the mixture stirred at room temp. for 0.5 h, KOBu-*t* (2.5 mmol) and *p*-nitrobenzaldehyde (0.75 mmol) added at 0°, the mixture stirred for 1 h at room temp., filtered, the solvent evaporated, the resulting crude residue dissolved in benzene (10 ml), 2,3-dichloro-4,5-dicyanobenzoquinone (0.6 mmol) added, and the mixture stirred at room temp. for 1 h → product. Y 85%. The procedure is compatible with both electron-rich and -poor ar. aldehydes, although yields are lower for *o*-subst. derivs. due to steric hindrance; in all cases, (E)-olefins are formed predominantly. Unsymmetrical dienes afford mixtures of regioisomers. F.e., **also styrenes** from aliphatic aldehydes (in moderate yield), s. G. Hilt, C. Hengst, J. Org. Chem. *2007*, *72* (19), 7337-42.

Bis(1,5-cyclooctadiene)nickel(0)/1,8-diazabicyclo[5.4.0]undec-7-ene *Ni(cod)$_2$/DBU*
N-Protected (E)-α,β-ethylene-β′-hydrazinocarboxylic acid esters from allenes ←
α-Alkylidene-γ-nickela-γ-lactones as intermediates

440.

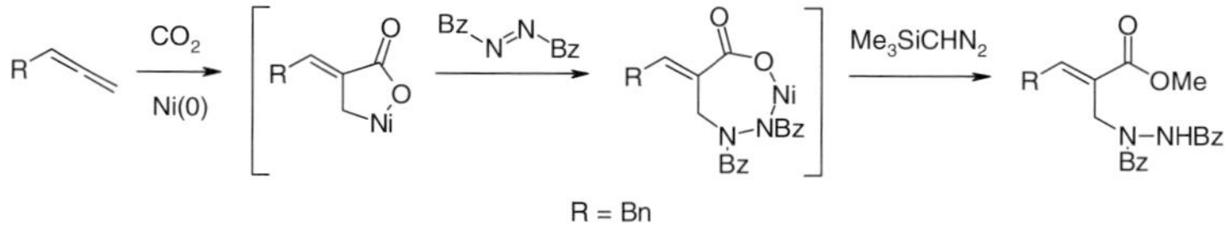

A stirred suspension of Ni(cod)$_2$ (0.3 mmol) in THF treated at 0° under N_2 with DBU (0.66 mmol), the mixture degassed then CO_2 introduced, a soln. of the startg. allene (0.33 mmol) in THF added dropwise, stirring continued at 0° for 2 h, a soln. of dibenzoyldiazene (0.33 mmol) in THF added, quenched after 2 h with 2 *M* aq. HCl, extracted with ethyl acetate, worked up, and the free acid converted to the methyl ester with trimethylsilyl(diazo)methane → product. Y 68% (98:2 mixture of regioisomers; E/Z 87:13). Ethers, esters and phthalimido groups remained unaffected. F.e. and conversion to **N-protected β′-amino-α,β-ethylenecarboxylic acid esters** (with SmI$_2$) s. M. Murakami, N. Ishida, T. Miura, Chem. Lett. *2007*, *36* (3), 476-7.

Bis(1,5-cyclooctadiene)nickel(0)/(S,S)-1,3-bis(1-mesitylpropyl)imidazol-2-ylidene ←
Siloxy-3(Z)-ethylenes from 1,3-dienes, aldehydes and organosilicon hydrides ←
Nickel-catalyzed asym. 3-component reaction

441.

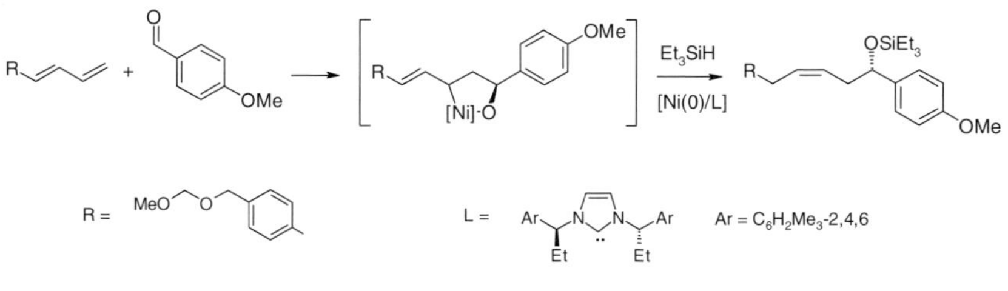

Ni(cod)$_2$ (0.072 mmol) and 0.072 mmol (S,S)-1,3-bis(1-mesitylpropyl)imidazolium chloride (previously dried *in vacuo* at 80° for 4 h) in THF stirred with 0.14 mmol Cs_2CO_3 at 0° for 1 h, 1.8 mmol triethylsilane added, followed by a soln. of 0.36 mmol each of the startg. 1,3-diene and aldehyde in the same solvent at 0°, stirred at 50° for 10 h, quenched with satd. aq. NH_4Cl at 0°, and worked up with column chromatographic purification on silica gel → (1S,3Z)-1-triethylsiloxy-5-(4-methoxymethoxymethylphenyl)-1-(4-methoxyphenyl)pent-3-ene. Y 83% (e.e. 71%). This asym. version of an established conversion is based on the use of a *chiral N-heterocyclic carbene* as ligand, which up to present has only found limited application. 1,4-Disubst. 1,3-butadienes gave chiral *anti*-products exclusively. F.e. and comparison of ligands s. Y. Sato, Y. Hinata, R. Seki, Y. Oonishi, N. Saito. Org. Lett. *2007*, *9* (26), 5597-9.

Bis(1,5-cyclooctadiene)nickel(0)/1,3-bis(2,6-diisopropylphenyl)imidazolium chloride/ potassium tert-*butoxide/titanium tetra*-tert-*butoxide* ←

Stereospecific [3+2]-cycloaddition of cyclopropanecarboxaldimines with α,β-ethyleneketones ○

442.

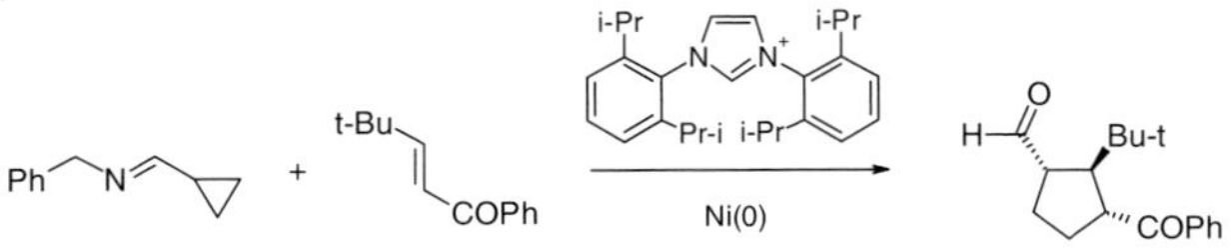

2-Subst. 3-acylcyclopentanecarboxaldehydes. A soln. of 10 mol% 1,3-bis(2,6-diisopropylphenyl)imidazolium chloride, 10 mol% KOBu-*t* and 10 mol% Ni(cod)$_2$ in toluene stirred 20-30 min, 1.5 eqs. startg. cyclopropanealdimine and 2 eqs. Ti(OBu-*t*)$_4$ in toluene added sequentially by syringe to the resulting red soln., 0.2 eqs. startg. enone in toluene added, heated to 90°, a further 0.8 eqs. enone in toluene added by syringe pump over 1.5-2 h, stirred 3-6 h, cooled, 10% HCl added, the mixture stirred for a further 20 min, and extracted with ether → (1S*,2R*,3R*)-3-benzoyl-2-*tert*-butylcyclopentanecarbaldehyde. Y 93% (d.r. >98:2). The scope of the reaction is broader than that with cyclopropyl ketones, being successful with aryl alkenyl enones (aryl is phenyl, naphthyl or furan-2-yl; alkenyl is propenyl, 3,3-dimethylbutenyl, styryl, 4-phenylbutadienyl, 1-octenyl or 2-methylpropenyl) with yields in the range 62-98% and diastereomeric ratios generally >7:3. The enone is the limiting reagent and is added gradually to avoid accumulation and reaction inhibition. No reaction occurs with enoates or alkyl alkenyl ketones, while in the absence of Ti(OBu-*t*)$_4$ reaction times are longer and yields lower. Key to the method is the intermediate generation of a metalloenamine, which (unlike the intermediate formed with cyclopropyl ketones) reacts by subsequent Michael addition *with the minimum of side-reactions*. F.e.s. L. Liu, J. Montgomery, Org. Lett. *2007*, *9* (20), 3885-7.

Bis(1,5-cyclooctadiene)nickel(0)/chiral 1,3-diaryl-Δ²-imidazolinium fluoroborate/ potassium tert-*butoxide* ←

Regiospecific asym. synthesis of siloxy-2-ethylenes C≡C → CH=C-C(OSi⩽)

from aldehydes, acetylene derivs. and organosilicon hydrides

Chiral 1,3-diarylimidazolidin-2-ylidenes as ligand

443.

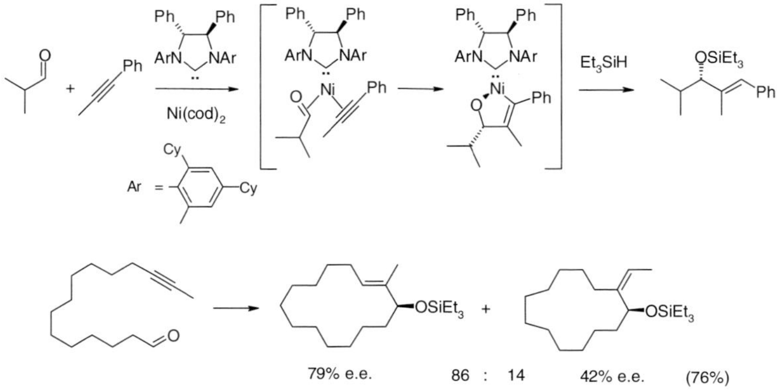

under mild conditions. A mixture of 10 mol% Ni(cod)$_2$, 10 mol% chiral 1,3-diaryl-Δ²-imidazolinium fluoroborate and 10 mol% KOBu-*t* stirred in THF at room temp. under inert atmosphere for 10 min, 2 eqs. triethylsilane added, stirring continued for 5 min at 0°, neat isobutanal and

1.2 eqs. 1-phenyl-1-propyne added sequentially at room temp., the mixture stirred until complete (30 min), quenched by stirring in air for 15 min, concentrated *in vacuo*, and purified by chromatography → (S)-1-phenyl-2-methyl-3-triethylsilyloxy-4-methylpent-1-ene. Y 86% (selectivity >19:1; e.e. 70%). This simple and rapid method is successful for aromatic and aliphatic (branched and unbranched) aldehydes reacting with mono- or di-substituted acetylenes giving yields of 64-99%, regioselectivities from 3:1 to >19:1 and enantiomeric excesses of 65-85%. High regioselectivities are obtained with alkyl(aryl)acetylenes and terminal alkynes by coupling predominantly at the less-congested carbon of the triple bond, while 13-pentadecynal underwent **intramolecular coupling** under these conditions to afford a 14-membered macrocycle as major product. F.e. incl. ligand development and synthesis s. M.R. Chaulagain, G.J. Sormunen, J. Montgomery, J. Am. Chem. Soc. *2007*, *129* (31), 9568-9.

Bis(1,5-cyclooctadiene)nickel(0)/tricyclopentylphosphine *$Ni(cod)_2/(c\text{-}C_5H_{11})_3P$*
(E)-2-Vinylpyridine N-oxides from pyridine N-oxides and acetylene derivs. H → C=CH
Regio- and stereo-specific hydroarylation under mild conditions

444. 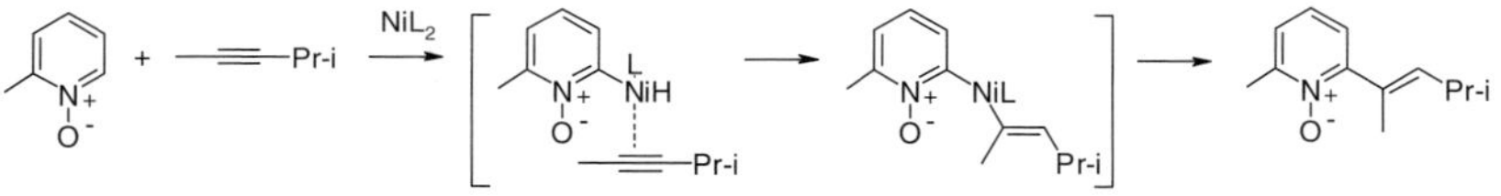

Startg. N-oxide (1 mmol) and 1.5 eqs. startg. alkyne added sequentially to 10 mol% each of $Ni(cod)_2$ and tricyclopentylphosphine in toluene (in a glove box), the reaction vial heated at 35° for 15 h, and worked up with flash chromatographic purification on silica gel → product. Y 56% (E/Z >99:1). The procedure is applicable to the hydroarylation of internal (not terminal!) acetylene derivs. with a wide range of pyridine and isoquinoline N-oxides, incl. carbalkoxy derivs., but addition of chloro-, bromo- and nitro-subst. pyridine oxides was sluggish. There was no reaction with the corresponding pyridines. Reaction is thought to involve initial formation of a pyridyl-(hydrido)nickel complex, followed by *cis*-hydronickelation of the alkyne and reductive elimination, the N-oxide group directing the metal to the proximal C_2-H bond. F.e. and N-deoxygenation s. K.S. Kanyiva, Y. Nakao, T. Hiyama, Angew. Chem. Int. Ed. Engl. *2007*, *46* (46), 8872-4.

Bis(1,5-cyclooctadiene)nickel(0)/1,4-bis(diphenylphosphino)butane *$Ni(cod)_2$/dppb*
(Z)-1-Arylseleno-1,4-dienes from terminal acetylene derivs. $C(SeAr)=C\text{-}CH_2CH=CH_2$
Regio- and stereo-specific synthesis with addition of three C-atoms

445. 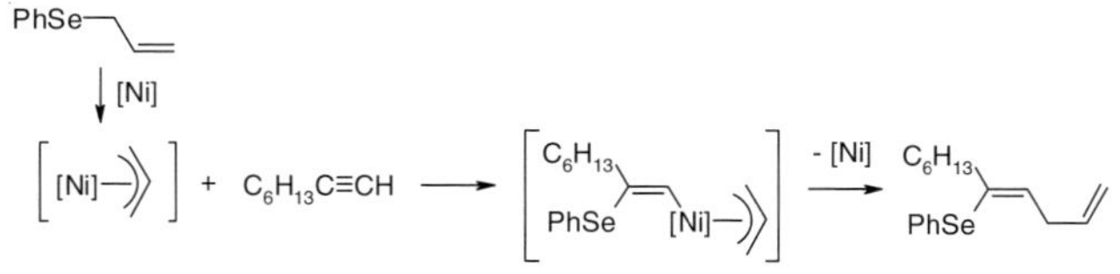

Allyl phenyl selenide and 2 eqs. oct-1-yne added to a mixture of 4 mol% $Ni(cod)_2$ and 4 mol% 1,4-bis(diphenylphosphino)butane in toluene under N_2, the resulting deep red soln. heated at 60° for 6 h, evaporated, and worked up with chromatographic purification on silica gel → product. Y 93% (89:11 mixture of regioisomers). Reaction is generally applicable to aliphatic and aromatic terminal acetylene derivs. (the latter with ca. 100% regioselectivity), leaving hydroxyl, chloro, cyano and methoxy groups unaffected. Hindered substrates, however, were less reactive. The key step appears be the *cis*-addition of an η^3-allylnickel complex across the triple bond. Palladium catalysts demonstrated low activity, while platinum complexes were inactive. F.e.s. K. Yamashita, H. Takeda, T. Kashiwabara, R. Hua, S. Shimada, M. Tanaka, Tetrahedron Lett. *2007*, *48* (38), 6655-9.

(Dimethoxyethane)nickel(II) bromide/zinc *(dme)NiBr$_2$/Zn*

Cyclooctatetraene ring from two terminal 1,6(7)-diyne molecules
Nickel(0)-catalyzed [2+2+2+2]-cycloaddition under mild conditions

446. 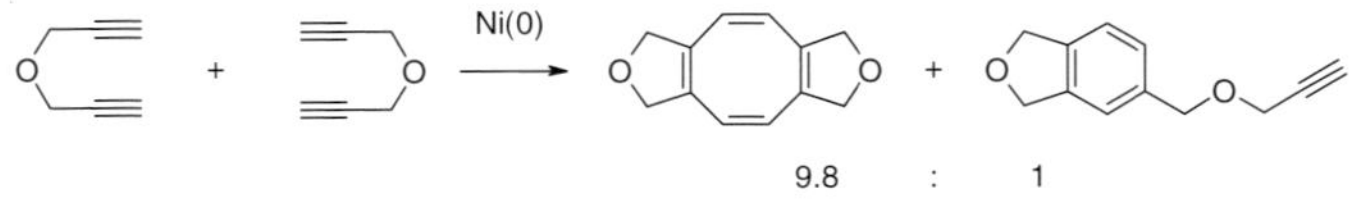

Currently there is no general strategy for controlling the regioselectivity of alkyne tetramerization, but the use of tethered alkynes offers a way of reducing the number of potential regioisomers that may be produced by this methodology. **E:** THF added via syringe through a septum to an oven-dried test-tube containing the startg. diyne, 20 mol% (dme)NiBr$_2$ and 2 eqs. Zn dust, 1 eq. water added in one portion, the mixture stirred and heated at 60° (with a colour change from purple to deep red over 20 min) until diyne was consumed (TLC; 4 h), the mixture cooled to room temp., filtered through Celite, washed with ether, the solvent removed *in vacuo*, and the residue purified by chromatography → cyclooctatetraene deriv. Y 84% (selectivity 9.8:1). The nickel complex is conveniently reduced *in situ* to afford a Ni(0)-complex which provides good selectivity over [2+2+2]-cycloaddition, generally affording a *single alkene positional isomer* in 59-97% yield (selectivity from 2.4:1 to >20:1). In initial experiments, the selectivity was enhanced by slow addition of the diyne and use of high catalyst loadings (up to 40 mol%). The method is tolerant of heteroatoms and ester, nitrile, ketone or aryl functionality and is successful with simple *unsubst. 1,6- and 1,7-diynes* (the latter giving a 66:25 mixture of alkene positional isomers). A single example using an equimolar mixture of two different 1,6-diynes resulted in preferential formation of the cross-product. F.e.s. P.A. Wender, J.P. Christy, J. Am. Chem. Soc. *2007*, *129* (44), 13402-3.

Nickel(II) perchlorate/chiral bis(Δ^2-oxazolines) *Ni(ClO$_4$)$_2$/box*

Asym. hetero-Diels-Alder reaction

with Cu(OTf)$_2$/chiral bis(Δ^2-oxazoline) complexes cf. *46*, 662s*51*; chiral 2-alkoxy-1-sulfonyl-1,2,3,4-tetrahydropyridines from α,β-ethylene-N-sulfonylketimines and enolethers with Ni(ClO$_4$)$_2$·6H$_2$O/(R,R)-4,6-dibenzofurandiyl-2,2′-bis(4-phenyl-Δ^2-oxazoline) s. J. Esquivias, R.G. Arrayás, J.C. Carretero, J. Am. Chem. Soc. *2007*, *129* (6), 1480-1.

Ruthenium complexes/microwaves *[Ru]/[\\\\]*

Addition of polyhalides to ethylene derivs. C=C → C(Hal)C(R$_{Hal}$)

under Ru-catalysis cf. *18*, 776s*71*; addition of CCl$_4$ under microwave enhancement s. Y. Borguet, A. Richel, S. Delfosse, A. Leclerc, L. Delaude, A. Demonceau, Tetrahedron Lett. *2007*, *48* (36), 6334-8; α,α-difluoro-γ-iodosulfones by addition of iododifluoromethyl phenyl sulfone in the presence of Et$_3$B/air, also one-pot reductive deiodination (with Bu$_3$SnH) or dehydroiodination (with DBU), s. Y. Li, J. Liu, L. Zhang, L. Zhu, J. Hu, J. Org. Chem. *2007*, *72* (15), 5824-7

(1,3,5-Cyclooctatriene)bis(dimethyl fumarate)ruthenium(0)/ 1-octanol *[Ru(η^6-cot)(η^2-dmfm)$_2$]/ROH*

Regiospecific [head-to-head] and stereospecific dimerization of styrenes 2 C=C → C=C-CHC

447. 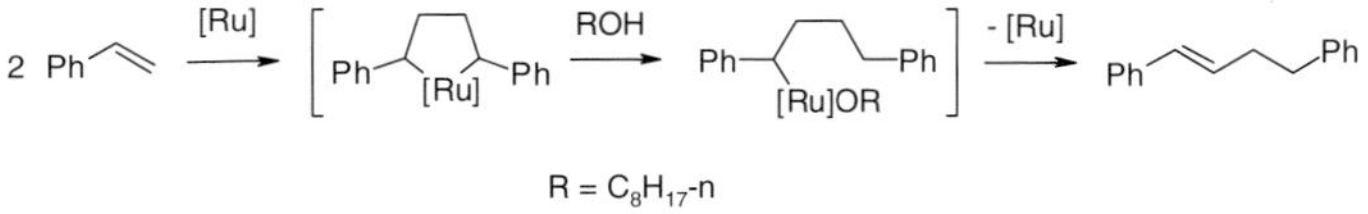

(E)-1,4-Diarylbut-1-enes. A mixture of styrene, 5 mol% [Ru(η^6-cot)(η^2-dmfm)$_2$] (where cot = 1,3,5-cyclooctatriene, and dmfm = dimethyl fumarate) and 25 mol% 1-octanol in mesitylene heated with stirring under argon at 80-110° for 24 h, cooled, and purified by distillation and flash column chromatography on silica gel → product. Y 53% (95% E). This unusual *head-to-head* dimerization possibly involves ring opening of a ruthenacyclopentane complex, facilitated by σ-bond metathesis with the primary alcohol as co-catalyst. F.e. and comparison of ruthenium catalysts, also selective codimerization of styrenes with ethylene, s. T. Kondo, D. Takagi, H. Tsujita, Y. Ura, K. Wada, T. Mitsudo, Angew. Chem. Int. Ed. Engl. *2007*, *46* (31), 5958-61.

Chloro(1,5-cyclooctadiene)(η⁵-pentamethylcyclopentadienyl)ruthenium(II) *Cp*RuCl(cod)*
[2+2]-Cycloaddition with acetylene derivs. □
to oxanorbornadienes s. *60*, 288s*72*

Carbonyl(chloro)hydridotris(triphenylphosphine)ruthenium(II) *[RuCl(CO)H(PPh₃)₃]*
β-Diketones from ethyleneketones and aldehydes CHO → CO-C(CH)CO
Regiospecific conversion under ruthenium hydride catalysis

448.

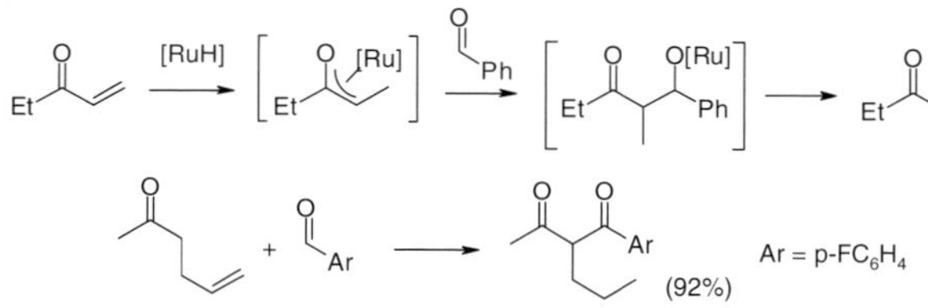

A mixture of ethyl vinyl ketone, 1.3 eqs. benzaldehyde and 10 mol% $[RuHCl(CO)(PPh_3)_3]$ in benzene stirred at reflux for 5 h under N_2, and worked up with column chromatographic purification on silica gel → 2-propanoylpropiophenone. Y 76%. The procedure is applicable to the addition of aliphatic, aromatic or α,β-ethylenic aldehydes to a variety of terminal, β-monosubst. and dialkyl-subst. enones, and equally facile with ketones *remotely-subst.* by an ethylene group **via double bond shift** along the chain. Reaction is thought to involve initial hydroruthenation of the ethylene residue, followed by aldol condensation of the generated ruthenium enolate then β-hydride elimination. There was no conversion, however, with phosphine-free catalysts. F.e.s. T. Fukuyama, T. Doi, S. Minamino, S. Omura, I. Ryu, Angew. Chem. Int. Ed. Engl. *2007*, *46* (29), 5559-61.

Acetoacetato(dicarbonyl)rhodium(I)/triphenylphosphine/L-proline/ sodium tetrahydridoborate ←
Asym. synthesis of 1,3-diols from ethylene derivs. and aldehydes ←
via hydroformylation-aldol condensation

449.

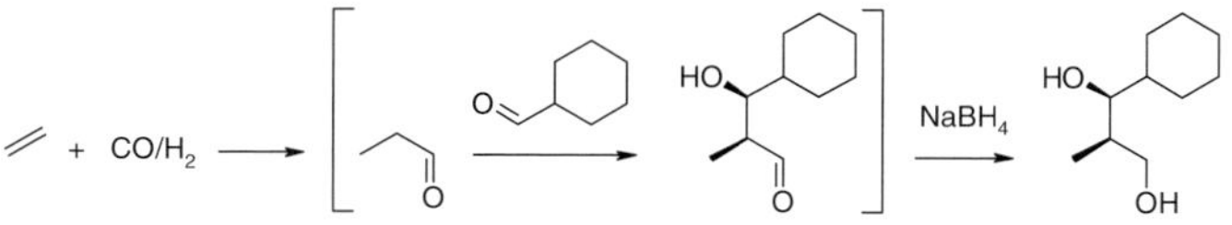

A soln. of cyclohexanecarboxaldehyde, 0.2 mol% $Rh(CO)_2(acac)$ and 6 mol% L-proline in DMF pressurized in an autoclave to 30 atm. with 1:1:1 ethylene/H_2/CO, stirred for 48 h *at 5°*, diluted with methanol, cooled to 0°, 2 eqs. $NaBH_4$ added portionwise, and worked up when reduction of the intermediate [unstable] β-hydroxyaldehyde complete → (2S,3R)-3-cyclohexyl-2-methyl-1,3-propanediol. Y 81% (based on conversion and after flash chromatography; d.r. 13:1; e.e. of major diastereoisomer 99%). A crucial factor was the choice of catalyst and ligand to control the hydroformylation rate relative to that of the subsequent aldol condensation to avoid accumulation of the donor aldehyde (thereby obviating its loss by hydrodimerization). With terminal ethylene derivs., however, 6-(diphenylphosphino)-2-pyridone was the optimum ligand. F.e., ligands and chiral proline derivs. s. O. Abillard, B. Breit, Adv. Synth. Catal. *2007*, *349* (11-12), 1891-5; f. ligands s. S. Chercheja, P. Eilbracht, ibid. 1897-905.

Bis(1,5-cyclooctadiene)rhodium(I) triflate/tri-2-furylphosphine *$[Rh(cod)_2]OTf/Fu_3P$*
***syn*-β-Sulfonylamino-** $C{=}NSO_2R \rightarrow C(NHSO_2R)C(CH)CO$
from α,β-ethylene-ketones and N-sulfonylimines
Stereospecific hydrogenative Mannich reaction

450. 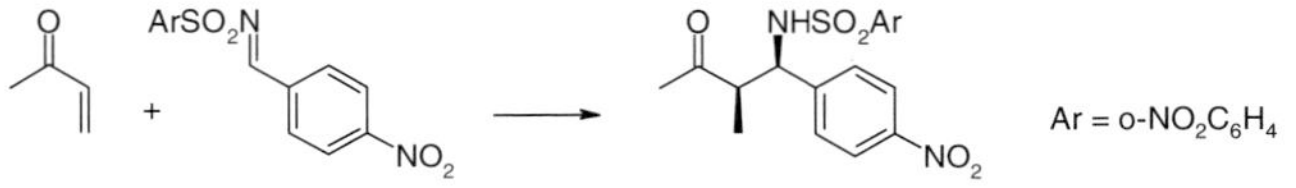

3 Eqs. methyl vinyl ketone added to a test-tube containing the startg. N-nosylimine, 5 mol% $[Rh(cod)_2]BF_4$ and 12 mol% tri-2-furylphosphine in methylene chloride under a balloon of hydrogen (1 atm.), stirred for 24 h, and worked up with column chromatographic purification on silica gel → product. Y 95% (d.r. 11:1). The level of *syn*-diastereoselectivity was lower with N-arylaldimines. F.e.s. S.A. Garner, M.J. Krische, J. Org. Chem. *2007*, *72* (15), 5843-6.

(1,5-Cyclooctadiene)[2,2′-bis(diphenylphosphino)biphenyl]rhodium(I) fluoroborate ← ○
Rhodium-catalyzed ring closures of diynes with α,β-unsatd. ketones
via carbonyl-directed activation of carbon-hydrogen bonds

451. 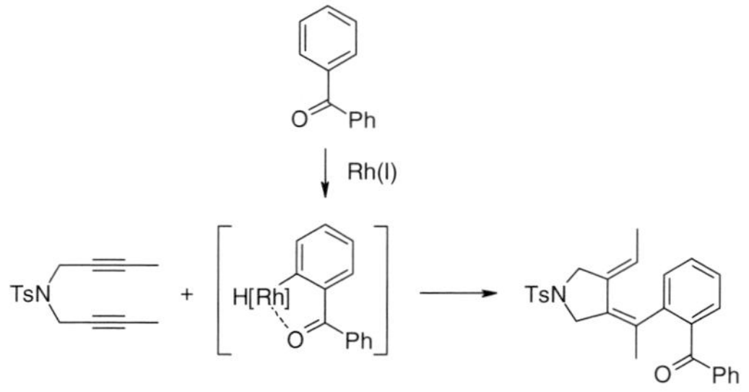

Exocyclic 1-(*o*-acylaryl)-1,3-dienes from aryl ketones. A soln. of 5 mol% (1,5-cyclooctadiene)[2,2′-bis(diphenylphosphino)biphenyl]rhodium(I) fluoroborate in methylene chloride stirred at room temp. under argon for 5 min, the mixture purged with H_2, stirring continued for 30 min, the solvent and H_2 removed under reduced pressure, argon introduced into the reaction flask, a soln. of 3 eqs. benzophenone in methylene chloride added, followed dropwise over 30 min by a soln. of the startg. diyne in the same solvent at room temp., stirred for a further 5 min, the solvent removed under reduced pressure, and the residue purified by TLC → product. Y 87%. Reaction is thought to involve initial carbonyl-directed *o*-activation by oxidative addition of rhodium(I) to the carbon-hydrogen bond, followed by hydrorhodation-intramolecular carbo-rhodation of diyne prior to reductive elimination of the catalyst. **Exocyclic 2,4,6-trienones** were obtained similarly **from α,β-ethyleneketones** via β-activation, while **asym. synthesis of exocyclic 2,6-dienones from enynes** and α,β-ethyleneketones was effected with an (S)-BINAP-based rhodium(I) complex. F.e.s. K. Tsuchikama, Y. Kuwata, Y. Tahara, Y. Yoshinami, T. Shibata, Org. Lett. *2007*, *9* (16), 3097-9.

Chloro(1,5-cyclooctadiene)rhodium(I) dimer *[Rh(cod)Cl]$_2$*

2-Acyl-3-vinylcycloalkanols
from 2-ene-1,n-diones and vinylzirconocene chlorides
via stereospecific rhodium-catalyzed 1,4-addition-intramolecular aldol condensation

452. 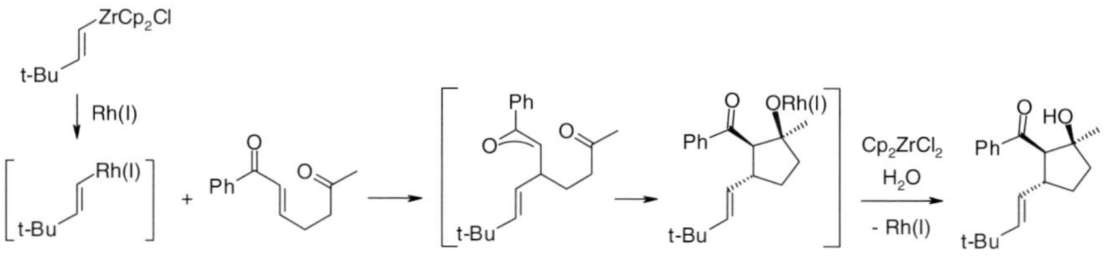

The startg. dienone and 2 mol% [RhCl(cod)]$_2$ in toluene added at 0° to a soln. of 5 eqs. startg. vinylzirconocene chloride in the same solvent under argon, stirred for 30 min, quenched with aq. $NaHCO_3$, and worked up with column chromatographic purification on silica gel → product. Y 88% (d.e. 85%). High diastereoselectivity (up to >95%) was recorded for the formation of 5- and 6-membered (not 7-membered) carbocyclic rings from unsatd. diketones, keto-esters or keto-amides. However, there was no reaction with alkylzirconocene chlorides. F.e., also ring closures of bis(enones) via double 1,4-addition, and from chiral [Oppolzer] sultam analogs with asym. induction (>95% d.e.) s. Y. Hanzawa, Y. Takebe, A. Saito, A. Kakuuchi, H. Fukaya, Tetrahedron Lett. *2007*, *48* (37), 6471-4.

Chloro(1,5-cyclooctadiene)rhodium(I) dimer/1,3-bis(diphenylphosphino)propane *[Rh(cod)Cl]$_2$/dppp*

3-Methylcyclopent-2-enones from spiropentanes
via regiospecific rhodium(I)-catalyzed carbonylation

453. 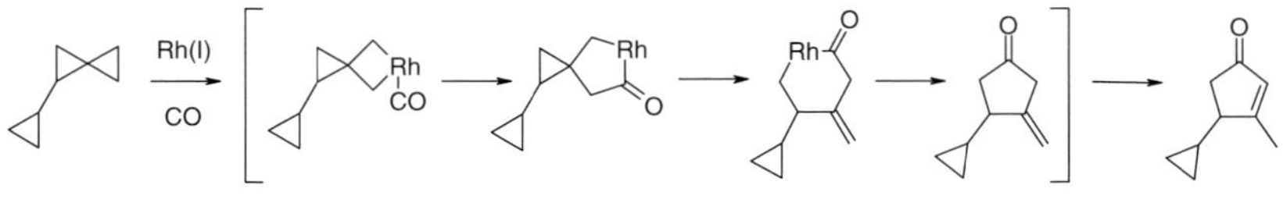

p-Xylene added to a mixture of 5 mol% [RhCl(cod)]$_2$ and 10 mol% 1,3-bis(diphenylphosphino)-propane under CO, the soln. stirred for 5 min at 130°, the startg. spiropentane added, the mixture stirred at 130° for 2.5 h, filtered through a pad of silica gel, washed with ethyl acetate, the filtrate evaporated *in vacuo*, and the residue purified by TLC → cyclopent-2-enone deriv. Y 82%. A series of 1,1- and 1,2-disubst. analogs reacted similarly in 56-82% yield (with retention of stereochemistry in the case of two cyclooctano-fused analogs), while 1-(benzyloxymethyl)spiropentane gave a mixture of 4- and 5-subst. 3-methylcyclopent-2-enones in 55 and 21% yield, respectively; a 1,1,2-trisubst. deriv. gave a mixture of the cyclopent-2-enone (Y 37%) and a diene (Y 18%). In one example, paraformaldehyde was used as carbonyl source, giving 81% yield, but required longer heating at 130°. F.e. incl. spiropentane preparation s. T. Matsuda, T. Tsuboi, M. Murakami, J. Am. Chem. Soc. *2007*, *129* (42), 12596-7.

Dicarbonyl(chloro)rhodium(I) dimer *$[Rh(CO)_2Cl]_2$*
cis-Fused 4-cyclooctenone ring from 1-cyclopropyl-1(E),n-dienes
via regiospecific carbonylation of the *cis*-fused rhodacyclooct-4-ene ring

454. 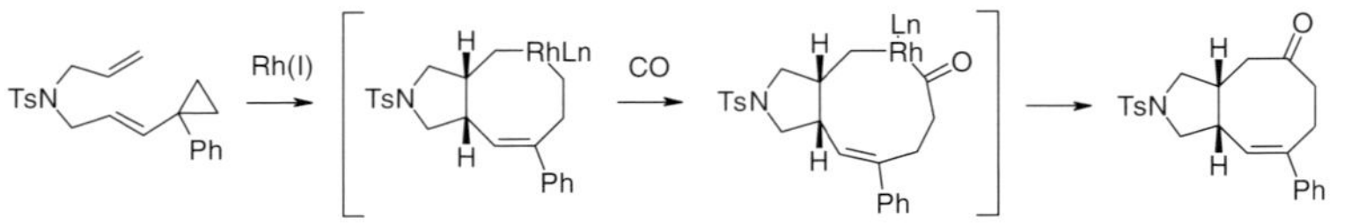

A soln. of startg. cyclopropyldiene in degassed dioxane added to 5 mol% *$[Rh(CO)_2Cl]_2$* in an oven-dried Schlenk flask under N_2, the soln. bubbled with CO/N_2 (0.2:0.8 atm.) for 5 min, the mixture stirred at 80° under CO/N_2 for 10 h (TLC), cooled to room temp., the mixture concentrated *in vacuo*, and the residue purified by chromatography → *cis*-6-phenyl-N-*p*-toluenesulfonyl-3-oxo-10-azabicyclo[6.3.0]undec-6-ene. Y 92%. The reaction is tolerant of tethers incorporating geminal diester, sulfonamide and ether functionality, with highest yields (71-92%) being obtained for the heteroatom-based tethers (29 and 70% for carbon-based tethers). The [5+2+1] pathway was shown computationally to be favoured over the [5+2] non-carbonylative route, although the cycloheptene-fused analog was isolated as the main product at a more elevated temp. (Y 59% at 110°). Other rhodium(I) catalysts were ineffective. F.e. incl. startg. m. preparation s. Y. Wang, J. Wang, J. Su, F. Huang, L. Jiao, Y. Liang, D. Yang, S. Zhang, P.A. Wender, Z.-X. Yu, J. Am. Chem. Soc. *2007*, *129* (33), 10060-1.

trans-*Carbonyl(chloro)bis(triphenylphosphine)rhodium(I)* *$RhCl(CO)(PPh_3)_2$*
7,8(E)-Di(alkylidene)bicyclo[4.4.0]dec-1-enes from 1,2,7,8-tetraenes and allenes

455. 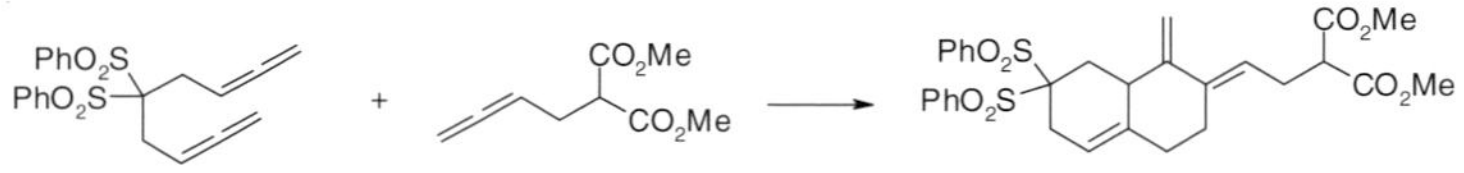

A degassed soln. of the startg. bis(allene) (0.25 mmol) in toluene added via syringe pump at 80° over ca. 10 h to a soln. of 5 mol% *trans*-$RhCl(CO)(PPh_3)_2$ and the startg. monoallene (0.625 mmol) in the same solvent, stirred until reaction complete (TLC monitoring), solvent removed, and the residue worked up with flash chromatographic purification on silica gel → 7-methylene-8-(3′,3′-bis(methoxycarbonyl)prop-1′(E)-ylidene)-4,4-bis(phenylsulfonyl)bicyclo[4.4.0]dec-1-ene. Y 46%. A variety of functionalized monoallenes participated in the conversion, e.g. phenylallenes, 3-allenealcohols, and protected 2- and 3-alleneamines. F.e. and subsequent Diels-Alder reaction to **steroid-like tetracyclics** s. P. Lu, S. Ma, Org. Lett. *2007*, *9* (25), 5319-21.

Chlorotris(triphenylphosphine)rhodium(I) *$RhCl(PPh_3)_3$*
Succinic acid amides C≡C → CH(CON<)CH(CON<)
from terminal acetylene derivs. by biscarbonylation

456. 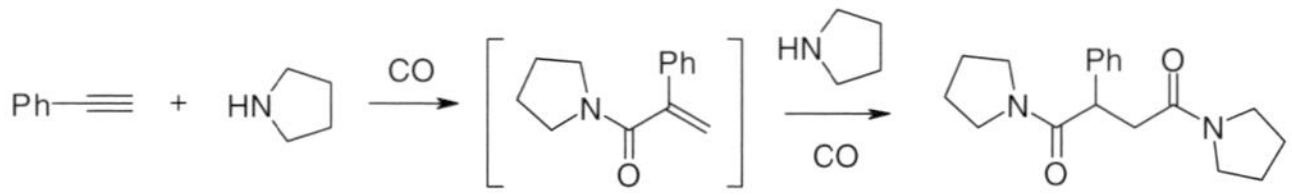

A mixture of phenylacetylene, 4 eqs. pyrrolidine, 5 mol% $RhCl(PPh_3)_3$ and toluene pressurized at room temp. with 1.5 MPa CO in an autoclave, heated at 120° with stirring for 15 h, cooled to room temp., CO released slowly, and the mixture worked up with column chromatographic purification → product. Y 70%. Reaction is applicable to aliphatic and aromatic terminal acetylene derivs., good yields being recorded for coupling of pyrrolidine and piperidine (not dibutylamine). F.e. and comparison of rhodium catalysts and solvents s. Q.F. Huang, R.M. Hua, Adv. Synth. Catal. *2007*, *349* (6), 849-52.

Dichloro(pentamethylcyclopentadienyl)rhodium(III) dimer/copper(II) acetate/air ←
Isocoumarins from arylcarboxylic acids and acetylene derivs. ○
Oxidative ring closure via *o*-rhodation

457. 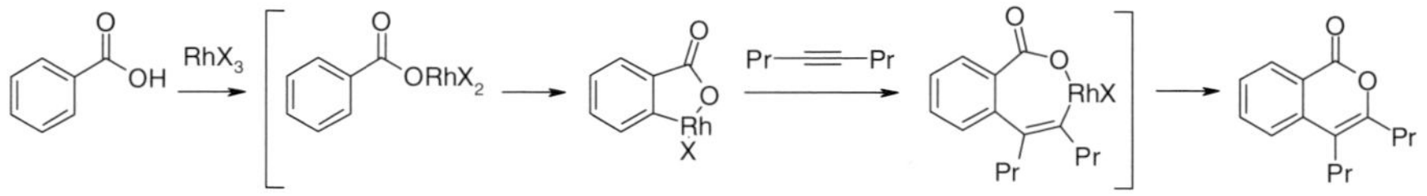

A soln. of benzoic acid, 1.2 eqs. oct-4-yne, 1 mol% $[(Cp^*)RhCl_2]_2$, 5 mol% $Cu(OAc)_2$-hydrate and 1-methylnaphthalene (internal standard) in DMF stirred under air at 120° for 2 h with exclusion of atmospheric moisture, the solvent removed under vacuum, and worked up with chromatographic purification on silica gel → product. Y 77%. This waste-free procedure (only water being eliminated) is generally applicable to the coupling of arylcarboxylic acids (incl. hindered substrates, and irrespective of the nature of ring substituents) with dialkyl- or diarylacetylenes as well as alkyl(aryl)acetylenes, which afford single regioisomers. Reaction appears to involve initial formation of a rhodium(III) benzoate, followed by *o*-rhodation to activate the *ortho*-position, then alkyne insertion and elimination of rhodium(I); the latter is then re-oxidized by copper(II) under air. **7-Vinylphthalides** were obtained similarly from two electron-deficient ethylene deriv. [acrylate] molecules. F.e. and comparison of catalysts s. K. Ueura, T. Satoh, M. Miura, Org. Lett. *2007*, *9* (7), 1407-9.

Tetrakis(triphenylphosphine)palladium(0) *$Pd(PPh_3)_4$*
2-Ethyleneamines from azomethines and acetylene derivs. ←
via silaaziridines and 1,2-azasilacyclopent-3-enes

458. 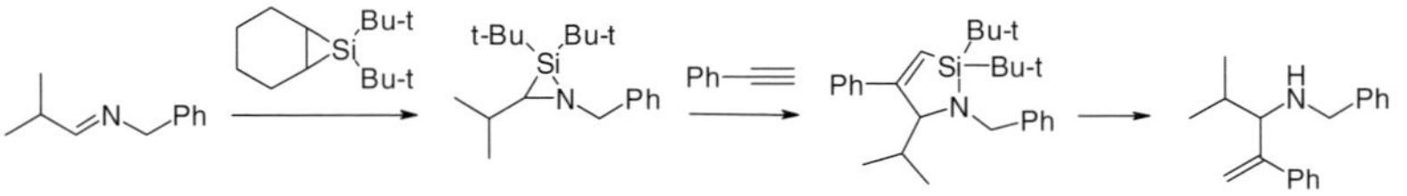

1 Mol% Ag-triflate added to a mixture of the startg. imine, 1 eq. of the startg. silacyclopropane and 20 mol% phenyltrimethylsilane in toluene, and stirred at 23° for 12 h → intermediate silaaziridine (Y 84%), added to 1.25 eqs. phenylacetylene and 1 mol% tetrakis(triphenylphosphine)-palladium(0) in toluene, heated to 80° for 8 h, cooled, and extracted → intermediate azasilacyclopentene (Y 90%), added to a mixture of 3 eqs. K-*tert*-butoxide and 6 eqs. tetra-*n*-butylammonium fluoride in toluene, and heated to 80° for 31 h → allylic amine (Y 72%). The reaction is successful with N-benzyl- and N-aryl-imines. F.e. incl. treatment of intermediate **silaaziridines** with electrophiles to afford other ring-expanded silicon-containing heterocycles s. Z. Nevárez, K.A. Woerpel, Org. Lett. *2007*, *9* (19), 3773-6.

Tetrakis(triphenylphosphine)palladium(0)/formic acid *$Pd(PPh_3)_4/HCOOH$*
2-(2-Alkylidenecyclopent-1-yl)furans from 2,4,9-enediynals ○
Reductive double ring closure

459. 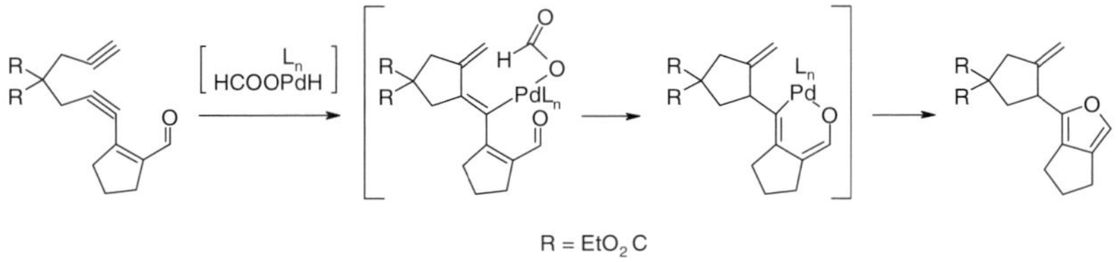

A soln. of the startg. enediynal, 3 mol% $Pd(PPh_3)_4$ and 1.5 eqs. formic acid in dry dioxane mixed at 0° under argon, stirred at 70° for 0.5 h (TLC monitoring), solvent removed under vacuum, and worked up with flash column chromatographic purification → product. Y 77%. Reaction appears to involve initial hydropalladation of the terminal alkyne residue, followed by intramolecular carbopalladation, decarboxylation, hydride transfer and finally reductive elimination. A nitrogen or oxygen tether between the two alkyne residues was also supported to give the corresponding **alkylideneheterocyclic analogs**. F.e. incl. carbalkoxymethylene analogs s. C.H. Oh, H.M. Park, D.I. Park, Org. Lett. *2007*, *9* (7), 1191-3.

*Dichlorobis(triphenylphosphine)palladium(II)/1,3-bis(diphenylphosphino)propane/ 1-*n*-butyl-3-methylimidazolium triflimide* ←

3,4-Dihydro-2-pyrone-5-carbonyl compds. ○
from terminal acetylene derivs. and β-ketocarbonyl compds.
Regiospecific cyclocarbonylation in ionic liquids

460.
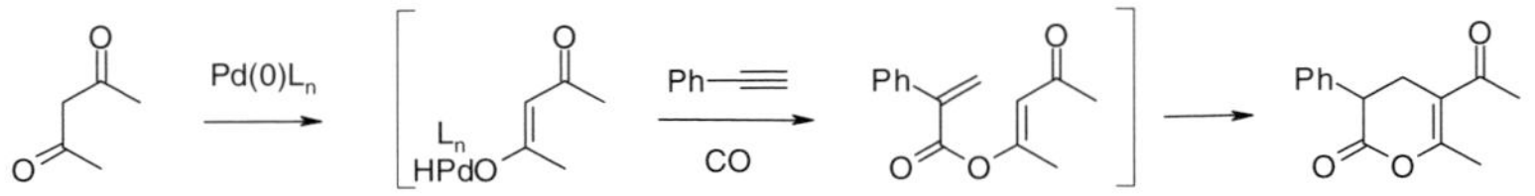

3 Mol% $PdCl_2(PPh_3)_2$, 6 mol% 1,3-bis(diphenylphosphino)propane, 1-*n*-butyl-3-methylimidazolium triflimide, 2 eqs. phenylacetylene and 2,4-pentanedione introduced in this order into an autoclave, the latter purged three times with CO, pressurized with 200 psi CO, heated to 110° for 22 h, excess of CO discharged at room temp., extracted with ether, and worked up with flash chromatographic purification on silica gel → product. Y 63%. The catalyst remains in the ionic liquid, and may be recycled at least five times with only a slight reduction in activity. Yields were low in conventional solvents, and there was no reaction with internal acetylene derivs. Reaction is presumed to involve regiospecific intramolecular acylpalladation through initial formation of a hydridopalladium enolate. F.e. incl. 5-carbalkoxy derivs., also with aliphatic acetylene derivs., s. Y. Li, Z. Yu, H. Alper, Org. Lett. *2007*, *9* (9), 1647-49.

Bis(1,5-cyclooctadiene)iridium(I) tetrakis[3,5-bis(trifluoromethyl)phenyl]borate/(R)-3,3′-dichloro-2,2′-dimethoxy-6,6′-bis(diphenylphosphino)biphenyl/triphenylacetic acid ←

(E)-2-Ethylenesulfonylamines C=NSO_2R → C($NHSO_2R$)C=CH
from acetylene derivs. and N-sulfonylimines
Regiospecific asym. hydrogenative coupling

461.
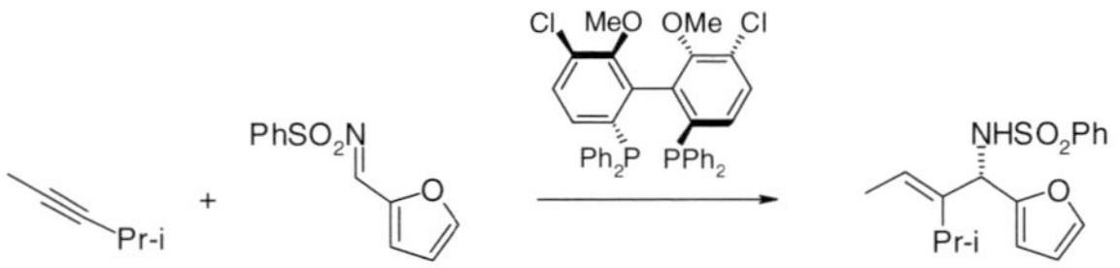

3 Eqs. startg. acetylene deriv. added to a soln. of 5 mol% bis(1,5-cyclooctadiene)iridium(I) tetrakis[3,5-bis(trifluoromethyl)phenyl]borate, 5 mol% (R)-3,3′-dichloro-2,2′-dimethoxy-6,6′-bis(diphenylphosphino)biphenyl and 5 mol% triphenylacetic acid in toluene, the system purged with argon, followed by H_2, and stirred at 60° under 1 atm. H_2 until TLC indicated completion of reaction → product. Y 80% (E/Z >99:1; e.e. 97%). The procedure is generally applicable to N-sulfonylimines derived from aromatic, heteroaromatic, α,β-ethylenic and aliphatic aldehydes with outstanding levels of (E)-selectivity and enantioselectivity, while unsym. acetylene derivs. afford high levels of regiospecificity (10:1). Reaction is presumed to involve initial formation of a chiral iridium(I) alkyne complex having iridacyclopropane character, followed by enantio-determining insertion of the imine to give the corresponding aza-iridacyclopentene prior to liberation of the product. F.e. and comparison of chiral ligands s. M.Y. Ngai, A. Barchuk, M.J. Krische, J. Am. Chem. Soc. *2007*, *129* (42), 12644-5; achiral procedure s. ibid. *129* (27), 8432-3; **chiral 5-sulfonylamino-1,3-dienes** by asym. hydrogenative coupling of two ethyne molecules (with the corresponding rhodium(I) complex) cf. E. Skucas, J.R. Kong, M.J. Krische, ibid. *129* (23), 7242-3.

1,5-Cyclooctadiene[2,2′-bis(diphenylphosphino)biphenyl]iridium(I) tetrakis[3,5-bis(tri-fluoromethyl)phenyl]borate/cesium carbonate ←

Regiospecific synthesis of 3-ethylene-*sec*-alcohols from prim. alcohols and allenes via iridium-catalyzed autotransfer-hydrogenation $CH_2OH \rightarrow CH(OH)C\text{-}C{=}C$

462. 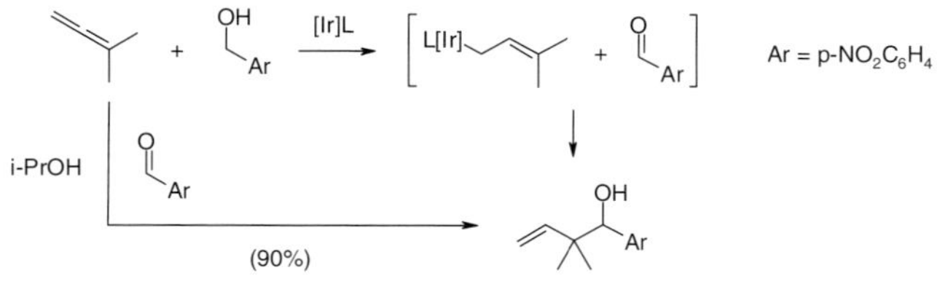

An **alternative to classical nucleophilic allylation** of aldehydes with allylmetallics is now at hand *from prim. alcohols* via iridium-catalyzed transfer-hydrogenation to form the aldehyde *in situ* in the presence of an allene, itself accepting the molecule of hydrogen to generate an intermediate allyliridium complex which adds to the aldehyde. **E:** 4 Eqs. dimethylallene added to a reaction tube containing a soln. of the startg. prim. alcohol, 5 mol% [Ir(cod)BIPHEP]BARF and 5 mol% Cs_2CO_3 in 1:1 dichloroethane/ethyl acetate, heated at 75° for 15 h, the mixture concentrated *in vacuo*, and worked up with flash column chromatographic purification on silica gel → product. Y 92%. This is the first such metal-catalyzed transfer-hydrogenation to provide products of carbonyl addition. The procedure is byproduct-free, notably circumventing over-reduction of the product, and applicable to reverse prenylation, crotylation and allylation *in the absence of a preformed allylmetallic.* However, the method appears to be limited to activated alcohols. F.e., **also directly from pre-formed aldehydes** (with 2 eqs. isopropanol as terminal reductant), s. J.F. Bower, E. Skucas, R.L. Patman, M.J. Krische, J. Am. Chem. Soc. *2007*, *129* (49), 15134-5; **from aldehydes by hydrogenative coupling** (with Li_2CO_3 as base) cf. E. Skucas, J.F. Bower, M.J. Krische, ibid. *129* (42), 12678-9.

Platinum(II) bis(phosphinous acid) complexes/acetic acid ←

Alkylidenecyclopropanation of strained bicyclic ethylene derivs. with terminal acetylene derivs. ▽

463. 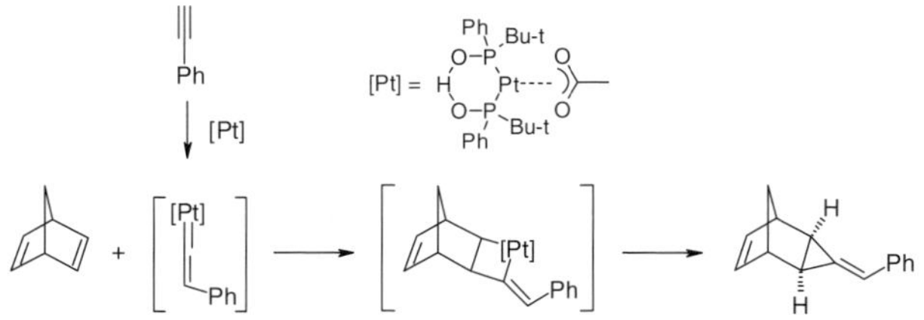

Benzylidenecyclopropane ring. A soln. of 0.5 eq. norbornadiene and 1 eq. phenylacetylene in dry, degassed toluene containing acetic acid added successively to 2.5 mol% platinum(II) bis(phosphinous acid) complex in dry, degassed toluene under argon, and stirred at 55° for 20 min → 3-benzylidenetricyclo[3.2.1.0^{2,4}]oct-6-ene. Y 81%. This is the first example of C-C bond formation with such air-stable and robust platinum(II) phosphinous acid complexes, yields being noticeably lower with the corresponding palladium complexes. Reaction, which is thought to involve initial [2+2]-cycloaddition with an intermediate vinylideneplatinum complex, is applicable to both all-carbon as well as diaza- and oxaza-bicyclic strained ethylene derivs. F.e.s. J. Bigeault, L. Giordano, I. de Riggi, Y. Gimbert, G. Buono, Org. Lett. *2007*, *9* (18), 3567-70.

Platinum(II) chloride *$PtCl_2$*
Synthesis of thioenol carbamates from propargyl thionocarbamates via transition metal-carbene complexes ←

464. 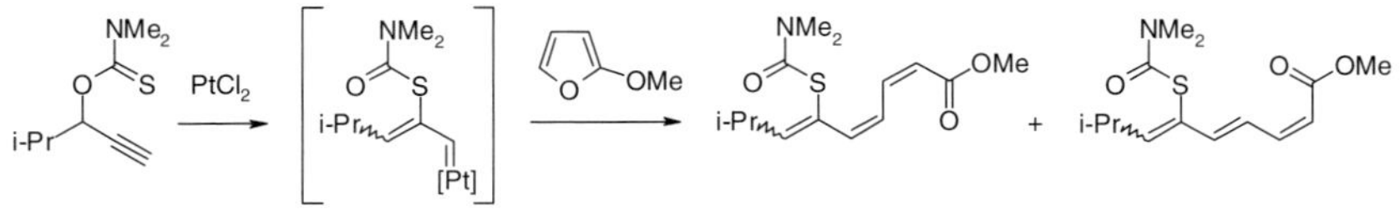

The startg. propargyl thiocarbamate and 2 eqs. 2-methoxyfuran in anhydrous THF added to 5 mol% $PtCl_2$ at room temp. under N_2, stirred at 50° for 0.5 h, cooled to room temp., evaporated under reduced pressure, and worked up with column chromatographic purification on silica gel → (2Z,4Z)-product (Y 52%) and (2Z,4E)-product (Y 38%). F.e. and via Ru-, Rh- or Au-carbenoids, also substitution with pyrroles, and synthesis of **1-cyclopropylthioenol carbamates** from ethylene derivs., s. Y. Ikeda, M. Murai, T. Abo, K. Miki, K. Ohe, Tetrahedron Lett. *2007*, *48* (38), 6651-4.

Potassium tetrachloroplatinate/silver triflate/trifluoroacetic acid ←
Double thienylation of propiolic acid or esters s. *59*, 311s*72* C≡C → CH_2CR_2

Rearrangement ∩

Hydrogen/Oxygen Type CC ∩ HO

Silver mesylate/organic base/carbon dioxide ←
Cationic gold complexes *[Au]*
Meyer-Schuster rearrangement C(OH)C≡C → C=CHCO
under Ru-catalysis cf. *31*, 665s*64*; with cationic gold catalysts, α,β-ethylenecarboxylic acid ethyl esters from ethoxyethynylcarbinols, s. S.S. Lopez, D.A. Engel, G.B. Dudley, Synlett *2007* (6), 949-53; with 10 mol% AgOMs and 0.2 eq. DBU (or DBN or *i*-Pr_2NEt) in formamide at 60° under 1 MPa CO_2, α,β-ethylene-ketones or esters, s. Y. Sugawara, W. Yamada, S. Yoshida, T. Ikeno, T. Yamada, J. Am. Chem. Soc. *2007, 129* (43), 12902-3.

Gold(III) chloride *$AuCl_3$*
3(2*H*)-Furanones from β,γ-acetylene-α-hydroxyketones ○
s. *72*, 466

(1,5-Cyclooctadiene)(hydroxy)rhodium(I) dimer/(R)-2,2′-bis(diphenylphosphino)-1,1′-binaphthyl ←
γ,δ-Acetyleneketones from 1,4-enyn-3-ols via asym. 1,3-(alk-1-ynyl) group migration ←

465. 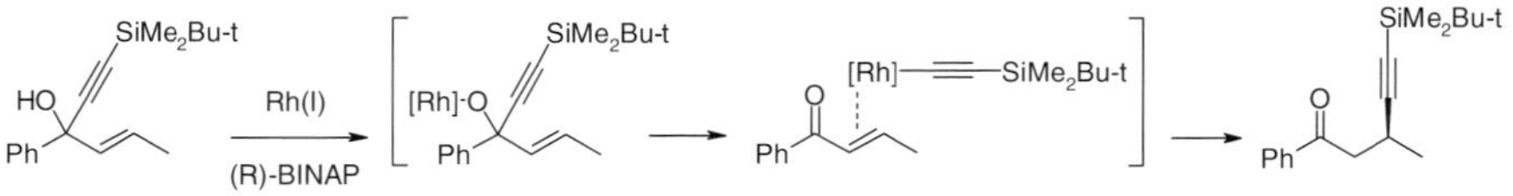

A mixture of 2.5 mol% $[Rh(OH)(cod)]_2$ and 1 mol% (R)-BINAP in toluene heated at 60° for 5 min, a soln. of the startg. enynol in toluene added, stirred for 3 h at the same temp., the soln. passed through a short column of silica with ethyl acetate as eluent, the solvent removed, and the residue purified by preparative TLC on silica gel → product. Y 88% (e.e. 94%). The procedure is a useful **alternative to asym. 1,4-addition of alk-1-ynylmetallic compds. to α,β-ethylene-ketones,** rearrangement being facilitated by retention of the alkynyl group within the sphere of the reaction centre by complexation with rhodium via β-alkynyl elimination. F.e. and rearrangement of cyclic 1,4-enyn-3-ols and 2-furylcarbinol derivs. s. T. Nishimura, T. Katoh, K. Takatsu, R. Shintani, T. Hayashi, J. Am. Chem. Soc. *2007*, *129* (46), 14158-9.

Platinum(II) chloride *$PtCl_2$*
3(2*H*)-Furanones from β,γ-acetylene-α-hydroxyketones ○

466. 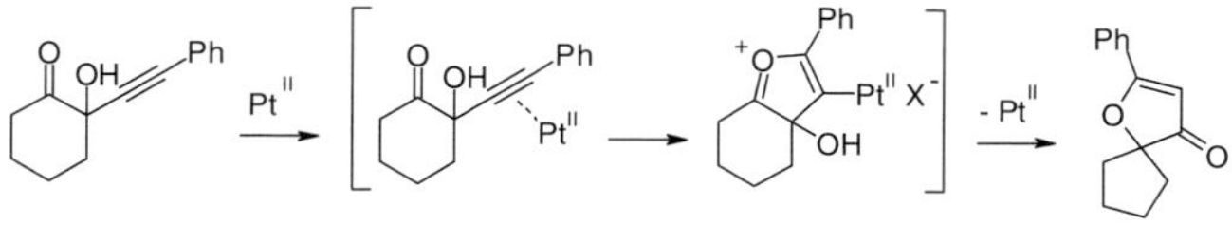

5 Mol% $PtCl_2$ added to a soln. of the startg. acetylene(hydroxy)ketone in toluene, the reaction vial sealed, the pale yellow mixture stirred at 80° for 90 min, concentrated, and worked up with flash chromatographic purification on silica → product. Y 93%. The procedure is generally applicable to cyclic or acyclic substrates possessing an aryl, hetaryl, alkyl, alkenyl, cycloalkyl or silyl group at the alkyne terminus, rearrangement taking place with ring contraction or by simple aryl or alkyl shift. $AuCl_3$ was also an effective catalyst but only with arylalkynyl derivs. An intermediate oxonium ion is presumed to trigger the irreversible 1,2-shift analogous to a formal α-ketol rearrangement. F.e. and comparison of catalysts, also stereospecificity, s. S.F. Kirsch, J.T. Binder, C. Liébert, H. Menz, Angew. Chem. Int. Ed. Engl. *2006*, *45* (35), 5878-80.

Hydrogen/Carbon Type CC ∩ HC

Lithium diisopropylamide/copper(I) phenylmercaptide i-*$Pr_2NLi/CuSPh$*
Anionic cycloisomerization of acetylene derivs. ○
cf. *51*, 337s*71*; 3*H*-azepine ring from *o*-acetylenealdimines by deprotonation with *i*-Pr_2NLi at low temp. then transmetalation with CuSPh s. V. Lyaskovskyy, K. Bergander, R. Fröhlich, E.-U. Würthwein, Org. Lett. *2007*, *9* (6), 1049-52.

Silver hexafluoroantimonate s.a. under $V(salen)Cl_2$ and $RhCl(PPh_3)_3$ *$AgSbF_6$*

*(Tri-*tert*-butylphosphine)gold(I) chloride/silver hexafluoroantimonate* (t-*$Bu_3P)AuCl/AgSbF_6$*
Cyclopentadienes from 1,2,4-trienes
Regiospecific cycloisomerization under mild conditions

467. 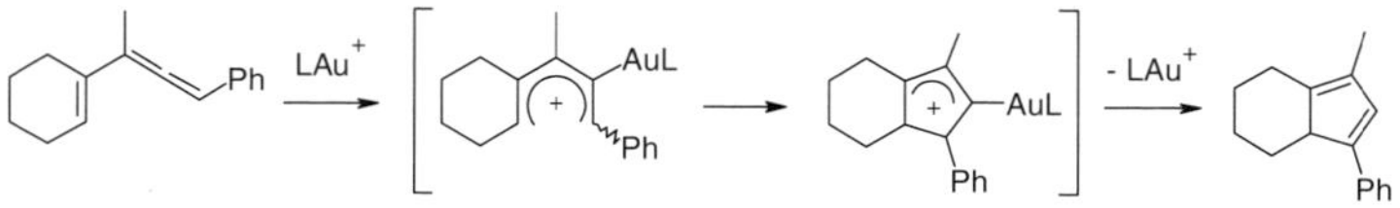

with gold(I) catalysis. 1 Mol% (*t*-Bu_3P)AuCl and methylene chloride added sequentially to 1 mol% $AgSbF_6$, stirred at room temp. for 5 min, cooled to 0°, startg. 1,2,4-triene (0.3 mmol scale) in methylene chloride added dropwise via cannula, stirring continued at 0° for 1 min, filtered, concentrated, and purified on a short column of silica gel → product. Y 96%. Yields and regioselectivities were high for the cyclization of 1,2,4-trienes possessing a phenyl, alkyl or subst. alkyl group at the terminal carbon of the allene, and acid-sensitive protective groups (THP and silyl ethers, isopropylidene acetals and N-carbo-*tert*-butoxyamines) remained intact. It is thought that reaction involves electrocyclization of a pentadienyl cation, the ultimate 1,2-hydrogen shift being consistent with the formation a single regioisomer. Other transition metal catalysts and Au(III) complexes were ineffective. F.e. and on the gram scale, also with ring expansion of a vinylidenecyclopentane analog, and intramolecular trapping of the generated cyclic carbocation with an appended hydroxyl group to give a tricyclic ether, s. J.H. Lee, F.D. Toste, Angew. Chem. Int. Ed. Engl. *2007*, *46* (6), 912-4; with $PtCl_2$ cf. H. Funami, H. Kusama, N. Iwasawa, ibid. 909-11.

*(2-Di-*tert*-butylphosphinobiphenyl)gold(I) chloride/silver triflate* ←
3-Acyl-2-pyrrolidones from β-ketocarboxylic acid N-allylamides ○
Stereospecific gold(I)-catalyzed cycloisomerization

468. 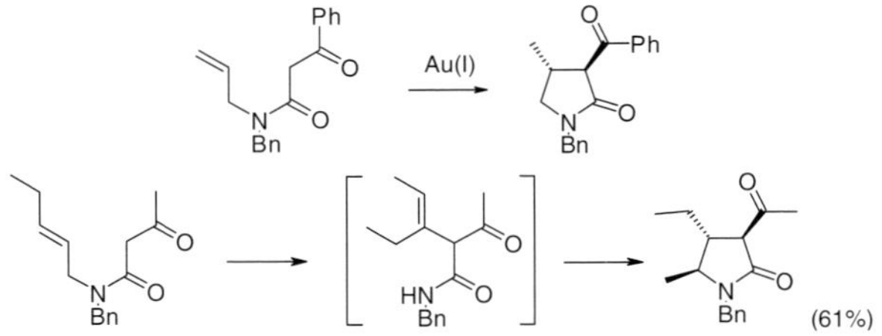

Startg. ketoamide added to a toluene soln. of 5 mol% (2-di-*tert*-butylphosphinobiphenyl)gold(I) chloride and 5 mol% AgOTf, the mixture stirred at 50° for 4 h, the solvent removed *in vacuo*, and the residue purified by chromatography → *trans*-3-benzoyl-1-benzyl-4-methyl-2-pyrrolidone. Y 99%. Eleven examples gave 90-99% yield of predominantly *exo-trig* cyclization products, incl. spirolactam derivs. (no *endo* cyclization being observed). Substitution at the internal alkene atom has no effect on the yield but the reaction is slower (12 h) as is a preparative scale reaction (5 g) using *1 mol%* catalyst, requiring 3 days to reach completion (Y 90%). Interestingly, terminal substitution (*cis* or *trans*) at the alkene moiety resulted in *exo*-cyclization **via stereospecific Claisen rearrangement-intramolecular hydroamination**. F.e. incl. reaction optimization s. C.-Y. Zhou, C.-M. Che, J. Am. Chem. Soc. *2007*, *129* (18), 5828-9.

Indium(III) or scandium(III) triflate *In(OTf)$_3$ or Sc(OTf)$_3$*
Nazarov cyclization s. *29*, 688s*72*

*Tri-*n*-butyltin hydride s. under Pd(OAc)$_2$* *Bu$_3$SnH*
*Tris(*p*-methoxyphenyl)phosphine/ethylene s. under [Rh(coe)$_2$Cl]$_2$* *Ar$_3$P/CH$_2$=CH$_2$*
2,3:2′,3′-Bis(methylenedioxy)-6,6′-bis(diphenylphosphino)biphenyl/ 1,2-cyclohexanedione s. under [Rh(cod)$_2$]BF$_4$ ←

Antimony pentafluoride *SbF$_5$*
1,2,3,4-Tetrahydro-1,2-benzothiazine 1,1-dioxides by cycloisomerization s. *72*, 287

Dichlorovanadium(IV) salen complexes/silver hexafluoroantimonate *V(salen)Cl$_2$/2AgSbF$_6$*
Nazarov cyclization
with $AlCl_3$ cf. *29*, 688s*67*; with 2 mol% V(salen)Cl_2 complex/2$AgSbF_6$, 2,3,4-trisubst. 2-cyclopentenone-5-carboxylic acid esters, s. I. Walz, A. Bertogg, A. Togni, Eur. J. Org. Chem. *2007* (16), 2650-8; with 3-15 mol% In(OTf)$_3$ or Sc(OTf)$_3$, 2,3-fused 2-cyclopentenones from dienones obtained by carbonylative coupling of lactam-, lactone- or thiolactone-derived vinyl triflates and vinylboronic acids, s. L. Bartali, P. Larini, A. Guarna, E.G. Occhiato, Synthesis *2007* (11), 1733-7.

*Acetonitrile[4-*tert*-butyl-2-(diisopropylphosphino)-1-methylimidazole]-(cyclopentadienyl)ruthenium(II) hexafluorophosphate* *[Ru(II)]*
Stereospecific ruthenium-catalyzed ethylene zipper ←

469. 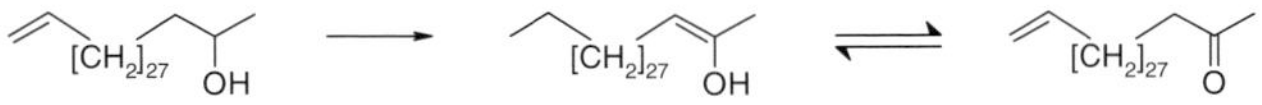

A catalyst has been discovered for **isomerization of** a variety of **heterofunctionalized ethylene derivs.** with movement of an unhindered alkene bond *over as many as 30 positions* along the alkyl chain: a record to date!! **E: Ketones from ethylenealcohols.** 30 Mol% acetonitrile[4-*tert*-butyl-2-(diisopropylphosphino)-1-methylimidazole](cyclopentadienyl)ruthenium(II) hexafluorophosphate added to a soln. of the startg. ethylenealcohol in *acetone-d$_6$*, heated at 70° for 3 days, and worked up → product. Y 81%. The hemilabile phosphinoimidazole serves not only as ligand, but also as an internal base promoting deprotonation of each intermediate ruthenium alkene complex along the chain before returning the proton onto the generated ruthenium allyl complex. In most instances the catalyst loading can be lowered to 2-5 mmol% for high-yielding conversions. F.e. incl. the formation of enoxysilanes from siloxyethylenes, and isomerization of N-protected ethyleneamines, s. D.B. Grotjahn, C.R. Larsen, J.L. Gustafson, R. Nair, A. Sharma, J. Am. Chem. Soc. *2007*, *129* (31), 9592-3.

Bis(1,5-cyclooctadiene)rhodium(I) fluoroborate/2,3:2′,3′-bis(methylenedioxy)-6,6′-bis(diphenylphosphino)biphenyl/1,2-cyclohexanedione ←

Stereospecific cycloisomerization of 1,6(7)-diynes

470. 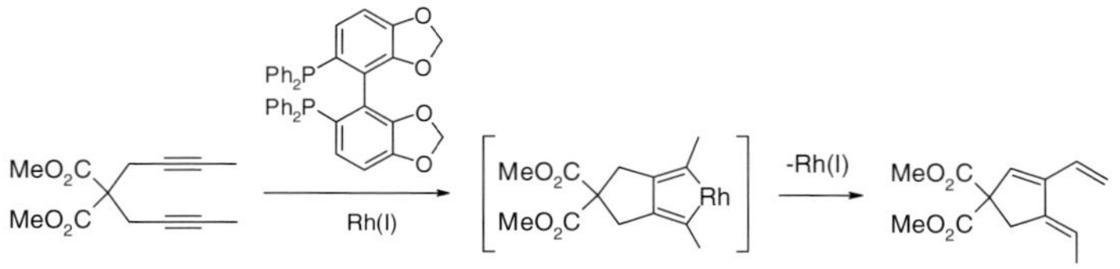

Cyclic cross-conjugated trienes. 10 Mol% SegPhos and 10 mol% bis(1,5-cyclooctadiene)-rhodium(I) fluoroborate stirred in methylene chloride for 5 min, H_2 introduced to the soln. in a Schlenk tube, stirred for 0.5 h, evaporated, the residue dissolved in methylene chloride, a soln. of malonate-derived 1,6-diyne and 1 eq. *1,2-cyclohexanedione* in methylene chloride added dropwise, stirred at 50° for 3 h, and the soln. concentrated → product. Y 85% (E/Z >20:1). SegPhos was the most suitable ligand tested but required the additional chelating 1,2-diketone ligand to promote cycloisomerization *via β-hydride elimination*, rather than the usual [2+2+2]-cycloaddition by coupling two molecules of the substrate. A similar yield was obtained with 5 mol% catalyst but with 2.5 mol% the reaction did not reach completion. The procedure was successful with 1,6-diynes derived from malonate and acetylacetone, and with 1,7-diynes derived from tetraethyl ethane-1,1,2,2-tetracarboxylate. Replacing the malonate moiety with N-tosyl, N-benzyl or oxygen afforded **pyrroles and furans** in low yield (21, 16 and <5% respectively). F.e. and comparison of ligands and diketones s. K. Tanaka, Y. Otake, M. Hirano, Org. Lett. *2007*, *9* (20), 3953-6.

*Chlorobis(cyclooctene)rhodium(I) dimer/tris(*p*-methoxyphenyl)phosphine/ethylene* ←
Chlorotris(triphenylphosphine)rhodium(I)/silver hexafluoroantimonate *$RhCl(PPh_3)_3/AgSbF_6$*

Cycloisomerization of cyclopropylidenealdehydes
via rhodium-catalyzed activation of the formyl carbon-hydrogen bond

471. 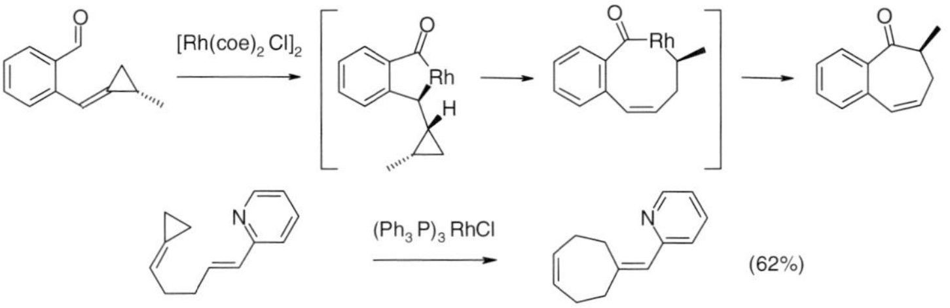

3*H*-Benzocycloheptadien-1(2*H*)-ones. Ethylene bubbled for 1 min through a mixture of the startg. chiral alkylidenecyclopropane, 5 mol% $[Rh(coe)_2Cl]_2$ and 20 mol% tris(*p*-methoxyphenyl)-phosphine in 1,2-dichloroethane contained in a Schlenk tube, the latter sealed then immersed into a preheated oil bath at 120°, stirred for 2.5 h, the mixture cooled, diluted with ether, filtered through a short pad of silica, the filtrate evaporated *in vacuo*, and the residue purified by chromatography → benzocycloheptadienone deriv. Y 73%. Six examples containing the pendant aldehyde gave 70-87% yield in the presence of ether and halo functionality, while substrates possessing a vinylpyridine trigger in place of the aldehyde gave **5-(2-pyridylmethylene)-cycloheptenes** in 53-62% yield with chlorotris(triphenylphosphine)rhodium(I)/$AgSbF_6$.

Cyclization of the aldehyde derivs. is thought to involve initial Rh insertion into the formyl C-H bond, as supported by deuterium labelling experiments. F.e. incl. preparation of the startg. m. s. C. Aïssa, A. Fürstner, J. Am. Chem. Soc. *2007*, *129* (48), 14836-7.

trans-*Carbonyl(chloro)bis(triphenylphosphine)rhodium(I)* *$RhCl(CO)(PPh_3)_2$*
(Z)-5-Alkylidene-4-vinyl-5,6-dihydro-2-pyrones ○
from α-allenecarboxylic acid propargyl esters
Intramolecular Alder-ene-type reaction

472. 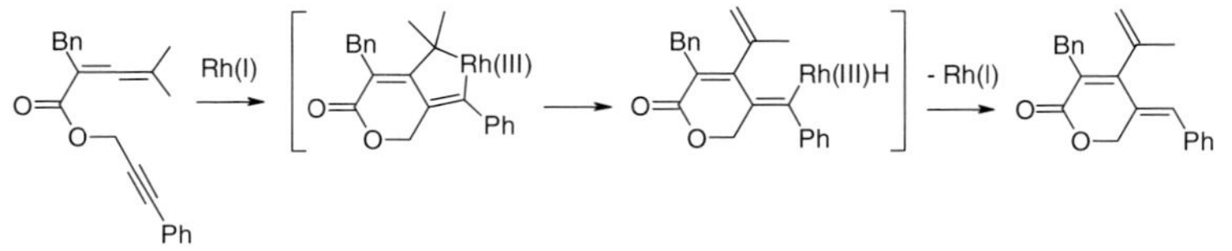

A soln. of 3′-phenylprop-2′-ynyl 2-benzyl-4-methylpenta-2,3-dienoate and 5 mol% *trans*-$RhCl(CO)(PPh_3)_2$ in dry toluene refluxed under argon for 15 h → (Z)-3-benzyl-5-benzylidene-4-(prop-1-en-2-yl)-5,6-dihydropyran-2-one. Y 67%. Reaction proceeds via an Alder-ene-type mechanism, involving initial cyclometalation, followed by β-hydride elimination to form an intermediate rhodium hydride complex prior to reductive elimination. With 4-phenyl-subst. allenoates [lacking a 4-substituent possessing a β-hydrogen], two molecules of the substrate unite via a Diels-Alder reaction to give symmetrical bi(pyrano)norbornenes. F.e.s. X. Jiang, S. Ma, J. Am. Chem. Soc. *2007*, *129* (37), 11600-7.

*Palladium(II) acetate/tri-*n*-butyltin hydride/triethylamine* ←
(Z)→(E)-Rearrangement of ethylene derivs. ←
with NBS/benzoyl peroxide cf. *31*, 714s*66*; of aryl- or alkyl-subst. alkenes with 5 mol% $Pd(OAc)_2$/Bu_3SnH/Et_3N in methylene chloride at reflux s. I.S. Kim, G.R. Dong, Y.H. Jung, J. Org. Chem. *2007*, *72* (14), 5424-6

Platinum(II) chloride *$PtCl_2$*
Cyclopentadienes from 1,2,4-trienes ○
Regiospecific cycloisomerization s. *72*, 467

Oxygen/Sulfur Type CC ∩ OS

1,3-Dimesitylimidazol-2-ylidenegold(I) chloride/silver hexafluoroantimonate ←
2,3-Dihydrobenzo[*b*]thiepin-4(5*H*)-ones from ar. γ,δ-acetylenesulfoxides ○
Gold(I)-catalyzed cycloisomerization via 1,4-S→C-oxygen shift

473. 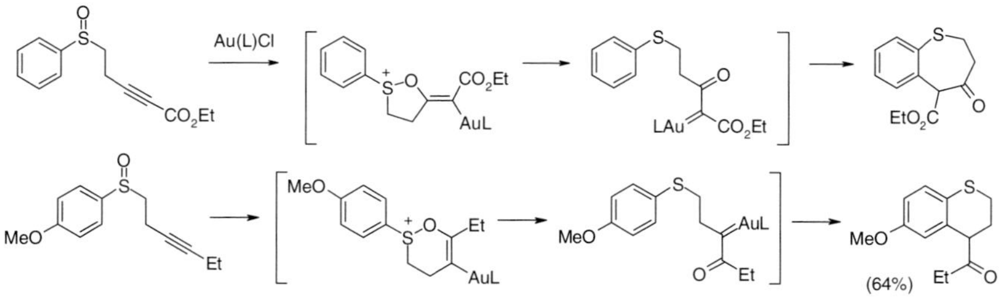

under mild conditions. A mixture of the startg. sulfoxide, 5 mol% (IMes)AuCl, 5 mol% $AgSbF_6$ and 5 Å molecular sieves stirred in methylene chloride at room temp. until reaction complete (TLC monitoring), and the product isolated directly by chromatography → benzo[*b*]thiepin-4-one deriv. Y 91%. The reaction proceeds smoothly for electron-poor and -rich aryl sulfoxides and is tolerant of substitution at the homopropargyl and propargyl positions (the latter requiring heating at 60° in dichloroethane) giving 63-91% yield in the presence of halo, cyclopropane and ether functionality. The terminal alkyne substituent can be H, aryl or an electron-withdrawing group, but with a terminal alkyl substitution the reaction took a different course affording a 4-acyl-thiochroman in 64% yield. The authors suggest the latter arises from formation of an intermediate gold complex via 6-*endo-dig* cyclization rather than a 5-*exo-trig* cyclization *en route* to the benzothiepinones. F.e.s. N.D. Shapiro, F.D. Toste, J. Am. Chem. Soc. *2007*, *129* (14), 4160-1.

Oxygen/Carbon Type — CC ∩ OC

Methyllithium/microwaves — *MeLi/[\\\\]*

4-Cycloheptenones from 1,6-enyn-3-ols
via stereospecific oxyanionic cycloisomerization-Claisen rearrangement

474.

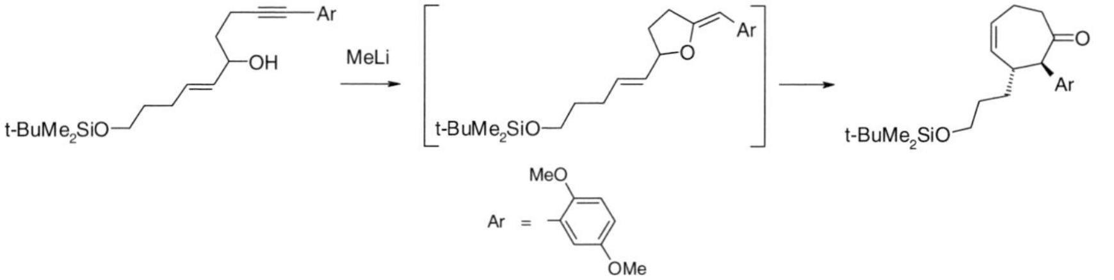

Microwave-assisted conversion. A mixture of the startg. enynol and 10 mol% methyllithium (as an ether soln.) in *phenetole* heated at 210° in a microwave oven for 45 min, the solvent removed *in vacuo*, and the residue purified directly by chromatography → *trans*-2,3-disubst. 4-cycloheptenone deriv. Y 81% (d.r. 85:15). The startg. enynols are prepared from readily available substrates and give yields of 60-82% in the presence of silyl ether and ether functionality, *trans*-disubst. products predominating (selectivity 67-92%) from terminally subst. alkyne derivs. Reaction is thought to involve an initial oxyanionic 5-*exo-dig* cyclization of the deprotonated substrate to the furan ring (isolated as sole product in the case of a terminal alkyne ester in 89% yield) and subsequent Claisen rearrangement. F.e. incl. startg. m. prepn. s. X. Li, R.E. Kyne, T.V. Ovaska, J. Org. Chem. *2007*, *72* (17), 6624-7.

Triethylenediamine/1,8-diazabicyclo[5.4.0]undec-7-ene/trimethylsilyl chloride ←

Terminal 1,4(E)-diene-2-carboxylic acids from acrylic acid allyl esters ∩
via DABCO-catalyzed Ireland-Claisen rearrangement under mild conditions

475.

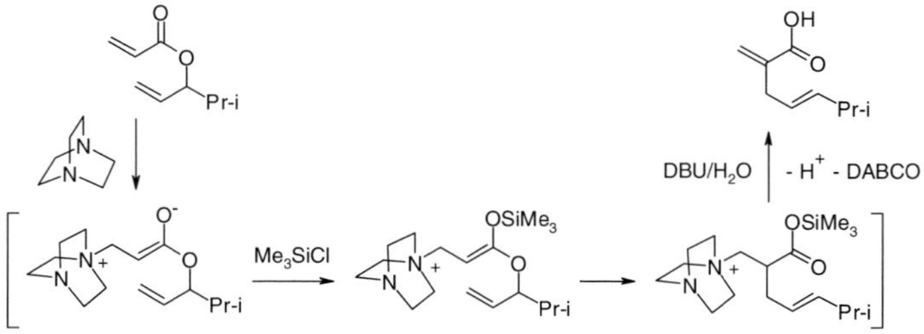

A mixture of the startg. allyl acrylate, 20 mol% DABCO, 3 eqs. trimethylsilyl chloride and 2 eqs. 1,8-diazabicyclo[5.4.0]undec-7-ene in acetonitrile refluxed for 7.5 h, volatiles removed under reduced pressure, the residue suspended in ether, stirred with 3 *M* HCl for a few min, and worked up with chromatographic purification, distillation or recrystallization → product. Y 86% (E/Z 99:1). The procedure is mild, inexpensive, based on manageable reagents (cf. triphenylphosphine), and the solvent does not require degassing; it is also applicable to a wide range of allyl acrylates derived from sec. or tert. alcohols, and tolerates alkyl, olefinic or aryl substitution of the allyl residue. Mechanistically, it is presumed that reaction is initiated by 1,4-addition of DABCO to give a zwitterionic species, which is silylated *in situ* prior to classical Ireland-Claisen rearrangement, and terminated by DBU-induced deprotonation with elimination of DABCO. F.e.s. Y.X. Li, Q.R. Wang, A. Goeke, G. Frater, Synlett *2007* (2), 288-92.

Gold(III) chloride/silver hexafluoroantimonate *$AuCl_3/AgSbF_6$*

Cyclic α,β-ethyleneketones from acetyleneketones ○

476.

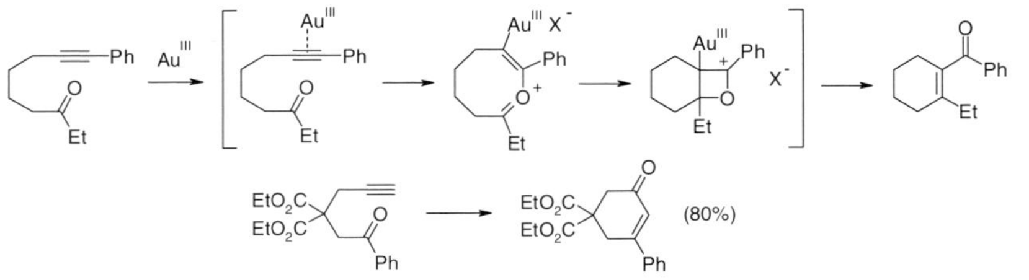

Cyclohex-1-enyl ketones. The startg. *internal* acetyleneketone added to a soln. of 2 mol% $AuCl_3$ (0.2 *M*) and 6 mol% $AgSbF_6$ in toluene (contained in a pressure vial) at room temp. under argon, stirred at 60° for 30 min, filtered through a short Florisil pad, concentrated, and worked up with column chromatographic purification on silica gel → product. Y 87%. The triple bond can be terminally substituted by alkyl or aryl groups (irrespective of the electronic nature of the ring substituent), while cyclopentanone and cyclooctanone analogs gave the corresponding condensed cyclohex-2-enyl ketones. Reaction is presumed to involve a formal intramolecular [2+2]-cycloaddition via initial activation of the alkyne residue by gold(III) and generation of an intermediate cyclic oxonium species. *Terminal* δ,ε-acetyleneketones, however, gave the corresponding **2-cyclohexenones** with $AuCl_3/AgOTf$. F.e. incl. 1-alkanoylcyclohexenes and bicyclic 2-cyclohexenones s. T. Jin, Y. Yamamoto, Org. Lett. *2007*, *9* (25), 5259-62.

Trimethylsilyl chloride s. under Triethylenediamine *Me_3SiCl*

Sulfated zirconia *ZrO_2/SO_4^{2-}*

Claisen rearrangement of allyl phenolethers ∩

under microwave irradiation on silica in the absence of solvent cf. 2, 621s*68*; with sulfated zirconia [UDCaT-5] as solid superacid in refluxing chlorobenzene s. G.D. Yadav, S.V. Lande, Synth. Commun. *2007*, *37* (6), 941-6.

Platinum(II) chloride *$PtCl_2$*

2-Oxabicyclo[3.3.0]oct-7-en-3-one ring ○

via stereospecific [3.3]-sigmatropic rearrangement-intramolecular [3+2]-cycloaddition

477.

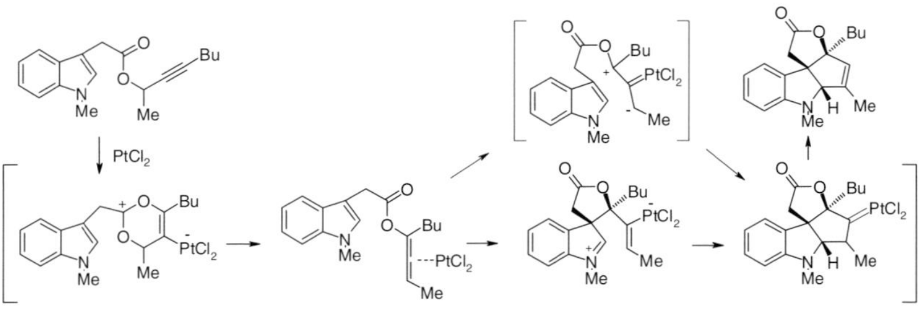

Indoline-fused derivs. While under Au(I)-catalysis [3.3]-sigmatropic rearrangement of propargylic indole-3-acetates is followed by intramolecular [2+2]-cycloaddition (*70*, 341), under $PtCl_2$ catalysis the [3+2]-cycloadduct is obtained. Such divergent selectivity between cationic Au(I) catalysts and $PtCl_2$ is rare. **E:** A soln. of startg. indole-3-acetate in toluene (0.05 *M*) treated with 5 mol% $PtCl_2$, then stirred under 1 atm. CO in a pre-heated oil-bath at 80° for 1.5 h (TLC monitoring), concentrated, and purified by flash column chromatography on silica → product. Y 94%. The efficiency of cyclopentene formation was substantially improved under CO (Y 57% at 81% conversion after 7 h otherwise). The method was highly effective with substrates bearing prim. alkyl groups at both ends of the propargyl moiety, and these groups may contain functionality such as olefin, protected hydroxyl or bromine. It is believed [3.3]-sigmatropic rearrangement leads to an acoxyallene, which may be further activated by the catalyst to yield an oxocarbenium ion, followed by electrophilic attack of the indole to afford a lactone, 5-*exo* ring closure to give a cyclopentane containing a Pt-carbenoid moiety, hydride migration and elimination of the Pt-catalyst; alternatively an intramolecular, concerted 1,3-dipolar cycloaddition of a mesomeric all-carbon 1,3-dipole would lead to the same Pt-carbenoid. F.e. incl. N-benzyl-derivs. and application to the synthesis of the tetracyclic core of vindolinine s. G. Zhang, V.J. Catalano, L. Zhang, J. Am. Chem. Soc. *2007*, *129* (37), 11358-9.

Nitrogen/Carbon Type CC ∩ NC

Sodium methoxide *NaOMe*

3-Aminothiophene ring closure ○

with KOH cf. *19*, 803s*61*; with NaOMe in 2-methoxyethanol, 2,5-diamino-4-(organothio)-thieno[2,3-*d*]pyrimidine-6-carbonyl compds., s. S. Drescher, D. Ramsbeck, D. Briel, B. Dobner, J. Heterocycl. Chem. *2007*, *44* (4), 757-63.

Potassium tert-*butoxide* *KOBu*-t

N-Protected 3-amino-2,4-pyrrolidiones from 2,5-piperazinediones
Regio- and stereo-specific ring contraction under mild conditions ○

478.

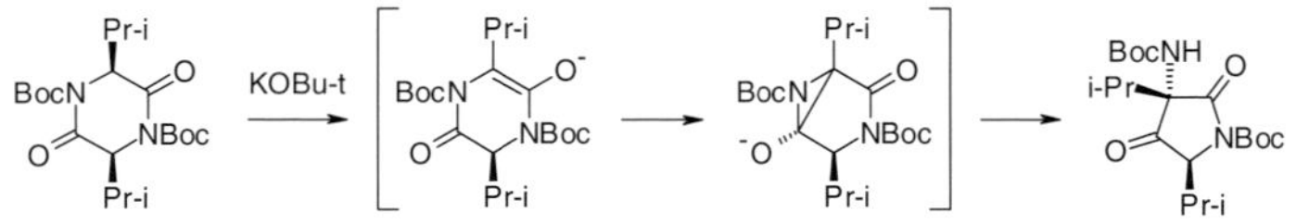

1.1 Eqs. *t*-BuOK added at room temp. to a soln. of the startg. Boc-diprotected cyclic dipeptide in dry THF, stirred for 12 h under argon, and worked up → product. Y 68% (d.e. >95%). Chiral mono-3-alkylated substrates reacted similarly **with retention of configuration,** possibly as a result of the initial formation of a kinetic enolate and the subsequent generation of an intermediate condensed aziridine. However, it does appear that the two electron-withdrawing N-Boc groups are critical for the conversion. A one-pot **alkylative ring contraction** was also effected **with asym. induction** to provide a series of chiral N-protected 3,5-disubst. 3-amino-2,4-pyrrolidiones. F.e.s. D. Farran, I. Parrot, J. Martinez, G. Dewynter, Angew. Chem. Int. Ed. Engl. *2007*, *46* (39), 7488-90.

*[Tris[*p*-(trifluoromethyl)phenyl]phosphine]gold(I) triflimide* *[(Ar$_3$P)Au]NTf$_2$*

2-Homoallyl-N-tosylpyrroles from N-allyl-N-tosyl-2,4-enyn-1-amines
via gold(I)-catalyzed aza-Claisen-type rearrangement ○

479.

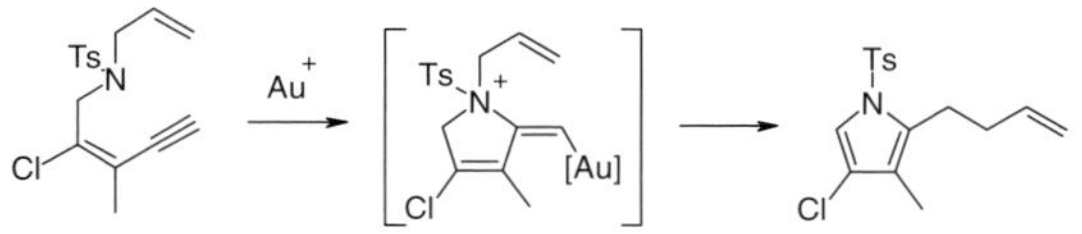

2 Mol% tris[*p*-(trifluoromethyl)phenyl]gold(I) triflimide added to a 0.1 *M* soln. of the startg. N-allyl-N-tosyl-2,4-enyn-1-amine in methylene chloride, stirred at room temp. for 30 min, evaporated, and worked up with column chromatographic purification on silica gel → product.

Y 88%. The procedure is mild and efficient, and notably applicable to the generation of **quaternary hydrocarbon groups** in the chain. Reaction is presumed to involve initial activation of the alkyne residue by gold(I), followed by intramolecular aminoauration and a concerted aza-Claisen-type rearrangement. However, there was no reaction with substrates bearing a terminal methyl or phenyl group on the alkyne residue. F.e. and comparison of catalysts s. F.M. Istrate, F. Gagosz, Org. Lett. *2007*, *9* (16), 3181-4.

Silica or Platinum(II) chloride *SiO_2 or PtO_2*
N-Subst. 2-homoallylindoles from tert. N-allyl-*o*-propargylanilines via 3-azonia-Cope rearrangement

480. 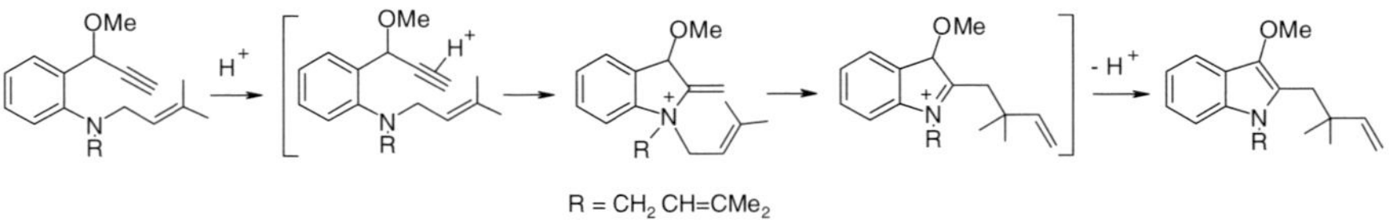

N-Subst. 3-alkoxy-2-homoallylindoles. A 0.025 *M* soln. of the startg. *o*-propargylaniline (0.23 mmol) in methylene chloride submitted to argon-bubbling for 15 min, 300 mg silica added, stirred for 6 h at room temp., and worked up with flash chromatographic purification → 2-(2,2-dimethylbut-3-enyl)-3-methoxy-1-(2-methylbut-2-enyl)-1*H*-indole. Y 81%. Reaction was smooth, versatile, simple and high-yielding, while the same reaction with 5 mol% $PtCl_2$ gave a significant amount of a regioisomer in some instances. Formally, this amounts to an intramolecular aminoallylation of the triple bond, which is presumed to take place via a charge-accelerated aza-Cope rearrangement after initial activation of the alkyne residue (by protonation or π-complexation in the case of platinum catalysis). F.e. incl. 3-acoxy-derivs. s. K. Cariou, B. Ronan, S. Mignani, L. Fensterbank, M. Malacria, Angew. Chem. Int. Ed. Engl. *2007*, *46* (11), 1881-4; indoles **from N-aryl-2-acetyleneamines via amino-Claisen rearrangement** with $[Rh(cod)_2]OTf$/dppp cf. A. Saito, A. Kanno, Y. Hanzawa, ibid. *46* (21), 3931-3.

Sulfur/Carbon Type CC ∩ SC

Gold(I) or gold(III) chloride *AuCl or $AuCl_3$*
2-(Alkylthio)indene ring by gold-catalyzed cycloisomerization with 1,2-alkylthio group migration via thiiranium ions

481. 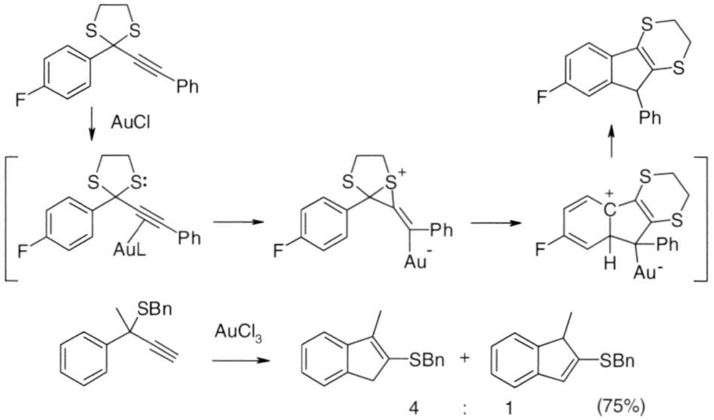

2,3-Dihydro-9*H*-1,4-dithiafluorenes. A mixture of the startg. propargylic dithioacetal and 5 mol% AuCl heated at 80° in dry toluene under N_2 for 1 h (TLC monitoring), the solvent removed *in vacuo*, and the residue purified by chromatography → 7-fluoro-9-phenyl-2,3-dihydro-9*H*-1,4-dithiafluorene. Y 91%. A series of propargylic dithioacetals gave the 9*H*-isomers as sole products in 85-95% yield within 1 h at 80°, while the 7-bromo-9-phenyl analog was obtained as a 4:1 mixture with the 8b*H*-isomer in 93% yield, and a furo-fused analog was obtained in only 43% yield after 36 h (with recovery of the startg. m. in 22% yield). Uncondensed **2-(alkylthio)-indenes** were obtained similarly **from 1-aryl-2-acetylenethioethers** (regioselectivity 3:1 to >20:1; Y 13-75%). Both AuCl and $AuCl_3$ were effective, while other catalysts and the use of additives gave lower yields and selectivity. F.e. incl. startg. m. prepn. and reaction optimization s. L. Peng, X. Zhang, S. Zhang, J. Wang, J. Org. Chem. *2007*, *72* (4), 1192-7.

Carbon/Carbon Type CC ∩ CC

Without additional reagents *w.a.r.*
Intramolecular [2+2]-cycloaddition with allenes □○
of 1,2-dien-n-ynes under microwave irradiation cf. *70*, 347; of 3-phenylsulfonyl-1,2-dien-n-ynes thermally in refluxing xylene (or mesitylene) s. C. Mukai, Y. Hara, Y. Miyashita, F. Inagaki, J. Org. Chem. *2007*, *72* (12), 4454-61; of 1,2-dien-7-ynes or 1,2,7-trienes in DMF s. H. Ohno, T. Mizutani, Y. Kadoh, A. Aso, K. Miyamura, N. Fujii, T. Tanaka, ibid. 4378-89.

Irradiation ///
Copper(I) triflate/butyl(trimethyl)ammonium triflimide/irradiation ←
Intramolecular [2+2]-cycloaddition
s. *22*, 761s*66*; *24*, 751s*66*; of substrates derived by Ugi 4-component condensation of ethylene-amines, ethylene(keto)carboxylic acids, oxo compds. and isonitriles (cf. *17*, 809s*67*), especially for the formation of the 3-α-carbamyl-3-azabicyclo[4.2.0]octan-4-one ring with the formation of up to five stereocenters but (in most cases) only two diastereomers, s. I. Akritopoulou-Zanze, A. Whitehead, J.E. Waters, R.F. Henry, S.W. Djuric, Org. Lett. *2007*, *9* (7), 1299-302; intra- or inter-molecular (cf. *22*, 705) [2+2]-photocycloadditions with $(CuOTf)_2$-benzene as catalyst in ionic liquid solvents, e.g. butyl(trimethyl)ammonium triflimide (but not imidazolium salts due to decomposition), s. C.K. Malik, M. Vaultier, S. Ghosh, Synthesis *2007* (8), 1247-50.

Chiral copper(II) bis(Δ^2-oxazoline) complex *Cu(II)-box*
Asym. Nazarov cyclization-Wagner-Meerwein rearrangement ○

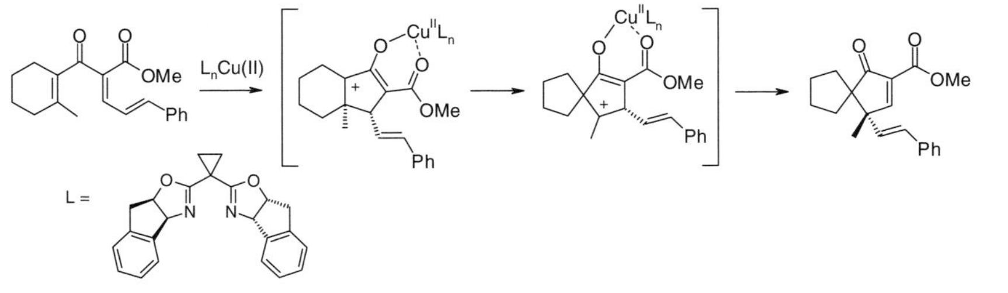

482.

Chiral 5-spiro-2-cyclopentenone-2-carboxylic acid esters. A soln. of the startg. dienone in methylene chloride stirred with 1 eq. chiral Cu(II)-bis(Δ^2-oxazoline) complex under argon at 25° for 20 h, quenched with aq. NH_4OH, and worked up → product. Y 96%. For substrates possessing an aryl or styryl group on the alkene residue, the Wagner-Meerwein shift onto the generated oxyallyl cation is followed by 1,2-aryl or 1,2-styryl migration, respectively, to provide **vicinal quaternary hydrocarbon groups.** However, hydride shift prevails with substrates substituted on the alkene residue by an alkyl group. The high chemoselectivity over the classical Nazarov route with 1 eq. of the copper(II) complex suggests that both carbonyl oxygens of the substrate are bonded to the metal in the transition state. Enantioselectivities were in the range 29-64%. F.e. and from cycloheptenyl derivs., also with $Cu(SbF_6)_2$, s. J. Huang, A.J. Frontier, J. Am. Chem. Soc. *2007*, *129* (26), 8060-1.

Chiral [2,3;2′,3′-bis(methylenedioxy)-6,6′-bis(diarylphosphino)biphenyl]-gold(I) chloride complex/silver fluoroborate ←
Alkylidenecyclobutane ring from 1,2,n-trienes □○
via gold(I)-catalyzed asym. intramolecular [2+2]-cycloaddition

483.

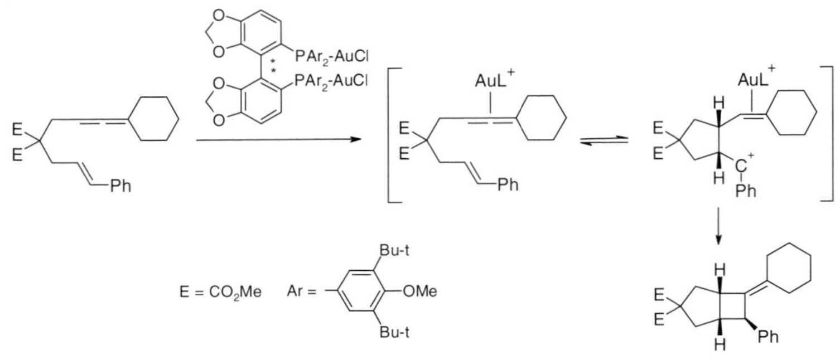

under mild conditions. A mixture of 3-5 mol% chiral [2,3;2′,3′-bis(methylenedioxy)-6,6′-bis-(diarylphosphino)biphenyl]gold(I) chloride complex and 3-6 mol% $AgBF_4$ stirred in methylene chloride at room temp. for 5 min, the soln. added to the startg. 1,2,n-triene in methylene chloride at room temp., the cloudy white mixture stirred until reaction complete (TLC monitoring), filtered through silica, the filtrate evaporated *in vacuo*, and the residue purified by chromatography → product. Y 86% (e.e. 95%). This asym. cycloaddition appears to be limited to substrates with a *gem*-diester tether (NTs in place of *gem*-diester gave only 54% e.e.) giving 76-92% yield and 92-7% enantiomeric excess. A series of achiral reactions utilizing $(Ph_3P)AuCl/AgBF_4$ as catalysts were more flexible giving 80-92% yield in the presence of ester, sulfone and additional alkene functionality, and were able to accommodate disubstitution at the distal position. The mechanism is thought to involve reversible formation of carbocationic intermediates which are trapped in the *cis* form in the absence of exogeneous nucleophiles (addition of methanol trapped the *trans* intermediate). F.e. incl. catalyst prepn. s. M.R. Luzung, P. Mauleón, F.D. Toste, J. Am. Chem. Soc. *2007*, *129* (41), 12402-3.

Mercury(II) triflate *$Hg(OTf)_2$*
5-Membered cyclic 1,2,4-trienes from 1,2-dien-7-ynes ○
Mercury(II)-catalyzed cycloisomerization

484.

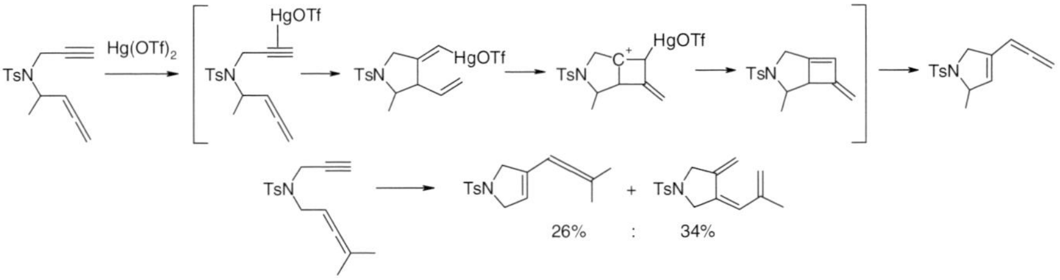

under mild conditions. A soln. of startg. allenyne in acetonitrile added slowly to 10 mol% *$Hg(OTf)_2$* in *acetonitrile*, the mixture stirred for 0.5 h at room temp., the reaction quenched with aq. $NaCl/NaHCO_3$, extracted with ether, the organic layer dried over $MgSO_4$, concentrated *in vacuo*, and the residue purified by chromatography → product. Y 72%. This is the first mercury-catalyzed cycloisomerization of allenynes and is most effective for terminally unsubst. derivs. affording yields of 59-79% within 1 h at room temp. A substrate with a single substituent at the terminus of the allene gave a yield of 48%, while disubstitution at this position produced cyclic trienes as major products (Y 34 and 46%). There was no reaction, however, with internal alkyne derivs., nor with substrates possessing an internal allene substituent, nor with 1,2,8-dienynes. Reaction, for which proposed mechanistic details are outlined, seems unique to $Hg(OTf)_2$ with no other catalyst tested affording the same products. F.e. incl. 1-allenylcyclopentenes s. S.H. Sim, S.I. Lee, J. Seo, Y.K. Chung, J. Org. Chem. *2007*, *72* (25), 9818-21.

Chiral N-condensed 2-(o-fluorophenyl)-1,3,2-oxazaborolidinium triflimide ←
Lewis acid-catalyzed asym. transannular Diels-Alder reaction ○

485. 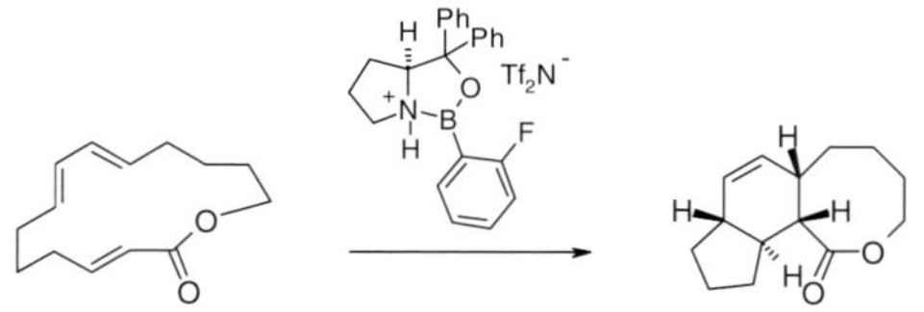

0.2 Eq. 1 *M* triflimide in toluene (freshly prepared) added dropwise to a soln. of chiral N-condensed 2-(*o*-fluorophenyl)-1,3,2-oxazaborolidine in toluene under N_2 at -20°, stirred for 30 min at the same temp., a soln. of the startg. macrocycle (previously azeotroped with benzene three times) in toluene added under N_2, the mixture immediately warmed to room temp., stirred for 20 h, triethylamine added, concentrated, and worked up with flash chromatographic purification → (3aR,5aS,11aS,11bS)-2,3,3a,5a,6,7,8,9,11a,11b-decahydro-1*H*-10-oxacycloocta[*e*]inden-11-one. Y 69% (d.r. >19:1; e.e. 90%). The procedure is highly enantioselective and *endo*-specific for the transannular Diels-Alder reaction of macrocyclic α,β-ethylene-ketones and -lactones possessing a (E,E)-1,3-diene residue, affording 5-6-7-, 6-6-7- and 7-6-7-membered tricyclic systems in good yield. There was no reaction, however, with macrocycles possessing a trisubst. 1,3-diene or (E,Z)-1,3-diene residue, and the *o*-fluorophenyl group in the chiral auxiliary is critical for high enantioselectivity. F.e. and comparison of chiral reagents, also application to the total synthesis of the sesquiterpene, 11,12-diacetoxydrimane, s. E.P. Balskus, E.N. Jacobsen, Science *2007*, *317* (5845), 1736-40.

Butyl(trimethyl)ammonium triflimide s. under CuOTf *$[BuMe_3N][NTf_2]$*

Ruthenium carbene [Grubbs′] complex *$PhCH{=}RuCl_2(PCy_3)_2$*
Ring-closing ene-yne metathesis
s. *50*, 443s*70*; chiral α-(3-vinyl-Δ^3-pyrrolin-1-yl)carboxylic acid esters (without requiring ethylene gas) s. Q. Yang, H. Alper, W.-J. Xiao, Org. Lett. *2007*, *9* (5), 769-71.

Tris(dibenzylideneacetone)dipalladium/o,o′-biphenylyl phosphoromonoamidites ←
Intramolecular [4+3]-cycloaddition with alkylidenecyclopropanes ○◐

486. 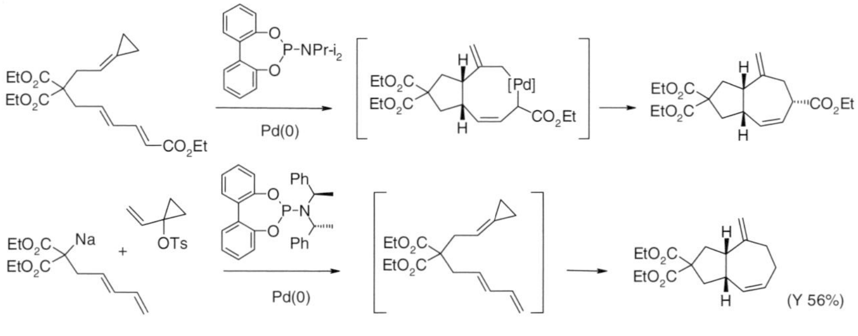

5-Methylenecycloheptene ring. Startg. octa-5,7-dienylidenecyclopropane added to a suspension of 6 mol% $Pd_2(dba)_3$ and 24 mol% phosphoramidite ligand in dry, freshly-distilled and deoxygenated dioxane under argon, refluxed for 3 h in a preheated bath, cooled to room temp., diluted with hexanes, filtered through a short pad of silica gel eluting with 10% ethyl acetate/ hexanes, and purified by flash chromatography → product. Y 65%. This novel, transition metal-catalyzed intramolecular [4+3]-cycloaddition is highly diastereoselective, generating three stereocenters. An asym. variant with a chiral phosphoramidite ligand afforded an enantiomeric excess of 47% (Y 73%) for this example. There is evidence for a concerted cycloaddition mechanism involving an intermediate palladacyclooctane. F.e. incl. substrates with oxa or aza groups in the tether, also from 1-vinylcyclopropyl tosylate and diethyl 2-((E)-penta-2,4-dienyl)-malonate Na-salt in a one-pot process [in the presence of the chiral ligand and 1,2-bis(diphenyl-phosphino)ethane], s. M. Gulías, J. Durán, F. López, L. Castedo, J.L. Mascareñas, J. Am. Chem. Soc. *2007*, *129* (37), 11026-7.

Exchange ↓↑

Hydrogen ↑ CC ↓↑ H

Electrolysis/sodium halide ⚡*/NaHal*

Sym. ethylene derivs. by oxidative dimerization 2 >CH₂ → >C=C<

tetraarylethylenes with $PhNO_2/K_2CO_3$ cf. *14*, 772; *trans*-α,β-dicyanostilbenes by electrolysis of arylacetonitriles in methanol containing Na-halide, e.g. NaBr, as mediator using an undivided cell s. M.N. Elinson, A.S. Dorofeev, S.K. Feducovich, P.A. Belyakov, G.I. Nikishin, Eur. J. Org. Chem. *2007* (18), 3023-7.

Microwaves s. under Pd(OCOCF₃)₂ [\\\\]

Potassium hydroxide/O-benzyl-N-(9-anthracenylmethyl)dihydrocinchonidinium bromide/ trichloroisocyanuric acid ←

Organocatalyzed asym. synthesis of chalcone epoxides from acetophenones and ar. aldehydes via Claisen-Schmidt condensation CHO →

487. 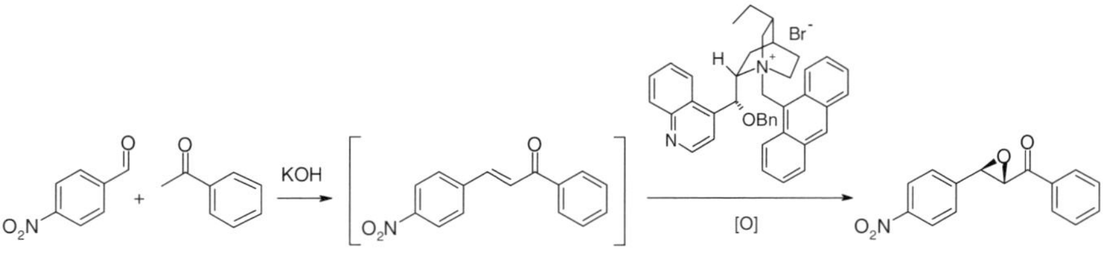

One-pot conversion under mild conditions. A suspension of equimolar amounts of acetophenone and *p*-nitrobenzaldehyde in 10% aq. KOH (6 eqs.) stirred at 25° for 12 h, toluene and 5 mol% O-benzyl-N-(9-anthracenylmethyl)dihydrocinchonidinium bromide added, cooled to 0°, treated with 1 eq. trichloroisocyanuric acid in one portion, stirred for a further 10 h at 0°, treated with water, and worked up with column chromatographic purification on silica gel → product. Y 68% (e.e. 96%)The procedure is highly enantioselective, but poor results were obtained with aliphatic ketones and aldehydes. F.e.s. Y. Wang, J. Ye, X. Liang, Adv. Synth. Catal. *2007*, *349* (7), 1033-6.

Sodium ethoxide s. under Et₃N *NaOEt*

n-*Butyllithium/zirconocene dichloride/quinones/copper(I) chloride* ←
Cross-conjugated enediynes 3 RC≡CH → RCH=C(C≡CR)$_2$
from three terminal acetylene deriv. molecules

488.

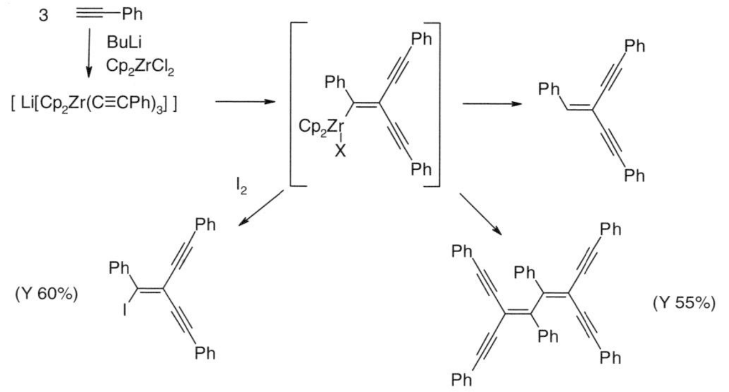

A soln. of phenylacetylene in THF treated at -78° under N_2 with 1.07 eqs. 1.6 *M* BuLi in hexane, stirred at room temp. for 1 h, treated at -78° with 0.33 eq. Cp_2ZrCl_2, stirred at room temp. for 12 h, the dark red soln. treated with 0.43 eq. tetrachlorobenzoquinone, stirred for a further 8 h, and quenched with 3 *N* HCl → product. Y 67%. The method is applicable to arylethynes, but aliphatic derivs. gave only a trace of product and 1-ethynylcyclohexene gave a 17% yield. The intermediate zirconoenediynes undergo coupling reactions with electrophiles, such as iodine, iodobenzene or allyl bromide [in the presence of CuCl and/or $Pd(PPh_3)_4$], or dimerization. F.e. and with DDQ s. C. Xi, Y. Liu, X. Yan, C. Chen, J. Organomet. Chem. *2007*, *692* (21), 4612-7.

*Lithium bis(trimethylsilyl)amide/biphenyl-2,2′-ylene(*p*-tolyl)bismuth bis(triflate)* ←
Oxidative dimerization of enolates 2 C(O)CH → C(O)C-C-C(O)

489.

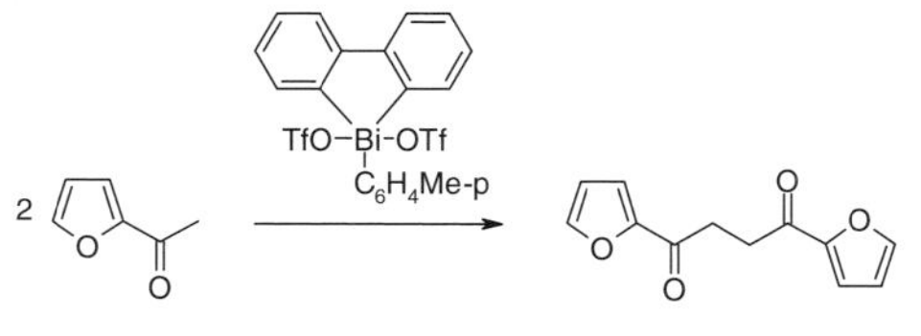

A soln. of 2-acetylfuran in THF treated with 0.95 eq. Li-bis(trimethylsilyl)amide at -78°, stirred for 30 min, allowed to warm to -45°, 0.45 eq. biphenyl-2,2′-ylene(*p*-tolyl)bismuth bis(triflate) added, and stirring continued for 3 h at the same temp. → product. Y 84%. The labile triflate group is critical to high yields, as is the tricyclic triarylbismuth skeleton which holds the two enolate residues in close proximity prior to coupling. The procedure is generally applicable to ketone, ester and thioester enolates. F.e.s. S. Imachi, T. Mukaiyama, Chem. Lett. *2007*, *36* (6), 718-9.

Cesium pivalate s. under Pd(OCOCF$_3$)$_2$ *CsOCOR*
Sodium halide s. under Electrolysis *NaHal*
Potassium iodide s. under PdI$_2$ *KI*

Triethylamine/sodium ethoxide/air *$Et_3N/NaOEt/O_2$*

***m*-Nitro-*o*-aminonitriles from 1,1-nitroethylene derivs. and alkylidenemalononitriles via Michael addition-oxidative cycloaromatization** ○

490. 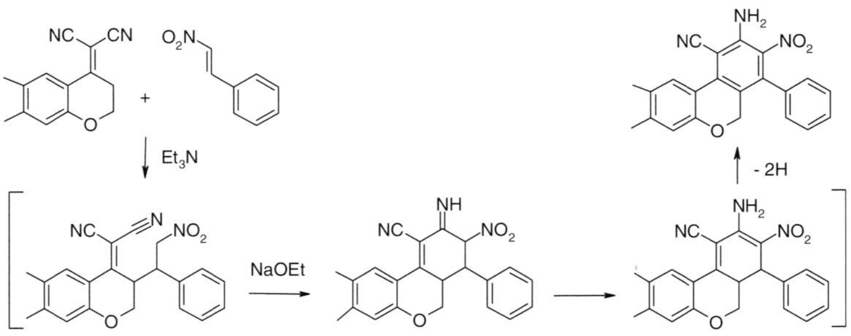

One-pot conversion. An equimolar mixture of β-nitrostyrene and the startg. alkylidenemalononitrile stirred with 10 mol% triethylamine in methylene chloride for 57 h (TLC monitoring), volatiles removed *in vacuo*, 1 eq. *Na-ethoxide* in *acetonitrile* added, the mixture heated under reflux for 3 h (TLC monitoring), the solvent removed *in vacuo*, and the residue purified by chromatography → 9-amino-2,3-dimethyl-8-nitro-7-phenyl-6*H*-benzo[*c*]chromene-10-carbonitrile. Y 64%. The reaction is successful with alkyl- and aryl-subst. nitroethylenes and alkylidenemalononitriles giving 61-79% yield (13 examples) of bi- and ter-phenyl products. In the absence of air or oxygen the reaction afforded complex mixtures of products. F.e.s. D. Xue, J. Li, Z.-T. Zhang, J.-G. Deng, J. Org. Chem. *2007*, *72* (14), 5443-5.

3-Nitropyridine s. under $Pd(OCOCF_3)_2$ ←
Copper(II) acetate s. under $Pd(OCOCF_3)_2$ and $PdCl_2$ *$Cu(OAc)_2$*
Copper(I) chloride s. under BuLi *CuCl*

Copper(I) bromide/oxygen *$CuBr/O_2$*

Tert. 2-nitroamines from tert. amines and aliphatic nitro compds. Catalytic aerobic cross-coupling H → $CHNO_2$

491. 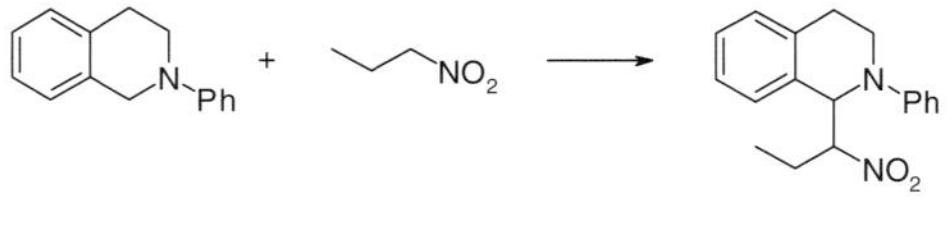

Oxidative alkylation of an sp^3 C-H bond adjacent to a nitrogen atom with CH-acidic compds., such as nitroalkanes, can be achieved aerobically under copper(I) catalysis. **E:** A mixture of 2-phenyltetrahydroquinoline (0.2 mmol), CuBr (5 mol%), distilled water (0.6 ml) and 1-nitropropane (0.4 mmol) stirred under O_2 at 60° for 16 h → product. Y 82%. Mechanistic studies, based on the amount of molecular oxygen consumed, indicate that the reaction proceeds via a Henry-type coupling between the nitroalkane and an initially-formed imine-type intermediate. F.e., **also with malonic acid diesters** in place of nitroalkanes, s. O. Baslé, C.-J. Li, Green Chem. *2007*, *9* (10), 1047-50.

Silver carbonate s. under $Pd(OAc)_2$ *Ag_2CO_3*

*Silver triflate or hexafluoroantimonate/2,6-di-*tert-*butyl-4-methylpyridine/air* ←
Pyrrolo[2,1-*b*]isoquinoline-4-carboxylic acid esters ○
from N-[*o*-(alk-1-ynyl)benzylidene]glycinates and acetylene derivs.
via 1,3-dipolar cycloaddition to isoquinolinium carbalkoxymethylids

492. 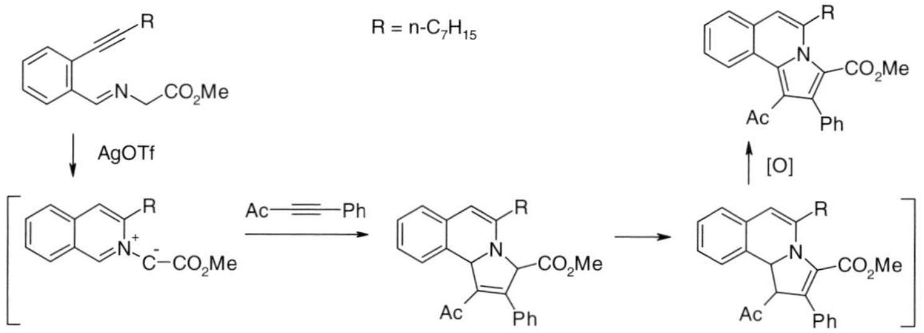

Startg. crude imine (prepared from the corresponding aldehyde), 1.2 eqs startg. acetyleneketone, 1 eq. 2,6-di-*tert*-butyl-4-methylpyridine, toluene and 10 mol% *AgOTf* added sequentially to a flask, the mixture heated to 60°, stirred for 6 h in the air, cooled to room. temp., filtered through silica gel, washed with ethyl acetate, the filtrates concentrated *in vacuo*, and the residue purified by chromatography → pyrroloisoquinoline deriv. Y 56% (regioselectivity 3:1). The conversion is catalyzed by metal salts (decomposition products being observed in the absence of catalyst), with AgOTf and $AgSbF_6$ giving the best results, but is limited to reactions of electron-poor acetylenes as dipolarophile (Y 28-83%). Highest yields were generally obtained with acetylenedicarboxylates, with non-symmetrical acetylenes giving lower yields with the predominant formation of *single* regioisomers. The reaction tolerates aryl, cyclopropane, propargyl ether, silyl and enyne functionality but reduced yields are obtained in the presence of an additional alkene group. Reaction is presumed to involve initial formation of an azomethinium ylid. F.e. incl. optimization s. S. Su, J.A. Porco, Jr, J. Am. Chem. Soc. *2007*, *129* (25), 7744-5.

Quinones s. under BuLi and $Pd(OAc)_2$ ←
Carboxylic acids s. under $Pd(OCOR)_2$ and $PdCl_2$ *RCOOH*

o-*Iodoxybenzoic acid* *$ArIO_2$*
N-Acyl-1,2,3,4-tetrahydroisoquinoline-1-carboxamides
by oxidative Ugi-type condensation

493. 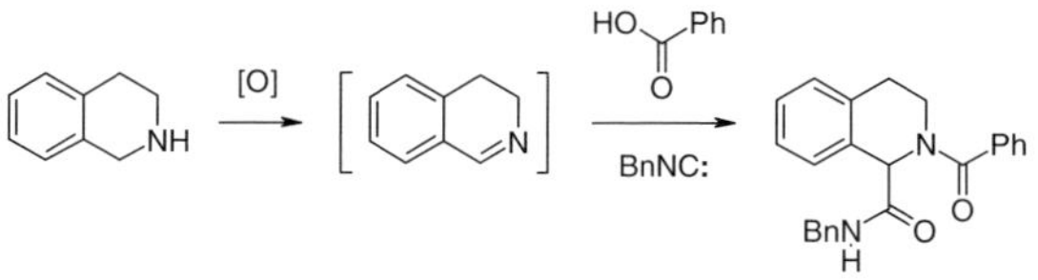

under mild conditions. 1.5 Eqs. benzoic acid, 1.5 eqs. 1,2,3,4-tetrahydroisoquinoline and 1 eq. benzyl isocyanide added successively to a suspension of 2 eqs. *o*-iodoxybenzoic acid in dry *THF* (0.5 *M*), heated at 60° for 6-72 h (TLC monitoring), cooled to room temp., hydrolyzed with 1 *N* HCl, stirred for a further 10 min, and worked up with chromatographic purification on silica gel → 2-benzoyl-N-benzyl-1,2,3,4-tetrahydroisoquinoline-1-carboxamide. Y 87%. The intermediate imine required for the Ugi condensation is generated *oxidatively* from the cyclic sec. amine, the acyclic sec. amine, dibenzylamine, reacting similarly but in low yield (32%). The carboxylic acid may be aliphatic, aromatic, or α,β-unsatd., and the isonitrile may be aliphatic, cycloalkyl, or benzylic. F.e. and in 19:1 DMSO/THF, also with N-Cbz-glycine and an α-isocyanocarboxamide, and application to the synthesis of β-carboline analogs, s. T. Ngouansavanh, J. Zhu, Angew. Chem. Int. Ed. Engl. *2007*, *46* (30), 5775-8.

Trichloroisocyanuric acid s. under KOH ←
Zirconocene dichloride s. under BuLi Cp_2ZrCl_2
Triethyl phosphite s. under $Rh_4(CO)_{12}$ $(EtO)_3P$
*Biphenyl-2,2′-ylene(*p*-tolyl)bismuth bis(triflate) s. under $LiN(SiMe_3)_2$* $\geqslant Bi(OTf)_2$
Oxygen s. under Et_3N, CuBr, AgOTf, $Pd(OAc)_2$ and PdI_2 O_2
Heteropolyacid s. under $Pd(OAc)_2$ ←
O-Benzyl-N-(9-anthracenylmethyl)dihydrocinchonidinium bromide s. under KOH ←

Bromo(pentacarbonyl)manganese $Mn(CO)_5Br$
Synthesis of *o*-subst. sec. benzyloxysilanes H → CH(OSi≤)R
from arenes, aldehydes and organosilicon hydrides via *o*-manganesation

494. 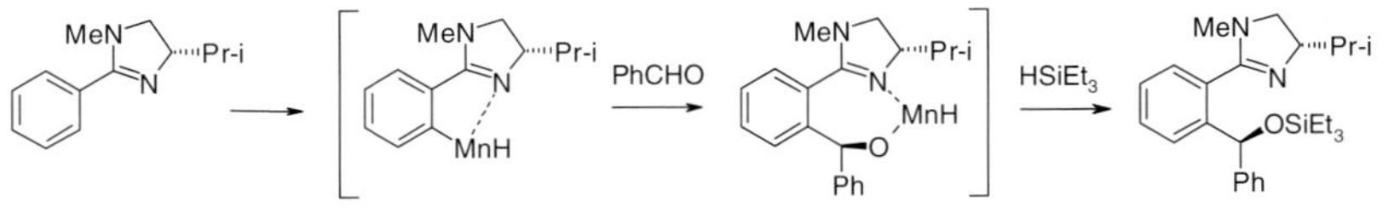

with asym. induction. A mixture of the startg. chiral imidazoline, 2 eqs. benzaldehyde, 2 eqs. triethylsilane and 5 mol% $Mn(CO)_5Br$ refluxed (110°) in toluene for 24 h, the mixture cooled, and the product isolated with chromatographic purification → 4-isopropyl-1-methyl-2-[phenyl-(triethylsiloxy)methyl]phenyl-4,5-dihydro-1*H*-imidazole. Y 80% (d.e. 95%). Replacing the isopropyl group with phenyl or benzyl in this example gave lower diastereoselectivity (60 and 30% d.e., respectively) as did the use of octanaldehyde in place of benzaldehyde (38% d.e.). A total of eleven chiral and achiral startg. ms. gave isolated yields of 48-87% using aromatic (electron-rich and -poor), heteroaromatic and aliphatic aldehydes. In the absence of triethylsilane the catalyst is not recycled and is required in stoichiometric amounts to afford the product. This ***o*-directed insertion of aldehydes into the carbon-hydrogen bond** is believed to be promoted by the polarity of the aryl-manganese bond formed by the initial oxidative addition of the metal. F.e.s. Y. Kuninobu, Y. Nishina, T. Takeuchi, K. Takai, Angew. Chem. Int. Ed. Engl. *2007*, *46* (34), 6518-20.

Dodecacarbonyltetrarhodium/triethyl phosphite $Rh_4(CO)_{12}/(EtO)_3P$
Maleimides from acetylene derivs. ○
via chelation-assisted rhodium-catalyzed biscarbonylation

495. 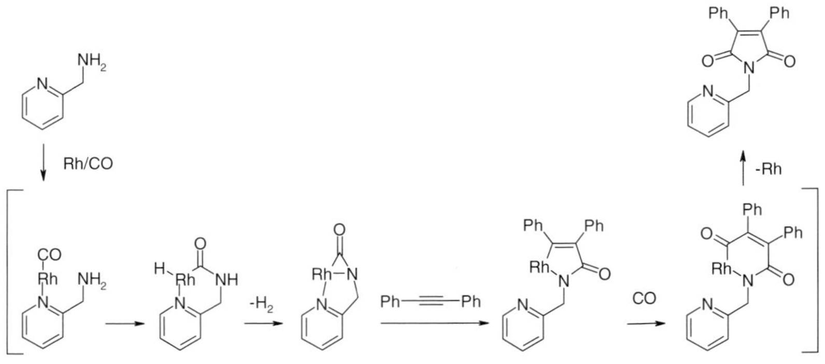

N-(2-Pyridylmethyl)maleimides. A mixture of 2 eqs. diphenylacetylene, 2-pyridylmethylamine, 2 mol% $Rh_4(CO)_{12}$ and 8 mol% triethyl phosphite in toluene heated in a stainless steel autoclave at 100° under 3 atm. CO for 20 h, cooled to room temp., the pressure released, the crude product washed out with ethyl acetate, the volatiles removed *in vacuo*, and the residue purified by chromatography → 1-(2-pyridylmethyl)-3,4-diphenylmaleimide. Y 52%. The proposed mechanism is tentative, but it does appear that initial coordination of rhodium to the pyridylmethylamine is critical to the process. Maleimide formation is achieved with internal alkynes at 100-120° (the catalyst decomposing at 130°) giving 42-52% yield (isolated yields), while the terminal alkyne, 1-octyne, gave only 18% yield. In preliminary experiments with 4-octyne, an α,β-unsatd. amide was also isolated (highest selectivity for maleimide formation being 6:1) and became the major product (20:3) with tris(*p*-trifluoromethylphenyl)phosphine as ligand and triphenyl phosphite as catalyst. Similar reactions with benzylamine afforded mixtures of succinimides and α,β-unsatd. amides. F.e.s. S. Inoue, Y. Fukumoto, N. Chatani, J. Org. Chem. *2007*, *72* (17), 6588-90.

*Palladium(II) acetate/silver carbonate/*p*-benzoquinone* ←
Palladium(II) acetate/heteropolyacid/acetic acid/oxygen ←
Palladium(II) trifluoroacetate/3-nitropyridine/copper(II) acetate/cesium pivalate/ pivalic acid/microwaves ←
Catalytic oxidative cross-coupling of unactivated arenes ←

496. 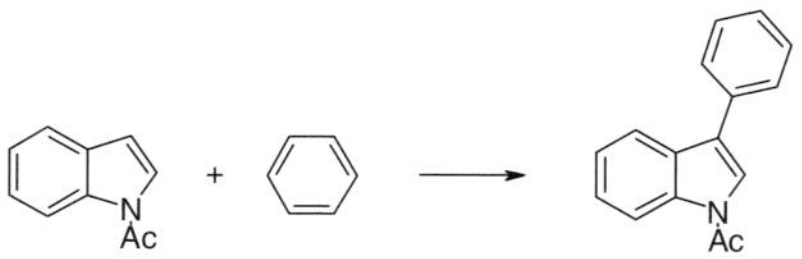

Regiospecific palladium-catalyzed 3-arylation of N-acylindoles. 10 Mol% $Pd(OCOCF_3)_2$, 10 mol% 3-nitropyridine, 3 eqs. $Cu(OAc)_2$ and 0.4 eq. Cs-pivalate weighed into a microwave tube, 0.295 *M* N-acetylindole in 4:1 benzene/pivalic acid added, the tube sealed, placed in a CEM Discover Microwave, subjected to microwave irradiation for 5 h at 140° (power setting 300 W; maximum pressure 250 psi), filtered over Celite, then washed with ethyl acetate, solvent removed, partitioned between ethyl acetate and satd. aq. NH_4Cl, stirred at room temp. for ca. 30 min, and worked up with column chromatographic purification on silica gel → 1-acetyl-3-phenylindole. Y 87%. Key to the procedure is the choice of a catalyst which metalates one arene molecule selectively with an associated *inversion of catalytic selectivity* for metalation of the other arene in the second phase of the catalytic cycle. In this way *homo-coupling is minimized.* The procedure is also waste-free [unlike classical arylations with arylmetallic compds. or aryl halides] and highly regioselective, although the N-acetyl group of the indole does appear critical (N-unsubst. indoles failing to react, while the N-methyl deriv. undergoes self-dimerization). The corresponding thermal conversion is much slower (48 h), and the additives (3-nitropyridine and cesium pivalate), although not critical, enhance the process by increasing the catalytic turnover. F.e.s. D.R. Stuart, K. Fagnou, *Science 2007*, *316* (5828), 1172-5; palladium-catalyzed *aerobic* arylation **of benzofurans** with 10 mol% $Pd(OAc)_2$ and $H_4PMo_{11}VO_{40}$ in acetic acid at 120° under 3 atm. O_2, also intramolecular 2-arylation of indoles (with added CuOAc in place of the heteropolyacid), s. T.A. Dwight, N.R. Rue, D. Charyk, R. Josselyn, B. DeBoef, Org. Lett. *2007*, *9* (16), 3137-9; regiospecific cross-coupling of 2-arylpyridine, 1-arylpyrazole and 2-arylpyrimidine with a range of arenes possessing electron-withdrawing or -donating groups with $Pd(OAc)_2/Ag_2CO_3/p$-benzoquinone s. K.L. Hull, M.S. Sandford, J. Am. Chem. Soc. *2007*, *129* (39), 11904-5.

Palladium(II) chloride/copper(II) acetate/acetic acid $PdCl_2/Cu(OAc)_2/AcOH$
Regiospecific oxidative vinylation H → C=C
of indoles using $Pd(OAc)_2/Cu(OAc)_2$, solvent-dependent regioselectivity, s. *69*, 369; *o*-vinylation of tert. benzylamines with electron-deficient ethylene derivs. using 5 mol% $PdCl_2$, 1 eq. $Cu(OAc)_2$ and 16 eqs. AcOH in trifluoroethanol at 85°, also 2-step or one-pot prepn. of 3-(*o*-tolyl)propanoic acid and its derivs. via hydrogenation over Pd-C, s. G. Cai, Y. Fu, Y. Li, X. Wan, Z. Shi, J. Am. Chem. Soc. *2007*, *129* (24), 7666-73.

Palladium(II) iodide/potassium iodide/oxygen *$PdI_2/KI/O_2$*

5-Carbamylmethylene-2-oxazolidones from terminal 2-acetyleneamines via palladium-catalyzed oxidative aminocarbonylation-cyclocarbonylation ○

497.

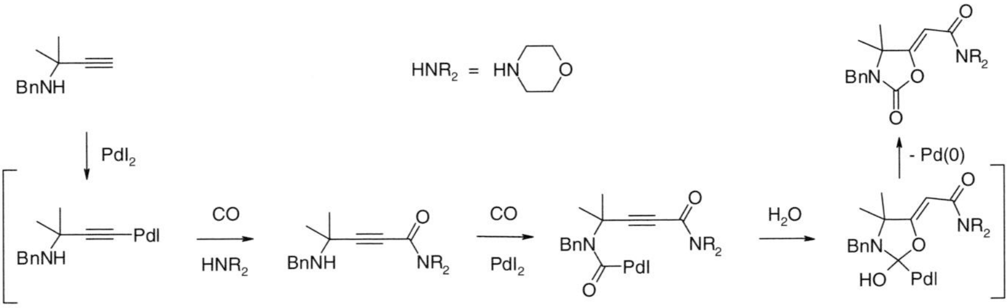

21.1 Mmol water added to a stainless-steel autoclave containing a soln. of 0.042 mmol PdI_2, 0.42 mmol KI, the startg. acetylene deriv. (4.2 mmol) and morpholine (21 mmol) in DME, the autoclave sealed, charged with CO (16 atm.) and air (up to 20 atm.) while stirring the mixture, heated at 100° with continued stirring for 5 h, cooled, and worked up with column chromatographic purification on neutral alumina → product. Y 94%. α-Monosubst. propargylamines gave the corresponding **Δ^4-2-oxazolone-5-acetic acid amides**. The procedure is applicable to both acyclic and [more rapidly] cyclic sec. amines. Reaction involves a dual catalytic cycle: in the first the propargylamine undergoes PdI_2-catalyzed oxidative aminocarbonylation, followed by nucleophilic addition of water, and entry into the second cycle for the cyclocarbonylation, this too being catalyzed by PdI_2. F.e.s. B. Gabriele, P. Plastina, G. Salerno, R. Mancuso, M. Costa, Org. Lett. *2007*, *9* (17), 3319-22.

Oxygen ↑ CC ↓↑ O

Without additional reagents *w.a.r.*

4-Iodotetrahydropyrans from 3-ethylenealcohols and ketones via Prins cyclization ○

4-Iodotetrahydro-2-spiropyrans – 4-Iodo-3,6-dihydro-2*H*-pyrans from 3-acetylenealcohols s. *72*, 294

3-Arylidene-5-sulfonyloxytetrahydropyrans

from 2-arylidenecyclopropylcarbinols, aldehydes and sulfonic acids – Regio- and stereo-specific Prins-type reaction s. *72*, 407

Stereospecific 4-component synthesis of 4-organothio-2-pyrrolidone-4-carboxylic acids

498.

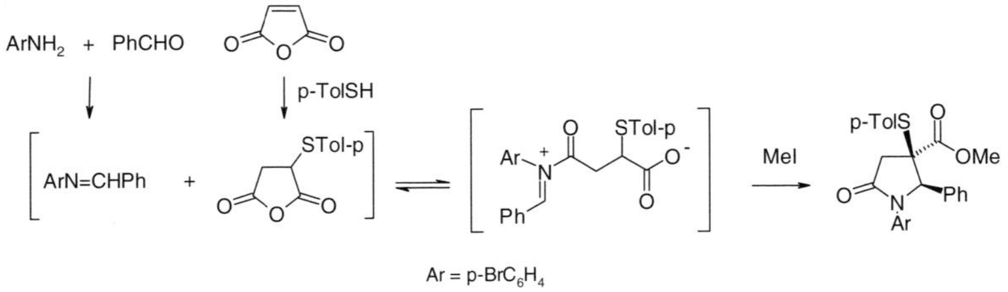

4-Bromoaniline added to a soln. of 1 eq. maleic anhydride, 1 eq. 4-mercaptotoluene and 1 eq. benzaldehyde in toluene, the mixture refluxed under argon in a Dean-Stark apparatus for 24 h, cooled, concentrated *in vacuo*, and the resulting acid isolated as the methyl ester by dissolving in acetone, treating with 4 eqs. anhydrous K_2CO_3 and methyl iodide at 0°, and stirring overnight at room temp. → pyrrolidone deriv. Y 88% (d.r. >95:5). The reaction is >95% diastereoselective for the 4,5-*syn* isomer and is versatile, accepting aliphatic and aromatic amines (incl. branched alkyl) and both aryl (incl. *o*-subst.) and alkyl thiols (the latter being less diastereoselective). The aldehyde component is limited to non-enolizable aldehydes due to enamide formation, but if the enolization may be suppressed the reaction is still successful (cyclopropanecarboxaldehyde giving a 64% yield). The use of benzaldehyde dimethyl acetal in place of the aldehyde, however, gave a reduced yield. The mechanism is thought to involve formation of an imine which reacts with *in situ*-generated 2-(organothio)succinic anhydride to give an adduct prior to ring closure. Ten examples gave 63-94% yields. F.e.s. J. Wei, J.T. Shaw, Org. Lett. *2007*, *9* (20), 4077-80.

3-Component synthesis of 3,5-isoxazolidione-4-acetic acid amides ○

499. 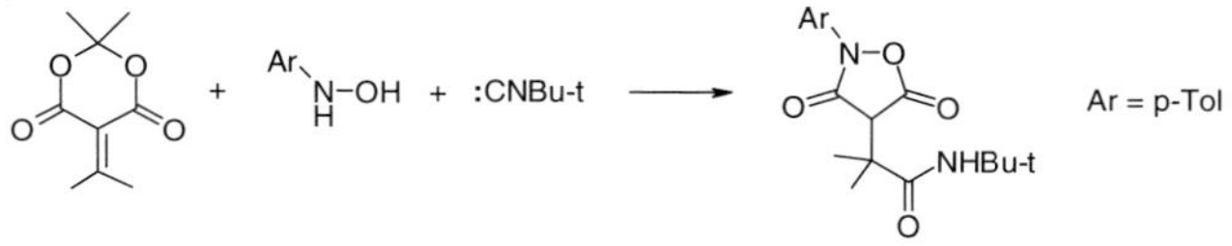

under mild, neutral conditions. A soln. of 1 eq. *tert*-butyl isocyanide in dry methylene chloride added at -5° over 10 min to a soln. of isopropylidene-Meldrum's acid and 1 eq. N-(*p*-tolyl)-hydroxylamine in the same solvent, allowed to warm to room temp., and stirred for a further 24 h → product. Y 91%. F.e.s. A. Habibi, L. Mousavifar, I. Yavari, M.R. Yazdanbakhsh, Monatsh. Chem. *2007*, *138* (6), 603-6.

4-Component synthesis of 1,2,3,6-tetrahydropyrimidine-4,5-dicarboxylic acid esters

500. 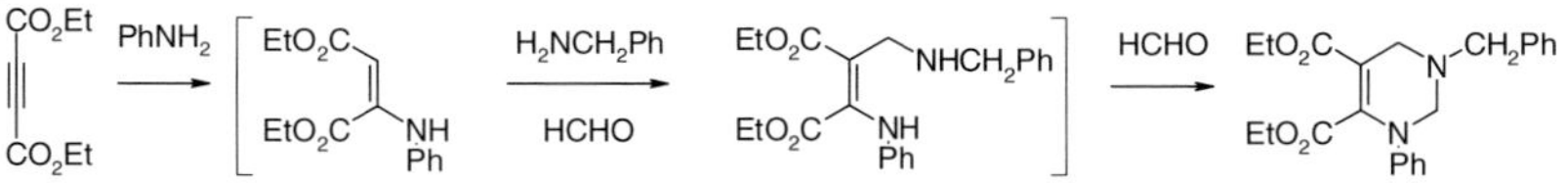

1 Eq. aniline, 1.1 eqs. benzylamine and 4 eqs. 35% formaldehyde added dropwise in this order to diethyl acetylenedicarboxylate in DMF, the mixture stirred at 100° for 3 h, cooled to room temp., diluted with water, and extracted with ether → diethyl 1-benzyl-3-phenyl-1,2,3,6-tetrahydro-pyrimidine-4,5-dicarboxylate. Y 93%. The method is rapid, simple in work-up, free of toxic by-products, and applicable to both aliphatic and aromatic prim. amines (irrespective of ring substituents). The reaction involves hydroamination of the alkyne moiety with the first amine, Mannich-type reaction with the second amine to afford the intermediate ene-1,3-diamine, and cyclodehydration with a second formaldehyde molecule. Interestingly, reversing the order of amine addition afforded the isomeric 1-phenyl-3-benzyl deriv. in this case in 86% yield. F.e.s. M. Zhang, H. Jiang, H. Liu, Q. Zhu, Org. Lett. *2007*, *9* (21), 4111-3.

Irradiation/triethylamine *⫽⫽/Et_3N*
Metal-free synthesis of crowded biaryls from arenes Ar-Ar′
and aryl sulfonates or phosphates s. *72*, 546

Microwaves [s.a. under $Cu(SO_4)_2$, CuCl, $AgNO_3$, $Zn(CN)_2$, ZnI_2 and $HClO_4$] [\\\\]
Knoevenagel condensation CO → C=C
under microwave irradiation in the absence of solvent – 5-Arylidenebarbituric acids s. *46*, 713s*72*

Biginelli synthesis ○
in water under microwave irradiation – 3,4-Dihydropyrimido[4,5-*d*]pyrimidine-2,4,7-triones s. *55*, 337s*72*

Potassium tert-*butoxide* *KOBu*-t

4,4-Disubst. cyclohexanone-2-carboxylic acid esters
from two α,β-ethylenecarboxylic acid ester molecules
Double Michael addition-Dieckmann condensation

501. 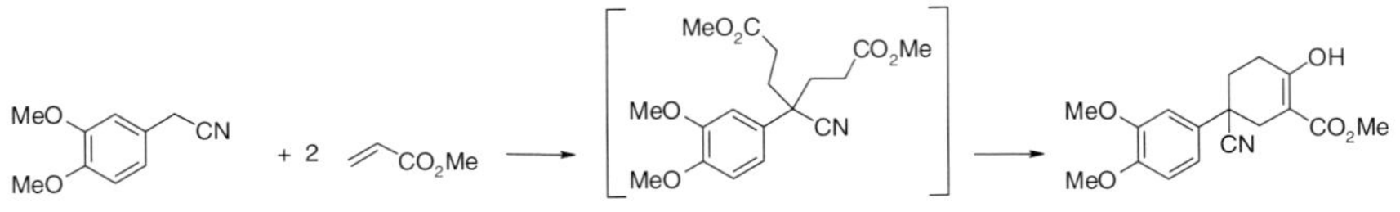

4-Aryl-4-cyanocyclohexanone-2-carboxylic acid esters. Solid KOBu-*t* (3 eqs.) added to a soln. of startg. arylacetonitrile (1 eq.) and methyl acrylate (2 eqs.) in THF, and stirred for 1 h at 25° → product. Y 90%. The procedure is applicable to a wide variety of substituted arylacetonitriles, including electron-rich, electron-deficient and sterically-hindered substrates, although both 3- and 4-nitrophenyl derivs. are unsuitable. Unlike previously-reported two-step methods, this **one-pot procedure** is safe, efficient and scalable. F.e., **also 4-arylcyclohexanone-2,4-dicarboxylic acid esters** with methyl arylacetates, and with methyl crotonate as the Michael acceptor, s. M.R. DeGraffenreid, S. Bennett, S. Caille, F. Gonzalez-Lopez de Turiso, R.W. Hungate, L.D. Julian, J.A. Kaizerman, D.L. McMinn, Y. Rew, D. Sun, X. Yan, J.P. Powers, J. Org. Chem. *2007*, *72* (19), 7455-8.

n-*Butyllithium s.a. under TiCl(OPr*-i*)$_3$* *BuLi*

n-*Butyllithium/morpholine/boron fluoride* *BuLi/O(CH$_2$CH$_2$)$_2$NH/BF$_3$*

Aryl α,β-acetyleneketones COOR → C(O)C≡C
from arylcarboxylic acid esters and terminal acetylene derivs.

502. 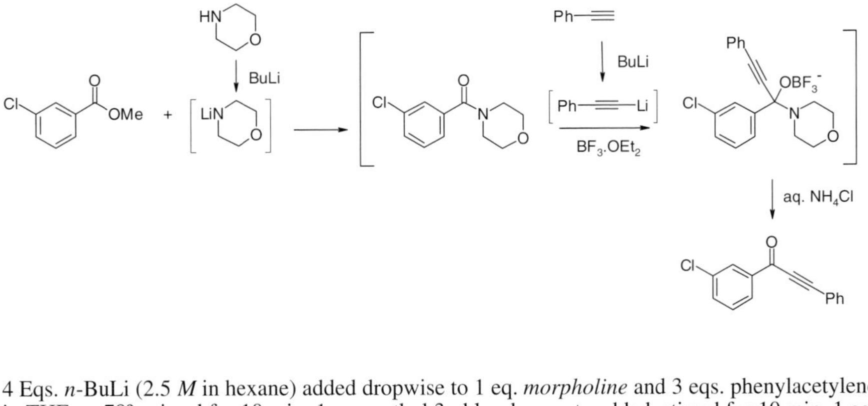

4 Eqs. *n*-BuLi (2.5 *M* in hexane) added dropwise to 1 eq. *morpholine* and 3 eqs. phenylacetylene in THF at -78°, stirred for 10 min, 1 eq. methyl 3-chlorobenzoate added, stirred for 10 min, 1 eq. $BF_3 \cdot OEt_2$ added, stirred for a further 30 min, quenched with satd. aq. NH_4Cl, extracted with ether, the organic layer dried over $MgSO_4$, concentrated *in vacuo*, and the residue purified by chromatography on silica gel → 3-chlorophenyl phenylethynyl ketone. Y 78%. Ten examples gave 78-87% yield using methyl, isopropyl, ethyl and *tert*-butyl esters in the presence of ether and halo functionality. From initial experiments, using ethyl benzoate in the absence of the Lewis acid, N-benzoylmorpholine was obtained in 91% yield. F.e.s. S.J. Yim, C.H. Kwon, D.K. Am, Tetrahedron Lett. *2007*, *48* (31), 5393-5.

n-Butyllithium/trimethylsilyl triflate/triethylamine *BuLi/Me$_3$SiOTf/Et$_3$N*
3-Alkoxy-4-hydroxypyridines from alkoxyallenes, nitriles and carboxylic acids ○
One-pot 3-component synthesis

503.

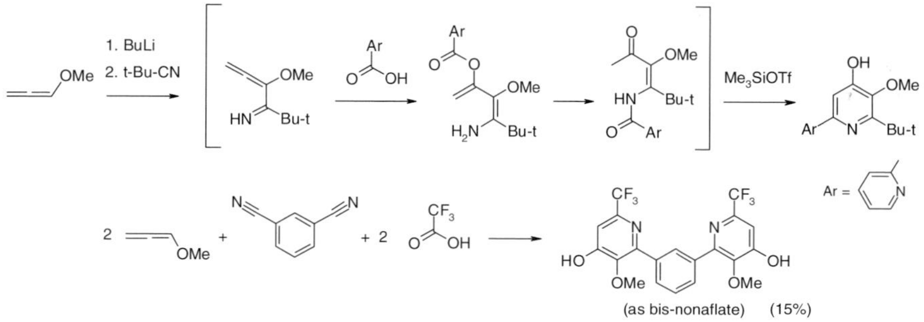

2.75 Eqs. *n*-butyllithium stirred with 3.5 eqs. methoxyallene in ether at -50 to -40° for 25 min, the soln. cooled to -78°, 1 eq. *tert*-butyl cyanide added, the mixture stirred for 4 h, 6.5 eqs. picolinic acid added, warmed to room temp. overnight, quenched with satd. aq. $NaHCO_3$, extracted with methylene chloride, the combined organic phases dried over Na_2SO_4, evaporated to dryness, the residue dissolved in methylene chloride, cooled to 0°, 3 eqs. triethylamine and 3 eqs. trimethylsilyl triflate added, the mixture refluxed for 3 days, quenched at room temp. with satd. aq. NH_4Cl, extracted with methylene chloride, dried over Na_2SO_4, the solvents removed *in vacuo*, the residue dissolved in methylene chloride, treated with acetic acid and water, the organic layer dried over Na_2SO_4, concentrated *in vacuo*, and the residue purified by chromatography → 2-*tert*-butyl-4-hydroxy-3-methoxy-6-(2-pyridyl)pyridine. Y 73%. Yields obtained are only fair to good (17-73%) but the method allows a convenient one-pot synthesis of a diverse range of **2,3,6-trisubst. 4-hydroxypyridines** with only one purification step by variation in the allenyl ether (methyl, benzyl, trimethylsilylethyl and 3-methoxybenzyl), carboxylic acid (acetic, benzoic and pivalic acids) and nitrile (alkyl, aryl and hetaryl) components. The use of dicyano-benzenes and -biphenyls allowed synthesis of some novel **dipyridyl derivs.** F.e. incl. coupling reactions of the products s. J. Dash, T. Lechel, H.-U. Reissig, Org. Lett. *2007*, *9* (26), 5541-4.

Sodium acetate s. under 2(S)-[Diphenyl(triethylsiloxy)methyl]pyrrolidine *NaOAc*
Sodium iodide s. under Me$_3$SiCl *NaI*
Chiral 2-aminoalcohols s. under Me$_2$Zn ←

2(S)-[Diphenyl(trimethylsiloxy)methyl]pyrrolidine ←
Isoxazolidine-4-carboxaldehydes
from α,β-ethylenealdehydes, hydroxylamines and aldehydes
Organocatalyzed asym. 3-component reaction

504.

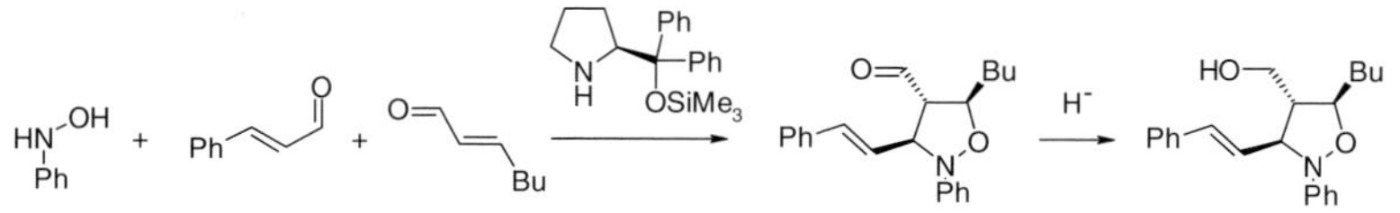

and reduction to chiral 4-(hydroxymethyl)isoxazolidines. 2 Eqs. phenylhydroxylamine added to 2 eqs. cinnamaldehyde in chloroform, the mixture stirred at room temp. for 1 h, 20 mol% 2(S)-[diphenyl(trimethylsiloxy)methyl]pyrrolidine and 1 eq. 2-heptenal added, the mixture stirred for 16 h, the crude product reduced *in situ* with $NaBH_4$ in methanol at 0° for 5 min, quenched with 1 *M* HCl, extracted with ethyl acetate, the extracts dried over Na_2SO_4, concentrated *in vacuo*, and the residue purified by chromatography on silica gel → (3R,4R,5R)-2-phenyl-3-styryl-4-

hydroxymethyl-5-*n*-butylisoxazolidine. Y 73% (d.r. >25:1; e.e. 99%). *in situ*-Reduction of the initially formed 4-formylisoxazolidine was required to eliminate epimerization on silica gel chromatography. Eleven examples using phenylhydroxylamine or its 4-chloro deriv. with ar., alkyl and styryl aldehydes as the variable aldehyde component gave 58-74% yield with d.r. >25:1 in each case and e.e. of 91-9%. Ester groups were tolerated on the enal component. Comparison with the literature allowed assignment of absolute stereochemistry in one case in keeping with efficient shielding of the *si*-face of the intermediate chiral iminium ion leading to stereoselective *re*-face addition. F.e.s. R. Rios, I. Ibrahem, J. Vesely, G.-L. Zhao, A. Cordova, Tetrahedron Lett. *2007*, *48* (32), 5701-5.

2(S)-[Diphenyl(triethylsiloxy)methyl]pyrrolidine/sodium acetate ←

Organocatalyzed asym. Michael addition-intramolecular aldol condensation O

505. 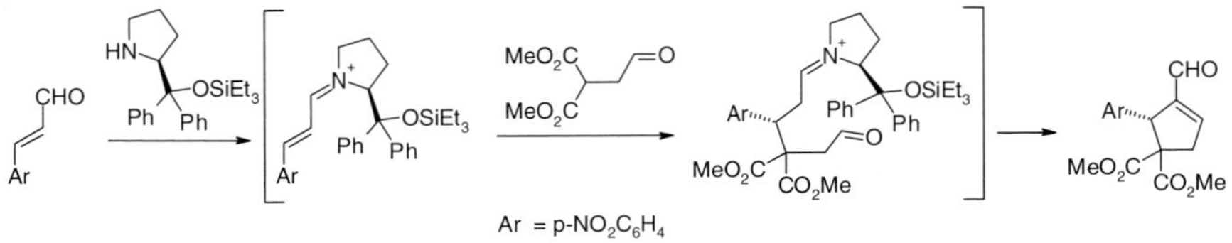

Chiral 4,4-di(carbalkoxy)cyclopent-1-enecarboxaldehydes. A mixture of 1.2 eqs. *p*-nitrocinnamaldehyde, the startg. dicarboxylic acid ester, 0.5 eq. NaOAc and 10 mol% 2(S)-[diphenyl-(triethylsiloxy)methyl]pyrrolidine in *dichloroethane* stirred at room temp. for 12 h, and worked up with column chromatographic purification on silica gel → product. Y 88% (e.e. 92%). Two new C-C bonds are created with high enantioselectivity (91-7%) from readily available substrates **under mild conditions,** reaction being facile with cinnamaldehydes and heterocyclic analogs irrespective of the electronic nature of ring substituents. Alternative condensation manifolds are minimized by the choice of a bulky organocatalyst and a bulky nucleophile. The intermediate formation of iminium ions from the enal are presumed to be central to the conversion. F.e. and comparison of organocatalyst and added bases s. J. Wang, H. Li, H. Xie, L. Zu, X. Shen, W. Wang, Angew. Chem. Int. Ed. Engl. *2007*, *46* (46), 9050-3; **chiral 5-nitrocyclohex-1-enecarboxaldehydes** from enals and γ-nitroketones with 2(S)-[diphenyl(trimethylsiloxy)methyl]pyrrolidine s. D. Enders, A.A. Narine, T.R. Benninghaus, G. Raabe, Synlett *2007* (11), 1667-70.

(S)-2-(5,6-Dimethylbenzimidazol-2-yl)pyrrolidine ←

3-Hydroxy-2(5*H*)-furanones [isotetronic acids]
from α-ketocarboxylic acids and aldehydes
via organocatalyzed asym. aldol condensation

506. 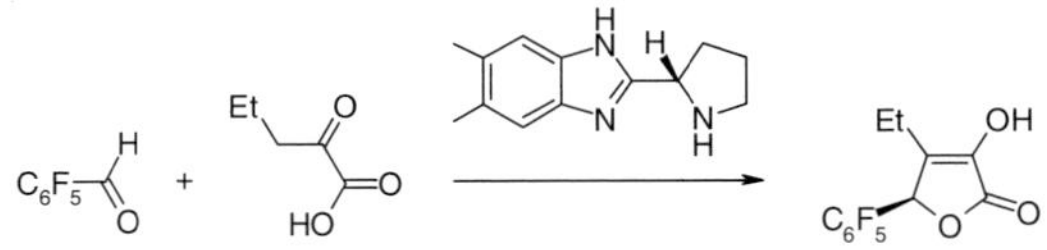

Equivalent amounts of the startg. α-ketocarboxylic acid and aldehyde in THF stirred at room temp. for 5 days with 10 mol% (S)-2-(5,6-dimethylbenzimidazol-2-yl)pyrrolidine, and worked up with column chromatographic purification on silica gel → product. Y 71% (e.e. 90%). The procedure is applicable to a variety of aliphatic or aromatic aldehydes, reaction being facilitated by electron-withdrawing groups in the aromatic ring. There was no reaction, however, with hindered pivaldehyde, nor with pyruvic acid as the keto-acid component, and both yield and enantioselectivity were low with L-proline as organocatalyst. Reaction may also be undertaken **in water** for condensations with viscous or liquid aldehydes, and enantioselectivity is generally higher. F.e.s. J.-M. Vincent, C. Margottin, M. Berlande, D. Cavagnat, T. Buffeteau, Y. Landais, Chem. Commun. *2007* (48), 4782-4.

Morpholine s. under BuLi *$O(CH_2CH_2)_2NH$*

Copper nanoparticles *Cu*

3-Component synthesis of 2-acetyleneamines CO → C(N<)C≡C

updates s. *66*, 353s*69-71*; with 15 mol% Cu-nanoparticles in acetonitrile at 100-10° s. M. Kidwai, V. Bansal, N.K. Mishra, A. Kumar, S. Mozumdar, Synlett *2007* (10), 1581-4; under microwave irradiation with CuCl in a minimal amount of DMF s. D. Khrustalev, G. Khamzina, S. Fazylov, A. Gazaliev, Russ. J. Gen. Chem. *2007*, *77* (5), 970; with 5 mol% $AgNO_3$ in the absence of solvent s. T. He, Z. Zha, C. Pan, Z. Wang, Synth. Commun. *2007*, *37* (5), 849-58.

Copper(II) nitrate *$Cu(NO_3)_2$*

Sym. 1,1-bis(indol-3-yl)alkanes from aldehydes s. *5*, 549s*72* CHO → $CHAr_2$

Copper(II) sulfate/camphorsulfonic acid/microwaves *$CuSO_4/RSO_3H$/[\\\\]*

3-Acylpyrrolidines from 2-ethylene-2′-aminoalcohols and aldehydes via stereospecific aza-Cope rearrangement-Mannich reaction ○

507. 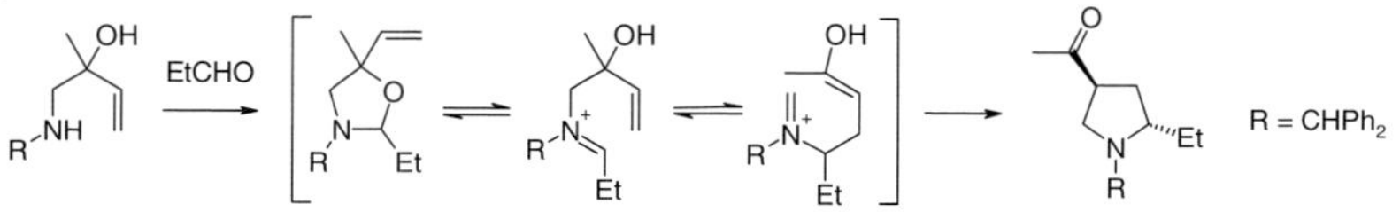

One-pot conversion under microwave enhancement. A mixture of the startg. aminoalcohol, 4 eqs. propionaldehyde, 2 eqs. $CuSO_4$ and 1 eq. camphorsulfonic acid in acetonitrile sealed in a vial, the mixture heated at 60° under microwave irradiation (200 W) for 2.5 h, cooled, diluted with ethyl acetate, the organic layer washed with aq. satd. $NaHCO_3$, dried over $MgSO_4$, concentrated *in vacuo*, and the residue purified by chromatography → *trans*-4-acetyl-1-benzhydryl-2-ethylpyrrolidine. Y 82% (*trans/cis* 8:1). Diastereoselectivity is favoured by the use of a bulky N-protecting group and lower reaction temp. (attributed to a decrease in the epimerization rate), with the use of microwave irradiation reducing the reaction time significantly compared with conventional heating (cyclization of the illustrated aminoalcohol with cinnamaldehyde under these conditions, however, reached only 70% conversion after irradiation for 6 h). Nine examples using propionaldehyde, cinnamaldehyde or ethyl glyoxylate as aldehyde component gave yields of 63-84% and *trans*-selectivities of 1:1 to 3:1 with N-benzyl protection, and 2:1 to 8:1 with N-benzhydryl protection, *without the need to isolate the intermediate oxazolidine*. F.e.s. B.F. Johnson, E.L. Marrero, W.A. Turley, H.A. Lindsay, Synlett *2007* (6), 893-6.

Copper(I) chloride/microwaves *CuCl/[\\\\]*

Silver nitrate/microwaves *$AgNO_3$/[\\\\]*

3-Component synthesis of 2-acetyleneamines CO → C(N<)C≡C

under microwave irradiation s. *66*, 353s*72*

Copper(I) iodide s. under $Pd(OAc)_2$ *CuI*

Silver triflate/proline *AgOTf/Pro-OH*

1-β-Keto-1,2-dihydroisoquinolines from *o*-acetylenealdehydes, ketones and amines ○

Regiospecific 3-component synthesis under dual organo- and Lewis acid-catalysis

508. 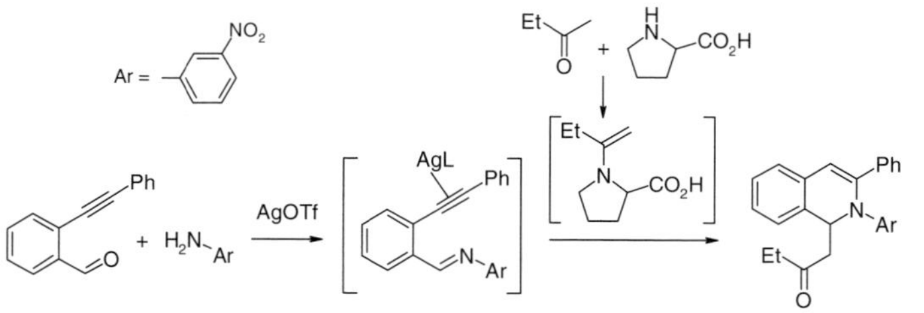

under mild conditions. A soln. of 2-(phenylethynyl)benzaldehyde, 1 eq. 3-nitroaniline, 5 eqs. 2-butanone, 10 mol% AgOTf and 10 mol% proline stirred at 50-60° under N_2 in ethanol for 2 h

(TLC), the solvent evaporated, the residue dissolved in water, extracted with ethyl acetate, dried over Na_2SO_4, the solvent removed *in vacuo*, and the residue purified by chromatography → 1-[2-(3-nitrophenyl)-3-phenyl-1,2-dihydroisoquinolin-1-yl]butan-2-one. Y 78% (single regio-isomer). Significantly, proline activates the ketone by intermediate formation of an enamine, which then undergoes cyclization with the *in situ*-generated *o*-acetylenealdimine via activation of the alkyne residue by the Lewis acid. Reaction is successful and usually complete in 2 h with cyclic and acyclic dialkyl or aralkyl ketones, giving a single isomer as sole or major product with subst. anilines (electronically diverse) as the amine component and phenyl or 4-methoxyphenyl terminated alkynes (Y 50-95%). Low yields were obtained, however, where the alkyne terminus was cyclopropyl (30%, 35%) or *n*-butyl (28%), and using benzylamine as amine component (30%), while a trimethylsilyl terminated alkyne suffered desilylation only in 12% yield. F.e. incl. reaction optimization s. Q. Ding, J. Wu, Org. Lett. *2007*, *9* (24), 4959-2.

Chloro(1,3-diarylimidazol-2-ylidene)gold(I) complexes/silver triflate *(NHC)AuCl/AgOTf*
Quinolines from *o*-aminoketones and terminal acetylene derivs. ○
Improved procedure with gold(I) catalysis under mild conditions

509. 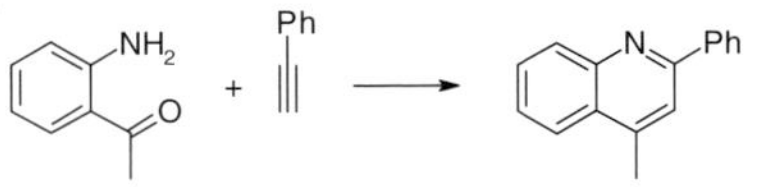

Gram-scale synthesis. 2-Aminoacetophenone (2.7 g) and 2 eqs. phenylacetylene added with stirring to a mixture of 3 mol% chloro[1,3-bis(2,6-diisopropylphenyl)imidazol-2-ylidene]gold(I) complex, 3 mol% AgOTf and 4 Å molecular sieves in toluene, the reaction vessel capped, stirred for 36 h at 100°, and worked up with flash chromatographic purification on silica gel → product. Y 93%. Yields are lower and/or reaction times significantly longer under harsher conditions by existing routes [e.g. with $Ru_3(CO)_{12}$ reaction requires heating at 150° for 12 h]. F.e., also under microwave irradiation, and 1,2-dihydroquinolines from anilines and two acetylene deriv. molecules, s. X.-Y. Liu, P. Ding, J.-S. Huang, C.-M. Che, Org. Lett. *2007*, *9* (14), 2645-8.

(Triphenylphosphine)gold(I) chloride/silver triflate *(Ph_3P)AuCl/AgOTf*
Friedel-Crafts alkylation with *o*-(alk-1-ynyl)benzoic acid esters H → R
s. *72*, 126

Sodium tetrachloroaurate *$NaAuCl_4$*
Gold-catalyzed 3-component synthesis of 1-aminoindolizines ○
from pyridine-2-carboxaldehydes

510. 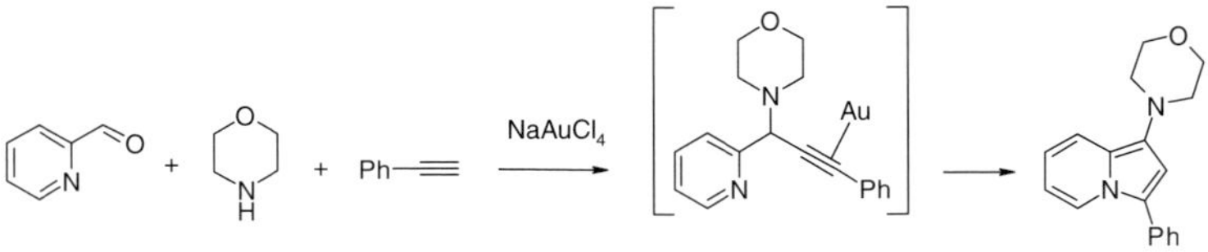

One-pot conversion in the absence of solvent. Pyridine-2-carboxaldehyde, 1.2 eqs. phenyl-acetylene, 1 mol% $NaAuCl_4 \cdot 2H_2O$ and 1.1 eqs. morpholine added successively to a Schlenk tube at room temp., the mixture stirred at 60° for 1-4 h until complete (TLC monitoring), and the crude product purified by flash chromatography → 3-phenyl-1-(morpholin-4-yl)indolizine. Y 96%. Reaction is high-yielding for the condensation of a series of pyridine- and quinoline-2-carboxaldehydes with mono-alkyl-subst. or electron-rich/poor mono-aryl-subst. acetylenes and sec. amines (prim. amines giving complex mixtures). Low yields (<32%) were obtained with arylamines but chiral *sec*-amino acids may be incorporated in high yield without loss of enantio-meric purity. Ester, ether and thioether functionality is well tolerated and the reaction may also be carried out **in water** at 60° but yields are generally lower. F.e.s. B. Yan, Y. Liu, Org. Lett. *2007*, *9* (21), 4323-6.

Magnesium/titanium tetraisopropoxide *Mg/Ti(OPr-i)$_4$*
Sec. perfluoroalkylcarbinols COOR → CH(OH)R$_f$
from carboxylic acid esters and perfluoroalkylmagnesium halides
Stable perfluoroalkyltitanate complexes as intermediates

511. 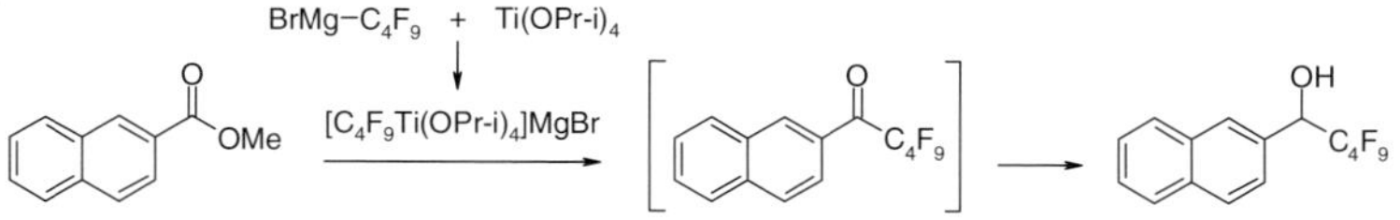

under mild conditions. 3.6 Eqs. phenylmagnesium bromide added to a soln. of 5.8 eqs. perfluorobutyl iodide in ether at -78°, the mixture stirred for 1 h, 3.6 eqs. Ti(IV)-isopropoxide added at -78°, the mixture stirred at room temp. for 15 min, 1 eq. methyl 2-naphthoate added, stirring continued for 2 h, the reaction quenched with 1 *M* HCl, extracted with ether, the combined organic layers washed with brine, dried over Na_2SO_4, the solvent removed *in vacuo*, and the residue purified by chromatography → 1-(2-naphthyl)-2,2,3,3,4,4,5,5,5-nonafluoropentan-1-ol. Y 99%. Methyl and phenyl benzoates and cinnamates react efficiently via stable perfluorotitanate complexes in high yield (72-100%), while aliphatic esters required 6 eqs. of the reagent to afford 41 and 51% yield. The reaction involves formation of an intermediate perfluoroalkyl ketone which suffers **Meerwein-Ponndorf-Verley-type** reduction to the sec. alcohol with an excess of Grignard reagent. F.e.s. K. Mikami, T. Murase, Y. Itoh, J. Am. Chem. Soc. *2007*, *129* (38), 11686-7.

Cyclopentylmagnesium chloride s. under TiCl(OPr-i)$_3$ *c-C$_5$H$_9$MgCl*

Dimethylzinc/chiral 2-aminoalcohols ←
Asym. 3-component synthesis of 2-acetyleneamines CO → C(N<)C≡C
with CuOTf/chiral bis(Δ^2-oxazolines) cf. *63*, 356; with Me_2Zn/(1R,2S)-2-[bis(4-methoxybenzyl)-amino]-1-phenylpropan-1-ol s. L. Zani, T. Eichhorn, C. Bolm, Chem. Eur. J. *2007*, *13* (9), 2587-600.

Bis(iodozincio)methane *$CH_2(ZnI)_2$*
Cyclopropanols from α-sulfonyloxyketones ▽

512. 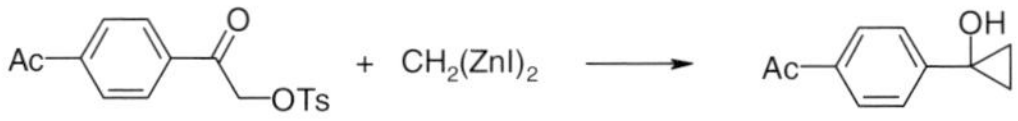

3 Eqs. bis(iodozincio)methane added to a soln. of the startg. α-tosyloxyketone in dry THF under argon at room temp., stirred for 2 h, and poured into satd. aq. NH_4Cl → product. Y 92%. F.e., also trapping the intermediate zinc cyclopropoxide by copper-catalyzed allylation to give the corresponding **δ,ε-ethyleneketones,** s. K. Nomura, S. Matsubara, Chem. Lett. *2007*, *36* (1), 164-5.

Zinc cyanide/palladium(II) acetate/polymer-based triphenylphosphine/microwaves ←
Ar. nitriles from aryl triflates OTf → CN
under microwave irradiation with $Pd(PPh_3)_4$ cf. *45*, 447s*65*; with $Pd(OAc)_2$/polymer-based triphenylphosphine s. R.R. Srivastava, A.J. Zych, D.M. Jenkins, H.-J. Wang, Z.-J. Chen, D.J. Fairfax, Synth. Commun. *2007*, *37* (3), 431-8.

Magnesium sulfate *$MgSO_4$*
Biginelli synthesis ○
cf. *55*, 337s*71*; 3,4-dihydro-2(1*H*)-pyrimi-dinones or -dinethiones with anhydrous $MgSO_4$ at room temp. in the absence of solvent s. D.M. Pore, U.V. Desai, T.S. Thopate, P.P. Wadgaonkar, Aust. J. Chem. *2007*, *60* (6), 435-8; with Dowex 50W in water at 70° s. C. Mukhopadhyay, A. Datta, B.K. Banik, Heterocycles *2007*, *71* (1), 181-8; with propanephosphonic acid anhydride in ethyl acetate at reflux s. F.L. Zumpe, M. Flüß, K. Schmitz. A. Lender, Tetrahedron Lett. *2007*, *48* (8), 1421-3; under microwave irradiation in the absence of solvent using a little perchloric acid, 4-aryl-derivs., s. C. Mukhopadhyay, A. Datta, B.K. Banik, J. Heterocycl. Chem. *2007*, *44* (4), 979-81; using ZnI_2 s. B. Liang, X. Wang, J.-X. Wang, Z. Du, Tetrahedron *2007*, *63* (9), 1981-6; 3,4-dihydropyrimido[4,5-*d*]pyrimidine-2,4,7-triones from barbituric acid in water under microwave irradiation s. M. Kidwai, K. Singhal, S. Kukreja, Z. Naturforsch. *2007*, *62B* (5), 732-6.

Zinc triflate *Zn(OTf)$_2$*
2-Vinylmaleimides from α-allenecarboxylic acid esters and isonitriles ○

513.

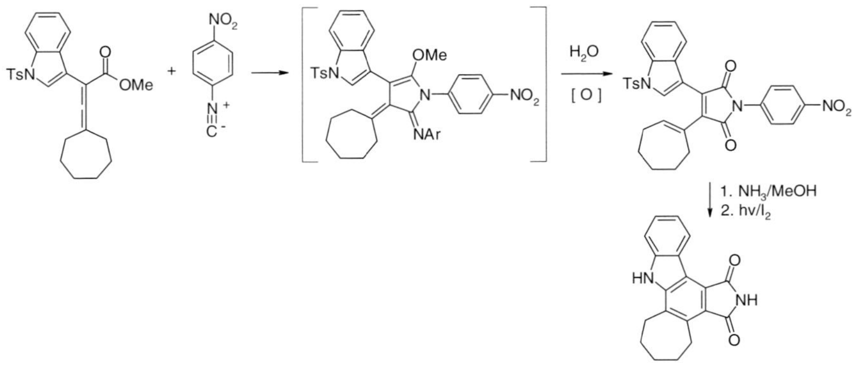

p-Nitrophenyl isocyanide and 3 mol% $Zn(OTf)_2$ added to a soln. of 1.2 eqs. startg. allenic ester in THF/water (ca. 20:1), stirred at 50° for 24 h, and evaporated → product. Y 58%. The formation of maleimides appears to be general, occurring with both electron-poor and -rich isonitriles in 54-85% yield. It is thought to involve initial attack of the central allene carbon by the nucleophilic isonitrile, followed by ring closure to give the corresponding 5-alkoxy-2-imino-3-alkylidene-2,3-dihydrofuran; oxidation then furnishes an intermediate hydroperoxide which fragments to the product. By using *p*-nitrophenyl isocyanide, an entry into the **carbazole ring** may be effected by removal of the N-protective group (with satd. methanolic ammonia), followed by classical photoinduced dehydrogenative cycloaromatization. F.e. and preparation of the startg. allenic esters s. Y. Li, H. Zou, J. Gong, J. Xiang, T. Luo, J. Quan, G. Wang, Z. Yang, Org. Lett. *2007*, *9* (20), 4057-60.

Magnesium bromide-etherate/ethyldiisopropylamine *$MgBr_2 \cdot OEt_2$/i-Pr_2NEt*
β-Diketones from ketones and carboxylic acid pentafluorophenyl esters H → COR
s. *72*, 540

Zinc chloride/silicon tetrachloride *$ZnCl_2$/$SiCl_4$*
β-Acylaminoketones from ketones, aldehydes and nitriles H → CH(NHCOR)R′
One-pot 3-component reaction via Lewis acid-catalyzed aldol-type condensation

514.

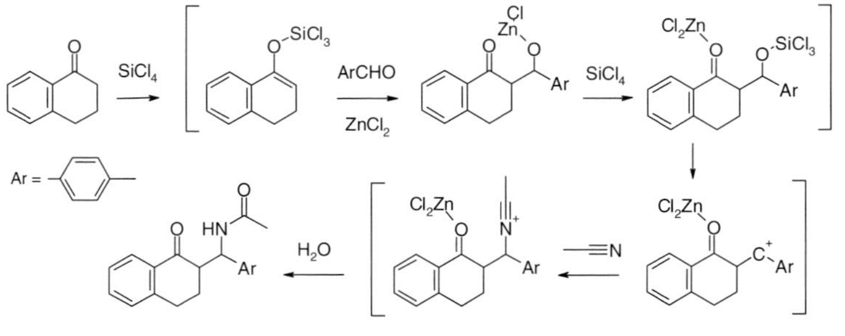

under mild conditions. Equimolar 4-methylbenzaldehyde, 1-tetralone and acetonitrile added to 2 eqs. anhydrous $ZnCl_2$ in methylene chloride, the mixture stirred at room temp. for 10 min, 4 eqs. $SiCl_4$ added, stirring continued with exclusion of moisture for 13 h (TLC), the mixture added to water, extracted with methylene chloride, the extracts dried over $MgSO_4$, evaporated *in vacuo*, and the residue purified by chromatography on silica → N-[α-(1-tetralon-2-yl)-4-methylbenzyl]-acetamide. Y 65%. Twenty-nine examples were prepared by reaction of ar. aldehydes (and 3-formylchromone) with acetophenones, 1-tetralone or 1-benzosuberone as ketone component, and phenylacetonitrile, acetonitrile, benzonitrile or ethyl cyanoacetate as the nitrile component giving 60-87% yield in the presence of halo, ether and ester functionality. The reaction may accommodate sterically-hindered aldehydes or ketones but if both are hindered the reaction affords enones only. α-Subst. enolizable ketones gave a single diastereomer but the stereochemistry was not established. F.e.s. T.A. Salama, S.S. Elmorsy, A.-G.M. Khalil, M.A. Ismail, Tetrahedron Lett. *2007*, *48* (35), 6199-203.

Zinc iodide/microwaves *ZnI_2/[\\\\]*
Biginelli synthesis ○
under microwave irradiation in the absence of solvent s. *55*, 337s*72*

Zeolite H-USY ←
1-Indanones from arenes and α,β-ethylenecarboxylic acids

515.
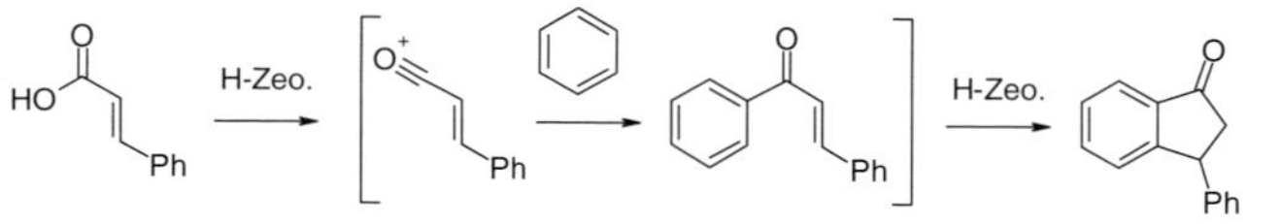

Cinnamic acid and 15 eqs. activated zeolite H-USY heated and magnetically stirred in benzene at 130° for 20 h in a sealed tube, cooled, heated and stirred with methanol at 80° for 5 h, cooled, filtered, and concentrated → 3-phenyl-1-indanone. Y 72%. The reaction is successful with cinnamic acid, 3-methylacrylic acid and 3,3-dimethylacrylic acid and also with toluene and *m*-xylene replacing benzene, but acrylic and 2-methylacrylic acids give low yields or decomposition products. The reaction requires a temperature of at least 130° and >10 molar excess of acidic sites to reach completion, and is presumed to involve the generation and trapping of an unsatd. ethyleneacylium ion by the arene. Use of a less acidic zeolite catalyst, however, produces mixtures of the corresponding 1-indanones and 1,1-diarylalkanes, the latter predominating with zeolite H-ZSM5 (cf. *72*, 623). F.e.s. S. Chassaing, M. Kumarraja, P. Pale, J. Sommer, Org. Lett. *2007*, *9* (20), 3889-92.

Boron fluoride (s.a. under BuLi) *BF_3*
Friedel-Crafts propargylation with propargyl trichloroacetimidates H→C-C≡C

516.
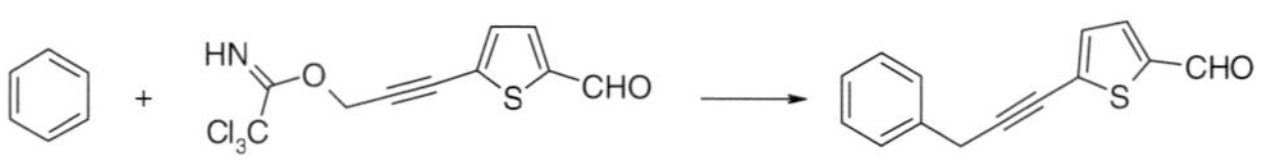

1,3-Diarylpropynes. Startg. trichloroacetimidate (0.387 mmol) and BF_3-etherate (30 mol%) added to benzene (3 ml), and the soln. stirred for 5 min at room temp. → product. Y 89%. The method is unsuccessful with substrates that stabilize intermediate propargyl cations; the presence of strong electron-donating substituents and trichloroacetimidates derived from secondary propargyl alcohols afford poor yields or alternative products. F.e., also 3-alkylprop-2-ynyl derivs., s. C. Li, J. Wang, J. Org. Chem. *2007*, *72* (19), 7431-4.

Ammonium fluoroborate s. under Chiral thiolate-bridged diruthenium complex *NH_4BF_4*

Indium(III) chloride *$InCl_3$*

1-Aryl-4-arylamino-1,2,5,6-tetrahydropyridine-3-carboxylic acid esters from β-ketocarboxylic acid esters and two prim. ar. amine and aldehyde molecules
One-pot 3-component [five molecule] reaction under mild conditions

517.

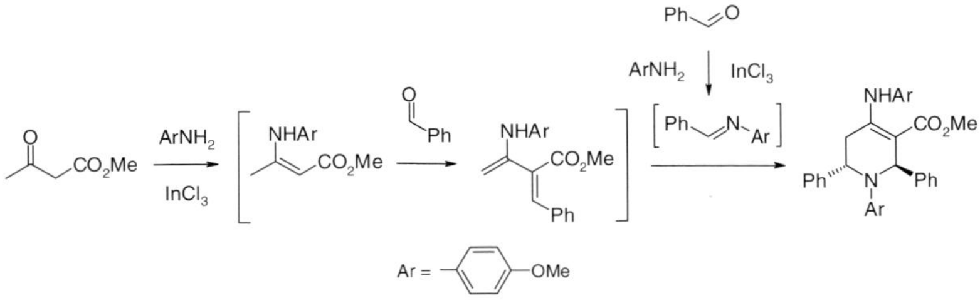

A mixture of methyl acetoacetate, 2 eqs. benzaldehyde, 2 eqs. 4-methoxyaniline and 0.33 eqs. $InCl_3$ stirred in acetonitrile at room temp. for 24 h, and the product isolated by filtration and washing with a little acetonitrile → *trans*-1-(4-methoxyphenyl)-2,6-diphenyl-3-carbomethoxy-4-(4-methoxyphenyl)amino-3-piperideine. Y 74%. A wide range of ar. aldehydes may be used in the reaction (13 examples; Y 16-74%) but aliphatic aldehydes give multiple products, while reactivity is reduced with electron-deficient ar. amines (4-chloroaniline requiring 7 days for a 52% yield and 4-nitroaniline being unreactive) and is sensitive to steric effects (examples using 2-methylbenzaldehyde gave 16 and 24% yield). Furthermore, some aniline-aldehyde combinations produced insoluble imine intermediates that would not react further, and partial hydrolysis of products to afford 4-hydroxy-derivs. was also observed in some cases. F.e.s. P.A. Clarke, A.V. Zaytzev, A.C. Whitwood, Tetrahedron Lett. *2007*, *48* (30), 5209-12.

Scandium(III) triflate *$Sc(OTf)_3$*

2,3-Disubst. quinolines from anilines, aldehydes and α,β-acetylenecarboxylic acid esters
Lewis acid-catalyzed 3-component reaction

518.

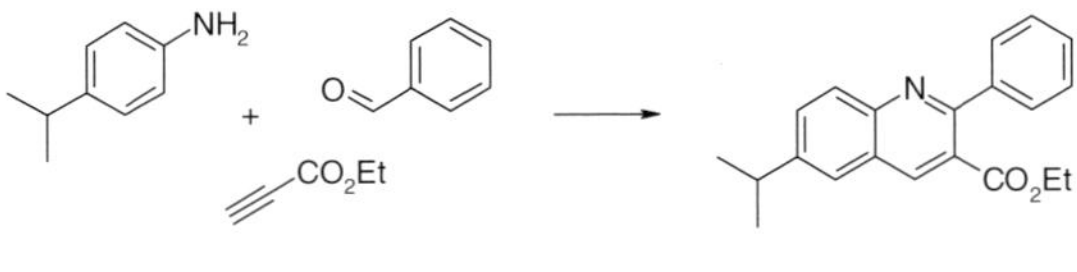

Benzaldehyde (0.54 mmol) and $Sc(OTf)_3$ (10 mol%) added to a soln. of *p*-isopropylaniline (0.5 mmol) in ethanol (5 ml), and the mixture refluxed for 24 h, before quenching with aq. $NaHCO_3$ → product. Y 79%. A variety of subst. anilines and benzaldehyde derivs. afforded the corresponding quinolines mainly in low to moderate yields. No significant trends were observed, with respect to product yield, with either electron-withdrawing or -donating substituents in the aniline *para* position. Use of ethyl glyoxylate in place of an ar. aldehyde in the reaction with *p*-toluidine successfully gave the equivalent 2,3-di(ethoxycarbonyl)quinoline in moderate yield (34%). F.e.s. S. Kikuchi, M. Iwai, S.-I. Fukuzawa, Synlett *2007* (17), 2639-42.

Scandium(III), lanthanum(III) or ytterbium(III) triflate *$Sc(OTf)_3$ or $La(OTf)_3$ or $Yb(OTf)_3$*

C-α-Alkylation with alcohols H → R

α-Allylation or -benzylation s. *22*, 782s*72*

Ytterbium(III) perfluorooctanesulfonate *$Yb(OSO_2R_f)_3$*

Knoevenagel condensation CO → C=C

cf. *46*, 713s*70*; with Yb(III)-perfluorooctanesulfonate in a fluorous 2-phase system (perfluorodecalin/toluene) s. W.B. Yi, Y.Q. Yin, C. Cai, Org. Prep. Proced. Int. *2007*, *39* (1), 71-5; 5-arylidene-barbituric acids under microwave irradiation in the absence of solvent or catalyst s. C.S. Reddy, A. Nagaraj, P. Jalapathi, Indian J. Chem. *2007*, *46B* (4), 660-3.

Benzophenone s. under $RuCl_2(DMSO)_4$ *Ph_2CO*
(R,R)-1,2-Bis(picolinamido)cyclohexane s. under $Mo(CO)_3$(cycloheptatriene) ←
Dowex 50W ←
Biginelli synthesis in water s. *55*, 337s*72* ○
Formic acid s. under $Pd(PPh_3)_4$ *HCOOH*
Acetic acid s. under $Pd(OAc)_2$ *AcOH*
Proline s. under AgOTf *Pro-OH*
Trifluoroacetic acid [s.a. under [RuCO(η^3-methallyl)dppf]SbF_6] *CF_3COOH*
Pictet-Spengler cyclization
with CF_3COOH cf. *8*, 823s*62*; as a 10% soln. in water at room temp., 1,2,3,4-tetrahydro-β-carbolines, stereospecificity, s. B. Saha, S. Sharma, D. Sawant, B. Kundu, Tetrahedron Lett. *2007*, *48* (8), 1379-83; by heating with 20 mol% perfluorooctanesulfonic acid (as combined Brønsted acid-surfactant) in water or (more quickly) in ca. 10% 1,1,1,3,3,3-hexafluoroisopropanol/water, 2-carbomethoxy-1,2,3,4-tetrahydroisoquinolines, s. A. Saito, J. Numaguchi, Y. Hanzawa, ibid. *48* (5), 835-9; with 3.5 eqs. Me_3SiCl/NaI in acetonitrile at room temp., 2-acylamino-1,2,3,4-tetrahydroisoquinolines, s. L. Zhang, X.-F. Zheng, G. Linn, K. Zhao, Synlett *2007* (3), 374-80.

Silica s.a Silica-based N-(4-chlorosulfonylbutyl)imidazolium triflate and Silica-sulfuric acid *SiO_2*
Titanium tetraisopropoxide s. under Mg and Bu_3P *$Ti(OPr\text{-}i)_4$*
Trimethylsilyl cyanide (s.a. under Me_3SiOTf) *Me_3SiCN*
Trimethylsilyl cyanide/sulfamic acid *Me_3SiCN/H_2NSO_3H*
α-Aminonitriles from aldehydes CO → C(N<)CN
uncatalyzed Strecker reaction with Me_3SiCN alone in acetonitrile cf. *52*, 449s*70*; *7*, 777s*70*; in the absence of solvent, self-catalysis by the amine, also [more slowly] from ketones, s. A. Baeza, C. Nájera, J.M. Sansano, Synthesis *2007* (8), 1230-4; with 10 mol% sulfamic acid in acetonitrile s. Z. Li, Y. Sun, X. Ren, P. Wei, Y. Shi, P. Ouyang, Synlett *2007* (5), 803-5.

Trimethylsilyl chloride [s.a. under $Hf(OTf)_4$] *Me_3SiCl*
2-Imino-3,4,7,8-hexahydropyrimido[5,4-*e*][1,3]thiazin-5(6*H*)-ones by 3-component [four molecule] double ring closure ○

519.

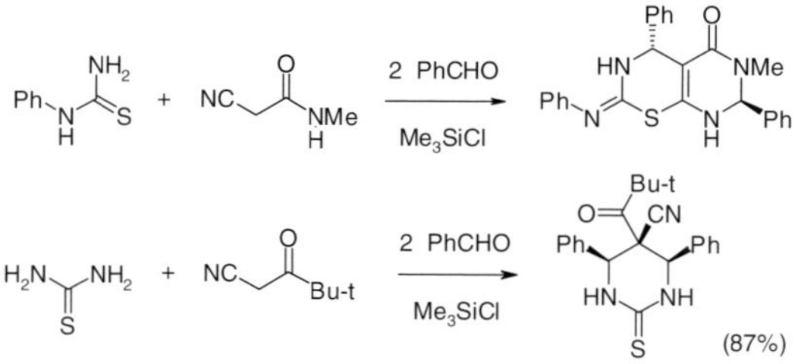

1 Eq. N-methylcyanoacetamide, 2 eqs. benzaldehyde, and 1-1.5 eqs. thiourea stirred in dry DMF, 4 eqs. trimethylsilyl chloride added dropwise, the soln. sonicated for 1 h, stirred for 2-3 days, poured into water, sonicated for a further 1 h, filtered, washed with isopropanol, and the filtrate evaporated → hexahydropyrimido[5,4-*e*][1,3]thiazin-5-one deriv. Y 75%. Highest yields (13 examples; Y 56-92%) were obtained with stoichiometric amounts of the three reactants. The benzaldehyde may possess *p*-chloro or *p*-methoxy groups, the thiourea may be unsubstituted or mono-N-alkylated (incl. aliphatic ether) or N-phenylated, and the cyanoacetamide may be unsubstituted or N-substituted (Me, Bn, *p*-chlorophenyl and aliphatic ether). Replacing the cyanoacetamide with cyanoketones affords **5-acyl-3,4,5,6-tetrahydropyrimidine-2(1*H*)-thiones** (4 examples; Y 85-96%). F.e.s. A.N. Shivanyuk, A.A. Tolmachev, Org. Lett. *2007*, *9* (21), 4215-8.

Trimethylsilyl chloride/sodium iodide *Me_3SiCl/NaI*
Pictet-Spengler cyclization s. *8*, 823s*72*

*Chlorotitanium triisopropoxide/cyclopentylmagnesium chloride/*n-*butyllithium* ←
1,4-Dienes from acetylene derivs. and 2-ethylenealcohols C(OH)C=CH → C=C-C-C=CH
Titanium(IV)-catalyzed regio- and stereo-specific cross-coupling

520. 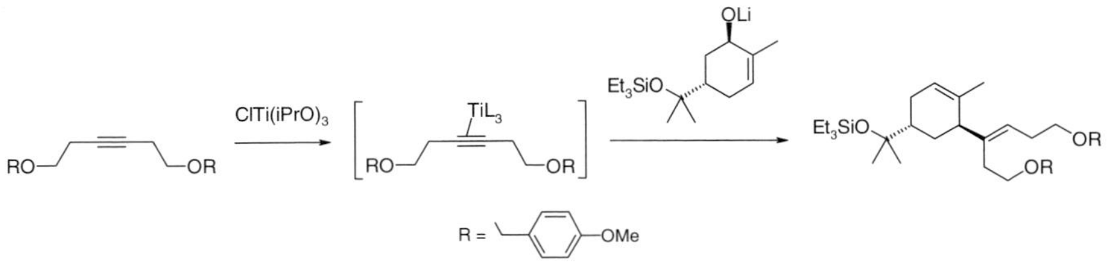

2 Eqs. $TiCl(OPr\text{-}i)_3$ added by gas-tight syringe to a stirred soln. of the startg. alkyne in toluene, the mixture cooled to -78°, treated dropwise with 4 eqs. cyclopentylmagnesium chloride in ether, warmed slowly to -35°, stirred for 1 h, cooled to -78°, a soln. of the startg. Li-allyloxide (prepared from the corresponding allyl alcohol and 1.1 eqs. *n*-BuLi at -78° to 0°) added dropwise via cannula to the black soln., the mixture warmed to 0° over 30 min, stirred at 0° for 1 h, quenched by addition of water and rapid stirring (grey precipitate), the aq. layer extracted with ether, the extracts dried over $MgSO_4$, concentrated *in vacuo*, and the residue purified by chromatography → 1,4-diene. Y 50% (d.r. 20:1). Initial generation of a titanium π-complex is followed by condensation with the lithium allyloxide with **allylic transposition,** high diastereoselectivity being recorded with both acyclic and cyclic allyl alcohols in the presence of silyl ethers, *p*-methoxybenzyl ethers and additional conjugated double bonds giving 40-79% yield (22 examples). Lowest yields are obtained with primary and sterically-hindered allylic alcohols while (Z)-disubst. allyl alcohols provide the highest stereoselectivity. F.e.s. F. Kolundzic, G.C. Micalizio, J. Am. Chem. Soc. *2007*, *129* (49), 15112-3.

Silicon tetrachloride s. under $ZnCl_2$ *$SiCl_4$*

Hafnium(IV) triflate *$Hf(OTf)_4$*
C-α-Alkylation with alcohols H → R
α-Allylation or -benzylation s. *22*, 782s*72*

Hafnium(IV) triflate/trimethylsilyl chloride *$Hf(OTf)_4/Me_3SiCl$*
Ar. amino(cyano)methylation with alkoxy(amino)acetonitriles $H \rightarrow CH(CN)NR_2$

521. 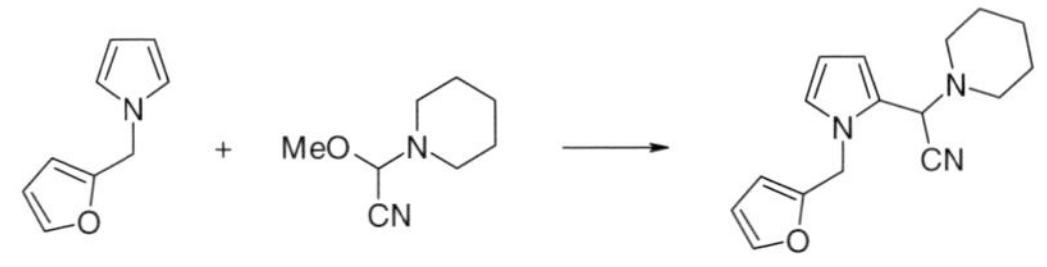

N,N-Disubst. arylglycine precursors. N-(2-Furanomethyl)pyrrole (0.5 mmol), startg. N,O-acetal (0.6 mmol) and *freshly-distilled* Me_3SiCl (0.6 mmol) added successively to methylene chloride (2 ml) under N_2, followed by $Hf(OTf)_4$ after 1 min, and the resulting thick suspension stirred vigorously at room temp. for 4 h → product. Y 85%. The procedure is successful for a variety of subst. pyrroles, indoles and furans and even for an electron-rich arene, N,N-dimethylaniline, which affords the *p*-subst. aminoalkyl derivative, albeit in moderate yield (38%). Substrates bearing electron-donating ring-substituents generally give higher yields than those bearing electron-withdrawing groups such as esters and halogens. Other N,O-acetals (including hydroxyglycine) are also successful, although no reaction occurs using an N-methylpiperazine analog, presumably due to preferential coordination of the hafnium catalyst with the *tert*-amino group. A plausible mechanism is proposed in which $Hf(OTf)_4$ coordinates with the acetal oxygen atom, giving rise to an intermediate iminium complex which is then attacked by the arene. The role of Me_3SiCl is not known, although it may trap the *in situ*-generated methoxy ion, helping to drive the equilibrium to completion. F.e.s. N. Sakai, J. Asano, Y. Shimano, T. Konakahara, Synlett *2007* (17), 2675-8.

Tri-n-*butylphosphine/titanium tetraisopropoxide* *$Bu_3P/Ti(OPr$-i$)_4$*
Electron-deficient 1,3-dienes from two aldehyde molecules via modified Baylis-Hillman reaction RCH═C(E)CH═CHR

522.
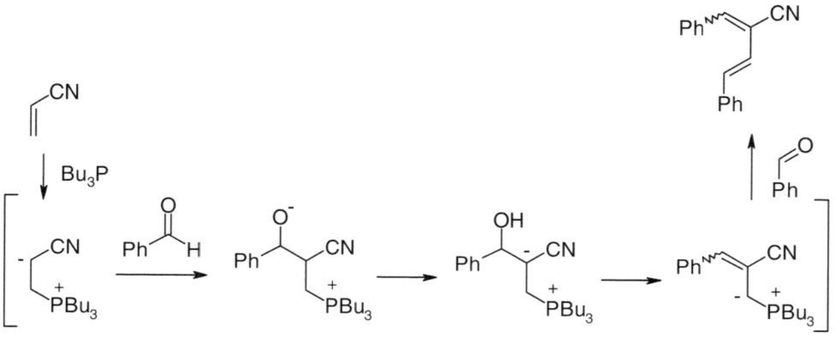

Tri-*n*-butylphosphine added to 0.8 eq. acrylonitrile, 1 eq. benzaldehyde and 1 eq. Ti(OPr-*i*)$_4$ in methylene chloride under N_2, and stirred for 15 h → product. Y 66% (E,Z/E,E 4:1). Reaction appears generally applicable to aliphatic and cycloaliphatic aldehydes, and yields were good with ar. aldehydes possessing electron-donating and neutral substituents. Yields were low, however, without the added Lewis acid, which inhibits undesirable Stetter reaction and benzoin condensation. The conversion is initiated by 1,4-addition of the phosphine (à la classical Baylis-Hillman reaction), followed by addition to the aldehyde; subsequent proton shift and dehydration lead to an unsatd. P-ylid which duly undergoes Wittig synthesis with a second molecule of the aldehyde to give the product. The (Z)-product predominated (E,Z/E,E being maximally 6:1). F.e. and with ethyl acrylate s. A. Palmelund, E.L. Myers, L.R. Tai, S. Tisserand, C.P. Butts, V.K. Aggarwal, Chem. Commun. *2007* (40), 4128-30.

2-Dicyclohexylphosphino-2′,4′,6′-triisopropylbiphenyl s. under Pd(OAc)$_2$ ←
Triphenylphosphine s. under Pd(OAc)$_2$ *Ph_3P*
Polymer-based triphenylphosphine s. under Zn(CN)$_2$ *Ⓟ-ArPPh$_2$*
1,3-Bis(diphenylphosphino)propane s. under Pd(OAc)$_2$ *dppp*
(S,S)-N,N′-Bis[2-(diphenylphosphino)-1-naphthoyl]-1,2-diaminocyclohexane s. under Pd$_2$(dba)$_3$ ←

Propanephosphonic acid anhydride *[PrP(O)O]$_3$*
Biginelli synthesis s. *55*, 337s*72* ○

Chiral 1,1′-binaphthyl-2,2′-diyl hydrogen phosphates/thiophenol/hydrogen chloride ←
Organocatalyzed asym. Pictet-Spengler cyclization via N-sulfenyliminium ions
s. 72, 579

Diammonium hydrogen phosphate *$(NH_4)_2O_2P(O)OH$*
Sym. 1,1-bis(indol-3-yl)alkanes from oxo compds. CO → CAr_2
update s. *5*, 549s*70,71*; in the absence of solvent with $(NH_4)_2O_2P(O)OH$ at 80° s. M. Dabiri, P. Salehi, M. Baghbanzadeh, Y. Vakilzadeh, S. Kiani, Monatsh. Chem. *2007*, *138* (6), 595-7; with sulfamic acid by grinding with a mortar and pestle s. L.T. An, F.Q. Qing, J.P. Zou, X.H. Lu, L.L. Zhang, Chin. J. Chem. *2007*, *25* (6), 822-7; with 10 mol% $MoO_2(acac)_2$ in water s. B. Banerjee, S.K. Mandal, S.C. Roy, Indian J. Chem. *2007*, *46B* (4), 669-73; from aldehydes with 10 mol% $Cu(NO_3)_2 \cdot 3H_2O$ in acetonitrile at room temp. s. A. Nasreen, R. Varala, S.R. Adapa, J. Heterocycl. Chem. *2007*, *44* (4), 983-7; with a [reusable] silica-immobilized N-(4-chlorosulfonylbutyl)imidazolium triflate as acidic ionic liquid in acetonitrile at room temp. s. H. Hagiwara, M. Sekifuji, T. Hoshi, K. Qiao, C. Yokoyama, Synlett *2007* (8), 1320-2; rapid method **from prim. alcohols** with

1.5 eqs. $Bi(NO_3)_3 \cdot 5H_2O$ in the absence of solvent cf. A.R. Khosropour, I. Mohammadpoor-Baltork, M.M. Khodaei, P. Ghanbary, Z. Naturforsch. *2007*, *62B* (4), 537-9; **sym. 1,1,1-tris(indol-3-yl)methanes** with ethyl orthoformate and 15 mol% I_2 in acetonitrile cf. Z.-H. Zhang, J. Lin, Synth. Commun. *2007*, *37* (2), 209-15.

Bismuth(III) nitrate *$Bi(NO_3)_3$*
Sym. 1,1-bis(indol-3-yl)alkanes from prim. alcohols $CH_2OH \rightarrow CHAr_2$
in the absence of solvent s. *5*, 549s*72*

Bismuth(III) triflate *$Bi(OTf)_3$*
C-α-Alkylation with alcohols $H \rightarrow R$
α-Allylation or -benzylation s. *22*, 782s*72*

Bismuth trichloride *$BiCl_3$*
Bismuth(III)-catalyzed nucleophilic substitution of 2-acetylenealcohols $OH \rightarrow Ar$
by (het)aryl s. *72*, 106

Oxygen s. under $Pd(OAc)_2$ *O_2*

*N-Tosylimidazole/triethylamine/tetra-*n*-butylammonium iodide* ←
Nitriles from alcohols $OH \rightarrow CN$

523. 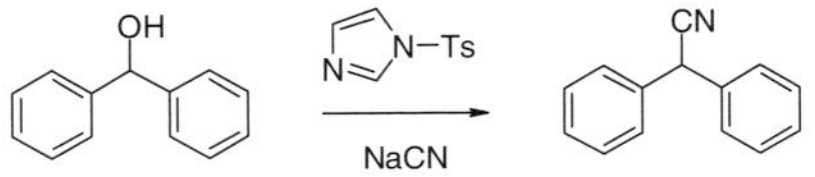

A mixture of benzhydryl alcohol, 1.2 eqs. N-tosylimidazole, 2 eqs. triethylamine, 2 eqs. NaCN and a catalytic amount of tetra-*n*-butylammonium iodide refluxed in *DMF* for 6 h (TLC), solvent removed *in vacuo*, the residue dissolved in chloroform, washed with water, dried over Na_2SO_4, evaporated *in vacuo*, and purified by chromatography on silica → diphenylacetonitrile. Y 71%. This convenient and efficient method is successful with prim., sec. and tert. aliphatic alcohols (incl. benzylic and allylic) in the presence of nitro, ether and imide functionality giving 51-94% yield (17 examples). Cholesterol was also converted under these conditions in 43% yield but a chiral alcohol suffered partial racemization. *No tosyl cyanide was observed even in trace amounts.* F.e. and optimization, also comparison with sulfonyl chlorides as *in situ*-sulfonating agents, s. M.N.S. Rad, A. Khalafi-Nezhad, S. Behrouz, M.A. Faghihi, Tetrahedron Lett. *2007*, *48* (38), 6779-84.

Trifluoromethanesulfonic anhydride s. under $Pd(OAc)_2$ *Tf_2O*

Silica-based N-(4-chlorosulfonylbutyl)imidazolium triflate ←
Sym. 1,1-bis(indol-3-yl)alkanes from aldehydes $CO \rightarrow CAr_2$
s. *5*, 549s*72*

Potassium hydrogen sulfate *$KHSO_4$*
Double ring closure via Biginelli-type synthesis ○

524. 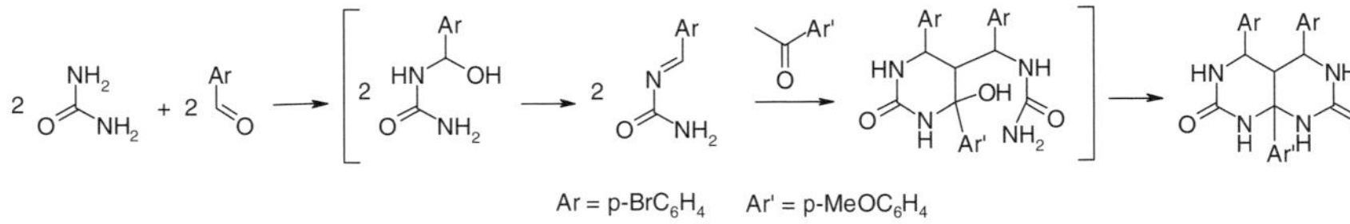

4,5,8a-Triarylhexahydropyrimido[4,5-*d*]pyrimidine-2,7(1*H*,3*H*)-diones. A mixture of the startg. ar. aldehyde, 1.5 eqs. urea, 0.5 eq. startg. ar. ketone and 0.5 eq. $KHSO_4$ in acetic acid refluxed with stirring at 80° for 1.5 h → 4,5-bis(4-bromophenyl)-8a-(4-methoxyphenyl)hexahydropyrimido[4,5-*d*]pyrimidine-2,7(1*H*,3*H*)-dione. Y 96%. The procedure is mild, high-yielding, broad in scope and simple to undertake. Reaction is thought to involve initial condensation of aldehyde and urea to give a 1,1-hydroxyurea, followed by its dehydration, sequential additions across the C=N bond, intramolecular cyclization and a further dehydration. Two equivalents of aldehyde and urea condensed similarly with one equivalent of acetone under the same conditions to give the corresponding **6,6′-diaryl-4,4′-spirobi[hexahydropyrimidine]-2,2′-diones**. F.e.s. F. Shi, R. Jia, X. Zhang, S. Tu, S. Yan, Y. Zhang, B. Jiang, J. Zhang, C. Yao, Synthesis *2007* (18), 2782-90.

Trimethylsilyl triflate s.a. under BuLi — *Me_3SiOTf*

Trimethylsilyl triflate/trimethylsilyl cyanide — *Me_3SiOTf/Me_3SiCN*
α-Arylaminonitriles from oxo compds. — CO → C(NHAr)CN
Metal-free 3-component Strecker synthesis

525. 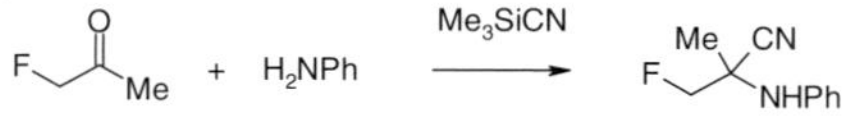

A mixture of the startg. fluoroketone, 0.67 eq. aniline, 1 eq. trimethylsilyl cyanide and 5 mol% trimethylsilyl triflate *in methylene chloride* stirred for 6 h in a sealed pressure tube *at room temp.*, and quenched with water → product. Y 84%. Significantly, the procedure does not require the presence of a metal catalyst (cf. *55*, 341s*70*), and is mild, clean, high-yielding (without chromatographic purification), and generally applicable to the reaction of both aldehydes *and* ketones, incl. fluorinated ketones. Aliphatic amines, however, gave a mixture of products, while benzophenone required a higher temperature and excess of trimethylsilyl cyanide. F.e.s. G.K.S. Prakash, C. Panja, C. Do, T. Mathew, G.A. Olah, Synlett *2007* (15), 2395-9.

Perfluorooctanesulfonic acid — *R_fSO_3H*
Pictet-Spengler cyclization ○
with a combined Brønsted acid-surfactant as catalyst – 2-Acylamino-1,2,3,4-tetrahydroisoquinolines s. *8*, 823s*72*

Camphorsulfonic acid s. under Copper(II) sulfate — *RSO_3H*

p-*Toluenesulfonic acid* — p-*TsOH*
3-Amino-5,6-dicyano-1,2-dihydropyrazines
from ketones, isonitriles and diaminomaleonitrile
One-pot 3-component synthesis

526. 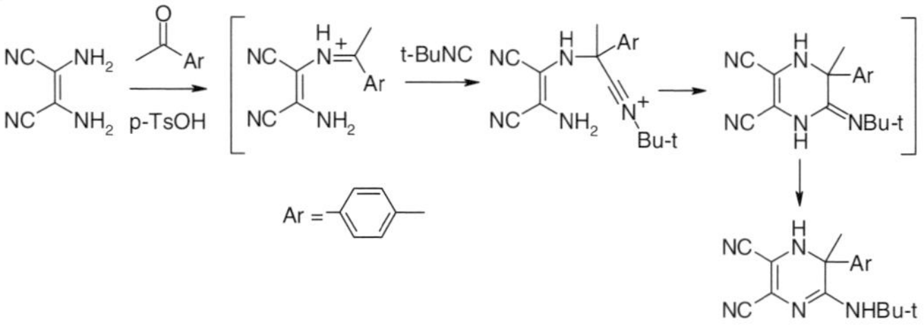

under mild conditions. *p*-Toluenesulfonic acid (5 mol%) added to an equimolar soln. of diaminomaleonitrile, 4-methylacetophenone, and *tert*-butyl isocyanide in ethanol, the mixture stirred *at room temp.* for 2 h (TLC monitoring), the product precipitated by addition of water, filtered off, and crystallized from acetone → 5-(*tert*-butylamino)-1,6-dihydro-6-methyl-6-(4-tolyl)pyrazine-2,3-dicarbonitrile. Y 82%. This rapid, convenient, simple and clean method is successful for coupling cyclic or acyclic dialkyl and aralkyl ketones with alkyl and aryl isocyanides giving yields of 80-98% (15 examples). Reaction with benzaldehyde in place of the ketone component, however, gave only the imine. F.e.s. A. Shaabani, A. Maleki, J. Moghimi-Rad, J. Org. Chem. *2007*, *72* (16), 6309-11.

*Tetra-*n*-butylammonium triflate s. under $Pd_2(dba)_3$* — *Bu_4NOTf*

Sulfamic acid (s.a. under Me_3SiCN) H_2NSO_3H
Sym. 1,1-bis(indol-3-yl)alkanes from oxo compds. CO → CAr_2
by grinding in the absence of solvent s. *5*, 549s*72*

Silica-sulfuric acid SiO_2-OSO_3H
3-Component Mannich reaction CHO → CH(NHR)C-CO
with $Zn(OTf)_2$ cf. *68*, 361s*70*; β-aminoketones with 0.04 mol% silica-sulfuric acid in ethanol s. H. Wu, Y. Shen, L. Fan, Y. Wan, P. Zhang, C. Chen, W. Wang, Tetrahedron *2007*, *63* (11), 2404-8; with 5 mol% iodine in excess ketone at room temp., fluorinated compds., s. M. Xia, Y. Lu, Synth. Commun. *2007*, *37* (5), 725-35; β-(acylamino)ketones with 25 mol% iodine in acetyl chloride/acetonitrile at 80-85° s. M.A. Pasha, V.P. Jayashankara, N. Ramachandraswamy, ibid. *37* (9), 1551-6.

Tricarbonyl(cycloheptatriene)molybdenum/(R,R)-N,N′-bis(picolinoyl)cyclohexane-1,2-diamine/sodium tert-*butoxide* ←
Regiospecific molybdenum-catalyzed asym. 3-allylation of 3-aryloxindoles H → C-C=C

527. 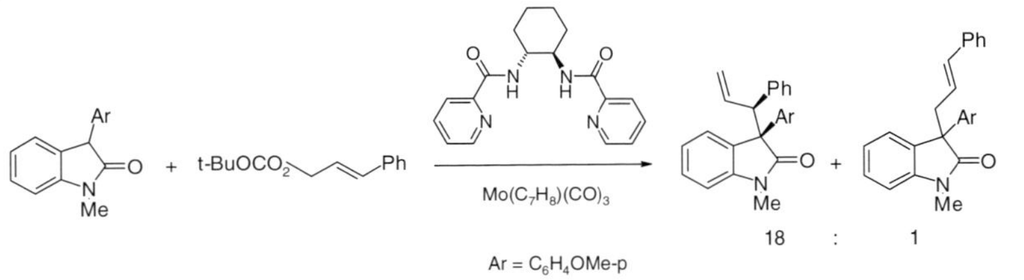

In a dry box 1.1 eqs. NaOBu-*t* added to 1-methyl-3-(4-methoxyphenyl)oxindole, the reaction tube sealed, removed from the dry box, degassed THF added at room temp., the mixture cooled to 0°, a soln. of the active molybdenum catalyst [freshly prepared from 15 mol% (R,R)-1,2-bis(picolinamido)cyclohexane and 10 mol% (η^3-cycloheptatriene)$Mo(CO)_3$ by heating at 60° for 3-5 min to give a deep purple colour] added via cannula, 1.2 eqs. *tert*-butyl cinnamyl carbonate added via syringe, the soln. stirred at 60° under argon until consumption of the oxindole complete (TLC; 2-16 h), the mixture concentrated *in vacuo*, and purified by chromatography → product. Y 92% (branched/linear selectivity 18:1; d.r. 8:1; e.e. 92%). Significantly, reaction provides *a quaternary carbon centre* at the oxindole 3-position which is difficult to generate by other routes, and is successful in the presence of silyl ether, nitrile, tosyl, Boc and MOM amine and halo functionality. Twenty-eight examples are given (Y 63-95%) where the 1- and 3-substituents on the oxindole and the terminal group of the allylic carbonate were systematically varied to assess the outcome on selectivity. In general, bond formation occurs at the more-substituted position of the π-allyl system (2.5:1 to 19:1) giving branched products *even with bulky substituents* but use of the smaller thienyl, thiazolyl and indolyl 3-substituents gives linear products only. Furthermore, the presence of electron-withdrawing groups on the 3-aryl group decreases the diastereoselectivity but enantioselectivity (e.e. 89-97%) is largely unaffected by electronic variations. F.e. incl. oxindole prepn. s. B.M. Trost, Y. Zhang, J. Am. Chem. Soc. *2007*, *129* (47), 14548-9.

Molybdenyl acetoacetonate $MoO_2(acac)_2$
Sym. 1,1-bis(indol-3-yl)alkanes from oxo compds. CO → CAr_2
in water s. *5*, 549s*72*

Iodine I_2
3-Component Mannich reaction CHO → CH(NHR)C-CO
Fluorinated β-aminoketones – Also β-(acylamino)ketones s. *68*, 361s*72*

Sym. 1,1,1-tris(indol-3-yl)methanes from indoles s. *5*, 549s*72* ←

Perchloric acid/microwaves $HClO_4$/
Biginelli synthesis ○
under microwave irradiation in the absence of solvent s. *55*, 337s*72*

Tetra-n-*butylammonium iodide s. under N-Tosylimidazole* *Bu_4NI*

Manganese(III) acetate *$Mn(OAc)_3$*
Radical-cascade reactions of Ugi adducts s. *72*, 625 ←

Iron(III) chloride *$FeCl_3$*
C-α-Alkylation with alcohols H → R
α-Allylation or -benzylation s. *22*, 782s72

Carbonyl(η³-methallyl)[1,1′-bis(diphenylphosphino)ferrocene]ruthenium(II) hexafluoroantimonate/trifluoroacetic acid ←
Furans from 2-acetylenealcohols and ketones in the absence of solvent ○

528. 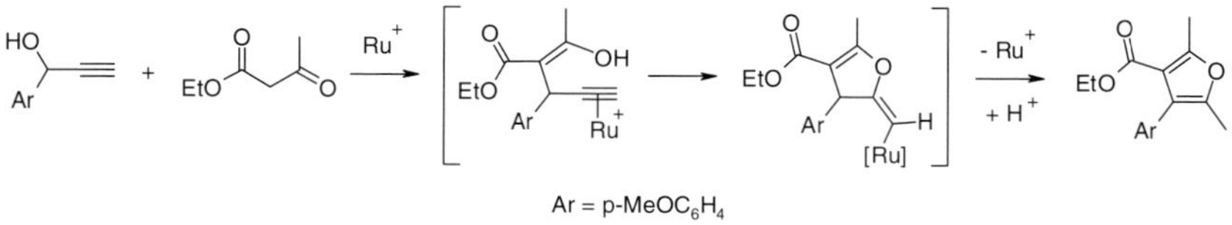

4-Arylfuran-3-carboxylic acid esters. A mixture of 1 eq. of the startg. aryl-subst. propargyl alcohol, 10 eqs. ethyl acetoacetate, 5 mol% carbonyl(η³-methallyl)[1,1′-bis(diphenylphosphino)-ferrocene]ruthenium(II) hexafluoroantimonate and 0.5 eqs. trifluoroacetic acid heated in a sealed tube under N_2 at 75° for 2 h → product. Y 94%. **Tetrasubst. furans** are thus obtained simply and efficiently in one pot from terminal or internal sec. 1-arylpropargyl alcohols and β-diketones or β-ketocarboxylic acid esters (though not *tert*-butyl acetoacetate), irrespective of the electronic nature of the substituents in the aromatic ring. Reaction is presumed to involve initial acid-catalyzed α-propargylation of the β-dicarbonyl compd., followed by the novel Ru-catalyzed cycloisomerization of the intermediate α-propargyl-β-dicarbonyl compd. via a ruthenium π-alkyne complex. F.e.s. V. Cadierno, J. Gimeno, N. Nebra, Adv. Synth. Catal. *2007*, *349* (3), 382-94.

Chiral thiolate-bridged diruthenium complex/ammonium fluoroborate ←
Asym. Friedel-Crafts propargylation with propargyl alcohols H → C-C≡C

529. 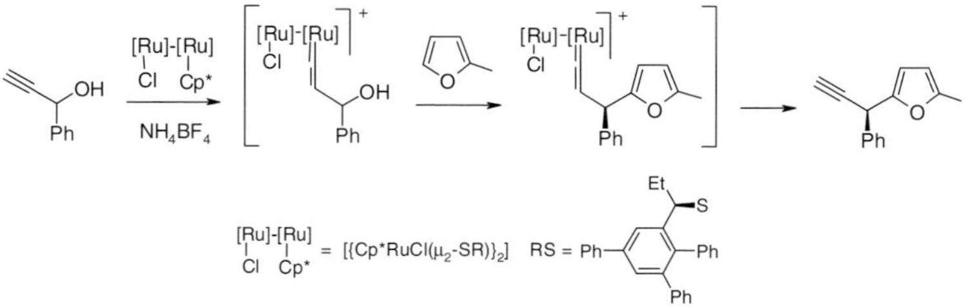

of furans. Dry THF added to 2.5 mol% $[Cp^*RuCl]_4$ and 5 mol% chiral disulfide under N_2, stirred at room temp. for 12 h, solvent removed *in vacuo*, 10 mol% NH_4BF_4 and dry dichloroethane added under N_2, stirring continued at room temp., 1-phenyl-2-propyn-1-ol and 10 eqs. 2-methylfuran added to the generated chiral thiolate-bridged diruthenium complex, heated at 60° for 3 h, and worked up with column chromatographic purification on silica gel → product. Y 75% (e.e. 77%). This is the first example of an asym. ar. propargylation with propargyl alcohols as electrophile, enantioselectivities being high (76-94%) for the coupling of variously ring-substituted 1-arylpropargyl alcohols with 2-alkylfurans and N,N-dimethylaniline. π-π-Interactions between the phenyl rings of the chiral ligand and the allenylidene moieties of the intermediate ruthenium allenylidene complex are thought to play an important role in achieving high asym. induction, and it is significant that chiral thiolate residues with less than three aryl rings afforded much reduced enantioselectivity. F.e. and comparison of chiral ligands s. H. Matsuzawa, Y. Miyake, Y. Nishibayashi, Angew. Chem. Int. Ed. Engl. *2007*, *46* (34), 6488-91.

Chloro(cyclopentadienyl)bis(triphenylphosphine)ruthenium(II)/ 1,8-diazabicyclo[5.4.0]undec-7-ene *$CpRuCl(PPh_3)_2$/DBU*

Benzene ring annelation of 2,4-enynals with cyclic ketones under dual catalysis ○

530. 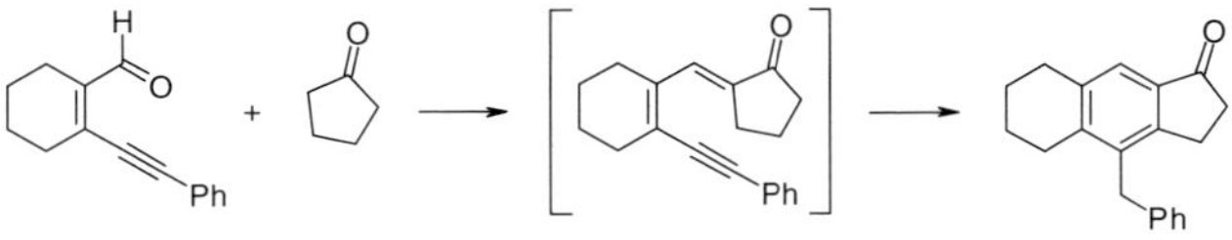

1-Indanones. 5 Mol% $CpRu(PPh_3)_2Cl$, 5 mol% DBU and the startg. 2,4-enynal added to 5 eqs. cyclopentanone, heated at 140° for 12 h, filtered through Celite, and eluted through a column of silica gel → product. Y 65%. The procedure is applicable to the synthesis of both 1-indanones **and 1-tetralones** from variously 5-subst. (by aryl, hetaryl, alkenyl or alkyl) cyclic and acyclic 2,4-enynals. Reaction is favoured by electron-withdrawing groups in the 5-aryl ring in keeping with the expectation that an increase in H-acidity of the intermediate α,β-ethyleneketone accelerates the conversion by deprotonation-reprotonation under the influence of the base. F.e. and comparison of catalysts, ligands and bases s. C.-W. Yang, R.-S. Liu, Tetrahedron Lett. *2007*, *48* (48), 5887-9.

Dichlorotetrakis(dimethyl sulfoxide)ruthenium(II)/potassium tert-*butoxide/benzophenone* ←

Oxidative Friedländer quinoline synthesis with *in situ*-generated ketones

531. 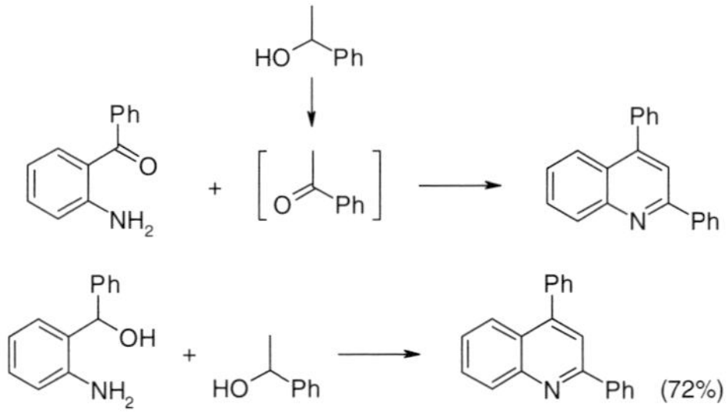

under mild conditions in the absence of solvent. Equivalent amounts of the startg. alcohol and KOBu-*t* added to a mixture of 1 mol% $RuCl_2(DMSO)_4$, 1 eq. benzophenone (as H-acceptor) and 1 eq. of the startg. *o*-aminoketone, heated at 100° for 48 h, quenched successively with satd. aq. NH_4Cl and ethyl acetate, filtered off, and worked up with column chromatographic purification on silica gel → 2,4-diphenylquinoline. Y 99%. The procedure is inexpensive, energy-efficient, high-yielding, and selective; furthermore, work-up and purification are easy, and there is no need for a strong acid catalyst. It is also applicable to a range of aliphatic and aromatic prim. or sec. alcohols, and unaffected by the electronic nature of substituents in the aromatic ring. It appears that the catalytic cycle involves initial Ru-catalyzed oxidation of the K-alkoxide to the ketone, followed by standard Friedländer condensation with the *o*-aminoketone, the eliminated ruthenium hydride being reoxidized with benzophenone. Quinolines were also obtained under the same conditions **from *o*-aminobenzyl alcohols and ketones or alcohols.** F.e. and comparison of bases s. R. Martinez, D.J. Damon, M. Yus, Eur. J. Org. Chem. *2007* (10), 1599-605.

Palladium(II) acetate s.a. under $Zn(CN)_2$ *$Pd(OAc)_2$*

Palladium(II) acetate/copper(I) iodide/triphenylphosphine/triethylamine
Fluorene ring from terminal acetylene derivs.
Palladium-catalyzed double ring closure via intramolecular activation of aromatic or vinylic carbon-hydrogen bonds

532.
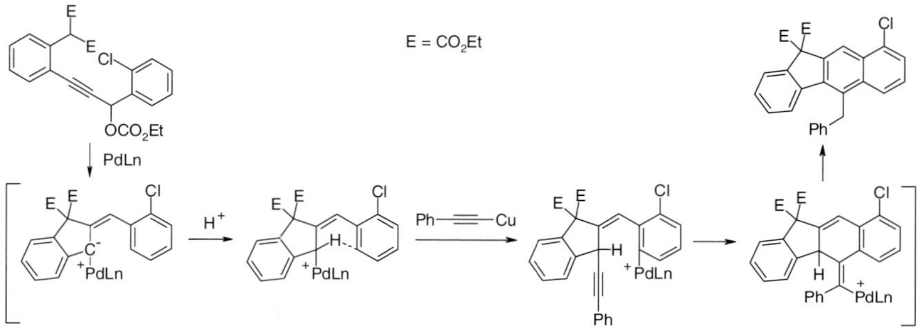

On attempting to prepare some alkynylindenes, a new **one-pot polyarene cyclization** has emerged via a carboannulation, coupling, C-H activation and C-C bond formation sequence. **E: 11*H*-Benzo[*b*]fluorenes.** 5 Eqs. *triethylamine*, 5 mol% *Pd(OAc)*$_2$, 10 mol% triphenylphosphine and 10 mol% CuI added sequentially to a mixture of 2 eqs. phenylacetylene and the startg. propargylic carbonate in *DMF*, the mixture stirred under argon at 60° for 2 h, cooled to room temp., quenched with satd. aq. NH_4Cl, extracted with ethyl acetate, the combined extracts washed with water and satd. brine, dried over Na_2SO_4, solvent removed *in vacuo*, and the residue purified by chromatography → tetracycle. Y 91%. Yields of 56-93% (14 examples) are reported in the presence of halo, ether and tetrahydropyranyl functionality, but are low (Y 30%) with SO_2Ph replacing one ester group or with a bulky *tert*-propargylic acetate (Y 56%), and are dramatically reduced in the absence of CuI. The proposed mechanism is supported by deuterium-labelling experiments. F.e. incl. simple fluorene derivs. via intramolecular activation of the ethylenic C-H bond, also heteroaromatic analogs and optimization, s. L.-N. Guo, X.-H. Duan, X.-Y. Liu, J. Hu, H.-P. Bi, Y.-M. Liang, Org. Lett. *2007*, *9* (26), 5425-8.

Palladium(II) acetate/2-dicyclohexylphosphino-2′,4′,6′-triisopropylbiphenyl/ cesium carbonate/acetic acid/oxygen
Carbazoles from aryl triflates and anilines
Palladium-catalyzed N-arylation-intramolecular oxidative biaryl coupling

533.
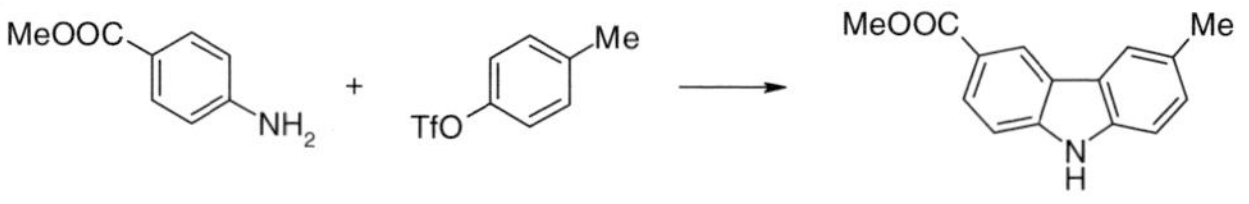

One-pot conversion. Toluene added to a flask containing 4-methylphenyl triflate, 1.1 eqs. 4-methoxycarbonylaniline, 10 mol% Pd(OAc)$_2$, 15 mol% 2-dicyclohexylphosphino-2′,4′,6′-triisopropylbiphenyl and 1.2 eqs. Cs_2CO_3 under argon, stirred at 100° for 1.5 h then at room temp. for 5 min, acetic acid added then an O_2 balloon connected to the flask, stirred again at 80° for 22.5 h, cooled, and worked up with flash chromatographic purification → product. Y 78%. The procedure is atom-economical and applicable to substrates possessing electron-withdrawing or -donating groups, while *m*-subst. substrates gave a *single regioisomer* by ring closure at the least congested site. There was no reaction, however, with aryl halides which interfered with the oxidative coupling. Significantly, both steps of the conversion are effected *by the same catalyst.* F.e. and in air (a little more slowly) s. T. Watanabe, S. Ueda, S. Inuki, S. Oishi, N. Fujii, H. Ohno, Chem. Commun. *2007* (48), 4516-8.

Palladium(II) acetate/1,3-bis(diphenylphosphino)propane/ethyldiisopropylamine ←
Arylcarboxylic acid derivs. from aryl triflates ArOTf → ArCO-
arylcarboxylic acid amides and esters by carbonylation using $Pd(OAc)_2/Ph_3P/Et_3N$ cf. *41*, 734; arylcarboxylic acids and esters using $Pd(OAc)_2$/dppp/*i*-Pr_2NEt, chiral phenylglycine derivs., s. J.B. Grimm, K.J. Wilson, D.J. Witter, Tetrahedron Lett. *2007*, *48* (26), 4509-13.

Palladium(II) acetate/1,3-bis(diphenylphosphino)propane/pyridine/ trifluoromethanesulfonic anhydride ←
Ar. aldehydes from phenols via aryl triflates ArOH → ArCHO
One-pot procedure by reductive carbonylation with synthesis gas

534.

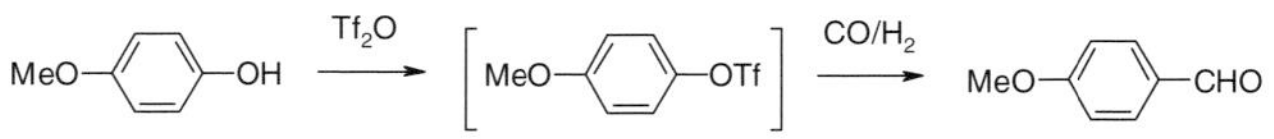

Pyridine added through a septum to 4-methoxyphenol in a Schlenk flask, cooled to 0°, ca. 1.1 eqs. triflic anhydride added dropwise, the resulting orange soln stirred overnight at room temp., the mixture combined with a soln. of 1.5 mol% $Pd(OAc)_2$, 2.25 mol% dppp, DMF and 2-butoxyethyl ether (as internal standard), the whole transferred to a Parr autoclave under argon, flushed three times with 1:1 CO/H_2, the pressure of the synthesis gas adjusted to 20 atm. at room temp., and heated at 100° for 24 h → 4-methoxybenzaldehyde. Y 44%. This is the first direct conversion of a phenol to an ar. aldehyde. A preliminary study was based on the catalytic conversion of preformed aryl triflates to ar. aldehydes under synthesis gas, a reaction favoured by electron-withdrawing groups in the aromatic ring, and hindered by *ortho* substituents. F.e.s. A. Brennführer, H. Neumann, M. Beller, Synlett *2007* (16), 2537-40.

*Tris(dibenzylideneacetone)dipalladium/(S,S)-N,N′-bis[2-(diphenylphosphino)-1-naphthoyl]cyclohexane-1,2-diamine/tetra-*n*-butylammonium triflate* ←
Regiospecific asym. C-allylation with 1,4-dien-3-ol carbonates ←
of Meldrum′s acid s. *72*, 135

Tetrakis(triphenylphosphine)palladium(0)/formic acid *Pd(PPh₃)₄/HCOOH*
2-(2-Methylenecycloalkyl)pyrroles from 2-ene-4,n-diynals and prim. amines ○
Reductive palladium-catalyzed double ring closure

535.

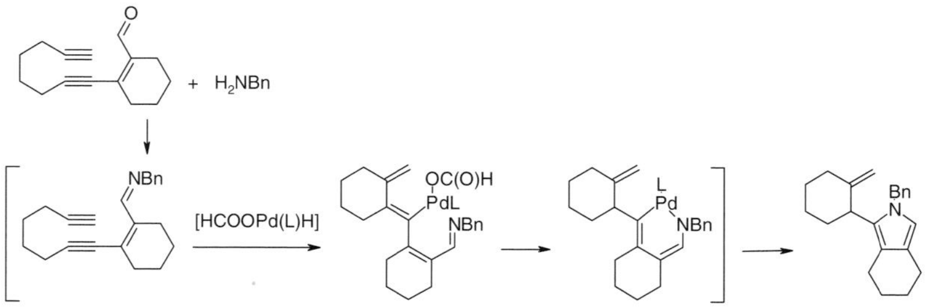

1.2 Eqs. benzylamine added via microsyringe to the startg. enediynal in dry dioxane contained in a test-tube, stirred at room temp. for 30 min, treated in succession with 3 mol% $Pd(PPh_3)_4$ and 2 eqs. formic acid under argon, stirring continued for 1 h at 60°, and worked up with flash chromatographic purification → product. Y 85%. The procedure is atom-economical, environmentally friendly, and applicable to the reaction of cycloalkenecarboxaldehyde derivs. with aliphatic amines, benzylamine or aniline. Reaction is thought to involve initial formation of the aldimine, followed by hydropalladation of the terminal alkyne residue and intramolecular carbopalladation; the introduced nitrogen atom then coordinates to the metal with simultaneous transfer of hydride, and reaction is terminated by decarboxylation and reductive elimination to complete the catalytic cycle. Nitrogen-tethered diyne analogs gave 3-methylenepyrrolidine derivs. F.e.s. C.H. Oh, W. Park, M. Kim, Synlett *2007* (9), 1411-5.

Chloro(1,5-cyclooctadiene)iridium(I) dimer *[Ir(cod)Cl]$_2$*
Iridium-catalyzed intramolecular Alder-ene reaction-Knoevenagel-type [Murahashi] condensation ◯

536. 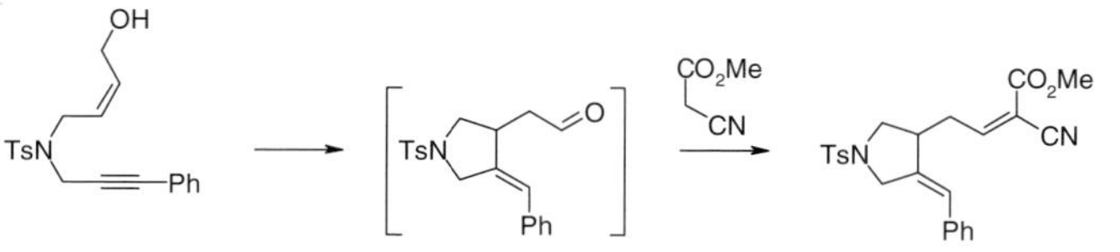

3(Z)-Alkylidene-4-homoallyl-1-tosylpyrrolidines. 1.01 Eqs. startg. allyl(propargyl)tosylamine added to a soln. of 4.4 mol% [Ir(cod)Cl]$_2$ in dry THF, stirred at room temp. for 24 h, 1.51 eqs. methyl cyanoacetate added, stirring continued for a further 24 h, and worked up with chromatographic purification on silica gel → product. Y 72%. The procedure is mild and applicable to substrates bearing electronically diverse aryl [not alkyl] groups. Both the initial cycloisomerization and the subsequent condensation are catalyzed by the iridium complex. F.e. and tetrahydrofuran analogs, also isolation of the intermediate s. M. Kummeter, C.M. Ruff, T.J.J. Muller, Synlett *2007* (5), 717-20.

Chloro-(cyclooctadiene)- or -bis(cyclooctene)-iridium(I) dimer/triphenylphosphine ←
C-α-Alkylation with alcohols H → R
using RuCl$_2$(dmso)/KOH cf. *22*, 782s*69*; base-free method using [Ir(cod)Cl]$_2$ or [Ir(coe)$_2$Cl]$_2$/Ph$_3$P s. M. Morita, Y. Obora, Y. Ishii, Chem. Commun. *2007* (27), 2850-2; α-allylation or -benzylation using 5 mol% Yb(OTf)$_3$ in nitromethane s. W. Huang, J. Wang, Q. Shen, X. Zhou, Tetrahedron Lett. *2007*, *48* (23), 3969-73; using 0.5 mol% La(III)-, Yb(III)-, Sc(III)- or Hf(IV)-triflate s. M. Noji, Y. Konno, K. Ishii, J. Org. Chem. *2007*, *72* (14), 5161-7; using 1 mol% Bi(OTf)$_3$ s. M. Rueping, B.J. Nachtsheim, A. Kuenkel, Org. Lett. *2007*, *9* (5), 825-8, using 10 mol% anhydrous FeCl$_3$ in methylene chloride at reflux s. U. Jana, S. Biswas, S. Maiti, ibid. *48* (23), 4065-9; mono-α-benzylation using 5 mol% FeCl$_3$·6H$_2$O in nitromethane, also with benzyl acetates (cf. *69*, 395), s. J. Kischel, K. Mertins, D. Michalik, A. Zapf, M. Beller, Adv. Synth. Catal. *2007*, *349* (6), 865-70.

Via intermediates *v.i.*
Asym. α-allylation of N-carbobenzoxyglycinamides via Eschenmoser-Claisen rearrangement H → C-C=C

537. 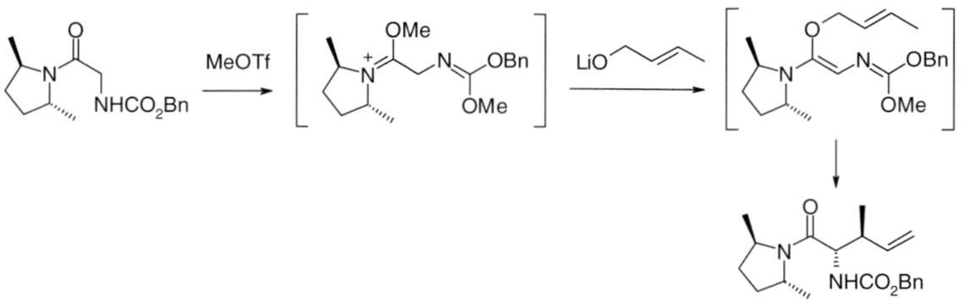

Chiral N-protected *anti*-α-amino-γ,δ-ethylenecarboxylic acid amides. Startg. chiral glycinamide, 2.2 eqs. methyl triflate and 2,6-di-*tert*-butylpyridine stirred under argon in dry methylene chloride for 4 h, the obtained Meerwein salt cooled to -35°, dissolved in methylene chloride (or toluene), 4 eqs. pre-formed Li-allyloxide soln. added by cannula, warmed to ambient temp. during 4 h, quenched with satd. aq. NH$_4$Cl, separated, and dried → product. Y 84.6% (d.e. 87%; *anti/syn* 98:2). Rearrangement occurs at room temp. for linear allyloxides while branched analogs required heating to 60° to effect the reaction with some loss of diastereoselectivity. Highest diastereoselectivities were obtained from *trans*-allyloxides, in keeping with the intermediate formation of (Z)-N,O-ketene acetals with a pseudochair conformation favouring formation of *anti* products. F.e., also removal of the chiral auxiliary via iodolactonization and zinc reduction to afford the corresponding chiral amino acids, s. H. Qu, X. Gu, Z. Liu, B.J. Min, V.J. Hruby, Org. Lett. *2007*, *9* (20), 3997-4000.

Nitrogen ↾ CC ⇃↾ N

Irradiation ///

Δ^2-Pyrazolines from tetrazoles and ethylene derivs. ←

Regiospecific photoactivated 1,3-dipolar cycloaddition under mild conditions

538. 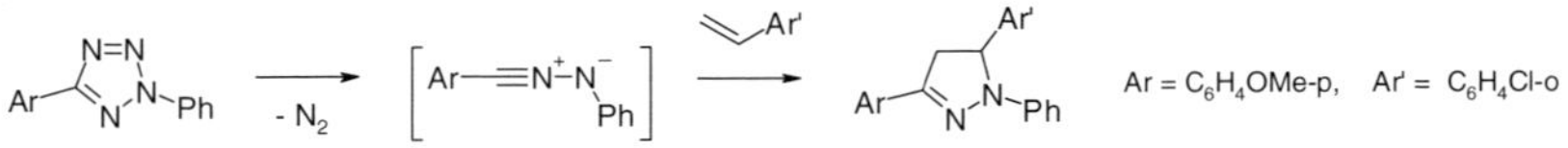

An ethyl acetate soln. of the startg. tetrazole and 2 eqs. dipolarophile irradiated with a hand-held 302-nm UV lamp for 2 h, and solvent and excess of dipolarophile removed under reduced pressure → 2-pyrazoline deriv. Y 100%. The method is mild, regioselective and successful for a range of 1,3-dipolarophiles (not electron-rich alkenes) in a range of solvents (incl. water) and is tolerant of many functional groups (MeO, CO_2Me, CN, CHO, CONHPh, Cl, Br) with yields generally >80%. Both electron-rich and electron-neutral tetrazoles are accommodated but the presence of a nitro substituent on either the N^2- or C^5-phenyl group slows the reaction considerably (only 10% complete after 2 h). The reaction is presumed to involve **nitrile imines as intermediates**. F.e. incl. details for preparation of startg. tetrazoles s. Y. Wang, C.I. Rivera Vera, Q. Lin, Org. Lett. *2007*, *9* (21), 4155-8.

Microwaves s. under $POCl_3$ [\\\\]

Sodium/alcohol/sulfuric acid *NaOR/H_2SO_4*

One-pot Fischer-type indole synthesis with aliphatic nitro compds. under mildly acidic conditions

539. 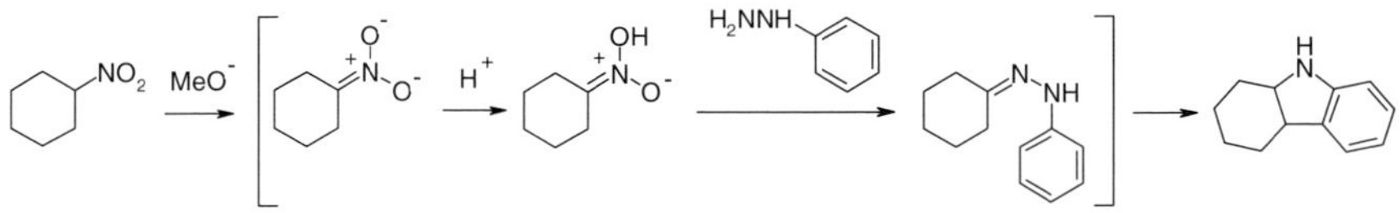

via Nef reaction. 1.1 Eqs. NaOMe in methanol (1 *M*) added under argon at room temp. to a soln. of 1.1 eqs. phenylhydrazine in the same solvent, nitrocyclohexane added, stirred at room temp. for 2 h, cooled to 0°, acidified with 2 eqs. H_2SO_4, heated at 90° for 3 h, cooled to 0°, the bulk of the methanol removed *in vacuo*, and the residue worked up with flash column chromatographic purification on silica gel → product. Y 72%. Reaction is applicable to both prim. and sec. nitro compds., and substituted phenylhydrazines are supported. Base converts the aliphatic nitro compds. into the corresponding nitronate anions which are efficiently protonated to give the corresponding ketone/oxime prior to Fischer condensation with the hydrazine. F.e.s. C.A. Simoneau, A.M. Strohl, B. Ganem, Tetrahedron Lett. *2007*, *48* (10), 1809-11.

Sodium methoxide/hydrogen chloride *NaOMe/HCl*

Ring closures with 2-dimethylaminomethylene-1,3-bis(dimethylimmonio)propane salts
cf. *47*, 318s*71*; *25*, 581s*71*; 3-acyl-5-formyl-2-pyridones from the diperchlorate salt and β-keto-carboxylic acid amides with added HCl to effect hydrolysis s. G. Thouraya, K. Jameleddine, H.A.M. Adnen, B.H. Béchir, Synth. Commun. *2007*, *37* (7), 1053-8.

Potassium tert-*butoxide* *KOBu*-t

N-Protected 3-amino-2,4-pyrrolidiones from 2,5-piperazinediones
Regiospecific alkylative ring contraction with asym. induction – Chiral N-protected 3,5-disubst. 3-amino-2,4-pyrrolidones s. *72*, 478

Sodium bis(trimethylsilyl)amide *$NaN(SiMe_3)_2$*
β-Ketocarboxylic acid allyl esters H → COOC-C=C
from ketones and N-carballyloxyimidazoles s. *72*, 112

*Sodium bromide s. under 5,10,15,20-Tetrakis(*p*-chlorophenyl)porphyrinato-iron(III) chloride* *NaBr*

Tetrakis(acetonitrile)copper(I) hexafluorophosphate/chiral tris(Δ^2-oxazolines) ←
Silver hexafluoroantimonate *$AgSbF_6$*
Cyclopropane ring from diazo compds. and ethylene derivs.
silylcyclopropane ring with $Cu(OTf)_2$ cf. *23*, 819s*71*; 1-arylcyclopropanecarboxylic acid esters with 10 mol% $AgSbF_6$ in refluxing methylene chloride (d.e. >94%), especially for hindered compds., s. J.L. Thompson, H.M.L. Davies, J. Am. Chem. Soc. *2007*, *129* (19), 6090-1; asym. cyclopropanation with $[Cu(MeCN)_4]PF_6$/chiral tris(Δ^2-oxazolines) (e.e. up to 95%) s. Z.-H. Xu, S.-N. Zhu, X.-L. Sun, Y. Tang, L.-X. Dai, Chem. Commun. *2007* (19), 1960-2; with chiral pentadentate Co(II)-salen complexes bearing an imidazolyl or pyridyl moiety as the fifth coordinating group, also asym. epoxidation with Mn(III) analogs/aq. H_2O_2 (cf. *46*, 106s*70*), s. H. Shitama, T. Katsuki, Chem. Eur. J. *2007*, *13* (17), 4849-58.

Copper(II) triflate/chiral 1,2-diamines ←
Asym. Michael addition of enacylamines to alkylidenemalonic acid esters C=C → CHC(R)
Chiral δ-ketomalonic acid esters s. *72*, 419

Magnesium bromide-etherate/ethyldiisopropylamine *$MgBr_2{\cdot}OEt_2$/i-Pr_2NEt*
β-Diketones from ketones and 1-acylbenzotriazoles H → COR

540. 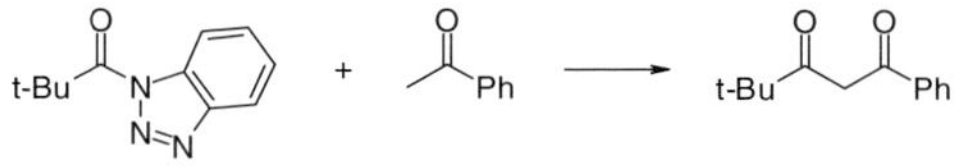

Acetophenone added dropwise via syringe to a stirred mixture of 1.2 eqs. N-pivaloylbenzotriazole and $MgBr_2{\cdot}OEt_2$ in methylene chloride, 1.27 eqs. *i*-Pr_2NEt added (the suspension changing from colorless to yellow), stirred for 2.5 h, then for a further 5 min with aq. HCl, extracted, and dried → the β-diketone. Y 99%. The method is simple (using reagent-grade solvent without further purification under air), efficient, and gives good yields with both ar. and aliphatic ketones,while tolerating MeO, $OSiMe_2Bu$-*t*, OCO_2Bu-*t* and ethenyl functions. Both pentafluorophenyl esters and N-acylbenzotriazoles are more efficient acylating agents than acyl chlorides and, being less sensitive to moisture, require no stringent reaction conditions. The method is based on a **soft enolization approach** wherein a weak base and a Lewis acid effect reversible deprotonation. F.e.s. D. Lim, F. Fang, G. Zhou, D.M. Coltart, Org. Lett. *2007*, *9* (21), 4139-42.

Triphenyl borate/chiral 3,3′-diphenyl-2,2′-bi-1-naphthol or 2,2′-diphenyl-3,3′-biphenanthrene-4,4′-diol ←
Indium(III) triflate *$In(OTf)_3$*
Boron fluoride *BF_3*
Aziridines from diazo compds.
2-carbalkoxy derivs. from azomethines with Rh(II)-exchanged clay cf. *53*, 387; effect of Lewis acid and solvent in the synthesis of aziridine-2-phosphonic acid esters, especially formation of *cis*-isomers from N-(diphenylmethyl)benzylimines with $In(OTf)_3$ in methylene chloride at 0°, s. R. Pellicciari, L. Amori, N. Kuznetsova, S. Zlotsky, A. Gioiello, Tetrahedron Lett. *2007*, *48* (28), 4911-4; asym. synthesis of *cis*-2-carbalkoxy-derivs. with $(PhO)_3B$/(S)-VAPOL or (S)-VANOL (e.e. up to 97%), also conversion to chiral β-amino esters, and racemic products with $BF_3{\cdot}OEt_2$, s. Z. Lu, Y. Zhang, W.D. Wulff, J. Am. Chem. Soc. *2007*, *129* (22), 7185-94.

Chiral ar. diols s. under Triphenyl borate ←
Chiral tris(Δ^2-oxazolines) s. under $[Cu(MeCN)_4]PF_6$ ←
*(S)-6,6′-Dimethoxy-2,2′-bis[bis(*m*-methoxyphenyl)phosphino]biphenyl s. under 5,10,15,20-Tetrakis(*p*-chlorophenyl)porphyrinatoiron(III) chloride* ←

Phosphorus oxide chloride/microwaves *POCl₃/[\\\\]*
Ring closures via Vilsmeier reaction ○
under microwave irradiation in DMF cf. *22*, 826s*71*; 3-formyl-4-chromanones from *o*-hydroxy-ketones in 4 eqs. DMF-$POCl_3$ s. P. Kamotra, A.K. Gupta, R. Gupta, Indian J. Chem. *2007*, *46B* (5), 866-8.

1-Butyl-3-methylimidazolium hydrogen sulfate *[bmim]HSO_4*
Fischer indole synthesis
in choline chloride·2$ZnCl_2$ cf. *61*, 357s*67*; in [bmim]HSO_4 as [recyclable] Brønsted acidic ionic liquid solvent/catalyst with high regioselectivity with unsym. ketones s. D.-Q. Xu, W.-L. Yang, S.-P. Luo, B.-T. Wang, J. Wu, Z.-Y. Xu, Eur. J. Org. Chem. *2007* (6), 1007-12.

Sulfuric acid s. under NaOR *H_2SO_4*
Hydrogen chloride s. under NaOMe *HCl*

*5,10,15,20-Tetrakis(*p*-chlorophenyl)porphyrinatoiron(III) chloride/(S)-6,6′-dimethoxy-2,2′-bis[bis(*m*-methoxyphenyl)phosphino]biphenyl/sodium bromide* ←
Asym. synthesis of α-allenecarboxylic acid esters from ketenes with addition of two C-atoms C=C=CHCOOR

541. 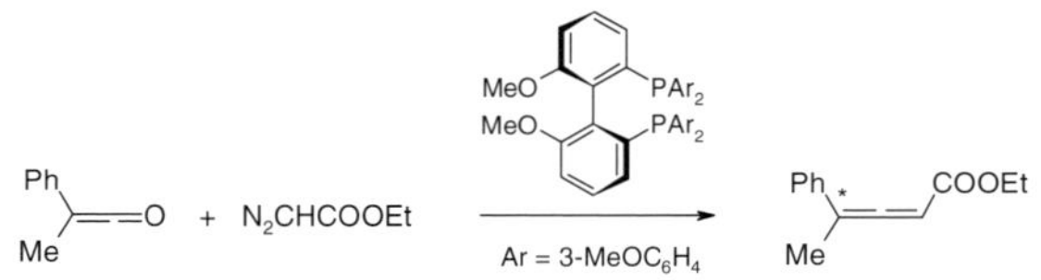

under mild, neutral conditions. 2 Eqs. ethyl diazoacetate added in four portions every 30 min to a stirred soln. of *0.5 mol%* 5,10,15,20-tetrakis(*p*-chlorophenyl)porphyrinatoiron(III) chloride and 1 eq. (S)-6,6′-dimethoxy-2,2′-bis[bis(*m*-methoxyphenyl)phosphino]biphenyl in toluene at room temp. under N_2, concentrated after ca. 6 h, treated with 1 eq. NaBr, taken up in THF, cooled to -65°, the startg. ketene added via syringe, stirring continued for 72 h, and the mixture worked up with column chromatographic purification → 4-phenylpenta-2,3-dienoic acid ethyl ester. Y 72% (e.e. 93%). The procedure was applied to the asym. synthesis of 4-alkyl-4-arylallenecarboxylic acid esters in high yield and with high enantioselectivity (93-8%), this being the development of preliminary syntheses of racemic mono- and di-subst. α-allenecarboxylic acid esters with the same iron catalyst and triphenylphosphine as ligand. Mechanistically, it is presumed that the key step in the reaction is the catalytic transfer of the carbethoxymethylene group from iron to *one* of the phosphorus atoms of the ligand to provide a reactive mono-P-ylid. The chiral di(phosphine) is readily recovered by reduction of the liberated phosphine oxide with $HSiCl_3$/dimethylaniline. F.e.s. C.-Y. Li, X.-B. Wang, X.-L. Sun, Y. Tang, J.-C. Zheng, Z.-H. Xu, Y.-G. Zhou, L.-X. Dai, J. Am. Chem. Soc. *2007*, *129* (6), 1494-5.

Iron(II) sulfate *$FeSO_4$*
2-Arylalkoxylamines C=C → C(Ar)C(ON<)
from ethylene derivs. and diazonium fluoroborates
Radical aryl(aminooxyl)ation

542. 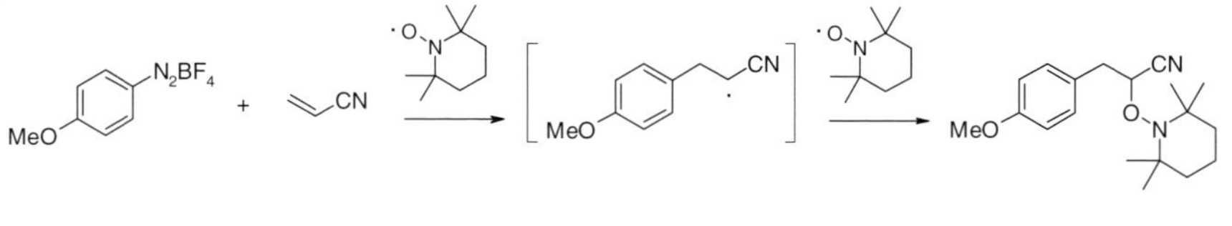

A soln. of *p*-methoxybenzenediazonium fluoroborate in DMSO/water (2:1) added over 10-15 min to a stirred and degassed mixture of 10 eqs. acrylonitrile, 3 eqs. $FeSO_4$·7H_2O and 2 eqs. 2,2,6,6-tetramethylpiperidine nitroxyl (TEMPO) in DMSO/water (1:2) under argon, stirred 10 min, 3 eqs. ascorbic acid added, stirred 10 min, diluted with water, and extracted with ether → 3-(4-methoxyphenyl)-2-(2,2,6,6-tetramethylpiperidin-1-yloxy)propionitrile. Y 84%. The use of the polar solvent reduces the reactivity of TEMPO, allowing the olefins to compete for the generated aryl radicals; TEMPO then traps the products of the addition step to complete the process. Examples are given using both electron-rich and -poor diazonium salts, but yields are variable (31-84%). Two cyclic olefins were *trans*-selective with diastereomeric ratios of 72:28 and 90:10. The addition of ascorbic acid to the reaction mixture prior to workup facilitates product isolation. F.e. and cleavage of the alkoxylamine bond with Zn/AcOH (cf. *39*, 351) s. M.R. Heinrich, A. Wetzel, M. Kirschstein, Org. Lett. *2007*, *9* (19), 3833-5.

Chiral cobalt(II) porphyrin complex ←
***trans*-Cyclopropanecarboxylic acid esters** C=C → ▽
from electron-deficient ethylene derivs. and diazoacetic acid esters
Cobalt(II)-catalyzed asym. conversion

543. 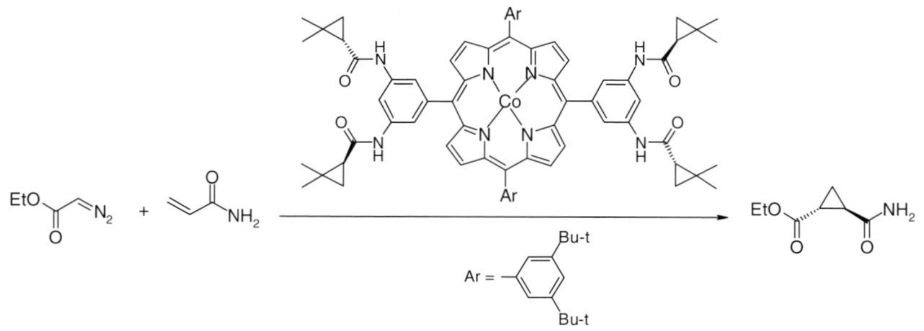

under mild conditions. A mixture of acrylamide and 1.2 eqs. ethyl diazoacetate stirred under N_2 with 1 mol% chiral cobalt(II) porphyrin complex in chlorobenzene at room temp. for 20 h → (1R,2R)-ethyl 2-aminocarbonylcyclopropanecarboxylate. Y 90% (d.r. 99:1; e.e. 90%). The reaction is successful for ethyl and *tert*-butyl diazoacetates reacting with vinyl esters, amides (incl. substituted derivs.), ketones and nitriles, and is tolerant of an α-methyl group giving 73-98% yield, 52-99% *trans*-selectivity and 61-97% e.e. Toluene is also a suitable solvent with use of chlorobenzene showing a dramatic improvement in low-yielding cases. Lowest selectivities were obtained with 2-methylacrylonitrile as substrate. F.e.s. Y. Chen, J.V. Ruppel, X.P. Zhang, J. Am. Chem. Soc. *2007*, *129* (40), 12074-5.

Chiral pentadentate cobalt(II) salen complexes *Co(II)-salen*
Asym. cyclopropanation with diazo compds. s. *23*, 819s*72*

Palladium(II) acetate *Pd(OAc)$_2$*
Heck-type arylation with diazonium salts ←
4-aryl-N-carbalkoxy-1,2,3,4-tetrahydro- from N-carbalkoxy-1,2,3,6-tetrahydro-pyridines cf. *38*, 793s*71*; chiral 4-aryl-1-carbalkoxy-Δ^3-pyrroline-2-carboxylic acid esters from N-carbomethoxy-Δ^3-dehydro-L-proline methyl ester via 2-alkoxy-4-arylproline derivs. (in refluxing methanol) or 2-hydroxy-analogs (in 6:3:1 water/acetonitrile/AcOH) s. K.P. da Silva, M.N. Godoi, C.R.D. Correia, Org. Lett. *2007*, *9* (15), 2815-8.

Palladium(II) acetate/methanol *Pd(OAc)$_2$/MeOH*
Sym. biaryls from arenediazonium fluoroborates 2 ArN_2BF_4 → Ar-Ar

544. 

p-Methoxybenzenediazonium fluoroborate (0.5 mmol), Pd(OAc)$_2$ (15 mol%) and methanol (1 ml) stirred under reflux until diazonium salt consumed (<1 h), then filtered through Celite →

product. Y 88%. Unlike other, commonly-used, biaryl coupling methods, the procedure is operationally simple, can be carried out with no special oxygen- or water-exclusion techniques and does not require the addition of a zero-valent metal or other terminal reductant, the role of which is likely performed by methanol (supported by the fact that Pd-black is formed during the course of the reaction). A number of substituents, including aromatic chlorine and bromine, are tolerated, although the corresponding *p*-cyano-substrate affords none of the required biaryl, possibly due to competing coordination of the cyano group with $Pd(OAc)_2$. By-product formation (including simple replacement of the diazonium group with hydrogen, methoxy or fluorine) is minimized by increasing the concentration of the reaction mass, in keeping with the proposed mechanism involving oxidative addition of the arenediazonium salt to low-valent palladium. F.e.s. M.K. Robinson, V.S. Kochurina, J.M. Hanna, Jr., Tetrahedron Lett. *2007*, *48* (43), 7687-90.

Halogen ↑ CC ↓↑ Hal

Without additional reagents *w.a.r.*

3-Component synthesis of 4-hydroxy-2-pyridone-5,6-dicarboxylic acid esters ○

545. 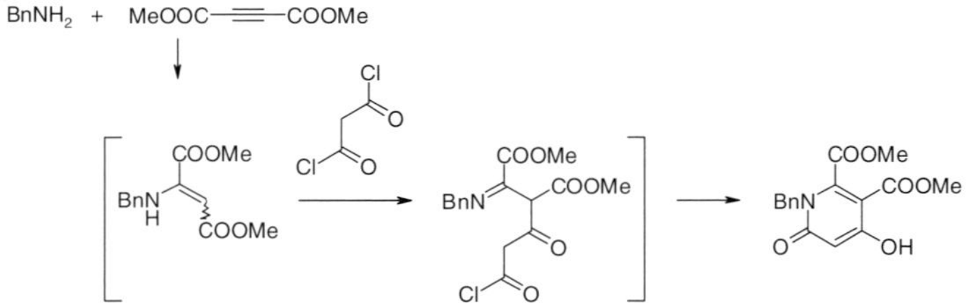

Malonyl dichloride added at room temp. to a soln. of 1.2 eqs. dimethyl acetylenedicarboxylate and 1 eq. benzylamine in methylene chloride, stirred for 24 h, and worked up with flash column chromatographic purification on silica gel → product. Y 85%. The procedure is simple, does not require any additives, and is generally applicable to alkyl- and benzyl-amines. It appears that reaction involves initial formation of an enaminone, followed by attack of malonyl dichloride prior to cyclization, elimination of HCl and keto-enol tautomerism. F.e.s. I. Yavari, S. Souri, Synlett *2007* (19), 2969-70.

Irradiation/triethylamine *⫽/Et_3N*

Metal-free synthesis of crowded biaryls from arenes Ar-Ar′

546. 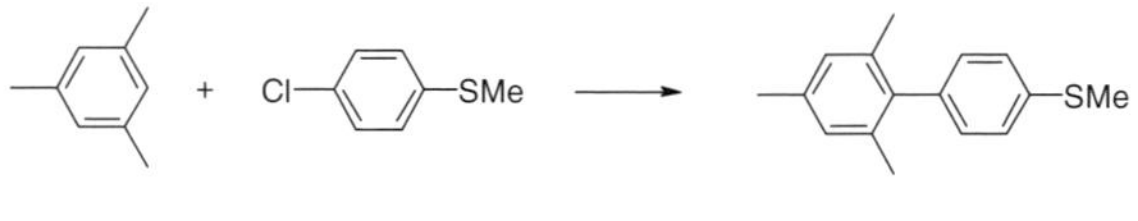

and ar. chlorides under irradiation. A 0.05 *M* soln. of the startg. aryl chloride (1.5 mmol), Et_3N (0.05 *M*) and mesitylene (1 *M*) in 2,2,2-trifluoroethanol purged with argon for 10 min in two quartz tubes, the latter closed with serum caps, irradiated with six 15 W phosphor-coated lamps (emission centred at 310 nm) for 8 h *at room temp.*, concentrated under reduced pressure, and worked up with column chromatographic purification → product. Y 70%. The procedure is atom-economical, mild, unaffected by moisture, and environmentally friendly as no metallic residues are in the waste. It is also broad in scope, being applicable to the coupling of methylated benzenes with ar. chlorides, aryl triflates, aryl mesylates or aryl phosphates possessing electron-donating groups in the aromatic ring (MeO, MeS, siloxy, amino, hydroxyl, methylenedioxy) at any position. F.e. incl. tetra-*ortho*-subst. biphenyls s. V. Dichiarante, M. Fagnoni, A. Albini, Angew. Chem. Int. Ed. Engl. *2007*, *46* (34), 6495-8.

Microwaves s. under CuI and Pd(OAc)$_2$ [\\\\\]

Potassium hydroxide/N-(3,4,5-trimethoxybenzyl)quinidinium bromide ←
Asym. synthesis of (Z)-ketimines from iminochlorides under phase transfer catalysis C(=N)Cl → C(=N)R

547. 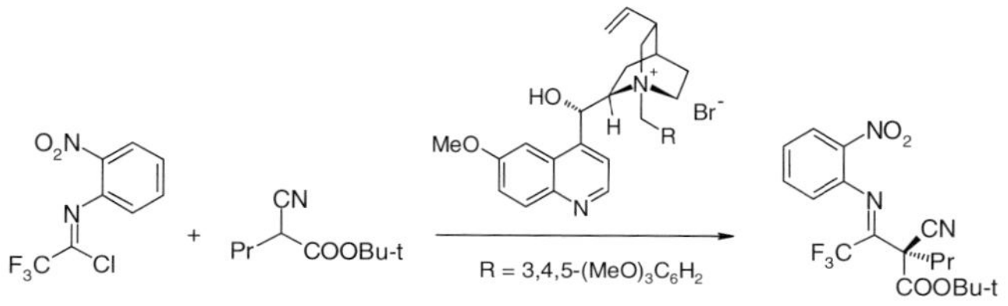

The first example of an asym. coupling of an iminohalide with an electrophile is reported. **E: Chiral α-subst. α-cyano-β-iminocarboxylic acid esters.** A soln. of the startg. α-cyanoacetate, 1.3 eqs. startg. iminohalide and 6 mol% N-(3,4,5-trimethoxybenzyl)quinidinium bromide in 7:1 *o*-xylene/chloroform stirred for a short time at room temp., cooled to -20°, a cold (-20°) soln. of 20% aq. KOH added, and stirred vigorously for 18 h → (S,Z)-*tert*-butyl 2-cyano-2-[2,2,2-trifluoro-1-(2-nitrophenylimino)ethyl]pentanoate. Y 76% (e.e. 86%). The procedure is simple, inexpensive, applicable to α-*prim*-alkyl- and α-*sec*-alkyl-subst. cyanoacetates (preferably *tert*-butyl esters), and notable in that **chiral quaternary hydrocarbon groups** can be generated with high enantioselectivity (up to 90%). However, it appears that an electron-withdrawing aryl group (preferably *o*-nitrophenyl) on nitrogen is essential for high enantioselectivity, which was low with α-aryl-α-cyanoacetates. F.e. and with carbethoxyacetimidoyl chloride, also comparison of other Cinchona derivs. as phase transfer catalysts, s. S. Santoro, T.B. Poulsen, K.A. Jørgensen, Chem. Commun. *2007* (48), 5155-7.

Potassium tert-*butoxide* *KOBu*-t
4-Component synthesis of 2-alkylthio-4(3*H*)-pyrimidinones ○

548. 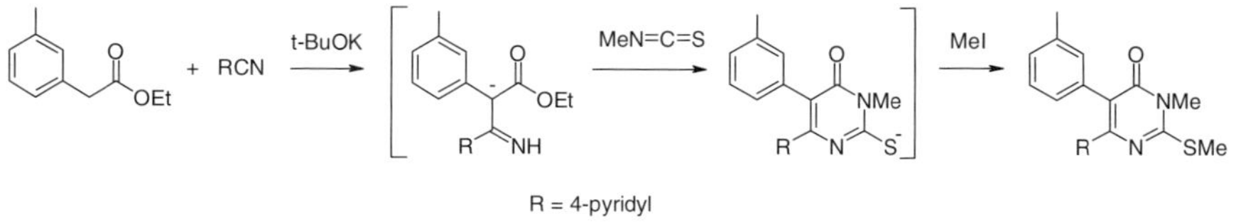

A soln. of ethyl 3-methylphenylacetate and ca. 1 eq. 4-cyanopyridine in DMF treated slowly with 1 eq. 1 *M* KOBu-*t* in *tert*-butanol, stirred at room temp. for 1.5 h, a soln. of 1.2 eqs. methyl isothiocyanate in DMF added below 25°, stirred at room temp. for 1 h, cooled to 0°, treated with 1 eq. methyl iodide below 5°, stirring continued at room temp. for 1 h, cooled to 0°, quenched with deionized water below 10°, and the suspension stirred again at room temp. for 16 h → product. Y 51%. Reaction appears limited to 3- or 4-cyano-pyridines and -quinolines, while 2-(4-fluorophenyl)acetates gave a complex mixture of products. Enolate addition to the nitrile is the initiating event. F.e.s. Y. Lu, T. Xiang, M.D. Bartberger, C. Bernard, T. Bostick, L. Huang, L. Liu, A. Siegmund, G. Sukay, G. Guo, M. Silva Elipe, W. Tormos, C. Dominguez, K. Koch, L.E. Burgess, T.C. Basil, P. Ibrahim, C. Hummel, Tetrahedron *2006*, *62* (50), 11714-23.

n-*Butyllithium/hexamethylphosphoramide* *BuLi/HMPA*

1-Acoxycyclopentadienes

from 1,4-diiodo-1,3-dienes and two carboxylic acid anhydride molecules ○

549.

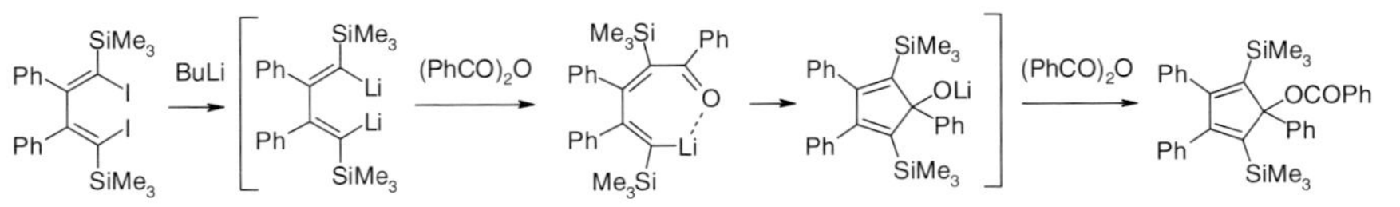

4 Eqs. 1.5 *M* *n*-BuLi in pentane added under N_2 at -78° to the startg. 1,4-diiodo-1,3-diene in dry ether, stirred for 1 h, treated with 4.4 eqs. HMPA, stirring continued for 20 min, 2.1 eqs. benzoic anhydride added, allowed to warm to room temp., quenched with water after a further 1 h, and worked up with column chromatographic purification on silica gel → product. Y 88%. Initial generation of the 1,4-dilithio-1,3-diene is followed by coupling with one molecule of the anhydride to give a stabilized lithioketone, which then undergoes ring closure prior to O-acylation with a second molecule of the anhydride. The latter may be aromatic or aliphatic, and the diiododiene may be alkyl-substituted. F.e. and with carboxylic acid chlorides, also reaction with DMF, s. C. Wang, G.L. Mao, Z.H. Wang, Z.F. Xi, Eur. J. Org. Chem. *2007* (8), 1267-73.

tert-*Butyllithium* t-*BuLi*

Methylarenes from ar. bromides via aryllithium compds. Br → Me

550.

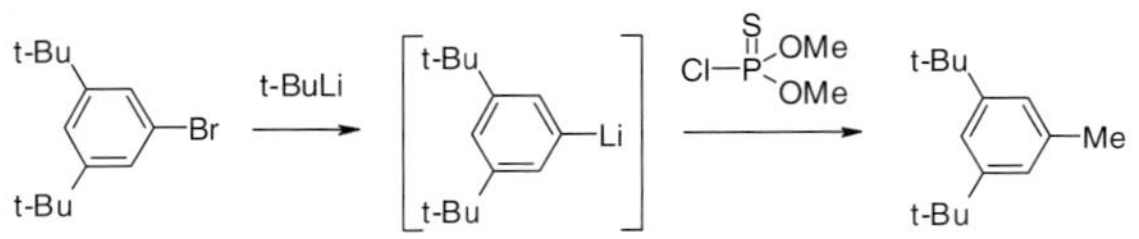

2.1 Eqs. *t*-BuLi in pentane added dropwise to a soln. of the startg. aryl bromide in dry THF at -78°, stirred for 1 h, 1.1 eqs. dimethyl thionophosphorochloridate added at the same temp., allowed to warm to room temp. over 4 h with continued stirring, quenched with water, and worked up with column chromatographic purification on silica gel → product. Y 80%. There was no reaction with the corresponding diethyl ester. F.e.s. N. Ribeiro, J. Maeda, Y. Ishida, Y. Kobayashi, K. Saigo, Chem. Lett. *2007*, *36* (12), 1412-3.

Metal-free synthesis of *o*-bromobiaryls from aryllithium compds. and *o*-dibromides Ar-Ar′

Benzynes as intermediates

551.

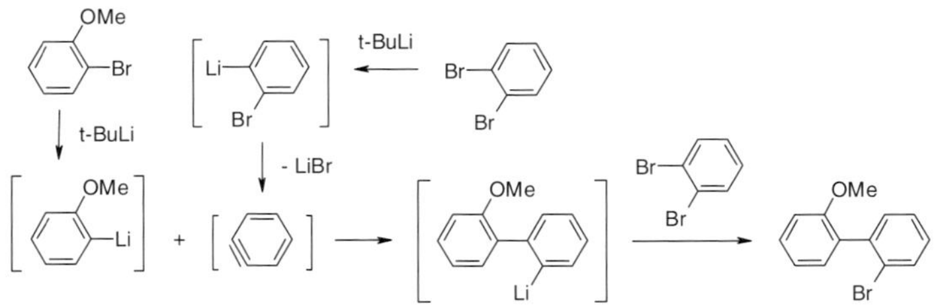

2-Bromoanisole added at -75° to a soln. of *2 eqs.* *t*-BuLi in 3:7 pentanes/THF, after 5 min the temp. increased to -25° for 15 min, a soln. of 1 eq. 1,2-dibromobenzene in THF added dropwise over 10 min, allowed to warm to room temp., water added, and the mixture worked up with column chromatographic purification on silica gel → 2′-bromo-2-methoxybiphenyl. Y 67%. One equivalent of the base is required for the initial formation of the aryllithium compd., and the second reacts with the *o*-dibromide to generate benzyne via elimination of LiBr from 2-bromo-phenyllithium; addition of the thermally less-sensitive aryllithium compd. ensues, followed by transfer of bromine from the *o*-dibromide. *o,o′*-Dibromo- and *o*-bromo-*o′*-iodo-biphenyls were obtained similarly **by homocoupling** of two molecules of 1,2-dibromo- and 1-bromo-2-iodo-benzene, respectively. The coupling is clean, inexpensive and notably suitable for preparing unsym. *o,o′*-di-, tri- and even tetra-subst. bromo- and iodo-biaryls on the gram scale. F.e. incl. the synthesis of 10,10′-dibromo-9,9′-biphenanthryl by homocoupling, and generation of the aryl-lithium compds. by direct lithiation of arenes, s. F.R. Leroux, L. Bonnafoux, C. Heiss, F. Colobert, D.A. Lanfranchi, Adv. Synth. Catal. *2007*, *349* (17-18), 2705-13.

Lithium diisopropylamide i-*Pr_2NLi*

α-Substitution of sec. nitriles with 1,2-asym. induction ←

552.

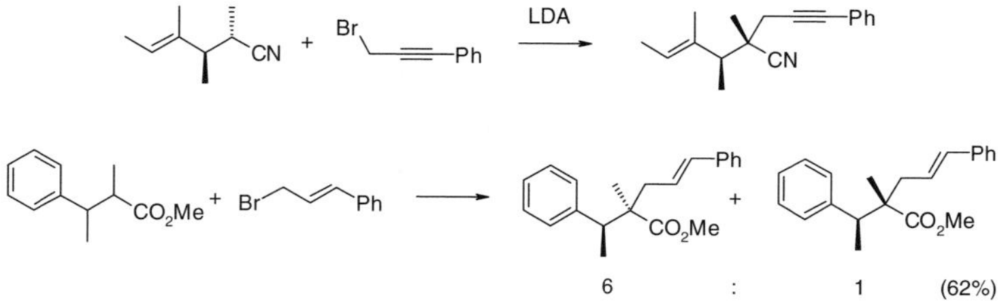

N-Metallonitriles as intermediates. Startg. nitrile added to a soln. of LDA (from 1.05 eqs. BuLi and 1.15 eqs. diisopropylamine in THF) at -78°, the mixture stirred for 50 min, 1.2 eqs. 1-phenyl-3-bromopropyne added, stirring continued for 3 h, the reaction quenched with satd. aq. NH_4Cl, extracted with ethyl acetate, the extracts dried over $MgSO_4$, concentrated *in vacuo*, and purified by radial chromatography → product. Y 79%. The proposed substitution model (supported by X-ray crystallography) suggests that the intermediate metalated nitrile adopts a conformation in which the smaller nitrile group occupies the more sterically-demanding environment, thereby forcing attack from the opposite face **with inversion of configuration.** Nine electrophiles (esters, carbonyl compds., thiosulfonates and alkyl halides) gave single diastereomers (apart from cyclohexanone) in 72-84% yield (methyl cyanoformate giving 58% yield and sterically demanding isopropyl iodide 52%). By contrast an ester analog gave a mixture of diastereomers under these conditions. F.e.s. F.F. Fleming, W. Liu, S. Ghosh, O.W. Steward, Angew. Chem. Int. Ed. Engl. *2007*, *46* (37), 7098-100.

Lithium bis(trimethylsilyl)amide *$LiN(SiMe_3)_2$*

C-α-Alkylation of N-acyl-2-iminooxazolidines with asym. induction H → R

553.

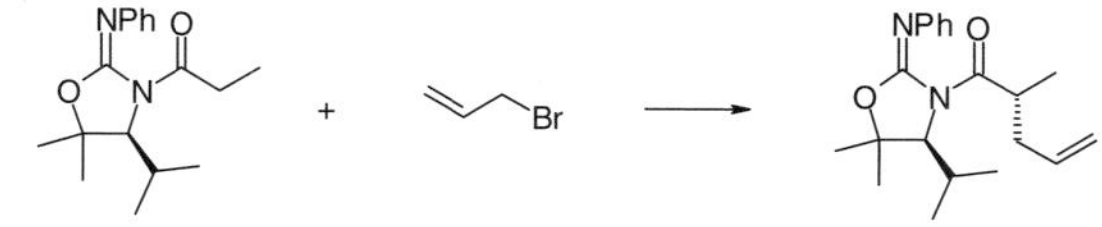

A soln. of $LiN(SiMe_3)_2$ in THF (1.0 *M*; 2-4 eqs.) added dropwise to a soln. of (S)-N-propionyl-5,5-dimethyl-4-isopropyl-2-(phenylimino)oxazolidine (0.1 g) [prepared in 63% yield by treatment of (S)-5,5-dimethyl-4-isopropyl-2-(phenylimino)oxazolidine with *t*-BuOK/propionyl chloride] in dry THF (4 ml) at -78° under N_2, stirred 30 min, allyl bromide (3-8 eqs.) added to the resulting Li-enolate, and stirred for 30 min at -78°, then a further 1 h at 0°, before quenching with satd. aq. NH_4Cl → (R)-alkylated product. Y 99% (d.e. >99%). Hydrolysis of the products with 2 *M* NaOH in refluxing dioxane affords the corresponding **chiral carboxylic acids** along with recovered chiral auxiliary. The 5,5-dimethyl subst. chiral auxiliary was shown to be superior to its 5,5-diphenyl- and previously-reported unsubst. analogs, affording alkylated products in high yields and excellent diastereoselectivities. F.e.s. T.N. Le, Q.P.B. Nguyen, J.N. Kim, T.H. Kim, Tetrahedron Lett. *2007*, *48* (44), 7834-7.

*Potassium carbonate/tetra-*n*-butylammonium iodide/chiral cyclic thioethers* ←

Asym. synthesis of N-tosylaziridines from N-tosylimines and bromides

554. 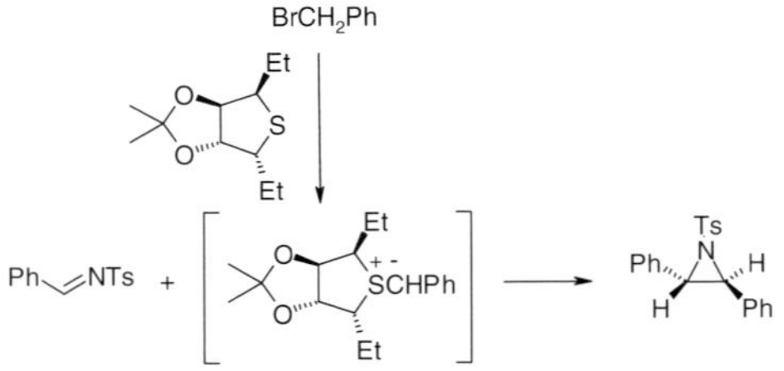

via chiral sulfonium ylids. A mixture of the chiral C_2-symmetric cyclic thioether (readily prepared from inexpensive L-tartaric acid), 1.5 eqs. benzyl bromide, 0.5 eq. startg. N-tosylimine, 1.5 eqs. K_2CO_3 and 5 mol% tetra-*n*-butylammonium iodide (as phase transfer catalyst) *in acetonitrile* stirred for 2 days at room temp. → product. Y 72% (*trans/cis* 75:25; e.e. 96%). Enantioselectivity was in the range 80-96%, and the *trans/cis* ratio reached a maximum at 93%. F.e.s. Y. Gui, S. Shen, H.-Y. Wang, Chem. Lett. *2007*, *36* (12), 1436-7.

Cesium carbonate *Cs_2CO_3*

Synthesis of α-(indan-1-yl)carbonyl compds.
from (E)-β-[*o*-(chloromethyl)aryl]-α,β-ethylenecarbonyl compds. and CH-acidic compds.
α-Benzylation-intramolecular Michael addition

555. 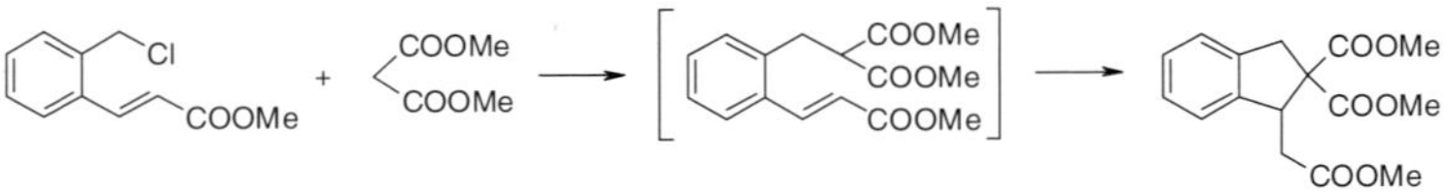

1,2,2-Trisubst. indans. 2 Eqs. dimethyl malonate and 3 eqs. Cs_2CO_3 added to a stirred soln. of the startg. chloroester (0.475 mmol) in DMF, stirred at room temp. for 3 h (TLC monitoring), and worked up with column chromatographic purification on silica gel → product. Y 81%. The procedure is applicable to CH-acidic compds. bearing ester, keto or sulfonyl groups (pK_a 10-15), but 1,1-disulfones required harsher conditions. 2-Spiroindan derivs. were also obtained from enoates and 1,3-indanedione (with added tetrabutylammonium bromide), but a mixture of products was formed with enones. F.e. and diastereoselectivity s. S.J. Gharpure, S.R.B. Reddy, U. Sanyal, Synlett *2007* (12), 1889-92.

Potassium phosphate s. under $Pd(OAc)_2$ *K_3PO_4*

Triethylamine *Et_3N*
Sym. 3-subst. 3-acoxy-1,4-diyne-1,5-dicarboxylic acid esters ←
from two propiolic acid ester and two carboxylic acid chloride molecules

556. 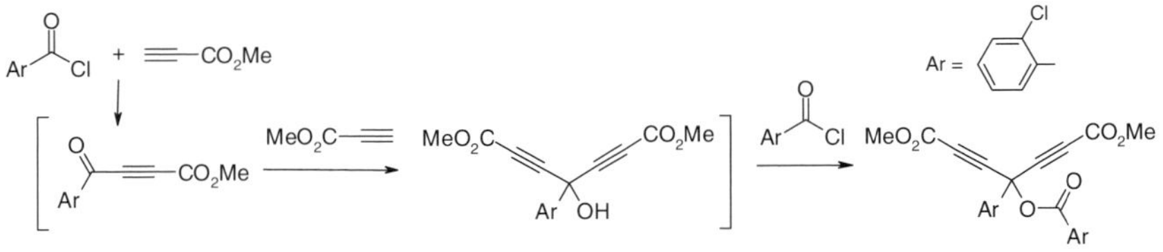

under mild conditions. 2-Chlorobenzoyl chloride and 2.2 eqs. methyl propiolate dissolved in methylene chloride, cooled to 0°, 1.05 eqs. triethylamine added, the mixture stirred at room temp. for 6 h, filtered, solvent removed *in vacuo*, and the product purified by chromatography on silica → dimethyl 4-(2-chlorobenzoyloxy)-4-(2-chlorophenyl)hepta-2,5-diynoate. Y 77%.

The method is successful with aromatic acid chlorides in 69-85% yield (9 examples), even with deactivated 4-methoxybenzoyl chloride (Y 34%). Aliphatic acid chlorides also participated in lower yield (4 examples; Y 28-56%), but here reaction is limited to substrates lacking α-hydrogens or those with low acidity as ketene formation is a likely side reaction. Other nucleophiles such as DABCO or tri-*n*-butylphosphine were ineffective triggers. F.e.s. D. Tejedor, S. López-Tosco, J. González-Platas, F. García-Tellado, J. Org. Chem. *2007*, *72* (14), 5454-6.

Triethylamine/1-butyl-3-methylimidazolium hexafluorophosphate *Et_3N/[bmim]PF_6*
Ethyldiisopropylamine i-*Pr_2NEt*
2-Azetidinones from azomethines and carboxylic acid chlorides
cf. *7*, 836; *trans*-3-acoxy-1-phenylthio-derivs. from N-phenylsulfenylimines with *i*-Pr_2NEt s. S. Coantic, D. Mouysset, S. Mignani, M. Tabart, L. Stella, Tetrahedron Lett. *2007*, *48* (24), 4301-3; library synthesis of ionic liquid-based *cis*-2-azetidinones with Et_3N in [bmim]PF_6 as ionic liquid solvent, and removal of the imidazolium moiety by transesterification s. X.-L. Tao, M. Lei, Y.-G. Wang, ibid. *48* (29), 5143-6.

2(S)-[Diphenyl(trimethylsiloxy)methyl]pyrrolidine/potassium carbonate
4-Formylcyclopentanone-2-carboxylic acid esters
from α,β-ethylenealdehydes and γ-bromo-β-ketocarboxylic acid esters
Organocatalyzed asym. Michael addition-intramolecular α-alkylation

557. 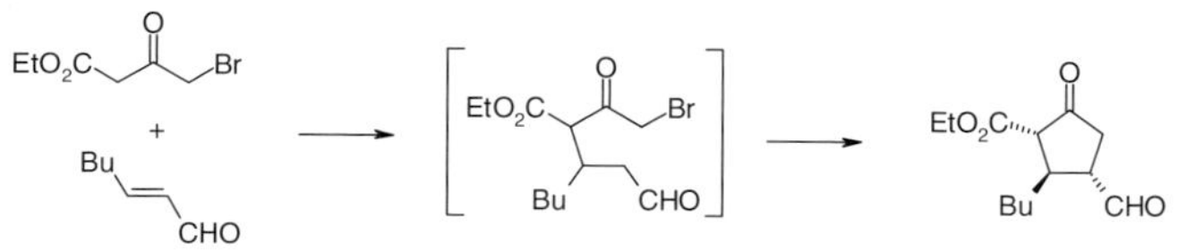

The startg. enal, 0.83 eq. K_2CO_3 and 0.83 eq. ethyl 4-bromo-3-oxobutanoate added to a stirred soln. of 20 mol% 2(S)-[diphenyl(trimethylsiloxy)methyl]pyrrolidine in chloroform, stirred vigorously for 14 h at room temp., and worked up with column chromatographic purification on silica gel → product. Y 88% (d.r. 8:1; e.e. 98%). Very high enantioselectivity (93-9%) was recorded, the diastereoselectivity increasing with decrease in temp. to 4°. It is presumed that reaction is initiated by iminium activation of the enal, followed by Michael addition of the keto-ester, intramolecular [5-*exo-tet*] α-alkylation and hydrolysis. F.e. and comparison of chiral bases, also reduction to the corresponding **chiral 2-hydroxy-4-(hydroxymethyl)cyclopentane-carboxylic acid esters,** s. R. Rios, J. Vesely, H. Sunden, I. Ibrahem, G.-L. Zhao, A. Cordova, Tetrahedron Lett. *2007*, *48* (33), 5835-9.

Copper(I) chloride s. under Cp_2ZrEt_2 and Cp_2ZrCl_2 *CuCl*
Copper(I) iodide s.a. under $Pd(PPh_3)_4$ *CuI*

Copper(I) iodide/potassium carbonate *CuI/K_2CO_3*
Benzofuran-3-carboxylic acid esters
from *o*-bromoiodides and β-ketocarboxylic acid esters

558. 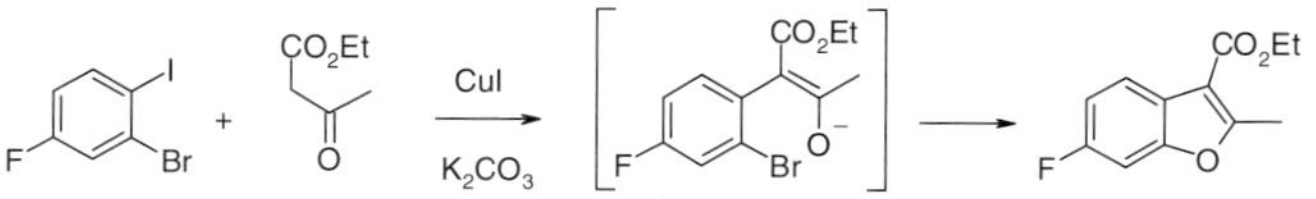

One-pot reaction. 2 Eqs. ethyl acetoacetate and THF added to a mixture of 10 mol% CuI, 3 eqs. K_2CO_3 and 1-iodo-2-bromo-4-fluorobenzene in a Schlenk tube under argon, the tube sealed, the mixture heated at 100° for 24 h, the cooled soln. diluted with ethyl acetate, washed with satd. brine, the organic layer dried over Na_2SO_4, concentrated *in vacuo*, and the residue purified by chromatography on silica → 2-methyl-3-carbethoxy-6-fluorobenzofuran. Y 80%. In this *domino reaction*, both the initial intermolecular coupling and the intramolecular ring closure are effected by the same catalyst. The method is successful for electronically diverse *o*-bromoiodobenzenes in the presence of chloro, fluoro, ether, silyl ether, nitro, acetal, ketone and ester functionality to give 70-88% yields of single regioisomers. Low yields were obtained, however, with a hydroxymethyl deriv. (32%, increased to 75% on protection as a silyl ether) and a nitro deriv. (52%, attributed to partial instability under these conditions). The reaction is also sensitive to steric hindrance at the γ-position of the keto ester, with ethyl isobutyrylacetate and methyl benzoylacetate affording 32% and 0% yield, respectively. A bis-benzofuran was obtained from 1,4-dibromo-2,5-diiodobenzene in 44% yield. F.e.s. B. Lu, B. Wang, Y. Zhang, D. Ma, J. Org. Chem. *2007*, *72* (14), 5337-41.

Copper(I) iodide/N-methylimidazole/potassium hexacyanoferrate ←
Hetar. nitriles from halides s. *67*, 419s*72* Hal → CN

*Copper(I) iodide/triphenylphosphine/tetra-*n*-butylammonium bromide/ potassium carbonate/microwaves* ←
Palladium-free Sonogashira coupling C≡CH → C≡CR
under microwave irradiation in water s. *66*, 384s*72*

Silver acetate s. under Pd(OAc)$_2$ *AgOAc*

Magnesium/manganese(II) chloride-lithium chloride/air *Mg/MnCl$_2$·2LiCl/O$_2$*
Manganese(II)-catalyzed homocoupling of Grignard compds. 2 RMgHal → R-R

559.

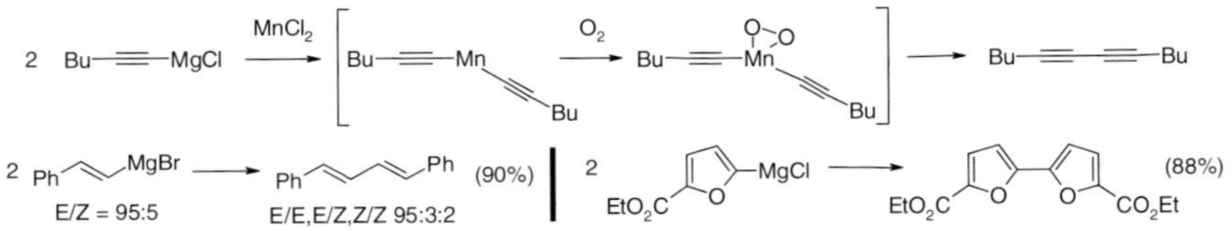

under mild conditions. 1-Hexynylmagnesium chloride added in one portion to 5 mol% $MnCl_2$·2LiCl in THF, dry air bubbled into the mixture via cannula for 45 min maintaining a temp. of <30°, the reaction quenched with 1 *M* HCl, the mixture extracted with ethyl acetate, the organic layer washed with brine, dried over $MgSO_4$, concentrated *in vacuo*, and the residue purified by chromatography on silica gel → 5,7-dodecyne. Y 91%. The reaction is successful with alkenes (giving (E)-coupled dienes), alkynes, aryl, hetaryl and benzyl Grignard reagents in the presence of ester, nitrile, nitro and ether functionality giving 75-95% yield (18 examples) and requires only atmospheric oxygen. $FeCl_3$ was also an effective catalyst but was limited (alkyne Grignard agents gave complex mixtures) and yields were generally lower (7 examples; 60-88%). F.e.s. G. Cahiez, A. Moyeux, J. Buendia, C. Duplais, J. Am. Chem. Soc. *2007*, *129* (45), 13788-9.

Magnesium/iron(III) fluoride/1,3-bis(2,6-isopropylphenyl)-Δ^2-imidazolinium chloride/ ethylmagnesium bromide ←
Biaryls from ar. chlorides and arylmagnesium bromides ArCl → Ar-Ar′
Iron-catalyzed cross-coupling avoiding homocoupling

560.

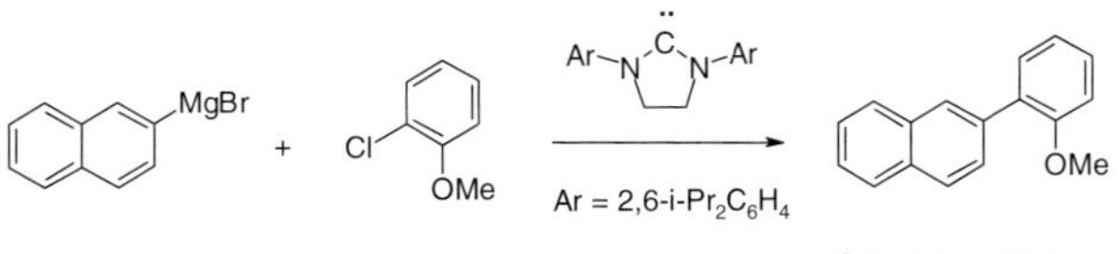

3 Mol% FeF_3·3H_2O and 9 mol% 1,3-bis(2,6-isopropylphenyl)-Δ^2-imidazolinium chloride added to 18 mol% ethylmagnesium bromide in THF at 0°, the mixture stirred at room temp. for 5 h, 1 eq. *o*-chloroanisole and 1.2 eqs. 2-naphthylmagnesium bromide added at 0°, heated at 60° for 24 h, cooled to room temp., quenched with aq. sodium potassium tartrate, extracted with hexane, the extracts filtered through Florisil (100-200 mesh), the solvent removed *in vacuo*, and the residue purified by chromatography → 2-(2-methoxyphenyl)naphthalene. Y 96%. The *fluoride anion* was shown to play a key role *in suppressing homocoupling* to a negligible level, affording cross-coupled biaryls and bihetaryls in 74-96% yield from ar. and hetar. chlorides (homocoupling being significant/predominant with phenyl bromide, iodide and triflate). Reaction is compatible with halo, ether, thioether, amine and acetal functionality and not affected by small amounts of

water (which may assist in solubilizing the reagents), but is slow in the absence of the *in situ*-generated N-heterocyclic carbene ligand. F.e. incl. reaction optimization s. T. Hatakeyama, M. Nakamura, J. Am. Chem. Soc. *2007*, *129* (32), 9844-5.

Zinc/dichlorotetrakis(pyridine)nickel(II)/chiral 2,6-bis(Δ²-oxazolin-2-yl)pyridines ←

Nickel-catalyzed radical cross-coupling of alkyl iodides with alkylzinc halides R-R′

561. 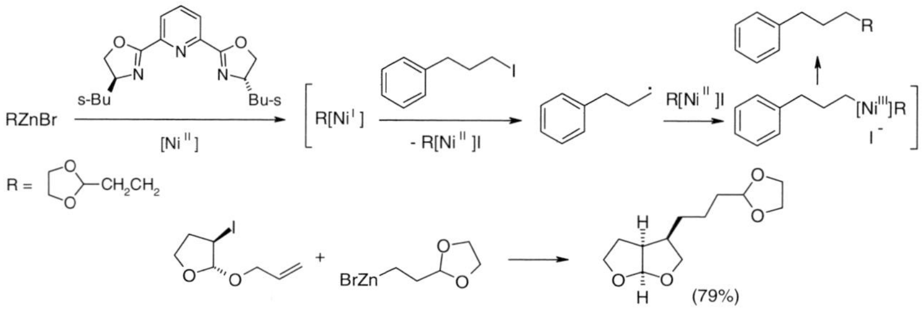

A soln. of the startg. alkyl iodide in dry THF added via syringe to a flask containing 10 mol% each of $Ni(py)_4Cl_2$ and (S)-(*i*-Bu)-Pybox under argon, stirred for 3-4 min at room temp., the resulting deep-blue soln. treated with 2 eqs. of the startg. alkylzinc bromide in THF (0.5 *M*), stirring continued for 5.5 h, a further 2 eqs. of the alkylzinc bromide soln. added, stirred for a further 5.5 h, and worked up with column chromatographic purification → product. Y 69%. The procedure is applicable to both *prim*- and *sec*-alkyl iodides and leaves a number of functional groups unaffected (e.g. ketones and nitriles). Mechanistically, it is likely that a Pybox-coordinated alkylnickel(I) complex is initially generated on reaction of the alkylzinc bromide with the nickel catalyst; this subsequently reacts with the alkyl iodide with **generation of an alkyl radical** and the liberated alkylnickel(II) iodide then couples with the latter to produce a mixed dialkylnickel(II) iodide prior to product formation by reductive elimination and regeneration of nickel(I). With ethyleneiodides, **stereospecific radical ring closure-cross-coupling** takes place if the rate of radical ring closure is faster than intermolecular cross-coupling with the alkylnickel(II) iodide. F.e.s. V.B. Phapale, E. Bunuel, M. Garcia-Inglesias, D.J. Cardenas, Angew. Chem. Int. Ed. Engl. *2007*, *46* (46), 8790-5.

Diethylzinc/methylene iodide Et_2Zn/CH_2I_2

β-Diketones from two bromomethyl ketone molecules 2 $ArCOCH_2Br$ → $ArCOCH_2COCH_2Ar$

Reformatsky-type dimerization with 1,2-aryl migration

562. 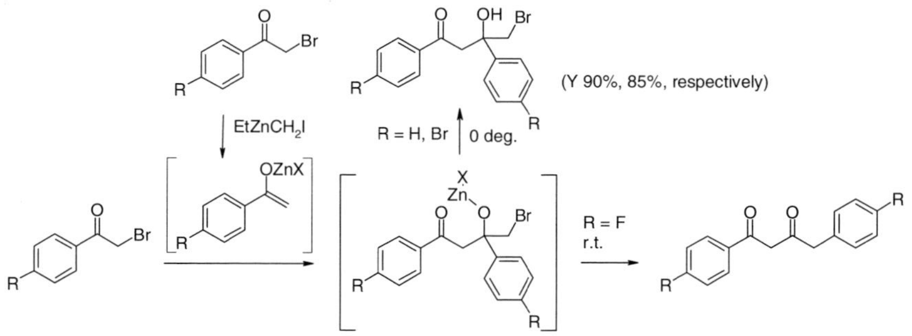

under mild conditions. 1 Eq. methylene iodide added dropwise via syringe to 1 eq. diethylzinc in methylene chloride at 0° under N_2, the white suspension stirred for 10 min, 0.5 eq. 4-fluorophenacyl bromide added, the mixture warmed to room temp., stirred for 6 h (TLC monitoring), quenched with satd. aq. NH_4Cl, extracted with ether, the organic layer washed with brine, dried over $MgSO_4$, concentrated *in vacuo*, and the residue purified by chromatography → 1,4-bis(4-fluorophenyl)butane-1,3-dione. Y 82%. The reaction is sensitive to temperature and substitution pattern but is successful with aryl and 2-thiazolyl ketones (single example) generally affording 60-85% yield. High yields are favoured by electron-releasing groups (at 0°), with lower yields (44 and 46%) being obtained at room temp. with electron-releasing *p*-bromo- or *p*-chloro- (but not *p*-fluoro-) as well as 3-methoxy-phenacyl bromide (Y 40%). At 0°, however, the intermediate γ-bromo-β-hydroxyketones were obtained in high yields (85-90%). The reaction was also tolerant of an α-phenyl substituent (Y 72%) but not α-propyl (Y 0%). Reaction with alkyl or cycloalkyl bromomethyl ketones took a different course, affording furans or disubst. cyclopropanols (*72*, 563). No reaction occurs with Et_2Zn alone. F.e.s. L. Li, P. Cai, D. Xu, Q. Guo, S. Xue, J. Org. Chem. *2007*, *72* (21), 8131-4.

Furans and 1,2-disubst. cyclopropanols from bromomethyl ketones under mild conditions ○▽

563.

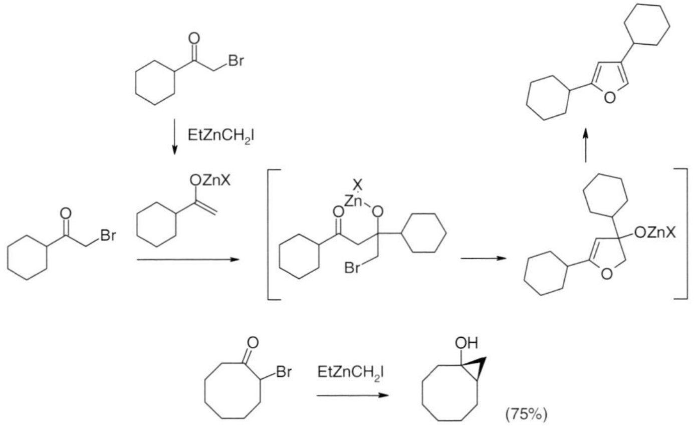

Furans. 1 Eq. methylene iodide added dropwise via syringe to 1 eq. diethylzinc in methylene chloride at 0° under N_2, the white suspension stirred for 10 min, 0.5 eq. 2-bromo-1-cyclohexyl-ethanone added, the mixture warmed to room temp., stirred for 5 h (TLC monitoring), quenched with satd. aq. NH_4Cl, extracted with ether, the organic layer washed with brine, dried over $MgSO_4$, concentrated *in vacuo*, and the residue purified by chromatography → 2,4-dicyclohexylfuran. Y 56%. The cycloheptyl analog reacted similarly (Y 45%), whereas less-sterically hindered bromomethyl ketones or those carrying an α-substituent afforded **cyclopropanols** via homologation (4 examples; Y 48-75%). There was no reaction with highly hindered *tert*-butyl bromomethyl ketone. Reaction with aryl and hetaryl bromomethyl ketones took a different course affording β-diketones (*72*, 562). No reaction occurs in either case with Et_2Zn alone. F.e.s. L. Li, P. Cai, D. Xu, Q. Guo, S. Xue, J. Org. Chem. *2007*, *72* (21), 8131-4.

Ethylmagnesium bromide s. under Mg and Cp_2ZrEt_2 *EtMgBr*
Zinc cyanide s. under $Pd(OAc)_2$ *$Zn(CN)_2$*
*Zinc chloride s. under 1-*n*-Butylpyridinium chloride-stannous chloride* *$ZnCl_2$*

Indium *In*

5-Methyl-Δ^2-isoxazolines from nitrile oxides ○

564. 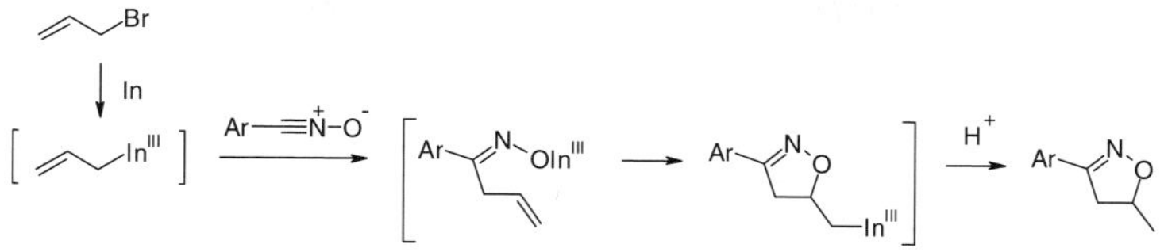

Ar = p-ClC_6H_4

A suspension of indium powder (10 mmol) and 1 eq. allyl bromide in 1:1 THF/water stirred at room temp. for 3 h until homogeneous, cooled to 0-5°, added dropwise over 5 min to a stirred soln. of 1 eq. *in situ*-generated *p*-chlorobenzonitrile oxide in THF at 0-5°, allowed to warm to room temp., stirred for 20 h at room temp., quenched with aq. NH_4Cl, and worked up with column chromatographic purification on silica gel → product. Y 84%. Reaction involves initial generation of allylindium bromide, followed by nucleophilic addition to the nitrile oxide prior to ring closure. F.e. and from *in situ*-generated nitrile oxides s. S.D. Sawant, P.P. Singh, N.A. Qazi, H.M.S. Kumar, Chem. Lett. *2007*, *36* (2), 296-7.

Triarylaluminum compds./palladium(II) acetate/tricyclohexylphosphine *$Ar_3Al/Pd(OAc)_2/Cy_3P$*
Biaryls from ar. halides and triarylaluminum compds. ArBr → Ar-Ar′
Palladium-catalyzed cross-coupling s. *72*, 610

Alumina s. under $Pd(OAc)_2$ *Al_2O_3*
1-Butyl-3-methylimidazolium salts s. under Et_3N *$[bmim]^+$*
1,3-Bis(2,6-isopropylphenyl)-Δ^2-imidazolinium chloride/ethylmagnesium bromide s. under Mg ←
Chiral 2,6-bis(Δ^2-oxazolin-2-yl)pyridines s. under Zn *Pybox*
Alcohol dehydrogenase s. under $Pd(OAc)_2$ ←
Trifluoroacetic acid/trifluoroacetic anhydride s. under $Pd(OAc)_2$ *$CF_3COOH/(CF_3CO)_2O$*
Chiral cyclic thioethers s. under K_2CO_3 ←
n-*Butyl bromide s. under $Pd(PPh_3)_4$* *BuBr*
Methylene iodide s. under Et_2Zn *CH_2I_2*
Polymethylhydrosiloxane s. under $Pd(OAc)_2$ *PMHS*

Diethylzirconocene/copper(I) chloride *$Cp_2ZrEt_2/CuCl$*
5-Hydroxy-Δ^3-2-pyrrolones ○
from acetylene derivs., isocyanates and carboxylic acid chlorides
Regiospecific 3-component synthesis via zirconacyclopentenes

565. 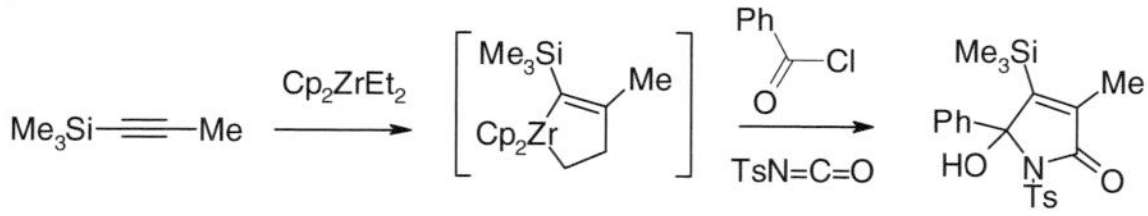

The startg. alkyne treated in the normal way with Cp_2ZrEt_2 in THF at 0° to generate the intermediate zirconacyclopentene, 1.2 eqs. N-tosylisocyanate added, the mixture stirred for 1 h at 50°, treated with 2 eqs. CuCl and the startg. acyl chloride at room temp., stirred for 3 h, and worked up with column chromatographic purification on silica gel → product. Y 68%. The procedure is applicable to the coupling of sym. or unsym. alkynes (the latter affording a *single* regioisomer) with aliphatic or aromatic acyl chlorides (incl. hindered pivaloyl chloride) and N-alkyl-, N-aryl- and

N-tosyl-isocyanates. However, there was no reaction with *tert*-butyl isocyanate. The yield was the same with 10 mol% CuCl after 24 h. F.e.s. K. Kanno, S. Ren, Y. Li, K. Nakajima, T. Takahashi, Tetrahedron Lett. *2007*, *48* (52), 9199-202.

Zirconocene dichloride/ethylmagnesium bromide/copper(I) chloride ←
2(Z),4-Enynols from acetylene derivs., aldehydes and α,β-acetylenebromides ←

566. 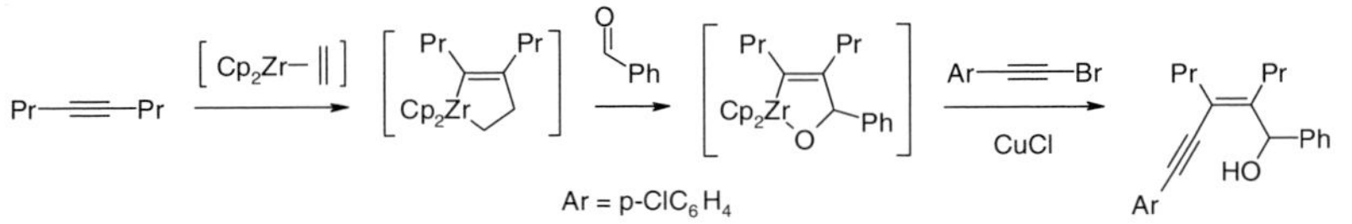

A soln. of 1.25 eqs. Cp_2ZrCl_2 in THF treated with 2.5 eqs. 1 *M* ethylmagnesium bromide in THF at -50°, stirred for 1 h, 1 eq. oct-4-yne added, warmed to 0°, stirred for 3 h, warmed to 50° over ca. 5 min, treated slowly with 1.5 eqs. benzaldehyde, stirring continued for 2 h, cooled to room temp., 0.1 eq. CuCl and 1.2 eqs. of 1-bromoethynyl-4-chlorobenzene added, stirred at room temp. overnight, quenched with 3 *N* aq. HCl, and worked up with chromatographic purification on neutral alumina → product. Y 79%. Reaction is applicable to the coupling of alkyl-, aryl- or cycloalkenyl-ethynyl bromides as electrophile, and the aldehyde may be aromatic or aliphatic (although yields were lower with the latter). Intermediate oxazirconacyclopentene ring formation is followed by Zr→Cu-transmetalation and quenching with the acetylenehalide. F.e. and quenching with iodine and proton s. S.H. Guo, H. Zhang, F.J. Song, Y.H. Liu, Tetrahedron *2007*, *63* (9), 2009-18.

*1-*n*-Butylpyridinium chloride-stannous chloride/zinc chloride* ←
4-Arylchromans from *o*-hydroxyaldehydes, activated arenes and β,γ-ethylenehalides ○

567. 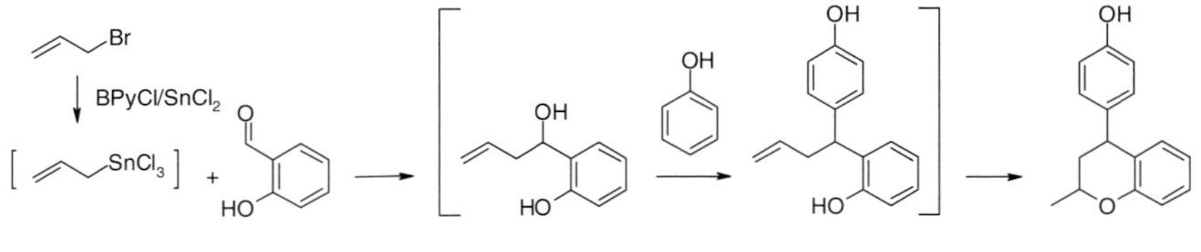

Lewis acidic ionic liquids as medium. A mixture of salicylaldehyde, 2 eqs. allyl bromide and 1.5 eqs. phenol in 1:2 *n*-butylpyridinium chloride/$SnCl_2 \cdot 2H_2O$ stirred at room temp. for 10 h, 1.5 eqs, $ZnCl_2$ added, stirred at 70° for 5 h, and worked up with flash column chromatographic purification → 2-methyl-4-(4′-hydroxyphenyl)chroman. Y 70%. Reaction is presumed to involve initial Barbier-type allylation of the aldehyde with *in situ*-generated allyltin(IV) trichloride, followed by Friedel-Crafts coupling with the phenol and $ZnCl_2$-catalyzed cycloisomerization to the product. β,γ-Ethylenechlorides were less reactive, but did participate after conversion to the corresponding β,γ-ethyleneiodides *in situ* using *n*-butylpyridinium iodide/$SnCl_2 \cdot 2H_2O$. Other Lewis acids and Brønsted acids were ineffective for the cycloisomerization. F.e., **also from ar. aldehydes** (with *p*-subst. phenols and allyl bromide), s. X.L. Zhao, L. Liu, Y.J. Chen, D. Wang, Synlett *2007* (9), 1357-64.

Tert. phosphines s. under Ar_3Al, $Pd(OAc)_2$ and $Pd_2(dba)_3$ *R_3P*
1,2,3,4-Tetrakis(diphenylphosphinomethyl)cyclopentane s. under $[(\eta^3\text{-allyl})PdCl]_2$ *Tedicyp*
Air s. under Mg *O_2*

Trimethylsilyl triflate/ethyldiisopropylamine *Me_3SiOTf/i-Pr_2NEt*
γ,δ-Ethylenecarboxylic acid amides ←
from 2-ethyleneamines and carboxylic acid chlorides
Stereospecific silyl-modified Bellus-Claisen rearrangement

568. 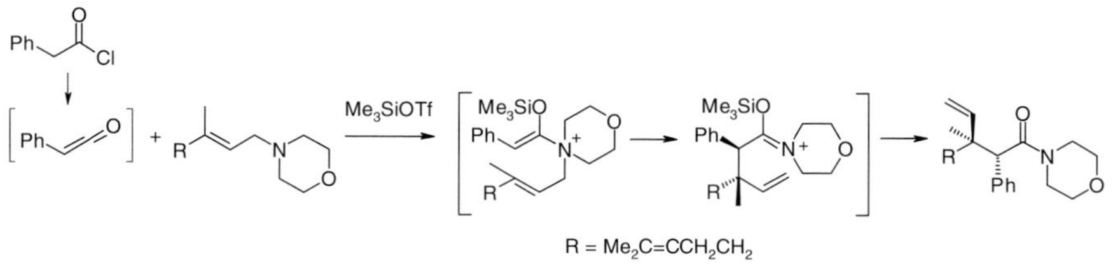

A soln. of N-geranylmorpholine in methylene chloride treated at room temp. with 1 eq. trimethylsilyl triflate followed by 1.5 eqs. ethyldiisopropylamine, the soln. stirred for 5 min, a soln. of 1.2 eqs. phenylacetyl chloride in methylene chloride added dropwise over 4 h, the mixture stirred for 30 min, then diluted with ether, and treated with 4.2 eqs. 1 *M* aq. NaOH for 10 min → product. Y 89% (d.r. 97:3). This *metal-free* procedure is applicable to a wide range of ketene precursors and allylamines, incl. those with trisubst. ethylene residues for the generation of quaternary carbon centres, and is potentially applicable to complex allylamines possessing heteroatomic substituents which may deactivate cationic Lewis acids, such as Ti(IV). The *anti*-stereochemistry contrasts with the established *syn*-stereochemistry of the Al-catalyzed conversion, and is in keeping with the intermediate formation of a 2-siloxy-3-azonia-1(Z),5-diene. F.e. and comparison with Ti(IV)-reagents s. D. Craig, N.P. King, D.M. Mountford, Chem. Commun. *2007* (10), 1077-9.

*Tetra-*n*-butylammonium fluoride s. under $PdCl_2(PPh_3)_2$* *Bu_4NF*
*Tetra-*n*-butylammonium bromide s. under CuI and $PdCl_2$* *Bu_4NBr*
*Tetra-*n*-butylammonium iodide s. under K_2CO_3* *Bu_4NI*
N-(3,4,5-Trimethoxybenzyl)quinidinium bromide s. under KOH ←
Manganese(II) chloride-lithium chloride s. under Mg *$MnCl_2 \cdot 2LiCl$*
Potassium hexacyanoferrate s. under CuI, Pd-C and $Pd(OAc)_2$ *$K_4Fe(CN)_6$*
1,1′-Bis(diphenylphosphino)ferrocene s. under $Pd(OAc)_2$ *dppf*
Iron(III) fluoride s. under Mg *FeF_3*
Dichlorotetrakis(pyridine)nickel(II) s. under Zn *$Ni(py)_4Cl_2$*

*Palladium-carbon/tri-*n*-butylamine/sodium carbonate/potassium hexacyanoferrate* ←
Ar. nitriles from halides Hal → CN
under Pd-catalysis s. *67*, 419s*72*

Palladium(II) acetate s.a. under Ar_3Al *$Pd(OAc)_2$*

Palladium(II) acetate/potassium carbonate *$Pd(OAc)_2/K_2CO_3$*
Palladium(II) acetate/potassium carbonate/microwaves ←
Palladium(II) acetate/alumina/potassium phosphate/microwaves ←
Heck arylation H → Ar
update s. *27*, 871s*71*; improved general methods for Pd-catalyzed Heck arylation under microwave irradiation, incl. cinnamaldehydes from acrolein diethyl acetal with $Pd(OAc)_2/K_2CO_3$ in DMF, s. A. Togninelli, H. Gevariya, M. Alongi, M. Botta, Tetrahedron Lett. *2007*, *48* (28), 4801-3; 2-hydroxy-3-vinyl-1,4-naphthoquinones with $Pd(OAc)_2/K_2CO_3$ in refluxing water s. A.L. Perez, G. Lamoureux, B.Y. Zhen-Wu, ibid. *48* (23), 3995-8; ligand- and solvent-free reaction with $Pd(OAc)_2/Al_2O_3/K_3PO_4$ under microwave irradiation s. L.-H. Du, Y.-G. Wang, Synth. Commun. *2007*, *37* (2), 217-22; with a sterically demanding dichloro-Pd(II)-bis(phosphinite) complex (derived from salicylaldehyde and 2-amino-3-hydroxypyridine)/Bu_3N s. D.D. Pathak, H. Maheswaran, K. Leon Prasanth, M. Lakshmi Kantam, Synlett *2007* (5), 757-60; of pent-4-en-

oates or allylmalonates without double bond migration with $[(\eta^3\text{-allylPdCl}]_2$/*cis,cis,cis*-1,2,3,4-tetrakis(diphenylphosphinomethyl)cyclopentane/NaOAc s. M. Lemhadri, A. Battace, T. Zair, H. Doucet, M. Santelli, J. Organomet. Chem. *2007*, *692* (11), 2270-81.

Palladium(II) acetate/sodium acetate *$Pd(OAc)_2$/NaOAc*

Palladium(II)-catalyzed 4-arylation of Δ^4-2-imidazolones H → Ar

569.

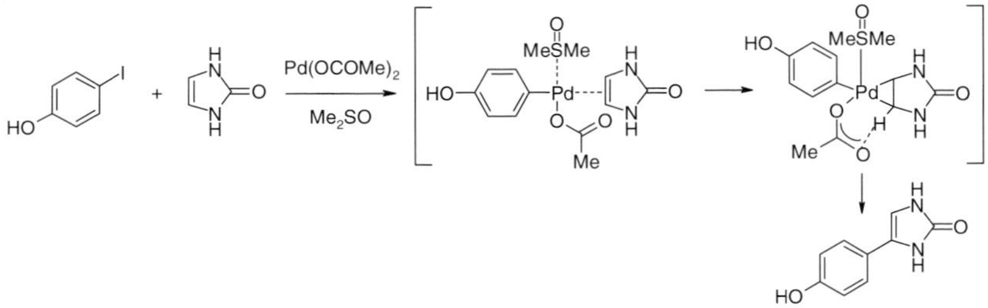

under mild conditions. 1.5 Eqs. 4-hydroxyphenyl iodide and degassed DMSO added to a mixture of 5 mol% $Pd(OAc)_2$, Δ^4-2-imidazolone, and 3 eqs. $NaOAc{\cdot}3H_2O$ under argon, the reaction vial purged with argon for 1 min, the mixture stirred in the dark at 80° for 6 h, the solvent removed *in vacuo*, and the residue purified by chromatography → 4-(4-hydroxyphenyl)-Δ^4-2-imidazolone. Y 85%. The method is successful with electron-rich and -poor aryl iodides (although the latter reacted more slowly), and is also efficient with electron-deficient aryl bromides after a prolonged reaction time (24-36 h), giving 52-87% yields (16 examples) in the presence of halo, ether, nitro, ketone and notably *alcohol* and *phenol* functionality. Substitution in the 2-position of the aryl iodide is tolerated as is dimethylation of the imidazolone nitrogens (single example; Y 93%). Mechanistic studies indicate that arylation most likely occurs via a C-H insertion pathway. F.e. incl. 4-cinnamyl derivs., also optimization, s. J. Lu, X. Tan, C. Chen, J. Am. Chem. Soc. *2007*, *129* (25), 7768-9.

Palladium(II) acetate/silver acetate/trifluoroacetic acid/trifluoroacetic anhydride ←

Phenanthridines from carboxylic acid anilides and ar. iodides ○

One-pot procedure via biaryl cross-coupling-intramolecular cyclodehydration

570.

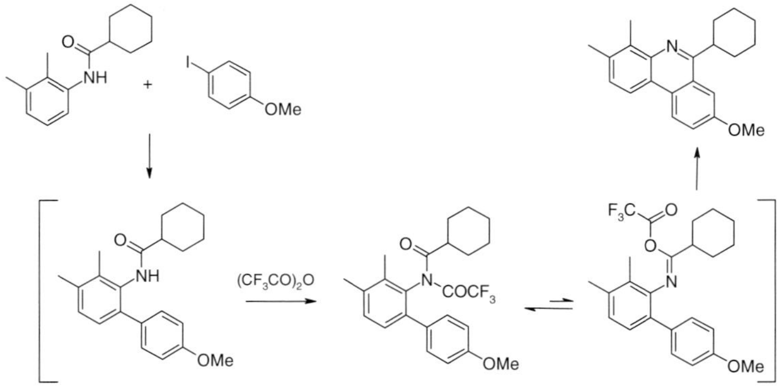

A soln. of N-(cyclohexanecarbonyl)-2,3-dimethylaniline (0.7 mmol), $Pd(OAc)_2$ (0.035 mmol), AgOAc (0.91 mmol) and 4-methoxyiodobenzene (2.1 mmol) in trifluoroacetic acid (0.5 ml) heated at 120° for 40 min, cooled to room temp., trifluoroacetic anhydride (1.4 mmol) added, and the mixture heated at 110° for 25 min → 6-cyclohexyl-8-methoxy-3,4-dimethylphenanthridine.

Y 99%. The reaction is successful for both electron-poor and -rich anilides and is tolerant of a wide variety of substituents, including bromo, chloro and trifluoromethoxy groups. The cyclization step proceeds via an initial N-trifluoracetylated deriv. (characterized by X-ray analysis), which is assumed to exist in equilibrium with its O-acylated analog in order for the cyclization to occur. In the absence of trifluoroacetic anhydride, cyclization is accompanied by partial hydrolysis of the N-trifluoracetylated amide by adventitious water. The procedure also represents a convenient method for the synthesis of **2-aryl-** or **2,6-diaryl-anilines**, since the intermediate cross-coupled biaryls can be readily isolated and deacylated by base hydrolysis with NaOH in ethanol. F.e.s. D. Shabashov, O. Daugulis, J. Org. Chem. *2007, 72* (20), 7720-5.

*Palladium(II) acetate/diadamantyl(*n*-butyl)phosphine/sodium carbonate/ potassium hexacyanoferrate* ←
Palladium(II) acetate/1,1′-bis(diphenylphosphino)ferrocene/polymethylhydrosiloxane/ zinc cyanide ←

Ar. nitriles from halides Hal → CN

under Pd-catalysis cf. *67*, 419(s*71*); from chlorides with diadamantyl(*n*-butyl)phosphine/Na_2CO_3/$K_4Fe(CN)_6$ in NMP at 140° s. T. Schareina, A. Zapf, W. Mägerlein, N. Müller, M. Beller, Tetrahedron Lett. *2007, 48* (7), 1087-90; from bromides in the presence of dppf/$Zn(CN)_2$ in 10:1 NMP/water under air at 80°, prevention of catalyst poisoning using a little polymethylhydrosiloxane, s. M.T. Martin, B. Liu, B.E. Cooley, Jr., J.F. Eaddy, ibid. *48* (14), 2555-7; with Pd-C/Bu_3N/Na_2CO_3/$K_4Fe(CN)_6$ in NMP at 140° s. Y.-Z. Zhu, C. Cai, Eur. J. Org. Chem. *2007* (15), 2401-4; hetar. nitriles from bromides with CuI/$K_4Fe(CN)_6$ in N-methylimidazole at 140° s. T. Schareina, A. Zapf, W. Mägerlein, N. Müller, M. Beller, Synlett *2007* (4), 555-8.

Palladium(II) acetate/tris(2,4,6-trimethoxyphenyl)phosphine/1,8-bis(dimethylamino)- naphthalene/alcohol dehydrogenase ←

(E)-3-Aryl-2-ethylenealcohols I → C=C-CH(OH)
from α,β-ethyleneketones and ar. iodides
One-pot Heck arylation-asym. enzymatic reduction

571. 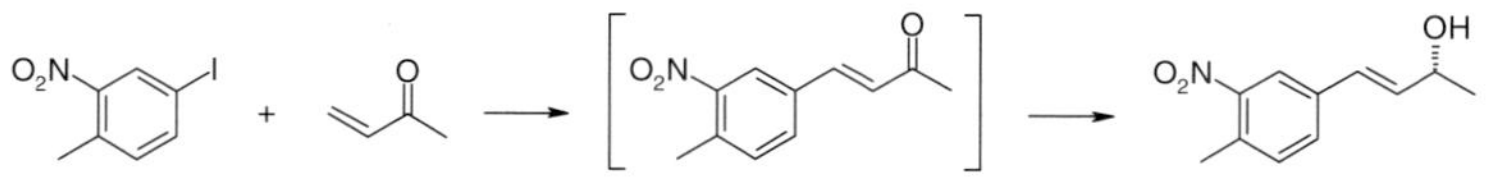

A mixture of the startg. aryl iodide (1 mmol), 3 eqs. 3-buten-2-one, 5 mol% $Pd(OAc)_2$, 1.1 eqs. Proton Sponge and 10 mol% tris(2,4,6-trimethoxyphenyl)phosphine warmed at 80° with stirring for 16 h, cooled to room temp., concentrated under reduced pressure, the residue suspended in 1:2 methoxyethanol/isopropanol, phosphate buffer (pH 7 containing 4.45 μmol NADP+ and 100 U *Lactobacillus brevis* alcohol dehydrogenase) added, allowed to react at 25° for 2 h, and worked up with column chromatographic purification on silica gel → (2R,3E)-4-(3-nitro-4-methylphenyl)-3-buten-2-ol. Y 80% (e.e. >99%). Significantly, enantioselectivities exceeded 99%, there was *no enzyme inhibition* by the palladium catalyst or any of the additives, and there was no need to purify the intermediate. Furthermore, the (S)-products were obtained (e.e. >99%) with the alcohol dehydrogenase from a *Thermoanaerobacter* species. F.e.s. S. Sgalla, G. Fabrizi, R. Cirilli, A. Macone, A. Bonamore, A. Boffi, S. Cacchi, Tetrahedron:Asym. *2007, 18* (23), 2791-6.

Tris(dibenzylideneacetone)dipalladium/2-dicyclohexylphosphino-2′,4′,6′-triisopropyl- biphenyl/lithium tert-*butoxide* ←

Styrenes from ar. halides and tosylhydrazones Hal → C=C

572. 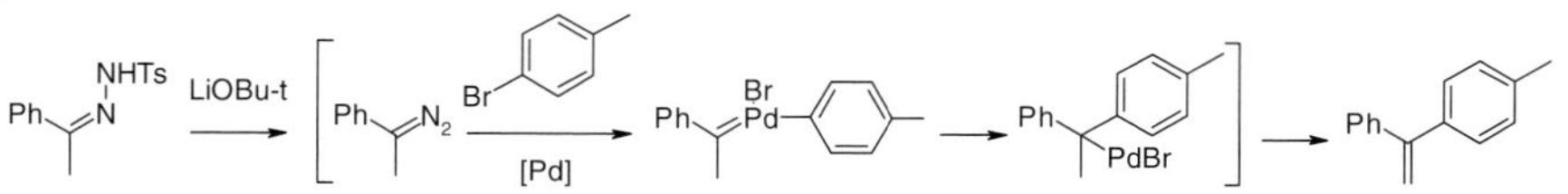

A soln. of the startg. tosylhydrazone, 2 mol% 2-dicyclohexylphosphino-2′,4′,6′-triisopropyl-biphenyl, 1 mol% $Pd_2(dba)_3$ and 2.2 eqs. LiOBu-*t* mixed in dioxane for 1 min under N_2 in a carousel tube, 1 eq. of the startg. aryl halide added, heated at 70° for 4 h with stirring under reflux, allowed to cool to room temp., and worked up → 1-phenyl-1-*p*-tolylethylene. Y 98%. Significantly, *no stoichiometric organometallic compds*. are required. The procedure is applicable to the synthesis of di- and tri-subst. styrenes by coupling aldehyde- or ketone-derived tosylhydrazones with aryl or hetaryl bromides or chlorides. *trans*-Products were obtained from tosylhydrazones derived from linear aldehydes, but the stereochemistry of the formed trisubst. styrenes seems to be directed by the size of the substituents. Nitriles, carbonyl groups and esters remained unaffected. It appears that reaction involves intermediate generation of a transient *arylpalladium carbene complex*. F.e. and regioselectivity s. J. Barluenga, P. Moriel, C. Valdés, F. Aznar, Angew. Chem. Int. Ed. Engl. *2007*, *46* (29), 5587-90.

*Tetrakis(triphenylphosphine)palladium(0)/triethylamine/*n*-butyl bromide* ←

Palladium(0)-catalyzed cross-coupling of unsatd. halides with ethyl diazoacetate Hal → C(=N_2)COOR

573. 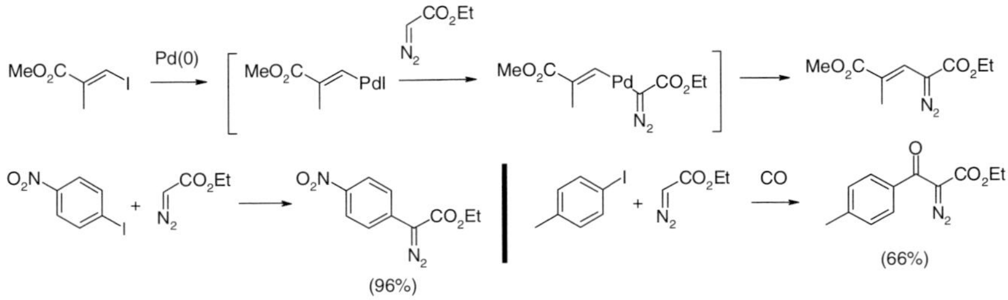

α,β-Ethylenediazo compds. from α,β-ethyleneiodides. 5 Mol% Pd(PPh_3)$_4$ added to a mixture of methyl 2-methyl-3-iodopropenoate, 2.5 eqs. ethyl diazoacetate, 1.5 eqs. triethylamine and 1 eq. 1-bromobutane in acetone, the mixture stirred at 35° under N_2 for 2 h, the solvent evaporated *in vacuo*, and the residue purified by chromatography → 5-ethyl 1-methyl (E)-4-diazo-2-methylpent-2-enedioate. Y 88%. The cross-coupling is successful with electron-deficient *vinyl iodides* (6 examples; 48-88%) using triethylamine as base, and with *aryl iodides* using DBU as base for preparing **aryl(diazo)acetic acid esters** (8 examples; Y 74-96%). However, low yields were obtained in the latter series with aryl bromides (the sterically-hindered aryl iodides, 2-chloro- and 2-bromo-iodobenzene, giving only traces of product) and with electron-rich aryl iodides (4-methoxyiodobenzene giving 39% yield). Furthermore, conducting the reaction with aryl iodides/triethylamine under a carbon monoxide atmosphere afforded **α-aroyl(diazo)acetic acid esters** (7 examples; Y 43-66%). No reaction occurs in the absence of the Pd(0) catalyst. F.e. incl. reaction optimization s. C. Peng, J. Cheng, J. Wang, J. Am. Chem. Soc. *2007*, *129* (28), 8708-9.

Tetrakis(triphenylphosphine)palladium(0)/copper(I) iodide/potassium carbonate ←

4-Benzylimidazoles from N-propargylamidines and ar. halides ○

574. 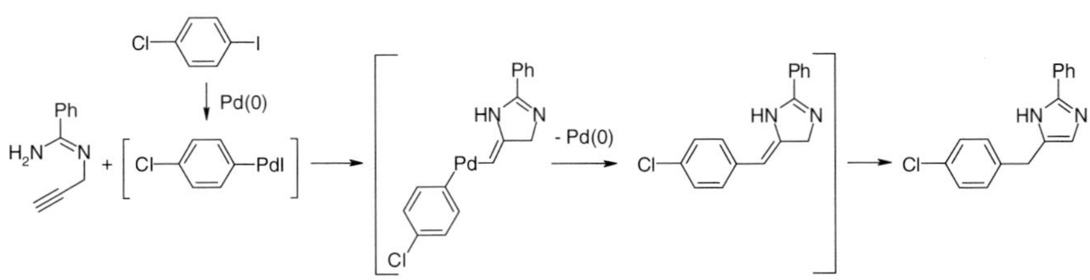

The startg. aryl halide added to a suspension of ca. 1 eq. N-propargylbenzamidine and 4.78 eqs. K_2CO_3 in dry DMF under N_2, 2 mol% Pd(PPh_3)$_4$ and ca. 4 mol% CuI added, stirred at 60° for 1 h (TLC monitoring), and worked up with flash chromatographic purification on silica gel →

product. Y 65%. Moderate-to-good yields (40-80%) were obtained with aryl iodides and 5-bromopyrimidine, yields being highest with aryl iodides possessing electron-withdrawing groups. Reaction may take place by initial oxidative addition of palladium(0) to the aryl-halogen bond, followed by activation of the alkyne residue by the generated arylpalladium species prior to intramolecular aminopalladation and reductive elimination. F.e.s. G. Abbiati, A. Arcadi, V. Canevari, E. Rossi, Tetrahedron Lett. *2007*, *48* (48), 8491-5.

Tetrakis(triphenylphosphine)palladium(0)/copper(I) iodide/ethyldiisopropylamine ←
Sonogashira coupling C≡CH → C≡CR
update s. *27*, 851s*71*; of N-hetar. halides with $Pd(PPh_3)_4$/CuI/*i*-Pr_2NEt *in water* s. N. Inoue, O. Sugimoto, K. Tanji, Heterocycles *2007*, *72* (Special issue), 665-71; improved general method for Pd-catalyzed coupling under microwave irradiation s. A. Togninelli, H. Gevariya, M. Alongi, M. Botta, Tetrahedron Lett. *2007*, *48* (28), 4801-3; palladium-free coupling of ar. iodides with CuI/Ph_3P/Bu_4NBr/K_2CO_3 under microwave irradiation (cf. *66*, 384s*67*) in water s. G. Chen, X. Zhu, J. Cai, Y. Wan, Synth. Commun. *2007*, *37* (8), 1355-61; Sonogashira (and Negishi) coupling with 3-halogeno-1,3-diene-2-carboxylic acid esters s. *72*, 295; Sonogashira coupling with 4-iodo-3(2*H*)-furanones s. *72*, 304.

Dendritic palladium phosphine complex-on-silica ←
Macrocyclic lactams from halogenamines by cyclocarbonylation ○

575.

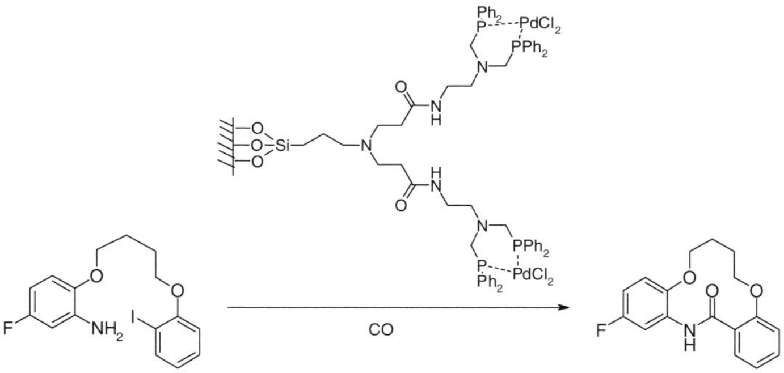

A mixture of the startg. aminohalide (1 mmol), 15 mg dendritic palladium phosphine complex-on-silica and 1.5 eqs. ethyldiisopropylamine in dry toluene flushed three times with CO in a glass-lined autoclave, the latter pressurized with CO to 100 psi, stirred at 80° for 22 h, cooled to room temp., excess CO released, filtered, and the filtrate worked up with column chromatographic purification on silica gel → product. Y 87%. The metallodendrimer is easy to retrieve and retains its high activity over several cycles; it is also very stable and can be stored at room temp. for several months. The procedure is suitable for the synthesis of dibenzo-fused 12- to 18-membered O-, N- and S-macrolactams, and leaves a number of functional groups unaffected (e.g. ketones, esters, ethers, fluorides, chlorides, and trifluoromethyl groups). Furthermore, the aniline ring may possess electron-withdrawing or -donating groups, but the corresponding aryl bromide was less reactive. F.e.s. S.-M. Lu, H. Alper, Chem. Eur. J. *2007*, *13* (20), 5908-16.

Palladium(II) chloride/sodium hydroxide *$PdCl_2$/NaOH*
*Dichlorobis(triphenylphosphine)palladium(II)/tetra-*n*-butylammonium fluoride* ←
Copper-free Sonogashira coupling C≡CH → C≡CR
update s. *63*, 411s*71*; rapid reaction with 2 mol% $PdCl_2$ under high pressure in 2 *M* aq. NaOH at high temp. in a microfluidic reactor s. H. Kawanami, K. Matsushima, M. Sato, Y. Ikushima, Angew. Chem. Int. Ed. Engl. *2007*, *46* (27), 5129-32; with $PdCl_2(PPh_3)_2$/Bu_4NF in the absence of solvent s. R. Sanz, M.P. Castroviejo, V. Guilarte, A. Pérez, F.J. Fañanás, J. Org. Chem. *2007*, *72* (14), 5113-8.

*Palladium(II) chloride/tetra-*n*-butylammonium bromide/sodium carbonate* ←
Stereospecific synthesis of electron-deficient 1,3-enynes C=C-C≡C
from ethylene derivs. and (E)-1,2-ethylene-1,2-diiodides
Catalysis with palladium(0) nanoparticles in water

576.

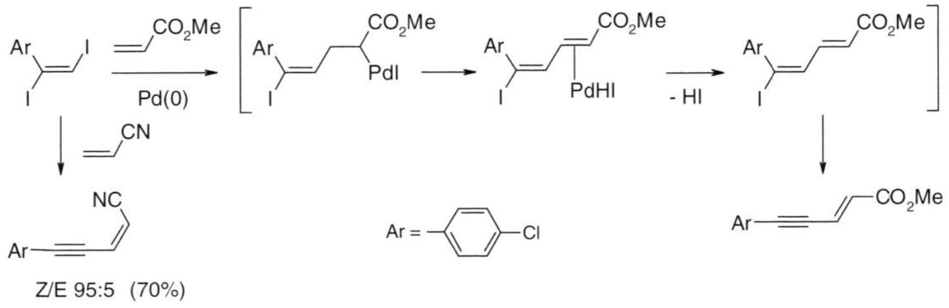

(E)-2,4-Enynecarboxylic acid esters. (E)-1,2-Diiodovinyl-4-chlorobenzene, 4 eqs. methyl acrylate and 4 eqs. Na_2CO_3 added to a stirred mixture of tetra-*n*-butylammonium bromide and 2 mol% $PdCl_2$ in water, heated at 80° for 6.5 h (TLC), the cooled mixture extracted with ether, the extracts washed with water and brine, dried over Na_2SO_4, the solvent removed *in vacuo*, and the residue purified by chromatography → methyl (E)-5-(4-chlorophenyl)pent-2-en-4-ynoate. Y 72% (selectivity 100%). A series of aryl- and alkyl-subst. (E)-1,2-diiodoalkenes reacted with acrylate esters to afford (E)-enynoates exclusively in 65-82% yield (11 examples), while reaction with acrylonitrile afforded **(Z)-2,4-enynenitriles** as major products in 70-72% yield (3 examples; selectivity 80-95%). This convenient and efficient method is compatible with ether, halo and methylenedioxy functionality, but produces only 30-40% yield with the corresponding 1,2-dibromoalkenes. The outlined mechanism is based on experimental and theoretical considerations. F.e. incl. reaction optimization s. B.C. Ranu, K. Chattopadhyay, Org. Lett. *2007*, *9* (12), 2409-12.

*Bis(η^3-allylpalladium chloride)/*cis,cis,cis-*1,2,3,4-tetrakis(diphenylphosphinomethyl)-cyclopentane/sodium acetate* ←
Heck arylation H → Ar
of pent-4-enoates or allylmalonates without double bond migration s. *27*, 871s*72*

Dichloropalladium(II) bis(phosphinite) complex/tributylamine ←
Heck arylation s. *27*, 871s*72*

Sulfur ↑ CC ↓↑ S

Sodium hydride *NaH*
Asym. synthesis of aziridines from imines and sulfonium salts 
chiral 1-sulfinylaziridines with *t*-BuOLi cf. *53*, 435s*68*; chiral *cis*-2-sulfinyl-1-tosylaziridines with NaH s. W.H. Midura, Tetrahedron Lett. *2007*, *48* (22), 3907-10.

*Potassium cyanide/(R,R,R)-2,6-bis[2,4-bis(4-trifluoromethylphenyl)naphth-1-yl]-4,4-dimethyl-4,5-dihydro-3*H*-4-azoniacyclohepta[2,1-*a*;3,4-*a′*]dinaphthalene iodide* ←
Potassium phosphate/chiral spirocyclic quaternary ammonium salt ←
Asym. synthesis of N-protected amines C(NHCO-)SO_2R → C(NHCO-)R′
from α-aminosulfones
chiral N-carbalkoxy-2-nitramines from aliphatic nitro compds. with KOH or CsOH/cinchonidinium salt cf. *70*, 431; chiral N-protected α-aminonitriles with KCN/(R,R,R)-2,6-bis[2,4-bis(4-trifluoromethylphenyl)naphth-1-yl]-4,4-dimethyl-4,5-dihydro-3*H*-4-azoniacyclohepta[2,1-*a*;3,4-*a′*]-dinaphthalene iodide (cf. *71*, 266) s. T. Ooi, Y. Uematsu, J. Fujimoto, K. Fukumoto, K. Maruoka, Tetrahedron Lett. *2007*, *48* (8), 1337-40; asym. vinylogous Mannich reaction of alkylidenemalononitriles with K_3PO_4/chiral spirocyclic quaternary ammonium salt s. B. Niess, K.A. Jørgensen, Chem. Commun. *2007* (16), 1620-2.

Sodium iodide s. under (S)-La-Li$_3$-(biphenyldiol)$_3$ *NaI*

Ethyldiisopropylamine/lithium perchlorate/O-trimethylsilylquinine ←
2-Alkylthio-4*H*-1,3-thiazin-6(5*H*)-ones ○
from 1,1-tosyldithiourethans and carboxylic acid chlorides
via asym. [4+2]-cycloaddition of ketenes to N-dithiocarbalkoxyimines

577.

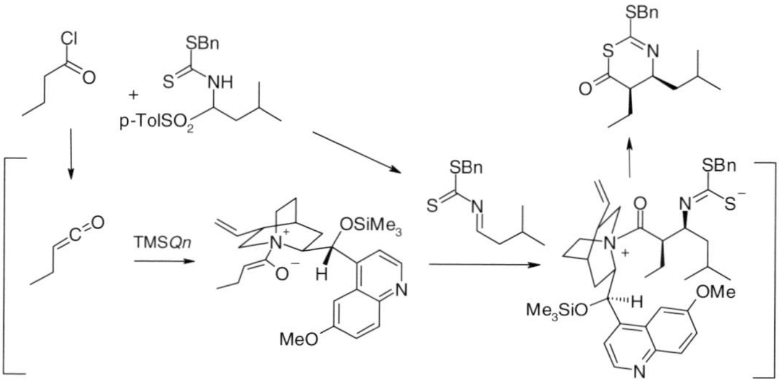

Ethyldiisopropylamine (3.5 eqs.), followed by $LiClO_4$ (1 eq.) in ether (1 ml), added to a soln. of O-trimethylsilylquinine (TMS*Qn*) (20 mol%) and the startg. dithiocarbamic acid ester (1 mmol) in methylene chloride (9 ml) at -78°, the mixture stirred 10 min, a soln. of butanoyl chloride (2 eqs.) in methylene chloride added via syringe pump over 2 h, the mixture stirred a further 10 h at -78°, filtered through a silica gel pad, and purified by column chromatography on silica gel → product. Y 72% (e.e. >98%; *cis/trans* >97:3). The procedure provides a useful alternative to the Mannich reaction for imines which are subject to thermodynamically-driven imine-to-enamine tautomerization under the basic reaction conditions required for enolate formation. Attempts to effect the cyclization using pregenerated imines or with carbamic acid esters were unsuccessful. Yields were generally moderate to good using aliphatic 1,1-tosyldithiourethans but were slightly lower using aryl analogs owing to competing β-lactam formation, a feature of which was unexpected *trans*-diastereoselectivity. F.e.s. X. Xu, K. Wang, S.G. Nelson, J. Am. Chem. Soc. *2007*, *129* (38), 11690-1.

Lanthanum(III) trilithium tris[(S)-6,6'-pentamethylenedioxybiphenyl-2,2'-diolate]/ sodium iodide/molecular sieves ←
Catalytic asym. cyclopropanation of α,β-ethyleneketones ▽

578.

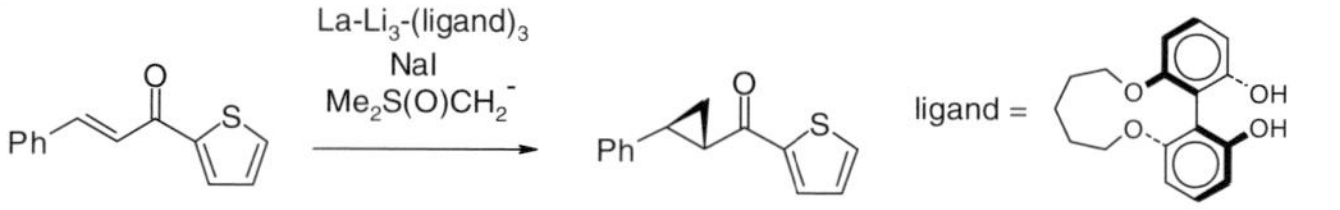

A *catalytic*, asym. Corey-Chaykovsky cyclopropanation of enones using a reactive ylid affords exclusively, under optimized conditions, **chiral *trans*-cyclopropyl ketones** in high yield and enantioselectivity. **E:** (S)-La-Li_3-(biphenyldiol)$_3$ (0.01 mmol; 0.1 *M* soln. in THF) and THF (320 μl) added to a flame-dried mixture of NaI (0.01 mmol) and 4 Å molecular sieves (20 mg) in a test tube, the mixture stirred at room temp. for 30 min (to dissolve NaI), cooled to -55°, dimethylsulfoxonium methylid (0.20 mmol; 0.881 *M* soln. in THF) and toluene (868 μl) added, stirred 10 min at -55°, startg. enone [0.20 mmol, as a soln. in THF/toluene (196 μl/244 μl)] added over 4 h using a syringe pump, aq. NH_4Cl and ethyl acetate added after 18 h, and the soln. warmed to room temp. → product. Y 94% (e.e. 94%). The beneficial effect of NaI on enantioselectivity was initially noticed as an artefact of the method of preparation of the ylid. Subsequent ESI-MS analysis and experimentation using La-Na_3-(biphenyldiol)$_3$, with or without added LiI, supported the idea that La-Li_2-Na-(biphenyldiol)$_3$ is the active species, formed by partial alkali metal exchange. F.e.s. H. Kakei, T. Sone, Y. Sohtome, S. Matsunaga, M. Shibasaki, J. Am. Chem. Soc. *2007, 129* (44), 13410-1.

Chiral spirocyclic quaternary ammonium salt s. under K_3PO_4 ←

*(S)-(-)-Indolin-2-yl-1*H*-tetrazole* ←

Asym. synthesis of cyclopropanes from ethylene derivs. ▽

and sulfonium salts cf. *63*, 413s*70*; and sulfonium ylids with 20 mol% (S)-(-)-indolin-2-yl-1*H*-tetrazole (e.e. 99%; d.e. up to 98%) s. A. Hartikka, P.I. Arvidsson, J. Org. Chem. *2007, 72* (15), 5874-7.

Lauroyl peroxide *$(RCOO)_2$*

Radical C-alkylation with xanthates H → R

of 5-membered 2-subst. heteroarenes cf. *48*, 819s*66*; 2-alkylation of 3-aryl- and 3-formyl-pyrroles s. O. Guadarrama-Morales, F. Méndez, L.D. Miranda, Tetrahedron Lett. *2007, 48* (26), 4515-8.

Chiral 1,1′-binaphthyl-2,2′-diyl hydrogen phosphates/thiophenol/hydrogen chloride ←

Organocatalyzed asym. Pictet-Spengler cyclization via N-sulfenyliminium ions ○

579.

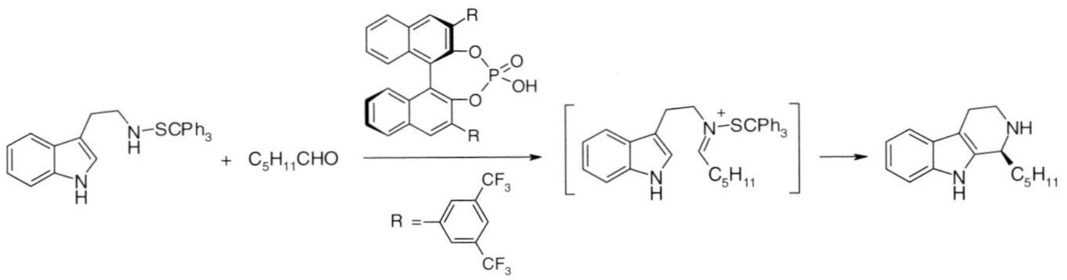

3 Eqs. startg. aldehyde added to a soln. of N-tritylsulfenyltryptamine and 5 mol% chiral cyclic 1,1′-binaphthyl-2,2′-diyl hydrogen phosphate *in toluene* containing 3 Å molecular sieves (to absorb liberated water) and 10 mol% 2,6-di-*tert*-butyl-4-methylphenol (to inhibit radical processes) at 0°, stirred at the same temp. for 2 h (TLC monitoring), the molecular sieves filtered off and washed with dioxane, 1.3 eqs. thiophenol added to the filtrate and washings, followed by 4 *M* HCl in dioxane, stirred overnight, the HCl salt filtered off, and the free base liberated with satd. aq. Na_2CO_3 and 25% ammonia (5:1 v/v) → product. Y 87% (e.e. 84%). The procedure is high-yielding, highly enantioselective (e.e. 72-87%), and more generally applicable and convenient than existing methods, notably those where the N-substituent has to be cleaved subsequently. Reaction is applicable to α-branched, α-unbranched (except acetaldehyde) and arylalkyl aldehydes, but ar. aldehydes reacted more slowly. Enantioselectivity was low with the corresponding N-2-nitrophenylsulfenyl deriv. F.e., comparison of organocatalysts, also reaction at the 3 g level, and with isolation of the intermediate N-sulfenyltetrahydro-β-carbolines s. M.J. Wanner, R.N.S. van der Haas, K.R. de Cuba, J.H. van Maarseveen, H. Hiemstra, Angew. Chem. Int. Ed. Engl. *2007, 46* (39), 7485-7.

Lithium perchlorate s. under i-*Pr_2NEt* *$LiClO_4$*

*(R,R,R)-2,6-Bis[2,4-bis(4-trifluoromethylphenyl)naphth-1-yl]-4,4-dimethyl-4,5-dihydro-3*H*-4-azoniacyclohepta[2,1-*a*;3,4-*a*′]dinaphthalene iodide s. under KCN* ←

Remaining Elements ↑ CC ↓↑ Rem

Without additional reagents *w.a.r.*

Cycloaddition-intramolecular cyclopropanation ○

with transition metal α,β-(*o*-styryl)acetylene(alkoxy)carbene complexes s. *72*, 638

Microwaves s. under K_2CO_3, $MgCl_2$, Pd-C and $Pd_2(dba)_3$ [\\\\]

Potassium tert-*butoxide* *KOBu*-t

Stilbenes from 1,3-dienes, 2-acetylenephosphonium salts and ar. aldehydes CHO → CH=C in one-pot via cobalt(I)-catalyzed Diels-Alder reaction and Wittig synthesis s. *72*, 439

n-*Butyllithium s.a. under 2(S)-[Bis[3,5-bis(trifluoromethyl)phenyl]trimethylsiloxymethyl]pyrrolidine* *BuLi*

n-*Butyllithium/sodium hydroxide* *BuLi/NaOH*

Nitriles from bromides Br → CH_2CN

Synthesis with addition of two C-atoms

580. 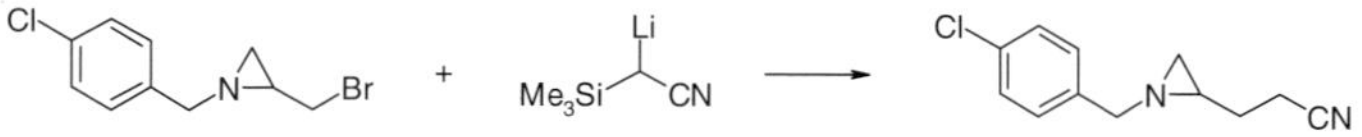

2-(2-Cyanoethyl)aziridines. A soln. of 1-(*p*-chlorophenyl)methyl-2-(bromomethyl)aziridine (5 mmol) in THF added to a soln. of α-lithiated trimethylsilylacetonitrile (1.5 eqs.) in THF at 0° (formed by treatment of trimethylsilylacetonitrile with *n*-BuLi in THF for 1 h at 0°), refluxed for 4 h, then poured into aq. NaOH (1 *N*) → product. Y 65%. The workup procedure cleaves the trimethylsilyl group from the initially-formed silylated aziridine intermediate. The method offers a general and convenient entry into 2-(2-cyanoethyl)aziridines, which constitute a novel and versatile class of synthetic building blocks. It should be noted that no reaction occurs using the corresponding trimethylsilylacetic acid esters under similar or modified conditions. F.e.s. M. D'hooghe, K. Vervisch, N. De Kimpe, J. Org. Chem. *2007*, *72* (19), 7329-32.

n-*Butyllithium/1,3-bis(2,6-diisopropylphenyl)imidazol-2-ylidene/hydrogen chloride* ←

(Z)-α-Alkylidene-β-hydroxyketones CH(OH)C(=C)CO

from 2-acetylene-1,1-hydroxysilanes and aldehydes

under mild N-heterocyclic carbene catalysis

581. 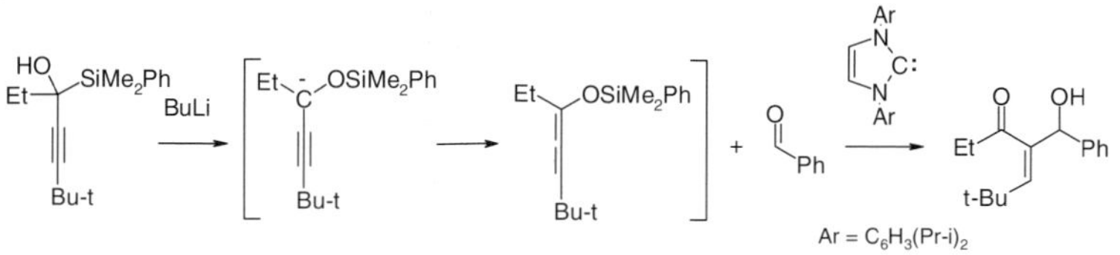

5 Mol% BuLi added to a THF soln. of the startg. 1-hydroxypropargylsilane under N_2 at 0°, the mixture stirred until conversion to the siloxyallene was complete (TLC), cooled to -78°, 2 eqs. benzaldehyde and 0.2 eqs. 1,3-bis(2,6-diisopropylphenyl)imidazol-2-ylidene added, the soln. stirred at -78° until the aldehyde consumed (TLC; 15 min to 18 h), filtered through silica gel with ether washing, the filtrate concentrated *in vacuo*, the residue dissolved in THF, treated with 1 *N* HCl, stirred at room temp. for 1-5 h, diluted with ether/water (2:1), the organic layer washed with satd. aq. $NaHCO_3$ and brine, dried over Na_2SO_4, concentrated *in vacuo*, and the residue purified by flash chromatography → (Z)-4-(α-hydroxybenzyl)-6,6-dimethylhept-4-en-3-one. Y 88% (Z/E 20:1). The silylpropargylic alcohol reacted readily with ar. (9 examples; Y 61-94%) and aliphatic (2 examples; Y 38 and 45%) aldehydes in the presence of halo, ether and silyl ether functionality affording the trisubst. (Z)-alkenes as major products (10:1 to 20:1). Unlike the standard Morita-Baylis-Hillman approach the method is suitable for preparing *β-subst. products* (*tert*-butyl, methyl, phenyl and silyl being tolerated). The role of the carbene was not certain and is under investigation, but other tested nucleophile sources were required in stoichiometric amounts to be effective. F.e.s. T.E. Reynolds, C.A. Stern, K.A. Scheidt, Org. Lett. *2007*, *9* (13), 2581-4.

Potassium carbonate/palladium(II) acetate/triphenylphosphine/microwaves ←
4-Component synthesis of 2,3,6-trisubst. imidazo[1,2-*a*]pyridines from 2-aminopyridines via sequential Suzuki coupling and heteroarylation ○

582. 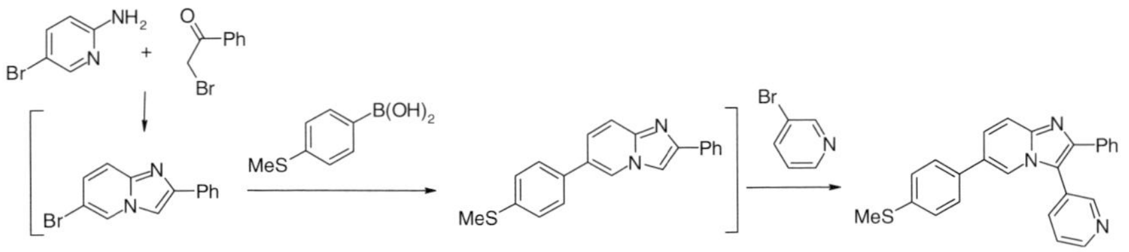

One-pot procedure under microwave enhancement. K_2CO_3 (2 eqs.) and α-bromoacetophenone (1.05 eqs.) added to a soln. of 2-amino-5-bromopyridine (0.1 g) in 1,4-dioxane/ethanol (2:1, 2 ml), the resulting mixture subjected to microwave irradiation with stirring in a sealed microwave tube under argon for 20 min at 150°, cooled, startg. arylboronic acid (1.2 eqs.), triphenylphosphine (0.2 eq.) and $Pd(OAc)_2$ (0.1 eq.) added, the sealed mixture irradiated at 150° for a further 20 min before addition, by syringe, of 3-bromopyridine (1.5 eqs.) to the cooled mixture, and irradiation for a further 2 h at 150° → product. Y 44%. Alternative one-pot, two-step procedures, involving heteroarylation and Suzuki coupling of pre-prepared imidazo[1,2-*a*]pyridine precursors, is also described. The development of a one-pot strategy was based on the observation that 6-chloro-substituents had survived previous heteroarylation reactions with similar substrates. Although the 6-chloro-substrate was found to be unsuitable for the two-step process (affording product mixtures resulting from incomplete reactions), the corresponding 6-bromo-deriv. was found to react cleanly and further optimizations were developed therefrom. F.e.s. J. Koubachi, S.E. Kazzouli, S. Berteina-Raboin, A. Mouaddib, G. Guillaumet, J. Org. Chem. *2007*, *72* (20), 7650-5.

Potassium fluoride s.a. under Pd(OAc)$_2$ *KF*

Potassium fluoride/18-crown-6 polyether *KF/crown*
***o*-Disubst. arenes from benzynes** ←
with $Pd(OAc)_2$/dppe/CsF cf. *67*, 255s*71*; *o*-chloro- or *o*-bromo-ketones from acyl chlorides or bromides with KF/18-crown-6, also ester analogs, s. H. Yoshida, Y. Mimura, J. Ohshita, A. Kunai, Chem. Commun. *2007* (23), 2405-7; *o*-cyanobenzhydryl nitriles or sulfones from two benzyne molecules and malononitrile or sulfonylacetonitriles, respectively, s. H. Yoshida, M. Watanabe, T. Morishita, J. Ohshita, A. Kunai, ibid. *2007* (15), 1505-7.

Potassium fluoride/18-crown-6 polyether *KF/crown*
Cesium fluoride [s.a. under Pd(OAc)$_2$ and Pd$_2$(dba)$_3$] *CsF*
***o*-Annelation of benzynes** ○
cf. *68*, 464s*71*; *70*, 463s*71*; N-unsubst. or 1-arylated indazoles from diazo compds. with KF/18-crown-6 or CsF, respectively, s. T. Jin, Y. Yamamoto, Angew. Chem. Int. Ed. Engl. *2007*, *46* (18), 3323-5; 9-(arylamino)anthracenes from imidazoles and three benzyne molecules with CsF s. C. Xie, Y. Zhang, Org. Lett. *2007*, *9* (5), 78-4; indolo[1,2-*f*]phenanthridines from N-(*o*-bromo-aryl)indoles with 2.5 mol% $Pd_2(dba)_3$, 5 mol% dppp and 3 eqs. CsF in 1:1 acetonitrile/toluene s. C. Xie, Y. Zhang, Z. Huang, P. Xu, J. Org. Chem. *2007*, *72* (14), 5431-4.

trans-*N,N'-Dimethylcyclohexane-1,2-diamine s. under NiCl$_2$(glyme)* ←

*2(S)-[Bis[3,5-bis(trifluoromethyl)phenyl]trimethylsiloxymethyl]pyrrolidine/*n*-butyllithium* ←
(E)-γ-Alkylthio-α,β-ethylenecarboxylic acid esters from aldehydes C(SR)CH=CHCOOR
Organocatalyzed asym. α-sulfenylation-Horner synthesis with addition of two C-atoms

583. 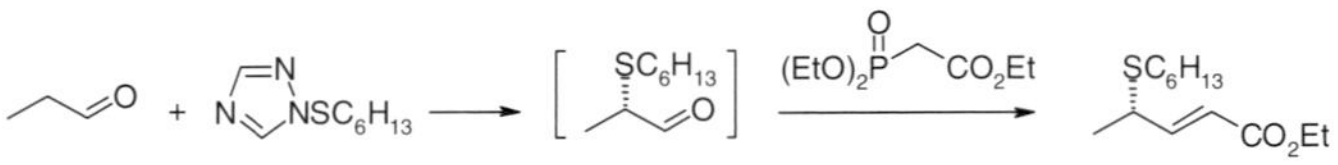

One-pot conversion. 1.3 Eqs. freshly prepared 1-hexylthio-1,2,4-triazole added to a soln. of 0.1 eq. 2(S)-[bis[3,5-bis(trifluoromethyl)phenyl]trimethylsiloxymethyl]pyrrolidine and 1 eq.

propional in dry toluene, stirred at room temp. for 5 h, cooled to -78°, a pre-mixed (at -78°) soln. of 1.5 eqs. *n*-BuLi and 1.5 eqs. triethyl phosphonoacetate in dry methylene chloride added, stirring continued at the same temp. for 1 h, quenched with satd. aq. NH_4Cl, and worked up with column chromatographic purification on silica → methyl (4S)-4-(hexylthio)pent-2(E)-enoate. Y 52% (e.e. 93%). The procedure is highly enantioselective (e.e. 88-93%), tolerant of the size of the β-substituent on the aldehyde (even *t*-Bu), and leaves benzyl, allyl and O-*tert*-butyldimethylsilyl groups unaffected. F.e. and conversion to **chiral (E)-β,γ-ethylene-α-sulfenamidocarboxylic acid esters** with complete transfer of chirality s. A. Armstrong, L. Challinor, J.H. Moir, Angew. Chem. Int. Ed. Engl. *2007*, *46* (28), 5369-72.

(-)-Sparteine s. under $PdCl_2$((-)-sparteine) ←

Copper(I) 3-methylsalicylate *RCOOCu*

Unsatd. ketones from thiolic acid esters and unsatd. boronic acids via copper-catalyzed aerobic cross-coupling COSR → COR′

584. 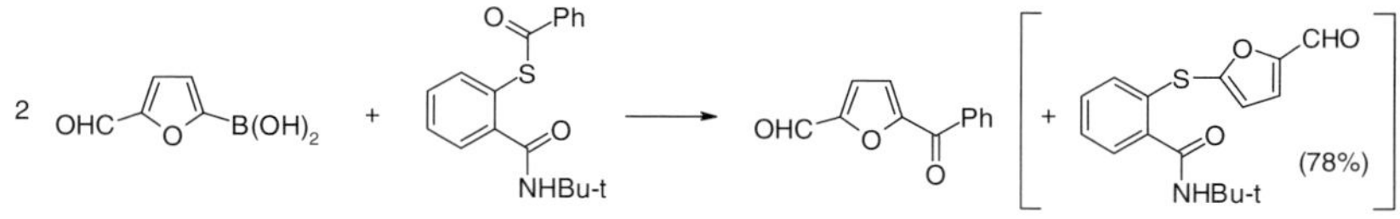

Aryl ketones. Copper(I) 3-methylsalicylate (0.013 mmol) and 5-formylfuran-2-boronic acid (0.63 mmol) added to S-2-(*tert*-butylcarbamoyl)phenyl benzothioate (0.25 mmol) in a 10 ml vial, DMF (2 ml) added, and the mixture stirred *in air* at 50° until completion (by TLC) → 5-benzoylfuran-2-carboxaldehyde. Y 91%. On the basis of control experiments and mechanistic studies, the authors suggest that the reaction proceeds via initial coordination of Cu(I) to the thiosalicylamide, followed by aerial oxidation to a novel Cu(II/III) species. This intermediate reacts with one equivalent of boronic acid to give the ketone and a Cu(II/III)-thiolate deriv., which then reacts with a second (sacrificial) equivalent of boronic acid, thereby regenerating Cu(I) and removing thiolate from the system. The procedure is carried out at neutral pH and avoids the need for Pd catalysts and stoichiometric quantities of Cu(I) reagents. Suitable substrates include aromatic, heteroaromatic and alkenyl boronic acids along with aromatic, aliphatic and α,β-unsaturated thiolic acid esters; cyclohexylboronic acid, the only aliphatic boronic acid explored, was unreactive. F.e.s. J. Villalobos, J. Srogl, L.S. Liebeskind, J. Am. Chem. Soc. *2007*, *129* (51), 15734-5.

Gold(I) chloride s. under $ReBr(CO)_3(thf)_2$ *AuCl*

Magnesium chloride/dichloro[1,1′-bis(diphenylphosphino)ferrocene]palladium(II)/ potassium carbonate/microwaves ←

4-Component synthesis of 3-amino-6-arylimidazo[1,2-*a*]pyridines via ring-closure-Suzuki coupling ○

585. 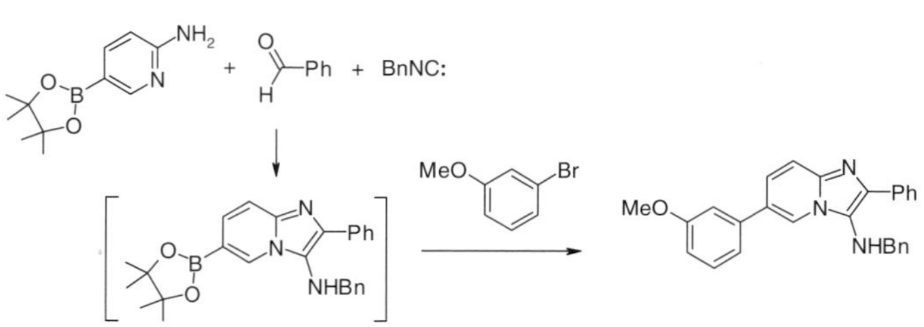

The classical 3-component synthesis of 3-aminoimidazo[1,2-*a*]pyridines (cf. *55*, 340) has been sequenced with Suzuki coupling **under microwave enhancement. E:** A mixture of 4 mol%

$MgCl_2$, 2-aminopyridine-5-boronic acid pinacol ester, and 1 eq. benzyl isocyanide in dry methanol sealed in a microwave reaction vessel under N_2, irradiated in a microwave reactor for 10 min at 160°, allowed to cool to room temp., the vessel opened, 2.5 eqs. K_2CO_3, 10 mol% $PdCl_2$(dppf) and 1 eq. startg. aryl halide added, purged with N_2, the vessel resealed, and irradiated again in the microwave reactor at 90° for 30 min → product. Y 58%. The procedure is efficient and rapid, and generally applicable to the coupling of aliphatic or aromatic aldehydes (incl. formaldehyde!) with aliphatic or aromatic isonitriles, although there was no reaction with the hindered 2,6-dimethylphenyl isocyanide. An electron-withdrawing and -donating group in the aromatic ring were both tolerated. F.e.s. E.F. DiMauro, J.M. Kennedy, J. Org. Chem. *2007*, *72* (3), 1013-6.

Chiral 2-phenyl-3-tosyl-1,3,2-oxazaborolid-5-one/isopropanol ←

(E)-α,β-Ethylene-δ-hydroxycarboxylic acid esters CHO → CH(OH)CH=CH-COOR
from aldehydes with addition of four C-atoms
Organocatalyzed asym. vinylogous aldol-type condensation

586. 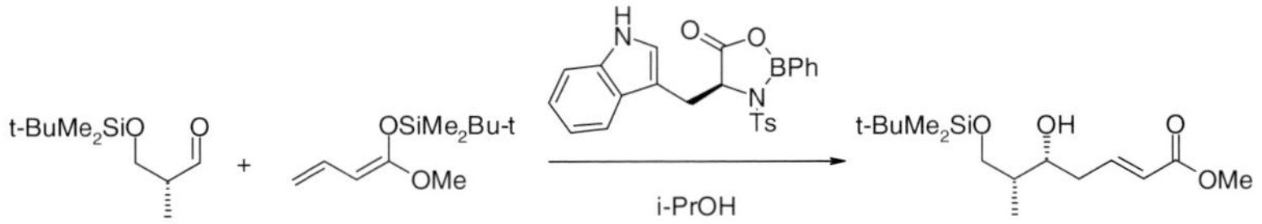

under mild conditions. 1 Eq. dichloro(phenyl)borane added to 1 eq. N-tosyl-L-tryptophan in methylene chloride, the mixture stirred at room temp. under argon for 1 h, concentrated *in vacuo*, the residue dissolved in *n*-butyronitrile, cooled to -78°, a soln. of 1.2 eqs. *tert*-butyl(1-methoxy-1,3-butadienyl-1-oxy)dimethylsilane, 1 eq. (R)-3-*tert*-butyldimethylsilyloxy-2-methylpropanal and 1.2 eqs. abs. isopropanol in *n*-butyronitrile added via syringe pump over 10 min, the mixture stirred at -78° for 3 h, quenched with satd. aq. $NaHCO_3$, the mixture extracted with ether, the extracts dried over $MgSO_4$, concentrated *in vacuo*, and the residue purified by chromatography → methyl (5R,6R)-7-(*tert*-butyldimethylsiloxy)-5-hydroxy-6-methylhept-2-enoate. Y 65% (e.e. 94%). A series of chiral and non-chiral alkyl, aryl and unsatd. aldehydes gave vinylogous aldol adducts regioselectively in 59-80% yield (e.e. 76-99%) with *tert*-butyldimethylsilyl as hydroxyl protecting group, while enantiomeric products were formed with a D-tryptophan-based catalyst (2 substrates gave 64 and 11% yield; 90 and 56% e.e.). Lower yields and selectivities were observed in both cases using 4-methoxybenzyl protection. Addition of isopropanol is required to suppress catalysis by cationic silyl species (which leads to racemic products). F.e. incl. optimization and prepn. of the startg. m. s. S. Simsek, M. Horzella, M. Kalesse, Org. Lett. *2007*, *9* (26), 5637-9.

Boric acid s. under [Rh(cod)(OH)]$_2$ *H_3BO_3*
Diethylaminopropylsilylated alumina s. under Pd(OAc)$_2$ ←

Aluminum chloride/ammonium chloride *$AlCl_3/NH_4Cl$*
Furans from carboxylic acid chlorides ○

587. 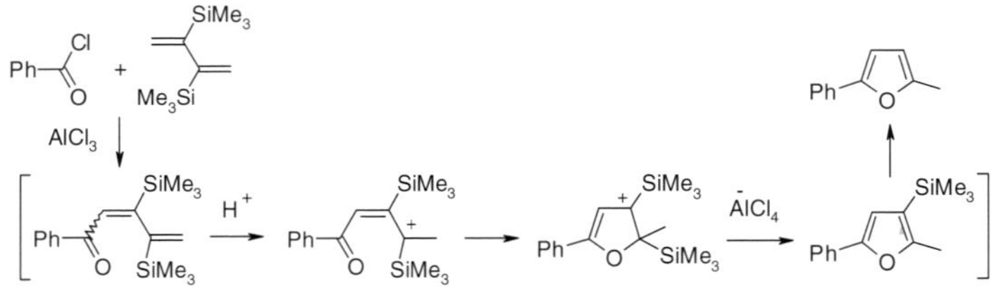

2,5-Disubst. furans. 1.5 Eqs. freshly distilled benzoyl chloride in methylene chloride added under N_2 to a stirred suspension of 1.5 eqs. $AlCl_3$ in the same solvent at 0°, when homogeneous the mixture treated slowly (over ca. 20 min) with a soln of 2,3-bis(trimethylsilyl)buta-1,3-diene in the same solvent (maintaining the temp. at 0°), and quenched with satd. aq. NH_4Cl (to desilylate the intermediate) → product. Y 63%. The intermediate **2,5-disubst. 3-silylfurans** were isolable after a shorter reaction time, and with certain acyl chlorides were obtained as the sole products, although they could be simply desilylated on subsequent treatment with BF_3-etherate. Reaction appears to involve deprotonation of an intermediate allyl cation, followed by re-protonation and cyclization. F.e.s. F. Babudri, S.R. Cicco, G.M. Farinola, L.C. Lopez, F. Naso, V. Pinto, Chem. Commun. *2007* (36), 3756-8.

Indium(III) chloride *$InCl_3$*

N-Carbalkoxy-3-ethyleneamines from α-(carbalkoxyamino)sulfones $SO_2R \rightarrow CH_2CH{=}CH_2$

Sakurai reaction with addition of three C-atoms via N-carbalkoxyimines

588. 

1.5 Eqs. allyltrimethylsilane added dropwise to a soln. of the startg. α-(carbalkoxyamino)sulfone and *5 mol%* $InCl_3$ in methylene chloride under N_2, and stirred for 9 h *at room temp.* → product. Y 87%. The procedure is mild and generally applicable to aliphatic, cycloaliphatic and aromatic substrates, as well as acid-sensitive furfuraldehyde and 2-naphthaldehyde, while electron-withdrawing and -donating substituents in the aromatic ring were tolerated. $TiCl_4$, $SnCl_4$ and $Zn(OTf)_2$ were only effective in stoichiometric amount, and $Bi(OTf)_3$ required refluxing with excess of the allylsilane. F.e.s. B. Das, K. Damodar, D. Saritha, N. Chowdhury, M. Krishnaiah, Tetrahedron Lett. *2007*, *48* (45), 7930-3.

Indium(III) bromide *$InBr_3$*

Indium(III)-catalyzed synthesis of carbonyl compds. from enoxysilanes C=C(OSi≤) → C(R)CO

589. 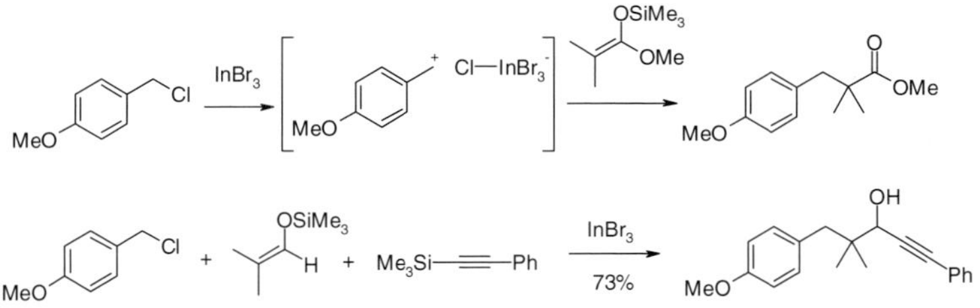

under mild conditions. 4-Methoxybenzyl chloride added to a mixture of 5 mol% $InBr_3$ and 1.5 eqs. dimethylketene methyl trimethylsilyl acetal in methylene chloride under N_2, stirred for 2 h, added to a mixture of ether and aq. $NaHCO_3$, extracted with ether, the extracts dried over $MgSO_4$, the solvent removed *in vacuo*, and the residue distilled → methyl 3-(4-methoxyphenyl)-2-dimethylpropanoate. Y 86%. The reaction is successful with enolates derived from ketones, esters, thioesters, amides and hindered aldehydes affording the corresponding carbonyl compds. from unactivated alkyl and benzyl chlorides in 47-99% yield in the presence of ether, halo and ester functionality. In a further development, products of aldehyde-derived silyl enolates were reacted *in situ* with alkenyl- or alkynyl-silanes to afford the corresponding **2-ethylene- and 2-acetylene-*sec*-alcohols by 3-component synthesis.** F.e.s. Y. Nishimoto, M. Yasuda, A. Baba, Org. Lett. *2007*, *9* (24), 4931-4.

*Ammonium cerium(IV) nitrate s. under (S,S)-5-Benzyl-2-*tert*-butyl-3-methyl-4-imidazolidonium trifluoroacetate* *CAN*

Scandium(III) triflate s. under o-*Iodoxybenzoic acid* *$Sc(OTf)_3$*

Lanthanum(III) triflate *La(OTf)₃*
3-Ethylene(trifluoroacetyl)amines C(NHCOCF₃)C-C=C
from N-phosphinylimines and 2-ethylenesilanes
Regiospecific synthesis with asym. induction by dual activation

590. 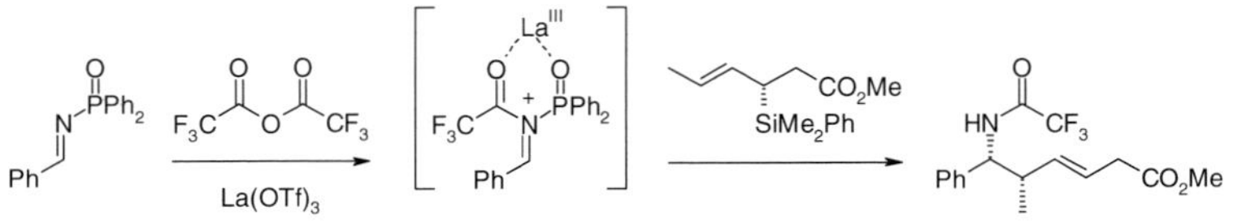

A mixture of 18 mol% hydrated $La(OTf)_3$, 1.8 eqs. trifluoroacetic anhydride and *ACS-grade* acetonitrile sonicated, added via syringe to a mixture of the startg. phosphinylimine and ca. 1 eq. startg. crotylsilane, vortexed until clear, and stirred at room temp. under argon for 3 h → (5S,6R)-methyl 5-methyl-6-phenyl-6-(2,2,2-trifluoroacetamido)hex-3(E)-enoate. Y 93% (d.r. >20:1 after column chromatography). Reaction is generally applicable to phosphinylimines derived from aromatic, α,β-ethylenic and cyclopropyl aldehydes, and tolerates both electron-donating and -withdrawing groups. There was no reaction, however, with aliphatic derivs. The isolation of N-trifluoroacetyl-3-ethyleneamines was quite unexpected, and argues in favour of dual activation of both the phosphinyl group and the anhydride by the Lewis acid, followed by acylation/hydration prior to nucleophilic addition. Friedel-Crafts addition of furan to the imine, and imino-ene reaction with methylenecyclopentane, were also effected under the same dual activation to give the corresponding trifluoroacetylamine derivs. F.e. and comparison of Lewis acids s. W.W. Ong, A.B. Beeler, S. Kesavan, J.S. Panek, J.A. Porco Jr., Angew. Chem. Int. Ed. Engl. *2007*, *46* (39), 7470-2.

Alcohols s. under Chiral 2-phenyl-3-tosyl-1,3,2-oxazaborolid-5-one, $NiCl_2$(glyme), Pd(OAc)₂(dppe) and $PdCl_2$((-)-sparteine) *ROH*
Polyethylene glycol s. under $Pd(OAc)_2$ *PEG*
18-Crown-6 polyether s. under KF *crown*

*(S,S)-5-Benzyl-2-*tert*-butyl-3-methyl-4-imidazolidonium trifluoroacetate/ ammonium cerium(IV) nitrate/sodium hydrogen carbonate* ←
Asym. syntheses with aldehydes ←
via organocatalyzed enamine activation-single electron transfer

591. 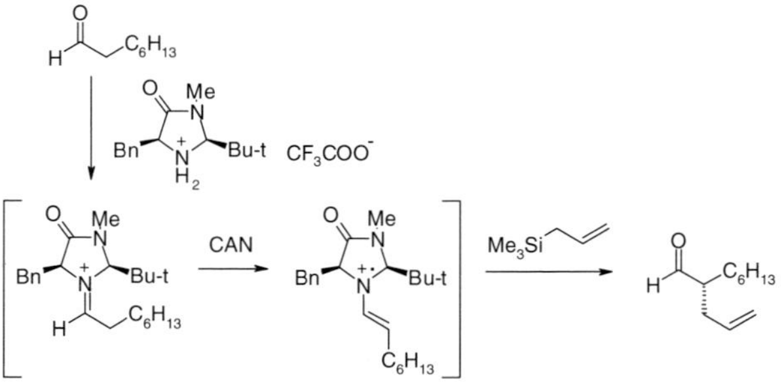

Asym. α-allylation with 2-ethylenesilanes. Dimethoxyethane added to an oven-dried flask containing 0.2 eq. (S,S)-5-benzyl-2-*tert*-butyl-3-methyl-4-imidazolidonium trifluoroacetate, 2.5 eqs. CAN and 1.5 eqs. oven-dried $NaHCO_3$, the suspension cooled to -50°, deoxygenated by stirring vigorously *in vacuo* for 3-5 min, the mixture back-filled with argon, degassed twice more, 2.5 eqs. allyltrimethylsilane added, followed by 1 eq. octanal, warmed to -20°, stirred for

24 h under argon, re-cooled to -50°, filtered, and worked up with forced-flow chromatographic purification → (R)-2-allyloctanal. Y 81% (e.e. 91%). This is the first development of organo-catalyzed asym. enamine activation by subsequent conversion to a *reactive radical cation* through single electron oxidation, thereby facilitating face-selective attack by a variety of *nucleophiles* with high enantioselectivity. F.e., also asym. α-pyrrol-2-ylation (with N-Boc-pyrrole), asym. synthesis of cyclic γ-chloraldehydes from ethylenealdehydes, and asym. synthesis of ε-alkoxy-ζ-nitrato-γ,δ-ethylenealdehydes (with 2-methoxy-3-phenyl-1-vinylcyclopropane), s. T.D. Beeson, A. Mastracchio, J.-B. Hong, K. Ashton, D.W.C. MacMillan, Science *2007*, *316* (5824), 582-5.

1,3-Bis(2,6-diisopropylphenyl)imidazol-2-ylidene s. under BuLi ←

o-*Iodoxybenzoic acid/scandium(III) triflate* *$ArIO_2/Sc(OTf)_3$*

δ,ε-Ethylene-β′-ketocarboxylic from β-hydroxy-α-methylenecarboxylic acid esters Synthesis with addition of three C-atoms via oxidative 1,4-addition ←

592.

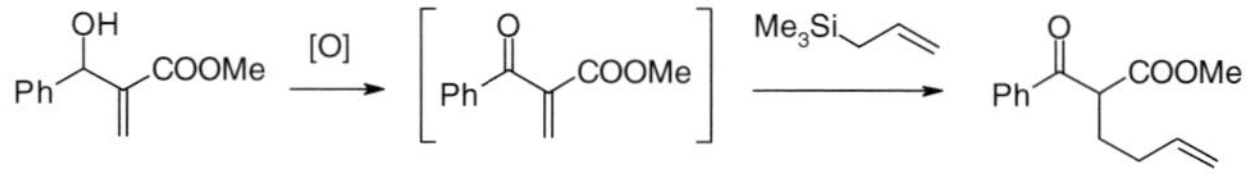

A mixture of the startg. Baylis-Hillman adduct and 1.2 eqs. *o*-iodoxybenzoic acid in acetonitrile stirred at room temp. until oxidation complete, 1.5 eqs. allyltrimethylsilane and 10 mol% $Sc(OTf)_3$ added sequentially, stirred at room temp., and worked up after 24 h → product. Y 84%. The procedure is simple, clean, mild, inexpensive and leaves various functionalities, e.g. halides, phenolethers, esters and alkenes, unaffected. Other hypervalent iodine reagents were less effective, and other oxidants (Oxone, CAN, MnO_2, $KBrO_3$) and Lewis acids ($InCl_3$, $InBr_3$, $CeCl_3$) were ineffective. Liberated *o*-iodobenzoic acid is recyclable. F.e.s. J.S. Yadav, B.V. Subba Reddy, A.P. Singh, A.K. Basak, Tetrahedron Lett. *2007*, *48* (42), 7546-8.

Tin(IV) chloride *$SnCl_4$*

Stereospecific Lewis acid-catalyzed Michael-type addition of 1-silyl-1,3-(dien)oxysilanes to 1,1-nitroethylene derivs. ←

593.

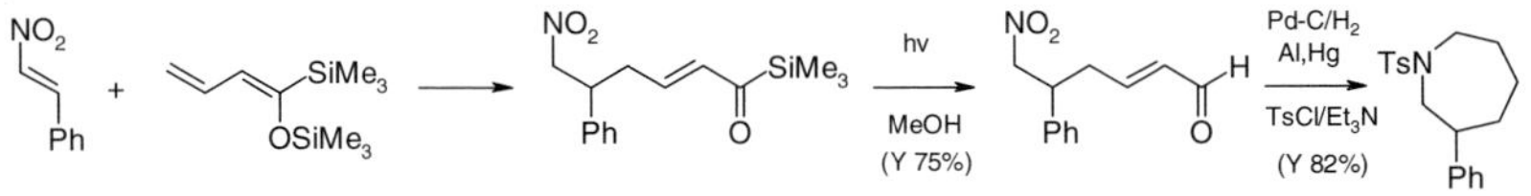

1.1 Eqs. 0.5*M* $SnCl_4$ in methylene chloride added dropwise via syringe under N_2 over 40 s to a cold (-70°) soln. of the startg. nitrostyrene and 1.2 eqs. of the startg. dienol silyl ether in the same solvent, stirring continued for 10 min at -70°, 4.9 eqs. 1 *M* Et_3N in methanol added rapidly with vigorous stirring, and quenched with water → product. Y 73%. cf. *42*, 899. The bulky silyl group at the 1-position of the dienol silyl ether prevents undesirable α-addition and Diels-Alder reaction. F.e. and comparison of Lewis acids, also conversion to **1-tosylhexahydroazepines** via photochemical desilylation and reductive ring closure, s. S.E. Denmark, M. Xie, J. Org. Chem. *2007*, *72* (18), 7050-3.

Tert. phosphines and di(phosphines) s. under K_2CO_3, $MoO_2Cl_2(dmf)_2$, $Rh(acac)(CO)_2$, $Pd(OAc)_2$ and $Pd_2(dba)_3$ ←

(R)-5,5′,6,6′,7,7′,8,8′-Octahydro-2,2′-bis(diphenylphosphino)-1,1′-binaphthyl s. under $[RhCl(ethylene)]_2$ *(R)-H_8-BINAP*

(R,R)-1-Benzyl-3,4-bis(diphenylphosphino)pyrrolidine s. under $Rh(acac)(coe)_2$ *(R,R)-DeguPhos*

Chiral phosphoromonoamidites s. under $Pd(dba)_2$ ←

Bismuth trichloride *$BiCl_3$*

Bismuth(III)-catalyzed nucleophilic substitution of 2-acetylenealcohols OH → C-C≡C
by allyl s. *72*, 106

*Oxygen s. under Diacetato((4R)-2-pyrid-2-yl-4-*tert*-butyl-Δ^2-oxazoline)palladium(II) and $PdCl_2$((-)-sparteine)* *O_2*

Hydrogen peroxide/sodium hydroxide *H_2O_2/NaOH*
3-Component synthesis of sec. amines CHO → CH(R)NHR′
from aldehydes, prim. amines and borinic acid chlorides

594. 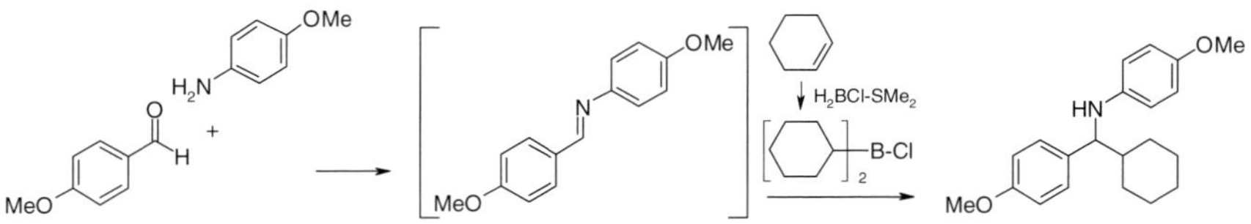

7.6 Eqs. cyclohexene added to a 1 *M* soln. of monochloroborane-dimethyl sulfide in methylene chloride at 0°, warmed to room temp., stirred for 3 h, *p*-methoxybenzaldehyde and 1 eq. *p*-methoxyaniline added, stirred at room temp. for a further 2 h, treated with 30 mol% H_2O_2, stirring continued for 30 min, 1 *M* NaOH added, and stirred again for 30 min → product. Y 66%. The borinyl chloride acts as a Lewis acid for the initial condensation to give the azomethine, then adds to the latter. Yields were lower, however, with di-*n*-hexylborinyl chloride. F.e. and with Oxone, urea-hydrogen peroxide or *m*-chloroperoxybenzoic acid as oxidant, also from the intermediate azomethines, s. M. Valpuesta, C. Muñoz, A. Diaz, R. Suau, G. Torres, Eur. J. Org. Chem. *2007* (27), 4467-70.

Molybdenyl chloride·bis(dimethylformamide)/triphenylphosphine *$MoO_2Cl_2(dmf)_2/Ph_3P$*
Indoles from *o*-nitroaldehydes and alkylidenephosphoranes ○
One-pot Wittig synthesis-molybdenum(VI)-catalyzed ring closure via *o*-nitrostyrenes

595. 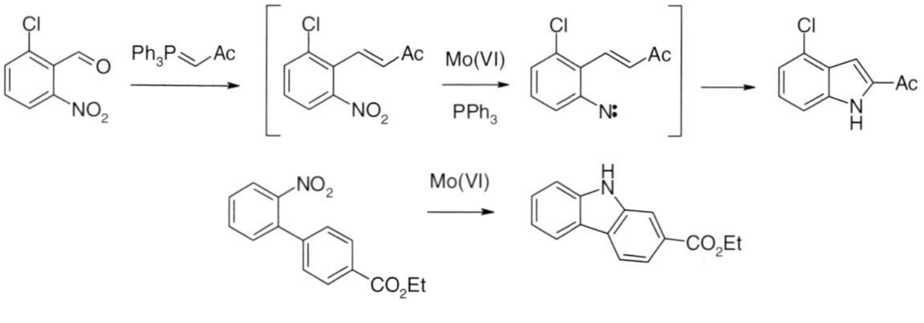

2-Acylindoles. 1 Eq. startg. alkylidenephosphorane added to 2-chloro-6-nitrobenzaldehyde in dry toluene at room temp., the mixture stirred for 1-2 h (GC-MS monitoring), 2.4 eqs. triphenylphosphine and 5 mol% $MoO_2Cl_2(dmf)_2$ added, refluxed with stirring under N_2 for 12 h, a second portion of catalyst added, reflux continued for a further 12 h, excess of triphenyl-phosphine oxidized by addition of dimethyl sulfoxide and 1 mol% catalyst with further refluxing and stirring for 30 min, the solvents removed *in vacuo*, and the residue purified by chromatography on silica → 2-acetyl-4-chloroindole. Y 93%. The method involves the use of commercially available benzaldehydes and alkylidenephosphoranes to generate styrene derivs. *in situ*, which undergo reductive cyclization **via nitrenes** in 45-93% yield in the presence of ether, halo and methylenedioxy functionality; low yields were obtained, however, for 5-chloro-2-carbethoxy- and -2-acetylindoles (15% and 39%, respectively). In additional examples, 2,3-disubst. indoles were prepared from pre-formed styrene derivs. in 64-84% yield (7 examples; phenyl, alkyl and carbethoxy substituents) and **3,8-disubst. carbazoles** (7 examples; methoxy, formyl, acetyl, fluoro and carbethoxy substituents) from the corresponding disubst. 2-nitrobiphenyls in 70-86% yield (3-hydroxy and 3-carboxy derivs. being obtained in 33 and 30% yield, respectively). Removal of excess of the reductant was also facilitated in several examples by use of polymer-bound triphenylphosphine. F.e. incl. **indole-2-carboxylic acid esters** s. R. Sanz, J. Escribano, M.R. Pedrosa, R. Aguado, F.J. Arnáiz, Adv. Synth. Catal. *2007*, *349* (4-5), 713-8.

*(2S,5S)-2-*tert*-Butyl-5-[(1-benzyl-1*H*-indol-3-yl)methyl]-3-methyl-4-imidazolidone hydrochloride/hydrogen fluoride/hydrogen chloride* ←
Organocatalyzed asym. 1,4-addition C=C → CHC(R)
of potassium trifluoro(hetaryl or vinyl)borates to α,β-ethylenealdehydes

596. 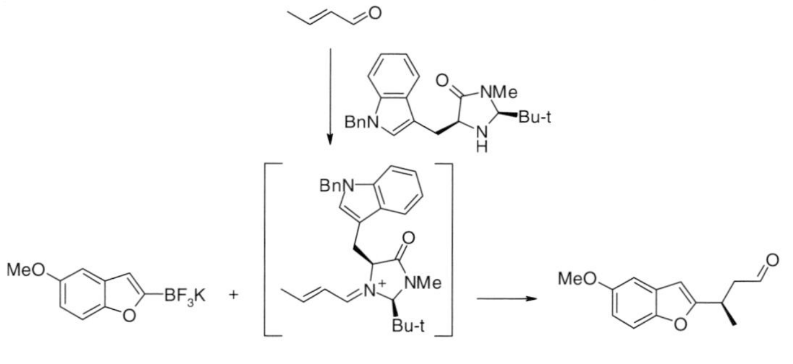

The use of preformed Vedejs-Molander potassium trifluoro(organo)borate salts enables π-neutral and π-deficient aryl and vinyl nucleophiles to undergo conjugate addition to a broad range of enals. In the case of the hetaryl examples, the method can provide complementary regiocontrol to that which is obtained from traditional Friedel-Crafts reactions. **E:** (2S,5S)-2-*tert*-Butyl-5-[(1-benzyl-1*H*-indol-3-yl)methyl]-3-methyl-4-imidazolidone (0.03 mmol) and HCl (0.03 mmol) added to HF (48 wt%; 0.15 mmol) and 1,2-dimethoxyethane (450 μl) in a plastic vial, the mixture cooled to -20°, crotonaldehyde (0.45 mmol) and K-2-(5-methoxybenzofuranyl)trifluoroborate (0.15 mmol) added, the mixture stirred at -20° for 24 h, then quenched with 1 *M* HCl (1.0 ml) and stirred with chloroform (1.5 ml) for 30 min → (R)-3-(5-methoxybenzofuran-2-yl)butanal. Y 94% (e.e. 92%). F.e.s. S. Lee, D.W.C. MacMillan, J. Am. Chem. Soc. *2007*, *129* (50), 15438-9.

Bis(cinchoninium bromides)/tetramethylammonium fluoride/tetrabutylammonium fluoride ←
2,2,2-Trifluoroalcohols from ketones CO → C(OH)CF_3
Asym. synthesis with addition of one C-atom

597. 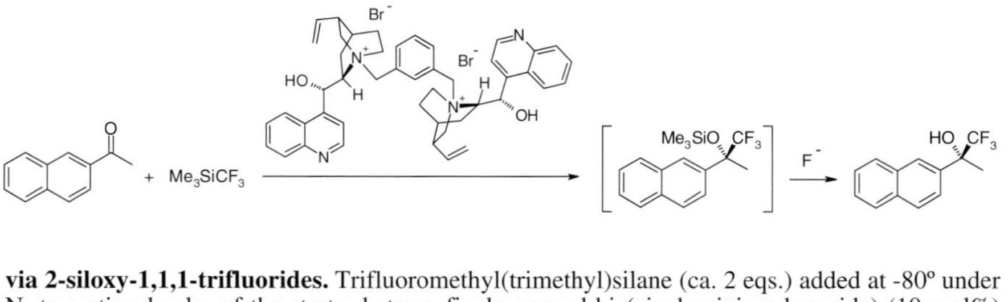

via 2-siloxy-1,1,1-trifluorides. Trifluoromethyl(trimethyl)silane (ca. 2 eqs.) added at -80° under N_2 to a stirred soln. of the startg. ketone, finely ground bis(cinchoninium bromide) (10 mol%) and tetramethylammonium fluoride (10 mol%) in 2:1 toluene/methylene chloride, stirred at the same temp. for 6 h, quenched with satd. aq. NH_4Cl, worked up, and the crude silyl ether treated with tetrabutylammonium fluoride (ca. 1 eq.) in THF at room temp. for 1 h → product. Y 87% (e.e. 85%). The procedure is simple, reliable, and generally applicable to acyclic and cyclic aryl ketones, affording products with high enantioselectivity (up to 94%). The choice of tetramethylammonium fluoride as fluoride ion source was critical. cf. *58*, 435. F.e. and comparison of cinchoninium salts s. S. Mizuta, N. Shibata, S. Akita, H. Fujimoto, S. Nakamura, T. Toru, Org. Lett. *2007*, *9* (18), 3707-10.

Manganese dioxide s. under $PdCl_2$((-)-sparteine) *MnO_2*

Bromotris(carbonyl)bis(tetrahydrofuran)rhenium(I)/gold(I) chloride *$ReBr(CO)_3(thf)_2$/AuCl*
Sym. 1,4-diynes from aldehydes and two silylacetylene molecules C≡C-CH(R)C≡C
Rhenium(I)/gold(I)-catalyzed coupling under mild conditions

598.

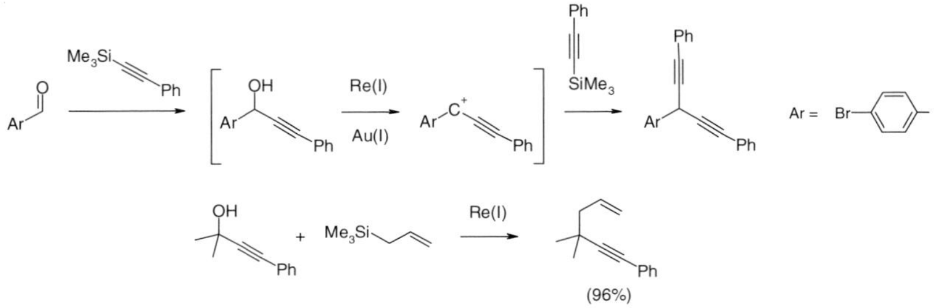

Sym. 3-aryl-1,4-diynes. A mixture of 4-bromobenzaldehyde, 4 eqs. trimethylsilylphenylacetylene, 2.5 mol% $ReBr(CO)_3(thf)_2$ and 5 mol% AuCl stirred in methylene chloride at room temp. for 3 h, the solvent removed *in vacuo*, and the residue purified by chromatography on silica → 1,5-diphenyl-3-(4-bromophenyl)-1,4-pentadiyne. Y 66%. Seven examples gave 43-79% yield in the presence of *bromo* and ether functionality with the lowest yield recorded for electron-rich 4-methoxybenzaldehyde while electron-poor 4-trifluoromethylbenzaldehyde gave a mixture of the diyne (33% yield) and mono-coupled propargylic alcohol (66% yield). In the absence of either catalyst the products were only obtained in trace amounts. A mechanism was proposed based on experimental evidence. In a further series of reactions, propargylic (incl. alkyl-subst.) and benzylic alcohols coupled with allyl- or phenylethynyl-silanes in the presence of the rhenium catalyst alone in 56-99% yield. F.e.s. Y. Kuninobu, E. Ishii, K. Takai, Angew. Chem. Int. Ed. Engl. *2007*, *46* (18), 3296-9.

1,1′-Bis(diphenylphosphino)ferrocene s. under $Pd(OAc)_2$ *dppf*

Tetrakis(triphenylphosphine)nickel(0)/potassium phosphate *$Ni(PPh_3)_4/K_3PO_4$*
Nickel(0)-catalyzed synthesis of α-aryl- Hal → Ar
from α-halogeno-carbonyl compds. and arylboronic acids

599.

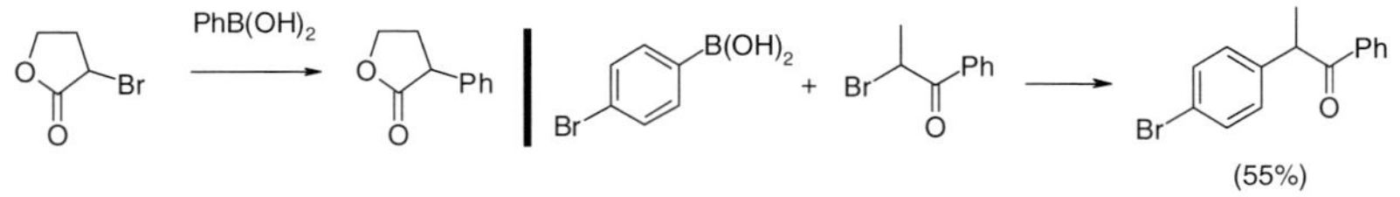

under mild conditions. In a glove box, phenylboronic acid, 5 mol% $Ni(PPh_3)_4$ and 2 eqs. *anhydrous* K_3PO_4 combined in a Schlenk tube, the latter capped with a rubber septum then removed from the glove box, backfilled with N_2, 0.5 eqs. 2-bromobutyrolactone and toluene added, the soln. stirred at 60° for 16 h, reaction quenched with 2 *M* HCl, extracted with ethyl acetate, the organic layers dried over Na_2SO_4, and purified by flash chromatography → 2-phenylbutyrolactone. Y 74%. This efficient and simple method is general for the coupling of readily available α-bromo-esters and -amides (free NH group being tolerated) and α-bromo- or α-chloroketones with chloro-, methyl-, methoxy- and notably *bromo*-boronic acids in 55-95% yield (30 examples) with the minimum (0-5%) formation of by-products (arising from, for example, dehydrobromination or debromination). Use of the *anhydrous phosphate as base* appears critical since no products could be isolated when the trihydrate was used. Furthermore, the procedure is complementary with related palladium-based routes, such as the Buchwald-Hartwig reaction, and advantageous in that regioselectivity can be controlled at the *most hindered site*. F.e. incl. reaction optimization s. C. Liu, C. He, W. Shi, M. Chen, A. Lei, Org. Lett. *2007*, *9* (26), 5601-4.

Dichloro(glyme)nickel(II)/potassium tert-*butoxide/isobutanol/*trans-*N,N′-dimethyl-cyclohexane-1,2-diamine* ←

Branched hydrocarbons RB< + R′Hal → R-R′

from B-alkyl-9-borabicyclo[3.3.1]nonanes and unactivated *sec*-alkyl halides

Nickel-catalyzed Suzuki coupling under mild conditions

600. 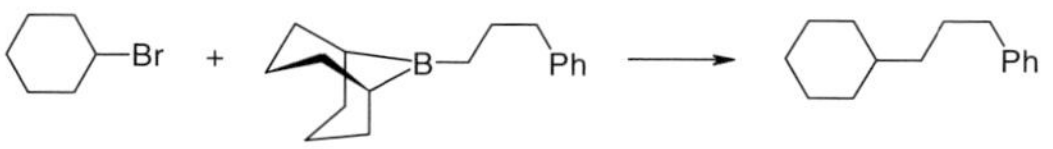

A ca. 1.5 *M* soln. of B-3-phenylpropyl-9-borabicyclo[3.3.1]nonane in dioxane added [in a glove box] to a mixture of 0.67 eqs. KOBu-*t* and 1.11 eqs. isobutanol contained in a vial, stirred vigorously at room temp. for at least 30 min in order to activate the borane, the mixture added to a soln. of 3.33 mol% $NiCl_2$·glyme and *trans*-N,N′-dimethylcyclohexane-1,2-diamine in dioxane, followed by 0.55 eqs. cyclohexyl bromide, stirred vigorously *at room temp.* for 24 h, filtered through a short pad of silica gel, and worked up with flash chromatographic purification → product. Y 72% (78% on the second run). This is the first example of a Suzuki coupling of an unactivated *sec*-alkyl bromide or iodide with an *alkyl*boron compd., involving a relatively simple 1,2-diamine as ligand (there being no reaction with bathophenanthroline, *trans*-2-amino-cyclohexanol or prolinol). The addition of KOBu-*t* and isobutanol is also critical, and a number of functional groups (ethers, carbamates and esters) remain unaffected. F.e. and with *prim*-bromides and -iodides s. B. Saito, G.C. Fu, J. Am. Chem. Soc. *2007*, *129* (31), 9602-3.

(1,5-Cyclooctadiene)rhodium(I) hydroxide dimer *[Rh(cod)(OH)]₂*

Stereospecific synthesis of 3-alkylideneoxindoles ○

from *o*-acetyleneisocyanates and unsatd. boronic acids

601. 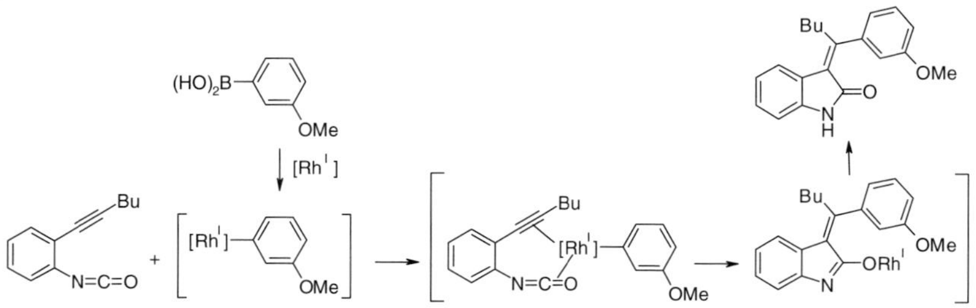

3(Z)-Arylideneoxindoles. A soln. of the startg. *o*-acetyleneisocyanate, 1.5 eqs. startg. arylboronic acid and 5 mol% $[Rh(cod)(OH)]_2$ in dry THF stirred at room temp. for 12 h under argon, quenched with water, and worked up with preparative thin layer chromatographic purification → product. Y 78%. The procedure is generally applicable to both terminal alkyne derivs. and those substituted by *prim*- and *sec*-alkyl-groups (not *tert*-butyl!), and leaves chloro, methoxy, cyano and ester groups unaffected. Reaction is thought to involve intermediate formation of a 3-alkylidene-indolenin-2-yloxyrhodium(I) species, although it is not clear whether a concerted or stepwise mechanism is implicated. F.e., **also 3-(alk-2-en-1-ylidene)oxindoles** from α,β-ethyleneboronic acids, s. T. Miura, Y. Takahashi, M. Murakami, Org. Lett. *2007*, *9* (24), 5075-7.

(1,5-Cyclooctadiene)rhodium(I) hydroxide dimer/boric acid *$[Rh(cod)(OH)]_2/H_3BO_3$*
Arylglyoxylic acid esters $B(OH)_2 \rightarrow COCOOR$
from arylboronic acids with addition of two C-atoms

602. 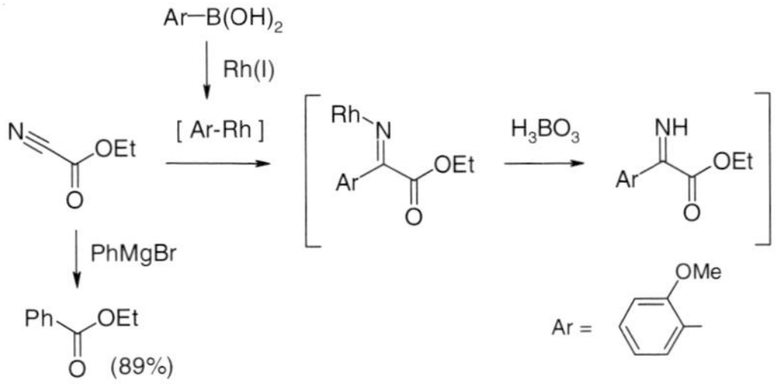

A mixture of 1.2 eqs. 2-methoxyphenylboronic acid, 2 eqs. boric acid, 2.5 mol% $[Rh(OH)(cod)]_2$ and ethyl cyanoformate in dioxane stirred under argon at room temp. for 30 min, heated at 60° for 3 h, the mixture cooled to room temp., diluted with ethyl acetate and 10% aq. citric acid, the aq. layer extracted with ethyl acetate, the extracts washed with water and brine, dried over Na_2SO_4, the solvent removed *in vacuo*, and the residue purified by TLC → ethyl 2-methoxybenzoylformate. Y 87%. Selective reaction *at the cyano group* of cyanoformates with organometallic reagents is rare (indeed, phenylmagnesium bromide and phenyllithium reacted at the ester moiety under these conditions to afford ethyl benzoate in 89 and 36% yield, respectively). Yields of 46-87% were recorded in the presence of halo, ester, ether, *o*-methyl and *formyl* groups and were lowest from arylboronic acids possessing electron-withdrawing *o*- and *p*-functionality, while 2-biphenylylboronic acid reacted in 50% yield. Little or no reaction occurred in the absence of boric acid which is thought to assist in the release of rhodium from the intermediate iminorhodium(I) species. F.e.s. H. Shimizu, M. Murakami, Chem. Commun. *2007* (27), 2855-7.

Acetoacetonatobis(cyclooctene)rhodium(I)/(R,R)-1-benzyl-3,4-bis(diphenylphosphino)pyrrolidine/triethylamine/potassium carbonate ←
Rhodium(I)-catalyzed asym. synthesis of N-protected mixed 1,1-diarylamines $SO_2R \rightarrow Ar$
from α-aryl-α-aminosulfones and arylboronic acids
N-Protected imines as intermediates

603. 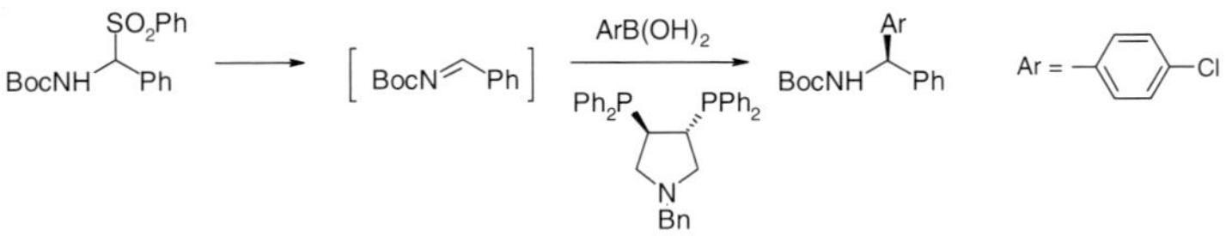

A premixed dioxane soln. of 5 mol% $Rh(acac)(coe)_2$ and 5.5 mol% (R,R)-DeguPhos added to the startg. N-Boc-protected α-aminosulfone, 2 eqs. 4-chlorophenylboronic acid, 4 Å molecular sieves and dioxane in a Schlenk flask, 1.5 eqs. triethylamine and 6 eqs. K_2CO_3 added sequentially, heated to 70° for 20 h, cooled to room temp., filtered through a pad of Celite, the Celite washed with ethyl acetate, the combined filtrates concentrated *in vacuo*, the residual oil dissolved in THF, shaken with PS-DEAM resin for 10 h, filtered, the resin washed with ethyl acetate, the combined filtrates concentrated *in vacuo*, and the residue purified by chromatography → (R)-product. Y 76% (e.e. 98%). The absolute configuration was determined, in this case, by comparison with the known amine. A series of derivs. was prepared by variation of the two aryl groups giving 51-76% yield and enantiomeric excesses generally >90% (although the 4-trifluoromethylphenyl deriv. gave 79% e.e.) in the presence of halo, ketone and ether functionality. Lower yields are obtained with electron-poor ar. rings but the reaction is tolerant of *o*-methyl substituents on the phenylboronic acid. F.e. incl. reaction optimization s. H. Nakagawa, J.C. Rech, R.W. Sindelar, J.A. Ellman, Org. Lett. *2007*, *9* (25), 5155-7.

Acetoacetonato(dicarbonyl)rhodium(I)/2,2′-bis(diphenylphosphino)-1,1′-binaphthyl *Rh(acac)(CO)$_2$/BINAP*

N-Protected *trans*-3-aroyl-4-hydrazinocyclopentenes from *meso*-2,3-diazabicyclo[2.2.1]hept-5-enes and arylboronic acids via rhodium(I)-catalyzed carbonylative generation of aroyl carbanion equivalents C

604.

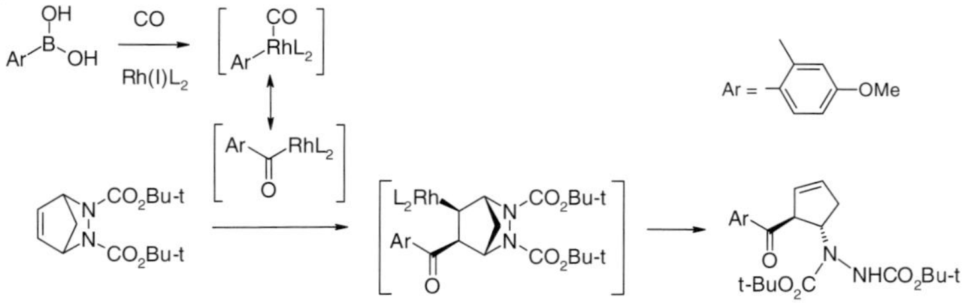

under mild conditions. 10 Mol% Rh(acac)(CO)$_2$ and 12 mol% rac-BINAP added to a vial, the latter sealed and flushed with CO gas (balloon), toluene and water (10:1) added, the mixture stirred at room temp. for 15 min, a soln. of 1 eq. 2,3-di(carbo-*tert*-butoxy)-2,3-diazabicyclo[2.2.1]-hept-5-ene and 1.2 eqs. 2-methyl-4-methoxyphenylboronic acid in toluene added, stirred at 20-25° for 16-20 h, ether added, the mixture filtered through a silica gel pad, washed with ether, the filtrate concentrated *in vacuo*, and purified by chromatography on silica → *trans*-N,N′-di(carbo-*tert*-butoxy)-N-[2-(2-methyl-4-methoxybenzoyl)]cyclopent-3-enyl)hydrazine. Y 91%. Despite *the low pressure of CO* none of the 2-arylated analog (the normal product in the absence of CO) was observed. The reaction is successful with arylboronic acids (with highest yields obtained from electron-rich or -neutral derivs.), and is unaffected by the nature of the N-protecting group (CO_2Et, Boc, Cbz or Troc) in the startg. hydrazine (19 examples; 38-96%) and can be accomplished with a number of neutral Rh(I) complexes. A styrylboronic acid also reacted successfully (Y 63%) but required slow addition to the biscarbamate to minimize competing dibenzylideneacetone formation. F.e. incl. prepn. of the startg. m. and reaction optimization s. F. Menard, C.F. Weise, M. Lautens, Org. Lett. *2007, 9* (26), 5365-7.

Bis(norbornadiene)rhodium(I) fluoroborate/P-chiral P,P′-di(alk-1-ynyl)-1,2-di(phosphines)/potassium hydroxide ←

Asym. 1,4-addition of arylboronic acids C=C → C(Ar)CH

to cyclic α,β-ethyleneketones s. *72*, 376

Chlorobis(ethylene)rhodium(I) dimer/(R)-5,5′,6,6′,7,7′,8,8′-octahydro-2,2′-bis(diphenylphosphino)-1,1′-binaphthyl/potassium hydroxide ←

Asym. synthesis of cyclic β′-aryl-β-amino-α,β-ethylenecarboxylic acid esters from cyano-α,β-ethylenecarboxylic acid esters and 9-aryl-9-borabicyclo[3.3.1]nonanes Asym. 1,4-addition-cycloisomerization ○

605.

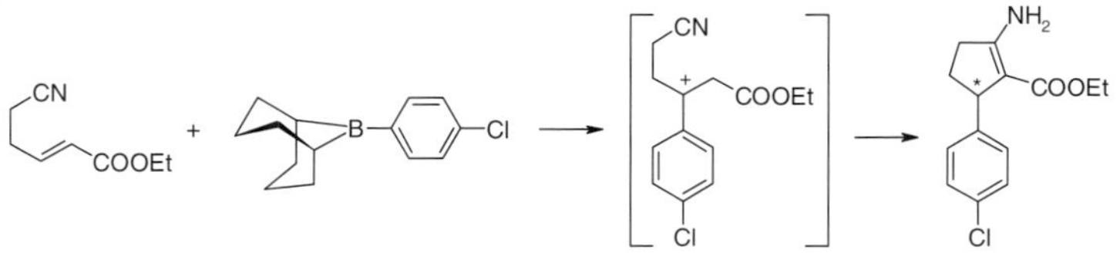

A mixture of the startg. cyano-subst. enoate, 5 mol% [RhCl(ethylene)]$_2$, 10 mol% (R)-5,5′,6,6′,7,7′,8,8′-octahydro-2,2′-bis(diphenylphosphino)-1,1′-binaphthyl, 0.5 eq. KOH and 2 eqs. startg. 9-aryl-9-borabicyclo[3.3.1]nonane in toluene stirred under N_2 at 70° for 12 h, cooled, water added, the aq. layer extracted, and the extracts worked up with thin-layer chromatographic purification → product. Y 70% (e.e. 95%). The procedure is applicable to the formation of 5- and 6-membered cyclic enaminoesters, and tolerates both electron-withdrawing and -donating groups in the aromatic ring. Reaction is presumed to involve intermediate formation of an (oxa-π-allyl)rhodium(I) complex. F.e. and racemic products without the chiral ligand s. T. Miura, T. Harumashi, M. Murakami, Org. Lett. *2007*, *9* (5), 741-3.

Chloro(1,5-cyclooctadiene)rhodium(I) dimer/sodium carbonate *[Rh(cod)Cl]$_2$/Na$_2$CO$_3$*
1-Indenones from *o*-bromoarylboronic acids and acetylene derivs. ○
Regioselective rhodium(I)-catalyzed carbonylative ring closure

606. 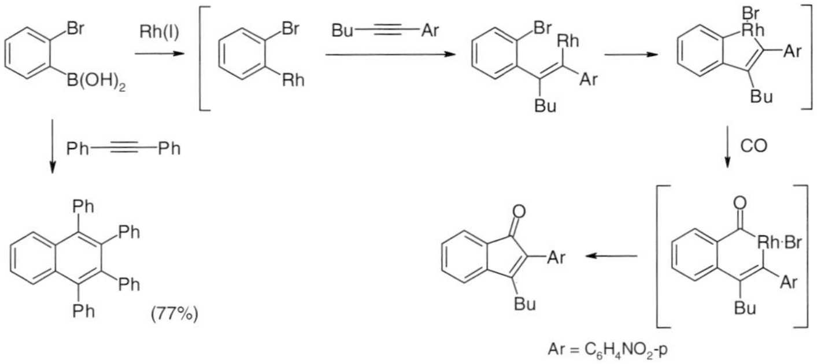

A soln. of 2 eqs. *Na$_2$CO$_3$*, 1.5 eqs. 2-bromophenylboronic acid, 2.5 mol% [RhCl(cod)]$_2$ and 1 eq. 1-(4-nitrophenyl)-1-hexyne in dioxane/water (10:1) purged of air, the reaction vessel filled with CO to 1 atm., heated at 80° for 20 h, cooled to room temp., volatiles removed *in vacuo*, and the residue purified by chromatography on silica → 2-(4-nitrophenyl)-3-*n*-butyl-1-indenone. Y 91% (10:1 mixture with regioisomer). High regioselectivity is favoured by electron-withdrawing or sterically hindered alkyne substituents, with SiMe$_3$ and COOR groups producing the corresponding 2-subst. indenones with up to 100% regioselectivity. There was no reaction, however, with terminal acetylenes. Yields are high with 2-bromophenylboronic acid (62-97%) and the method tolerates other halogeno, ether and trifluoromethyl substituents (14 examples; Y 47-92%). In some cases, **1,2,3,4-tetrasubst. naphthalenes** were also detected in low yield by non-carbonylative ring closure of two molecules of the alkyne and these became major products (2 examples; Y 76 and 77%) when the carbon monoxide atmosphere was replaced with nitrogen. F.e. and from *o*-bromophenyl(trimethylsilyl)acetylenes and phenylboronic acid s. Y. Harada, J. Nakanishi, H. Fujihara, M. Tobisu, Y. Fukumoto, N. Chatani, J. Am. Chem. Soc. *2007*, *129* (17), 5766-71.

Chloro(norbornadiene)rhodium(I) dimer/potassium hydroxide *[Rh(nbd)Cl]$_2$/KOH*
Stereoselective rhodium(I)-catalyzed synthesis C(OH)C=C=C(Ar)
of 4-aryl-2-allenealcohols from 3-acetyleneepoxides and arylboronic acids

607. 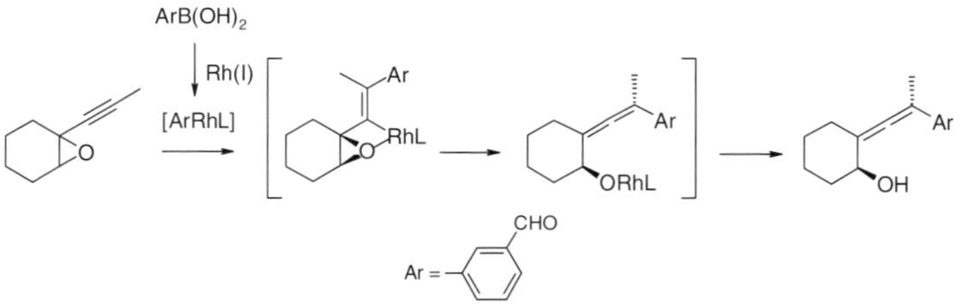

under mild conditions. A mixture of 5 mol% [Rh(nbd)Cl]$_2$, 1.5 eqs. 3-formylphenylboronic acid, 0.6 eqs. KOH and 1 eq. startg. alkynyloxirane stirred in THF under argon at room temp. for 2 h, quenched with water, extracted with ethyl acetate, the extracts washed with brine, dried over

$MgSO_4$, the solvent removed *in vacuo*, and the residue purified by TLC → allenol deriv. Y 72% (*syn/anti* 96:4). The method is successful with a range of electronically and sterically diverse arylboronic acids (and 2-thienylboronic acid) giving 61-86% yield and *syn*-selectivities of 94-9% in the presence of *formyl*, ether, halo and silyl ether functionality and without loss of enantiopurity with chiral substrates. Lower selectivities were obtained, however, with hindered *o*-tolylboronic acid (Y 83%; selectivity 83%) and with a terminal alkynyloxirane (Y 19%; selectivity 83%). It was also noted that MeMgCl reacted analogously but with lower diastereoselectivity (Y 65-86%; selectivity 57-77%). F.e.s. T. Miura, M. Shimada, S.-Y. Ku, T. Tamai, M. Murakami, Angew. Chem. Int. Ed. Engl. *2007*, *46* (37), 7101-3.

Palladium-carbon/microwaves *Pd-C/[\\\\]*

β-Arylcarboxylic acid esters $CH_2NH \rightarrow CH_2CH_2COOR$

from benzylamines via ar. aldehydes

One-pot synthesis with addition of two C-atoms via Wittig synthesis-hydrogenation

608. 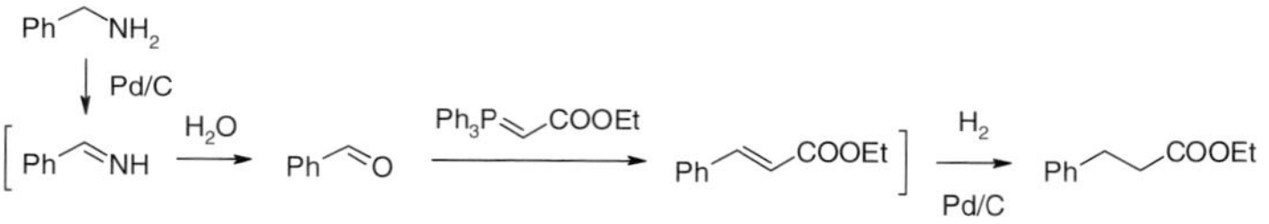

A mixture of benzylamine and 5 mol% Pd-C *in PEG-400* contained in a pressure-resistant glass ampoule irradiated in a microwave oven at 100 W and 170° for 3 h, cooled to room temp., stirred with added water at the same temp. for 30 min, 1.5 eqs. ethyl triphenylphosphoranylideneacetate added, stirring continued for 12 h at room temp., a gas balloon fitted to the reaction vessel, and the mixture hydrogenated (1 atm.) for 8 h → product. Y 60%. Reaction involves initial palladium-catalyzed generation of an ar. aldehyde *in situ*, followed by Wittig synthesis and hydrogenation of the resulting enoate *with the same palladium catalyst*. The procedure is applicable to a variety of prim. and sec. benzylamines, as well as 2-naphthylmethylamine and 4-aminomethylpyridine, but aliphatic amines gave low yields (18-26%). F.e.s. S. Chandrasekhar, G.P.K. Reddy, C. Nagesh, C.R. Reddy, Tetrahedron Lett. *2007*, *48* (7), 1269-71.

Palladium-carbon/sodium phosphate *Pd-C/Na_3PO_4*

Suzuki biaryl coupling Ar-Ar′

cf. *37*, 902s*70*; with Pd-C/Na_3PO_4, ligand-free coupling in 1:1 water/isopropanol, also *trans*-stilbenes from styrylboronic acids, s. T. Maegawa, Y. Kitamura, S. Sako, T. Udzu, A. Sakurai, A. Tanaka, Y. Kobayashi, K. Endo, U. Bora, T. Kurita, A. Kozaki, Y. Monguchi, H. Sajiki, Chem. Eur. J. *2007*, *13* (20), 5937-43; of aryl or hetaryl chlorides with $Pd_2(dba)_3$/2-aryl-1-phosphino-indoles/K_3PO_4 s. C.M. So, C.P. Lau, F.Y. Kwong, Org. Lett. *2007*, *9* (15), 2795-8; of ar. bromides, iodides or chlorides with the water-soluble, air-stable catalyst derived from 1:2 $Pd(OAc)_2$/1,1,3,3-tetramethyl-2-*n*-butylguanidine in 3:2 water/ethanol containing K_2CO_3 at room temp. under air (TON up to 850,000) s. S. Li, Y. Lin, J. Cao, S. Zhang, J. Org. Chem. *2007*, *72* (11), 4067-72; of chiral (4-bromophenyl)- or (3,5-dibromophenyl)-glycidyl ethers under Buchwald conditions [$Pd_2(dba)_3 \cdot C_6H_6$/S-Phos/Cs_2CO_3] *en route* to C_2-symmetric bis(Δ^2-oxazolines) s. X. Cattoën, M.A. Pericàs, ibid. *72* (9), 3253-8; with $Pd(OAc)_2$ immobilized in diethylaminopropylsilylated alumina pores with the aid of [bmim]PF_6 in 50% aq. ethanol containing K_2CO_3, allowing recycling 5 times and TONs of 2,000,000 (TOF 30,000 h^{-1}) s. H. Hagiwara, K.H. Ko, T. Hoshi, T. Suzuki, Chem. Commun. *2007* (27), 2838-40; of ar. chlorides with Na_2PdCl_4 and water-soluble, sulfonated imidazol-2-ylidene or Δ^2-imidazolin-2-ylidene ligands in aq. KOH s. C. Fleckenstein, S. Roy, S. Leuthäußer, H. Plenio, ibid. 2870-2; of (het)ar. bromides or chlorides with $PdCl_2$(adamantyl isocyanide)$_2$/Cs_2CO_3 in refluxing dioxane s. D. Villemin, A. Jullien, N. Bar, Tetrahedron Lett. *2007*, *48* (24), 4191-3; sym. 2,5-diaryl-3,4-dibromo- and sym. or unsym. tetraaryl-thiophenes with $Pd(PPh_3)_4$/K_3PO_4 in toluene/water s. T.T. Dang, N. Rasool, T.T. Dang, H. Reinke, P. Langer, ibid. *48* (5), 845-7.

Palladium(II) acetate s.a. under K_2CO_3 *$Pd(OAc)_2$*

Palladium(II) acetate/sodium hydroxide/polyethylene glycol *$Pd(OAc)_2$/NaOH/PEG*
Palladium(II) acetate/potassium hydroxide *$Pd(OAc)_2$/KOH*
Biaryls from ar. halides and aryl(trialkoxy)silanes Ar-Ar′
cf. *57*, 439s*71*; with $Pd(OAc)_2$ (1.8 mol%) and NaOH in water/PEG-2000 s. S. Shi, Y. Zhang, J. Org. Chem. *2007*, *72* (15), 5927-30; from hexaarylcyclosiloxanes with $Pd(OAc)_2$ (5 mol%) and KOH in dioxane/water, compatibility with base-sensitive groups such as acetyl or nitro, s. M. Endo, T. Sakurai, S. Ojima, T. Katayama, M. Unno, H. Matsumoto, S. Kowase, H. Sano, M. Kosugi, K. Fugami, Synlett *2007* (5), 749-52.

Palladium(II) acetate/sodium hydroxide/polyethylene glycol *$Pd(OAc)_2$/NaOH/PEG*
(E)-1,2-Diarylethylenes ArCH=CHAr′
from two different ar. halide molecules with insertion of two C-atoms
One-pot ligand-free Hiyama-Heck reaction in water

609. 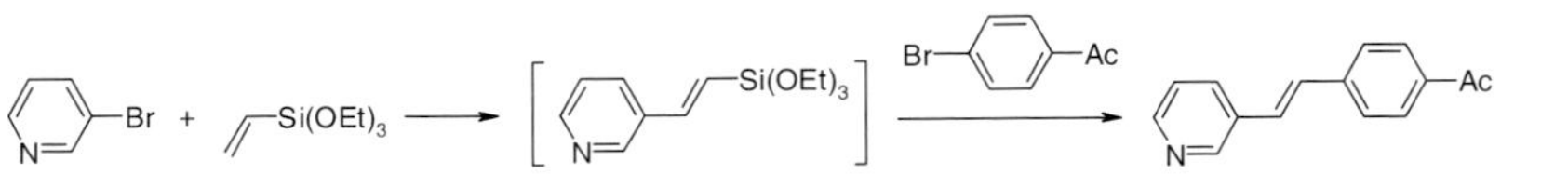

at low catalyst loading. 3.54 Mmol triethoxy(vinyl)silane and 5.35 mmol 0.5 *M* aq. NaOH containing 2.14 g PEG 2000 vigorously stirred for 5 min at room temp., 1.78 mmol 3-bromopyridine and *0.1 mol%* $Pd(OAc)_2$ added, stirring continued for 3 h at 140°, 1.78 mmol 4-bromoacetophenone and a further *0.1 mol%* $Pd(OAc)_2$ in 20% wt/wt aq. PEG-2000 added at room temp., heated again at 140° for 6 h, and worked up → (E)-4-[2-(3-pyridyl)vinyl]acetophenone. Y 87%. Reaction is carried out in air, and most products are solid so that they can be isolated by simple filtration. The aq. medium is essential since conventional organic solvents would be too volatile at the high temperature required. PEG reduces significantly the proportion of the vinylsilane, which would otherwise be required in large excess to prevent formation of undesired Heck products. For the formation of (E)-4,4′-diacetylstilbene, reaction is efficient at catalyst loadings as low as 0.005 *mol%*. F.e. incl. **(E)-stilbenes** possessing either electron-donating or -withdrawing groups s. Á. Gordillo, E. de Jesús, C. López-Mardomingo, Chem. Commun. *2007* (39), 4056-8.

Diethylaminopropylsilylated alumina-based palladium(II) acetate/potassium carbonate ←
Suzuki biaryl coupling s. *37*, 902s*72* Ar-Ar′

Palladium(II) acetate/tricyclohexylphosphine *$Pd(OAc)_2$/Cy_3P*
Biaryls from ar. halides and triarylaluminum compds. ArBr → Ar-Ar′
Palladium-catalyzed cross-coupling

610. 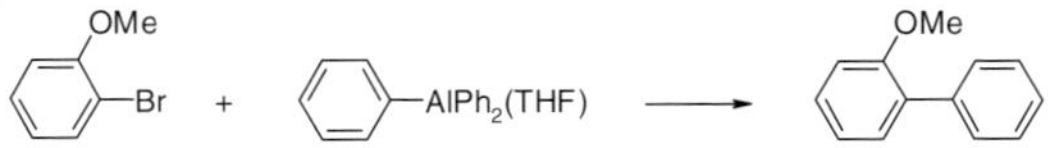

Triphenylaluminum(THF) (1.3 mmol) and *o*-bromoanisole (1 mmol) in DME (2 ml) added to a soln. of $Pd(OAc)_2$ (0.01 mmol) and Cy_3P (0.02 mmol) in DME (2 ml), and the mixture stirred at room temp. for 3 h → product. Y 92%. A variety of electron-withdrawing and -donating substituents are tolerated, although *p*-nitrophenyl bromide affords the corresponding biphenyl in only moderate yield (65%), presumably due to a competing affinity between the nitro group and the aluminum reagent. The reaction is also successful using aryl chlorides (although higher temperatures and longer reaction times are required) or more bulky triarylaluminum reagents [$AlAr_3$(THF): Ar = 2,4,6-trimethylphenyl, 2-naphthyl or 4-trimethylsilylphenyl]. In the most hindered case (Ar = 2,4,6-trimethylphenyl), the reactions proceed more smoothly when $AlAr_3$(THF), $Pd(OAc)_2$ and PCy_3 are preheated in toluene at 100° for 1 h, suggesting that formation of the active Pd(0) species is the rate-determining step. F.e.s. S.-L. Ku, X.-P. Hui, C.-A. Chen, Y.-Y. Kuo, H.-M. Gau, Chem. Commun. *2007* (37), 3847-9.

Palladium(II) acetate/2-(dicyclohexylphosphino)-2′,6′-dimethoxy-biphenyl/potassium phosphate *$Pd(OAc)_2/Cy_2PAr/K_3PO_4$*

Heteroarene-condensed 2-subst. pyrroles ○

from *o*-amino-*m*-(2,2-dichlorovinyl)heteroarenes and unsatd. boronic acids

611.

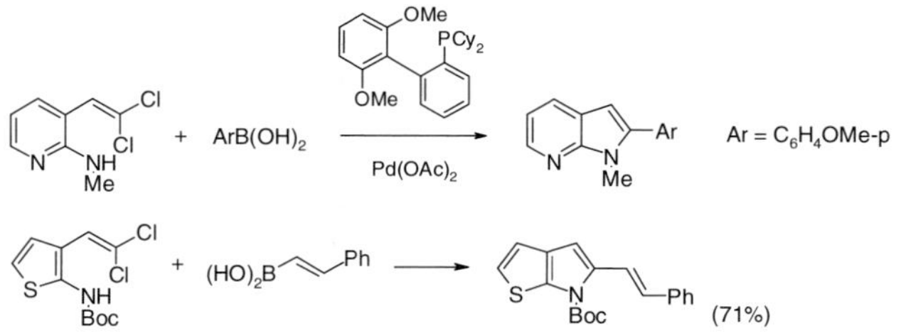

Bicyclic 2-arylpyrroles. A soln. of 10 mol% $Pd(OAc)_2$ and 20 mol% 2-(dicyclohexylphosphino)-2′,6′-dimethoxybiphenyl in toluene added to a mixture of N-methyl-3-(2,2-dichlorovinyl)pyridin-2-ylamine, 1.5 eqs. 4-methoxyphenylboronic acid and 3 eqs. $K_3PO_4{\cdot}H_2O$, the mixture heated at 100° for 2 h, cooled to room temp., quenched with satd. aq. $NaHCO_3$, extracted with ether, the extracts dried over Na_2SO_4, concentrated *in vacuo*, and the residue purified by chromatography → 1-methyl-2-(4-methoxyphenyl)pyrrolo[2,3-*b*]pyridine. Y 83%. By modification of the basic method, reaction was adapted to prepare 5-, 6- and the illustrated 7-aza-indoles in 55-96% yield (24 examples), while 4-azaindoles could only be obtained by protection of the pyridine nitrogen as the N-oxide (2 examples; Y 75 and 97%). The reaction was also extended to the preparation of thienopyrroles (9 examples; Y 66-81%). The reactions cannot be accomplished with the corresponding free amines due to catalyst-poisoning by the pyrrolyl NH group of the products, but reaction was possible with benzyl- or Boc-protected amines. F.e. incl. prepn. of the startg. *gem*-dichlorovinyl-deriv. from the corresponding aldehydes, also 2-vinyl-derivs. from α,β-ethylene-boronic acids, s. Y.-Q. Fang, J. Yuen, M. Lautens, J. Org. Chem. *2007*, *72* (14), 5152-60.

*Palladium(II) acetate/tri-*o*-tolylphosphine/cesium fluoride* *$Pd(OAc)_2/(o\text{-}Tol)_3P/CsF$*

***o*-Vinylbiaryls from ar. iodides, benzynes and electron-deficient ethylene derivs.** ←

Palladium-catalyzed 3-component synthesis via Heck arylation

612.

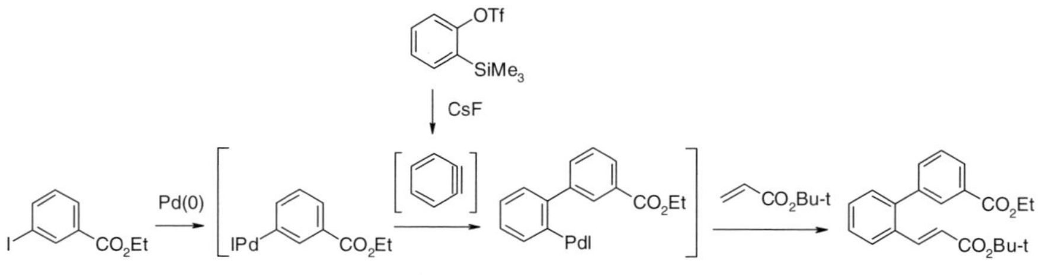

under mild conditions. 1.5 Eqs. 3-carbethoxyphenyl iodide added to a stirred suspension of 5 mol% $Pd(OAc)_2$, 10 mol% tri-*o*-tolylphosphine and 6 eqs. CsF in dry acetonitrile under N_2 in a carousel tube, the mixture heated at 45°, 1 eq. *tert*-butyl acrylate added followed by 2 eqs. 2-(tri-methylsilyl)phenyl triflate in two equal portions at 0 and 1 h, heating continued at 45° for a further 3 h, the mixture diluted with methylene chloride, filtered through a short plug of silica, the filtrate concentrated *in vacuo*, and the residue purified by chromatography on silica → ethyl 2′-((E)-2-*tert*-butoxycarbonylvinyl)biphenyl-3-carboxylate. Y 70%. The method is general and can be applied to the coupling of a range of aryl iodides (yields being highest with *m*- and *p*-halo and other electron-deficient derivs.) with methylenedioxy- and methyl-subst. or unsubst. arynes and with α,β-ethylene-carboxylic acid esters, -ketones, and -amides and also 4-methoxystyrene as Heck acceptors, giving 38-91% yield in the presence of nitro, halo, nitrile, ester and trifluoromethyl functionality. Key to the reaction appears to be the availability of sufficient aryne to intercept the initial arylpalladium complex while avoiding aryne trimerization. Products arising from palladium migration or 2-component coupling were also insignificant. F.e. incl. reaction optimization s. J.L. Henderson, A.S. Edwards, M.F. Greaney, Org. Lett. *2007*, *9* (26), 5589-92.

Palladium(II) acetate/1,1′-bis(diphenylphosphino)ferrocene/ potassium fluoride *Pd(OAc)$_2$/dppf/KF*

Diarylmethanes from arylstannanes and benzyl halides $ArCH_2Ar'$

613. 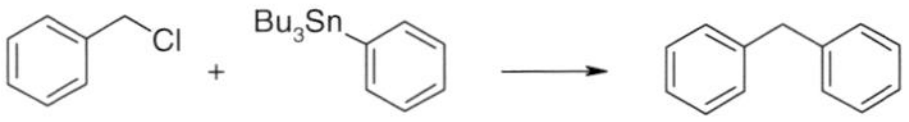

Stille coupling. Benzyl chloride, tributyl(phenyl)stannane and dioxane added to a Schlenk flask charged with 4 mol% Pd(OAc)$_2$, 8 mol% 1,1′-bis(diphenylphosphino)ferrocene and 4 eqs. KF under argon, and stirred at 80° for 3 h → diphenylmethane. Y 94%. Reaction is generally applicable in high yield with benzyl bromides and chlorides under mild conditions (cf. *64*, 463s*67*), being facilitated by an inexpensive system at low catalyst loading. Yields were also high without the base, but required a long reaction time. F.e. and comparison of P-ligands, also competitive studies in the presence of arylboronic acids, s. T.Z. Nichele, A.L. Monteiro, Tetrahedron Lett. *2007*, *48* (42), 7472-5.

Bis(dibenzylideneacetone)palladium(0)/chiral phosphoromonoamidites ←

3-Methylenepyrrolidines from imines ○

Asym. [3+2]-cycloaddition with a trimethylenemethane equivalent

614. 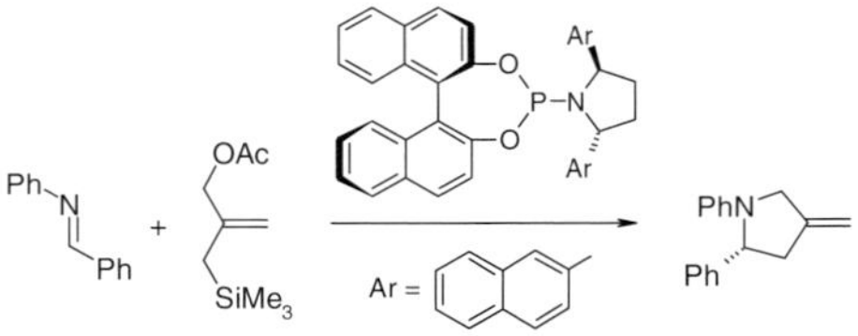

A soln. of 5 mol% Pd(dba)$_2$ and 10 mol% of the chiral phosphoromonoamidite in toluene stirred for 2 min at room temp. under argon, a soln. of the startg. imine in the same solvent added via cannula, followed by 1.2 eqs. 3-acetoxy-2-trimethylsilylmethyl-1-propene, stirring continued for 4 h at 45° (with GC monitoring), and worked up with flash column chromatographic purification on silica gel → product. Y 98% (e.e. 87%). High yields and enantioselectivities (82-93%) were recorded for reaction with N-carbo-*tert*-butoxy-aldimines and N-aryl-imines derived from aryl or hetaryl aldehydes (irrespective of the nature of ring substituents), but results were poor with N-tosyl-, N-benzyl-, N-phosphinyl-, N-Fmoc- and N-Cbz-derivs. F.e. and comparison of chiral phosphoromonoamidites s. B.M. Trost, S.M. Silverman, J.P. Stambuli, J. Am. Chem. Soc. *2007*, *129* (43), 12398-9.

Tris(dibenzylideneacetone)dipalladium(0)/tricyclohexylphosphine/ethyldiisopropylamine/ microwaves ←

Suzuki-type coupling with N-vinylpyridinium salts $B(OH)_2 \rightarrow C{=}C$

615. 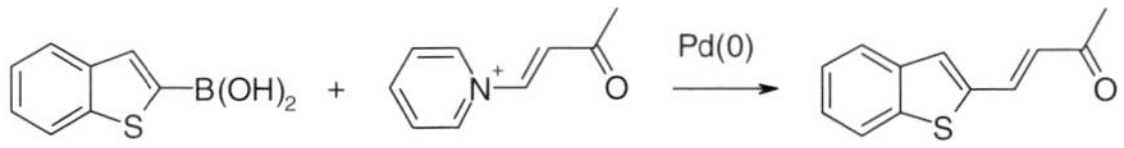

under microwave enhancement. 2.2 Eqs. ethyldiisopropylamine and startg. boronic acid added to a THF soln. of (E)-1-(3-oxobut-1-enyl)pyridinium fluoroborate, 5 mol% Pd$_2$(dba)$_3$ and 20 mol% tricyclohexylphosphine in a Biotage microwave vial, the vial sealed and heated at 150° in a microwave reactor for 12.5 min, cooled, the contents concentrated *in vacuo,* and the residue purified by flash chromatography → 2-(3-oxo-1-butenyl)benzothiazole. Y 76%. The pyridinium salts are crystalline, air-stable and non-hygroscopic, and are thus useful alternatives to the conventional vinyl halides and enol triflates. The method is successful with electronically-diverse aryl-, hetaryl- and vinyl-boronic acids reacting with alkyl or aryl 2-pyridiniovinyl ketones giving 60-92% yield in the presence of vinyl, nitro, trifluoromethyl and ether functionality. The selected Pd(0) catalyst gave the highest conversions but the conditions remain largely unoptimized. Attempted reaction with N-vinylammonium salts gave lower yields (35%) of coupled products, while reaction failed with the use of tetraphenylborate salts or by replacing tricyclohexylphosphine with triphenylphosphine. F.e.s. K.R. Buszek, N. Brown, Org. Lett. *2007*, *9* (4), 707-10.

Tris(dibenzylideneacetone)dipalladium/2-aryl-1-phosphinoindoles/potassium phosphate ←
Tris(dibenzylideneacetone)dipalladium-benzene/2-dicyclohexylphosphino-2′,6′-dimethoxy-biphenyl/cesium carbonate ←
Suzuki biaryl coupling s. *37*, 902s*72* Ar-Ar′

Tris(dibenzylideneacetone)dipalladium/1,3-bis(diphenylphosphino)-propane/cesium fluoride *$Pd_2(dba)_3$/dppp/CsF*
***o*-Annelation of benzynes** ○
Indolo[1,2-*f*]phenanthridines from N-(*o*-bromoaryl)indoles s. *68*, 464s*72*; *70*, 463s*72*

*Diacetato((4R)-2-pyrid-2-yl-4-*tert*-butyl-Δ^2-oxazoline)palladium(II)/oxygen* ←
Oxidative asym. Heck-type arylation $C{=}C\text{-}CH_3 \rightarrow C(Ar)C{=}CH_2$
of acyclic electron-deficient ethylene derivs. with arylboronic acids

616. 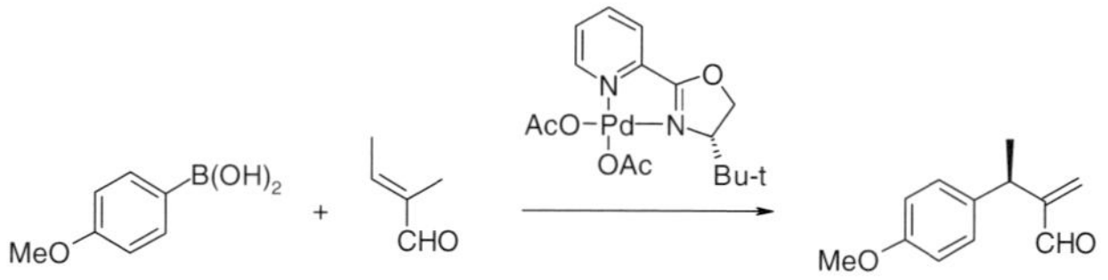

Chiral β-aryl-α-methylenealdehydes. *trans*-2-Methyl-2-butenal added to 1.1 eqs. 4-methoxyphenylboronic acid and 5 mol% diacetato((4R)-2-pyrid-2-yl-4-*tert*-butyl-Δ^2-oxazoline)-palladium(II) in DMF under O_2, stirred at room temp. for 16 h, diluted with ethyl acetate, washed with water, and separated → product. Y 67% (e.e. 73%). Initial screening identified chiral N,N-bidentate ligands as most effective (phosphine ligands promoting side-reactions) with the highest enantioselectivities achieved using the *preformed* Pd(II)-N,N-ligand complex. The latter was only stable over a few days and was generally freshly made. The reaction is applicable to coupling arylboronic acids (aryl = phenyl, 4-methoxyphenyl, 4-dimethyl-aminophenyl, 2-naphthyl and 6-methoxy-2-naphthyl) with 2-methyl-2-butenal and methyl 2-methylbutenoate in 67-79% yields (e.e. 62-75%), the latter yielding the corresponding **chiral β-aryl-α-methylenecarboxylic acid esters.** In most cases (R)-stereochemistry was determined by conversion to known compounds. F.e. incl. preparation of the Pd(II)-ligand complex s. K.S. Yoo, C.P. Park, C.H. Yoon, S. Sakaguchi, J. O′Neill, K.W. Jung, Org. Lett. *2007*, *9* (20), 3933-5.

*Bis(1,1,3,3-tetramethyl-2-*n*-butylguanidine)palladium(II) acetate/potassium carbonate* ←
Suzuki biaryl coupling s. *37*, 902s*72* Ar-Ar′

*Bis(tri-*tert*-butylphosphine)palladium(0)* *Pd(PBu-t$_3$)$_2$*
Stille coupling of triarylphosphonio-subst. enestannanes Sn⩽ → R
s. *17*, 117s*72*

Tetrakis(triphenylphosphine)palladium(0)/potassium phosphate *$Pd(PPh_3)_4/K_3PO_4$*
Suzuki biaryl coupling Ar-Ar′
Sym. 2,5-diaryl-3,4-dibromo- and sym. or unsym. tetraaryl-thiophenes s. *37*, 902s*72*

Diacetato[1,2-bis(diphenylphosphino)ethane]palladium(II)/methanol *$Pd(OAc)_2(dppe)/MeOH$*
1-Indenones from *o*-carbalkoxyboronic acids and acetylene derivs.

617. 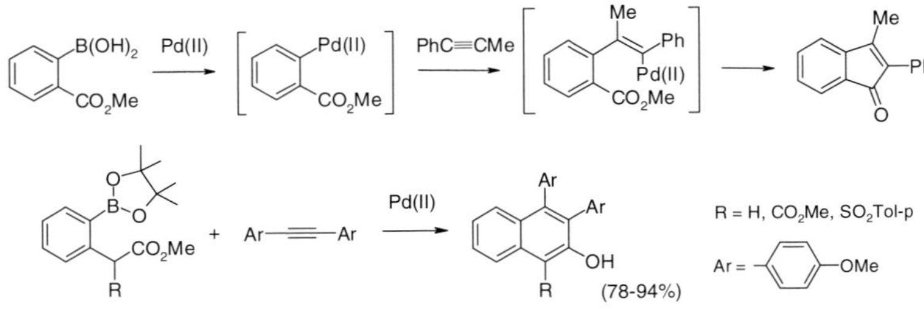

Dry methanol added to a mixture of 1-phenylprop-1-yne, 1.2 eqs. *o*-carbomethoxyphenylboronic acid and 5.4 mol% $Pd(OAc)_2$(dppe) under argon, agitated for 24 h, 2.3 eqs. *N,N-diethanolamino-methyl-polystyrene* in THF added (to remove excess of the boronic acid), the mixture agitated for 2 h, filtered, washed with chloroform, and the filtrate evaporated → 1-indenone deriv. Y 83% (5:1 mixture of regioisomers). The method is successful with aliphatic acetylenes, cyclododecyne, and diarylacetylenes containing electron-withdrawing or -releasing groups at the *para*-position (6 examples; 58-98% yields). The reaction failed, however, with alkynyl esters, alkynylsilanes and terminal acetylenes while the dialkyne, 1,4-bis(*tert*-butyl)butadiyne, gave only a 31% yield. The anticipated isocoumarins were obtained in trace amounts in some experiments. Annulation with the homologous boronic acid esters [e.g. methyl [2-(4,4,5,5-tetramethyl-1,3,2-dioxaborolan-2-yl)phenyl]acetate] afforded **2-naphthols** (3 examples; Y 78-94%). F.e.s. H. Tsukamoto, Y. Kondo, Org. Lett. *2007*, *9* (21), 4227-30.

Bis(aquo)[(R)-2,2′-bis(diphenylphosphino)-1,1′-binaphthyl]palladium(II) bis(triflate)
Inden-1-ols from *o*-oxoboronic acids and acetylene derivs.
Regiospecific asym. [3+2]-annelation

618. 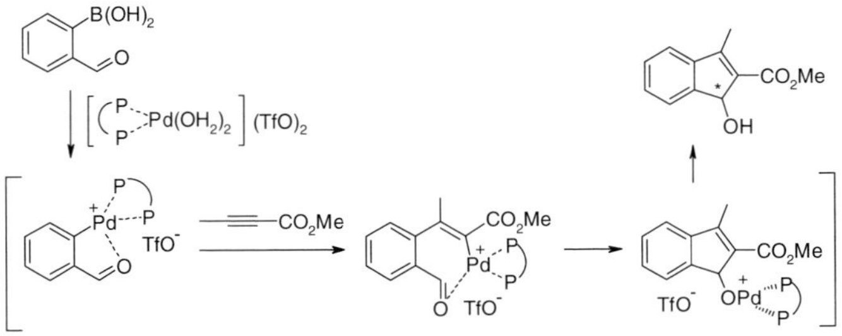

1.2 Eqs. *o*-formylphenylboronic acid, 3 mol% bis(aquo)[(R)-2,2′-bis(diphenylphosphino)-1,1′-binaphthyl]palladium(II) bis(triflate) and 1 eq. methyl 2-butynoate added in this order to a Schlenk tube under N_2, stirred at 20° for 12 h (TLC monitoring), concentrated under reduced pressure, and worked up with flash column chromatographic purification → product. Y 92% (e.e. 91%). The procedure is high yielding and highly enantioselective, but appears to be limited to α,β-acetylenecarboxylic acid esters and bis(methoxymethyl)ethyne. Reaction is thought to involve intermediate formation of a vinylpalladium species, which undergoes nucleophilic addition to the *o*-carbonyl group prior to protonolysis and regeneration of the catalyst. F.e. and under optimization with palladium(II) triflate·$2H_2O$/(S)-2,3;2′,3′-bis(isopropylidenedioxy)-6,6′-bis(di-phenylphosphino)biphenyl, also prepn. of 2-methoxymethyl derivs., s. M. Yang, X. Zhang, X. Lu, Org. Lett. *2007*, *9* (24), 5131-3.

Sodium tetrachloropalladate(II)/water-soluble, sulfonated imidazol-2-ylidene or Δ^2-imidazolin-2-ylidene/potassium hydroxide ←
Bis(adamantyl isocyanide)dichloropalladium(II)/cesium carbonate ←
Suzuki biaryl coupling s. *37*, 902s*72* Ar-Ar′

Dichloro((-)-sparteine)palladium(II)/(-)-sparteine/isopropanol/oxygen/manganese dioxide ←
Alkylarenes from styrenes and unsatd. stannanes C=C → C(R)CH
Regiospecific palladium-catalyzed reductive coupling under aerobic conditions

619. 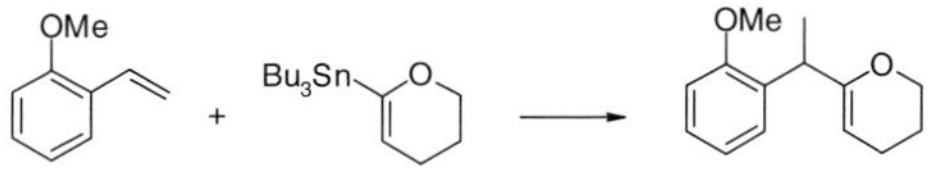

Allylarenes. 2-Vinylanisole (1 mmol) and tributyl(3,4-dihydro-2*H*-pyran-6-yl)stannane (1.5 mmol) added by syringe to a mixture of MnO_2 (0.75 mmol), isopropanol (4.3 ml) and (-)-sparteine (0.4 mmol; 0.2 ml of a 2 *M* soln. in isopropanol) in an *oven-dried 100 ml thick-wall glass pressure vessel* under O_2, the vessel pressurized to 25 psi O_2, evacuated and repressurized three times, then the mixture stirred *vigorously* for 20 min at room temp., Pd[(-)-sparteine]Cl_2 (0.035 mmol) added, the vessel repressurized to 25 psi O_2, the resulting mixture stirred *vigorously* for 5 min at room temp. then 18 h at 60° before workup with aq. NaOH and purification using alumina gel flash chromatography → 6-[1-(2-methoxyphenyl)ethyl]-3,4-dihydro-2*H*-pyran. Y 63%. High regioselectivity (>25:1) is observed in all cases, using a variety of subst. styrenes with either aryl- or vinyl-stannanes. Yields are minimally affected by the electronic nature of the arylstannanes and no Stille-type coupling products were observed with chlorinated substrates. Mechanistic studies, using deuterium-labelled isopropanol (*i*-PrOD and Me_2CDOH), support the hypothesis that the alkene inserts into an intermediate Pd-hydride species, formed via alcohol oxidation. Both the size of the reaction vessel and the elevated pressure are important in increasing the mass transport of O_2 and avoiding decomposition of the catalyst. The competing oxidative Heck-type reaction (affording the alternative regioisomer) is minimized by exclusion of Cu(II) salts, instead using (-)-sparteine to enhance reaction rates and improve catalyst stability. In the absence of MnO_2 (which acts as H_2O_2 scavenger), the reaction is inhibited by (-)-sparteine-N-oxide, formed by oxidation of (-)-sparteine with H_2O_2, itself a product of O_2-reduction. F.e.s. K.M. Gligorich, S.A. Cummings, M.S. Sigman, J. Am. Chem. Soc. *2007*, *129* (46) 14193-5.

Dichloro[1,1′-bis(diphenylphosphino)ferrocene]palladium(II) s. under $MgCl_2$ *$PdCl_2$(dppf)*

Cyclooctadiene(hydroxy)iridium(I) dimer/triethylamine *[Ir(cod)(OH)]$_2$/Et_3N*
3-Vinyl-1-indanols from *o*-oxoboronic acids and 1,3-dienes ○
Regio- and stereo-specific [3+2]-annelation

620. 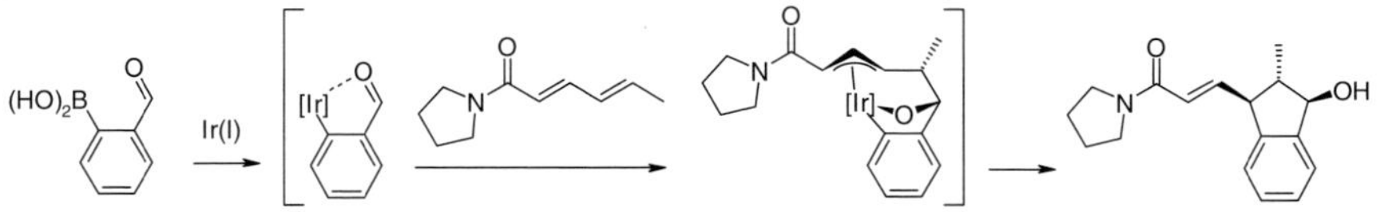

2.5 Eqs. triethylamine and the startg. diene added successively to a soln. of 1.67 mol% [Ir(cod)(OH)]$_2$ and 2 eqs. 2-formylphenylboronic acid in toluene, heated at 80° for 6 h, quenched with water, and worked up with flash column chromatographic purification on silica gel → product. Y 92%. The 1,3-diene may be substituted by an electron-withdrawing or -donating group, and cycloaddition takes place predominantly at the *more* electron-rich double bond of the diene to give the 1,2-*trans*- and 1,3-*cis*-product with, in some instances, perfect regioselectivity. However, there was no reaction with monoenes, e.g. styrene and 1-octene. Key to the reaction is the formation of an intermediate iridium(III) π-allyl complex, which is consistent with the stereochemistry, but it is presumed that annelation does not involve conjugate addition of the initially formed *o*-oxoarylindium species. F.e.s. T. Nishimura, Y. Yasuhara, T. Hayashi, J. Am. Chem. Soc. *2007*, *129* (24), 7506-7.

Carbon ↑ CC ↓↑ C

Potassium hydroxide or carbonate or Triethylamine *KOH or K_2CO_3 or Et_3N*
Benzene ring from 2-pyrones ←
cf. *22*, 877s*71*; 4,5-dihydronaphtho[2,1-*b*]furans or 2-hydroxy-6-methoxydibenzofurans from 6,7-dihydro-5*H*-benzofuran-4-one or 7-methoxybenzofuran-3-one, respectively, with KOH s. M. Dixit, R. Raghunandan, B. Kumar, P.R. Maulik, A. Goel, Tetrahedron *2007*, *63* (7), 1610-6; 4-arylthio-3-hydroxyphthalates from 4-arylthio-3-hydroxy-2-pyrones and acetylenedicarboxylates with Et_3N s. T. Komiyama, Y. Takaguchi, S. Tsuboi, Synth. Commun. *2007*, *37* (4), 531-4; 2-amino-3-aminomethyl-5-methylthio- or -5-*sec*-amino-biphenyl-4-carbonitriles from Boc-diprotected 1,3-diamino-2-propanone with K_2CO_3, followed by N-deprotection with TFA, s. Farhanullah, F. Samrin, V.J. Ram, Tetrahedron Lett. *2007*, *48* (18) 3187-90; 5-*sec*-amino-4-cyano-2,3,6,7-tetrahydrophenanthro[4,3-*b*]thiophenes from 4-*sec*-amino-3-cyano-5,6-dihydro-2*H*-benzo[*h*]chromen-2-ones and tetrahydrothiophen-3-one with KOH s. R. Pratap, V.J. Ram, ibid. *48* (27), 4715-8; 5-arylimidazo[4,5-*b*]pyrazin-2-ones from 7-acetyl-1,3-dimethyllumazine by ring transformation-ring contraction with KOH s. R. Pratap, A.D. Roy, R. Roy, V.J. Ram, ibid. *48* (7), 1281-5.

Organolithium compds. (or Magnesium) *RLi (or Mg)*
Synthesis of carboxylic acid esters from cyanoformic acid esters s. *72*, 602 R'COOR

N-Methylimidazole ←
o-Aminonitriles from acetylene derivs. and α,β-ethylenenitriles ○

621.

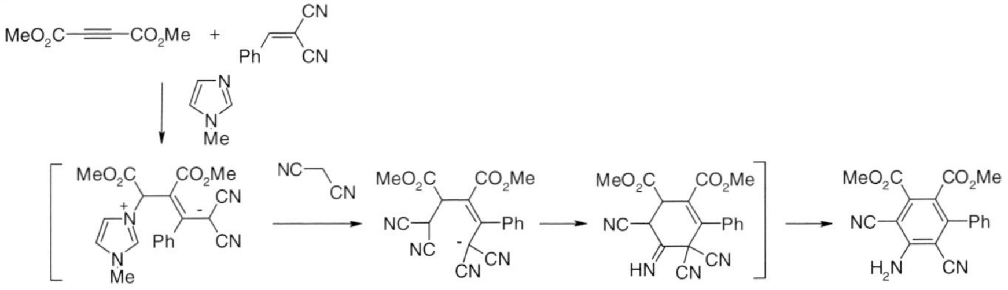

4-Amino-3,5-dicyanophthalic acid esters. A soln. of dimethyl acetylenedicarboxylate in dry methylene chloride added dropwise at -10° over 5 min to a soln. of benzylidenemalononitrile and 0.2 eq. N-methylimidazole in the same solvent, stirred at 25° for 25 min, 1 eq. malononitrile in dry methylene chloride added, stirred again at 25° for 2 h, and worked up → product. Y 95%. The procedure is mild, clean, simple, cost-effective, relatively short, high-yielding, and regioselective, leaving alkoxycarbonyl, cyano, aryl and amino functions untouched. Reaction involves intermediate formation of a series of zwitterions, the last of which undergoes reaction with malononitrile with subsequent intramolecular cyclization and aromatization with loss of HCN. F.e.s. M. Adib, B. Mohammadi, M. Mahdavi, A. Abbasi, M.R. Kesheh, Synlett *2007* (16), 2497-500.

Copper(I) bromide/trichloroacetic acid *$CuBr/Cl_3CCO_2H$*
Copper(I)-catalyzed α,α-diphenylation of β-diketones with benzyne $CH_2 \rightarrow CPh_2$

622.

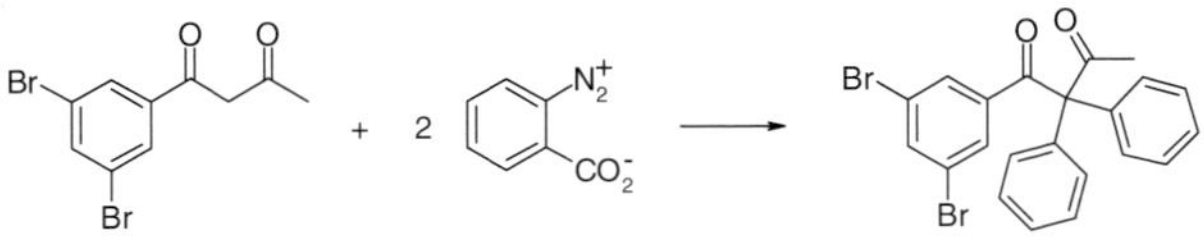

A suspension of benzenediazonium-2-carboxylate (4 mmol) in 1,2-dichloroethane (30 ml) added at 60° to a soln. of startg. 1,3-dione (2 mmol), trichloroacetic acid (5 mol%) and CuBr (5

mol%) in the same solvent (5 ml) over 1.5 h (a colour change from light yellow to deep brown being observed) → product. Y 90%. The general applicability of the reaction is demonstrated using a variety of substituted β-diketones, although yields are moderate (72-75%) with *o*-subst. 1-phenyl-1,3-diones (presumably due to steric hindrance) and relatively low (60-70%) with subst. 1-phenyl-1,3-diones bearing a strong electron-donating group. The expected benzyne insertion products were not observed. F.e.s. Y.-Y. Yang, W.-G. Shou, Y.-G. Wang, Tetrahedron Lett. *2007*, *48* (46), 8163-5.

Magnesium s. under RLi *Mg*

Zeolite H-ZSM5 *zeolite*

1,1-Diarylalkanes from cinnamic acids and arenes C=C-COOH → C(Ar)CH_2

Decarboxylative arylation

623. 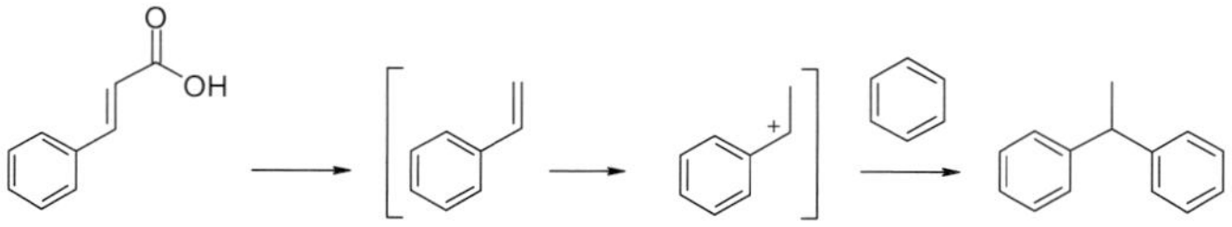

Cinnamic acid and 3 eqs. activated H-ZSM5 zeolite heated and magnetically stirred in benzene at 130° for 20 h in a sealed tube, cooled, heated and stirred with methanol at 80° for 5 h, cooled, filtered, and concentrated →1,1-diphenylethane. Y 78%. Toluene and *m*-xylene can replace benzene but the reaction is not successful with acrylic acid derivs. in place of cinnamic acid. The reaction requires temperatures of at least 130° and 3 or more molar excess of acidic sites to reach completion, and is presumed to involve initial decarboxylation, followed by protonation and quenching of the resulting carbonium ion by the arene. 1-Indanones, however, predominated with zeolite H-USY (cf. *72*, 515). F.e.s. S. Chassaing, M. Kumarraja, P. Pale, J. Sommer, Org. Lett. *2007*, *9* (20), 3889-92.

Chiral epiquinine-ureas ←

Organocatalyzed decarboxylative asym. Michael addition of monothiolmalonic acid S-monoesters C=C → CHC-COSR

624. 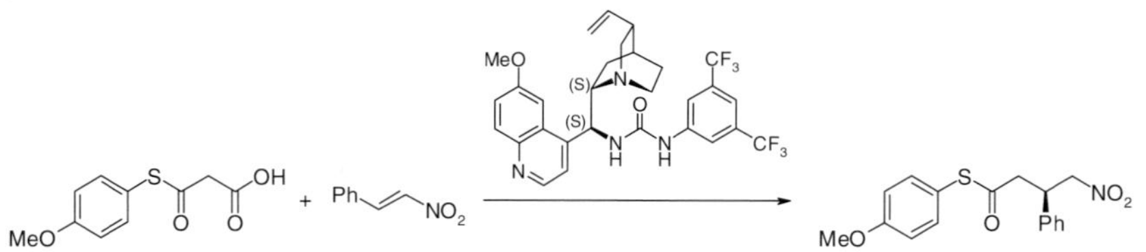

to 1,1-nitroethylene derivs. A mixture of the startg. nitroolefin, 1.34 eqs. startg. monothiolmalonic acid S-monoester and 20 mol% chiral epiquinine-urea deriv. in THF stirred in a capped vial for 24 h → product. Y 94% (e.e. 63%; 99% after recrystallization). Significantly, no metal ion is required: the basic site of the bifunctional catalyst serves to deprotonate the monothiolomalonate, while the urea residue coordinates with the thiocarbonyl group to present a chiral environment for the 1,4-addition. Interestingly, enantioselectivity was higher *in ethyl vinyl ether* although chemical yields were significantly lower. F.e. and comparison of urea- and thiourea-functionalized cinchona alkaloids, also with aliphatic nitroolefins, s. J. Lubkoll, H. Wennemers, Angew. Chem. Int. Ed. Engl. *2007, 46* (36), 6841-4.

Trichloroacetic acid s. under CuBr *Cl_3CCO_2H*

*Tri-*n*-butyltin methoxide s. under $PdCl_2(PTol_3)_2$* *Bu_3SnOMe*

Manganese(III) acetate *Mn(OAc)$_3$*

Syntheses with α-aryl-α-(N-allyl-N-acylamino)carboxylic acid amides via oxidative radical 1,4-aryl migration

625. 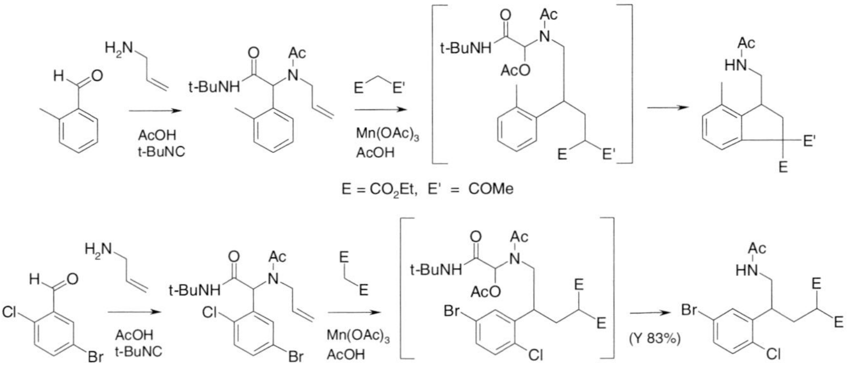

and indan ring closure. 1 Eq. allylamine, 1 eq. acetic acid and 1 eq. *tert*-butyl isocyanide added sequentially to a soln. of *o*-methylbenzaldehyde in methanol, and stirred until reaction complete by TLC → Ugi adduct (Y 100%), dissolved in 1 eq. 0.3 *M* acetic acid, 4 eqs. ethyl acetoacetate and 4.5 eqs. $Mn(OAc)_3 \cdot 2H_2O$ added, heated at 90° under argon, stirred until the dark brown colour of $Mn(OAc)_3$ disappeared and reaction complete by TLC, evaporated, the residue extracted with methylene chloride and satd. aq. NH_4Cl, and the organic extracts dried and evaporated → indan deriv. (Y 63%). The intermediate Ugi adducts are generally formed quantitatively and used without purification giving overall yields of 40-63%. Use of phenylacetic acid in place of acetic acid gave incomplete conversion (60%), and in this case purification of the adduct raised the final yield from 40% to 69%. The method appears to require an *o*-subst. benzaldehyde since adducts derived from 4-methoxy- or 4-chloro-benzaldehyde gave complex mixtures in the final step, as did pyridine 2-carboxaldehyde. Cyclohexyl isocyanide can replace *tert*-butyl isocyanide. Reaction with diethyl malonate was terminated before the indan ring closure, affording **δ-(acylamino)-malonic acid esters** from aldehydes **via 1,4-radical aryl migration** (four examples; Y 66-83%); these could be converted to the corresponding 3-α-(acylamino)indan-1,1-dicarboxylates by further treatment with $Mn(OAc)_3$. F.e.s. L. El Kaim, L. Grimaud, E. Vieu, Org. Lett. *2007*, *9* (21), 4171-3.

Alkylidene(dichloro)ruthenium(II) imidazolidin-2-ylidene complex ←

Cross-metathesis of ethylene derivs. C=C

with Grubbs′ (or Hoveyda′s) second generation complex cf. *49*, 932s*70,71*; inter- and intra-molecular metathesis of ethylenenitriles, enhanced reaction rates and yields with Ru-catalysts obtained by treating the second generation Grubbs′ catalyst with *o*-subst. pyridines, especially $(H_2IMes)(2\text{-}MeC_5H_4N)(Cl)_2Ru{=}CHPh$, s. W. Zhang, R. Zhang, R. He, Tetrahedron Lett. *2007*, *48* (24), 4203-5.

*Dichlorobis(tri-*o*-tolylphosphine)palladium(II)/tri-*n*-butyltin methoxide* ←

Aryl ketones from ar. bromides and enolesters Br → COCH

626. 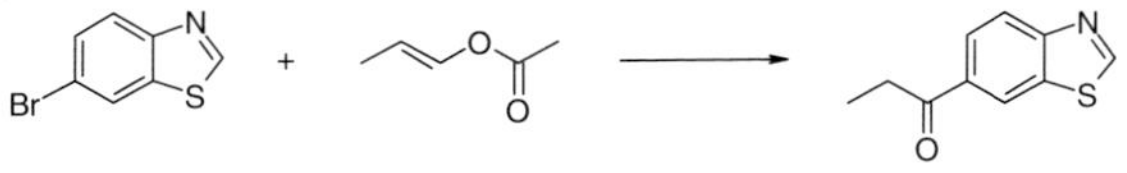

2 Eqs. prop-1-enyl acetate then 5 mol% dichlorobis(tri-*o*-tolylphosphine)palladium(II) added to a soln. of 6-bromobenzothiazole and 2 eqs. tri-*n*-butyltin methoxide in dry DMSO, the mixture

heated at 100° under N_2 for 14 h, cooled to room temp., diluted with ethyl acetate, treated with 4 *M* aq. KF, and filtered through Celite → 6-propionylbenzothiazole. Y 56%. This is the first example of such a palladium-catalyzed synthesis of aryl ketones, which is applicable to coupling aryl bromides with methyl-, dimethyl- and benzyl-subst. vinylic acetates. Ar. bromides possessing electron-withdrawing groups, however, gave low yields, as did 3-bromopyridine, while 2-bromopyridine gave 2,2′-bipyridyl. The anticipated α-arylation was not observed, and there was no reaction in the absence of catalyst or tri-*n*-butyltin methoxide. Initial mechanistic studies indicate that reaction proceeds via addition of the aryl group to an intermediate ketene. F.e.s. M. Jean, J. Renault, P. Uriac, M. Capet, P. van de Weghe, Org. Lett. *2007*, *9* (18), 3623-5.

Elimination ⥮

Hydrogen ↑ CC ⥮ H

Irradiation s. under Iodine ⫻

Copper(I) acetate s. under Pd(OAc)₂ *CuOAc*

Activated carbon/oxygen *C/O₂*

Imidazoles from Δ²-imidazolines CHCH → C=C

with Magtrieve cf. *4*, 316s*71*; by heating with activated carbon in xylene under O_2, 2-aryl-derivs., and applications of the substrates and products as ligands in Heck reactions, s. S. Haneda, A. Okui, C. Ueba, M. Hayashi, Tetrahedron *2007*, *63* (11), 2414-7.

Polymer-based methyl trityl ether s. under [Pt(dppe)](BF₄)₂ *Ⓟ-ArCPh₂OMe*

Bromotrichloromethane/1,8-diazabicyclo[5.4.0]undec-7-ene *BrCCl₃/DBU*

Bis(azoles) from bis(azolines) s. *72*, 270

N-Bromosuccinimide *NBS*

Dehydrogenation of N-heterocyclics

2*H*-pyrroles from pyrrolidines with DDQ cf. *10*, 639s*65*; pyrrole ring with NBS, 5,6-dihydropyrrolo[2,1-*a*]isoquinolines, s. J. Tóth, L. Váradi, A. Dancsó, G. Blaskó, L. Tõke, M. Nyerges, Synlett *2007* (8), 1259-63

Tin(IV) chloride or antimony pentachloride/triethylamine *SnCl₄ or SbCl₅/Et₃N*

Aromatization by dehydrogenation ←

with activated carbon/O_2 cf. *33*, 910s*66*; ar. amines from 1-aminocyclohexenes with $SnCl_4$ (or $SbCl_5$)/Et_3N in methylene chloride at room temp. s. M.A. Bigdeli, A. Rahmati, H. Abbasi-Ghadim, G.H. Mahdavinia, Tetrahedron Lett. *2007*, *48* (26), 4575-8.

Oxygen s. under Activated carbon and Pd(OAc)₂ *O₂*

Iodine/irradiation *I₂/⫻*

Carbazole ring ○

by photoinduced dehydrogenative cycloaromatization s. *72*, 513

Palladium(II) acetate/copper(I) acetate/acetic acid/oxygen ←

Catalytic oxidative intramolecular 2-arylation of indoles s. *72*, 496

[1,2-Bis(diphenylphosphino)ethane]platinum(II) fluoroborate/polymer-based methyl trityl ether ←

Regio- and stereo-specific oxidative polyenol cyclization

627.

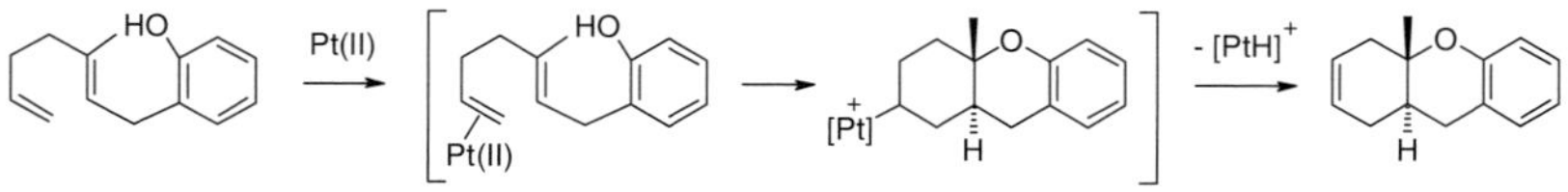

Catalyst turnover by hydride abstraction is the key step in a novel platinum(II)-catalyzed polyenol cyclization which, unlike the corresponding palladium(II) conversion, proceeds with high diastereo- and regio-selectivity. **E:** 2.2 Eqs. $AgBF_4$ added to a soln. of 10 mol% (dppe)PtI_2 in nitroethane, stirred in the dark for 1 h, 21 eqs. polystyrene-supported methyl trityl ether and 10 eqs. of the startg. polyenol added, stirred in the dark for 8 h, and worked up → product. Y 73%. A single diastereoisomer with a *trans*-fused junction is formed following β-hydride elimination of an intermediate cycloalkyl[1,2-bis(diphenylphosphino)ethane]platinum cation, the liberated platinum(I) species [(dppe)PtH^+] being oxidized to the original platinum(II) state by the trityl cation to continue the cycle. F.e. and homogeneous procedure with added diphenylamine s. C.A. Mullen, M.R. Gagné, J. Am. Chem. Soc. *2007, 129* (39), 11880-1.

Oxygen ↑ CC ⥣ O

Sodium hydroxide *NaOH*

4(1*H*)-Quinolones from *o*-acylaminoketones s. *28*, 903s*72* ○

Chloro(triphenylphosphine)gold(I)/silver triflate *$(Ph_3P)AuCl/AgOTf$*

Gold(I)-catalyzed Friedel-Crafts alkylation with *o*-(alk-1-ynyl)benzoic acid alkyl esters
4-Metalloisobenzopyrylium ions as intermediates

628. 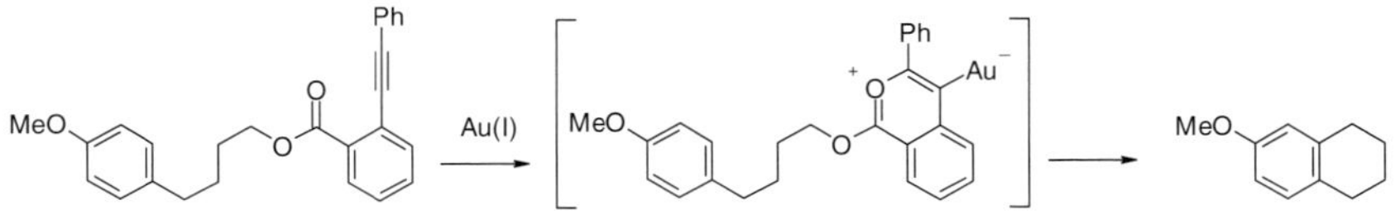

Intramolecular Friedel-Crafts alkylation. 4-(4-Methoxyphenyl)butyl 2-(phenylethynyl)-benzoate added to a mixture of 5 mol% $(Ph_3P)AuCl$ and 5 mol% AgOTf in 1,2-dichloroethane under argon, the mixture heated at 100° for 2 h, cooled, satd. aq. $NaHCO_3$ added, the mixture extracted with ether, the extracts washed with brine, dried, and evaporated → 6-methoxytetralin. Y 81%. In the absence of the methoxy group the reaction is slower giving a 70% yield after 7 h. A series of 2-alkyl-5-methylfurans were prepared similarly by **intermolecular Friedel-Crafts alkylation** in chlorobenzene at room temp. Yields were generally >70% but the cinnamyl 2-(phenyl-ethynyl)benzoate gave a 7:3 mixture of 5-(3/1-phenylallyl) derivs. in 61% yield. The reaction is assumed to involve initial activation of the alkyne residue with Au(I), followed by ring closure to a 4-metalloisobenzopyrylium ion prior to Friedel-Crafts alkylation. F.e. and O-alkylation of alcohols (*72*, 126) s. N. Asao, H. Aikawa, S. Tago, K. Umetsu, Org. Lett. *2007*, *9* (21), 4299-302.

Formic acid s. under $(CF_3SO_2)_2O$ *HCOOH*

Titanium tetrachloride/N,N′-carbonyldiimidazole *$TiCl_4/[Im]_2CO$*

Phenanthrene-9,10-diones from *o*-biphenylylglyoxylic acids via their imidazolides

629. 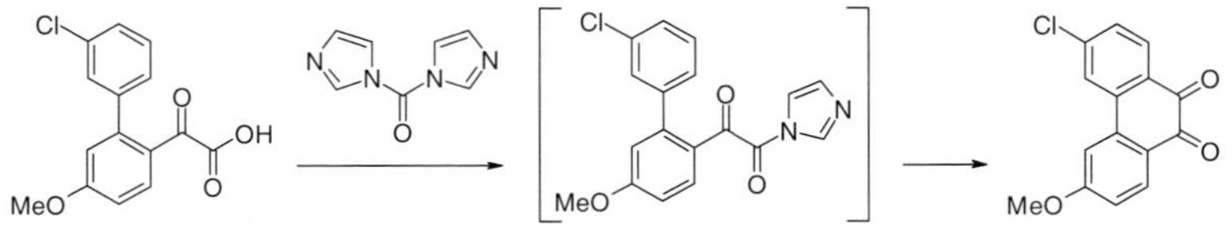

One-pot procedure. 1.4 Eqs. N,N′-carbonyldiimidazole added in portions to a soln. of the startg. biphenylylglyoxylic acid in methylene chloride, stirred for 15 h, the mixture added to 3 eqs. $TiCl_4$ in methylene chloride at 0-5° over 1 h, stirred at 0° for 80 min, quenched with cold methanolic 6 *M* HCl, stirred at room temp. for 3 h, filtered, washed, and dried → phenanthrene-9,10-dione deriv. Y 85%. Attempts to effect Friedel-Crafts cyclization of the free acid directly were ineffective as were attempts to prepare an acid chloride. It was possible, however, to generate the acid chloride *in situ* and prepare the corresponding benzotriazolide, which cyclized in lower yield, but there were reservations about the use of benzotriazole for large-scale synthesis.

The illustrated method is scalable to 20 Kg without appreciable loss in yield or regioselectivity (2% of the regioisomer being formed in the illustrated case). The cyclization is sensitive to substituents in the reacting ring with electron-releasing *m*-substituents (Cl or F) producing yields of 70-85%, while *p*-MeO and *m*-Me gave lower yields, and substrates with electron-deficient rings (e.g. *m*-CHO) failed to react. F.e. incl. formation of the startg. biaryls via Suzuki-Miyaura coupling s. N. Yoshikawa, A. Doyle, L. Tan, J.A. Murry, A. Akao, M. Kawasaki, K. Sato, Org. Lett. *2007*, *9* (21), 4103-6.

Phosphorus pentoxide/methanesulfonic acid *$P_2O_5/MeSO_3H$*

4(1*H*)-Quinolones from β-arylamino-α,β-ethylenecarboxylic acid esters ○

thermally cf. *27*, 943s*36*; with Eaton's reagent (P_2O_5 in $MeSO_3H$) s. D. Zewge, C. Chen, C. Deer, P.G. Dormer, D.L. Hughes, J. Org. Chem. *2007*, *72* (11), 4276-9.

*Trifluoromethanesulfonic anhydride/2,6-di-*tert*-butyl-4-methylpyridine/diacetato-bis(triphenylphosphine)palladium(II)/tri-*n*-butylamine/formic acid* ←

Methylene compds. from aldehydes via enol triflates CHCHO → C=CH_2

630. 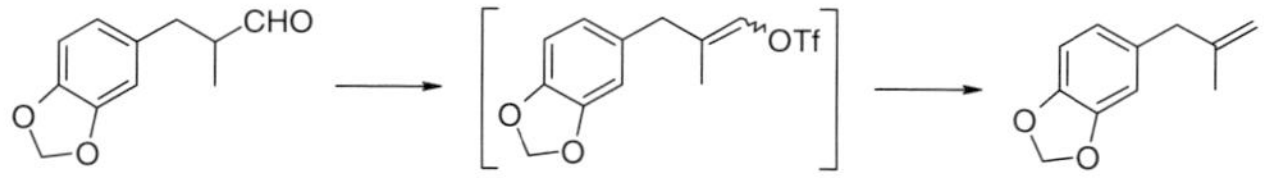

One-pot conversion. Triflic anhydride (1.1 mmol) added to a stirred soln. of startg. aldehyde (1 mmol) and 2,6-di-*tert*-butyl-4-methylpyridine (1.2 mmol) in 1,2-dichloroethane (4 ml) under argon, the resulting soln. heated at 70° for 1 h, cooled to room temp., filtered through a short silica gel plug, concentrated under reduced pressure, then DMF (2 ml), $Pd(OAc)_2(PPh_3)_2$ (5 mol%), tributylamine (3 mmol) and formic acid (2 mmol) added, and the mixture stirred at 35° for 30 min → product. Y 63%. Notably, in this and similar examples, little or no double bond isomerization to form the styryl derivative is observed. The method is suitable for both linear and α-branched aldehydes, affording the corresponding olefins in moderate to good yields. Although olefins had previously been prepared by reduction of ketone enol triflates [for example, s. S. Danishefsky et al., J. Am. Chem. Soc. *2006*, *128* (2), 426-7], this is the first report of the strategy being applied to the reduction of aldehydes. F.e.s. S.K. Pandey, A.E. Greene, J.-F. Poisson, J. Org. Chem. *2007*, *72* (20) 7769-70.

Palladium-carbon/cesium carbonate *$Pd\text{-}C/Cs_2CO_3$*

Regioselective synthesis of 1-vinylideneindans from *o*-subst. 3-arylpropargyl carbonates ○

631. 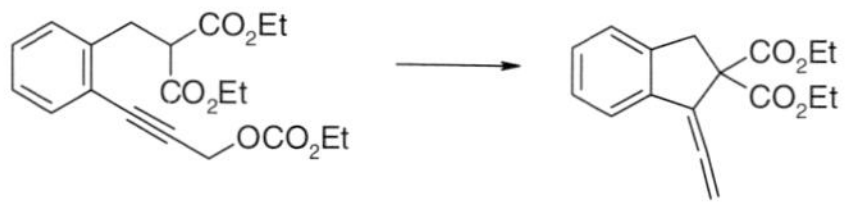

1-Vinylideneindan-2,2-dicarboxylic acid esters. 2 Eqs. Cs_2CO_3 added to a soln. of the startg. propargylic carbonate in DMF, stirred 1 min, *0.5 mol%* Pd-C added, stirred at room temp. *in air* for 6 h, quenched with satd. aq. NH_4Cl, and extracted with ethyl acetate → product. Y 96%. The reaction is successful with the inexpensive palladium catalyst (lacking a phosphine ligand!) at room temperature and does not require an inert atmosphere. The malonate moiety can be replaced by acetonylacetone, bis(benzoyl)methane or bis(diethyl phosphonato)methane. Carbonates of primary, secondary and tertiary propargylic alcohols participated in the cyclization, but yields were lowest with bulky tertiary alcohols. Other solvents (THF and toluene) and bases (K_2CO_3) were less effective. An intermediate allenylpalladium species is likely involved in the process. F.e. and bis(cyclization), also with tetrakis(triphenylphosphine)palladium and tris(dibenzylideneacetone)dipalladium, s. H.-P. Bi, X.-Y. Liu, F.-R. Gou, L.-N. Guo, X.-H. Duan, Y.-M. Liang, Org. Lett. *2007*, *9* (18), 3527-9.

Diacetatobis(triphenylphosphine)palladium(II) s. under $(CF_3SO_2)_2O$ *$Pd(OAc)_2(PPh_3)_2$*

Nitrogen ↑ CC ⇑ N

*Dirhodium(II) tetrakis(N-phthaloyl-*L-tert-*leucinate)* ←
Intramolecular insertion of carbenes into carbon-hydrogen bonds ○
with $Rh(OAc)_2$ cf. *38*, 954s*70*; 2-arylcyclopentanones from α-aryl-α-diazoketones with dirhodium(II) tetrakis(N-phthaloyl-L-*tert*-leucinate) s. D.F. Taber, W. Tian, J. Org. Chem. *2007*, *72* (9), 3207-10.

Halogen ↑ CC ⇑ Hal

Microwaves s. under 1-Methyl-3-pentylimidazolium fluoroborate [\\\\]

Potassium tert-*butoxide* t-*BuOK*
Cyanocyclopropanes from γ-bromonitriles ▽

632. 

2-(Dibenzylaminomethyl)-1-cyanocyclopropanes. KOBu-*t* (1.5 eqs.) added to a soln. of 5-(dibenzylamino)-4-bromopentanenitrile (2 mmol) in dry THF (15 ml), and stirred at room temp. for 5 h → product. Y 75% (*cis/trans* 35:65). This is the first report of the use of γ-halogeno-δ-aminopentanenitriles as substrates for a 3-*exo-tet* cyclization towards this class of compound. F.e.s. M. D'hooghe, K. Vervisch, N. De Kimpe, J. Org. Chem. *2007*, *72* (19), 7329-32.

Samarium diiodide SmI_2
Ethylene derivs. from 1,2-acoxyhalides C(OAc)C(Hal) → C=C
(Z)-2-Ethylenesilanes s. *57*, 477s*72*

(E)-1,1-Nitroethylene derivs. from 1-nitro-1,2-bromohydrins C(OH)C(Hal) → C=C
s. *59*, 484s*72*

*1,3-Dimesitylimidazolin-2-ylidene s. under Tris[di(*p-*methoxybenzylidene)acetone]-dipalladium* ←

1-Methyl-3-pentylimidazolium fluoroborate/microwaves *[pmim]*BF_4/[\\\\]
Ethylene derivs. from 1,2-acoxyhalides C(OAc)C(Hal) → C=C
cf. *53*, 487s*65*; (E)-ethylene derivs. from 1,2-acoxybromides under microwave irradiation in 1-methyl-3-pentylimidazolium fluoroborate (as ionic liquid solvent) containing a little water, also mono- or di-debromination of α,α-dibromoketones (cf. *57*, 44s*69*), s. B.C. Ranu, K. Chattopadhyay, R. Jana, Tetrahedron *2007*, *63* (1), 155-9; (Z)-2-ethylene- from 2-acoxy-3-chloro-silanes with SmI_2 (cf. *57*, 477s*69*) s. J.M. Concellón, H. Rodríguez-Solla, C. Simal, C. Gómez, Synlett *2007* (1), 75-8.

Palladium(II) acetate/triphenylphosphine/triethylamine *Pd(OAc)*$_2$*/Ph*$_3$*P/Et*$_3$*N*
2-Cyclopentenones from 5-bromo-1,4-dien-3-ols ○
via oxidative intramolecular Heck vinylation

633. 

A soln. of the startg. bromodienol, 5 mol% $Pd(OAc)_2$, 0.5 eq. triphenylphosphine and 1.2 eqs. triethylamine in acetonitrile degassed with argon, and heated at 80° for 8-10 h → product. Y 65%. This is part of a two-step synthesis **from α,β-ethylene-β-bromoaldehydes** via Grignard addition of vinylmagnesium bromide. F.e.s. D. Ray, S. Paul, S. Brahma, J.K. Ray, Tetrahedron Lett. *2007*, *48* (45), 8005-8.

*Tris[bis(*p*-methoxybenzylidene)acetone]dipalladium/1,3-dimesitylimidazol-2-ylidene/ cesium carbonate or potassium phosphate* ←
Intramolecular Heck alkylation with unactivated alkyl bromides or chlorides ○

634.

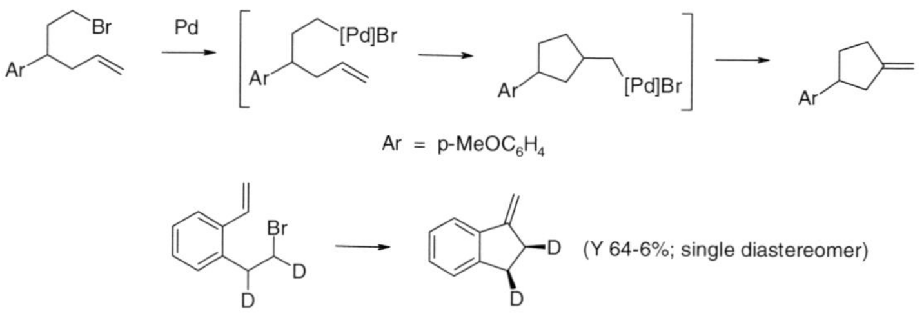

Palladium-catalyzed intramolecular Heck reactions of unactivated β-hydrogen-containing alkyl halides may now be achieved by judicious choice of catalyst and N-heterocyclic carbene ligand, enabling the creation of an array of 5-membered rings. **E: Methylenecyclopentanes.** 5 Mol% Pd_2(MeO-dba)$_3$, 20 mol% SIMes·HBF_4, 20 mol% KOBu-*t*, and 1.1 eqs. Cs_2CO_3 added to an oven-dried Schlenk flask under argon, stirred for 5 min, dry acetonitrile added, stirred for 10 min, startg. ethylenebromide added via a gas-tight syringe to the resultant orange-brown mixture, the flask sealed, the mixture stirred at 65° for 16 h, cooled to room temp., passed though a pad of silica gel, washing with ether, concentrated under vacuum, and purified by column chromatography (10% w/w $AgNO_3$ on SiO_2) → product. Y 82% (average of 2 experiments). The same product was obtained from the corresponding chloride in 72% yield by using K_3PO_4 instead of Cs_2CO_3 and heating at 100° in N-methyl-2-pyrrolidone. The method is compatible with aryl, hetaryl, alkyl and ester substituents and is also applicable to 1-methyleneindan formation. The formation of 2,3-dideuterio-1-methyleneindan as a single diastereomer is indicative of the formation of an alkylpalladium intermediate (rather than the radical pathway proposed for other metal-catalyzed alkyl-Heck-like reactions). This alkylpalladium species must undergo intramolecular insertion more rapidly than β-hydride elimination, while the cyclized palladium species must favour β-hydride elimination over insertion. F.e.s. L. Firmansjah, G.C. Fu, J. Am. Chem. Soc. *2007*, *129* (37), 11340-1.

Sulfur ↑ CC ⥣ S

Potassium tert-*butoxide* *KOBu*-t
Ramberg-Bäcklund rearrangement $CHSO_2C(Hal) \rightarrow C{=}C$
with KOH cf. *8*, 942; with KOBu-*t*, 1-methylenesugars, s. X. Zhu, Y. Jin, J. Wickham, J. Org. Chem. *2007*, *72* (7), 2670-3.

Iron(III) acetoacetonate/methylmagnesium iodide *Fe(acac)$_3$/MeMgI*
Geoselective synthesis of ethylene derivs. from mercaptals and oxo compds. via 2-hydroxythioethers C(OH)C(SR) → C=C

635.

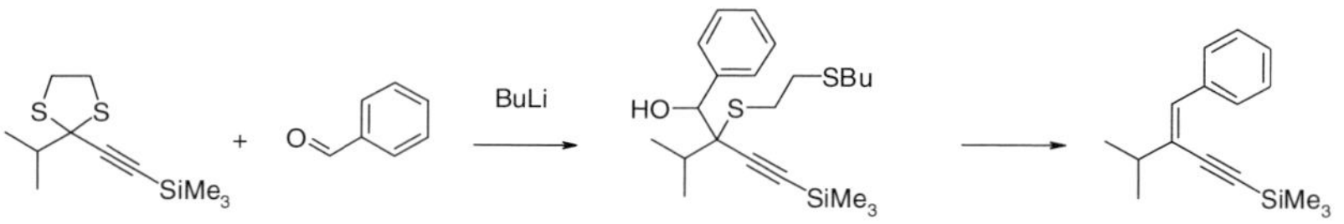

(Z)-1,3-Enynes. 2-Isopropyl-2-trimethylsilylethynyl-1,3-dithiolane added to 1.1 eqs. of *n*-butyllithium (2.5 *M* in hexane) in THF at -78° under N_2, stirred 1 h, 1 eq. benzaldehyde in THF added dropwise, stirred 2-8 h, quenched with satd. aq. NH_4Cl, and extracted with ether → intermediate homopropargylic alcohol (Y 84%; d.r. 7.1:1), dissolved in toluene with 1 eq. Fe(III)-acetoacetonate, treated dropwise with 5 eqs. of methylmagnesium iodide in ether, warmed to room temp., refluxed for 24-48 h, cooled, and quenched with satd. aq. NH_4Cl → 1-isopropyl-1-trimethylsilylethynyl-2-phenylethene (Y 74%; Z/E 20:1). Excess of methylmagnesium iodide is required to drive the reaction to completion. Other alkyl-substituted propargylic dithioacetals were also (Z)-selective, whereas dithioacetals derived from acetophenone or benzaldehyde were (E)-selective. Dithioacetals derived from dialkyl ketones, however, gave a complex mixture of products. F.e. incl. **tri- and tetra-subst. styrenes** s. L.-F. Huang, C.-W. Chen, T.-Y. Luh, Org. Lett. *2007*, *9* (18), 3663-5.

Remaining Elements ↑ CC ⇑ Rem

Silver fluoroborate *$AgBF_4$*

Tricyclics by interrupted Nazarov cyclization with appended aryl groups ○

636. 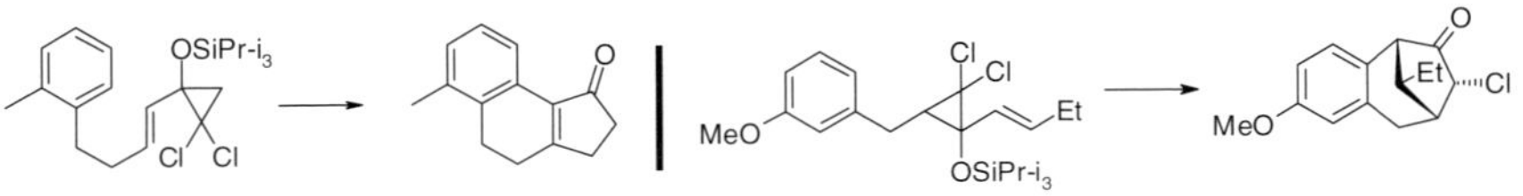

4,5-Dihydrobenz[*g*]indan-1-ones. 1 Eq. $AgBF_4$ added to (1E)-1,1-dichloro-2-[4-(2-methylphenyl)-1-buten-1-yl]-2-triisopropylsilyloxycyclopropane in acetonitrile, refluxed 12 h, cooled to room temp., filtered through silica, and purified by flash chromatography → 6-methyl-2,3,4,5-tetrahydrocyclopenta[*a*]napthalen-1-one. Y 79%. The intermediate siloxycyclopentadienyl cation is trapped by the pendant phenyl group to afford the tricyclic product. Reaction is successful with phenyls containing both electron-releasing and -withdrawing groups but yields are variable (39-80%). The analog with a pendant furyl group gave only decomposition products while the presence of a methyl alpha to the cyclopropane moiety gave a tricyclic product only with a pendant electron-rich 4-methoxyphenyl group. The conventional Nazarov product was obtained from the phenyl deriv. itself (Y 97%). Benzylic substrates unexpectedly led to **6-chlorobenzo[*b*]bicyclo[3.2.1]oct-2-en-7-ones** stereospecifically (4 examples; Y 57-81%). F.e. and preparation of the starting *gem*-dichloropropanes by dichlorocyclopropanation of the corresponding siloxydiene s. T.N. Grant, F.G. West, Org. Lett. *2007*, *9* (19), 3789-92.

Boron fluoride *BF_3*

Bicyclic 2-ethyleneethers from (enesilyl)lactolides
via regio- and stereo-specific intramolecular quenching of cyclic oxocarbenium ions

637. 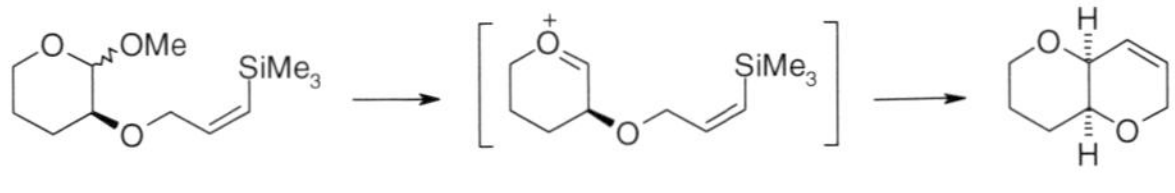

Enesilyl residues may be used in the same manner as allylsilyl groups (cf. *46*, 984s*66*) for intramolecular quenching of *in situ*-generated oxocarbenium ions to yield a range of ***cis*-2,(n+3)-dioxabicyclo[n.4.0]alk-(n+5)-enes. E:** 2.5 Eqs. BF_3-etherate added dropwise to a stirred soln. of the startg. lactolide in dry methylene chloride, cooled to 0°, stirring continued with cooling, allowed to warm to room temp., and quenched with satd. aq. $NaHCO_3$ when TLC indicated completion of reaction → (4aS*,8aS*)-2,3,4,4a,6,8a-hexahydropyrano[3,2-*b*]pyran. Y 75%. The procedure is rapid, clean and affords a single stereoisomer. F.e. incl. carbohydrate derivs. s. S.P. Fearnley, P. Lory, Org. Lett. *2007*, *9* (18), 3507-10.

Carbon ↑ CC ⥣ C

Without additional reagents *w.a.r.*
Syntheses via cycloaddition
to transition metal α,β-(*o*-styryl)acetylene(alkoxy)carbene complexes

638.

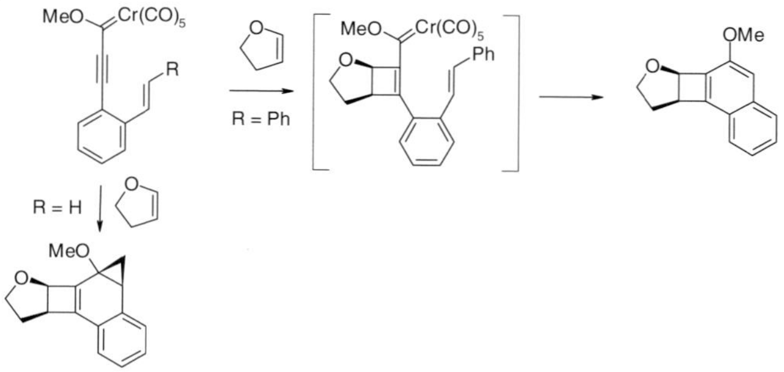

2-Naphthol ether ring via intramolecular metathesis. The startg. Fischer carbene complex stirred with 5 eqs. of 2,3-dihydrofuran in dry THF until reaction complete, then the intermediate [2+2]-cycloadduct heated in a sealed tube at 90° for 2 h → product. Y 74%. The method is successful with both tungsten and chromium carbene complexes utilizing enolethers, nitrones or 1,3-dienes for the initial cycloaddition in 52-81% yields. The reaction is controlled by the *o*-ethenyl substituent, giving metathesis products where the substituent is Me or Ph, while *o*-vinyl derivs. underwent **cycloaddition-intramolecular cyclopropanation** (spontaneously at room temp.). F.e. and preparation of the startg. m. s. J. Barluenga, F. Andina, F. Aznar, C. Valdés, Org. Lett. *2007, 9* (21), 4143-6.

Alkylidene(dichloro)ruthenium(II) imidazolidin-2-ylidene complex ←
Alkylidene(dichloro)ruthenium(II) imidazolidin-2-ylidene complex/trifluoroacetic acid ←
Ring closing metathesis
with Grubbs′ complex cf. *49*, 985s*70*; chiral cyclic ethylene derivs. using *p*-menthane-3-carboxaldehyde as chiral auxiliary followed by cleavage via ring-closing metathesis with dichloro(1,3-dimesitylimidazolidin-2-ylidene)(tricyclohexylphosphine)(benzylidene)ruthenium(II) (e.e. up to 98%) s. L. Boisvert, F. Beaumier, C. Spino, Can. J. Chem. *2006*, *84* (10), 1290-3; 2-cyanonaphthalenes from β-(*o*-allylaryl)-β-hydroxy-α-methylenenitriles s. P.-Y. Chen, H.-M. Chen, L.-Y. Chen, J.-Y. Tzeng, J.-C. Tsai, P.-C. Chi, S.-R. Li, E.-C.Wang, Tetrahedron *2007*, *63* (13), 2824-8; functionalized cyclic ethylene derivs. with benzylidene(chloro)(tricyclohexylphosphine)-ruthenium(II) L-prolinate s. J.S.M. Samec, R.H. Grubbs, Chem. Commun. *2007* (27), 2826-8; furans from 5-alkoxy-3-oxa-1,6-dienes with dichloro(1,3-dimesitylimidazolidin-2-ylidene)(2-isopropyloxybenzylidene)ruthenium(II)/trifluoroacetic acid s. T.J. Donohoe, L.P. Fishlock, A.R. Lacy, P.A. Procopiou, Org. Lett. *2007*, *9* (6), 953-6.

Tris(dibenzylideneacetone)dipalladium(0)/2,2′-bis(diphenylphosphino)-1,1′-binaphthyl *$Pd_2(dba)_3$/rac-BINAP*
Decarboxylative intramolecular benzylic allylation of nitrophenylacetic acid allyl esters ←

639.

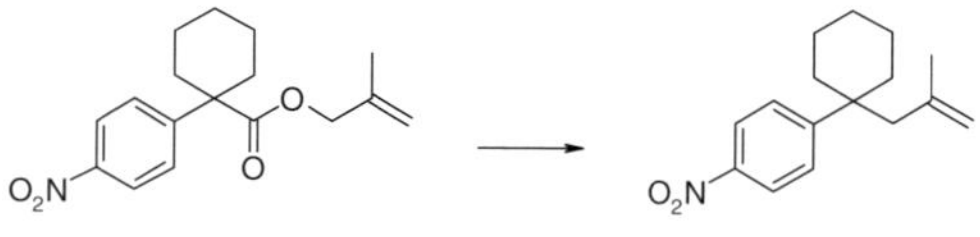

$Pd_2(dba)_3$ (0.0075 mmol) and *rac*-BINAP (0.015 mmol) added to a soln. of the startg. ester (0.3 mmol) in toluene (5 mmol) in a dried Schlenk tube under argon, the soln. heated at 110° for 1-3 h

(TLC monitoring), cooled to room temp., concentrated *in vacuo,* and purified directly via column chromatography → product. Y 94%. The reaction is applicable to α-monosubst. *o*- and *p*-nitrophenylacetic acid esters and α,α-disubst. *p*-nitrophenylacetic acid esters, the more highly-subst. *o*-nitrophenylacetic acid esters being unreactive due to steric effects. Esters and ketones are tolerated as are aromatic methoxy substituents. Alkyl- and terminal-subst. allyl esters reacted smoothly only when *o,p*-dinitrophenyl substrates were employed, using $Pd(PPh_3)_4$ as the preferred catalyst. For the analogous mononitrophenyl substrates, it was reasoned that the intermediate benzylic anion is too basic to undergo coupling, instead affording elimination products. F.e.s. S.R. Waetzig, J.A. Tunge, J. Am. Chem. Soc. *2007*, *129* (48), 14860-1.

Resolutions

(s.a. Subject Index under Resolution)

Without additional reagents *w.a.r.*
Determination of absolute configuration of sec. alcohols and prim. amines ←
via NMR of chiral silyl derivs. s. *72*, 1

Sodium hydroxide/N-(4-trifluoromethylbenzyl)cinchoninium bromide ←
Kinetic resolution of 2-acyl-1-benzyl-1-cyano-1,2-dihydroisoquinolines via phase transfer-catalyzed asym. hydrolysis ←

640.

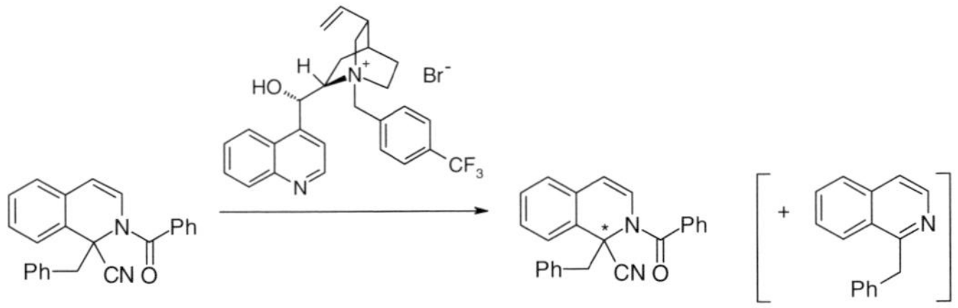

A stirred soln. of racemic 2-benzoyl-1-benzyl-1,2-dihydroisoquinoline-1-carbonitrile in methylene chloride treated with 10 mol% N-(4-trifluoromethylbenzyl)cinchoninium bromide in a small glass vial, the latter closed, cooled to 5°, cold (5°) 10 *N* NaOH added, stirred vigorously for 26 h at the same temp., and worked up → 1-benzylisoquinoline (Y not cited) and recovered startg. m. (Y 56%; e.e. 54%; conversion 43%). This is a useful alternative to asym. 1-alkylation of Reissert compds. Reaction is thought to take place via initial asym. hydrolytic N-deacylation through a chiral quaternary ammonium 1,1,1-aminodiolate, followed by a fast retro-Strecker reaction with loss of HCN where aromatization is the driving force. F.e.s. K. Frisch, K.A. Jørgensen, Org. Biomol. Chem. *2007, 5* (18), 2966-74.

Copper(II) triflate/chiral bis(Δ^2-oxazolines)/N-bromosuccinimide/potassium carbonate ←
α-Hydroxyketones from glycols by asym. oxidation CHOH → CO
with desymmetrization s. *72*, 137

Silver perchlorate/chiral di(phosphine) ←
Intramolecular hydroamination ○
of N-carbobenzoxyalleneamines with dynamic kinetic resolution s. *72*, 185

Dehydrogenase ←
Homogeneous hydrogenation CHO → CH_2OH
of α-subst. α-arylaldehydes with dynamic kinetic resolution s. *72*, 21

Lipase ←
Enzymatic resolution of 2-acoxythiocyanates s. *72*, 345 ←

Monooxygenases ←
Kinetic resolution of ketones by Baeyer-Villiger oxidation s. *44*, 114s*72* ←

Novozyme-433/palladium nanoparticles ←
Carboxylic acid amides from esters COOR → CON<
with dynamic kinetic resolution s. *45*, 210s*72*

Proline *Pro-OH*
Organocatalyzed aldol condensation ←
with dynamic kinetic resolution s. *58*, 245s72

Benzoic anhydrides/(+)-benzotetramisole or tetramisole/ethyldiisopropylamine ←
Kinetic resolution of sec. benzyl alcohols by esterification OH → OAc
with 3-phenylpropionic acid s. *72*, 101

Chiral (η^6-arene)chloro(N-sulfonyl-1,2-ethylenediamine)ruthenium(II) complex/ acetone/potassium hydroxide/chiral ruthenium(II) Δ^2-oxazolin-2-ylferrocenyl-phosphine complex/sodium isopropoxide/isopropanol ←
(Benzene)dichlororuthenium(II) dimer/(R)-2,2′-bis(diphenylphosphino)-1,1′-binaphthyl/ (R,R)-1,2-diphenylethylene-1,2-diamine/cyclohexanone/potassium tert-*butoxide* ←
Deracemization of sec. alcohols ←
via ruthenium-catalyzed oxidation-asym. hydrogenation – (S)-Benzyl from (R)-benzyl alcohols s. *72*, 84

Dichloro[2,2′-bis(di-3,5-xylylphosphino)-1,1′-binaphthyl]-((R,R)-1,2-di-phenyl-1,2-ethylenediamine)ruthenium(II)/potassium tert-*butoxide* *[Ru(II)]/*t*-BuOK*
Homogeneous hydrogenation CHO → CH_2OH
of α-subst. α-arylaldehydes with dynamic kinetic resolution s. *72*, 21

Chiral [7,7-bis(diarylphosphino)-1,1′-spirobiindane](1,2-diamine)(dichloro)-ruthenium(II)/potassium tert-*butoxide* ←
Cyclic *cis*-2-*tert*-aminoalcohols from α-*tert*-aminoketones CO → CHOH
Asym. homogeneous hydrogenation with dynamic kinetic resolution s. *72*, 22

Palladium nanoparticles s. under Novozyme-433 *Pd*

Reviews

This is a collection of reviews published
largely from May 2007 to April 2008
arranged in the following sections:

1 Asymmetric Catalysis – chiral ligands, asymmetric organocatalysis and hydrogenation, resolution

Recent advances in developing new axially **chiral phosphine ligands** for asymmetric catalysis, Y.-M. Li, F.-Y. Kwong, W.-Y. Yu, A.S.C. Chan, Coord. Chem. Rev. *2007, 251* (17-20), 2119-44; synthesis of chiral heterocyclic phosphines for application in asymmetric catalysis, J. Holz, M.N. Gensow, O. Zayas, A. Boerner, Curr. Org. Chem. *2007, 11* (1), 61-106; dendrimeric phosphines in asymmetric catalysis, A.-M. Caminade, P. Servin, R. Laurent, J.-P. Majoral, Chem. Soc. Rev. *2008, 37* (1), 56-67; combinatorial transition-metal catalysis: **mixing monodentate ligands** to control enantio-, diastereo-, and regioselectivity, M.T. Reetz, Angew. Chem. Int. Ed. Engl. *2008, 47* (14), 2556-88; synergistic effect of binary component ligands in chiral catalyst library engineering for enantioselective reactions, K. Ding, Chem. Commun. *2008* (8), 909-21; recent progress in asymmetric catalysis using **carbohydrate-based ligands**, M. Dieguez, C. Claver, O. Pamies, Eur. J. Org. Chem. *2007* (28), 4621-34; **chiral sulfur ligands** for asymmetric catalysis, M. Mellah, A. Voituriez, E. Schulz, Chem. Rev. *2007, 107* (11), 5133-209; new developments with electrophilic **chiral selenium reagents**, D.M. Browne, T. Wirth, Curr. Org. Chem. *2006, 10* (15), 1893-903; enantioselective synthesis mediated by catalytic chiral organoselenium compounds, A.L. Braga, D.S. Ludtke, F. Vargas, ibid. 1921-38; **Si-chiral silanes** in asymmetric catalysis, M. Oestreich, Russ. Chem. Rev. *2007, 76* (11), 1629-43; non camphor-based **sultam chiral auxiliaries**, A.W.M. Lee, W.H. Chan, S.-J. Zhang, H.-K. Zhang, Curr. Org. Chem. *2007, 11* (2), 213-28; chiral ligands with an isoborneol-10-sulfonamide structure: a ten-year odyssey, D.J. Ramon, M. Yus, Synlett *2007* (15), 2309-20; CBS **oxazaborolidines**, versatile catalysts for asymmetric synthesis, R.T. Stemmler, ibid. (6), 997-8; development of highly enantioselective oxazaborolidinone catalysts for the reactions of acyclic α,β-unsatd. ketones, T. Harada, T. Kusukawa, ibid. (12), 1823-35; 1,5-asymmetric induction in boron-mediated aldol condensation of β-oxygenated methyl ketones, L.C. Dias, A.M. Aguilar, Chem. Soc. Rev. *2008, 37* (3), 451-69; **C_2 and C_1 symmetry of chiral auxiliaries** in catalytic reactions on metal complexes, V.A. Pavlov, Tetrahedron *2008, 64* (7), 1147-79; asymmetric catalysis of metal complexes with non-planar **ONNO ligands: salen, salalen and salan**, K. Matsumoto, B. Saito, T. Katsuki, Chem. Commun. *2007* (35), 3619-27; **asymmetric organocatalysis**, H. Pellissier, Tetrahedron *2007, 63* (38), 9267-331; do we need asymmetric organocatalysis?, P.I. Dalko, Chimia *2007, 61* (5), 213-8; organocatalysis in the asymmetric synthesis of nitrogen-containing compounds – How and why, L. Bernardi, F. Fini, M. Fochi, A. Ricci, ibid. 224-31; enantioselective organocatalysis **using SOMO activation**, T.D. Beeson, A. Mastracchio, J.-B. Hong, K. Ashton, D.W.C. MacMillan, Science *2007, 316* (5824), 582-5; enantioselective direct aldol condensation: the blossoming of modern organocatalysis, G. Guillena, C. Nájera, D.J. Ramon, Tetrahedron:Asym. *2007, 18* (19), 2249-93; **peptides as catalysts** for asymmetric aldol condensation, H. Wennemers, Chimia *2007, 61* (5), 276-8; **organocatalytic asymmetric Mannich reactions**: new methodology, catalyst design, and synthetic applications, A. Ting, S.E. Schaus, Eur. J. Org. Chem. *2007* (35), 5797-815; organocatalyzed Mannich reactions, J.M.M. Verkade, L.J.C. van Hemert, P.J. L.M. Quaedflieg, F.P.J.T. Rutjes, Chem. Soc. Rev. *2008, 37* (1), 29-41; organocatalytic enantioselective **Michael and hetero-Michael addition**, J.L. Vicario, D. Badia, L. Carrillo, Synthesis *2007* (14), 2065-92; a practical synthesis of a key chiral drug intermediate via asymmetric organocatalysis, M. Seki, Synlett *2008* (2), 164-76; chiral amines as organocatalysts for asymmetric Michael addition to nitroolefins and vinyl sulfones **via enamine activation**, S. Sulzer-Mosse, A. Alexakis, Chem. Commun. *2007* (30), 3123-35; basic instinct: design, synthesis and evaluation of **(+)-sparteine** surrogates for asymmetric synthesis, P. O'Brien, Chem. Commun. *2008* (6), 655-67; **(S)- and (R)-5-pyrrolidin-2-yl-1*H*-tetrazoles** – enantiomeric organocatalysts of broad utility in organic synthesis, D.A. Longbottom, V. Franckevicius, S.V. Ley, Chimia *2007, 61* (5), 247-56; the **aminothiourea**-catalyzed asymmetric nucleophilic reactions, Y. Takemoto, H. Miyabe, ibid. 269-75; **mixed oxazoline-carbenes** as stereodirecting ligands for asymmetric catalysis, L.H. Gade, S. Bellemin-Laponnaz, Coord. Chem. Rev. *2007, 251* (5-6), 718-25; **chiral tertiary diamines** in asymmetric synthesis, J.-C. Kizirian, Chem. Rev. *2008, 108* (1), 140-205; metal complexes incorporating monoanionic **bisoxazolinate ligands** – Synthesis, structures, reactivity and applications in asymmetric catalysis, S. Dagorne, S. Bellemin-Laponnaz, A. Maisse-Francois, Eur. J. Inorg. Chem. *2007* (7), 913-25; developing **chiral ligands for asymmetric hydrogenation**, W. Zhang, Y. Chi, X. Zhang, Acc. Chem. Res. *2007, 40* (12), 1278-90; asymmetric hydrogenation of heteroaromatic compounds, Y.-G. Zhou, ibid. 1357-66; continuous asymmetric hydrogenation, L. Greiner, S. Laue, J. Woeltinger, A. Liese, Topics Curr. Chem. *2007, 276,* 111-24; **BINOL-derived phosphoramidites** in asymmetric hydrogenation: can the presence of a functionality in the amino group influence the catalytic outcome?, L. Eberhardt, D. Armspach, J. Harrowfield, D. Matt, Chem. Soc. Rev. *2008, 37* (4), 839-64; iridium-catalyzed asymmetric hydrogenation of olefins, S.J. Roseblade, A. Pfaltz, Acc. Chem. Res. *2007, 40* (12), 1402-11; aqueous-phase **asymmetric transfer-hydrogenation** of ketones – a greener approach to chiral alcohols, X.F. Wu, J.L. Xiao, Chem. Commun. *2007* (24), 2449-66; selected recent developments in the enantio-selective reduction of imines by asymmetric transfer-hydrogenation, P. Roszkowski, Z. Czarnocki, Mini-Rev. Org. Chem. *2007, 4* (3), 190-200; asymmetric organocatalytic reductions with **dihydropyridines as mediators**, S.J. Connon, Org. Biomol. Chem. *2007, 5* (21), 3407-17; progress in coenzyme NADH model compds. and asymmetric reduction of benzoylformate, N.-X. Wang, J. Zhao, Synlett *2007* (18), 2785-91; chiral ligands for practical rhodium-catalyzed asymmetric hydroformylation, J. Klosin, C.R. Landis, Acc. Chem. Res. *2007, 40* (12), 1251-9; **chiral phosphine-olefin ligands** for rhodium-catalyzed asymmetric 1,4-addition, W.L. Duan, H. Iwamura, R. Shintani, T. Hayashi, J. Am. Chem. Soc. *2007, 129* (7), 2130-8; a green chemistry approach to asymmetric catalysis: **solvent-free and highly concentrated asymmetric reactions**, P.J. Walsh, H. Li, C.A. de Parrodi, Chem. Rev. *2007, 107* (6), 2503-45; asymmetric catalysis at **chiral metal surfaces**, T. Mallat, E. Orglmeister, A. Baiker, ibid. (11), 4863-90; recent advances in **asymmetric phase-transfer catalysis**, T. Ooi, K. Maruoka, Angew. Chem. Int. Ed. Engl. *2007, 46* (23), 4222-66; design of C_2-symmetric chiral phase-transfer catalysts for practical asymmetric synthesis, K. Maruoka, Chimia *2007, 61* (5), 263-8; **asymmetric catalysis in aqueous media**, S. Kobayashi, Pure Applied Chem. *2007, 79* (2), 235-45; strategies in **optical resolution**: a practical guide, F. Faigl, E. Fogassy, M. Nógrádi, E. Pálovics, J. Schindler, Tetrahedron:Asym. *2008, 19* (5), 519-36; advantages of structural similarities of the reactants in optical resolution processes, F. Faigl, J. Schindler, E. Fogassy, Topics Curr. Chem. *2007, 269,* 133-57; racemization, optical resolution and crystallization-induced asymmetric transformation of amino acids and pharmaceutical intermediates, R. Yoshioka, ibid. 83-132; metal complex-catalyzed hydrolysis and alcoholysis of organic substrates and their application to **kinetic resolution**, M. Tokunaga, H. Aoyama, J. Kiyosu, Y. Shirogane, T. Iwasawa, Y. Obora, Y. Tsuji, J. Organomet. Chem. *2007, 692* (1-3), 472-80; recent developments in **dynamic kinetic resolution**, H. Pellissier, Tetrahedron *2008, 64* (8), 1563-601; application of electronic circular dichroism in configurational and **conformational analysis** of organic compounds, N. Berova, L. Di Bari, G. Pescitelli, Chem. Soc. Rev. *2007, 36* (6), 914-31.

2 Transition Metal Catalysis – methods, ligands, Groups 8-10, Re, Cu, Ag, Au, Zr ...

(for transition metal catalysis with ferrocenylpolyphosphines as ligand s. Section 2b; copper-facilitated synthesis of heterocyclics or palladium-catalyzed functionalization of heterocyclics s. Section 6; palladium-catalyzed natural product synthesis s. Section 7; catalytic arylation s under Section 9; transition metal catalysis with organometallic reagents s. Section 16; palladium-catalyzed cascade reactions s. Section 25; and microwave-assisted transition metal catalysis s. Section 29)

Hydrogen-mediated C-C bond formation – A broad new concept in catalytic C-C coupling, M.Y. Ngai, J.R. Kong, M.J. Krische, J. Org. Chem. *2007, 72* (4), 1063-72; enantiomerically enriched allylic alcohols and allylic amines via asymmetric hydrogen-mediated C-C bond formation: asymmetric carbonyl and imine vinylation, E. Skucas, M.-Y. Ngai, V. Komanduri, M.J. Krische, Acc. Chem. Res. *2007, 40* (12), 1394-401; unusual influence of the structures of transition metal complexes on catalytic **C-S and C-Se bond formation** under homogeneous and heterogeneous conditions, I.P. Beletskaya, V.P. Ananikov, Eur. J. Org. Chem. *2007* (21), 3431-44; transition metal-catalyzed carbochalcogenation of alkynes, H. Kuniyasu, N. Kambe, Chem. Lett. *2006, 35* (12), 1320-5; transition metal-catalyzed **oxidative carbonylation** of amines to ureas, D.J. Diaz, A.K. Darko, L. McElwee-White, Eur. J. Org. Chem. *2007* (27), 4453-65; **oxidative cyclization of dienes and polyenes** mediated by transition metal-oxo species, V. Piccialli, Synthesis *2007* (17), 2585-607; more than bystanders: the effect of olefins on transition metal-catalyzed **cross-coupling**, J.B. Johnson, T. Rovis, Angew. Chem. Int. Ed. Engl. *2008, 47* (5), 840-71; direct transition metal-catalyzed **functionalization of heteroaromatics**, I.V. Seregin, V. Gevorgyan, Chem. Soc. Rev. *2007, 36* (7), 1173-93; organic transformations on σ-aryl organometallic complexes, M. Gagliardo, D.J.M. Snelders, P.A. Chase, R.J.M. Klein Gebbink, G.P.M. van Klink, G. van Koten, Angew. Chem. Int. Ed. Engl. *2007, 46* (45), 8558-73; modularly designed transition metal PNP and PCP **pincer complexes** based on aminophosphines: synthesis and catalytic applications, D. Benito-Garagorri, K. Kirchner, Acc. Chem. Res. *2008, 41* (2), 201-13; catalysts based on **palladium dendrimers**, R. Andres, E. De Jesus, J.C. Flores, New J. Chem. *2007, 31* (7), 1161-91; oxime **palladacycles** revisited: stone-stable complexes nonetheless very active catalysts, E. Alacid, D.A. Alonso, L. Botella, C. Nájera, M.C. Pacheco, Chem. Record *2006, 6* (3), 117-32; **Suzuki coupling** of alkylboronic acid derivs. or alkyltrifluoroborates with aryl, alkenyl or alkyl halides and triflates, H. Doucet, Eur. J. Org. Chem. *2008* (12), 2013-30; non-conventional methodologies for transition metal-catalyzed C-C coupling: Part 2: the Suzuki reaction, F. Alonso, I.P. Beletskaya, M. Yus, Tetrahedron *2008, 64* (14), 3047-101; metal unsaturation and ligand hemi-lability in Suzuki coupling, Acc. Chem. Res. *2007, 40* (8), 676-84; monomolecular nanosized and heterogenized palladium catalysts for the **Heck reaction**, Coord. Chem. Rev. *2007, 251* (9-10), 1281-93; silica-supported palladium catalysts for Heck reactions, V. Polshettiwar, A. Molnar, Tetrahedron *2007, 63* (30), 6949-76; stereoselective carbonyl allylation by **umpolung of allylpalladium(II) complexes**, G. Zanoni, A. Pontiroli, A. Marchetti, G. Vidari, Eur. J. Org. Chem. *2007* (22), 3599-611; procedures for and possible mechanisms of **palladium-catalyzed allylations** of primary and secondary amines with allyl alcohols, J. Muzart, ibid. (19), 3077-89; aldehydes from palladium-catalyzed **oxidation of terminal olefins**, J. Muzart, Tetrahedron *2007, 63* (32), 7505-21; **addition to alkynes** of E-E and E-H bonds catalyzed by Pd, Pt, and Ni complexes (E = S, Se), I.P. Beletskaya, V.P. Ananikov, Pure Applied Chem. *2007, 79* (6), 1041-56; **ruthenium N-heterocyclic carbene complexes** – friendly catalytic tools for manifold chemical transformations, V. Dragutan, I. Dragutan, L. Delaude, A. Demonceau, Coord. Chem. Rev. *2007, 251* (5-6), 765-94; non-metathesis reactions of ruthenium carbene catalysts and their application to the synthesis of nitrogen-containing heterocycles, M. Arisawa, Y. Terada, K. Takahashi, M. Nakagawa, A. Nishida, Chem. Record *2007, 7* (4), 238-53; catalytic transformations of terminal alkynes by cationic **ruthenium tris(1-pyrazolyl)borate complexes**: versatile chemistry via catalytic allenylidene, vinylidene and π-alkyne intermediates, R.-S. Liu, Synlett *2008* (6), 801-12; catalytic hydrogenation of polar organic functionalities based on ruthenium-mediated heterolytic dihydrogen cleavage, M. Ito, T. Ikariya, Chem. Commun. *2007* (48), 5134-42; on inventing catalytic reactions via **ruthena- or rhoda-cyclic intermediates** for atom economy, T. Kondo, Synlett *2008* (5), 629-44; electron transfer, in **organonickel complexes** of α-diimines – versatile redox catalysts for C-C or C-P coupling reactions, A. Klein, Y.H. Budnikova, O.G. Sinyashin, J. Organomet. Chem. *2007, 692* (15), 3156-66; electrophilic activation of alkenes by **platinum(II):** so much more than a slow version of palladium(II), A.R. Chianese, S.J. Lee, M.R. Gagne, Angew. Chem. Int. Ed. Engl. *2007, 46* (22), 4042-59; platinum-catalyzed hydrofunctionalization of unactivated alkenes with carbon, nitrogen and oxygen nucleophiles, C. Liu, C.F. Bender, X. Han, R.A. Widenhoefer, Chem. Commun. *2007* (35), 3607-18; is **osmylation** always preferring the richest double bond? A. Francais, O. Bedel, A. Haudrechy, Tetrahedron *2008, 64* (11), 2495-524; osmium and palladium: complementary metals in alkene activation and oxidation, S.D.R. Christie, A.D. Warrington, Synthesis *2008* (9), 1325-41; recent developments of **rhenium**-catalyzed synthesis, R. Hua, J.-L. Jiang, Curr. Org. Synth. *2007, 42* (2), 151-74; **copper(I) N-heterocyclic carbene complexes** in homogeneous catalysis, S. Díez-González, S.P. Nolan, Synlett *2007* (14), 2158-67; Cu-, Ni-, and Pd-mediated homocoupling in biaryl syntheses: the **Ullmann reaction,** T.D. Nelson, R.D. Crouch, Org. Reactions *63*, 265-556; preparation and application of **silver(I) N-heterocyclic carbene complexes**, I.J.B. Lin, C.S. Vasam, Coord. Chem. Rev. *2007, 251* (5-6), 642-70; the organic chemistry of **silver acetylides**, U. Halbes-Letinois, J.-M. Weibel, P. Pale, Chem. Soc. Rev. *2007, 36* (5), 759-69; **gold-catalyzed reactions**, A.S.K. Hashmi, Chem. Rev. *2007, 107* (7), 3180-211; relativistic effects in homogeneous gold catalysis, D.J. Gorin, F.D. Toste, Nature *2007, 446* (7134), 395-403; recent advances in synthesis of heterocyclics and carbocyclics via homogeneous gold catalysis. Part 1. Heteroatom addition and hydroarylation of alkynes, allenes and alkenes, H.C. Shen, Tetrahedron *2008, 64* (18), 3885-903; copper, silver, and gold complexes in **hydrosilylation** reactions, S. Díez-González, S.P. Nolan, Acc. Chem. Res. *2008, 41* (2), 349-58; nanocrystalline gold and gold palladium alloy catalysts for chemical synthesis, G.J. Hutchings, Chem. Commun. *2008* (10), 1148-64; catalytic carbophilic activation: catalysis by platinum and gold π-acids, A. Fuerstner, P.W. Davies, Angew. Chem. Int. Ed. Engl. *2007, 46* (19), 3410-49; cross-coupling-elimination reactions mediated or catalyzed by **zirconium complexes**: a valuable tool in synthesis, F.J. Fañanás, F. Rodríguez, Eur. J. Org. Chem. *2008* (8), 1315-29; modular N,O-chelating ligands – **Group-4 amidate complexes** for catalytic hydroamination, A.V. Lee, L.L. Schafer, Eur. J. Inorg. Chem. *2007* (16), 2243-55; recent advances in **organo-thorium and -uranium catalysis**, T. Andrea, M.S. Eisen, Chem. Soc. Rev. *2008, 37* (3), 550-67.

2a **Metathesis** of ethylene and acetylene derivatives

Sustainable concepts in olefin metathesis, H. Clavier, K. Grela, A. Kirschning, M. Mauduit, S.P. Nolan, Angew. Chem. Int. Ed. Engl. *2007, 46* (36), 6786-801; evolution and applications of second-generation ruthenium olefin metathesis catalysts, Y. Schrodi, R.L. Pederson, Aldrichimica Acta *2007, 40* (2), 45-52; microwave-assisted olefin metathesis, Y. Coquerel, J. Rodriguez, Eur. J. Org. Chem. *2008* (7), 1125-32; preparation of ruthenium N-heterocyclic carbene complexes and their catalytic activity in metathesis, E. Colacino, J. Martinez, F. Lamaty, Coord. Chem. Rev. *2007, 251* (5-6), 726-64; metathesis of heteroatom-subst. olefins and alkynes – current scope and limitations, P. Van de Weghe, P. Bisseret, N. Blanchard, J. Eustache, J. Organomet. Chem. *2006, 691* (24-25), 5078-108; ruthenium vinyl carbene intermediates in **ene-yne metathesis**, Coord. Chem. Rev. *2007, 251* (5-6), 671-701; **ring-closing metathesis** of heteroatom-subst. dienes, R.C.D. Brown, V. Satcharoen, Heterocycles *2006, 70,* 705-36; preparation of nitrogen-containing heterocycles using ring-closing metathesis and its application to natural product synthesis, M. Arisawa, A. Nishida, M. Nakagawa, J. Organomet. Chem. *2006, 691* (24-25), 5109-21; preparation of cyclic molecules bearing 'strained' olefins using olefin metathesis, S.K. Collins, ibid. 5122-8; olefin ring-closing metathesis as a powerful tool in drug discovery and development – potent macrocyclic inhibitors of the hepatitis C virus NS3 protease, X.J. Wang, N. Haddad, X.D. Wei, J.H. Xu, L. Zhang, ibid. 5163-71; preparation of trialkyl-subst. olefins by ruthenium-catalyzed **cross-metathesis**, T. Netscher, ibid. 5155-62; synthesis of diverse polycyclics via catalytic metathesis, S. Kotha, K. Lahiri, Synlett *2007* (18), 2767-84; a resourceful new strategy in organic synthesis – tandem and stepwise metathesis/non-metathesis catalytic processes, V. Dragutan, I. Dragutan, J. Organomet. Chem. *2006, 691* (24-25), 5129-47.

2b **Catalyst – supports, scavenging, loading, recovery ...**

Modified mesoporous materials as palladium scavengers and catalyst supports, C.M. Crudden, K. McEleney, S.L. MacQuarrie, A. Blanc, M. Sateesh, J.D. Webb, Pure Applied Chem. *2007, 79* (2), 247-60; developments and applications of supported liquid-phase catalysts, F. Zhao, S. Fujita, M. Arai, Curr. Org. Chem. *2006, 10* (13), 1681-95; recent progress in the **immobilization** of catalysts for selective oxidation in the liquid phase, L. Alaerts, J. Wahlen, P.A. Jacobs, D.E. De Vos, Chem. Commun. *2008* (15), 1727-37; **ultra-low catalyst loading** as a concept in economical and sustainable modern chemistry – the contribution of ferrocenylpolyphosphine ligands, J.-C. Hierso, M. Beaupérin, P. Meunier, Eur. J. Inorg. Chem. *2007* (24), 3767-80; efficient **catalyst reuse** by simple dissolution in non-conventional media, C.A.M. Afonso, L.C. Branco, N.R. Candeias, P.M.P. Gois, N.M.T. Lourenco, N.M.M. Mateus, J.O.N. Rosa, Chem. Commun. *2007* (26), 2669-79.

2c **Activation of Carbon-Hydrogen Bonds ...**

Reactions of C-H bonds in water, C.I. Herrerias, X. Yao, Z. Li, C.-J. Li, Chem. Rev. *2007, 107* (6) 2546-62; direct sp^3 C-H bond activation adjacent to nitrogen in heterocycles, K.R. Campos, Chem. Soc. Rev. *2007, 36* (7), 1069-84; catalytic coupling of sp^2 and sp hybridized C-H bonds with vinylmetalloid compounds, B. Marciniec, Acc. Chem. Res. *2007, 40* (10), 943-52; metal-organic cooperative catalysis in C-H and C-C bond activation and its concurrent recovery, Y.J. Park, J.-W. Park, C.-H. Jun, ibid. *2008, 41* (2), 222-34; activation and functionalization of benzylic derivs. by palladium catalysis, B. Liégault, J.-L. Renaud, C. Bruneau, Chem. Soc. Rev. *2008, 37* (2), 290-9; **benzylic activation** and stereochemical control in reactions of tricarbonyl-(arene)chromium complexes, M. Uemura, Org. Reactions *67,* 217-658.

3 **Lewis and Brønsted Acid/Base Catalysis, solid acids ...**

Recent developments of Lewis acid catalysis: **lanthanide catalysts** with long perfluoro-chains, Y. Ma, L. Wang, J. Shao, H. Tian, Curr. Org. Chem. *2007, 11* (6), 559-76; **co-operative effects** of Lewis acids with transition metals for organic synthesis, C. Wang, Z. Xi, Chem. Commun. *2008* (4), 425-37; s.a. Chem. Soc. Rev. *2007, 36* (9), 1395-406; **Lewis base catalysis** in synthesis, S.E. Denmark, G.L. Beutner, Angew. Chem. Int. Ed. Engl. *2008, 47* (9), 1560-638; **stronger Brønsted acids**, T. Akiyama, Chem. Rev. *2007, 107* (12), 5744-58; **super-Brønsted acids**, H. Yamamoto, M.B. Boxer, Chimia *2007, 61* (5), 279-81; a link between reactivity and local structure in acid catalysis on **zeolites**, A. Bhan, E. Iglesia, Acc. Chem. Res. *2008, 41* (4), 559-67.

4 **N-Heterocyclic Carbenes as Organocatalysts and Ligands**

(for asymmetric organocatalysis s. Section 1 and for ruthenium, copper(I) and silver(I) N-heterocyclic carbene complexes s. Section 2)

Organocatalysis by N-heterocyclic carbenes, D. Enders, O. Niemeier, A. Henseler, Chem. Rev. *2007, 107,* (12), 5606-55; s.a. N. Marion, S. Diéz-González, S.P. Nolan, Angew. Chem. Int. Ed. Engl. *2007, 46* (17), 2988-3000; anionic tethered N-heterocyclic carbene chemistry, S.T. Liddle, I.S. Edworthy, P.L. Arnold, Chem. Soc. Rev. *2007, 36* (11), 1732-44; heterocyclic carbenes: synthesis and coordination chemistry, F.E. Hahn, M.C. Jahnke, Angew. Chem. Int. Ed. Engl. *2008, 47* (17), 3122-72; abnormal N-heterocyclic carbenes, P.L. Arnold, S. Pearson, Coord. Chem. Rev. *2007, 251* (5-6), 596-609; supported N-heterocyclic carbene complexes in catalysis, W.J. Sommer, M. Weck, ibid. 860-73; metal-mediated asymmetric alkylation using chiral N-heterocyclic carbenes derived from chiral amines, R.E. Douthwaite, ibid. 702-17.

5 **Biocatalysis**

(for enzymatic synthesis of carbohydrates and glycoproteins s. Sections 10 and 11, respectively)

New biocatalysts mimicking oxidative hemoproteins – Hemoabzymes, R. Ricoux, Q. Raffy, J.P. Mahy, C. R. Chim. *2007, 10* (8), 684-702; biocatalysis applied to the synthesis of pheromones, J. Aleu, A.J. Bustillo, R. Hernández-Galán, I.G. Collado, Curr. Org. Chem. *2007, 11* (8), 693-705; **microbial arene oxidations,** R.A. Johnson, Org. Reactions *63*, 117-264; zwitterionic glycosidase inhibitors – Salacinol and related analogs, S. Mohan, B.M. Pinto, Carbohydr. Res. *2007, 342* (12-13), 1551-80; vegetables as chemical reagents, G.A. Cordell, T.L.G. Lemos, F.J.Q. Monte, M.C. de Mattos, J. Nat. Prod. *2007, 70* (3), 478-92.

6 **Heterocyclic Chemistry – general, O-, S-, N-heterocyclics, condensed heterocyclics**

(for heterocyclics by ring-closing metathesis, polymer-based cycloaddition, multi-component synthesis, and domino synthesis s. Sections 2a, 19, 24 and 25, respectively, and via N-vinylic phosphazenes s. Section 14; for indole synthesis s.a. under Section 9; for synthesis of fluorinated heterocyclics s. Section 15; for 1,2,3-triazole ['click'] synthesis s. Section 19; for pyridine ring by cycloaddition s. Section 19; and for chromenes, chromans and coumarins by tandem routes s. Section 25)

Synthesis of **heterocyclics by copper-facilitated addition of heteroatoms** to alkenes, alkynes and arenes, S.R. Chemler, P.H. Fuller, Chem. Soc. Rev. *2007, 36* (7), 1153-60**;** **by carbenoid transfer** reactions, V.F. Ferreira, Curr. Org. Chem. *2007, 11* (2), 177-93; convenient and general **microwave-assisted protocols** for the expedient synthesis of heterocyclics, W.D. Shipe, F. Yang, Z. Zhao, S.E. Wolkenberg, M.B. Nolt, C.W. Lindsley, Heterocycles *2006, 70,* 655-89; microwave-assisted heterocyclic chemistry, E. Suna, I. Mutule, Topics Curr. Chem. *2006, 266,* 49-101; utilities of some **carbon nucleophiles in heterocyclic synthesis**, H.M.F. Madkour, A.S.H. Elgazwy, Curr. Org. Chem. *2007, 11* (10), 853-908; flexible elaboration of **halogenoheterocyclics:** metalated species as key intermediates, M. Schlosser, Synlett *2007* (20), 3096-102; regioselective (site-selective) functionalization of unsaturated halogeno-N-, -O- and -S-heterocyclics by Pd-catalyzed cross-coupling and direct arylation, I.J.S. Fairlamb, Chem. Soc. Rev. *2007, 36* (7), 1036-45; advances in the synthesis and biological perspectives of **benz-annulated medium-ring heterocyclics**, T.P. Majhi, B. Achari, P. Chattopadhyay, Heterocycles *2007, 71* (5), 1011-52; one-pot synthesis of heterocyclics initiated by chemoselective addition to β-acyl substituted unsaturated aldehydes with nucleophilic tin complexes, I. Shibata, H. Kato, M. Yasuda, A. Baba, J. Organomet. Chem. *2007, 692* (1-3), 604-19; synthesis of **6-membered O-heterocyclics** through C-O bond formation, I. Larrosa, P. Romea, F. Urpi, Tetrahedron *2008, 64* (12), 2683-723; regiodivergent **epoxide ring opening**: a concept in stereoselective catalysis beyond classical kinetic resolutions and desymmetrizations, A. Gansauer, C.-A. Fan, F. Keller, P. Karbaum, Chem. Eur. J. *2007, 13* (29), 8084-90; rather exotic types of cyclic peroxides: **heteroatom dioxiranes**, N. Sawwan, A. Greer, Chem. Rev. *2007, 107* (7), 3247-85; recent advances in the chemistry of **2-(2-oxoalkylidene)tetrahydrofurans**, E. Bellur, H. Feist, P. Langer, Tetrahedron *2007, 63* (45), 10865-88; synthesis and reactions of **2-alkylidene-thiiranes and -thietanes**, S.K. Richardson, A.R. Howell, Synthesis *2007* (18), 2755-78; **thiophene 1,1-dioxides** as unique building blocks in modern synthesis and materials chemistry, A.M. Moiseev, E.S. Balenkova, V.G. Nenajdenko, Russ. Chem. Rev. *2006, 75* (12), 1015-48; **thiafulvenes and thiafulvalenes** in organic chemistry – Synthesis and chemical reactions, S.E. Zayed, M.E.D. Hassan, R. Ragab, Phosphorus Sulfur Silicon Relat. Elem. *2007, 182* (9), 1945-2007; **1,2,3,4,5-pentathiepins and 1,2,3,4,5-pentathie-panes**, L.S. Konstantinova, S.A. Amelichev, O.A. Rakitin, Russ. Chem. Rev. *2007, 76* (3), 195-211; new **Te-heterocyclics**, T. Chivers, J. Konu, J.S. Ritch, M.C. Copsey, D.J. Eisler, H.M. Tuononen, J. Organomet. Chem. *2007, 692* (13), 2658-68; synthesis and reactivity of C-heteroatom-subst. **aziridines**, G.S. Singh, M. D'Hooghe, N. de Kimpe,

Chem. Rev. *2007, 107* (5), 2080-135; synthesis of **2-azetidinones** by the Kinugasa reaction, R. Pal, S.C. Ghosh, K. Chandra, A. Basak, Synlett *2007* (15), 2321-30; 2-azetidinones as versatile building blocks for the stereoselective synthesis of non-2-azetidinone products, B. Alcaide, P. Almendros, C. Aragoncillo, Chem. Rev. *2007, 107* (11), 4437-92; synthesis of highly functionalized **pyrroles**: a challenge in regioselectivity and chemical reactivity, C. Schmuk, D. Rupprecht, Synthesis *2007* (20), 3095-110; advances in the chemistry of **dipyrrins** and their complexes, T.E. Wood, A. Thompson, Chem. Rev. *2007, 107* (5), 1831-61; intramolecular aminopalladation of alkenes as the key step to **pyrrolidines** and related heterocyclics, A. Minatti, K. Muniz, Chem. Soc. Rev. *2007, 36* (7), 1142-52; synthesis and biological activity of vicinal diaryl-subst. **imidazoles**, F. Bellina, S. Cauteruccio, R. Rossi, Tetrahedron *2007, 63* (22), 4571-624; **pyridine elaboration** through organometallic intermediates: regiochemical control and completeness, M. Schlosser, F. Mongin, Chem. Soc. Rev. *2007, 36* (7), 1161-72; the Bohlmann-Rahtz pyridine synthesis: from discovery to applications, M.C. Bagley, C. Glover, E.A. Merritt, Synlett *2007* (16), 2459-82; the chemistry of 3-cyanopyridine-2(1*H*)-chalcogenones, V.P. Litvinov, Russ. Chem. Rev. *2006, 75* (7), 577-99; **diketopiperazines**: biological activity and synthesis, M.B. Martins, I. Carvalho, Tetrahedron *2007, 63* (40), 9923-32; the **Biginelli dihydropyrimidine synthesis,** C.O. Kappe, A. Stadler, Org. Reactions *63,* 1-116; use of polyfunctionalized **pyridazines** as reactive species for building chemical diversity, J.J. Bourguignon, S. Oumouch, M. Schmitt, Curr. Org. Chem. *2006, 10* (3), 277-95; functionalization of **diazines and benzo derivs.** through deprotonated intermediates, F. Chevallier, F. Mongin, Chem. Soc. Rev. *2008, 37* (3), 595-609; the diverse chemistry of **5(4*H*)-oxazolones**, J.S. Fisk, R.A. Mosey, J.J. Tepe, Chem. Soc. Rev. *2007, 36* (9), 1432-40; **2-oxazolidone ring**, a popular framework in synthetic organic chemistry: Part 1. The construction of the 2-oxazolidone ring, G. Zappia, E. Gacs-Baitz, G. Delle Monache, D. Misiti, L. Nevola, B. Botta, Curr. Org. Synth. *2007, 4* (1), 81-135; synthesis and functionalization of **indoles** through rhodium catalyzed reactions, S. Patil, R. Patil, ibid. (2), 201-22; the Witkop-Winterfeldt oxidation of **indoles**, M. Mentel, R. Breinbauer, Curr. Org. Chem. *2007, 11* (2), 159-76; synthesis and reactivity of 4-, 5- and 6-**azaindoles**, F. Popowycz, J.-Y. Merour, B. Joseph, Tetrahedron *2007, 63* (36), 8689-707; organometallic methods for the synthesis and functionalization of azaindoles, J.J. Song, J.T. Reeves, F. Gallou, Z. Tan, N.K. Yee, C.H. Senanayake, Chem. Soc. Rev. *2007, 36* (7), 1120-32; synthesis of **bridged and non-bridged N-heterocyclics** by cyclocondensation of bis(silyl enol ethers), P. Langer, Eur. J. Org. Chem. *2007* (14), 2233-8; new catalytic reactions of **oxa- and aza-bicyclic alkenes**, D.K. Rayabarapu, C.-H. Cheng, Acc. Chem. Res. *2007, 40* (10), 971-83; **pyrrolo[2,1-*b*]thiazoles**, A.V. Tverdokhlebov, Heterocycles *2007, 71* (4), 761-98; the chemistry of **indoloindoles,** S.A. Samsoniya, M.V. Trapaidze, Russ. Chem. Rev. *2007, 76* (4), 313-26; pyrazolo[4,3-*e*][1,2,4]triazolo[1,5-*c*]pyrimidine template: organic and medicinal chemistry approach, P.G. Baraldi, M.A. Tabrizi, R. Romagnoli, H. El-Kashef, D. Preti, A. Bovero, F. Fruttarolo, M. Gordaliza, P.A. Borea, Curr. Org. Chem. *2006, 10* (3), 259-75; syntheses of isotopically **labelled xanthines**, K.M. Cable, J. Label. Compd. Radiopharm. *2007, 50* (5-6), 509-10; modern aspects of the chemistry of **protoporphyrin IX**, V.Y. Pavlov, Russ. J. Org. Chem. *2007, 43* (1), 1-34; recent advances on the synthesis and chemistry of **carbaporphyrins** and related porphyrinoid systems, T.D. Lash, Eur. J. Org. Chem. *2007* (33), 5461-81; **porphycenes**: synthesis and derivatives, D. Sánchez-García, J.L. Sessler, Chem. Soc. Rev. *2008, 37* (1), 215-32; **corrole**-based applications, I. Aviv, Z. Gross, Chem. Commun. *2007* (20), 1987-99; the chemistry of **calixpyrroles**, V.K. Jain, H.C. Mandalia, Heterocycles *2007, 71* (6), 1261-314; N-confused calix[4]pyrroles, P. Anzenbacher, R. Nishiyabu, M.A. Palacios, Coord. Chem. Rev. *2006, 250* (23-24), 2929-38; bifunctional octasubst. **phthalocyanines** – Synthesis and properties, G.P. Shaposhnikov, V.E. Maizlish, V.P. Kulinich, Russ. J. Gen. Chem. *2007, 77* (1), 138-46; functionalization of phthalocyanines and subphthalocyanines by transition metal-catalyzed reactions, M. Martínez-Díaz, Q. Victoria, M. Quintiliani, T. Torres, Synlett *2008* (1), 1-20.

7 Natural Product Synthesis – general, alkaloids, terpenes, bioactive molecules, drug design

(for natural product synthesis by ring-closing metathesis s. Section 2a; and for glycopeptide antibiotics s. Section 11)

Natural products: synthetic aspects, J.C. Morris, G.M. Nicholas, A.J.M. Phillips, Nat. Prod. Rep. *2007, 24* (1), 87-108; microwave-assisted natural product synthesis, P. Appukkuttan, E. Van der Eycken, Topics Curr. Chem. *2006, 266,* 1-47; photocycloaddition in natural product synthesis, J. Iriondo-Alberdi, M.F. Greaney, Eur. J. Org. Chem. *2007* (29), 4801-15; chemical versatility of natural product assembly lines, C.T. Walsh, Acc. Chem. Res. *2008, 41* (1), 4-10; total synthesis approaches to natural product derivatives based on the combination of chemical synthesis and metabolic engineering, A. Kirschning, F. Taft, T. Knobloch, Org. Biomol. Chem. *2007, 5* (20), 3245-59; recent progress on the total synthesis of natural products in China, Nat. Prod. Rep. *2006, 23* (5), 772-88; Meldrum's acid and related compds. in the synthesis of natural products and analogs, A.S. Ivanov, Chem. Soc. Rev. *2008, 37* (4), 789-811; cascading pericyclic reactions – Building complex carbon frameworks for natural product synthesis, S. Arns, L. Barriault, Chem. Commun. *2007* (22), 2211-21; recent progress in the synthesis of bioactive polycyclic natural products, M. Miyashita, Pure Applied Chem. *2007, 79* (4), 651-65; synthesis of natural products and analogs using Pd-catalyzed transformations, L.F. Tietze, T. Kinzel, ibid. 629-50; the biology and chemistry of the zoanthamine **alkaloids**, D.C. Behenna, J.L. Stockdill, B.M. Stoltz, Angew. Chem. Int. Ed. Engl. *2008, 47* (13), 2365-86; synthesis of subst. 4-azaisoindoles – New tacrine analogs, H. Bekolo, G. Kirsch, Can. J. Chem. *2007, 85* (1), 1-6; β-phenylethylamines and the isoquinoline alkaloids, Nat. Prod. Rep. *2006, 23* (3), 444-63; imidazole, oxazole and thiazole alkaloids, Z. Jin, ibid. 464-96; total synthesis of complex cyclotryptamine alkaloids: stereocontrolled construction of quaternary carbon stereocenters, Angew. Chem. Int. Ed. Engl. *2007, 46* (29), 5488-508; new strategies for the synthesis of hexahydropyrroloindole alkaloids inspired by biosynthetic hypotheses, M.A. Schmidt, M. Movassaghi, Synlett *2008* (3), 313-24; quinoline, quinazoline and acridone alkaloids, J.P. Michael, Nat. Prod. Rep. *2007, 24* (1), 223-46; indolizidine and quinolizidine alkaloids, ibid. 191-222; some recent advances in the synthesis of polycyclic imidazole-containing **marine natural products**, S.M. Weinreb, ibid. (5), 931-48; synthesis of furopyranylpyrimidine natural products, J.D. More, Org. Prep. Proced. Int. *2007, 39* (2), 107-34; synthetic studies directed towards the histrionicotoxin family of alkaloids, A. Sinclair, R.A. Stockman, Nat. Prod. Rep. *2007, 24* (2), 298-326; recent advances in the synthesis of nicotine and its derivs., F.F. Wagner, D.L. Comins, Tetrahedron *2007, 63* (34), 8065-82; pyrrolidinyl-spirooxindole natural products as inspirations for the development of potential therapeutic agents, C.V. Galliford, K.A. Scheidt, Angew. Chem. Int. Ed. Engl. *2007, 46* (46), 8748-58; the pursuit of palau'amine, M. Koeck, A. Grube, I.B. Seiple, P.S. Baran, ibid. *46* (35), 6586-94; aza-annulation strategies in alkaloid total synthesis, A.K. Lawrence, K. Gademann, Synthesis *2008* (3), 331-51; dipyrrinones – constituents of the pigments of life, S.E. Boiadjiev, D.A. Lightner, Org. Prep. Proced. Int. *2006, 38* (4), 347-99; recent advances in the stereoselective synthesis of *trans*-3,4-disubst. piperidines: application to (-)-paroxetine, C. De Risi, G. Fanton, G.P. Pollini, C. Trapella, F. Valente, V. Zanirato, Tetrahedron:Asym. *2008, 19* (2), 131-55; stereoselective synthesis of cytoxazone and its analogs, A. Grajewska, M.D. Rozwadowska, ibid. *2007, 18* (7), 803-13; strategies to bypass the **taxol** problem: enantioselective cascade catalysis, a new approach for the efficient construction of molecular complexity, A.M. Walji, D.W.C. MacMillan, Synlett *2007* (10), 1477-89; paclitaxel from primary taxanes – A perspective on creative invention in organozirconium chemistry, B. Ganem, R.R. Franke, J. Org. Chem. *2007, 72* (11), 3981-7; synthesis and biological activity of **12-aza-epothilones** (azathilones): non-natural natural products with potent anti-proliferative activity, F. Feyen, J. Gertsch, M. Wartmann, K.H. Altmann, Chimia *2007, 61* (4), 143-6; fifteen years of biological and synthetic studies of decarestrictine family, V.B. Riatto, R.A. Pilli, M.M. Victor, Tetrahedron *2008, 64* (10), 2279-300; **vitamin B_{12}** and α-ribonucleosides, T. Chandra, K.L. Brown, ibid. *64* (1),

9-38; chemistry and biology of the aeruginosin family of serine protease inhibitors, K. Ersmark, J.R. Del Valle, S. Hanessian, Angew. Chem. Int. Ed. Engl. *2008, 47* (7), 1202-23; natural **sesquiterpenoids**, B.M. Fraga, Nat. Prod. Rep. *2006, 23* (6), 943-72; triterpenoids, J.D. Connolly, R.A. Hill, ibid. *2007, 24* (2), 465-86; total syntheses and synthetic studies of spongiane **diterpenes**, Tetrahedron *2008, 64* (3), 445-67; recent developments in the synthesis of **prostaglandins** and analogs, S. Das, S. Chandrasekhar, J.S. Yadav, R. Gree, Chem. Rev. *2007, 107* (7), 3286-337; **(+)-discodermolide**: total synthesis, construction of novel analogs, and biological evaluation, A.B. Smith III, B.S. Freeze, Tetrahedron *2008, 64* (2), 261-98; discovery and combinatorial synthesis of **fungal metabolites** beauveriolides, novel antiatherosclerotic agents, H. Tomoda, T. Doi, Acc. Chem. Res. *2008, 41* (1), 32-9; synthesis of biologically active guaianolides with a *trans*-annulated lactone moiety, A. Schall, O. Reiser, Eur. J. Org. Chem. *2008* (14), 2353-64; chemistry of epoxyquinols A, B, and C, and epoxytwinol A, M. Shoji, Y. Hayashi, Eur. J. Org. Chem. *2007* (23), 3783-800; advances in the synthesis of aryltetralin **lignan lactones**, J.D. Sellars, P.G. Steel, ibid. 3815-28; the conquest of vinigrol – Creativity, frustrations, and hope, G. Tessier, L. Barriault, Org. Prep. Proced. Int. *2007, 39* (4), 311-53; organocatalyzed **synthesis of drugs** and bioactive natural products, R.M. de Figueiredo, Eur. J. Org. Chem. *2007* (16), 2575-600; applications of total synthesis toward the discovery of clinically useful anticancer agents, R.M. Wilson, S.J. Danishefsky, Chem. Soc. Rev. *2007, 36* (8), 1207-26; unsym. ureas: synthetic methodologies and applications in drug design, Org. Prep. Proced. Int. *2007, 39* (4), 355-83; function-oriented synthesis, step economy, and drug design, P.A. Wender, V.A. Verma, T.J. Paxton, T.H. Pillow, Acc. Chem. Res. *2008, 41* (1), 40-9; development of environmentally benign organometallic catalysis for drug discovery and its application, M. Arisawa, Chem. Pharm. Bull. *2007, 55* (8), 1099-118; synthesis of antimycins, Y.Q. Yang, Y.K. Wu, Org. Prep. Proced. Int. *2007, 39* (2), 135-52; synthesis of platensimycin, K. Tiefenbacher, J. Mulzer, Angew. Chem. Int. Ed. Engl. *2008, 47* (14), 2548-55.

8 Carbocyclic Chemistry

(for multicomponent synthesis of carbocyclics s. Section 24; by ring-closing metathesis s. Section 2a; and from carbohydrates s. Section 10)

Cyclopropenes and cyclopropanes: transition metal chemistry, M. Rubin, M. Rubina, V. Gevorgyan, Chem. Rev. *2007, 107* (7), 3117-79; titanium-mediated synthesis of primary cyclopropylamines from nitriles and Grignard reagents, P. Bertus, J. Szymoniak, Synlett *2007* (9), 1346-56; natural occurrence, syntheses, and applications of cyclopropyl-containing (R)-amino acids. Part 1. 1-Aminocyclopropanecarboxylic acid and other 2,3-methanoamino acids, F. Brackmann, A. De Meijere, Chem. Rev. *2007, 107* (11), 4493-537; natural occurrence, syntheses, and applications of cyclopropyl-containing (R)-amino acids. Part 2. 3,4- And 4,5-methanoamino acids. ibid. 4538-83; metal-mediated synthesis of subst. **cyclooctatetraenes**, C. Wang, Z. Xi, Chem. Commun. *2007* (48), 5119-33; synthetic organic chemistry based on **small-ring compounds**, H. Nemoto, Chem. Pharm. Bull. *2007, 55* (7), 961-74; allylsilanes in the synthesis of **three-to-seven membered rings**: the silylcuprate strategy, A. Barbero, F.J. Pulido, M.C. Sañudo, Beilstein J. Org. Chem. *2007* (*3*), No.16.

9 Aromatic Chemistry – arylation, [de]aromatization, ar. substitution

(for microbial oxidation of arenes s. Section 5)

Recent advances in intermolecular **direct arylation**, L.-C. Campeau, D.R. Stuart, K. Fagnou, Aldrichimica Acta *2007, 40* (2), 35-41; recent developments in catalytic aryl coupling, M. Catellani, E. Motti, N. Della Ca, R. Ferraccioli, Eur. J. Org. Chem. *2007* (25), 4153-65; catalytic direct arylation of heteroaromatic compounds, T. Satoh, M. Miura, Chem. Lett. *2007, 36* (2), 200-5; recent advances in direct arylation via palladium-catalyzed aromatic C-H activation, B.-J. Li, S.-D. Yang, Z.-J. Shi, Synlett *2008* (7), 949-57; catalytic arylation with challenging substrates: from air-stable HASPO preligands to indole syntheses and C-H-bond functionalizations, L. Ackermann, ibid. *2007* (4), 507-26; palladium-catalyzed arylation for the synthesis of **polyarenes**, S. Pascual, P. de Mendoza, A.M. Echavarren, Org. Biomol. Chem. *2007, 5* (17), 2727-34; palladium- and copper-catalyzed **arylation in the presence of water**, with a focus on carbon-heteroatom bond formation, M. Carril, R. SanMartin, E. Dominguez, Chem. Soc. Rev. *2008, 37* (4), 639-47; aryl cation chemistry as an emerging tool for **metal-free arylation**, V. Dichiarante, M. Fagnoni, Synlett *2008* (6), 787-800; oxidative **dearomatization** of phenols: why, how and what for?, S. Quideau, L. Pouysegu, D. Deffieux, ibid. (4), 467-95; nucleophilic dearomatizing (D_NAr) reactions of aromatic C,H-systems. A mature paradigm in synthesis, F. López Ortiz, M.J. Iglesias, I. Fernández, I.C.M. Andujar Sánchez, G. Ruiz Gómez, Chem. Rev. *2007, 107* (5), 1580-691; selective hydrogenation of aromatic chloronitro compds., X. Wang, M. Liang, J. Zhang, Y. Wang, Curr. Org. Chem. *2007, 11* (3), 299-314; recent advances in the **benzannulation** of substituted 3-alkoxycarbonyl-3,5-hexadienoic acids and 3-alkoxycarbonylhex-3-en-5-ynoic acids to polysubstituted aromatics, S. Serra, C. Fuganti, E. Brenna, Chem. Eur. J. *2007, 13* (24), 6782-91; oxidative **amidation of phenols** through the use of hypervalent iodine reagents: developments and applications, M.A. Ciufolini, N.A. Braun, S. Canesi, M. Ousmer, J. Chang, D. Chai, Synthesis *2007* (24), 3759-72; *o*-aminomethyl derivatives of phenols. Part 1. **Benzylamines**: properties, structure, synthesis and purification, K. Bujnowski, A. Adamczyk, L. Synoradzki, Org. Prep. Proced. Int. *2007, 39* (2), 153-84.

10 Carbohydrate Chemistry, Oligosaccharides

(for carbohydrate ligands for asymmetric catalysis s. Section 1; glycopeptides s. Section 11; radical modification of carbohydrates s. Section 17; and use of protective groups in oligosaccharide synthesis s. Section 26)

Recent developments in preparative **enzymatic synthesis of carbohydrates**, A.S. Rowan, C.J. Hamilton, Nat. Prod. Rep. *2006, 23* (3), 412-43; going to extremes: 'super'-armed glycosyl donors in **glycosylation**, H.H. Jensen, C.M. Pedersen, M. Bols, Chem. Eur. J. *2007, 13* (27), 7576-82; glycosylation **on polymer supports**, S. Bufali, P.H. Seeberger, Org. Reactions *68,* 303-522; glycosylation **with sulfoxides and sulfinates** as donors or promoters, D. Crich, L.B.L. Lim, ibid. *64,* 115-252; transformation of **glycals into 2,3-unsatd. glycosyl derivs.**, R.J. Ferrier, O.A. Zubkov, ibid. *62,* 569-736; recent trends in the synthesis of O-glycosides of 2-amino-2-deoxysugars, A.F.G. Bongat, A.V. Demchenko, Carbohydr. Res. *2007, 342* (3-4), 374-406; recent advances in the glycosylation of sphingosines and ceramides, J.A. Morales-Serna, O. Boutureira, Y. Diaz, M.I. Matheu, S. Castillón, ibid. *342* (12-13), 1595-612; carbohydrate chemistry in the total synthesis of **saponins**, B. Yu, Y. Zhang, P. Tang, Eur. J. Org. Chem. *2007* (31), 5145-61; vinyl sulfone-modified carbohydrates: an inconspicuous group of chiral building blocks, T. Pathak, Tetrahedron *2008, 64* (17), 3605-28; designer **aminoglycosides**: the race to develop improved antibiotics and compds. for the treatment of human genetic diseases, M. Hainrichson, I. Nudelman. T. Baasov, Org. Biomol. Chem. *2008, 6* (2), 227-39; **glycomimetics** and scaffolds from carbohydrate building blocks, P.V. Murphy, Eur. J. Org. Chem. *2007* (25), 4177-87; **sialic acid** and structural analogs: stereoselective syntheses, I. Hemeon, A.J. Bennet, Synthesis *2007* (13), 1899-926; exocyclic olefinic carbohydrate derivs. to functionalized carbocyclic compds., J. Zhou, G. Wang, L.-H. Zhang, X.-S. Ye, Curr. Org. Chem. *2006, 10* (6), 625-42; synthesis and conformational and biological aspects of **carbasugars**, O. Arjona, A.M. Gomez, J.C. Lopez, J. Plumet, Chem. Rev. *2007, 107* (5), 1919-2036; nucleophilic substitution reactions of **2-phenylthio-subst. carbohydrate acetals** and related systems: episulfonium ions *vs*. oxocarbenium ions as reactive intermediates, M.G. Beaver, S.B. Billings, K.A. Woerpel, Eur. J. Org. Chem. *2008* (5), 771-81; recent studies on reaction pathways and applications of sugar orthoesters in synthesis of **oligosaccharides**, F.Z. Kong, Carbohydr. Res. *2007, 342* (3-4), 345-73; automated oligosaccharide synthesis, P.H. Seeberger, Chem. Soc. Rev. *2008, 37* (1), 19-28; oligosaccharide assembly by one-pot multi-step strategy, Y.H. Wang, X.S. Ye, L.H. Zhang, Org. Biomol. Chem.

2007, 5 (14), 2189-200; uronic acids in oligosaccharide synthesis, L.J. van den Bos, J.D.C. Codee, R.E.J.N. Litjens, Eur. J. Org. Chem. *2007* (24), 3963-76; non-conventional glycosidic linkages: syntheses and structures of **thiooligosaccharides** and carbohydrates with three-bond glycosidic connections, L. Szilagyi, O. Varela, Curr. Org. Chem. *2006, 10* (14), 1745-70; *in vitro* synthesis of **artificial polysaccharides** by glycosidases and glycosynthases, M. Faijes, A. Planas, Carbohydr. Res. *2007, 342* (12-13), 1581-94.

11 Amino Acid Chemistry, Peptides, Glycopeptides, Proteins

(for cyclopropyl-containing amino acids s. Section 8; radical modification of amino acids s. Section 17)

Catalytic asymmetric synthesis of **(R)-amino acids**, C. Nájera, J.M. Sanano, Chem. Rev. *2007, 107* (11), 4584-671; design and synthesis of **chiral α,α-disubst. amino acids** and conformational study of their oligopeptides, M. Tanaka, Chem. Pharm. Bull. *2007, 55* (3), 349-58; two highly effective phase-transfer catalysts for the enantioselective synthesis of α-amino acid derivs., B. Lygo, D.J. Beaumont, Chimia *2007, 61* (5), 257-62; recent advances in the synthesis of **unnatural (R)-amino acids**, A. Perdih, M.S. Dolenc, Curr. Org. Chem. *2007, 11* (9), 801-32; synthesis of **bicyclic peptides,** T. Karskela, P. Virta, H. Lonnberg, Curr. Org. Synth. *2006, 3* (3), 283-311; synthetic approaches for total synthesis of **proteins and protein-like macromolecules** of branched architecture, S. Papas, C. Strongylis, V. Tsikaris, *2006, 10* (14), 1727-44; histidinoalanine: a cross-linking amino acid, C.M. Taylor, W. Wang, Tetrahedron *2007, 63* (37), 9033-47; **phosphopeptides**: synthesis, analysis, outlook and limitations, G.K. Toth, Z. Kele, G. Varadi, Curr. Org. Chem. *2007, 11* (5), 409-26; synthesis of **peptidomimetics** based on pyranose and polyhydroxylated piperidine scaffolds, Curr. Org. Synth. *2006, 3* (4), 403-37; synthesis of **glycopeptides**, C. Haase, O. Seitz, Topics Curr. Chem. *2007, 267,* 1-36; a compendium of sugar amino acids: scaffolds, peptide- and glyco-mimetics, M.D.P. Risseeuw, M. Overhand, G.W.J. Fleet, M.I. Simone, Tetrahedron:Asym. *2007, 18* (17), 2001-10; enzymatic synthesis of glycopeptides and **glycoproteins**, D.A. Thayer, C.-H. Wong, Topics Curr. Chem. *2007, 267,* 37-63; chemo-enzymatic synthesis of glycoproteins, C.S. Bennett, C.-H. Wong, Chem. Soc. Rev. *2007, 36* (8), 1227-38; sugar-assisted ligation for the synthesis of glycopeptides, A. Brik, C.-H. Wong, Chem. Eur. J. *2007, 13* (20), 5670-5; synthesis and application of glycopeptide and glycoprotein mimetics, D. Specker, V. Wittmann, Topics Curr. Chem. *2007, 267,* 65-107; synopsis of structural, biosynthetic and chemical aspects of **glycopeptide antibiotics**, F. Wolter, S. Schoof, R.D. Suessmuth, Topics Curr. Chem. *2007, 267,* 143-85; use of solid-phase synthesis techniques for the preparation of **peptide-metal complex conjugates**, G. Dirscherl, B. Konig, Eur. J. Org. Chem. *2008* (4), 597-634.

12 Nucleosides, Nucleotides, Oligonucleotides

Recent approaches in the synthesis of conformationally restricted **nucleoside analogs**, C. Mathé, C. Périgaud, Eur. J. Org. Chem. *2008* (9), 1489-505; synthesis and properties of psico-nucleosides, S.K. Richardson, A.R. Howell, R. Taboada, Org. Prep. Proced. Int. *2006, 38* (2), 101-76; synthesis of 2′,3′-didehydro-2′,3′-dideoxynucleosides via nucleoside route, C. Len, D. Postel, Curr. Org. Synth. *2006, 3* (3), 261-81; nucleoside and **oligonucleoside boranophosphates:** chemistry and properties, P. Li, Z.A. Sergueeva, M. Dobrikov, B.R. Shaw, Chem. Rev. *2007, 107* (11), 4746-96; zinc(II) complexes as hydrolytic catalysts of phosphate diester cleavage: from model substrates to nucleic acids, New J. Chem. *2007, 31* (6), 800-17; recent highlights in **modified oligonucleotide chemistry**, A.J.A. Cobb, Org. Biomol. Chem. *2007, 5* (20), 3260-75; cycloaddition as a method for **oligonucleotide conjugation**, D. Graham, A. Enright, Curr. Org. Synth. *2006, 3* (1), 9-17; nucleoside-, nucleotide- and oligonucleotide-based amphiphiles: successful marriage of **nucleic acids with lipids**, A. Gissot, M. Camplo, M.W. Grinstaff, P. Barthélémy, ibid. *2008, 6* (8), 1324-33.

13 Steroid Chemistry

Enantioselective synthesis of steroids, A.-S. Chapelon, D. Moraleda, R. Rodriguez, C. Ollivier, M. Santelli, Tetrahedron *2007, 63* (47), 11511-616; homologation of estrone or estradiol: appending a carbon substituent into the 3-hydroxyestra-1,3,5-triene skeleton, S. Braune, S. Deb, T. Hase, K. Wahala, Curr. Org. Chem. *2007, 11* (4), 383-405; synthesis of modified steroids in the androstane and androstene series, E. Stephan, Org. Prep. Proced. Int. *2006, 38* (3), 217-305.

14 Phosphorus Chemistry

(for chiral phosphines as ligands for asymmetric catalysis s. Section 1; for phosphines as ligands s.a. under Section 2)

Synthesis, structure and properties of **crowded triarylphosphines**, S. Sasaki, M. Yoshifuji, Curr. Org. Chem. *2007, 11* (1), 17-31; new concepts in multidentate ligand chemistry: effects of multidentarity on catalytic and spectroscopic properties of **ferrocenylpolyphosphines**, J.-C. Hierso, R. Smaliy, R. Amardeil, P. Meunier, Chem. Soc. Rev. *2007, 36* (11), 1754-69; synthesis of tertiary and secondary phosphines and their applications, M.T. Honaker, J.M. Hovland, R.N. Salvatore, Curr. Org. Synth. *2007, 4* (1), 31-45; **P-chiral phosphorus**, R. Engel, J.I. Rizzo, Curr. Org. Chem. *2006, 10* (18), 2393-405; metal-catalyzed asymmetric synthesis of P-chiral phosphines, D.S. Glueck, Synlett *2007* (17), 2627-34; of P-chiral phosphorus(III) compds., A. Grabulosa, J. Granell, G. Muller, Coord. Chem. Rev. *2007, 251* (1-2), 25-90; synthesis of **phosphorus ligands** from zirconium reagents: a useful approach, M. Zablocka, J.-P. Majoral, Curr. Org. Chem. *2007, 11* (1), 49-60; **alkyl phosphinites**: versatile synthetic intermediates for dehydration condensation reactions, T. Mukaiyama, H. Yamabe, Chem. Lett. *2007, 36* (1), 2-7; the chemistry of **phosphinic and phosphinous acid derivs.** containing *t*-butyl group as a single bulky substituent: synthetic, mechanistic and stereochemical aspects, J. Drabowicz, W. Kudelska, A. Lopusinski, A. Zajac, Curr. Org. Chem. *2007, 11* (1), 3-15; syntheses and properties of **phosphinoselenoic chlorides**, acids and their salts, T. Murai, T. Kimura, ibid. *2006, 10* (15), 1963-73; **phosphoallenes** for building organophosphorus derivs., V.K. Brel, Heteroatom. Chem. *2006, 17* (6), 547-56; **chiral P-macrocyclics**, S. Cherenok, J.-P. Dutasta, V. Kalchenko, Curr. Org. Chem. *2006, 10* (18), 2307-31; recent advances in the chemistry of anellated azaphospholes, R.K. Bansal, N. Gupta, S.K. Kumawat, ibid. *2007, 11* (1), 33-47; synthesis of arylenephosphamacrocyclics using tri- and penta-valent phosphorus compds., E.E. Nifantyev, P.V. Slitikov, E.N. Rasadkina, Russ. Chem. Rev. *2007, 76,* (4), 327-38; N-vinylic **phosphazenes**: a useful tool for the synthesis of acyclic and heterocyclic compounds, F. Palacios, D. Aparicio, G. Rubiales, C. Alonso, J.M. de los Santos, Curr. Org. Chem. *2006, 10* (18), 2371-92; fragmentations and rearrangements of **(R)-hydroxyiminoalkylphosphonates**, E. Breuer, ibid. 2357-70.

15 Organofluorine and Fluorous Chemistry

(for asymmetric fluorination s. Section 21)

Understanding organofluorine chemistry: an introduction to the C-F bond, D. O'Hagan, Chem. Soc. Rev. *2008, 37* (2), 308-21; fluorine **in medicinal chemistry**, S. Purser, P.R. Moore, S. Swallow, V. Gouverneur, ibid. 320-30; selective **fluoroalkylations** with fluorinated sulfones, sulfoxides and sulfides, K.G.S. Prakash, J. Hu, Acc. Chem. Rev. *2007, 40* (10), 921-30; strategies for nucleophilic, electrophilic and radical **trifluoromethylation**, J.-A. Ma, J. Fluorine Chem. *2007, 128* (9), 975-96; recent advances in the chemistry of α-unsatd. trifluoromethyl ketones, S.V. Druzhinin, E.S. Balenkova, V.G. Nenajdenko, Tetrahedron *2007, 63* (33), 7753-808; the synthesis of **fluorinated heteroaromatics.** Part 2. Five-membered rings with two heteroatoms, A. Pace, S. Buscemi, N. Vivona, Org. Prep. Proced. Int. *2007, 39* (1), 1-70; fluorinated benzazoles and benzazines, E.V. Nosova, N.N. Mochul'skaya, S.K. Kotovskaya, Heteroatom. Chem. *2006, 17* (6), 579-94; β-fluorinated porphyrins and related compounds: an overview, J. Leroy, A. Bondon, Eur. J. Org. Chem. *2008* (3), 417-33; recent advances in practice and theory of **polyfluoroarene hydrodehalogenation**, V.D. Shteingaits, J. Fluorine Chem. *2007, 128* (7), 797-805; microwave-enhanced **high-speed fluorous synthesis**, W. Zhang, Topics Curr. Chem. *2006, 266,* 145-66.

16 Syntheses with Organometallics, Metal Carbenoids, Carbonyl Complexes; Metalation

(for benzylic activation with arene(tricarbonyl)chromium complexes s. Section 2c; natural product synthesis based on organozirconium chemistry s. Section 7)

Organometallic reactions under transition-metal catalysis that have revolutionized organic synthesis, E. Negishi, Bull. Chem. Soc. Jpn. *2007, 80* (2), 233-57; **1-lithio-1,3-dienes** – useful building blocks for cyclic compds., Z.F. Xi, ibid. *2007, 80* (6), 1021-32; **naphthyl-lithium and -magnesium compounds** in organic synthesis, A.F. Pozharskii, O.V. Ryabtsova, Russ. Chem. Rev. *2006, 75* (8), 709-36; **heavy Grignard reagents**: challenges and possibilities of **alkaline-earth arylmetal compds**., M. Westerhausen, M. Gaertner, R. Fischer, J. Langer, L. Yu, M. Reiher, Chem. Eur. J. *2007, 13* (22), 6292-306; transition metal-catalyzed carbon-silicon bond formation using chlorosilanes promoted by Grignard reagents, J. Terao, N. Kambe, Chem. Record *2007, 7* (1), 57-67; asymmetric cross-coupling of **organozinc reagents** with anhydrides: the controlled formation of C-C bonds through the nucleophilic interception of metalacycles, J.B. Johnson, T. Rovis, Acc. Chem. Res. *2008, 41* (2), 327-38; design *versus* discovery in synthetic applications of **organoalanes,** S. Woodward, Russ. Chem. Rev. *2007, 76* (10), 1490-500; trichlorogallium and **trialkylgalliums** in synthesis, M. Yamaguchi, Y. Nishimura, Chem. Commun. *2008* (1), 35-48; deprotonative **metalation using ate compounds** – synergy, synthesis, and structure building, R.E. Mulvey, F. Mongin, M. Uchiyama, Y. Kondo, Angew. Chem. Int. Ed. Engl. *2007, 46* (21), 3802-24; recent advances in the chemistry of **magnesium carbenoids**, T. Satoh, Chem. Soc. Rev. *2007, 36* (10), 1561-72; **titanium carbene complexes** as useful tools in synthesis, T. Takeda, Chem. Record *2007, 7* (1), 24-36; titanium carbenoid reagents for converting carbonyl groups into alkenes, R.C. Hartley, J. Li, C.A. Main, G.J. McKiernan, Tetrahedron *2007, 63* (23), 4825-64; **Fischer carbene complexes**: beautiful playgrounds to study single electron transfer (SET) reactions, M.A. Sierra, M. Gómez-Gallego, R. Martinez-Álvarez, Chem. Eur. J. *2007, 13* (3), 736-44; intermolecular reactions of electron-rich heterocyclics with **copper and rhodium carbenoids**, H.M.L. Davies, S.J. Hedley, Chem. Soc. Rev. *2007, 36* (7), 1109-19; **arene(tricarbonyl)chromium complexes** in synthesis, M. Rosillo, G. Dominguez, J. Perez-Castells, ibid. (10), 1589-604; **addition of organochromium reagents** to carbonyl compounds, K. Takai, Org. Reactions *64,* 353-612; tetrahydronaphthalene-1,4-dione and its tricarbonylchromium complex, Á.E. García, E.P. Kündig, T. Lomberget, R. Bragg, C. Poulard, G. Bernardinelli, Chimia *2007, 61* (4), 169-71.

17 Radical Chemistry, radical cations

(for radical photoaddition of tert. amines s. Section 28)

Radical functionalization catalyzed by N-hydroxyphthalimide, F. Recupero, C. Punta, Chem. Rev. *2007, 107* (9), 3800-42; modification of amino acids, peptides, and carbohydrates through radical chemistry, Topics Curr. Chem. *2006, 264,* 135-62; organosilicon peroxides: radicals and rearrangements, Tetrahedron *2007, 63* (42), 10385-405; the degenerative **radical transfer of xanthates** and related derivatives: an unusually powerful tool for the creation of carbon-carbon bonds, B. Quiclet-Sire, S.Z. Zard, Topics Curr. Chem. *2006, 264,* 201-36; the genesis of the reversible radical addition-fragmentation-transfer of thiocarbonylthio derivs. from the **Barton-McCombie deoxygenation**: a brief account and some mechanistic observations, S.Z. Zard, Austr.

J. Chem. *2006, 59* (10), 663-8; phenylselenoethers as precursors of acyclic free radicals: creating **tertiary and quaternary centers** using free radical-based intermediates, B. Cardinal-David, J.F. Brazeau, I.A. Katsoulis, Curr. Org. Chem. *2006, 10* (15), 1939-61; strategies for nucleophilic, electrophilic and **radical trifluoromethylation**, J.A. Ma, D. Cahard, J. Fluorine Chem. *2007, 128* (9), 975-96; catalysis of tin-mediated radical chain reactions by benzeneselenol, Acc. Chem. Rev. *2007, 40* (6), 453-63; **tin-free radical carbon-carbon bond formation** based on scission of alkylsulfonyl radicals, S. Kim, S. Kim, Bull. Chem. Soc. Jpn. *2007, 80* (5), 809-22; synthesis of azabicyclics using **nitrogen-directed radical rearrangements**, D.M. Hodgson, L.H. Winning, Org. Biomol. Chem. *2007, 5* (19), 3071-82; **alcohols and water as reducing agents** in radical reactions, D. Pozzi, P. Renaud, Chimia *2007, 61* (4), 151-4; generation of **alkene radical cations** by heterolysis of α-subst. radicals: mechanism, stereochemistry, and applications, D. Crich, F. Brebion, D.-H. Suk, Topics Curr. Chem. *2006, 263,* 1-38.

18 Functional Group Chemistry, synthesis, applications, leaving groups, equivalents

(for reactions of ethylene and acetylene derivs. s.a. under Sections 2 and 2a; for radical synthesis of tert. and quaternary centres s. Section 17)

Carboxylic acids as substrates in homogeneous catalysis, L.J. Gooßen, N. Rodríguez, K. Gooßen, Angew. Chem. Int. Ed. Engl. *2008, 47* (17), 3100-20; methods for the synthesis of **C^{13}-labelled carboxylic acids and esters**, A.C. Jordan, L.C. Axiford, J.R. Harding, Y. O'Connell, T.J. Simpson, C.L. Willis, J. Label. Compd. Radiopharm. *2007, 50* (5-6), 338-41; reactions of **mercaptans**, I.V. Koval, Russ. J. Org. Chem. *2007, 43* (3), 319-46; the chemistry of **sulfinyl- and sulfonyl-α,β-ethyleneketones,** G.V. Nenajdenko, A.L. Krasovskiy, E.S. Balenkova, Tetrahedron *2007, 63* (51), 12481-539; recent advances and perspectives in the chemistry of **sulfenic acids**, M.C. Aversa, A. Barattucci, P. Bonaccorsi, P. Giannetto, Curr. Org. Chem. *2007, 11* (12), 1034-52; synthesis and applications of chalcogenoamides: **carboxylic acid thio-, seleno- and telluro-amides**, M. Koketsu, H. Ishihara, Curr. Org. Synth. *2007, 4* (1), 15-29; asymmetric synthesis of **organoselenium compds**., C. Zhu, Y. Huang, Curr. Org. Chem. *2006, 10* (15), 1905-20; condensations of carboxylic acid thioamides with acetylenecarboxylic acid derivs., N.A. Danilkina, L.E. Mikhailov, B.A. Ivin, Russ. J. Org. Chem. *2006, 42* (6), 783-814; divinyl sulfide and its selenium and tellurium analogs as starting materials for the preparation of polyfunctional alkyl, aromatic, and heteroaromatic **vinyl chalcogenides**, G.M. Gavrilova, S.V. Amosova, Heteroatom. Chem. *2006, 17* (6), 491-8; the chemistry of **hydrazonates**, A.S. Shawali, S.M. Sherif, Curr. Org. Chem. *2007, 11* (9), 773-99; **nitrile ylids**: generation, properties and synthetic applications, C. Escolano, M.D. Duque, S. Vázquez, ibid. *11* (9), 741-72; **nitroso and azo compounds** in modern organic synthesis – late blooming but very rich, H. Yamamoto, M. Kawasaki, Bull. Chem. Soc. Jpn. *2007, 80* (4), 595-607; **potassium organotrifluoroborates**: new perspectives in organic synthesis, S. Darses, J.-P. Genet, Chem. Rev. *2008, 108* (1), 288-325; the synthesis and reactions of **betaines** formed in reactions of tertiary phosphines with unsaturated carboxylic acids and their derivatives, V.I. Galkin, Y.V. Bakhtiyarova, R.I. Galkina, R.A. Cherkasov, Heteroatom. Chem. *2006, 17* (6), 557-66; catalytic asymmetric syntheses of **chiral α-branched amines,** S. Bräse, T. Baumann, S. Dahmen, H. Vogt, Chem. Commun. *2007* (19), 1881-90; various approaches for the synthesis of **carbamates**, D. Chaturvedi, N. Mishra, V. Mishra, Curr. Org. Synth. *2007, 4* (3), 308-20; **enacylamines and enurethans** as nucleophiles in stereoselective C–C and C–N bond-forming reactions, R. Matsubara, S. Kobayashi, Acc. Chem. Res. *2008, 41* (2), 292-301; highly efficient one- or two-step sequences for the synthesis of fine chemicals from versatile **nitroalkanes,** R. Ballini, A. Palmieri, P. Righi, Tetrahedron *2007, 63* (49), 12099-121; catalytic asymmetric synthesis of **quaternary hydrocarbon groups**, P.G. Cozzi, R. Hilgraf, N. Zimmermann, Eur. J. Org. Chem. *2007* (36), 5969-94; synthesis of tetrasubst. carbons by self-assembled poly rare earth metal complexes, M. Shibasaki, M. Kanai, Org. Biomol. Chem. *2007, 5* (13), 2027-39; stereocontrolled synthesis of tetrasubst. **ethylene derivs**., A.B. Flynn, W.W. Ogilvie, Chem. Rev. *2007,*

107 (11), 4698-745; regio- and stereo-selective synthesis of tri- and tetra-subst. ethylene derivs. by introduction of CO_2 and alkylzinc reagents into acetylene derivs., M. Mori, Eur. J. Org. Chem. *2007* (30), 4981-93; synthesis, and biological activity of unnatural **enediyynes** and related analogs equipped with pH-dependent or photo-triggering devices, M. Kar, A. Basak, Chem. Rev. *2007, 107* (7), 2861-90; reactions of linear **push-pull enamines** at the β′-position, A.N. Kostyuk, D.M. Volochnyuk, D.A. Dmitriy, Synthesis *2008,* (2), 161-84; Group 14 and 15 **heteroallenes**, E:C:C and E:C:E′, J. Escudie, H. Ranaivonjatova, Organometallics *2007, 26* (7), 1542-59; α-substituted organic peroxides: synthetic strategies for a biologically important class of *gem*-**dihydroperoxide and perketal derivs.,** K. Zmitek, M. Zupan, J. Iskra, Org. Biomol. Chem. *2007, 5* (24), 3895-908; the conversion of prim. or sec. alcohols with nonaflyl fluoride into their corresponding inverted **fluorides**, H. Vorbrüggen, Synthesis *2008* (8), 1165-74; a bulky silylene generated under mild conditions: its application to the synthesis of **organosilicon compounds**., N. Takeda, N. Tokitoh, Synlett *2007* (16), 2483-91; catalytic reactions involving the **cleavage of C≡N and C≡C bonds**, M. Tobisu, N. Chatani, Chem. Soc. Rev. *2008, 37* (2), 300-7; **α-acetylenehalides and diynols,** A.N. Volkov, K.A. Volkova, Russ. J. Org. Chem. *2007, 43* (2), 161-9; **halogenovinylic aldehydes**: useful tools in organic synthesis, S. Brahma, K.J. Ray, Tetrahedron *2008, 64* (13), 2883-96; synthesis and applications of **2′-oxy-2-halogenamines,** M. D′hooghe, N. De Kimpe, ibid. *64* (15), 3275-85; stepwise two-electron-transfer **reduction of cyclic ethers and lactones** with alkalide $K^-,K^+(15\text{-Crown-}5)_2$, Z. Grobelny, A. Stolarzewicz, A. Maercker, Curr. Org. Chem. *2007, 11* (13), 1126-34; synthesis and cleavage of lactones and **thiolactones**: applications, Z. Paryzek, I. Skiera, Org. Prep. Proced. Int. *2007, 39* (3), 263-96**;** synthetic approaches to naturally-occurring **10-membered lactones**, H.M.C. Ferraz, F.I. Bombonato, L.S. Longo Jr., Synthesis *2007* (21), 3261-85; toward catalyst economy: a programmed approach to the synthesis of **bicyclic lactones and lactams**, C.D. Edlin, J. Faulkner, D. Fengas, M. Helliwell, C.K. Knight, D. House et al., J. Organomet. Chem. *2006, 691* (24-25), 5375-82; asymmetric synthesis of styryllactones, M. Monodon, J.P. Gesson, Curr. Org. Synth. *2006, 3* (1), 41-75; recent advances in heterolytic **nucleofugal leaving groups**, S.D. Lepore, D. Mondal, Tetrahedron, *2007, 63* (24), 5103-22; aldehyde *N,N*-dialkylhydrazones as neutral **acyl carbanion equivalents**: umpolung of the imine reactivity, R. Brehme, D. Enders, R. Fernandez, J.M. Lassaletta, Eur. J. Org. Chem. *2007* (34), 5629-60; small and versatile – formyl anion and dianion equivalents, A. Kirschning, C. Kujat, S. Luiken, E. Schaumann, ibid. (15), 2387-400.

19 Cycloaddition

(for photocycloaddition s. Sections 7 and 28)

Catalytic **azide-alkyne cycloaddition**: reactivity and applications, P. Wu, V.V. Folkin, Aldrichimica Acta *2007, 40* (1), 7-17; the growing applications of **'click' chemistry**, J.E. Moses, A.D. Moorhouse, Chem. Soc. Rev. *2007, 36* (8), 1249-62; click chemistry – what′s in a name? triazole synthesis and beyond, M.V. Gil, M.J. Arévalo, O. López, Synthesis *2007* (11), 1589-620; peptidomimetics via Cu-catalyzed azide-alkyne cycloaddition, Y.L. Angell, K. Burgess, ibid. *36* (8), 1674-89; the rise of azide-alkyne 1,3-dipolar 'click' cycloaddition and its application to polymer science and surface modification, R.A. Evans, Aust. J. Chem. *2007, 60* (6), 384-95; **intramolecular 1,3-dipolar cycloaddition** in targeted synthesis, V. Nair, T.D. Suja, Tetrahedron *2007, 63* (50), 12247-75; asymmetric catalysis of **transannular Diels-Alder reaction**, E.P. Balskus, E.N. Jacobsen, Science *2007, 317* (5845), 1736-40; asymmetric **hetero-Diels-Alder reaction** of Danishefsky′s and Brassard′s dienes with aldehydes, L. Lin, X. Liu, X. Feng, Synlett *2007* (14), 2147-57; catalytic, asymmetric inverse electron-demand hetero Diels-Alder reactions of *o*-benzoquinone derivs. and ketene enolates, D. Paull, J. Wolfer, J.W. Grebinski, A. Weatherwax, T. Lectka, Chimia *2007, 61* (5), 240-6; **Diels-Alder reactions of imino dienophiles,** G.R. Heintzelman, I.R. Meigh, Y.R. Mahajan, S.M. Weinreb, Org. Reactions *65,* 141-600; **cycloaddition of oxidopyrylium**

species in synthesis, V. Singh, U.M. Krishna, Vikrant, G.K. Trivedi, Tetrahedron *2008, 64* (16), 3405-28; **[8+2]-cycloaddition** in synthesis, V. Nair, K.G. Abhilash, K. Gopalakrishnan, Synlett *2008* (3), 301-12; cationic Rh(I)/BINAP-type di(phosphine) complexes: versatile new catalysts for highly chemo-, regio-, and enantio-selective **[2+2+2]-cycloaddition**, K. Tanaka, ibid. *2007* (13), 1977-93; the fascinating construction of pyridine ring systems by transition metal-catalyzed [2+2+2]-cycloaddition, B. Heller, M. Hapke, Chem. Soc. Rev. *2007, 36* (7), 1085-94; **cotrimerizations of acetylenic derivs.,** N. Agenet, O. Buisine, F. Slowinski, V. Gandon, C. Aubert, M. Malacria, Org. Reactions *68,* 1-302; advances in **solid-phase cycloaddition** for heterocyclic synthesis, L. Feliu, P. Vera-Luque, F. Albericio, M. Alvarez, J. Comb. Chem. *2007, 9* (4), 521-65.

20 Rearrangement, Isomerization, Cycloisomerization, Migration

(for radical rearrangements s. Section 17)

Investigations of the **[2.3]-sigmatropic rearrangement** of vinylaziridines and allylic amines, P. Somfai, O. Panknin, Synlett *2007* (8), 1190-202; [2.3]-sigmatropic rearrangement of allylic sulfur compds., M. Reggelin, Topics Curr. Chem. *2007, 275,* 1-65; of propargyl and allenyl systems, S. Braverman, M. Cherkinsky, ibid. 67-101; sulfur participation in **[3.3]-sigmatropic rearrangement**, R. Fernandez de la Pradilla, M. Tortosa, A. Viso, ibid. 103-29; catalysis of the **Claisen rearrangement**, K.C. Majumdar, S. Alam, B. Chattopadhyay, Tetrahedron *2008, 64* (4), 597-643; the **allylic trihaloacetimidate rearrangement,** L.E. Overman, N.E. Carpenter, Org. Reactions *66,* 1-108; **Smiles rearrangement** and the Julia-Kocienski olefination, K. Plesniak, A. Zarecki, J. Wicha, Topics Curr. Chem. *2007, 275,* 163-250; development and industrialization of the vapour-phase **Beckmann rearrangement**, Y. Izumi, H. Ichihashi, Y. Shimazu, M. Kitamura, H. Sato, Bull. Chem. Soc. Jpn. *2007, 80* (7), 1280-7; evolution, development and personal experience in studies of the **allyl cyanate-to-isocyanate rearrangement**, Y. Ichikawa, Synlett *2007* (19), 2927-36; thione-thiol rearrangement: **Miyazaki-Newman-Kwart rearrangement** and others, C. Zonta, O. De Lucchi, R. Volpicelli, L. Cotarca, Topics Curr. Chem. *2007, 275,* 131-61; mechanism and application of the Newman-Kwart O→S-rearrangement of aryl thiocarbamates, G.C. Lloyd-Jones, J.D. Moseley, J.S. Renny, Synthesis *2008* (5), 661-89; recent advances in the **semipinacol rearrangement** of (R)-epoxyalcohols and related compds., T.J. Snape, Chem. Soc. Rev. *2007, 36* (11), 1823-42; **the α-hydroxyketone (α-ketol) and related rearrangements,** L.A. Paquette, J.E. Hofferberth, Org. Reactions *62,* 477-568; **1,3-O→C-rearrangement**: opportunities for stereoselective synthesis, C.G. Nasveschuk, T. Rovis, Org. Biomol. Chem. *2008, 6* (2), 240-54; molecular rearrangements through **thermal 1,3-carbon shifts**, J.E. Baldwin, P.A. Leber, ibid. *6* (1), 36-47; **1,2-sulfur migration**, A.W. Sromek, V. Gevorgyan, Topics Curr. Chem. *2007, 274,* 77-124; **1,3-sulfur migration**: mechanism and synthetic utility, ibid. 125-71; the **halogen dance**, M. Schunarch, M. Spina, A.F. Khan, M. Mihovilovic, P. Stanetty, Chem. Soc. Rev. *2007, 36* (7), 1046-57; s.a. M.V. Nora de Souza, Curr. Org. Chem. *2007, 11* (7), 637-46; transition metal complexes as catalysts of **double bond migration** in O-allyl systems, N. Kuznik, S. Krompiec, Coord. Chem. Rev. *2007, 251* (1-2), 222-33; ruthenium-catalyzed **isomerization of allylic and propargylic alcohols** in aq. and organic media: applications in synthesis, V. Cadierno, P. Crochet, J. Jose, Synlett *2008* (8), 1105-24; polyyne synthesis using **carbene/carbenoid rearrangements,** W.A. Chalifoux, R.R. Tykwinski, Chem. Record *2006, 6* (4), 169-82; advances **in metallotropic 1,3-shift** of alkynyl carbenoids, D. Lee, M. Kim, Org. Biomol. Chem. *2007, 5* (21), 3418-27.

21 Name Reactions, Standard Transformations

(for asymmetric organocatalyzed aldol, Mannich and Michael reactions, and asymmetric hydrogenation, s. Section 1; transition-metal catalyzed conversions (Suzuki, Heck, allylation, hydrosilylation, hydroamination...) s. Section 2; Barton-McCombie deoxygenation s. Section 17; Diels-Alder reactions s. Section 19, and for classical

rearrangements s. Section 20)

The **Baylis-Hillman reaction** – A novel source of attraction, opportunities, and challenges in synthetic chemistry, D. Basavaiah, K.V. Rao, R.J. Reddy, Chem. Soc. Rev. *2007, 36* (10), 1581-8; advances in the Baylis-Hillman reaction-assisted synthesis of cyclic frameworks, V. Singh, S. Batra, Tetrahedron *2008, 64* (20), 4511-74; the **aza-Baylis-Hillman reaction** and applications, Y.-L. Shi, M. Shi, Eur. J. Org. Chem. *2007 (18),* 2905-16; **asymmetric Baylis-Hillman and aza-Baylis-Hillman reactions**, G. Masson, C. Housseman, J. Zhu, Angew. Chem. Int. Ed. Engl. *2007, 46* (25), 4614-28; **Prins reaction:** advances and applications, I.M. Pastor, M. Yus, Curr. Org. Chem. *2007, 11* (10), 925-57; the **Passerini reaction**, L. Banfi, R. Riva, Org. Reactions *65,* 1-140; **Pictet-Spengler reaction:** efficient carbon-carbon bond formation in heterocyclic synthesis, S.W. Youn, Org. Prep. Proced. Int. *2006, 38* (6), 505-91; recent advances in the **Pummerer reaction**, S. Akai, Y. Kita, Topics Curr. Chem. *2007, 274,* 35-76; **Baeyer-Villiger oxidation** on heterogeneous catalysts, C. Jimenez-Sanchidrian, J.R. Ruiz, Tetrahedron *2008, 64* (9), 2011-26; recent advances in the catalytic **asymmetric Henry reaction**, C. Palomo, M. Oiarbide, A. Laso, Eur. J. Org. Chem. *2007* (16), 2561-74; new approaches to **asymmetric fluorination** – Cinchona alkaloid combinations and chiral ligands/metal complexes, N. Shibata, T.S. Ishimaru, S. Nakamura, T. Toru, J. Fluorine Chem. *2007, 128* (5), 469-83; **electrophilic fluorination** with N–F reagents, J. Baudoux, D. Cahard, Org. Reactions *69,* 347-672; catalytic **asymmetric diboration, disilation and silaboration**: new opportunities for asymmetric synthesis, H.E. Burks, J.P. Morken, Chem. Commun. *2007* (45), 4717-25; **asymmetric alkylation of imines**, D. Ferraris, Tetrahedron *2007, 63* (39), 9581-97; **addition of allyl, allenyl, and propargylstannanes** to aldehydes and imines, B.W. Gung, Org. Reactions *64,* 1-114; modern **amination reactions,** M. Kienle, S.R. Dubbaka, K. Brade, P. Knochel, Eur. J. Org. Chem. *2007* (25), 4166-76; catalytic **hydroamination** of acetylene derivs., R. Severin, S. Doye, Chem. Soc. Rev. *2007, 36* (9), 1407-20; recent development of regio- and stereo-selective **halogenamination** of ethylene derivs., G. Li, S.R.S. Saibau Kotti, C. Timmons, Eur. J. Org. Chem. *2007* (17), 2745-58; renaissance of **deuteriation**, J. Atzrodt, V. Derdau, T. Fey, J. Zimmermann, Angew. Chem. Int. Ed. Engl. *2007, 46* (41), 7744-65; Lewis acid- and Brønsted acid-catalyzed **allylboration** of carbonyl compds.: from discovery to mechanism and applications, D.G. Hall, Russ. Chem. Rev. *2007* (11), 1644-55; creation of quaternary stereocenters in carbonyl allylation reactions, I. Marek, G. Sklute, Chem. Commun. *2007* (17), 1683-91; asymmetric **nucleophilic allylation** of carbonyl compds., M. Lombardo, C. Trombini, Chem. Rev. *2007, 107* (9), 3843-79; the **Ramberg-Bäcklund reaction,** R.J.K. Taylor, G. Casy, Org. Reactions *62,* 357-476; **α-hydroxylation** of enolates and silyl enol ethers, B.-C. Chen, P. Zhou, F.A. Davis, E. Ciganek, ibid. *62,* 1-356; **asymmetric dihydroxylation** of alkenes, M.C. Noe, M.A. Letavic, S.L. Snow, ibid. *66,* 109-626; **catalytic asymmetric aldol condensation,** E.M. Carreira, A. Fettes, C. Martl, ibid. *67,* 1-216.

22 Reagents and Auxiliaries

(for chiral reagents s. Section 1, N-heterocyclic carbenes s. Section 4, radical reagents s. Section 17, and use of CAN for deprotection s. Section 26)

Zinc bromide: an overview of its solution chemistry and recent applications, R.A. Gossage, Curr. Org. Chem. *2006, 10* (9), 923-36; **lithium hydrido(amino)borates**, P.P. Saikia, Synlett *2007* (6), 995-6; **CAN**: a versatile single electron oxidant, V. Nair, A. Deepthi, Chem. Rev. *2007, 107* (5), 1862-91; **zirconium tetrachloride**, G. Smitha, S. Chandrasekhar, C.S. Reddy, Synthesis *2008* (6), 829-55; functionalization of **polyoxometalates**: towards advanced applications in catalysis and materials science, A. Proust, R. Thouvenot, P. Gouzerh, Chem. Commun. *2008* (16), 1837-52; taking up the cudgels for **perchlorates** – uses and applications in organic reactions under mild conditions, G. Bartoli, M. Locatelli, P. Melchiorre, L. Sambri, Eur. J. Org. Chem. *2007* (13), 2037-49; **manganese(III) acetate**, A.S. Demir, M. Emrullahoglu, Curr. Org. Synth. *2007, 4* (3), 321-51; recent advances in **indium**-promoted reactions, J. Auge, N.

Lubin-Germain, J. Uziel, Synthesis *2007* (12), 1739-64; catalytic oxidation by molecular oxygen and **hydrogen peroxide** by multistep electron transfer – A biomimetic approach, J. Piera, J.-E. Bäckvall, Angew. Chem. Int. Ed. Engl. *2008, 47* (19), 3506-23; heterocyclic **hydroperoxides** in selective oxidations, F.G. Gelalcha, Chem. Rev. *107* (7), 3338-61; **dioxirane** oxidation of compounds other than alkenes, W. Adam, C.-G. Zhao, K. Jakka, Org. Reactions *69,* 1-346; versatile use of **CO_2** in synthesis of carbamates, D. Chaturvedi, S. Ray, Curr. Org. Chem. *2007, 11* (11), 987-98; a new organic chemistry of **SO_2**, P. Vogel, M. Turks, L. Bouchez, D. Markovia, A. Varela-Álvarez, J.Á. Sordo, Acc. Chem. Res. *2007, 40* (10), 931-42; applications of **N-chlorosuccinimide**, W.M. Golebiewski, M. Gucma, Synthesis *2007* (23), 3599-619; ***tert*-butyl-N-chlorocyanamide** as reagent, V. Kumar, M.P. Kaushik, Synlett *2007* (19), 2937-51; **hexadecyltributylphosphonium azide** for the synthesis of unusual azides, K. Banert, Synthesis *2007* (22), 3431-46; use of **Lawesson's reagent**, T. Ozturk, E. Ertas, O. Mert, Chem. Rev. *2007, 107* (11), 5210-78; amphiphilic **cyclodextrins** – advances in synthesis and supramolecular chemistry, F. Sallas, R. Darcy, Eur. J. Org. Chem. *2008* (6), 957-69; **cucurbit[n]uril**: a new molecule in host-guest chemistry, N.J. Wheate, Austr. J. Chem. *2006, 59* (5), 354; applications of **menthol**, H. Oertling, A. Reckziegel, H. Surburg, H.-J. Bertram, Chem. Rev. *2007, 107* (5), 2136-64; **citronellal** in synthesis, E.J. Lenardao, G.V. Botteselle, F. De Azambuja, G. Perin, R.G. Jacob, Heterocycles *2007, 63* (29), 6671-712.

23 Polymer-based Synthesis

(for glycosylation on polymer supports s. Section 10, and solid-phase cycloaddition s. Section 19)

Applications of **polymeric reagents**, H. Salimi, A. Rahimi, A. Pourjavadi, Monatsh. Chem. *2007, 138* (4), 363-79; solid-supported reagents and **catch-and-release techniques**, A. Solinas, M. Taddei, Synthesis *2007* (16), 2409-53; applications of phosphine-functionalized polymers, M. Guino, K.K. Hii, Chem. Soc. Rev. *2007, 36* (4), 608-17; electrons as reagent in solid-phase synthesis, M. Mentel, R. Breinbauer, Eur. J. Org. Chem. *2007* (26), 4283-92.

24 Multicomponent Synthesis

Multicomponent syntheses of heterocyclics under transition metal catalysis, D.M. D'Souza, T.J.J. Müller, Chem. Soc. Rev. *2007, 36* (7), 1095-108; chemodifferentiating ABB′ multicomponent reactions: privileged building blocks, D. Tejedor, F. García-Tellado, ibid. *36* (3), 484-91; recent developments in 3-component Grignard-Barbier-type reactions, V.V. Kouznetsov, L.Y. Vargas Méndez, Synthesis *2008* (4), 491-506; multicomponent reactions of carbonyl compds. and derivs. of cyanoacetic acid: synthesis of carbo- and hetero-cyclics, A.M. Shestopalov, A.A. Shestopalov, L. Rodinovskaya, Synthesis *2008* (1), 1-25.

25 Tandem, Sequenced, Domino, Cascade Reaction

(for cascade pericyclic reactions s. Section 7)

The synthesis of chromenes, chromans, coumarins and related heterocycles via tandem reactions of salicylic aldehydes or salicylic imines with α,β-unsaturated compounds, Y.L. Shi, M. Shi, Org. Biomol. Chem. *2007, 5* (10), 1499-504; the domino way to heterocyclics, A. Padwa, S.K. Bur, Heterocycles *2007, 63* (25), 5341-78; controlling competing pathways in palladium-catalyzed tandem/domino reactions of hindered Grignard reagents with 1,2-dihaloarenes and 2-haloaryl tosylates, Q.S. Hu, Synlett *2007* (9), 1331-45; domino reactions of 1,3-bis(silyl enol ethers) with 4-(trialkylsiloxy)-benzopyrylium triflates, P. Langer, Synlett *2007* (7), 1016-25; palladium-catalyzed cascade reactions of highly activated olefins, N.T. Patil, Y. Yamamoto, ibid. (13), 1994-2005.

26 **Protective Group Chemistry**

Mild and neutral deprotection catalyzed by CAN, N. Maulide, J.-C. Vanherck, A. Gautier, I.E. Marko, Acc. Chem. Res. *2007, 40* (6), 381-92; overcoming the demons of protective groups with amphoteric molecules, A.K. Yudin, R. Hili, Chem. Eur. J. *2007, 13* (23), 6538-42; the use of cyclic bifunctional protecting groups in oligosaccharide synthesis – An overview, R.E.J.N. Litjens, L.J. van den Bos, J.D.C. Codee, H.S. Overkleeft, G.A. van der Marel, Carbohydr. Res. *2007, 342* (3-4), 419-29; the fascination with 1,2-diacetals, S.V. Ley, A. Polara, J. Org. Chem. *2007, 72* (16), 5943-59.

27 **Media: Ionic Liquids, Water, ... Green Chemistry**

(for asymmetric catalysis, C-H bond activation, catalytic arylation, and microwave-assisted syntheses in water s. Sections 1, 2c, 9, and 29, respectively)

Understanding **ionic liquids** at the molecular level: facts, problems, and controversies, H. Weingärtner, Angew. Chem. Int. Ed. Engl. *2008, 47* (4), 654-70; protic ionic liquids: properties and applications, T.L. Greaves, C.J. Drummond, Chem. Rev. *2008, 108* (1), 206-37; applications of ionic liquids **in the chemical industry**, N.V. Plechkova, K.R. Seddon, Chem. Soc. Rev. *2008, 37* (1), 123-50; catalysis in ionic liquids, V.I. Parvulescu, C. Hardacre, Chem. Rev. *2007, 107* (6), 2615-65; recent advances in the synthesis and application of **chiral ionic liquids**, A. Winkel, P.V.G. Reddy, R. Wilhelm, Synthesis *2008* (7), 999-1016; advances in chiral ionic liquids derived from natural amino acids, X. Chen, X. Li, A. Hu, F. Wang, Tetrahedron:Asym. *2008, 19* (1), 1-14; asymmetric synthesis in ionic liquids, A. Paczal, A. Kotschy, Monatsh. Chem. *2007, 138* (11), 1115-23; sustainable processes employing ionic liquids for secondary alcohols separation, R. Bogel-Lukasik, ibid. 1137-44; Brønsted acids in ionic liquids: fundamentals, organic reactions, and comparisons, K.E. Johnson, R.M. Pagni, J. Bartmess, ibid. 1077-101; synthesis of ionic liquids using non conventional activation methods: an overview, J.-M. Lévêque, J. Estager, M. Draye, G. Cravotto, L. Boffa, W. Bonrath, ibid. 1103-13; the **phosphorus aspects of green chemistry**: the use of quaternary phosphonium salts and 1,3-dialkylimidazolium hexafluorophosphates, G. Keglevich, Z. Baan, I. Hermecz, T. Novak, I.L. Odinets, Curr. Org. Chem. *2007, 11* (1), 107-26; **fluorinated alcohols** as solvents, cosolvents and additives in homogeneous catalysis, I.A. Shuklov, N.V. Dubrovina, A. Boerner, Synthesis *2007* (19), 2925-43.

28 **Photochemistry**

Photochemical reactions as key steps in synthesis, N. Hoffmann, Chem. Rev. *2008, 108* (3), 1052-103; photocatalysis: a promising route for 21st century organic chemistry, G. Palmisano, V. Augugliaro, M. Pagliaro, L. Palmisano, Chem. Commun. *2007* (33), 3425-37; **photocatalysis** for the formation of the C-C bond, M. Fagnoni, D. Dondi, D. Ravelli, A. Albini, Chem. Rev. *2007, 107* (6), 2725-56; rearrangement and cyclization reactions by **photo-induced electron transfer**: new approaches for the synthesis of polycyclics, P.A. Waske, N.T. Tzvetkov, J. Mattay, Synlett *2007* (5), 669-85; photochemical transformations of pyridinium salts – Mechanistic studies and applications in synthesis, T. Damiano, D. Morton, A. Nelson, Org. Biomol. Chem. *2007, 5* (17), 2735-52; efficient radical addition of tert. amines to alkenes using photochemical electron transfer, N. Hoffmann, S. Bertrand, S. Marinkovic, J. Pesch, Pure Applied Chem. *2006, 78* (12), 2227-46; photolysis of carbonyl compounds, W.M. Horspool, Photochem. *2007* (36), 9-22; photochemistry of alkenes, alkynes and related compounds, ibid. 55-90; photochemistry of aromatic compounds, A. Gilbert, ibid. 91-132; **photo-oxidation and photo-reduction**, N.W.A. Geraghty, ibid. 133-204; **photoelimination**, I.R. Dunkin, ibid. 205-31; enone cycloadditions and rearrangements: photoreactions of dienones and quinones, W.M. Horspool, ibid. 23-54; **photooxygenation** of non-aromatic heterocycles, M.R. Iesce, F. Carmola, M. Rubino, Curr. Org. Chem. *2007, 11* (12), 1053-75.

29 Methodology: Microwave Chemistry, Microchemistry ...

(for microwave-assisted metathesis, and heterocyclic and natural product synthesis, s. Sections 6, 7 and 2a, respectively)

Microwaves in green and sustainable chemistry, C.R. Strauss, R.S. Varma, Topics Curr. Chem. *2006, 266,* 199-231; microwave-assisted synthesis in water, D. Dallinger, C.O. Kappe, Chem. Rev. *2007, 107* (6), 2563-91; microwave-assisted and metal-catalyzed coupling, P. Nilsson, K. Olofsson, M. Larhed, Topics Curr. Chem. *2006, 266,* 103-44; recent developments in microwave-assisted, transition metal-catalyzed C–C and C–N bond formation, P. Appukkuttan, E. Van der Eycken, Eur. J. Org. Chem. *2008* (7), 1133-55; greener approaches to synthesis using **microreactor technology**, B.P. Mason, K.E. Price, J.L. Steinbacher, A.R. Bogdan, D.T. McQuade, Chem. Rev. *2007, 107* (6), 2300-18; microreactors as tools for the chemist – The chemists′ round-bottomed flask of the 21st Century, K. Geyer, J.D.C. Codee, P.H. Seeberger, Chem. Eur. J. *2006, 12* (33), 8434-42; adventures in inner space: microflow systems for the chemist, T. Fukuyama, M.T. Rahman, M. Sato, I. Ryu, Synlett *2008* (2), 151-63; **continuous flow reactors** – a tool for the modern synthetic chemist, C. Wiles, P. Watts, Eur. J. Org. Chem. *2008* (10), 1655-71.

Index to Volume 72

As in previous volumes, reactions are indexed from both the starting material and product aspects, e.g. '**Azides** startg. m.f. amines' and '**Amines** from azides'. Nomenclature for complex functions can be located under the 'special s.' sub-entry, e.g. '**Carboxylic acids** special s. aminocarboxylic acids' or by consulting the Formula Index of Complex Functional Groups (Volume *48*, p. 471).

Hydrogenated and functionalized ring systems are indexed by the conventional reversal, e.g. '**Pyridines, aryl-**', the only important exception to the rule being alkylideneisocyclics which are indexed as such, e.g. '**Alkylidenecyclopentanes**'.

As from Volume *51*, '**Epoxides**' has been used in place of 'Oxido compds.'; '**Thiiranes**' in place of 'Sulfido compds.'; '**Diels-Alder reaction**' in place of 'Diene synthesis'; and '**Benzo[*b*]thiophenes**' in place of 'Thianaphthenes'. From this volume 'Biaryls' is used instead of 'Diaryls'.

References to abstracts in this volume are in the format **72,** 234. An entry such as '**Suzuki biaryl coupling**, update **37,** 902s**72**' refers to the indexing of a supplementary reference, which must be followed up via the Supplementary References section (p. 449), from which the page number on which the reference is located may be found.

Supplementary References in Volume 72

No.	Suppl. Vol.	Ref. Page

Volume 1

335	*72*, 135
376	*72*, 150
391	*72*, 133
492	*72*, 39

Volume 2

151	*72*, 47
312	*72*, 105
314	*72*, 107
580	*72*, 224
621	*72*, 288

Volume 3

115	*72*, 47
470	*72*, 185

Volume 4

86	*72*, 28 (2)
154	*72*, 47
212	*72*, 40
316	*72*, 369
337	*72*, 126
513	*72*, 187
566	*72*, 208

Volume 5

3	*72*, 1
12	*72*, 4
40	*72*, 12
101	*72*, 38
122	*72*, 48
192	*72*, 82
202	*72*, 79
216	*72*, 89
346	*72*, 156
549	*72*, 313

Volume 6

311	*72*, 99
367	*72*, 113
540	*72*, 172

Volume 7

281	*72*, 80
564	*72*, 179
634	*72*, 210
670	*72*, 223
836	*72*, 331

Volume 8

404	*72*, 120
411	*72*, 121
695	*72*, 214
823	*72*, 311
942	*72*, 373

Volume 9

41	*72*, 8

Volume 10

220	*72*, 45
257	*72*, 79
295	*72*, 130
394	*72*, 167
451	*72*, 181
639	*72*, 369

Volume 11

67	*72*, 8
171	*72*, 42
489	*72*, 149
536	*72*, 111

Volume 12

42	*72*, 9
55	*72*, 8
329	*72*, 95
465	*72*, 139
690	*72*, 212

Volume 13

192	*72*, 48
287	*72*, 80
361	*72*, 108
371	*72*, 110
561	*72*, 178
631	*72*, 213
896	*72*, 205

Volume 14

99	*72*, 29
181	*72*, 48
194	*72*, 53
366	*72*, 104
772	*72*, 294
982	*72*, 25

No.	Suppl. Vol.	Ref. Page

Volume 15

38	72, 14	
177	72, 60	
332	72, 31	
432	72, 181	

Volume 16

58	72, 104	
142	72, 37	
236	72, 70	
285	72, 82	
326	72, 98	
331	72, 98	
735	72, 251	
746	72, 270	

Volume 17

117	72, 32	
153	72, 39	
198	72, 251	
219	72, 65	
598	72, 179, 183	
705	72, 221	
809	72, 291	

Volume 18

307	72, 96	
373	72, 111	
561	72, 172	
608	72, 185	
689	72, 208	
704	72, 218	
776	72, 274	

Volume 19

93	72, 3	
125	72, 38	
211	72, 62	
253	72, 99	
354	72, 103	
481	72, 134	
490	72, 146	
563	72, 98	
674	72, 139	
734	72, 229	
803	72, 289	

Volume 20

51	72, 11	
119	72, 64	
373	72, 162	
417	72, 183	

Volume 21

9	72, 5	
78	72, 26	
121	72, 36	
186	72, 60	
192	72, 61	
244	72, 136	
543	72, 98	

Volume 22

44	72, 136	
163	72, 60	
191	72, 4	
286	72, 103	
761	72, 291	
782	72, 321	
826	72, 324	
877	72, 366	

Volume 23

6	72, 1	
13	72, 5	
15	72, 7	
51	72, 9, 14	
66	72, 25	
81	72, 29	
139	72, 47	
182	72, 61	
234	72, 101	
383	72, 98	
423	72, 133	
553	72, 183	
630	72, 212	
819	72, 323	

Volume 24

9	72, 3	
85	72, 31	
128	72, 47	
149	72, 53	
240	72, 71	
555	72, 178	
751	72, 291	

Volume 25

15	72, 9	
80	72, 38	
123	72, 62	
129	72, 66	
346	72, 165	
417	72, 192	
458	72, 212	
581	72, 322	
649	72, 162	

Volume 26

6	72, 5	
24	72, 9	
52	72, 16	
331	72, 133	
668	72, 232	

Volume 27

57	72, 24	
184	72, 92	
401	72, 141	
607	72, 204	

No.	Suppl. Vol.	Ref. Page

Volume 27 continued

851	72, 341
871	72, 337
943	72, 371

Volume 28

13	72, 5
15	72, 7
45	72, 26
107	72, 54
113	72, 56
141	72, 69
267	72, 107
437	72, 165
617	72, 111
753	72, 138
903	72, 152

Volume 29

9	72, 4
42	72, 19
53	72, 27
65	72, 30
184	72, 68
188	72, 73
415	72, 43
688	72, 284

Volume 30

5	72, 74
63	72, 44
72	72, 47
340	72, 169
365	72, 186
374	72, 190

Volume 31

57	72, 29
135	72, 62
159	72, 66
170	72, 28
185	72, 78
325	72, 122
373	72, 142
501	72, 186
522	72, 208
665	72, 282
714	72, 286

Volume 32

no references

Volume 33

21	72, 6
30	72, 155
43	72, 11
130	72, 56
200	72, 79
477	72, 179
487	72, 182
501	72, 190
540	72, 204
542	72, 203
593	72, 222
603	72, 226
910	72, 369

Volume 34

55	72, 83
114	72, 54
129	72, 58
172	72, 78
249	72, 99
285	72, 111
360	72, 141
417	72, 156
610	72, 230

Volume 35

57	72, 38
74	72, 53
85	72, 61
182	72, 94
312	72, 165
454	72, 264

Volume 36

148	72, 62
198	72, 81
270	72, 104
301	72, 119
476	72, 183
670	72, 18

Volume 37

81	72, 39
195	72, 85
220	72, 91
236	72, 97
264	72, 99
364	72, 150
446	72, 179
519	72, 221
646	72, 251
837	72, 225
902	72, 359

Volume 38

17	72, 10, 19
252	72, 103
461	72, 173
529	72, 209
793	72, 325
954	72, 372
998	72, 73

No.	Suppl. Vol.	Ref. Page

Volume 39

97	72, 43
104	72, 74
107	72, 48
189	72, 83
406	72, 161
458	72, 182
578	72, 243
593	72, 240
895	72, 239

Volume 40

127	72, 86
176	72, 108

Volume 41

38	72, 12
101	72, 40
104	72, 42
118	72, 46
166	72, 43
263	72, 102
377	72, 154
457	72, 173
493	72, 194
584	72, 207
734	72, 320

Volume 42

6	72, 5
11	72, 74
58	72, 30
127	72, 50
235	72, 96
420	72, 166
456	72, 202
507	72, 201
521	72, 207
522	72, 208
616	72, 232

Volume 43

249	72, 110
271	72, 126
402	72, 173
456	72, 174
458	72, 197
516	72, 218
520	72, 221
535	72, 224
564	72, 234
625	72, 177

Volume 44

8	72, 3
33	72, 29
83	72, 37
114	72, 53
211	72, 48, 88
214	72, 88
245	72, 101
249	72, 101
301	72, 96
521	72, 222

Volume 45

23	72, 22
35	72, 32
120	72, 94
150	72, 106
210	72, 134
255	72, 39
347	72, 222
350	72, 222
447	72, 307

Volume 46

42	72, 15, 17
47	72, 19
90	72, 37
106	72, 38, 323
172	72, 78
192	72, 82
228	72, 88
321	72, 136
530	72, 220
662	72, 258, 274
713	72, 310

Volume 47

35	72, 14
48	72, 25
106	72, 197
114	72, 58
146	72, 78
192	72, 96
248	72, 118
254	72, 121
270	72, 130
337	72, 152
390	72, 168
468	72, 78, 191
470	72, 191
479	72, 194
488	72, 218
495	72, 66
563	72, 216, 224

Volume 48

7	72, 4
86	72, 36
97	72, 41
104	72, 43
120	72, 56
170	72, 74
294	72, 126
367	72, 156
417	72, 172
434	72, 180
442	72, 183
529	72, 211
549	72, 218
658	72, 262

No.	Suppl. Vol.	Ref. Page

Volume 48 continued

819	72, 344

Volume 49

8	72, 5
15	72, 5
42	72, 13
91	72, 41
99	72, 37
105	72, 49
107	72, 50
117	72, 65
254	72, 108
284	72, 130
398	72, 173
508	72, 221
591	72, 239
644	72, 260
932	72, 368
985	72, 375

Volume 50

19	72, 17
56	72, 50
123	72, 95
151	72, 110
182	72, 141
183	72, 131
214	72, 154
230	72, 158
307	72, 202
343	72, 218
380	72, 244
443	72, 293

Volume 51

32	72, 26
45	72, 37
55	72, 46
65	72, 55
72	72, 65
98	72, 86
171	72, 156
204	72, 180
271	72, 232
337	72, 283

Volume 52

108	72, 84
119	72, 92, 93
147	72, 141
171	72, 156
207	72, 189
297	72, 262
449	72, 311

Volume 53

3	72, 5
59	72, 48
109	72, 97
116	72, 102
301	72, 263
387	72, 323
435	72, 342
487	72, 372

Volume 54

62	72, 66
79	72, 69
83	72, 72
90	72, 80
111	72, 97
143	72, 130
221	72, 207

Volume 55

49	72, 37
166	72, 85, 158
180	72, 178
227	72, 225
229	72, 218
337	72, 307
340	72, 347
341	72, 315

Volume 56

4	72, 1
21	72, 5
28	72, 93
33	72, 18
39	72, 27
63	72, 46
97	72, 80
136	72, 124
201	72, 191
242	72, 231

Volume 57

22	72, 15
44	72, 372
105	72, 88
439	72, 360

Volume 58

7	72, 5
32	72, 21
76	72, 70
78	72, 69 (2)
103	72, 94
194	72, 197
236	72, 233
245	72, 236
259	72, 248

Volume 59

66	72, 82
84	72, 95
92	72, 73
103	72, 84
134	72, 108

No.	Suppl. Vol.	Ref. Page

Volume 59 continued

173	*72*, 155
204	*72*, 194
219	*72*, 215
234	*72*, 222, 228
301	*72*, 253
311	*72*, 269

Volume 60

45	*72*, 35
55	*72*, 43
74	*72*, 62
123	*72*, 108
139	*72*, 141
219	*72*, 229
288	*72*, 270

Volume 61

7	*72*, 5
71	*72*, 57
79	*72*, 88
100	*72*, 158
134	*72*, 108
225	*72*, 220
357	*72*, 324

Volume 62

31	*72*, 19
117	*72*, 84
135	*72*, 96
146	*72*, 105
168	*72*, 146
171	*72*, 151
181	*72*, 162
203	*72*, 190
204	*72*, 191
230	*72*, 220
250	*72*, 230
272	*72*, 249
282	*72*, 256

Volume 63

97	*72*, 72
102	*72*, 69
114	*72*, 83
147	*72*, 118
191	*72*, 169
253	*72*, 239
266	*72*, 249
356	*72*, 307
377	*72*, 229
411	*72*, 341
413	*72*, 344

Volume 64

63	*72*, 47
66	*72*, 49
71	*72*, 55
107	*72*, 80
127	*72*, 95
137	*72*, 110
141	*72*, 115
219	*72*, 228
463	*72*, 362

Volume 65

14	*72*, 14
21	*72*, 17
66	*72*, 51
93	*72*, 69
99	*72*, 75
106	*72*, 79
110	*72*, 81
144	*72*, 119
197	*72*, 183
199	*72*, 182
208	*72*, 196
230	*72*, 225
249	*72*, 233
253	*72*, 239

Volume 66

33	*72*, 18
42	*72*, 28
57	*72*, 41
99	*72*, 66
169	*72*, 128
182	*72*, 137
224	*72*, 174
233	*72*, 186
290	*72*, 244
353	*72*, 305

Volume 67

57	*72*, 46
101	*72*, 85
113	*72*, 94
116	*72*, 96
118	*72*, 97
119	*72*, 97
127	*72*, 104
144	*72*, 117
185	*72*, 154
233	*72*, 203
255	*72*, 346
301	*72*, 254
419	*72*, 339

Volume 68

58	*72*, 108
64	*72*, 49
74	*72*, 57
90	*72*, 66
138	*72*, 100
178	*72*, 146
204	*72*, 179
220	*72*, 196
224	*72*, 74
230	*72*, 207
235	*72*, 223
239	*72*, 215, 220
245	*72*, 222
259	*72*, 237
272	*72*, 249